Simultaneous Localization and Mapping for Mobile Robots:

Introduction and Methods

Juan-Antonio Fernández-Madrigal
Universidad de Málaga, Spain

José Luis Blanco Claraco
Universidad de Málaga, Spain

A volume in the Advances in Computational Intelligence and Robotics (ACIR) Book Series

Managing Director: Lindsay Johnston
Book Production Manager: Jennifer Romanchak
Publishing Systems Analyst: Adrienne Freeland
Managing Editor: Joel Gamon
Development Editor: Hannah Abelbeck
Assistant Acquisitions Editor: Kayla Wolfe
Typesetter: Lisandro Gonzalez
Cover Design: Nick Newcomer

Published in the United States of America by
Information Science Reference (an imprint of IGI Global)
701 E. Chocolate Avenue
Hershey PA 17033
Tel: 717-533-8845
Fax: 717-533-8661
E-mail: cust@igi-global.com
Web site: http://www.igi-global.com

Library of Congress Cataloging-in-Publication Data

Fernandez-Madrigal, Juan-Antonio, 1970-
Simultaneous localization and mapping for mobile robots: introduction and methods / by Juan-Antonio Fernandez-Madrigal and Josi Luis Blanco Claraco.
p. cm.
Includes bibliographical references and index.
Summary: "This book investigates the complexities of the theory of probabilistic localization and mapping of mobile robots as well as providing the most current and concrete developments"-- Provided by publisher.
ISBN 978-1-4666-2104-6 (hardcover) -- ISBN 978-1-4666-2105-3 (ebook) -- ISBN 978-1-4666-2106-0 (print & perpetual access) 1. Mobile robots. 2. Geographical positions. 3. Localization theory. I. Blanco Claraco, Josi Luis, 1981- II. Title.
TJ211.415.F474 2013
629.8'932--dc23
2012015952

This book is published in the IGI Global book series Advances in Computational Intelligence and Robotics (ACIR) Book Series (ISSN: 2327-0411; eISSN: 2327-042X)

British Cataloguing in Publication Data
A Cataloguing in Publication record for this book is available from the British Library.

Advances in Computational Intelligence and Robotics (ACIR) Book Series

ISSN: 2327-0411
EISSN: 2327-042X

Mission

While intelligence is traditionally a term applied to humans and human cognition, technology has progressed in such a way to allow for the development of intelligent systems able to simulate many human traits. With this new era of simulated and artificial intelligence, much research is needed in order to continue to advance the field and also to evaluate the ethical and societal concerns of the existence of artificial life and machine learning.

The **Advances in Computational Intelligence and Robotics (ACIR) Book Series** encourages scholarly discourse on all topics pertaining to evolutionary computing, artificial life, computational intelligence, machine learning, and robotics. ACIR presents the latest research being conducted on diverse topics in intelligence technologies with the goal of advancing knowledge and applications in this rapidly evolving field.

Coverage

- Adaptive & Complex Systems
- Agent Technologies
- Artificial Intelligence
- Cognitive Informatics
- Computational Intelligence
- Natural Language Processing
- Neural Networks
- Pattern Recognition
- Robotics
- Synthetic Emotions

IGI Global is currently accepting manuscripts for publication within this series. To submit a proposal for a volume in this series, please contact our Acquisition Editors at Acquisitions@igi-global.com or visit: http://www.igi-global.com/publish/.

The Advances in Computational Intelligence and Robotics (ACIR) Book Series (ISSN 2327-0411) is published by IGI Global, 701 E. Chocolate Avenue, Hershey, PA 17033-1240, USA, www.igi-global.com. This series is composed of titles available for purchase individually; each title is edited to be contextually exclusive from any other title within the series. For pricing and ordering information please visit http://www.igi-global.com/book-series/advances-computational-intelligence-robotics-acir/73674. Postmaster: Send all address changes to above address.

Titles in this Series

For a list of additional titles in this series, please visit: www.igi-global.com

Intelligent Technologies and Techniques for Pervasive Computing
Kostas Kolomvatsos (University of Athens, Greece) Christos Anagnostopoulos (Ionian University, Greece) and Stathes Hadjiefthymiades (University of Athens, Greece)
Information Science Reference • copyright 2013 • 349pp • H/C (ISBN: 9781466640382) • US $195.00 (our price)

Mobile Ad Hoc Robots and Wireless Robotic Systems Design and Implementation
Raul Aquino Santos (University of Colima, Mexico) Omar Lengerke (Universidad Autónoma de Bucaramanga, Colombia) and Arthur Edwards-Block (University of Colima, Mexico)
Information Science Reference • copyright 2013 • 347pp • H/C (ISBN: 9781466626584) • US $190.00 (our price)

Intelligent Planning for Mobile Robotics Algorithmic Approaches
Ritu Tiwari (ABV – Indian Institute of Information, India) Anupam Shukla (ABV – Indian Institute of Information, India) and Rahul Kala (School of Systems Engineering, University of Reading, UK)
Information Science Reference • copyright 2013 • 320pp • H/C (ISBN: 9781466620742) • US $195.00 (our price)

Simultaneous Localization and Mapping for Mobile Robots Introduction and Methods
Juan-Antonio Fernández-Madrigal (Universidad de Málaga, Spain) and José Luis Blanco Claraco (Universidad de Málaga, Spain)
Information Science Reference • copyright 2013 • 497pp • H/C (ISBN: 9781466621046) • US $195.00 (our price)

Prototyping of Robotic Systems Applications of Design and Implementation
Tarek Sobh (University of Bridgeport, USA) and Xingguo Xiong (University of Bridgeport, USA)
Information Science Reference • copyright 2012 • 321pp • H/C (ISBN: 9781466601765) • US $195.00 (our price)

Cross-Disciplinary Applications of Artificial Intelligence and Pattern Recognition Advancing Technologies
Vijay Kumar Mago (Simon Fraser University, Canada) and Nitin Bhatia (DAV College, India)
Information Science Reference • copyright 2012 • 784pp • H/C (ISBN: 9781613504291) • US $195.00 (our price)

Handbook of Research on Ambient Intelligence and Smart Environments Trends and Perspectives
Nak-Young Chong (Japan Advanced Institute of Science and Technology, Japan) and Fulvio Mastrogiovanni (University of Genova, Italy)
Information Science Reference • copyright 2011 • 770pp • H/C (ISBN: 9781616928575) • US $265.00 (our price)

Particle Swarm Optimization and Intelligence Advances and Applications
Konstantinos E. Parsopoulos (University of Ioannina, Greece) and Michael N. Vrahatis (University of Patras, Greece)
Information Science Reference • copyright 2010 • 328pp • H/C (ISBN: 9781615206667) • US $180.00 (our price)

www.igi-global.com

701 E. Chocolate Ave., Hershey, PA 17033
Order online at www.igi-global.com or call 717-533-8845 x100
To place a standing order for titles released in this series, contact: cust@igi-global.com
Mon-Fri 8:00 am - 5:00 pm (est) or fax 24 hours a day 717-533-8661

Table of Contents

Foreword ix

Preface xi

Acknowledgment xiii

Section 1
The Foundations of Mobile Robot Localization and Mapping

Chapter 1
Introduction 1
Chapter Guideline 1
1. Overview 1
2. Taxonomies for the Problems 6
3. Historical Overview 14
4. Organization of the Book 21

Chapter 2
Robotic Bases 28
Chapter Guideline 28
1. Introduction 28
2. Turning Machines into Robots: Actuators 30
3. How Does the World Look to a Robot? Sensors 35
4. Proprioceptive Sensors: Inertial Sensors 36
5. Exteroceptive Sensors: Contact and Very Short-Range Sensors 39
6. Exteroceptive Sensors: Single-Direction Rangefinders 40
7. Exteroceptive Sensors: Two-Dimensional Rangefinders 45
8. Exteroceptive Sensors: Three-Dimensional Range Sensors 48
9. Exteroceptive Sensors: Range-Only Sensors 49
10. Exteroceptive Sensors: Imaging Sensors 51
11. Exteroceptive Sensors: Air analysis Sensors 51
12. Environmental Sensors: Absolute Positioning Devices 52
13. Energy Supply 54

Chapter 3
Probabilistic Bases 60
Chapter Guideline 60
1. Introduction 61
2. Historical Overview 64
3. Probability Spaces 69
4. Random Variables 73
5. The Shape of Uncertainty 78
6. Summarizing Uncertainty 80
7. Multivariate Probability 87
8. Transforming Random Variables 91
9. Conditional Probability 96
10. Graphical Models 99

Chapter 4
Statistical Bases 110
Chapter Guideline 110
1. Introduction 111
2. In Between Probability and Statistics 112
3. Estimators 120
4. Properties of Estimators: Use of the Sample 120
5. Properties of Estimators: Convergence to the Actual Value(s) 123
6. Properties of Estimators: Uncertainty (Variance) of the Estimator 125
7. Constructing Estimators: Classical Estimators 126
8. Constructing Estimators: Bayesian Estimators 129
9. Estimating Dynamic Processes 132

Section 2
Mobile Robot Localization

Chapter 5
Robot Motion Models 140
Chapter Guideline 140
1. Introduction 141
2. Constant Velocity Model 143
3. Holonomic Model with a Direction and a Distance 151
4. Non-Holonomic Model with Two Wheel Encoders 156
5. Non-Holonomic Model with One Angular and One Wheel Encoder 160
6. A Black Box Uncertainty Model for Commercial Robots 166
7. An Alternative Model: The No-Motion Motion Model 169
8. Improvements of the Basic Kinematic Models 171

Chapter 6
Sensor Models 174
Chapter Guideline 174
1. Introduction 175
2. The Beam Model and Ray-Casting 176
3. Feature Sensors: Probabilistic Models 179
4. Feature Sensors: Data Association 185
5. "Map" Sensors 193

Chapter 7
Mobile Robot Localization with Recursive Bayesian Filters 203
Chapter Guideline 203
1. Introduction 204
2. Parametric Filters for Localization 207
3. Non-Parametric Filters for Localization 234

Section 3
Mapping the Environment of Mobile Robots

Chapter 8
Maps for Mobile Robots: Types and Constructions 254
Chapter Guideline 254
1. Introduction 255
2. Explicit Representations of the Spatial Environment of a Mobile Robot 256
3. Bayesian Estimation of Grid Maps 269
4. Bayesian Estimation of Landmark Maps: General Approach 274
5. Bayesian Estimation of Landmark Maps: Range-Bearing Sensors 278
6. Bayesian Estimation of Landmark Maps: Bearing-Only Sensors 280
7. Bayesian Estimation of Landmark Maps: Range-Only Sensors 286
8. Other Map Building algorithms 290

Chapter 9
The Bayesian Approach to SLAM 298
Chapter Guideline 298
1. Introduction 299
2. On-Line SLAM: The Classical EKF Solution 306
3. Full SLAM: The Basic RBPF Solution 318
4. Full SLAM: Improved RBPF Solutions 325

Chapter 10
Advanced SLAM Techniques 336
Chapter Guideline 336
1. Introduction 337
2. Estimation as an Optimization Problem: The Topology of the State-Space 341
3. Graph SLAM: Introduction 348
4. Graph SLAM: Optimizing on Manifolds 353
5. Visual SLAM with Bundle-adjustment 364
6. Towards Lifelong SLAM 379

Appendix A: Common SE(2) and SE(3) Geometric Operations 390
1. About Geometric Operations and their Notation 390
2. Operations with SE(2) Poses 394
3. Operations with SE(3) Poses 396

Appendix B: Resampling Algorithms 407
1. Review of Resampling Algorithms 407
2. Comparison of the Different Methods 410

Appendix C: Generation of Pseudo-Random Numbers 412
1. Sampling from a Uniform Distribution 413
2. Sampling from a 1-Dimensional Gaussian 414
3. Sampling from an N-Dimensional Gaussian 419

Appendix D: Manifold Maps for SO(n) and SE(n) 424
1. Operator Definitions 424
2. Lie Groups and Lie Algebras 425
3. Exponential and Logarithm Maps 428
4. Pseudo-Exponential and Pseudo-Logarithm Maps 433
5. About Derivatives of Pose Matrices 435
6. Some Useful Jacobians 437

Appendix E: Basic Calculus and algebra Concepts 443
1. Basic Matrix Algebra 443
2. The Matrix Inversion Lemma 447
3. Cholesky Decomposition 448
4. The Gaussian Canonical Form 450
5. Jacobian and Hessian of a Function 453
6. Taylor Series Expansions 455

Appendix F: Table Notation 458

Compilation of References 460

About the Authors 477

Index 478

Foreword

There are theoretical experts and experimental experts, and often little overlap between the two. Dr. Juan Antonio Fernández Madrigal and Dr. Jose Luis Blanco Claraco are prominent examples of both: they have contributed novel concepts and these concepts have been rigorously tested in extensive real world experiments. I had the distinct pleasure of meeting Jose at Oxford University in the fall of 2007. We had both just joined the Mobile Robotics Group, Jose as a visiting scientist and me as a postdoc. As ever, Oxford was packed with world leaders in robotics and vision—and, in particular, the sub-fields of structure from motion and Simultaneous Localization And Mapping (SLAM). Even among such distinguished company, Jose's contributions are impressive. In his work, one finds efficient, elegant algorithms and robust real-time systems that work on live data.

The area of simultaneous localization and mapping is vast—for decades researchers have recognized SLAM as a fundamental prerequisite to capable autonomous robotics, and have built many theories and systems towards its solution. This present volume represents a monumental undertaking and in itself testifies to the breadth of the authors' experience. It takes the reader through the highlights of the field, providing sufficient historical context and theoretical foundation for the uninitiated to engage in and master this exciting topic. The authors begin with a taxonomy, dividing the problem along axes for spatial knowledge representation, the structure and dynamics of the scene, the availability of prior knowledge, and the types of sensors and actuators the robot has. They then go on to introduce a variety of robots available on the market, their sensing and actuation capabilities, and discuss the varied tasks these platforms are designed to accomplish.

Having introduced the problem and the hardware involved, the authors then dive into tools from probability and statistics (to complement this, they also offer a rich appendix, which helps make the book stand-alone and broadly accessible). In recent years, these tools have been remarkably helpful in building principled autonomous robot systems that actually work in the real world. It has been said that computer vision is estimation theory applied to images, and that SLAM is estimation theory applied to robot sensor data. Indeed, today we find probabilistic estimation theory at the heart of most perception problems. Starting with probability theory, the authors have distilled the core mathematical foundations needed to understand the topics of autonomous localization and mapping.

The problem of SLAM is often factored into two halves: first, solve the localization problem, and then solve the mapping problem. Due to the inherent uncertainty present in any real system, such a factored approach can lead to inconsistencies in a robot internal world-model. However, from a pedagogical point of view, it is favorable to approach SLAM by first discussing localization as a separate and distinct problem. This book takes that route and uses localization to introduce motion models, sensor models, and Bayesian filtering—all core concepts needed to understand the broader picture.

The third section of this book addresses mapping. There are many kinds of "maps" out there. Some scientists will argue that any state saving machine constructs a crude map. Others will argue that the internal representation must somehow "look" like the geometry we see. Generally, the kinds of maps one builds will depend entirely on the anticipated robot task and the sensors at hand. Vision-based maps look nothing like laser-based maps, and geometric maps are different from "appearance"-based maps. Having understood the chapters on mapping, the reader will know how to apply the right mapping tool and sensor suite to robotic mapping problems they may face in the field.

The book concludes with advanced topics in SLAM and directions for future research. The authors note that the problem of long-term autonomy and lifelong learning are attracting increased attention. Robots now routinely operate without human intervention for short periods of time, and a few systems have demonstrated operation over much longer periods. The state-of-the-art in mapping and localization systems has shown convincing results on large-scale environments. Three key lessons learned by the community and discussed in this book include: 1) the importance of properly modeling uncertainty; 2) using graphical, relative manifold representations; and 3) using scalable place recognition techniques. While these lessons are valuable, there are many challenges left to solve. The final chapter crystallizes and identifies the key issues and challenges we face as robotic systems are tasked to operate in increasingly large-scale environments and over long periods of time.

The techniques and algorithms presented in this book are at the heart of mobile robot perception. The authors are both expert theoreticians and experimentalists—they have much to offer and have worked hard to make this text complete and accessible. Having mastered the material in this book, the reader will be well positioned to contribute their own experience and knowledge to the growing field of mobile robotic localization and mapping, and help usher in the era of useful, long-term autonomy for mobile robots.

Gabe Sibley
George Washington University, USA

Preface

Today, robots face a similar challenge to what occurred to many members of human societies of the First World in the last century: they are trying to make their way out from a profitable and well-known position in the industry—mainly as robotic manipulators—to land into a much more unpredictable and undefined place in the service sector where they will have to work side by side with humans; from taking the role of humans at work to live *with* humans all the time; from the nuts and bolts of mechanics to the more ethereal challenges of understanding their place in our world. Mobile robots have left behind their cousins, the manipulator arms, along this way, for our world is much more dynamic, large, and complex than anything a fixed arm could handle.

From the first prototypes resembling home appliances in the 1960s to the present commercially-available humanoids that seem to have jumped out from a *manga* TV series, mobile robots have struggled to freely move among us efficiently and safely. When we look at them today, it is not difficult to imagine how they would interact with people if they had only part of the capabilities claimed by their manufacturing companies, in how many applications they might be employed, and in all the ways they could help us in our daily lives.

The general public would probably be surprised by the actual limitations of these robots. Amazing as they look (and as they truly are, from a scientific perspective), we would do better in remembering that it was only during the last two decades that robots were endowed with the first consistent and successful theory of localization and mapping, which are the two basic operations that underlie any task we could devise for any practical robot: knowing *where it is* within its environment and figuring out what *that environment looks like*. Today, these two fundamental problems cannot be considered to be completely solved for every practical situation yet, in spite of the remarkable scientific corpus developed around them. This book aims at introducing that corpus to the reader. More concretely, we focus on mobile robot localization and mapping approaches that rely on the theory of probability and statistics.

The theory involved in probabilistic localization and mapping methods can become quite cumbersome, in accordance with the importance and quality of the obtained results. Books and papers exploring those complexities are easy to find, but they may be difficult to grasp for those who are not active researchers in the area and do not have a solid background in mathematics. Furthermore, most of the material is quite scattered among journals, books, and conference papers, and in many occasions is addressed from the diverse—and often confusing—terminologies of very different disciplines. Since mobile robots have begun to get out of research labs and into the hands of the general public, we believe it is now time to offer a comprehensive introduction to these subjects that is appropriate for a wider audience than traditional scientific literature, and that gathers in a single place the fundamental concepts needed for fully understanding the problems, whatever area of science they come from.

From the perspective of two authors with many years of experience researching and teaching in this field, we have aimed this goal in the gentlest possible way, while still doing it rigorously. In particular, we have focused on three aspects: firstly, on explaining and justifying most deductions that are involved in the relevant parts of the theory, including step-by-step demonstrations that are typically obviated in specialized literature; secondly, on including the probabilistic, statistical, and robotic bases that other texts take for granted—even after saying otherwise; and thirdly, on providing a glimpse of the historical development of the covered theories and methods, not intending to offer an exhaustive historical timeline but a sufficient background. Our purpose is that the interested reader can really understand the treated issues in scope and depth, instead of just presenting powerful and sophisticated mathematical tools with obscure inner workings.

The book has been designed to be useful for practitioners, graduate and postgraduate students, and researchers mostly interested in a reference guide. No previous knowledge on probability and statistics is required—although it would speed up the reading, since two entire chapters are devoted to providing that background! Also, the prerequisites in physics, calculus, and algebra have been kept to the necessary minimum; alas, self-containment is just an ideal in any finite work these days. Thus, we have had to assume that the reader has the most elemental knowledge of those three disciplines—we provide, in Appendix E, some reinforcement on concepts that are especially important for the understanding of the problems.

This book is structured in three sections. The one that possibly makes this text more distinctive in its kind is section 1, which collects for the reader the robotic, probabilistic, and statistical backgrounds required for a good comprehension of the rest. Sections 2 and 3 follow the logical development of the main problems addressed in the book: localization and mapping, respectively. This organization is intended for both a sequential reading and for an easy selection of material for reference or teaching.

The first idea about writing this book came from the class notes by the first author for a postgraduate course on mobile robotics. Their main contents, and therefore a substantial number of concepts and explanations currently in the book, have been used for that purpose during several years; they should also be amenable for teaching in more introductory courses. In this use, a professor could choose to drop the first part if the mathematical background is assumed for the students, something that will depend on the academic context of the subject. The book also introduces some advanced issues in Simultaneous Localization And Mapping (SLAM) and many recent developments, mainly coming from the experience and continuous work of the second author during his PhD thesis and beyond.

Overall, we expect our book to serve as the starting point of a fascinating journey into this field, by setting the foundations of further detailed and thorough studies. Working in probabilistic robotics can certainly be tough, but we can assure you—this much we know—that it can be highly rewarding too. Our ultimate hope is that this text provides you with most of the tools needed to open a well-marked track into the jungle of probabilistic localization and mapping.

Juan-Antonio Fernández-Madrigal
Universidad de Málaga, Spain

José Luis Blanco Claraco
Universidad de Málaga, Spain

April 2012

Acknowledgment

The present book is the result of several years of continuous work, not only directly on the text, but also on the classes on probabilistic localization and mapping that gave rise to it, on continuing with our research in the area, and on documentation about topics not related to its main corpus, such as the history of mathematics or navigation. All of this has been benefited by the advice and aid of our supporters, colleagues, and students, but the final result would have not been possible at all without the intervention of the editors at IGI Global, the anonymous reviewers of the book, and other people who kindly offered to review early versions—we must thank especially Francisco Ángel Moreno and Eduardo Fernández for that.

The first author wishes to particularly thank several groups of students of the Master on Mechatronics Engineering of the University of Málaga, coordinated by Prof. Alfonso García-Cerezo, who provided continuous feedback to the class notes and lessons on which part of this book is based. Without the insightful and motivational questions and comments of Juan Carlos Aznar, Mariano Jaimez, Ángel Martínez, Antonio Menchero, Andrés San Millán, and many others, he would be less confident in the didactic value of relevant chapters of the book. He cannot forget either the very first students that were exposed to our stuff, especially Eduardo Fernández, Ana Gago, Javier G. Monroy, and Raúl Ruiz, who not only suffered beta versions of parts of the text, but, even now, continue actively doing research on these topics, which means that the probability that they liked the experience was strictly greater than zero.

Concerning the particular case of the methods for localization and mapping implemented by the first author on the *LEGO™ Mindstorms NXT* robots, Dr. Ana Cruz has had an invaluable role due to her pioneering efforts on the use of this robotic platform for educational purposes. Some results shown in this book have been obtained with the robots she has funded through the 2008-2010 educational research project entitled "Innovation in Engineering Control Subjects through Lego Mindstorms NXT Robots" (code PIE-008), through the *Escuela Superior de Ingeniería Informática,* and also through the System Engineering and Automation Dpt., all of them in the University of Málaga.

The second author would like to express his gratitude to the numerous researchers whom he had the luck of meeting in conferences and workshops all over the world, not only for the fun moments, but also for the inspiring talks and discussions which have always had the same effect: a continuous renewal of his motivation for continuing working hard in this exciting area. In particular, he wishes to thank Dr. Paul Newman for supervising his visit to his research lab in Oxford, an experience that enriched and widened the author's perspectives on many technical and theoretical aspects of mobile robotics. Gabe Sibley deserves a double special mention here: first, for kindly writing the foreword of the book, and second, for his suggestions that put the author on the "right track" of looking at many estimation problems in robotics as sparse, least-squares problems.

In a more practical context, he wants to thank all the researchers, from our lab in Málaga or elsewhere, who have contributed to the *Mobile Robot Programming Toolkit (MRPT)* in one way or another, either coding or providing patches and bug reports. They all have helped improve the reliability of a tool which has proven invaluable during the preparation of many graphs and results presented in this text. Special thanks go to Antonio J. Ortiz de Galisteo for his enthusiastic work in the early versions of MRPT and to Pablo Moreno Olalla for his gigantic contributions to the mathematical and geometry modules, from which some equations of Appendix A have been taken.

Both of us have developed most of our research career within the Machine Perception and Intelligence Robotics group (MAPIR), which has proven to be a fertile context for invaluable discussions and feedback on the topics at hand and, at the same time, has provided us with diverse perspectives for each problem, ranging from computer perception to artificial cognition, which have permeated our personal visions over the years. We both wish to thank the group's permanent members, Dr. Vicente Arévalo, Dr. Ana Cruz, and Dr. Cipriano Galindo, all the PhD students of the group, and also our guest researchers, who have contributed to our work in invaluable ways, particularly Prof. Alessandro Saffiotti, Assoc. Prof. Achim Lilienthal, and Assoc. Prof. Amy Loufti, from the AASS Research Center of the Örebro University (Sweden). Special thanks must go at this point to Prof. Javier González-Jiménez, the efficient lead researcher of the group; furthermore, he has been PhD advisor for both authors; thus, without his trust and constancy, we would not have started our research careers at all.

The authors also wish to thank the public institutions that contributed funding, especially the *Junta de Andalucía* (regional government), which, through a research project, allowed the first author to extend his research on probabilistic robotics, in spite of not being directly related to localization and mapping, and also allowed the second author to work on gas mapping for mobile robots. Both lines of study have had relevant benefits to this book. In addition, several national research projects funded by the Spanish Government and European research projects funded by the EU have provided invaluable support during all these years.

Some of the images that illustrate the text are from a number of researchers and companies who have all willingly granted us permissions for their inclusion here. Therefore, we sincerely thank all the owners for their unselfish contributions (and also the father- and mother-in-law of the first author, who obtained, cleaned, and photographed two astragali for us!). Likewise, our gratitude goes to those researchers who publicly released robotic datasets or source code of their own works, since such contents have also helped enrich the book with more demonstrations of the practical utility of the discussed topics.

Last but not least, our thanks must go to our families. They always helped and motivated us throughout all the years as students and, later on, during the tough (and satisfactory) times in our academic careers. Our warmest thank you is for our wives, Ana and María, who have both unconditionally supported us during the uncountable hours of preparation of this book without the least complaint.

Section 1
The Foundations of Mobile Robot Localization and Mapping

Chapter 1
Introduction

ABSTRACT

In this first chapter of the book, the authors provide an overview of the problems of mobile robot localization and mapping, including taxonomies over the axes of the representation of spatial knowledge, the structure and dynamics of the environment, the sensory apparatus of the robot, the motor apparatus of the robot, and the previous knowledge. They also provide a brief historical timeline and fundamental concepts. The goal is to provide the reader with a roadmap of the problems and also of the book, in order to allow her or him to choose the best way of approaching the text and also an appropriate understanding of the main limitations of the existing methods.

CHAPTER GUIDELINE

- You will learn:
    - The reasons why robotic manipulators have been much more successful in practical applications than mobile robots up to now, and the role of the problems of localization and mapping in this distinction.
    - A little on the history of mobile robotics.
    - How the rest of the book is organized and how to employ it depending on your purposes.
- Provided tools:
    - A taxonomy for the classification of the problems of localization and mapping, aimed at identifying the adequate approach for each situation.
- Relation to other chapters:
    - This chapter provides an overview of all subsequent chapters and their organization.

1. OVERVIEW

Robotics is a complex and fascinating discipline, not only for scientists. The idea of creating artifacts that perform autonomously and intelligently has appeared from time to time throughout the history of humankind in a variety of ways, ranging from the fearsome mythological form of the Golem of the early Judaism (Idel, 1990) to the inoffensive and very amusing automata of Jacques de Vaucanson in the 18th century (Landes, 2007), until it was definitely solidified in the last six decades of our era. In the transition from the 19th to the 20th century, science-fiction literature was already a

DOI: 10.4018/978-1-4666-2104-6.ch001

popular melting pot for these kinds of scientific and technologic fascinations, and therefore had a particular role in our modern conception of the discipline. In 1921, the Czech writer Karel Capek made use of the suggestion of his brother Josef to call "robota" the imaginary human-like creatures of his play "R.U.R. Rossum's Universal Robots" (Capek, 1921)—the word "robota" stands for "serf labor" in a few Eastern European languages. Twenty years later, Isaac Asimov introduced the term "robotics" in one of his science-fiction stories (Asimov, 1941) and developed in a number of further stories and novels many logical questions that rational robotic slaves would pose when working among humans. That seminal meaning for "robot," that is, a machine that imitates actions performed by a living being, particularly the hard work that people do not like or want to do, has survived through history without much change: virtually all the robots created by mankind, including the fictional ones that have become part of our collective memory across the centuries, have contributed—in reality or imaginarily—to the relief of some portion of our physical work.

However, the robotic inventions that have actually contributed the most to that relief are far from having human aspect, unlike in most fictional stories. Instead, mechanical arms fixed at some location, the so called *robotic manipulators* (Niku, 2010), have been the kings of all the real robotic creatures—see Figure 1. From the first arms built around the middle of the 20th century (Devol, 1961;

Figure 1. A programmable universal machine for assembly (PUMA manipulator arm), used for virtual reality research at NASA. This industrial robot was developed in 1978 by Unimation and General Motors based on the designs of Victor Scheinman, the creator of the Stanford Arm (NASA Ames Research Center, Mountain View, California).

Scheinman, 1969) to parallel manipulators or the ones used recently in space applications (Stoll, Letschnik, Walter, Artigas, Kremer, Preusche, & Hirzinger, 2009; Mamen, 2003), these robots have effectively substituted repetitive and hazardous human labor. They have been employed mainly in the industry, where, due to their flexibility for adaptation to variable production conditions with a relative low cost have been one of the keys for achieving the current development of modern societies.

Robotic manipulators became a success in such a short time not only for their utility, but also for the tractability of their design (Craig, 1989; Selig, 1992; Sandler, 1999; Kurfess, 2005), which follows from some implicit, but important, assumptions:

- Power sources do not significantly limit their morphology, behavior, or autonomy: a robotic arm can assume an unlimited supply of energy at every moment, disregarding unavoidable power shortages or other sporadic incidents external to the robotic system.
- Their working environment is basically invariant, of very limited extension, and known in advanced by the robot designers. Therefore, a nearly complete planning of the activities of the arm can be carried out before the robot moves into action.
- Not only the environment is static and small, but also the physical properties of the objects that the robotic arm has to manipulate are known in detail before operation; that is necessary in order to include the robot as part of an efficient production system. In addition, there are restrictions in the possible spatial location of those objects with respect to the robot.
- The state space of the arm—the angular position of all its joints plus their corresponding dynamics—is restricted to a bounded volume.

In contrast, designing a robot that moves on its own throughout the world while executing some orders (see Figure 2) is a completely different story (Cox, Wilfong, & Lozano-Pérez, 1990; Siegwart & Nourbakhsh, 2004). Problems multi-

Figure 2. Some of the mobile robots developed and used in the machine perception and intelligent robotics group (MAPIR) of the system engineering and automation Dpt. of the University of Málaga, where the authors have implemented a number of systems for localization and mapping, among other robotic problems.

ply in number and complexity, quickly becoming mathematically intractable. To begin with, we have the fundamental issue of power supply. A mobile robot, just like fixed manipulators, should be able of performing autonomously (i.e., without external intervention) during long periods of time operation; but these robots has to carry their own power source on board, and any mobile platform has restrictions regarding the maximum weight of its payload. Closing the loop, power can only be increased at the expense of larger and heavier power supplies. This issue is analogous to the one that limits the size and autonomy of space ships—which led to the invention of multistage rockets—or, for the matter, the one confronted by all kinds of automobiles. It is not the only chicken or the egg dilemma that we will find within mobile robotics. If you have thought of the robot going periodically and autonomously to a recharging station (i.e., finding it, approaching it, using it), please read the following issues.

A mobile robot must, by definition, change its location in order to perform it tasks, which tears down most of the other benefits enjoyed by robotic manipulators. Free motion makes the current environment of the robot vary, even largely, from one moment to another. Occasionally, engineers can build a model of the whole workspace, but in most practical applications, it is common having to face at least partially unknown regions. It is difficult to plan any complex sequence of actions with limited knowledge of their effects in all the particular scenarios that the robot may encounter: possibilities explode exponentially with time, and thus planning must be stopped regularly to perceive, analyze, and correct the results. Sensors become crucial devices for mobile robots precisely for that reason. The process of transforming sensory data into useful information, i.e., *perception*, arises then with all its complexity.

Motion also leads to the possibility of operating in spaces that are potentially unlimited in size (that is one of the main advantages of mobile robots, after all!). The issues that can arise from this are obvious for a programmable machine that has finite memory and computational capabilities—which, again, can only be increased at the expense of more size and weight, feeding again the beast of the power supply. Therefore, approximations may have to be done even for problems otherwise solvable. Furthermore, aside from computational limitations, we find the issue of merging all the information gathered at different locations of a large environment in order to get a fairly complete representation of it, which is a prerequisite for planning future tasks.

Finally, the objects, people and other robots or machines that a mobile robot is to interact with during its performance can also be considered as part of its environment, and as such they have a relevant influence in not having a complete, static, small model of the physical world. Unexpected obstacles with arbitrary shapes and unknown dynamics are one of the common problems that a real mobile robot will have to face without much previous knowledge. If those obstacles are intelligent beings, understanding and predicting their behavior becomes especially challenging.

In summary, the effects of including the capability of free motion in a robot lead to an important increase in dimensionality, size, and complexity of nearly all the issues of robotics at very different levels, from mechanics to electronics to computer science to artificial intelligence. Most of them have no satisfactory solution yet—albeit many proposals—, and that is the reason why today we do not see mobile robots everywhere in the real world, especially operating reliably during long periods of time. It is also the reason why mobile robotics remains such a fascinating discipline. It will probably be for quite a while.

In spite of all these difficulties, there is one basic problem of mobile robotics that has reached a reasonable maturity in the last decades, up to the point of being considered satisfactorily solved in many practical situations. It has to do with mobile robot navigation, that is, taking the robot from one point to another. Maybe the reader who is unfamiliar with mobile robotics is surprised by the apparent triviality of the problem. Six decades just to make the robot go well from one place to another, the most basic of the tasks of a *mobile* robot? For now, we hope that the previous rationale, besides the fact that a book like this one is needed just to introduce two of the main basic aspects of navigation, may convince him or her that solving this problem is actually a big deal. Nevertheless, robotics is a highly active area of research, and, apart from these problems, many other advances are succeeding concerning other issues (Siciliano & Khatib, 2005).

Let us focus on navigation then. It is obvious that a mobile robot must be able to navigate, that is, to change its location autonomously, since it performs operations that are distributed over space. It is also obvious, and much less striking at first sight, that in order to move to a target position the robot needs to figure out the spatial relation existing between that target and its current location. That target-robot relation is of the uttermost importance: it is needed for deciding actions, for example the motion of the robot actuators (Krzysztof, 2006) that reduce the distance to the target. Being able to estimate that relation is a basic, unavoidable step; it will be required for any navigation process.

The target-robot relation can be expressed in a variety of forms, mainly quantitative (geometrical, e.g. two points in a Euclidean coordinate reference system) or qualitative (a set of logical relations between places, such as "to the right of" or "farther than"), even in a mixture of both. How it is represented defines in turn the methods the robot may use to navigate. Scientific literature on mobile robot navigation algorithms has been quite prolific since the 1980s (Lozano-Pérez & Wesley, 1979; Thorpe, 1984; Crowley, 1985; Zheng & Zheng, 1994; Borenstein, Everett, & Feng, 1996; Siegwart & Nourbakhsh, 2004; Minguez & Montano, 2008; Blanco, González, & Fernández-Madrigal, 2008), but we are especially interested here just in the mentioned aspect: the relation existing between the target and the robot, and how it is acquired in order to navigate, rather than in the manner of performing that navigation.

Estimating the target-robot relation in a previously unmodeled environment can become a tough problem, which is clear when we realize that it involves handling sensory data reflecting unknown elements of the environment, as seen from unknown positions. It is worth mentioning that a robot can navigate without any representation—implicit or explicit—of the environment, but in practice such an approach limits the navigation autonomy to only small distances: sooner or later, the robot finds difficulties locating the target or itself since errors accumulate if it tracks its current position solely through odometry (Borenstein, 1998). Thus, at some point mobile robots have to use information from their environment to navigate.

And that is how we arrive at the main issues we are interested in: using *known* environmental information, that is, relations existing among environment elements, for estimating the actual position of the robot is the problem called *localization* in mobile robotics—less popular in the specialized literature is the term *positioning*—to which the second part of this book is entirely dedicated; it is one of the aspects of navigation that we labeled above as "satisfactorily solved in practical situations." Today, most mobile robots can localize themselves and their targets automatically and accurately as long as enough environment and sensory information is available, although the mathematical tools used in the problem are far from simple—that, in turn, is the reason for the size of the first part of the book: we need to provide the reader with those foundations.

On the other hand, estimating the *unknown* spatial relations that exist between environment elements in order to build an internal representation suitable for robot navigation is the problem called *mapping*. The third and last part of the book addresses this problem, which can also be considered as satisfactorily solved in some practical situations as the localization problem, something that we could have not claimed so categorically just fifteen years ago.

Now consider the following fact: if the robot has to estimate the direct relation existing between its current location and the target and uses environmental information for that (a bunch of spatial relations existing between elements of the workspace), that information must be known beforehand. However, in contrast, if the environmental information is to be estimated, it has to be with respect to the robot location (i.e., from the robot "subjective" point of view), thus the robot location must be calculated before building the map. We already warned the reader that there was more than one chicken or the egg problem in mobile robotics.

One solution to this particular one is to estimate concurrently both the map and the location of the robot. This is called *simultaneous localization and mapping*—also *concurrent mapping and localization*—, widely known pin the scientific community by the SLAM acronym. Chapter 8 in the last part of this book deals with the problem of building maps assuming that localization has been solved somehow. Subsequent chapters address SLAM. Naturally, SLAM is a much more difficult problem than localization, and the techniques devised up to now are not as universal or efficient as those applicable when we deal with localization and mapping separately. In particular, computational efficiency is still an issue on which many efforts are currently focused.

Due to the complexity and diversity of the methods available today for solving mobile robot localization and mapping, we firstly set up a particularly broad taxonomy in section 2, as a road map that clarifies why so many approaches exist and which factors we have to take into account for choosing one. For completeness, section 3 develops a brief historical overview of both problems. This introductory chapter will end with section 4, which provides the reader with a brief summary of the rest of the book.

2. TAXONOMIES FOR THE PROBLEMS

There are many methods available today to cope with robotic localization, mapping, and SLAM. Even if we focus only on those approaches based on probability theory and statistics (the ones explained in this text), the diversity of solutions is quite important. That is due to a number of factors: different ways of achieving computational efficiency, available sensors and actuators, characteristics of the environment, number of robots that work together, etc. Thus, it is convenient to put some order in the big picture before studying the details. We will examine the problems from five points of view in this section, summarized in Table 1.

The Representation of Spatial Knowledge

We have mentioned this perspective previously: the spatial relations involved in our problems can be stated quantitatively, qualitatively or in a mixture of both flavors.

Quantitative localization and mapping becomes the problem of determining the geometrical robot *pose* and constructing geometrical maps of the environment. A *pose*, a word related to *posture*, is a combined representation of the position and orientation of an object in space—one of its first uses in mobile robotics was in Moravec (1980). Often, robot poses are defined with two real numbers, the x and y coordinates of the center of motion of the robot in a given reference frame, for workspaces where the vertical dimension is irrelevant, plus one more real value for the orientation: an angle with respect to some of the reference axes, which indicates the robot heading. Regarding mapping, there is a variety of geometrical maps, which can represent empty regions of space, obstacles, poses of distinctive elements, etc. We will study the most important quantitative and qualitative representations for maps in chapter 8.

Table 1. Main axes of a broad taxonomy for the problems of mobile robot localization and mapping

Axis	Possible Values	Rationale
Representation of spatial knowledge	Quantitative (metrical) / Qualitative (topological) / Hybrid	Quantitative representations can be accurate and practical for small environments; qualitative are suitable for artificial cognition and large-scale scenarios.
Structure and dynamics of the environment	Structured / Unstructured	Structured environments are easier to deal with; unstructured environments are harder but a source of uncountable practical applications for mobile robots.
	Modified / Unmodified	When possible, modifying the environment before the robot operates in it improves localization and mapping.
	Dynamic / Static	Most localization and mapping methods can deal with (mostly) static environments only.
Sensory apparatus of the robot	Proprioceptive / Exteroceptive / Environmental	The robot can sense its own motion (cheap but autistic), its motion with respect to the external environment (it needs to process complex data), or obtain the motion observed by an environmental sensor (easy to use but usually out of its control), respectively.
Motor apparatus of the robot	Passive / Active	Most localization and mapping methods decouple motion from estimation, that is, perform passive estimation of location and/ or map without making motion decisions; by coupling both of them, i.e., by performing actions that affect sensor perceptions actively, the problems increase their difficulty but we can obtain better results, for instance, a quicker convergence to a good estimation.
	Off-line processing / On-line processing	Processing motion and perceptual data off-line permits the ability to obtain better estimates of both location and maps; processing data on-line enables the active paradigm but reduces accuracy.
Previous knowledge of the robot	Tracking / Awakening / Kidnapping / SLAM	Probabilistic methods permit the inclusion of previous knowledge in the estimation process. In the localization problem we can know where we are initially (tracking), we can be wrong about that (kidnapping), or may not have any information on location at all (awakening). If we do not have any previous information on both map and location we are confronted with the SLAM problem.

Metrical—the most common adjective in robotics—localization and mapping is required for machines to operate accurately, although it is not so useful for providing information to humans. Actually, it seems that we, human beings, do not rely on these kinds of representations as the basis of our *cognitive maps*, the "body of knowledge of a large-scale environment that is acquired by integrating observations gathered over time and is used to plan routes and determine the relative positions of places" (Kuipers, 1983). In spite of that, most modern practical solutions to mobile robot localization and mapping, including many large-scale approaches, are purely geometrical, and robotic engineers has achieved great success under this perspective in the last decades.

Qualitatively speaking, our localization and mapping problems look quite different. They consist of determining spatial relations that are insensitive to metrical deformations, i.e., elastic. A well-known form of such representation of the environment in mobile robotics is the so-called *topological map*: a graph with nodes representing places/objects/etc. and arcs representing relations—one of the seminal forms of the term in mobile robotics is the *topological network of paths and places* by Kuipers (1977). There are few existing solutions for purely topological localization or SLAM in mobile robotics, and many were explored before the main breakthrough in probabilistic, metrical SLAM—an exception is the modern probabilistic approach to pure topological mapping in Ranganathan, Menegatti, and Dellaert (2006). However, there has been a growing interest in the last years in studying topological approaches again, now that the metrical ones are mature. This follows from the suitability of qualitative representations for mimicking cognitive processes in computational machines (categorization, deduction, decision making, etc.) and for interacting with humans, a task more difficult to address with purely geometrical settings (Jefferies & Yeap, 2008). Notice that the two perspectives, quantitative and qualitative, have a clear duality: qualitative representations seem to be the basic building block when a human learns the structure of space, and are not too accurate—but serve well for us to perform operations—while the opposite holds for quantitative approaches.

An obvious way of taking advantage of both the qualitative and quantitative perspectives is to mix topological representations with geometrical data, not necessarily in that order. Actually, the so-called *hybrid metrical-topological maps* have become one of the keys to solving large-scale scenarios in SLAM efficiently (Blanco, 2009), where the growth of dimensionality and memory requirements of the problem render pure quantitative methods as inappropriate. In addition, promising results of linking other types of qualitative information, such as semantics, to traditional ones have been reported in recent scientific literature (Galindo, Fernández-Madrigal, González-Jiménez, & Saffiotti, 2008). Such a high-level hybridization allows the robot to cope with tasks that not only require some kind of artificial cognition (e.g., mission planning) but also accurate spatial operation (e.g., avoiding obstacles), being also an interesting way of filling the gap still existing between "pure" robotics and "pure" artificial intelligence.

The Structure and Dynamics of the Environment

Apart from the way of representing spatial relations, quantitatively or qualitatively, we can distinguish different kinds of environments, and this has an important influence on the methods that are available to address localization and mapping and also on the difficulty of these problems. The main axes that we can set across this point of view are: structured/unstructured, modified/unmodified, and dynamic/static—see Figure 3.

First of all, the fixed parts of the robot workspace—those that we usually represent in a map—may be composed of mostly regular and big planes, like in human indoor environments, or, on the contrary, may have irregular and small surfaces, like most outdoor scenarios. The former, called *structured* scenarios in mobile robotics—since at least the mid 1980s (Waldron, 1985)—present important advantages to localization and mapping, due to different reasons:

- Each part of a structured environment that is to be represented in the map can be distinguished by most robotic sensors:

Figure 3. Some of our mobile robots confronting different environments. (a) Roadbot, an outdoor robotic vehicle, whose development has been supported by the company SACYR S.A.U. for building accurate 3D reconstructions of the surface of roads. (b) The robotic wheelchair SENA, moving in a dynamic environment cluttered by people. (c) Picasso, a mobile robot, navigating in a structured, indoor, static environment before it unfortunately died in a fire. (d) The service robot Sancho demonstrates its capabilities and gives a speech in a public event celebrated in the Science Park of Granada (Spain).

frontiers are clean-cut and sharp, while visible surfaces have a uniform aspect, which produces less variations in the data gathered by sensors. Also, those elements that people need to detect—walls, doors, switches, etc.—are usually colored, textured or shaped to facilitate their detection. Consequently, the different elements of structured environments are easier to recognize by a localization/mapping robot than the ones of unstructured spaces.

- The possible spatial relations existing between elements of a structured environment are highly constrained—for example, walls in buildings are typically all perpendicular to each other. This information can be used by the robot to reduce the computational cost and to increase the accuracy and certainty of its estimates on the environment structure.
- There exist few possible different shapes, usually polygons of no more than four

sides. Again, the automatic detection and recognition of objects with those characteristics is much more tractable than a more general problem of artificial perception.

- The vertical dimension of the space can be safely ignored in most cases, since fixed surfaces extend orthogonally and without variation in their shape from the floor to the ceiling. This reduces the dimensionality, and therefore the computational cost, of localization and, particularly, of mapping.
- We can assume for certain that there is enough free space for robot motion in all the areas of interest for the application, and that the floor is flat enough for the robot to move without great effort—remember the power supply dilemma mentioned at the beginning of this chapter. This is very useful for motion planning and navigation tasks.

However, structured environments are not exempt from problems:

- The robot has to deal with ambiguities: there may be portions of the environment that look similar, such as corridors or offices. This would cause important issues in SLAM, especially when getting back to some previously visited place: how to distinguish it, using only data gathered locally by sensors, from a hypothetical new place that might look the same?
- Structured environments are designed and built by and for human beings, and people do not have the same way of interacting with the environment as robots. For a mobile robot, it becomes a challenge to open a door using its handle, to call an elevator, or to come down the stairs. Regarding perception, most common sensors detect distances to obstacles or light intensity and color, not the shapes of walls and of other complex objects. Sensory rich sensors like cameras provide so much raw information that interpreting it accurately in real-time is quite a challenge—even assuming that the problem of computer vision was solved!
- Indoor spaces are densely populated by people, which in case the robot has no human-machine interaction capabilities, become mobile parts of the environment of rather limited utility for localization and mapping. Therefore, our shapes must be detected and filtered out from those processes, which is far from being trivial. Most localization and mapping methods deal with this issue by considering people, as well as other mobile and unpredictable objects, as part of the "noise" that comes "embedded" into the data gathered by sensors. That works surprisingly well as long as the disturbances represent only a small fraction of the magnitudes of the sensory input. In the case that the presence of people exceeds the possibilities of the robot to deal with them as noise, mapping and localization can become almost impossible. That kind of environment is not rare (imagine a museum, or an office building at rush hour), and they are called *cluttered* environments—the term appears as early as in Moravec (1980). Obviously, cluttered environments require specialized approaches.
- Finally, the simplification of the vertical dimension is not always possible (buildings have several floors and the robot may not be confined to one), but taking it into account in every moment leads to the mentioned increase in computational cost, without being useful most of the time.

Most of these disadvantages are also present in unstructured environments. Outdoor scenarios are composed of many elements that result ambiguous to a robot perception system: trees, rocks, the sea floor. Obviously, unstructured environments have not been formed to serve any robot (or any human), and thus a machine will have difficulties reacting in and perceiving them. Of course, the vertical dimension is a must in unstructured environments. Actually, all the problems explained with regard to structured environments can render worse in the unstructured case; furthermore, the advantages that we have also mentioned about the former are mostly absent in the latter. The conclusion is that unstructured spaces are still a huge challenge for any mobile robot.

In some applications, it is possible to force, up to a certain degree, the "structuredness" of the environment. For instance, we may include artificial landmarks for facilitating the mapping or localization processes, or set up signaling devices that communicate their relative positions to the robot, or even limit the presence of people or other undesirable obstacles in the robot operation areas. These *modified* environments (or *engineered* environments) have several advantages:

- They allow the mobile robot to cope with tasks otherwise intractable.
- They may reduce (importantly) the cost of implementing the system.
- They increase safety and accuracy during operation.

Regretfully, there are many applications that need to keep the workspace unmodified, for instance those intended for unlimited, large sized or natural spaces. The typical engineered environment can be found in industrial factories.

The third axis to study in this section has to do with the dynamics of the environment. Most methods for mobile robot localization and mapping are designed to work with static workspaces: those with elements that do not vary in shape or position over time. Strictly speaking, almost no practical case involves a completely static environment: there will always be objects or persons moving around. As we commented above, a common approximation to that situation is to consider dynamic elements as noise in the perceptual data. When this is not possible, specialized methods for detecting or ignoring these elements will have to be employed. These include sophisticated recognition algorithms, dynamic models of the moving objects and prediction techniques; mobile robot localization and mapping in highly dynamic environments is still an intense research area, on which we will not focus in this text—consult, for instance, (Fox, Burgard, & Thrun, 1999; Andreasson, Treptow, & Ducket, 2007; Zhao, Chiba, Shibasaki, Shao, Cui, & Zha, 2009).

The Sensory Apparatus of the Robot

Sensors are the entry point for information on the robot environment, thus they determine in many aspects the available methods for localization and mapping and the quality of their results. Since part of chapter 2 will be devoted to robotic sensors, we will describe this perspective here broadly.

Let us distinguish three classes of sensors that the mobile robot can have access to, namely *proprioceptive*, *exteroceptive*, and *environmental*.

Proprioceptive sensors are the ones that measure changes occurring *within* the robot itself. A paradigmatic analogue in humans would be the vestibular system of our inner ear. In the case of a mobile platform, they mainly sense the motion of the robot body: accelerations, velocities, and changes in orientation. The most common and classical ones are odometers, which estimate the accumulated displacement of the robot from a previous location by just considering the movements of its actuators. Other popular devices in this category are inertial sensors, which measure instantaneous accelerations and rotational velocities. Proprioceptive sensors, in general, are subject to noise, which in addition to the incremental nature of their measurements leads to the accumulation of large errors in the midterm. These errors must be fixed whenever information from the outside is available; otherwise, the robot will be able to track its own movements only for short paths (the success of the Wii game console in spite of its cheap inertial sensors is due, technically, to the fact that they do not have to deal with long accumulated movements). Today, MEMS technology (MicroElectroMechanical Systems) is providing encapsulated sensors massively and at a low cost (Liu, 2011), thus several of these devices can be easily installed throughout a robot body to instrument it.

Exteroceptive sensors are also located in the robot, but, unlike the formers, acquire data about events occurring *outside* the robot. They can be used to reset the accumulated errors of proprioceptive sensors, although can also be used alone. An immensely popular family of such sensors in the context of this book are range-bearing sensors—usually based on the time of flight of ultrasonic or light waves—which measure the distance and orientation to solid obstacles, as we will see in chapter 2. They are suitable for building maps and for locating the robot in them, that is, to estimate the spatial relationships that we discussed in section 1. Different models of these sensors can widely vary in their static parameters: resolution, operating range, sensitivity, etc. The type of environment has an important influence on their choice: neither a laser scanner is appropriate when there are glass walls, nor ultrasound devices to detect very elastic or flexible obstacles. Budget limitation is another relevant factor: a laser scanner is more expensive than an ultrasonic sensor but provides way higher accuracy. Cameras are the other most popular exteroceptive sensor in current mobile robotics for SLAM, although they demand a much higher computational cost than range-bearing devices.

Finally, we can also include in our taxonomy those sensors that are not located in the robot body, what we have called environmental sensors. They provide the robot with spatial information, typically information for localization, which is *gathered actively from the outside*. For example, active beacons placed at fixed, well-known points of the workspace, which can serve to locate the robot through triangulation; also Global Navigation Satellite Systems (GNSS), of which GPS is the best current exponent, or any other device that can provide the robot with clues about its location. In the last years, there has also been an increasing interest in using communication signals for localization, due to their important presence in modern structured environments (e.g., Wifi access points). Notice that environmental sensors imply modified/engineered environments, something that is not always possible but, if possible, may be well worth due to the gains in localization and mapping.

A last few words on the sensory capabilities of a mobile robot. It turns out that it is possible to increase them by adding a special class of environmental sensors that we indeed control, without engineering the environment: using more than one robot. If a team of mobile robots includes communications, we have actually a "larger" robot with an enhanced perception apparatus, which can be useful not only to access several regions of the same environment faster (and maybe better if the sensory apparatus of each robot is highly heterogeneous), but also to improve the localization of a given robot by using the perceptions of the others. They become, in this particular sense, *environmental moving* sensors, with the advantage of being under the control of the multirobot system: they can be located at will in interesting positions in order to improve the overall localization and mapping process. Nevertheless, designing a multirobot architecture has its own issues, mainly concerning coordination and information fusion, that are outside of the scope of this book—you may start with a glance to Schultz and Parker (2010).

The Motor Apparatus of the Robot

In addition to sensors, the actuators of the robot constitute an important factor in localization and mapping. Actually, a mobile robot can completely decouple its actuation from its perception, which is, as explained before in this chapter, a common approach found in the localization and mapping literature. This separation allows us to examine the sensory data collected from an arbitrary robot navigation, even to process them off-line, in which case the data is what we call a *robotic data set* for localization and mapping. In this way, we can employ a much higher computational power than that available on the robot to obtain high quality estimates of the map and of the robot poses over time. This action-perception decoupling is a form of reductionism, and reductionism has led to brilliant discoveries in the history of science.

Of course, all robotic applications are not suitable for examining and processing sensory information *after* the robot navigates. For these cases, there exist outstanding methods to carry out localization and mapping in real-time, that is, on-line. In this regard, particle filters (to be introduced in chapter 7), have interesting characteristics to dynamically adapt their complexity to the available computational resources. However, you will see later that many on-line processes also work under a reductionist approach.

Neither on-line nor off-line methods are usually designed for taking advantage of the actuation capabilities of the robot, even when the utility of interlacing both perception and actuation seems obvious: a robot that decides its movements not only for reaching a desired target but for acquiring better perceptions or reducing its motion uncertainty can improve the results of its localization and mapping processes. For example, if the robot has a large uncertainty about its location and faces the possibility of either exploring an unknown area or going back to a previously visited space, it can obtain a greater improvement in localization and mapping by returning than by entering the unknown. The same is valid if it can decide to go either to a place with many distinctive elements or to a uniform or ambiguous region. For the case of sensors with a somewhat reduced field of view, like cameras, the decision on visiting one place or another may become an issue of at what direction to head the camera next, even if the robot itself does not move—this is the basic idea behind *active vision* techniques (Davison, 1998). Clearly, some motions imply a greater benefit for localization and mapping than others, therefore it can be really useful to couple motion/path planning with the SLAM process in order to optimize the results, as long as that is compatible with the rest of goals of the robot.

Including actions in localization and mapping is called *active exploration* in the literature. Active exploration (or more particularly, active localization, active mapping or active SLAM) requires decision making techniques which represent a large body of work on its own—for a review of the origins and evolution of that term, and some algorithms, you can consult the last chapter of (Thrun, Burgard, & Fox, 2005). Nevertheless, it is worth noting that active exploration methods are often based on the same probabilistic concepts introduced in this text.

The Previous Knowledge

The final axis of our taxonomy is concerned with the knowledge that the robot already has about its location and the environment before starting the localization and mapping processes. As we shall explain in chapter 4, considering previous knowledge about the state of a dynamic system is a necessary premise for modern probabilistic methods to estimate the current state of that system, since they are strongly based on the Bayesian paradigm, where the prior information is an essential part. We can distinguish three main cases regarding this previous knowledge—following the terminology of Fox, Burgard, and Thrun (1999).

Firstly, the robot may know the map of its environment and its starting spatial relation with that map (its location) exactly. In this situation, mapping is obviously not an issue, and localization becomes a *tracking problem* (also called *local localization*): the robot location is only to be updated as it navigates. Although this instance of localization may be unsuitable for long navigations or large-scale spaces, it is an important starting point for a good understanding of more sophisticated methods, and that is the reason why it is the subject of the second part of the book (chapters 5 to 7). An especial situation happens when the robot has previous knowledge about its location and map but it is suddenly moved to a different place without notification, or, maybe more realistically, its location estimation diverges and becomes completely wrong. This is called the *kidnapped robot problem*. It cannot be solved well by classical tracking methods, thus we should employ specific variations specially modified in order to deal with large and sudden changes of the robot location.

The second case of interest is when the robot has no information at all about its initial location within an environment whose structure is known. Again, mapping is not needed. This case, called the *robot awakening problem* or *global localization*, is more complicated than tracking. Due to the large initial uncertainty, the robot has often to move around first, in order to gather enough sensory data to disambiguate all its potential (usually uncountable) locations. In turn, in the most common approaches this involves the expensive cost of tracking each of those potential location hypotheses. Robot awakening is a very common scenario, and the basic methods to deal with it are explained in this book. The problem becomes particularly difficult for large-scale environments, where specialized techniques are required for achieving a tractable solution (we will deal with that in chapter 10).

The third case would be when the map is unknown but the localization of the robot is known. This is the mapping-only problem, explained in chapter 8, but it is rarely found alone in practice: obviously, if the robot is localized it must be *with respect to some map*. It is mostly considered a sub-problem that arises whenever neither the map nor the location are known, as explained next.

In the last possible case the robot does not have any previous knowledge about its location or map; then we find the *SLAM* problem: simultaneous localization and mapping, which is dealt with in chapters 9 and 10. If the environment is only known partially before navigating, we can include that previous knowledge in the initial map and work from there, but the problem would be similar.

3. HISTORICAL OVERVIEW

As the reader has noticed in the previous section, the problems of mobile robot localization and mapping have a rich structure, that is, the existing solutions are varied and complex. In this section, we will take a very broad glimpse of the path that it took to develop those solutions. Since the most recent developments will be reviewed in detail along further chapters, we will focus here instead on the long-term evolution. We believe that this perspective, unusually found in the scientific literature, can be enriching for the reader precisely due to that reason. Actually, the history of mobile robot localization and mapping is the history of how humankind got machines that automatically perform tasks as ancient as humanity, not only to alleviate the hard work of doing them manually, but also to open new ways of carrying out exploration more efficiently and accurately, and obtaining new knowledge.

Maybe exaggerating the perspective, we could start by going back before history, to the Middle and Upper Paleolithic era, to find the first of our ancestors that developed language and culture and thus needed to make some kind of spatial references explicit (Bagrow & Skelton, 2009). In that age, they lived as hunter-gatherers: they had to localize their prey and food, and move as these resources moved or got exhausted (a tracking problem!). The nomadic displacements were mostly based on natural landmarks in the landscape—a type of environment representation that mobile robots use today—and their map-making reduced to representing small areas, probably due to a lack of abstraction and generalization skills: they focused on survival. Unfortunately, prehistoric maps are subject to interpretation from our modern perspective, and in many cases their consideration as "maps" may be controversial.

It was in the Neolithic era that, in spite of becoming sedentary, some people were released from the pressure of immediate survival due to the division of work, and had time and resources to explore unknown spaces at an unprecedented scale. They did it for diverse reasons that demonstrate their highly evolved cognitive skills: curiosity, ambition and the need to acquire more power over other people (Diamond, 2005). They built specialized tools for that. Regarding truly maps, that is, graphic representations or models of spatial concepts, we find in the second millennium B.C. urban images and depictions of cultivated lands (see Figure 4), which implies an important gap from the ancient nomadic societies and the first civilizations. It is unknown when map-making became a recognized craft after that, but we can find an extraordinary development of cartography around the first centuries of our era, by Greeks and Romans.

There are two events of special interest for us in the ancient history of travel and exploration, both a few millennia B.C.: the invention of the wheel—the oldest known wooden wheel, found in Slovenia and dating back to about 5000 B.C. (Prešeren, 2004), corresponds to a single-axle cart—and the consistent development of sea travel by Egyptians, then Phoenicians, Greeks, and Romans in the area of the Mediterranean sea, and Chinese, Polynesian, and Mesoamerican people in the Atlantic and Pacific oceans. Some recent archaeological remains found at the time of writing this book have provided a much older evidence of maritime navigation, dating back to the Middle Paleolithic era, at around 130,000 B.C., but they were relatively short routes in the Mediterranean (Zorich, 2011).

The wheel was one of the bases for trading and thus the first big impulse to explore more and more distant places. Over the centuries, it became not only an efficient actuator, but a sensor (the odometer, that will be found in mobile robots in chapter 2), as described around 20 B.C. (Sleeswyk, 1981). Maritime navigators, on the other hand,

Figure 4. Petroglyph map (rock drawing) from the late Bronze Age, about 1900-1200 B.C., discovered at Giagidhe (also known as Plaz D'Ort), in Valcamonica, one of the largest valleys of the central Alps, eastern Lombardy. It also shows two anthropomorphic figures at the bottom. The reader will find along the book how similar is this map to some metrical maps built by modern mobile robots (© 1985, Ausilio Priuli. Used with permission.).

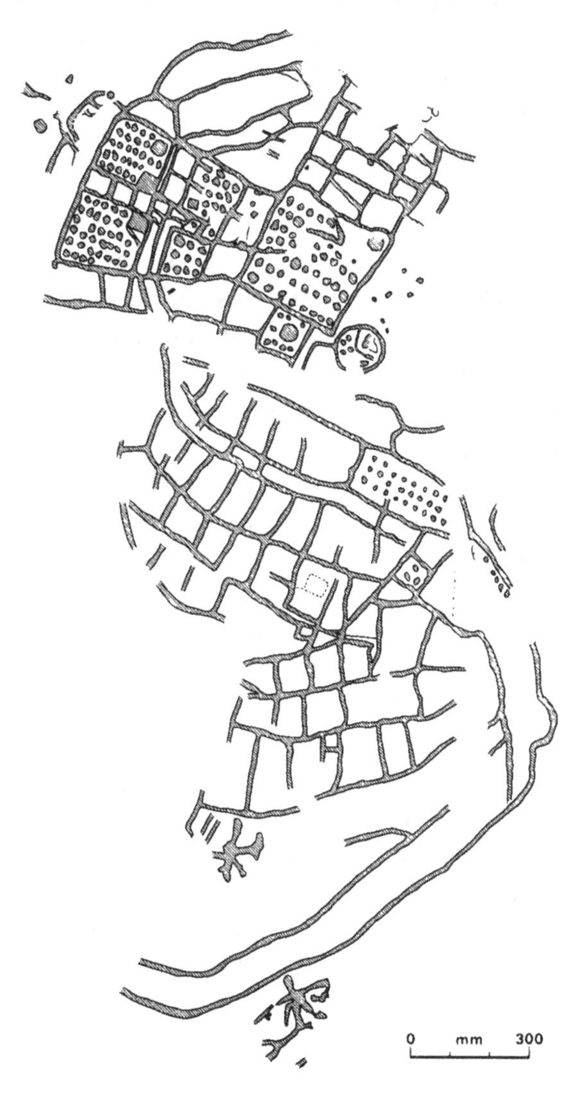

were required to position themselves with respect to other objects in large-scale spaces, thus they began by localizing their ships with respect to the coast and then went to the open sea with the help of stars. In time, they invented a bunch of remarkable, practical techniques for relative localization and mapping and for the "absolute" positioning with respect to celestial objects and on the surface of the Earth (Bowditch, 2010). Maritime navigation has been developing almost without interruption in the history of mankind. It has provided modern robotics with concepts as common as *dead-reckoning* (a generalization of odometry), triangulation, orientation (*bearing*), or distances to other objects as means for positioning. Therefore, maritime navigation serves us as a link to the present: it was the main agent that extended methods from the Middle Ages into the Modern era with its Renaissance, great explorations and the scientific revolution, and then into our Contemporary era with its Industrial Revolution and the Information Age.

It was during the Industrial Revolution that machines began to be a fundamental part of production processes, particularly cloth and textiles (Jenkins, 2003). They were not robotic manipulators or mobile devices, hence there was no need for localization and mapping: they were just automata intended to increase the productivity. Releasing people from hard work was not a real concern at the time, as the employment of children in those production processes demonstrates. Before the first manipulators arrived there was time enough in the 20th century for inventing the word "robot" and refining the concept of human-like machines, which, as commented in the introduction, had been present in mythology since Hebrews, Greeks, and in ancient China, introducing it into the collective memory as an actually realizable idea. There was also time for creating the first computers, in the middle of World War 2 (Ceruzzi, 2003), and cybernetics, right after that (Wiener, 1965). That led to truly programmable machines that imitated the homeostatic processes by which nature is able to maintain itself in stable states.

World War 2 and the post-war period came with an explosion of engineering machinery and new control and computer theoretical developments. We already find examples of machines that self-localize during the war, for instance the V-1 and V-2 flying bombs (Zaloga & Laurier, 2005), which carried a set of inertial sensors—several gyrocompasses and a pendulum—as a guidance system and also an anemometer for the integration of wind speed to obtain distance to target (a kind of odometer). Although the lack of programmability of these machines was a serious limiting factor, their basic settings could be changed before launching.

Shortly after WW2, from 1948 to 1949, the neurophysiologist and engineer William Grey Walter created the first well-known mobile robots, which he called *Machina speculatrix*. He aimed to prove the possibility of obtaining complex behaviors from simple actuator-sensor connections based on the connections of neurons (Walter, 1963). These robots, also called *The Walter's Tortoises*, were able to find a spot where to recharge their batteries, a sort of localization, but without using any explicit representation of the environment or being programmable at all, since they were based on hard-wired analog circuitry. The idea of avoiding the explicit representation of the environment can be thought of as a very preliminary precursor of the "environment is its best own model" paradigm, proposed by Rodney Brooks a few decades afterwards (Brooks, 1991).

In 1961, the mass production of consumer goods was at one of its peaks in the 20th century. The first robotic manipulator for industrial applications, called *Unimate*, was built for the assembly factories of General Motors in order to manufacture TV picture tubes, and was closely followed by others (Shimon, 1999). Of special relevance were the different institutes and research laboratories created in Stanford (California, USA) during the following years, which, among other outstanding developments, integrated for the first time a computer into a robotic arm, creating the Stanford Arm. From 1960 to 1961, also in Stanford, a mechanical engineering graduate named James L. Adams built the Stanford Cart, a mobile vehicle intended for studying the possible visual teleoperation of a robot on the surface of the Moon (Adams, 1961). Further refinements of the idea were implemented in the next year, but the teleoperation project vanished when the U.S. focused on the mission of sending directly humans to our satellite.

The 1961-1970 decade was also important for the development of artificial intelligence. In those times, the expectations of the discipline were largely overstated. Many researchers thought that the advent of intelligent machines able to read, speak, act, and understand was a matter of just a few more decades. Stanley Kubrick's movie, *2001: A Space Odyssey*, released in 1968, is a clear reflection of the so-called golden age of hard science-fiction—the most scientific subgenre—in literature. If all those dreams had come true, a book like this, about the rather "simple" problem of knowing only where a robot is within a certain area of space, would have been written many years ago... by a robot.

As a sign of the times, the first mobile robot that used an explicit representation of its environment and was able to execute tasks and doing reasoning with it was Shakey (Nilsson, 1988), which served for experimentation from 1966 to 1972 in the Artificial Intelligence Center at the Stanford Research Institute (see Figure 5). Sensors that are common today were already used in Shakey: range-finders, cameras, contact sensors... Shakey employed logical descriptions of indoor, structured spaces and objects that were based on predicate calculus for a good integration with its task planner. The representation was mostly filled in by engineers before experiments, but Shakey was also able to recognize lines and simple elements in camera images and modify its internal representation of the world with that new information during execution.

During the seventies, the invention of the microprocessor and new developments on expert systems kept the (today) classic artificial intelligence hopes growing—the *Star Wars* movie, featuring probably the two most popular intelligent robots of all time, was released in 1971. Again, in Stanford, Hans Moravec worked in his PhD thesis on the vision-based autonomous navigation of a renewed version of the Stanford Cart (Moravec, 1980). The robot was finally able to extract spatial relations between the elements of the environment projected on the image and its own position, and to plan and execute trajectories—slowly—based on that information.

Personal computers, the first ones providing a reasonable amount of computational power to the public in a reduced space, arrived with the IBM PC in 1981. The eighties, right in the transition to the Information Age, had begun. They were years of important changes and convulsions all over the world, which were of course translated into the robotics arena, where intelligent robots were a clear goal to achieve: machines with so complex internal architectures that the embodiment of autonomous, cognitive behaviors should have become a reality quickly. During that decade, mobile robotics flourished at the same time that classic AI found its main limitations. It was time to see a clash of paradigms in both areas: the deliberative vs. the behavior-based approaches in mobile robotics ran along the discussions between symbolic and subsymbolic AI (neural networks became popular in the mid eighties). Deliberative robot control architectures, which actually had been created with Shakey, were the direct reflection of classic artificial intelligence methods, those based on explicit—symbolic—representations of the environment, actions, and cognitive processes. They were based on a perceive-think-act working sequence. The reactive or behavior-based control architectures, on the other hand, emerged to substitute the thinking part by direct connections between sensors and actuators (they were hard-wired in the first prototypes, like in Walter's tortoises), and was firstly aimed to recreate more modest "intelligences," such as those of insects.

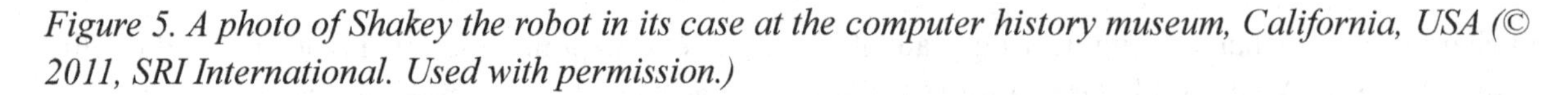

Figure 5. A photo of Shakey the robot in its case at the computer history museum, California, USA (© 2011, SRI International. Used with permission.)

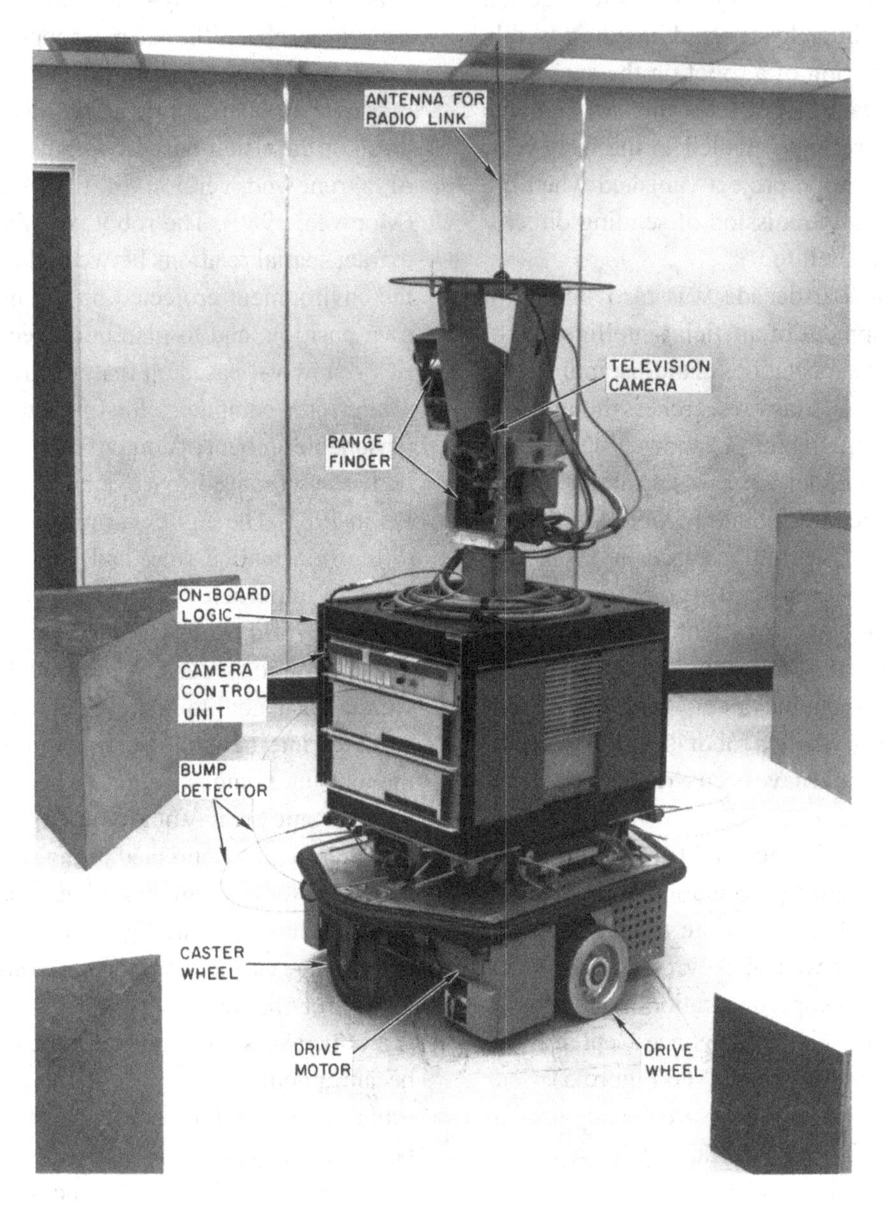

As a matter of fact, this new approach coincided with the beginning of the so-called *Nouvelle AI*, in contrast with the *Good Old Fashioned AI* (GOFAI). The former demonstrated the close relationship existing between AI and robotics, which is not always as solid as it should be (Nilsson, 1998). Behavior-based robots were aimed to gather from the environment the information that they needed at each moment without constructing any explicit, long-term model of anything. Up to date, none of these two broad AI projects alone has been able to replicate successfully human intelligence on a machine, and the current focus is on how to integrate both, as claimed in the field of *embodied cognition* (Anderson, 2003; Casacuberta, Ayala, & Vallverdú, 2011). Furthermore, the idea firstly mentioned by Alan Turing of making robots to develop like children (Turing, 1948) began to evolve at that time towards the modern paradigm of *developmental robotics* (Lee, Meng, & Chao, 2007; Lungarella, Metta, Pfeifer, & Sandini, 2003).

Concerning localization and mapping, in the transition between the seventies and the eighties, a new framework for large-scale robotic mapping was proposed (Kuipers, 1977) based on some of the experiments reported by Jean Piaget in psychology about the human cognitive map and its development in children (Piaget, 1948). The discovering of Piaget that children *first* learn topologies and *then* metrics was completely new in robotic mapping at the time. Kuipers studied comprehensively the problem under the perspective of classic AI, through thought-experiments and in some real robots (Kuipers, 2008). His model comprises a number of ontological levels that include topological, causal, and metrical maps.

It is surprising that, in spite of the remarkable developments of the eighties, as late as in the mid nineties the robotics community was still well aware of the difficulty of the localization and mapping problems in practical robots. The success had been noticeable in both tasks in the realms of civil and military maritime and aviation applications, but in the robotics arena, things appeared considerably hard. Among the reasons for that, the automatic extraction of sensory data and its correlation with existing information (maps) was recognized as one of the cornerstones. Concerning localization, Borenstein *et al.* stated (Borenstein, Everett, & Feng, 1996): "To date there is no truly elegant solution for the problem." Automatic mapping of real environments was not quite different.

There was a fundamental issue underlying the problem of interpreting and using sensory data in real robots, and it had already been recognized, for instance by Engelson and McDermott (1992): that of dealing with noisy sensor measurements or, more generally, with uncertainty. The point was that no clear, explicit, and comprehensive proposal for that had been reported yet. Earliest approaches to mobile robot localization and mapping extracted information from noisy data by minimization or approximation, being quite sensitive to noise and therefore not much robust. These uncertainty-unaware methods, mainly implemented during the eighties, included four basic ways of localization: *dead-reckoning*, *active beacons*, *passive beacons* (or environmental landmarks), and *map matching*. Dead-reckoning, already mentioned as one of the maritime techniques for estimating location, is the mathematical approach for estimating the accumulated distance from a given origin point, that is, to estimate the current position based on a past one and the theoretical behavior of the system. It can be implemented mainly based on odometric—motion of actuators—and inertial—motion of the robot body—measurements. Another method for approximating the actual robot location is by triangulation or minimization of the real distance of the robot to a set of active beacons given actual, noisy distance measurements. Distinctive, passive elements of the environments (*landmarks*) can also be used with this technique, but at the expense of the harder problem of their recognition. Finally, if the robot has a map of the entire environment and some sensor that provides a more limited map of its surroundings, the geometrical rigid transformation that minimizes the perception error when positioning the latter onto the former can also serve for localization.

All these methods were useful, and still form part of modern approaches, but they do not provide either an estimate of the quality of the result or more than one hypothesis about the robot location, which are essential issues in the robot awakening and kidnapped robot problems, for instance. This follows from not taking into account the uncertainty in the process, thus losing precious information. Surprisingly, the bases for coping with that had already been introduced decades ago, in particular the use in engineering of probability theory and statistics (Tabak, 2004). The Kalman Filter, a statistical algorithm that can be derived from the Bayes' rule (that in turn dates back to the middle of the 18th century), is able to estimate the position—or state—of a moving object by fusing a number of noisy observations, and it was invented in the early sixties (Kalman, 1960) and applied successfully to space and military control problems such as the guidance and navigation of missiles and spacecrafts. Why wasn't such success occurring in robotics just two decades afterwards? Undoubtedly, there were some specific factors that prevented a quick application into robotic localization and mapping. We may guess now that the lack of reasonably cheap and small sensors (and actuators) for robots, the little importance attached to the information that could be recovered from sensory noise (likely due to the lack of suitable sensors for robots to experiment with), the insufficient computational power available at that time and, finally, the lack of theoretical intersection between probability/statistics and robotics up to then, were among these factors.

Anyway, it was not until the late eighties and early nineties that the fundamental issue of how to deal with uncertainty in robotics was treated explicitly and started to produce practical results. Alberto Elfes finished his PhD thesis on occupancy grids (Elfes, 1990), a type of metrical map originally intended to accommodate measurements gathered by a kind of devices celebrated at the time for its low cost: ultrasonic sensors. Occupancy grids are now the best-known, oldest metric representation of the environment of a mobile robot that includes explicitly a measure of the uncertainty. In addition, by the end of the eighties and early nineties, the first approaches to the localization and the SLAM problems that reported an explicit treatment of uncertainty through probability and statistics were proposed (Smith & Cheeseman, 1986; Smith, Self, & Cheeseman, 1990; Leonard & Durrant-Whyte, 1991). Furthermore, more computational power in less space was available by the time, and sensor and actuator technologies were improving significantly (Everett, 1995); for instance, laser scanners arrived to robotics to provide highly accurate range measurements. At last, the way to go was wide open.

Therefore, the decades of 1990 and 2000 came with an immense effort in research on mobile robot localization and mapping. Apart from Kalman Filter derivatives, other statistical approaches like Monte Carlo methods served for coping with multi-hypothesis, non-linear dynamics and non-Gaussian models of uncertainty. In the late nineties and early two thousands Monte Carlo methods were applied to mobile robot global localization, obtaining a remarkable success in real applications (Thrun, Fox, Burgard, & Dellaert, 2000; Montemerlo & Thrun, 2007). Many researchers achieved important results in SLAM, in particular with real problems that had been really hard up to then, such as unmanned vehicles, mapping of mines, museum guiding or even robotic large-scale races—it is remarkable in this aspect the first DARPA Grand Challenge contest of the US Department of Defense, consisting in travelling more than one hundred kilometers with a driverless vehicle (Hoffmann, Tomlin, Montemerlo, & Thrun, 2007). At the end of the 20th century machines seemed to be able to solve autonomously the same problems that our ancestors had tackled when exploring the world thousands of years ago with the aid of probability and statistics (Thrun, Burgard, & Fox, 2005; Stachniss, 2009).

After the first decade of the 2000s, research in mobile robot localization and mapping is still highly active. In spite of the existence of a number of methods, mostly statistical filters, the scientific community is still looking to resolve issues concerning computational efficiency, large-scale SLAM, the use of new, more powerful sensors (range-finders are losing popularity in favor of computer vision and 3D sensors), and the integration of these quantitative results into complex robotic architectures that also reason about their states, goals, and experiences. Since statistics fits so well with numerical data, most localization and mapping developments of the last decades have focused on metrical representations of spatial relations, while pure qualitative approaches were kind of forgotten since the nineties until recent years. However, the scene is changing again, not only

Figure 6. The map resulting from a large-scale localization and mapping experiment conducted with our robot SENA in the technology complex of the University of Málaga. A hybrid metrical-topological representation of space is used.

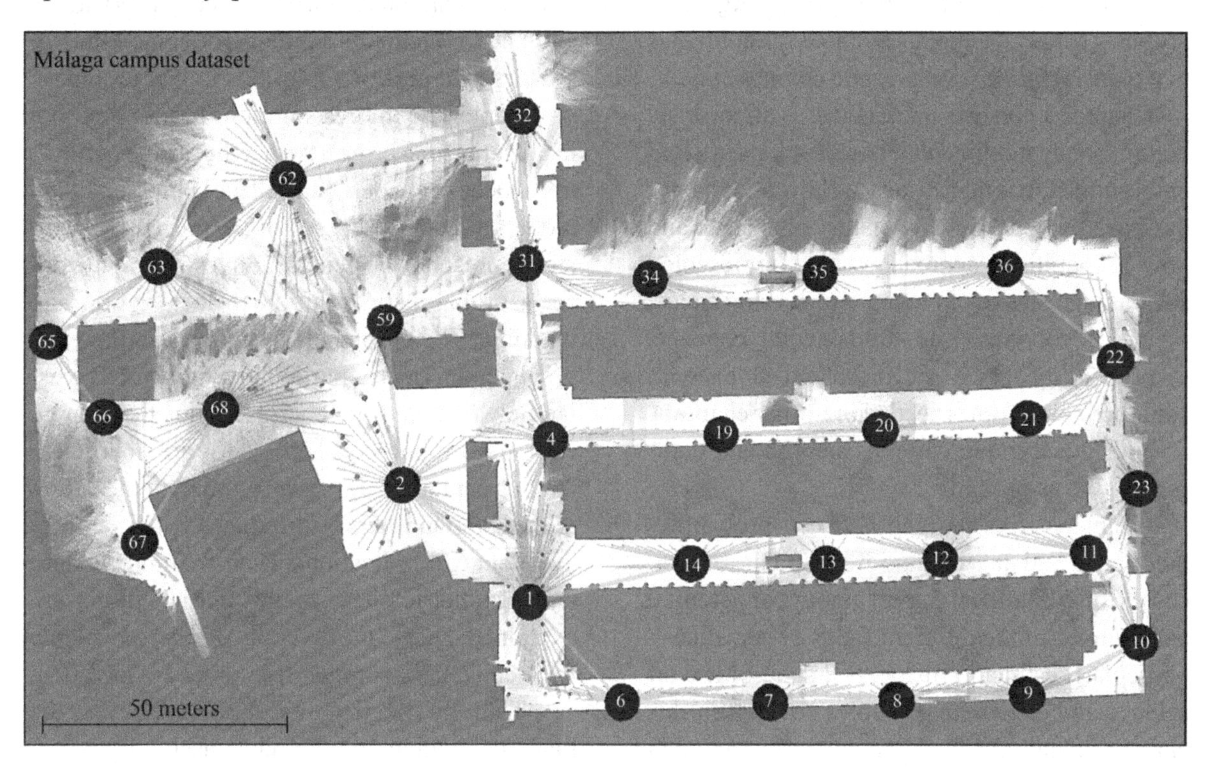

in merging these two paradigms (see for instance Figure 6), but also in exploring alternatives to model uncertainty that diverge from probability and statistics.

4. ORGANIZATION OF THE BOOK

The main text of this book is an elaboration of material coming from two main sources: the classes on probabilistic localization and mapping for mobile robots lectured by the first author in the Mechatronic Engineering doctoral program of the University of Málaga during the last years and the work on the PhD thesis of the second author on SLAM (co-supervised by the former). One of the main goals of this book has been to integrate all of this in a manner as self-contained as possible to serve as an introductory text in the subject. For that purpose, we have divided the contents into three separate parts. The first one includes the foundations of probability, statistics, and mobile robotics needed for the rest, and do not deal with the localization and mapping problems. The second part is devoted to localization. The third part deals with mapping and SLAM. Different parts of the book can be selected for different objectives, like teaching or research reference. For instance, the first part can be skipped if the mathematical and robotic bases are known. In any case, the book structure is also prepared for sequential reading.

The contents of the chapters are as follows.

Chapter 2 provides a detailed review of the most common electro-mechanical components found in state-of-the-art mobile robots, emphasizing practical aspects such as weight and size, power consumption and performance trade-offs. Sensors and actuators, in particular, are stated as the hardware basis for coping with localization and mapping, and thus, specialized sections are devoted to them. The described devices range from low-cost sensors/actuators suitable for hobbyists to expensive professional-grade components.

Chapter 3 presents uncertainty as an intrinsic feature of any mobile robot that develops in a real environment. It is then discussed how uncertainty has been treated along the history of science and how probabilistic approaches have represented such a huge success in many engineering fields, including robotics. The fundamental concepts of probability theory are discussed along with some advanced topics needed in further chapters, following a learning curve as smooth and comprehensive as possible.

Chapter 4 fills the gap between probability theory and real data coming from stochastic processes, highlighting the great amount of potential applications of the different fields of statistics—particularly estimation theory—in state-of-the-art science and engineering. Topics covered in this chapter include the fundamental tools needed in probabilistic robotics: probabilistic convergence, theory of estimators, hypothesis tests, etc. Special stress is on recursive Bayesian estimators, due to their central role in the problems of probabilistic robot localization and mapping.

Chapter 5 (the first of section 2) is dedicated to robot motion. It explores some of the reasons why any real robot cannot move as perfectly as planned, thus demanding a probabilistic model of the robot actions—mainly, its movements. Especial emphasis is put on the most common ground wheeled robots, although other configurations (including non-robotic ones) with more degrees of freedom, such as arbitrarily moving hand-held sensors or aerial vehicles, are also mentioned. The best-known approximate probabilistic models for robot motion are provided and justified.

Chapter 6 complements chapter 5 with a study of robot sensory models. It explains common mathematical models of sensors, stressing their differences and effects in further estimation techniques, in particular whether they are parametrical or not. It also points out the existence of the association problem between observations and known elements of maps for some kinds of sensors, and presents solutions to that problem.

Chapter 7 explores solutions to mobile robot localization that bring together the recursive filters introduced in chapter 4 and all the components and models already discussed in the preceding chapters. It presents the general, Bayesian framework for a probabilistic solution to localization and mapping. The problem is formally described as a graphical model (in particular a dynamic Bayesian network) and the characteristics that can be exploited to approach it efficiently are elaborated. Among parametric Bayesian estimators, the family of the Kalman filters is introduced with examples and practical applications. Then, the more modern non-parametric filters, mainly particle filters, are explained. Due to the diversity of filters available for localization, comparative tables are included.

Chapter 8 (the first of section 3) describes the kinds of mathematical models usable by a mobile robot to represent its spatial reality, and the reasons by which some of them are more useful than others depending on the task to be carried out. The most common metric, topological and hybrid map representations are described from an introductory viewpoint, emphasizing their limitations and advantages for the localization and mapping problems. It then addresses the problem of how to update or build a map from the robot raw sensory data, assuming known robot positions, a situation that becomes an intrinsic feature of some SLAM filters. Since the process greatly depends on the kind of map and sensors, the most common combinations of both are treated.

Chapter 9 deals with the situation arising when neither the environment nor the exact localization of a mobile robot are known, that is, when we face the hard problem of SLAM. It reviews the most common solutions to that problem found in literature, especially those based on statistical estimation. Both parametric and non-parametric

Figure 7. An overview of the dependencies between the different chapters of this text

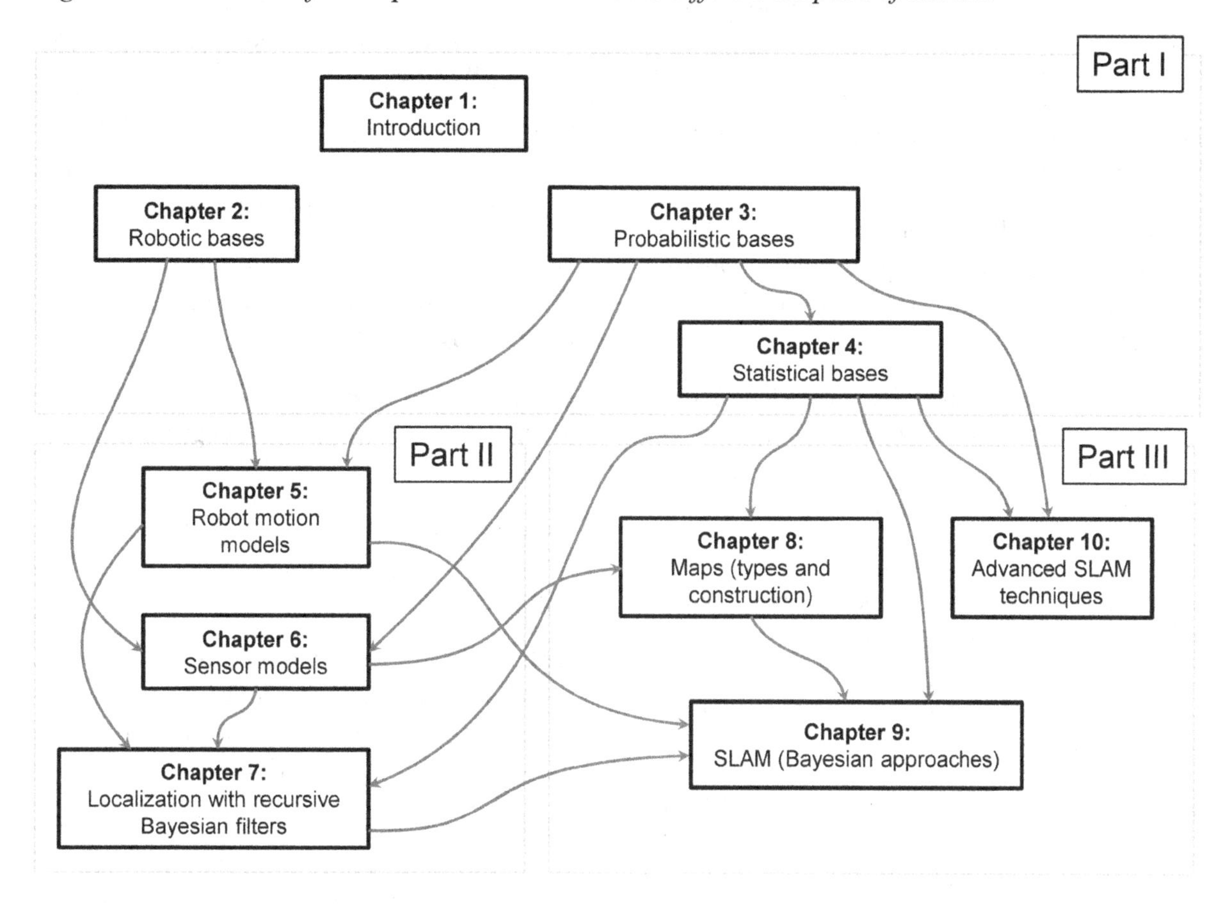

filters are explained as practical solutions to this problem, including analysis of their advantages and weaknesses that must be both taken into account in order to design a robust SLAM system. Complete examples and algorithms for these filters are included.

Being that SLAM is a very active research field, especially in its probabilistic formulation, and the wish of the authors for this book to serve as a first step in the study of the discipline, chapter 10 is devoted to providing an overview of emerging paradigms that are appearing as outstanding the traditional approaches in scalability or efficiency, such as hierarchical sub-mapping, or hybrid metric-topological map models. Other techniques not based on Bayesian filtering, such as iterative sparse least-squares optimization, are also introduced due to their efficiency and increasing popularity. This chapter has been devised as the conclusion part of the book, reviewing the major trends in the field (see Figure 7).

REFERENCES

Adams, J. L. (1961). *Remote control* with *long transmission delays.* (Doctoral Dissertation). Stanford University. Palo Alto, CA.

Anderson, M. L. (2003). Embodied cognition: A field guide. *Artificial Intelligence*, *149*, 91–130. doi:10.1016/S0004-3702(03)00054-7

Andreasson, H., Treptow, A., & Duckett, T. (2007). Self-localization in non-stationary environments using omni-directional vision. *Robotics and Autonomous Systems*, *55*(7), 541–551. doi:10.1016/j.robot.2007.02.002

Asimov, I. (1941, May). Liar!. *Astounding Science Fiction.*

Bagrow, L., & Skelton, R. A. (2009). *History of cartography* (2nd ed.). Piscataway, NJ: Transactions Publishers.

Blanco, J. L. (2009). *Contributions to localization, mapping and navigation in mobile robotics.* (Doctoral Dissertation). University of Málaga. Malaga, Spain.

Blanco, J. L., González, J., & Fernández-Madrigal, J. A. (2008). Extending obstacle avoidance methods through multiple parameter-space transformations. *Autonomous Robots, 24*(1), 29–48. doi:10.1007/s10514-007-9062-7

Borenstein, J. (1998). Experimental results from internal odometry error correction with the OmniMate mobile robot. *IEEE Transactions on Robotics and Automation, 14*(6), 963–969. doi:10.1109/70.736779

Borenstein, J., Everett, B., & Feng, L. (1996). *Navigating mobile robots: Systems and techniques*. Wellesley, MA: A. K. Peters, Ltd.

Bowditch, N. (2010). *The American practical navigator*. Arcata, CA: Paradise Cay Publications.

Brooks, R. A. (1991). Intelligence without representation. *Artificial Intelligence*, 47.

Capek, K. (1921). *RUR (Rossum's universal robots)*. New York, NY: Penguin Books.

Casacuberta, D., Ayala, S., & Vallverdú, J. (2011). Embodying cognition: A morphological perspective. In *Machine Learning: Concepts, Methodologies, Tools and Applications*. Hershey, PA: IGI Global. doi:10.4018/978-1-60960-818-7.ch707

Ceruzzi, P. E. (2003). *A history of modern computing*. Cambridge, MA: MIT Press.

Cox, I. J., Wilfong, G. T., & Lozano-Pérez, T. (1990). *Autonomous robot vehicles*. Berlin, Germany: Springer. doi:10.1007/978-1-4613-8997-2

Craig, J. J. (1989). *Introduction to robotics mechanics and control* (2nd ed.). Reading, MA: Addison-Wesley.

Crowley, J. L. (1985). Navigation for an intelligent mobile robot. *IEEE Journal on Robotics and Automation, 1*(1). doi:10.1109/JRA.1985.1087002

Davison, A. J. (1998). *Mobile robot navigation using active vision.* (Doctoral Dissertation). University of Oxford. Oxford, UK.

Devol, G. C. (1961). *Programmed article transfer*. US Patent no. 2988237. Washington, DC: US Patent Office.

Diamond, J. (2005). *Guns, germs, and steel: The fates of human societies*. New York, NY: W. W. Norton and Co.

Elfes, A. (1990). *Occupancy grids: A stochastic spatial representation for active robot perception*. Paper presented at the 6th Conference on Uncertainty and AI. Cambridge, MA.

Engelson, S. P., & McDermott, D. V. (1992). Error correction in mobile robot map learning. In *Proceedings of the International Conference on Robotics and Automation*. IEEE.

Everett, H. R. (1995). *Sensors for mobile robots*. Boca Raton, FL: CRC Press.

Fox, D., Burgard, W., & Thrun, S. (1999). Markov localization for mobile robots in dynamic environments. *Journal of Artificial Intelligence Research, 11*, 391–427.

Galindo, C., Fernández-Madrigal, J. A., González-Jiménez, J., & Saffiotti, A. (2008). Robot task planning using semantic maps. *Robotics and Autonomous Systems, 56*(11), 955–966. doi:10.1016/j.robot.2008.08.007

Herman, W. A. (1923). *Founders of oceanography and their work: An introduction to the science of the sea*. London, UK: Edward Arnold & Co. Retrieved Mar 1, 2012, from http://www.archive.org/details/foundersofoceano1923herd

Hoffmann, G. M., Tomlin, C. J., Montemerlo, M., & Thrun, S. (2007). Autonomous automobile trajectory tracking for off-road driving: Controller design, experimental validation and racing. In *Proceedings of the American Control Conference (ACC)*, (pp. 2296-2301). ACC.

Idel, M. (1990). *Golem: Jewish magical and mystical traditions on the artificial anthropoid.* Albany, NY: State University of New York Press.

Jefferies, M., & Yeap, W. K. (Eds.). (2008). *Robotics and cognitive approaches to spatial mapping.* Berlin, Germany: Springer. doi:10.1007/978-3-540-75388-9

Jenkins, D. (2003). The western wool textile industry in the nineteenth century. In *The Cambridge History of Western Textiles.* Cambridge, UK: Cambridge University Press.

Kalman, R. E. (1960). A new approach to linear filtering and prediction problems. *Transactions of the ASME: Journal of Basic Engineering, Series D, 82*, 35–45. doi:10.1115/1.3662552

Krzysztof, K. (Ed.). (2006). Robot motion and control: Recent developments. *Lecture Notes in Control and Information Sciences, 335.*

Kuipers, B. J. (1977). *Representing knowledge of large-scale space.* (Doctoral Dissertation). Massachusetts Institute of Technology. Cambridge, MA.

Kuipers, B. J. (1983). *The cognitive map: could it have been any other way? Spatial Orientation: Theory, Research, and Application* (pp. 345–359). New York, NY: Plenum Press.

Kuipers, B. J. (2008). *An intellectual history of the spatial semantic hierarchy.* Berlin, Germany: Springer. doi:10.1007/978-3-540-75388-9_15

Kurfess, T. R. (Ed.). (2005). *Robotics and automation handbook.* Boca Raton, FL: CRC Press.

Landes, J. B. (2007). The anatomy of artificial life: An eighteenth-century perspective. In Riskin, J. (Ed.), *Genesis Redux: Essays in the History and Philosophy of Artificial Life.* Chicago, IL: The University of Chicago Press.

Lee, M. H., Meng, Q., & Chao, F. (2007). Developmental learning for autonomous robots. *Robotics and Autonomous Systems, 55*, 750–759. doi:10.1016/j.robot.2007.05.002

Leonard, J. J., & Durrant-Whyte, H. F. (1991). Mobile robot localization by tracking geometric beacons. *IEEE Transactions on Robotics and Automation, 7*(3). doi:10.1109/70.88147

Liu, C. (2011). *Foundations of MEMS* (2nd ed.). Upper Saddle River, NJ: Prentice Hall.

Lozano-Pérez, T., & Wesley, M. A. (1979). An algorithm for planning collision-free paths among obstacles. *Communications of the ACM, 22*, 560–570. doi:10.1145/359156.359164

Lungarella, M., Metta, G., Pfeifer, R., & Sandini, G. (2003). Developmental robotics: A survey. *Connection Science, 15*(4), 151–190. doi:10.1080/09540090310001655110

Mamen, R. (2003). Applying space technologies for human benefit: The Canadian experience and global trends. In *Proceedings of the International Conference on Recent Advances in Space Technologies (RAST 2003).* RAST.

Minguez, J., & Montano, L. (2008). Extending collision avoidance methods to consider the vehicle shape, kinematics, and dynamics of a mobile robot. *IEEE Transactions on Robotics, 25*(2), 367–381. doi:10.1109/TRO.2009.2011526

Montemerlo, M., & Thrun, S. (2007). *FastSLAM: A scalable method for the simultaneous localization and mapping problem in robotics.* Berlin, Germany: Springer.

Moravec, H. (1980). *Obstacle avoidance and navigation in the real world by a seeing robot rover.* (Doctoral Dissertation). Stanford University. Palo Alto, CA. Retrieved Mar 1, 2012, from http://www.frc.ri.cmu.edu/~hpm/hpm.pubs.html

Niku, S. B. (2010). *Introduction to robotics: Analysis, control, applications*. New York, NY: John Wiley & Sons.

Nilsson, N. (1988). *Shakey the robot. Tech. Note 323*. Palo Alto, CA: Artificial Intelligence Center, SRI International.

Nilsson, N. (1998). *Artificial intelligence: A new synthesis*. San Francisco, CA: Morgan Kaufmann Publishers Inc.

Piaget, J. (1948). *The child's conception of space*. London, UK: Routledge.

Prešeren, P. (Ed.). (2004). *World's oldest wheel*. Slovenia News.

Ranganathan, A., Menegatti, E., & Dellaert, F. (2006). Bayesian inference in the space of topological maps. *IEEE Transactions on Robotics, 22*(1), 92–107. doi:10.1109/TRO.2005.861457

Sandler, B. Z. (1999). *Robotics: Designing the mechanisms for automated machinery* (2nd ed.). New York, NY: Academic Press.

Scheinman, V. (1969). *Design of a computer controlled manipulator. Tech Report*. Palo Alto, CA: University of Stanford.

Schultz, A. C., & Parker, L. E. (Eds.). (2010). *Multi-robot systems: From swarms to intelligent automata*. Berlin, Germany: Springer.

Selig, J. M. (1992). *Introductory robotics*. Upper Saddle River, NJ: Prentice-Hall.

Shimon, Y. N. (1999). *Handbook of industrial robotics* (2nd ed.). New York, NY: John Wiley & Sons.

Siciliano, B., & Khatib, O. (Eds.). (2008). *Handbook of robotics*. Berlin, Germany: Springer. doi:10.1007/978-3-540-30301-5

Siegwart, R., & Nourbakhsh, I. R. (2004). *Introduction to autonomous mobile robots: Intelligent robotics and autonomous agents*. Cambridge, MA: MIT Press.

Sleeswyk, A. W. (1981, October). Vitruvius' odometer. *Scientific American*.

Smith, R. C., & Cheeseman, P. (1986). On the representation and estimation of spatial uncertainty. *The International Journal of Robotics Research, 5*(4), 56–68. doi:10.1177/027836498600500404

Smith, R. C., Self, M., & Cheeseman, P. (1990). Estimating uncertain spatial relationships in robotics. *Autonomous Robot Vehicles, 1*, 167–193. doi:10.1007/978-1-4613-8997-2_14

Stachniss, C. (2009). *Robotic mapping and exploration*. Berlin, Germany: Springer.

Stoll, E., Letschnik, J., Walter, U., Artigas, J., Kremer, P., Preusche, C., & Hirzinger, G. (2009). On-orbit servicing. *IEEE Robotics & Automation Magazine, 6*(4), 29–33. doi:10.1109/MRA.2009.934819

Tabak, J. (2004). *Probability and statistics: The science of uncertainty*. New York, NY: Facts on File.

Thorpe, C. E. (1984). *Path relaxation: Path planning for a mobile robot.* Technical Report CMU-RI-TR-84-5. Pittsburgh, PA: Carnegie Mellon University.

Thrun, S. (2001). Is robotics going statistics? The field of probabilistic robotics. *Communications of the ACM, 45*(3), 1–8.

Thrun, S., Burgard, W., & Fox, D. (2005). *Probabilistic robotics*. Cambridge, MA: MIT Press.

Thrun, S., Fox, D., Burgard, W., & Dellaert, F. (2000). Robust Monte Carlo localization for mobile robots. *Artificial Intelligence*, *128*(1-2).

Turing, A. (1948). *Intelligent machinery.* Retrieved Feb 1, 2012, from http://www.turingarchive.org/browse.php/C/11

Waldron, K. J. (1985). Mobility and controllability characteristics of mobile robotic platforms. In *Proceedings of the IEEE International Conference on Robotics and Automation*, (pp. 237-243). IEEE Press.

Walter, W. G. (1963). *The living brain*. New York, NY: W. W. Norton and Co.

Wiener, N. (1965). *Cybernetics or the control and communication in the animal and the machine*. Cambridge, MA: MIT Press. doi:10.1037/13140-000

Zaloga, S., & Laurier, J. (2005). *V-1 flying bomb 1942-52: Hitler's infamous 'doodlebug*. London, UK: Osprey Publishing.

Zhao, H., Chiba, M., Shibasaki, R., Shao, X., Cui, J., & Zha, H. (2009). A laser-scanner-based approach toward driving safety and traffic data collection. *IEEE Transactions on Intelligent Transportation Systems*, *10*(3), 534–546. doi:10.1109/TITS.2009.2026450

Zheng, Y. F., & Zheng, Y. F. (1994). *Recent trends in mobile robots*. New York, NY: World Scientific Pub Co Inc.

Zorich, Z. (2011). Paleolithic tools - Plakias, Crete. *Archaeology Magazine, 64*(1).

Chapter 2
Robotic Bases

ABSTRACT

This is the second chapter of the first section. It presents the mechanical and physical foundations of mobile robots that are needed for a complete understanding of the concepts of further chapters, such as sensor and motion models. It provides a detailed review of the most common electro-mechanical components found in state-of-the-art mobile robots, emphasizing practical aspects, such as weight and size, power consumption, and performance trade-offs. Sensors and actuators, in particular, are stated as the hardware basis for coping with localization and mapping, and thus, specialized sections are devoted to them. The described devices range from low-cost sensors/actuators suitable for hobbyists to expensive professional-grade components.

CHAPTER GUIDELINE

- You will learn:
  - The most common types of robot locomotion systems.
  - A classification of sensors by the type of information they provide.
  - The physical principles behind advanced robotic sensors.
- Provided tools:
  - A glimpse at existing commercial robots and sensors.
- Relation to other chapters:
  - Motion models for each kinematic configuration are covered in chapter 5.
  - Probabilistic models for each sensor will be discussed in chapter 6.

DOI: 10.4018/978-1-4666-2104-6.ch002

1. INTRODUCTION

As established in the previous chapter, this book mainly focuses on *mobile* robots, which are indeed more interesting and have a greater potential in the service sector than static robots or robotic arms. Unfortunately, their advantages only come at the cost of an increased complexity, in both software and hardware. The first step to cope with that complexity is to provide the bases related to the mechanics and electronics of this kind of robots.

In this chapter, readers not familiarized with mobile robots will find an accessible, while thorough, overview of typical engineering designs and technologies employed in current research labs to build autonomous robots. Those already familiar with robotic sensors may also find it motivating to go through all the introduced technologies, ranging from those intended for the hobbyists to the latest developments.

A mobile robot must be able to interact with its environment through a set of *basic actions*. What we mean here by basic action is the execution of some set of minimal operations—in the sense that they do not invoke other actions—which are very close (and coupled) to the hardware of the robot. The set of basic actions of a mobile robot often comprises only one capability: to move around. For instance, in the case of an intelligent vehicle aimed at transporting a payload from one point to another, this capability of moving permits the design of a complete control architecture, capable of accomplishing uncountable specific tasks. More complex robots are equipped in such a way that they can also do other things, for instance grasp and manipulate objects by themselves, which exponentially increases their potential applications. In fact, moving around and manipulating objects are the two most common robot actions today, followed by the ability to communicate with humans and other machines. Just think of how many real-life tasks can be decomposed into sequences of the moving-manipulating pair of actions. Section 2 will review these and other less common robot actions, and how they are implemented in the hardware of current robots.

From the very first instant that a robot interacts with its environment, by moving itself or by manipulating other objects, there appears the need to gain some feedback from the world. For example, if a robot grasps a cup in a kitchen with the intention of taking it somewhere else, it absolutely needs to figure out whether the grasping was achieved at the intended points of the object. As another example closer to the scope of this book, when a wheeled robot intends to walk down a corridor in a straight path it needs to sense the world in order to compare the planned and the actual trajectories to avoid clashing with the walls.

Several reasons exist behind this fundamental need for feedback. First, the world may change while the robot performs its actions, which makes it necessary to constantly reevaluate the planned behavior. Secondly, the world—including the mechanical parts of the robot—will not always behave as the robot expects. In most cases the mathematical models of the robot-world interactions are only approximations (just to mention one example: wheel-ground friction while moving may lead to complex slippage behaviors), while in other cases the effects of actions must be treated as if they were random, either due to the lack of information (after trying to open a door a robot may find it was locked) or to the intractable complexity of an exact mathematical model (where all the billiard balls will end up after striking the cue ball?).

It is important at this point to stress the existence of several paradigms regarding the way feedback information is employed in automated systems. In classic control theory (Doyle, Francis, & Tannenbaum, 1992) the ultimate purpose of the system is to actuate such that a given property remains as close as possible to a desired reference level—for example, the temperature of a boiler or the speed of a motor. Solutions based on this classic control theory have been proposed to mobile robotics problems, and thus we can find in the literature examples such as *visual servoing* (Espiau, Chaumette, & Rives, 1993) or *motion control* (Quinlan & Khatib, 1993; Desai, Ostrowski, & Kumar, 1998). Unfortunately, many practical robotic tasks cannot be attacked by means of such a direct and well-studied approach. To cite one key issue central to this book: finding out its own position within the environment is a prerequisite for a mobile robot to perform a wide range of tasks—e.g. how could a robot head office 20.3 if the current robot location is unknown? As it will be explained in chapters 6 to 9, estimating the location of a robot often involves the use of intricate data fusion and statistical inference algorithms, hence localization cannot be modeled under the perspective of classic control theory.

All in all, a mobile robot absolutely needs to be equipped with sensors for gathering this essential feedback while carrying out its mission through its actuators. From section 3 on, we will

Table 1. A non-exhaustive list of companies that market the most common types of robots and robotic sensors at present. Products from these companies and concrete product names are mentioned throughout this chapter.

Robots	Wheeled mobile robots	• Adept Mobile Robots (formerly Mobile Robotics Inc.) • iRobot • KUKA • LEGO • Robotnik • Robosoft • Willow Garage
	Biped robots	• Aldebaran Robotics • Honda Motor Company, Ltd. • Pal Robotics
Sensors	Inertial	• Analog Devices Inc. • xSens Technologies B.V.
	1D ranging and very short ranging	• LEGO • Parallax Inc. • Robot Electronics (formerly Devantech)
	2D ranging	• HOKUYO Automatic Co. • RIEGL GmbH • SICK AG • Velodyne
	3D ranging	• Fotonic • Mesa Imaging AG • Microsoft / Primesense

review a variety of available sensor families, each targeted at each particular sensing need. Table 1 provides a summary of most companies mentioned in this chapter, classified by the kind of devices they market.

After discussing why having both actuators and sensors is a prerequisite for a machine to be considered a mobile robot we will address another fundamental component of chief practical importance: their power supply. Section 13 ends this chapter by exploring the different technological solutions to the problem of power storage on this these robots.

2. TURNING MACHINES INTO ROBOTS: ACTUATORS

As defined above, robot actions are any of the different ways in which a robot can interact with the external world, i.e., with its environment. Since the main concern of this book is mobile robotics, the action of moving is the only one that will be treated throughout the text. A robot, however, may have many other capabilities, such as grasping and manipulating objects, communicating with other machines or speaking to human users. Since all those are somewhat orthogonal to localization and mapping, we will not focus on them in this text.

The particular hardware implementation of a robot capacity for mobility does not affect much the applicable localization and mapping techniques, disregarding the natural division between 2D-only and 3D-capable methods. In the following, we classify robots, in a non-exhaustive way, by their locomotion types and stress the concepts about those systems that may be useful in later chapters, where we will study the probabilistic motion models required for addressing localization and mapping. Some of the diversity in actuation is captured in Figure 1.

Figure 1. Some examples of the diversity found among mobile robots. (a) Three quadrotors, from the Flying Machine Arena, fly in formation (© 2010 Raymond Oung. Used with permission). (b) A hexapod robot, a popular instance of multi-legged robots, built by the first author for his MSc thesis in 1994. (c) Our wheeled robot SANCHO, built upon the commercial mobile base Pioneer™ 3-DX from Mobile Robotics Inc. (Mobile Robotics, 2011). (d, e) The wheeled and biped humanoid robots REEM™ and REEM-B™, respectively, from Pal Robotics (Pal Robotics, 2011) (© 2011 PAL Robotics. Used with permission).

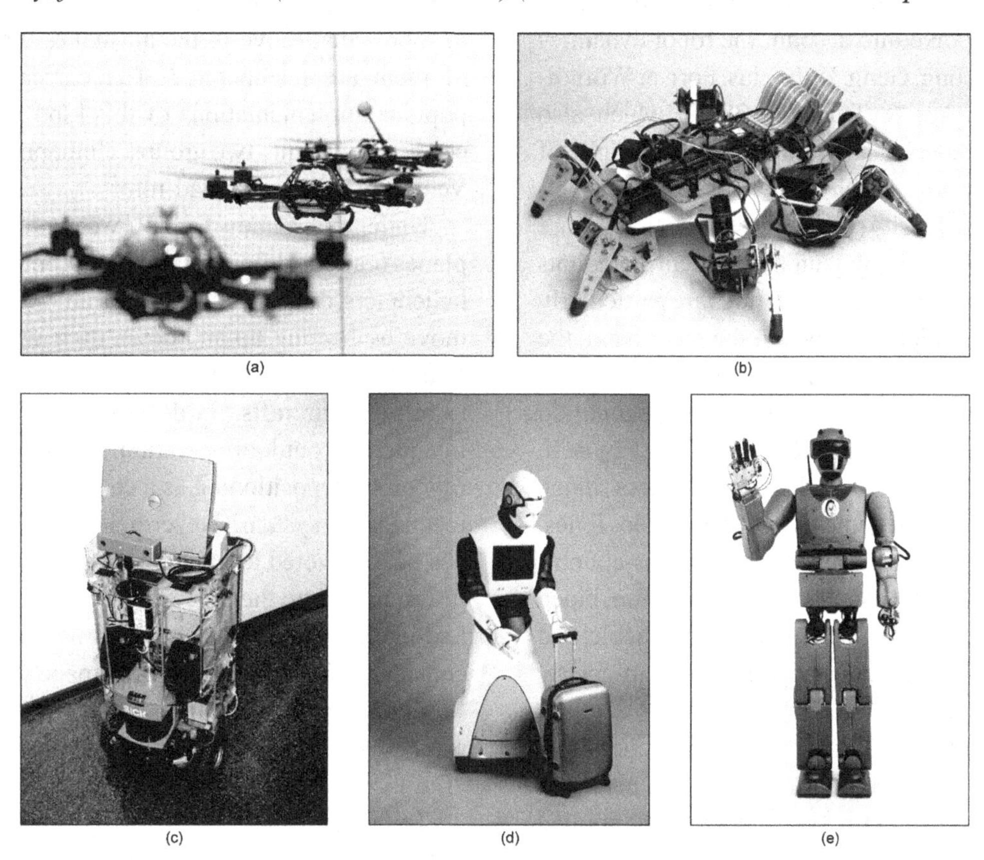

Legged Robots

In the last decades of the 20th century, there was a considerable interest in legged robots, including one-leg jumper robots, rudimentary bipeds, and multi-legged machines (four or more legs). At present, the research community is more focused on biped robots than in the other configurations, although six and eight-legged configurations are still popular among students and hobbyists. The reader interested in the history of legged robot research can refer to Raibert and Tello (1986).

Regarding biped robots, nowadays they account for only a tiny fraction of the existing mobile robots, mainly due to the several challenges for stability that represents the usage of only two articulated legs. We must state the difference between the terms *biped* and *humanoid*, the latter being applied to any robot with an upper body resembling a human being. Humanoids are sometimes bipeds, like the famous Honda™'s Asimo (Hirose & Takenaka, 2001), but there exist other human-like robots that move on wheeled platforms—like NASA's Robonaut (Aldridge, et al., 2002) and Willow Garage's PR2™ (Willow Garage, 2012)—or some that even do not move at all (Endo, et al., 2009).

Efficient and secure control of agile walking or running biped robots is thus still an active area of research. A clear indicator of that activity is the existence of specialized annual conferences focused solely on this topic. Unlike the case of wheeled robots, designers of biped robots must absolutely take into account the robot dynamics (Manoonpong, Geng, Kulvicius, Porr, & Wörgötter, 2007; Aoi & Tsuchiya, 2011), which also imposes especial restrictions on the weight of onboard systems. Both equipment weight and battery energy density (please refer to section 13) are limiting factors for the autonomy of current robots of all kinds, but weight limitations are particularly relevant for biped robots. On the other hand, the main advantage of bipeds is their capability to traverse uneven terrains the way we human do (Huang, Yokoi, Kajita, Kaneko, Arai, Koyachi, & Tanie, 2001), thus their workspaces range from unmodified homes (Ciocarlie, Hsiao, Jones, Chitta, Rusu, & Sucan, 2008) to cross-country zones (Boston Dynamics, 2010). In practice, biped technology is not yet as consolidated, reliable, and energy-efficient as other approaches—particularly wheeled robots.

Concerning the aspects of biped robots that influence localization or mapping, the most relevant one is that 2D or 3D methods can be applied depending on the physical placement of the sensors in the robot body. For instance, the work (Tellez, Ferro, Mora, Pinyol, & Faconti, 2008) describes the humanoid robot Reem-B™, whose feet are equipped with planar laser scanners (see section 6) which allow the usage of well-established 2D localization and mapping algorithms. Other humanoid robots perform vision-based localization and mapping by means of single or stereo cameras mounted on their heads, which requires employing full 3D methods (Davison, Reid, Molton, & Stasse, 2007) to track the sway of the robot head as it walks.

Flying Robots

While weight is an important design factor in biped and other types of robots, it truly becomes the primary design constraint for flying robots. In addition, the complexity of achieving autonomy in robots that move in the air makes that many of them are teleoperated. The currently most popular implementations of this kind of robots can be divided into two groups: Unmanned Aerial Vehicles (UAV) and quad-rotors.

Under the name of UAV, we typically find planes (called "drones" in military applications), helicopters, dirigibles and even ornithopters—that move by beating up and down their wings—all of them in a wide range of sizes, from miniature to full-size aircrafts. By design, these UAVs are intended for outdoor operation, hence they often rely on GPS positioning as a central part of their localization system. Nevertheless, much work has been devoted to integrate visual and inertial information into the process of UAV localization and maneuvering control, which naturally must consider a full 3D vehicle workspace (Saripalli, Montgomery, & Sukhatme, 2003; Ollero & Merino, 2004; Caballero, Merino, Ferruz, & Ollero, 2009).

Today, an especial case of flying robots are quad-rotors, which are gaining popularity in indoor scenarios since their set of four individually-controlled rotors allows a quick and precise control of its trajectory with great stability (Michael, Mellinger, Lindsey, & Kumar, 2010). Due to their typical small sizes some of these robots are also called *Micro Air Vehicles* (MAVs). It has been proposed to use 2D sensors—such as planar laser scanners—onboard these vehicles in order to localize, build maps, and navigate within indoor structured environments using 2D-only techniques (Morris, Dryanovski, & Xiao, 2010).

As will be discussed in chapter 5, the availability of an approximate measure of how much the robot has moved at each time simplifies the problem of mapping, by means of what is called a *motion model*. However, one aspect that applies equally to all types of flying robots is the absence of a mechanical, reliable odometry, unlike in grounded legged or wheeled robots. This lack can be alleviated with a combination of inertial sensors (see section 8) and monocular or stereo cameras (Nistér, Naroditsky, & Bergen, 2006).

Submarine Robots

The problem of localization for a submarine robot is cumbersome, mainly for the unfeasibility of using the most common robotic sensors under water. In particular, these vehicles move in a three-dimensional workspace, like aerial robots, but in this case, there is no GPS-like system readily available due to the impossibility of underwater radio-wave transmission in microwave frequencies. In fact, practical usage of subsea radio waves is limited to a few kilohertz. In addition, locomotion is harder in these environments and consumes more energy.

The work in Yuh (2000) presents an interesting survey of existing Autonomous Underwater Vehicles (AUVs), including an analysis of the dynamics of this kind of robots. AUVs are equipped with a set of thrusters that can be controlled independently in order to pilot the vehicle movement. Dead reckoning can be achieved with either Doppler Velocity Log (DVL), inertial sensors or a combination of both (Larsen, 2000).

Wheeled Robots

At present, using wheels is the preferred way of locomotion for most terrestrial mobile robots: it requires less energy, the actuators are cheaper, stability is not an issue, and wheels serve well for most scenarios, as long as the floor surfaces are smooth enough. Different designs for wheeled robots exist depending on the number and physical placement of the *drive wheels* (those receiving torque from a motor) and *casters wheels* (undriven passive wheels). Typically, these robots rely on electric brushed DC motors as actuators for their simplicity, power capability, and price. Carefully controlling the speed of such motors is the lowest-level stage of robot motion control and navigation, and requires its own feedback sensors, which conform with what we have called in chapter 1 the vehicle *odometry*.

A wheeled-robot *odometry* consists of sensors that measure how much one or two wheels have rotated between two instants of time (see Figure 2), which allows the system to obtain a short-term estimation of its location in space based on the traveled distance, i.e. to perform dead-reckoning. In some kinematic designs, the orientation of some wheel may also be measured. The exact way all these pieces of information are put together to estimate the instantaneous vehicle motion will be discussed below, since it depends on the particular kinematic design of the robot.

The utility of odometry is twofold: for dead reckoning, and also as the input to the low-level closed-loop control of the vehicle velocity. Odometry is most commonly implemented by means of dual-channel tachometers, or phase-quadrature optical encoders (Borenstein, Everett, & Feng, 1996), which consist of disks mechanically attached to the wheel axis and which have been perforated following a pattern of holes along two circular paths of different radii. A pair of photodetectors reads out these patterns as the disk turns, being the angle that a wheel has turned proportional to the number of counted holes. Both patterns in the disk are exactly the same but shifted one half of the hole length with respect to the other (i.e. an offset of "90 degrees" if the pattern period corresponds to 360 degrees), so this kind of rotational encoder makes possible to also determine the direction of rotation.

We introduce now the most popular kinematic designs for wheeled robots (Borenstein, Everett, & Feng, 1996), together with their possible instrumentations for providing odometry. The kinematic equations for each system will be derived in chapter 5.

Figure 2. The different kinematic configurations of wheeled robots discussed in the text, along with their corresponding odometry instrumentations

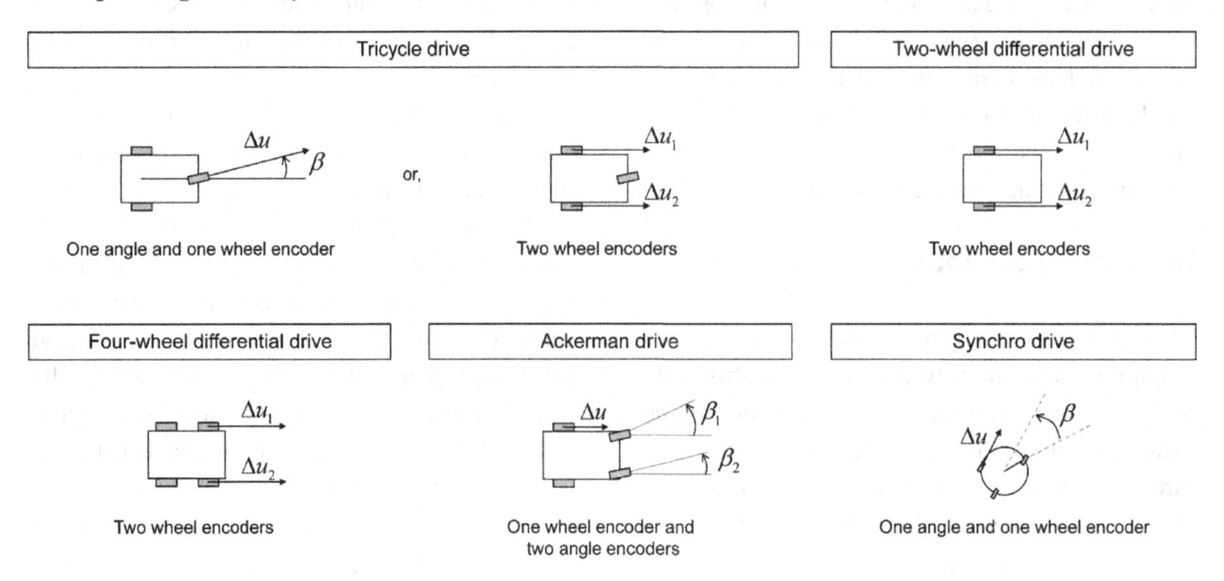

- **Differential drive.** In this design, there exists at least one drive wheel at each side of the robot, each with its own motor. If there are only two driven wheels, rear or front casters must be installed to stabilize the mobile platform. Another possibility is to place two drive wheels at each side instead of one, leading to a four drive wheels differential robot. Commercial examples of these designs are the Pioneer™ P3-DX and 3-AT robotic bases, respectively, both manufactured by Mobile Robotics Inc. (Mobile Robots, 2011). Notice that the lack of a steering mechanism means that the four-wheel drive version will always suffer from drive-induced wheel slippage: since all four wheels are fixed and aligned with the front-rear axis, they exert a resistance force whenever the robot turns. On the other hand, in comparison to the tricycle and Ackerman designs—see below—differential drive has two advantages: (1) a minimum turning radius of zero for the robot, and (2) not needing a steering wheel, which would require its own actuator and exhibit a strong resistance to be turned while the vehicle is still.
- **The tricycle model.** This design comprises only one drive wheel, which is also the steering wheel. Two passive wheels in the opposite part of the robot complete this model. Regarding its odometry instrumentation, there are two possible approaches. In the first one, the steering-drive wheel is equipped with both rotational and angle encoders, while the passive wheels are not instrumented. Alternatively, only the passive wheels can be instrumented with rotational encoders. Both alternatives lead to exactly the same differential-drive odometry equations, although this model imposes a minimum turning radius, which is avoided with differential-drive.
- **Ackerman steering.** This is the model found in all automobiles. Here, the steering angles of two front wheels are slightly different such that the extended axes of all four wheels intersect at a common point, as illustrated in Figure 2. It will be shown (in chapter 5 section 5) that this model is equivalent to a tricycle model for some equivalent steering angle. An Ackerman vehicle is, however, more stable than a tricycle vehicle in outdoor environments,

thanks to the extra front wheel. In comparison to the differential four-wheel drive, it also has the advantage of avoiding geometry-induced slippage while turning. On the other hand, Ackerman steering presents a reduced maneuverability due to the existence of a minimum, non-zero vehicle-turning radius.

- **Synchro drive.** In this case there exist three or more wheels, all of them being both steering and drive wheels simultaneously. All the wheels are mechanically (e.g. through a chain or belt) or electronically coupled such that they align to define the motion direction. The main advantage of this approach is obtaining a robot capable of moving in any direction at any instant, as long as time is available to allow the wheels to align in the required direction first. An example of a commercial robot equipped with synchro drive was the B21™ robot, manufactured by Real World Interface Inc. (2011)—now known as iRobot.

Regarding the aspects of wheeled robots that influence localization or mapping, the most remarkable is that two-dimensional simultaneous localization and mapping (SLAM) methods—those that consider the robot moving on a plane—are often applicable to this kind of robots if two-dimensional sensors are also employed. Another particularity is that odometry is an exception in the way sensors are treated in probabilistic approaches: it will not be considered an *observation* of the external world, but a robot *action* instead. At first sight, it is common to find this distinctive treatment shocking and confusing since, after all, odometry readings are nothing else than readings from a robot sensor (i.e. rotational and/or angular encoders). However, it is easy to see that odometry actually measures, with great accuracy, how much each wheel has turned, which in turn determines where the robot *intended* to move to (its *action*). Odometry is only inaccurately and approximately related to where the robot actually ends up moving to. Modeling robot actions from odometry readings will be further explored in chapter 5 in the context of probabilistic Bayesian approaches to localization.

3. HOW DOES THE WORLD LOOK TO A ROBOT? SENSORS

Choosing which sensors a robot will be equipped with is a crucial stage in its design. Among other implications, this election absolutely defines the set of applicable methods for localization and mapping.

Thinking of a simile with the biological world, it is obvious that the senses of a given animal must determine the models it builds up about the external world. For our case as humans, the preponderant sense for localization and mapping is vision, which allows us to build 3D representations—of a kind not well known yet—of our environment and recognize a known place quite easily by identifying colors, shapes or the arrangement of individual objects. The olfactory, auditory and other perceivable characteristics of the world seem secondary for us (although they are not for many other animals), and that is why we do not account for them while thinking of how to drive into downtown. In short: visual features and three-dimensional models are the dominant world representational elements for us humans, but it must be made clear that this is neither the only way of representing the world nor even the *best* model in all situations, since in the end the model must fulfill the principle of being useful for its intended purposes.

The biological world displays a vast repertory of senses evolved to detect properties of the world quite different than those we are used to, and which indeed shape whatever world models their owners might have. Well-known examples are the capability sharks have revealed of detecting extremely weak electric fields (such as those produced by the muscles of preys) and the echo-localization capabilities of bats, whales, and dolphins. Probably, the perceptions these animals have are quite different than ours, one point that enforces the idea that our 3D (mainly visual-based) models are not the only possible or best representations of the world.

It becomes natural to extend this discussion to artificial autonomous agents such as mobile robots, whose models of their environment can be anything that is useful to their assigned tasks and their sensors. For the aims of this book, the main task to be solved is being able to localize itself within the environment without getting lost.

As explained in chapter 1, a first classification of robot sensors includes the following separate groups: proprioceptive and exteroceptive. The first one groups those devices in charge of detecting the internal state of the robot (*proprioception*), such as the current position of its wheels, its orientation—much like the internal ear in humans—or the different joint angles in the kinematic chain of a robotic arm. The second class of sensors is aimed at gathering information from the external world (*exteroception*), and is the main source of data for localization and mapping. In chapter 1, we also distinguished within this class a third one: *environmental sensors*, which are exteroceptive sensors that are located outside the robot body.

It is clear that building a model of the world (or *mapping*) with a certain quality requires the use of exteroceptive sensors. Proprioceptive sensors can contribute information, which makes the process easier and more robust, but they alone render insufficient in practice. Each type of exteroceptive sensor provides the robot with a measure of some given property of the elements in the environment, e.g. their spatial location or their color. It is worth highlighting at this point that, even if a sensor is able to take a measurement related to the physical placement of an external object, it does not immediately follow that we will be able to estimate the object position from one reading. For instance, in some cases the dimensionality of the sensor readings is smaller than the minimum required for unambiguously locating the object in space. In other words, the reconstruction of the sensed objects from the readings may have a *gauge freedom* (one or more free degrees-of-freedom). Accordingly, we present in Table 2 a classification of some of the sensors discussed in this chapter, attending to whether the spatial coordinates of sensed objects are *observable* (we can reconstruct them without ambiguity) or *partially observable* (there exists at least one free degree-of-freedom). In this latter case reconstruction is still possible thanks to the progressive accumulation of evidence within probabilistic filters, as we will see in chapter 8.

The following sections present a review of common sensors of both kinds typically employed for the instrumentation of mobile robots (Borenstein, Everett, & Feng, 1996). Notice that chapters 5 and 6 will provide more details on how to handle the data provided by these sensors within a probabilistic framework. If the reader is surprised of odometry not being included in this list of sensors, please refer to chapter 5 section 1, where we expose the reasons to treat odometry as an *action* rather than a robot *observation*.

4. PROPRIOCEPTIVE SENSORS: INERTIAL SENSORS

Under the category of inertial sensors, we find a kind of device fundamentally different from all others that will be addressed later on. Instead of performing measurements about the external world and its properties, inertial sensors are *proprioceptive*: when installed on a robot, they measure the first or second order derivatives of its pose (remember that the robot *pose* consists of its position and orientation). Note that we are assuming a modeling of mobile robots as single rigid bodies. Robots that can be represented by multibody models could be equipped with several

Table 2. A classification of some of the sensors described in the text regarding the observability of the spatial coordinates of sensed objects from just one sensor reading

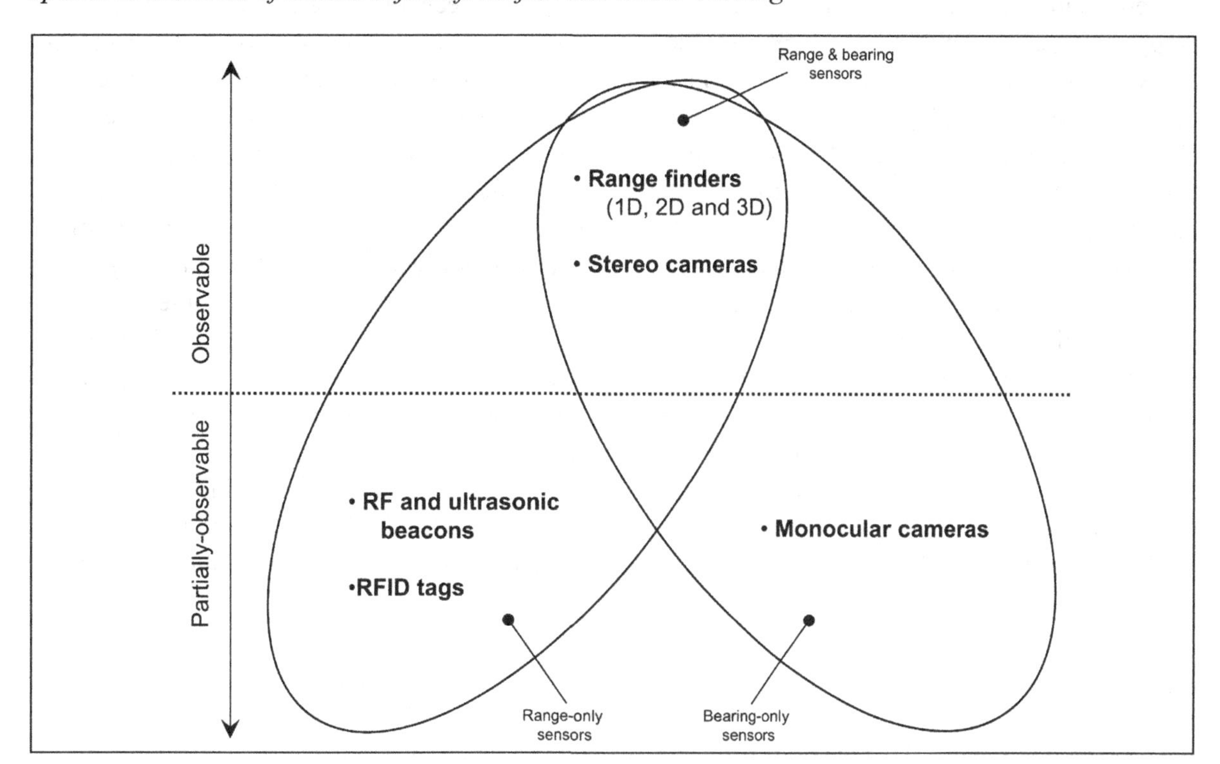

inertial sensors in their different parts, but this issue is far out of the scope of this text.

It is significant that inertial sensors only measure the derivatives of the robot pose instead of its absolute value. In the long-term, this implies that inertial sensors alone cannot be used to track the position of a robot accurately, and some kind of sensor fusion with exteroceptive ones becomes mandatory—an issue also discussed in chapter 5 section 8.

We can find two main families of inertial sensors: gyroscopes and accelerometers. Although some commercial devices integrate both, it is illustrative to describe them separately.

Firstly, gyroscopes typically employed in mobile robotics are devices capable of measuring changes in orientation, or more specifically, the orientation first-order derivatives, i.e. angular speeds. Notice that for robots moving in 3D this implies detecting the change in orientation for three degrees of freedom (i.e. rotations around three orthogonal axes). At present, MEMS technology has evolved enough to provide inexpensive gyroscopes weighting only a few grams and with an extremely limited power consumption. Doubtless, these reasons have made MEMS gyroscopes quite popular for equipping mobile robots.

Another relevant technology is fiber optic gyroscopes, where a laser device excites one end of a coil of optical fiber while its phase is measured at the other end. Einstein's special and general relativity theories postulate that the speed of light remains fixed no matter how complex is the motion state of the reference frame, that is, if the laser beam enters the fiber coil while the entire device rotates, the propagation velocity of the laser will not change with respect to its rest state; but while the speed of light does not change, the distance to be traversed along the coil may change depending on the rotation direction, thus leading to a phase shift measurable by advanced optoelectronics. This can be regarded as a sort of Doppler effect. This family of gyroscopes provides an exceptional accuracy, but although they have been occasionally employed in mobile robotics (Borenstein, Everett, & Feng, 1996), their excessive price restricts their application to avionics and other highly demanding fields.

Secondly, we have accelerometers. These devices measure the *proper acceleration* of the rigid body they are attached to. Again, we need to invoke here Einstein's General Relativity to define proper acceleration as that felt by an object in an inertial frame of reference, that is, in free fall. This concept is fundamental for interpreting data from accelerometers—see Figure 3b. The reader is probably familiar with the concept of (Newtonian) *acceleration*, the one employed in classical mechanics. For example, we know that gravity exerts a constant downwards acceleration to objects in free fall. After the object hits the ground and remains still, that acceleration becomes zero. It can be easily seen that this concept of acceleration in fact depends on the chosen frame of reference (in this example, the one of an external observer). On the other hand, *proper acceleration*, as measured by accelerometers, can be shown to equal zero when objects are in free fall, and becomes the value of gravity when objects are standing still (for instance, on the ground). In other words: proper acceleration equals the acceleration in any given coordinate frame with respect to free-fall, that is, subtracting the acceleration components from gravitational sources.

Figure 3. (a) Inertial sensors, like accelerometers, rely on the concept of proper acceleration. The figures illustrate a mass standing on a surface and in free fall, along with the corresponding acceleration in the coordinate frame (a_c) and the body proper acceleration (a_p). (b) A complete IMU system, the Xsens MTi™. (c) A MEMS gyroscope of the ADXRS™ family in its millimeter-sized package (manufactured by Analog Devices, Inc.).

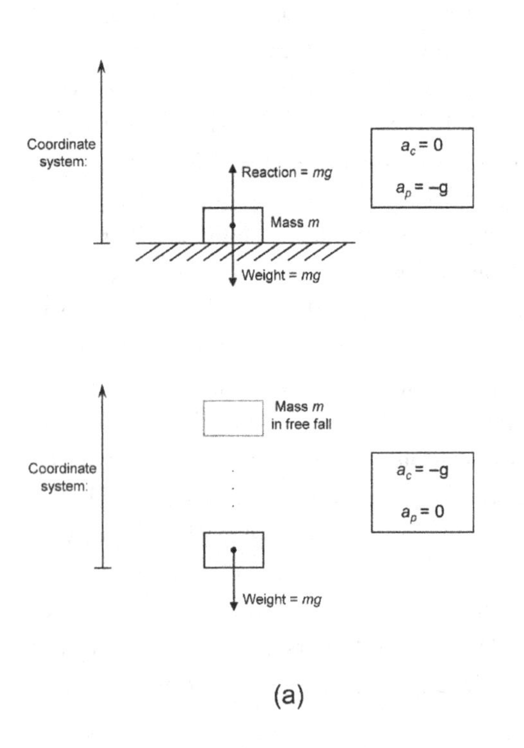

(a)

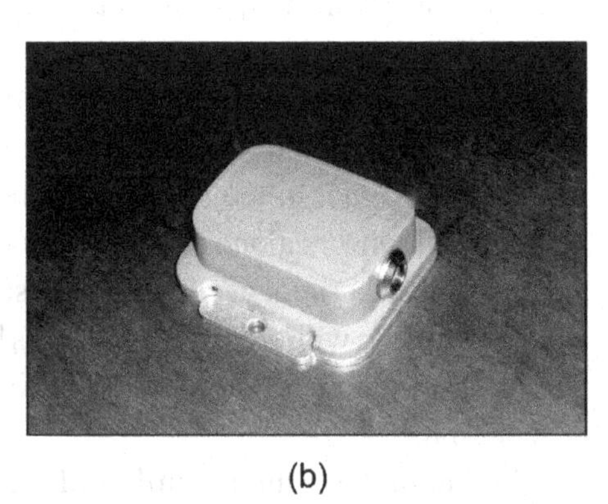

(b)

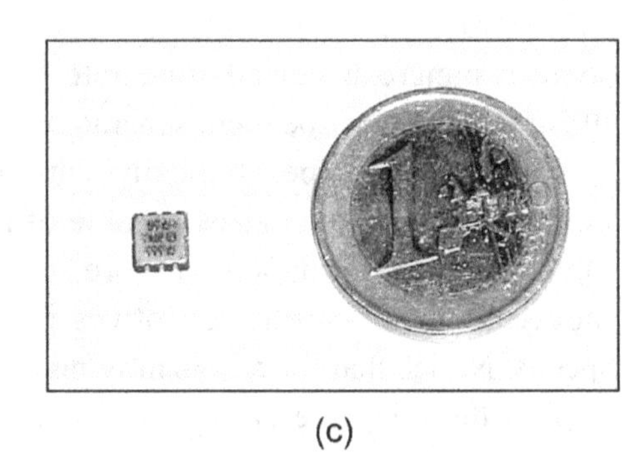

(c)

After clarifying this point, it can now be understood why accelerometers are used to detect tilt angles (with respect to downwards gravity) even for stationary objects with an externally observed null acceleration: assuming a fixed, known direction and magnitude for the gravity acceleration, we can determine the angle of any accelerometer with the downwards direction by computing the dot product of the two vector magnitudes, since the dot product of two vectors is proportional to the angle between them.

Regarding the technology employed for designing accelerometers, at present most commercial accelerometers for robotics are MEMS devices due to their low price, low weight, low power consumption and competent accuracy. Furthermore, one can find in the market inertial modules such as the xSens MTi™, shown in Figure 3a, which integrates gyroscopes, accelerometers and the required software to fuse both sensor data into one improved dead-reckoning estimate. Affordable accelerometers and gyroscopes as those in Figure 3c can be also purchased directly from companies such as Analog Devices for mounting into prototypes or mature electronic designs.

5. EXTEROCEPTIVE SENSORS: CONTACT AND VERY SHORT-RANGE SENSORS

Among the simplest exteroceptive robotic sensors, we find *bumpers*, or contact sensors. These devices are nothing else but an arbitrarily shaped contact surface mechanically attached to a push-button switch, such that a binary contact/no-contact signal can be read by the robot software. Several bumpers are often installed forming a ring of bumpers covering the entire robot perimeter. Despite their little helpfulness for localization and mapping, these simple sensors reveal themselves as extremely useful in obstacle avoidance and safe navigation, where they are employed as the last-chance detectors for unexpected obstacles. Examples of commercial robots equipped with bumpers are most mobile bases from Mobile Robotics Inc., or the educational NXT™ robots from LEGO MindStorms series (Lego, 2011).

While bumpers require physical contact for object detection, a small improvement can be achieved through contactless reflectivity sensors. Just slightly more complex than bumpers, these devices comprise an infrared (IR) emitting LED and a phototransistor that provides an electrical output voltage proportional to the intensity of the IR light reflected by any nearby object. The high sensitivity of this voltage to the wavelength of the reflected light makes these sensors very appropriate for detecting changes in color, hence they are often called "line following" sensors, after one of their most common applications. A commercially available representative device is Parallax's QTI™ reflectivity sensor (Parallax, 2011), which can be purchased at the time of writing this book by less than \$3.00; the equivalent ready-to-use sensor for LEGO's NXT robots is also available for less than \$20.00.

Both bumpers and short-range reflectivity sensors are quite popular in hobbyist robotics because of their low prices and easy interface to custom electronics. In principle, they could be used (together with some form of odometry) for localization or even map building by means of 2D grid maps and the techniques described from chapter 8 on. Such grid maps would represent the occupancy (from bumpers activation or detected color) of every portion of the environment. However, the short sensing range of these sensors would make the overall localization accuracy heavily reliant on odometry; hence such a sensor-minimalistic approach has not ever been reported in the scientific literature, to the best of our knowledge (see Figure 4).

Figure 4. One possible configuration of a LEGO MindStorms NXT™ robot, equipped with a light sensing device and a sonar prepared for sensing short distances—pointing downward and forward, respectively. In this setup, the robot is programmed by students of several engineering programs in the University of Málaga (Cruz-Martin, Fernandez-Madrigal, Galindo, Gonzalez-Jimenez, Stockmans-Daou, & Blanco, 2012).

6. EXTEROCEPTIVE SENSORS: SINGLE-DIRECTION RANGEFINDERS

Bumpers and short-range reflectivity sensors can be regarded as quite rudimentary since they can only detect objects that physically touch or are very close to touching the robot. We describe now the most common and mature technologies capable of detecting the presence of objects *at distance*. Doing so while simultaneously obtaining an estimate of that object distance is called *ranging*, and sensors capable of doing so are called *rangefinders* or *range* sensors. For now, it is enough to discuss about sensors providing one single measurement in a predetermined direction. Although this provides useful data for localization and mapping, we will see later on that 2D and even 3D extensions of these sensors are much more popular in the robotics community, for they provide richer information; naturally, single-direction sensors are much more inexpensive alternatives, posing an interesting choice for many hobbyists or low-cost designs. When the sensor has no associated directionality, we speak of a *range-only* devices, which are totally different sensors addressed more in depth in section 9.

There exist three main technological approaches for ranging in current commercially-available sensors: triangulation and two variants of Time-of-Flight (ToF). We review them below and describe some representative devices of each kind.

Triangulation-Based Proximity Sensors

This family of sensors relies on a geometrical approach to range sensing. An IR laser LED source illuminates the surface to be measured, conveniently focused by optical lens into a small brilliant spot—low-cost devices employ common LEDs instead of lasers, reducing the attainable measuring range. The projected light spot is detected by a separate part of the device, an electronic photoreceptor, which measures the incidence angle of the light reflected in the target object. Denoting that incident angle by α and the baseline (distance) between the emitter and receiver, which is known by design, by b, it follows from elementary geometry that the measured range r is obtained as:

$$\tan\alpha = \frac{r}{b} \quad \rightarrow \quad r = b\,\tan\alpha \tag{1}$$

In practice, however, directly measuring the angle of incidence of a light source is not straightforward. A solution based on optical lens is employed in Sharp™'s GP2D120/GP2Y-0A21YK0F IR proximity sensors (Sharp, 2011), quite popular due to their easy interfacing and low price. The idea, sketched in Figure 5c, consists of bending the reflected rays with lens such that they hit an array of photoreceptors at a location,

Figure 5. (a, b) Outer and internal views, respectively, of a SHARP™ GP2D120 infrared sensor. (c) A schematic illustration of the working principle behind triangulation-based IR rangefinders.

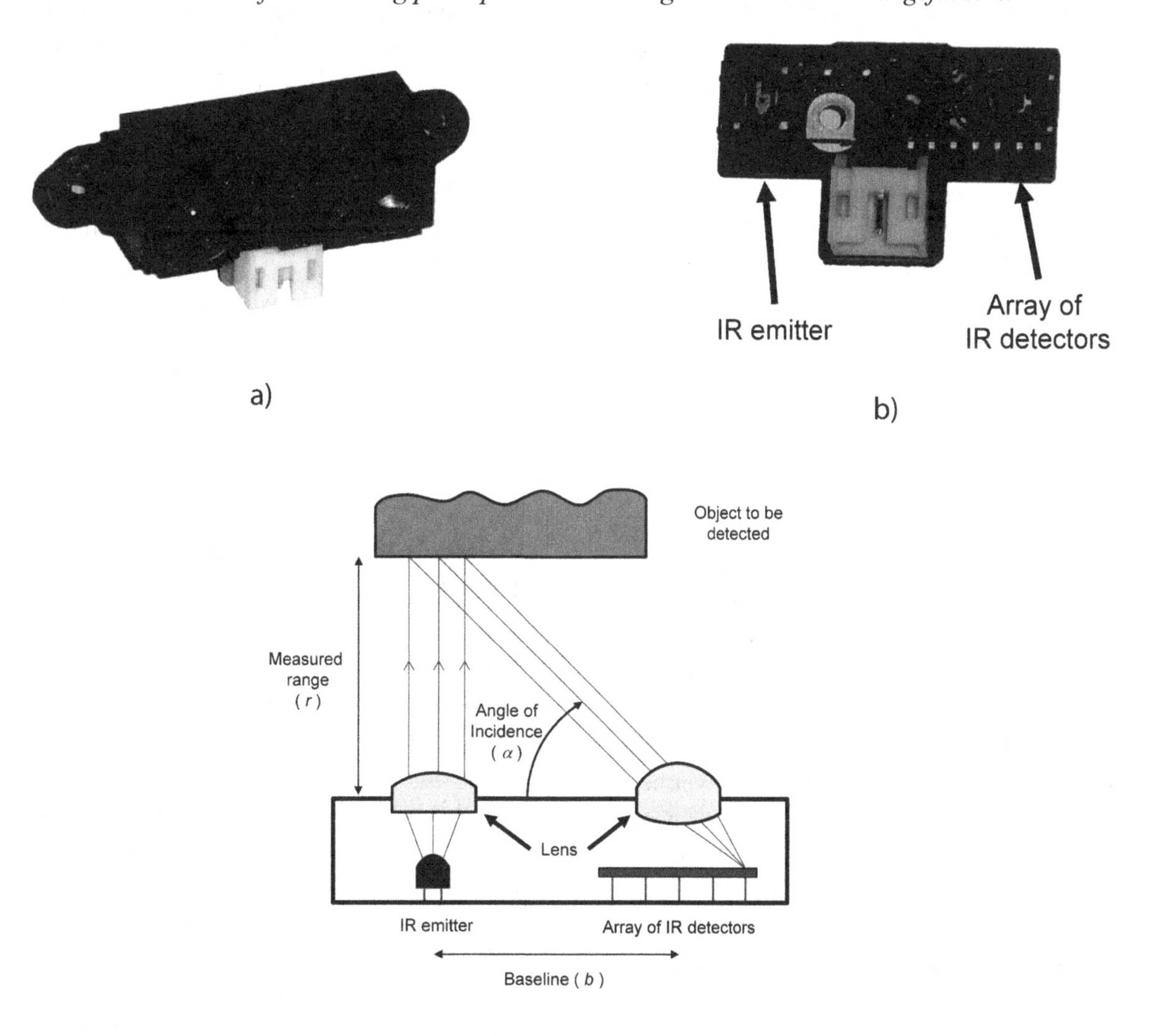

which becomes a function of the incident angle. Calibration tables of position-to-angles are then used to estimate this angle, which in turn is used via equation (1) to output the estimated distance to the sensed object.

This family of sensors is often used for obstacle and stair detection for their relatively good reliability at short ranges—up to one meter, approximately.

Pulsed-Signal Time of Flight (P-ToF)

The idea underlying this family of sensors is simply that of emitting a pulse of energy, then waiting for the received echo and finally estimating the round trip distance through the measured delay. This idea is the same approach exploited in RADAR detection systems. In the field of robotic sensors, the pulse can be of either acoustic, radio, or infrared light waves. Denoting the round trip delay as t and the wave speed of propagation as v, it turns out that the distance r to the object being detected can be obtained as (see Figure 6a):

$$t = \frac{2r}{v} \quad \rightarrow \quad r = \frac{t\ v}{2} \tag{2}$$

The most inexpensive sensors in this category are ultrasonic ranging devices, or sonars, such as the popular Devantech's SRF04—the company is now named Robot Electronics (2011). These devices comprise separate ultrasonic transceivers for emission and reception of short 40KHz pulses, allowing the detection of obstacles up to 3 meters in a *Field Of View* (FOV) of approximately 30 degrees. Such a large FOV is the outcome of the combined transmitter-receiver beam patterns, and can be an advantage if our goal is to detect any nearby obstacle… or it can be a challenge if we want to build accurate maps or localize the robot precisely. Due to the low speed of sound waves in air (340m/s), the maximum measuring rate is limited by the time taken by the wave traversing the round trip back to the sensor. For a maximum range of 4 meters, it turns out that the sensor must wait for the potential echo up to 12ms—which is a lot for a modern computing system—hence giving us a maximum firing rate of about 80Hz. In practice, crosstalk effects worsen the situation, even drastically decimating this rate, as we will discuss below.

The equivalent approach with laser beams instead of ultrasound waves has been extensively exploited commercially by surveying, automation, and robotic sensor manufacturers. Especially in surveying, these devices remain fixed during each measure in order to average the measure of several individual laser pulses and thus increase the system accuracy. The challenge of this kind of technology is the availability of very high-speed electronics, since resolving distances with a precision of one millimeter requires resolving sub-picosecond signals. It was not until the second part of the 1990 decade that such technology was matured enough to be commercially available at reasonable prices.

Maybe surprisingly, while unidirectional laser rangefinders are of little utility in mobile robotic localization and mapping due their negligible FOV, their two-dimensional versions, discussed in section 6, are one of most interesting sensors for the scope of this book.

Continuous-Wave Time of Flight (C-ToF)

There exists an alternative approach to ranging that does not require the emission of energy pulses, hence avoiding the potential inaccuracies in detecting the rising edge of a received echo.

In this scheme, an emitter (typically an array of laser or ordinary LEDs) radiates a continuous carrier signal modulated in amplitude with a pure sinusoidal tone in the order of a few tens of megahertz. This signal is reflected by the target object to be surveyed and is received by the appropriate receptor (typically an individual or

Figure 6. An illustration of the working principles of (a) P-ToF and (b) C-ToF, respectively. Notice how the former measures the delay between a fixed part of an amplitude-modulated pulse between its emission and its reception. On the other hand, C-ToF measures the phase delay between the modulating signal (the envelope of the real signal transmitted to the media), transmission, and reception.

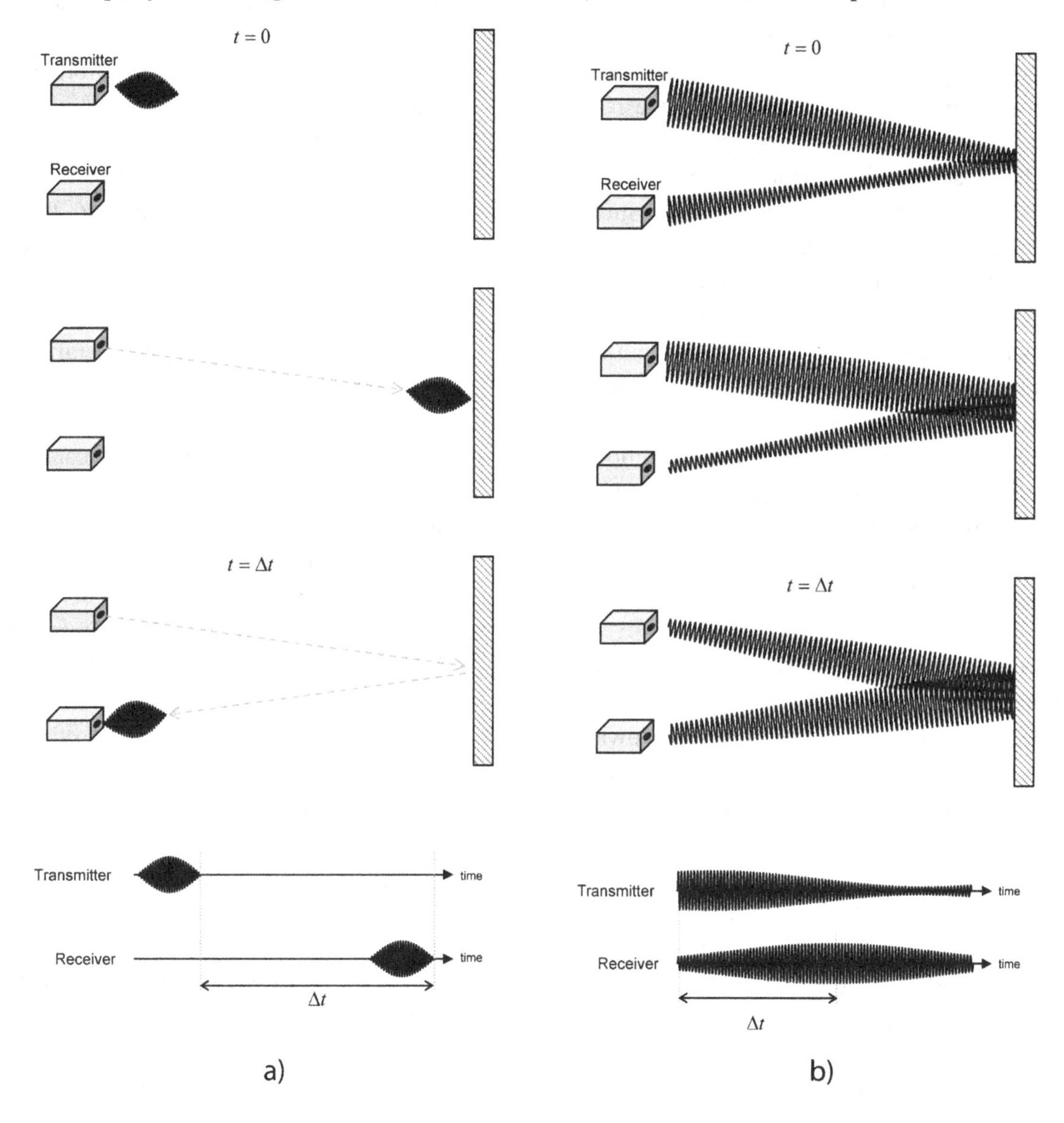

an array of phototransistors), which selectively amplifies the bandwidth of interest. At this point, the amplitude-modulated signal is demodulated to retrieve the original superposed sinusoidal tone, whose *phase* is finally compared to that of the emitted signal. It can be easily shown that the phase shift $\Delta\varphi$ is related to the wavelength λ of the modulating tone (not to be confused with the *carrier* wavelength, several orders of magnitude shorter) and the distance to the target obstacle r, as follows (see Figure 6b):

$$\Delta\varphi = 2r\frac{2\pi}{\lambda} \quad \rightarrow \quad r = \frac{\Delta\varphi\ \lambda}{4\pi} \tag{3}$$

One of the main drawbacks of this approach is revealed by the equation above itself: since the phase change $\Delta\varphi$ is by definition restricted to the range $[0,2\pi]$ (radians), the sensor would suffer from *phase aliasing*, that is, from not being able to discriminate between $\Delta\varphi$ and $\Delta\varphi+k2\pi$ for any positive integer value of k. In terms of ranges, this means that $\lambda/2$ determines the maximum distance that the sensor should measure, since any larger range would be ambiguously reported as a much shorter distance due to aliasing. Still, in practice, some state-of-the-art receivers implement smart solutions to this problem by deducing the integer number of wavelengths (the k above) from the attenuation of the reflected wave, thus effectively extending the sensing range beyond $\lambda/2$. This C-ToF technology is not common in unidirectional rangefinders, but is gaining popularity among 3D range finders, which we address in section 7.

Final Remarks

It is worth mentioning that all the three approaches to ranging explained in this section (triangulation, P-ToF, and C-ToF) may suffer from the following shortcomings, which affect their proper operation to a larger or lesser extent depending on the specific sensor technology:

- **Variations in the transmission speed of the medium.** This problem only affects significantly sonar sensors, since the speed of sound in the air depends on the temperature. For an operating temperature 10ºC off the nominal temperature for which the sensor was calibrated, readings can exhibit a 1.7% error.
- **Crosstalk.** In the case of multiple ranging sensors operating simultaneously in the same environment, caution must be paid to avoid each sensor interfering with the readings of the rest. The problem is much more relevant again in the case of ultrasonic sensors (sonars), due to their larger FOV and the low speed of sound. A solution to completely avoid the risk of crosstalk between sonars placed at different parts of a robot body is firing them in sequential order and waiting time enough for each echo to fade out before firing the next sonar. More elaborated solutions include using multiple frequencies for each sonar to make possible filtering each sonar signal out of the rest (Martínez, González, & Martínez, 1997), although that requires customized sensory electronics.
- **Multipath propagation.** A ranging sensor cannot only interfere with the operation of other similar devices, but with itself. The fact is that sharp corners and other adverse geometries of the environment can lead an emitted beam to be received as a series of multiple echoes, each having traversed a different distance and hence reporting a different range measurement. This problem, illustrated in Figure 7b, affects both P-ToF and C-ToF approaches and most commercially-available sensors. The only way to alleviate the problem is trying to avoid measuring under such adverse geometries as much as possible.

To complete the description of rangefinder technologies, it is appropriate to review their present relevance in mobile robotics. We first find the group of inexpensive devices such as IR triangulation-based sensors and P-ToF sonars. These are quite widespread among mobile robots and probably will remain being popular due to their easy interfacing, the lack of complex post-processing, and a reasonable ratio of sensed information quality versus price. On the other hand, P-ToF and C-ToF laser rangefinders are the ones almost uniquely employed today in mobile robotics as elements upon which to build more complex sensors like 2D or 3D scanners. Alone, unidirectional laser rangefinders only have practical utility in flying robots or in robotic space missions

Figure 7. An illustration of the (a) crosstalk and (b) multipath problems in ranging devices

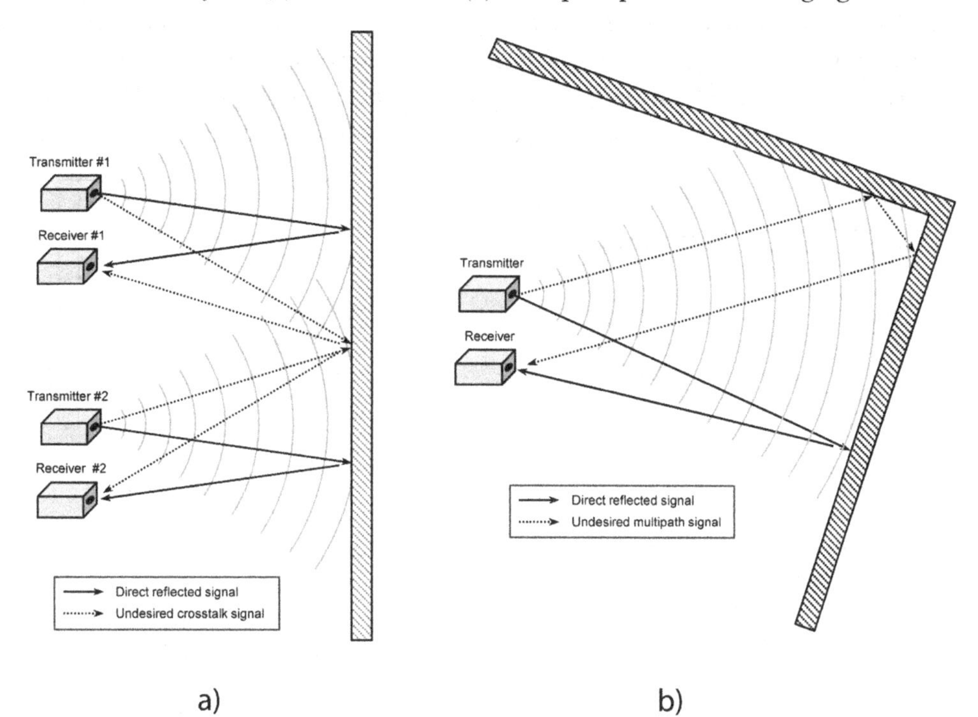

that need to accomplish automated landing: the altimetry achieved by pointing the sensor downwards is the essential input to perform an accurate landing control. Commercial devices suitable to these applications can be found at the time of writing this book at prices starting at $500 for maximum ranges of about 300m.

7. EXTEROCEPTIVE SENSORS: TWO-DIMENSIONAL RANGEFINDERS

The unidirectional rangefinders described previously have the desirable property of providing contactless range measurements. For some variants (namely, IR triangulation and P-ToF sonars) this is achieved by low cost hardware, which is the reason they are among the most popular devices employed in hobbyist robotics. However, localizing a robot from just one or a few unidirectional rangefinders is a complicated problem due to the very little information provided by such individual ranges. Stated differently: the information of one single range is very likely to be ambiguous, since it could correspond to a large number of potential positions and orientations of the robot within its environment. Simultaneously observing two ranges in different directions instead of just one would provide more information about the potential positions of the robot. Adding more and more simultaneous ranges increasingly improve the capability of successfully solving robot localization and mapping.

Following this idea, some manufacturers design two-dimensional rangefinders, which provide hundreds of ranges gathered from different directions, also called *range scans*. As shown in Figure 8, the principle of operation of these LIDAR sensors (after *Light Detection and Ranging*) is attaching a rotating mirror to a unidirectional laser rangefinder, typically based on the P-ToF technique described above. Quickly reading ranges as the mirror spins produces the apparent illusion

Figure 8. (a) The basic working principle of a common 2D laser rangefinder, where a rotating mirror allows one single laser range finder to scan in a variety of directions. (b) This top view shows an example of the points sampled in an environment during 2D laser scanning. (c) The defect known as "phantom points" emerges when the laser beam is wide enough as to being reflected by different parts of the environment simultaneously. Typically, this occurs while scanning sharp corners as illustrated here.

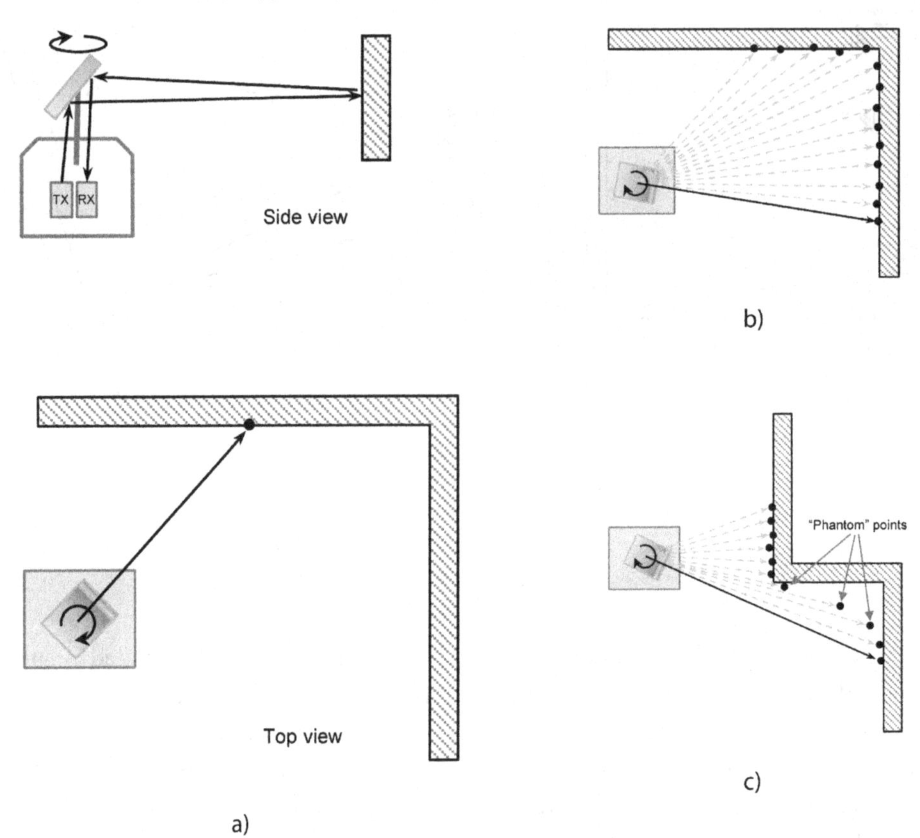

of scanning an entire 2D plane of the environment at once. Although each range is acquired at a different instant, modern scanners spin fast enough (15-100Hz) for allowing us to assume the simultaneity of all ranges without any significant errors, particularly in slow-moving robots.

This important family of devices is of chief importance in this book, since it includes archetypical sensors employed in a large fraction of all localization and SLAM research reported in the literature until the end of the 2000s, when cameras and other sensors started to gain popularity in mainstream localization and mapping research. It is easy to understand why so much research has focused on range scanners if we review their advantages:

- Typical accuracy (how close measures are from reality) is in the order of a few centimeters, with precision (how repeatable the sensor is) below one centimeter for most models.
- Laser scanners provide a dense set of range measurements that are spread over a large Field Of View (FOV). Most scanners offer 180 degrees FOV, while a few reach complete 360 degrees coverage.
- The sequence of range measurements evenly spaced at some given angular resolution can be easily converted into a 2D point cloud accurately representing a "plan" of the environment, easy to inspect and understand by a human.

In short, these sensors allow obtaining a quite accurate 2D representation of large environments almost instantly—as we will see in further chapters, that is still far from having solved the mapping problem, though. At present, even after an important development in computer vision techniques for 3D localization and SLAM, some well-established methods using laser scans are still of a huge practical utility and are widely employed as reliable solutions for both commercial and research mobile robots. The techniques developed with these sensors are the topic of interest for chapters in sections 2 and 3 of this book.

While designing a mobile robot equipped with laser scanning capabilities, it must be decided which of the commercially-available devices fit better the design goals. Extensive and quantitative characterizations of existing scanners have been reported in the literature (Lee & Ehsani, 2008; Kneip, Tache, Caprari, & Siegwart, 2009). For convenience, we next provide the reader with a short summary of some of the most common models, along with their strengths and weaknesses. Please note that the field has been experiencing a very quick development in the last years, thus this list may become incomplete and outdated in 2-3 years, if not before. In addition, it does not pretend to be complete but just to illustrate a representative sample of the devices that were currently in the market at a time when the mobile robot localization and mapping problems were quite mature.

- *HOKUYO Automatic Co.* (Hokuyo, 2011) manufactures a family of 2D scanners which, at present, are among the cheapest ($300-$3,000), lighter (below 0.2kg) and less energy demanding (below 3W) of all available. Therefore, these devices are very well suited for mobile robots, especially those of small dimensions or with energy-consumption restrictions. The main weaknesses of these scanners, in comparison to other more expensive devices, are: (1) a shorter maximum range (from 4m to 30m, depending on the model), (2) a lower accuracy in measured distances (from 1cm to 4cm), and (3) a higher probability of detecting “phantom points”—refer to Figure 7.
- *SICK AG* (Sick, 2011) marketed some of the first laser scanners that became widespread in the mobile robotics community, namely the family of sensors *SICK LMS™ 200*. These devices provide a tradeoff in price, weight and accuracy between HOKUYO’s and the next two companies’ scanners. Due to the weight and power consumption of most common SICK scanners, they are not well suited for lightweight vehicles, but can be perfectly installed into any mid- or large-sized mobile robot.
- RIEGL GmbH (Riegl, 2011) ships high-accuracy laser scanners aimed at surveying tasks. These scanners are typically more expensive than the models discussed above. For example, the RIEGL VZ-400™ scanner (reported in the SLAM literature) provides 42,000 range measures per second up to a distance of 300m with an accuracy of 5mm. Due to their weight and size, these scanners are typically operated manually or mounted atop a modified car.
- Velodyne (2011) offers a new type of scanner different from all those described above, where only one laser beam is simultaneously emitted in every direction. As opposed to the *single-beam* sensors, the *Velodyne HDL-64 S2™* can be considered as a *multi-beam* rangefinder, where the rotating mirror is replaced by a spinning structure, which holds 64 independent laser transceivers, each oriented in a different elevation angle. When the sensor spins at 15Hz the effect is that of 64 individual laser scanners in one, multiplying the amount of sensed points up to 1.8 million points per second. Such scanners can

be actually considered hybrids between 2D and 3D scanners, which we address in section 7.

The relevance of 2D laser scanners in mobile robotics cannot be fully appreciated without reviewing some of the robotic tasks they have been used for:

- Localization, mapping and Simultaneous Localization And Mapping (SLAM): Being this the topic of the present text, these are the applications we are most interested in. Different aspects of localization and SLAM with laser scanners are addressed in chapters 6 to 8. Note that 2D scanners only allow building 2D or semi-2D maps (see chapter 8), but in theory they can be used to localize a robot in a full 3D workspace—for instance, by using a technique known as particle filtering and introduced in chapter 7 for localization with a 3D robot pose and a 3D map.
- Some of the major public robotic events in the last years were the Grand Challenge and the Urban Challenge, both promoted by DARPA. In both events, contesters designed self-driven cars that had to traverse complex tracks with static and dynamic obstacles. All contesters heavily relied on laser scanners for instrumenting their cars.
- Reconstruction of structures and buildings for Architecture and Civil Engineering is another of the fields where 2D laser scanners find an interesting application. In those cases, they are typically mounted on a tilt-base to provide a kind of 3D scanner.
- Obstacle avoidance in mobile robotics, including humanoid robots, has also typically relied on laser scanners due to their high reliability and wide field of view, greater than those of other alternatives like cameras.

8. EXTEROCEPTIVE SENSORS: THREE-DIMENSIONAL RANGE SENSORS

We have dealt with two kinds of exteroceptive range sensors: simple rangefinders capable of measuring in just one direction, and mechanically improved versions, which rely on the very same principles than the former but provide the illusion of scanning a 2D plane by means of quickly spinning a mirror or the sensor itself.

Now, we define three-dimensional rangefinders as those based on a different principle: instead of using one (or a small set of) basic rangefinders and point it into different directions, we could employ more evolved techniques capable of measuring thousands to hundreds of thousands ranges at once, typically in the form of a 2D "depth image." This way of looking at the range information—as an image—is the reason behind calling these sensors "3D cameras." Notice that in these devices, all ranges are really measured simultaneously, unlike in 2D scanners. Achieving this requires using one of the following technologies, which are present in current most common 3D scanners:

- **ToF Cameras:** This family of sensors relies on the C-ToF principle described in section 5, that is, they emit an amplitude-modulated continuous luminous signal. The reflected signal is then captured by a two-dimensional array of sensors much like the CMOS imaging sensor of common video cameras. The difference is that each pixel is attached a high-bandwidth circuitry capable of receiving and processing the modulating signal, typically in the range of the few tens of megahertzs. Thus, these devices can be seen as monochrome cameras but with bandwidth orders of magnitude above that of a common camera, plus the extra electronics for processing that high frequency information.

Interestingly, the received signal for each pixel can be low-pass filtered and output as a normal intensity image; hence, these cameras typically provide in parallel both a range and an intensity image. Examples of such devices currently in the market are SwissRanger's SR4000™ (Mesa, 2011), providing 176x144 pixels at 50Hz, or Fotonic's C40™, featuring 160x120 images at up to 75Hz. In both cases the maximum measurable range is about 5m. Trying to measure larger distances will introduce aliasing errors, a phenomenon inherent to the C-ToF technology. At present, the price of these cameras is prohibitive to the hobbyists, falling in the range of the thousands of dollars.

- **Structured-light Cameras:** This represents an alternative approach to ranging that has not been commented yet in this text. Structured-light cameras comprise two parts: a common camera (either color or monochrome) and some kind of light emitter, which projects a known light pattern onto the objects being measured. The underlying idea is to resolve the depth ambiguity for each image pixel by matching it with its corresponding location in the projected pattern. Although this idea has been around for decades, the launch in 2010 of Microsoft's Kinect™ using *PrimeSense*™'s technology, at a very competitive price due to its mass production, has suddenly renewed the interest in this technology (Microsoft, 2011).

Regarding how to employ these sensors for localization and SLAM, it can be said that it is an active research field at present. An important distinction can be made between those devices where both the intensity and the depth cameras share the same focal point, and those where they are physically separated (e.g. in Microsoft's Kinect™). Independently of the concurrence or not of these two focal points, they can be used to find the color that corresponds to each 3D point sensed by the depth sensor. Using the range information the other way around (that is, finding the range that corresponds to each pixel in the intensity image) is only feasible without approximations when both focal points coincide. In practice, therefore, it is recommended to start with depth information, generate the 3D coordinates of each sensed point and then project them into pixel coordinates relative to the intensity camera in order to find the color that corresponds to each point. In general, it cannot be assured that all the pixels in the intensity image will have a valid sensed range; neither that all the sensed 3D points have a valid intensity or color, since their projection may fall out of the field of view of the intensity camera (see Figure 9).

9. EXTEROCEPTIVE SENSORS: RANGE-ONLY SENSORS

We already reviewed sensors capable of measuring one single distance in section 5. Although we did not make it explicit there, the act of performing the measuring along a defined direction provides each range reading with a direction or *bearing* information. In this section, we now describe sensors measuring one single range but, in contrast with single-directional range finders, these ones provide no extra information about the direction to the measured target.

Technologies employed in these range-only sensors are basically two: sound or radio waves, employed for operating underwater or in the air, respectively. Subsea operation involves acoustic ranging devices that communicate with a set of active transponders placed at fixed locations (Newman & Leonard, 2003). A commercial example of such a device is Sonardyne's AvTrak2™ acoustic instrument (Sonardyne, 2011), reportedly performing an accuracy and precision of tens of centimeters.

Figure 9. (a) Microsoft Kinect™ sensor, equipped with an IR laser pattern projector, a common color camera and an IR camera. (b) The color channel serves as a common imaging sensor, while the matching of the IR image with the expected projected pattern allows the computation of a dense depth image, as shown in (c). Combining both, depth and color data, and after a careful calibration, accurate 3D colored point clouds can be easily computed. These images have been generated with the Mobile Robot Programming Toolkit, an open-source C++ bunch of libraries for mobile robot programming developed by the second author (MRPT, 2011). They come from Kinect™ data taken from the dataset reported in (Sturm, et al., 2011).

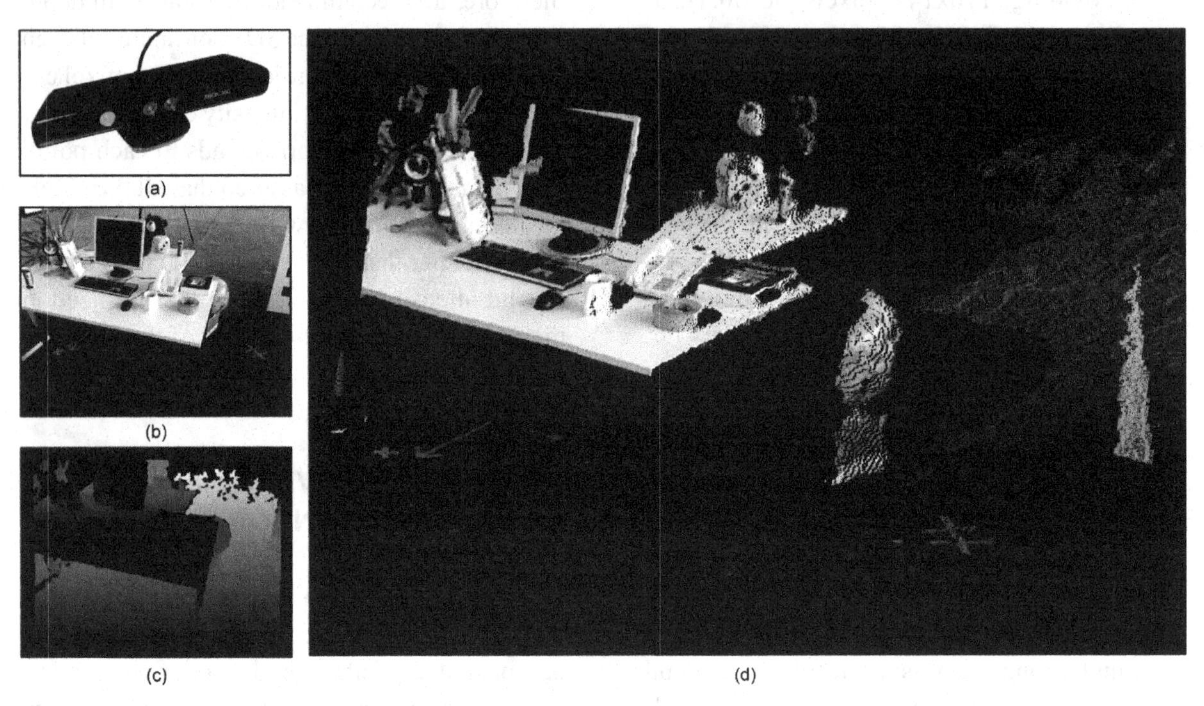

For the rest of ground or aerial scenarios, radio-based range-only sensors are the natural choice. A recent development in this field is the introduction of *Ultra Wide Band* (UWB) range finders. The key idea underlying this technology, shared by modern cell phones and WiFi standards, is to emit very low level of radio power but simultaneously occupying a huge bandwidth, from 500Mhz to more than 3Ghz (Gezici, Tian, Giannakis, Kobayashi, Molisch, Poor, & Sahinoglu, 2005). The ability of these mobile devices to work without *Line Of Sight* (LOS) to the fixed beacons or transceivers is a valuable advantage, especially in indoor scenarios where signal can often propagate through walls, without significant losses, up to tens of meters.

Despite the ambiguity of readings provided by range-only sensors, that is, the fact that the detected beacon or transceiver could be anywhere around within a given approximate distance, they have been demonstrated suitable for both localization (González, Blanco, Galindo, Ortiz-de-Galisteo, Fernández-Madrigal, Moreno, & Martínez, 2009) and SLAM (Blanco, Fernández-Madrigal, & González, 2008). We will come back to this kind of sensors in chapter 8 section 7 when discussing how to build maps with them.

10. EXTEROCEPTIVE SENSORS: IMAGING SENSORS

As stated earlier in this chapter, vision is the primary exteroceptive sense for us humans, and that is not without reasons: images provide an immense amount of information about the structure of the three-dimensional environment, even if we only process one intensity channel (monochrome cameras). Hence, there exist flourishing research fields on computer vision, structure from motion and vision-based SLAM, all getting increasingly interleaved over time.

Vision-based localization and SLAM makes use of very specific techniques, as detecting interest points, tracking, occlusion handling, feature descriptors, etc. which are far out of the scope of this text focused on the general approaches to localization and SLAM.

Therefore, for what we are concerned here a camera can be seen as a *bearing-only* sensor. That is, when projecting the surrounding three-dimensional world into a two-dimensional array of pixels, we can assign each pixel one direction (bearing) but have a completely unknown depth. Thus, mathematically speaking, the structure of the world is not directly *observable* from one image—recall Table 2. This is exactly the dual situation of the one found with range-only sensors described above, and also the reason why performing SLAM with these two families of sensors is much more challenging than with range-bearing sensors, like laser scanners.

A solution for making easier the application of computer vision is inspired in the human eyes and consists of the arrangement of two independent camera devices into a *stereo pair*, or stereo camera. Since the distance between both sensors is close enough with respect to the sensed objects, the two images are similar enough to make easy the determination of correspondences between elements that appear on both images. From those correspondences, and given that the distance between cameras (named the *baseline*) is fixed and known, it becomes easy to determine the range of each sensed object. In this way, stereo cameras are no longer bearing-only sensors. One can find other less common camera arrangements, such as trinocular sensors (an extension of stereo cameras with an additional extra sensor) or circular camera rigs for sticking panoramic views.

11. EXTEROCEPTIVE SENSORS: AIR ANALYSIS SENSORS

Among the best known, non range/bearing exteroceptive sensors for mobile robots, those that analyze the air have been the subject of research recently. Sensing the properties of the air, such as the concentration of certain chemicals and the airflow is a really complicated matter. Although works in this line were pioneered back in the 1990s (Ishida, Suetsugu, Nakamoto, & Moriizumi, 1994), research on gas sensing has remained marginal in mobile robotics.

One disposes of two main set of sensors when dealing with airflows and gas concentrations: electronic anemometers and gas detectors. Within the former we find accurate three-dimensional wind direction detectors based on arrays of acoustic transceivers, as shown in Figure 10a. The underlying working principle of these devices is the change in the transmission delay of a sound signal due to variations in the wind speed within the sensed volume. Measuring wind flows provides valuable information for predicting and understanding how gas pockets move and spread throughout an environment. The second set of sensors is that of gas detectors, employed to monitor the instantaneous concentration of some target gas (or family of chemicals) at the probing location. It has been demonstrated that, in spite of the inherent deficiencies and hurdles of such sensors, they are usable for building maps of gas concentration (Lilienthal & Duckett, 2004). In order to illustrate real-world designs with the abovementioned sensors, please refer to the designs in Figure 10.

Figure 10. (a) Mobile robot equipped with gas sensors (e-nose) and a three-dimensional sonic anemometer (courtesy of the Mobile Robotics and Olfaction Lab, AASS Research Centre, Örebro). (b) Our Rhodon mobile robot, here displaying a multi-chamber e-nose (prototype developed by the MAPIR lab, University of Málaga).

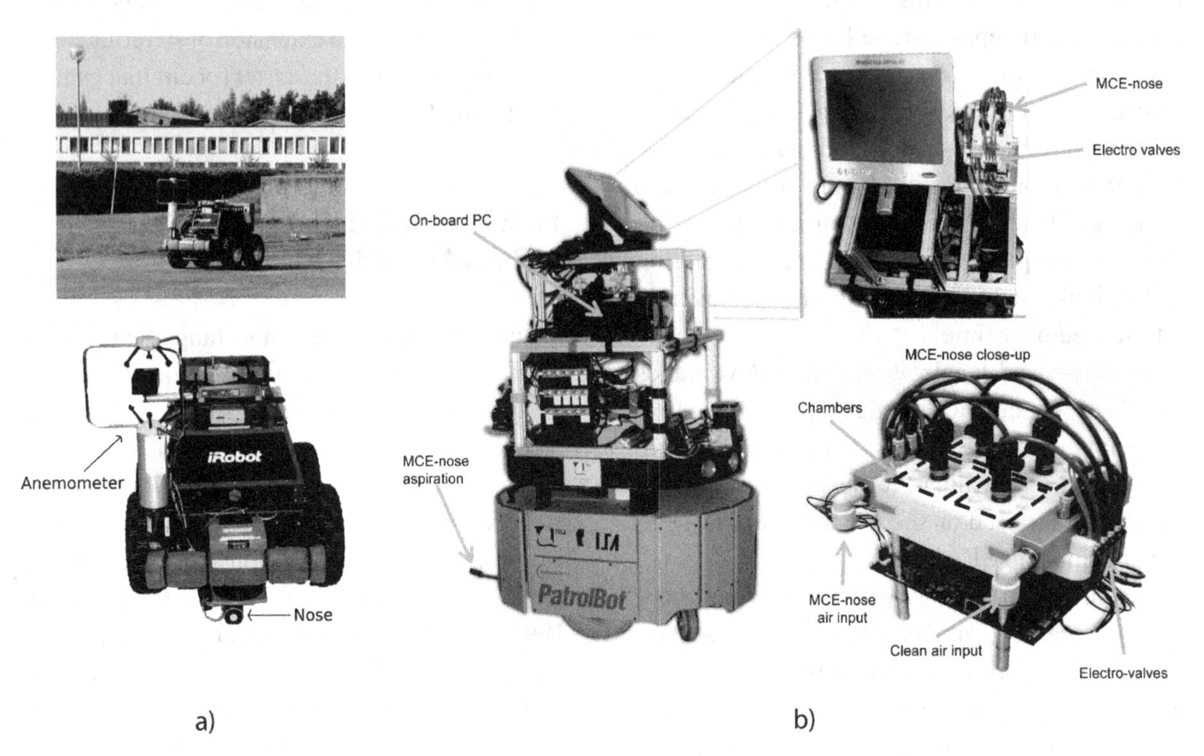

12. ENVIRONMENTAL SENSORS: ABSOLUTE POSITIONING DEVICES

Up to this point all described sensors are either exteroceptive located at the robot body (detecting and measuring distances or other properties about the objects in the environment) or proprioceptive (detecting the own robot motions from its inside). It must be noticed that all those sensors share one property: measurements are always dependent on the relative position and velocity of the robot within its environment. For example, a robot in a room equipped with a laser rangefinder or a sonar pointing at one of the walls will measure exactly the same if the whole room is translated or rotated arbitrarily as a whole (as a single rigid body). A gyroscope, as noted above, measures the derivatives of the robot heading, thus any constant offset in the robot heading is unobservable.

In contrast to this situation, there exist a few technologies, which allow the robot to measure the absolute coordinates it is located at—obviously by considering a broader environment with respect to which the measurements are still relative, or, in other words, by placing part of the sensory system outside the robot, what we called *environmental sensors* in chapter 1. Leaving apart sensors that provide an absolute position with respect to some physical element that fills the space and not the space itself—e.g., compasses to detect the magnetic field in the environment, which have been used recently for mapping indoor scenarios (Gozick, Subbu, Dantu, & Maeshiro, 2011)—the most common absolute positioning sensors are satellite-based systems. At present, three satellite networks coexist: *Global Positioning System* (USA, operating since 1994), *GLONASS* (Russia, working since 1995), and *Galileo* (European Union, expected to be functional by 2014). China also started its own global satellite networks, named Beidou-2, still in an early stage, but expected to work by 2020.

Global positioning sensors receive weak microwave signals from each satellite, which are enhanced by means of advanced signal processing techniques even when the noise levels are above the signal strength, and then estimate the time delays (at the speed of light) for those signals to travel the distance from each satellite. Using this information, and given an almanac of precise satellite orbits, receivers can obtain its location anywhere on the surface of the Earth by triangulation. This solution is called *Global Navigation Satellite Systems* (GNSS), though they are often named simply "GPS receivers" after the most popular positioning network. Modern GNSS receivers are designed to receive from all the three networks, thus measurements from each one enhance those from the others, improving the overall achievable accuracy up to 2-8 meters in favorable circumstances. It is worth mentioning that GNSS-like systems, usually useless inside buildings, have been adapted for indoor usage by placing dedicated radio beacons at fixed known locations. These arrangements can be also considered absolute positioning devices, although is seldom used in mobile robotics.

If better positioning accuracy is required for some application, there exists an extension to GNSS receivers called *Real-Time Kinematics* (RTK), capable of centimeter-level accuracy. The idea is that a ground station with a known precise coordinates can derive the error in the ranges obtained from each satellite and then transmit this corrections to the receiver for it to improve its own calculations. This is a mature technology, but deserves specific (and expensive) receivers and the availability of a nearby base station compatible with the receiver. At present, some countries support national public RTK networks, thus end users may not need to purchase their own base stations.

There exists another sensor, which could be considered capable of absolute positioning but only under special circumstances. We refer to accelerometers, already classified above as inertial, relative positioning sensors. To understand how they could also provide global positioning information, imagine a robot standing still and equipped with three accelerometers aligned in orthogonal directions. If we try to estimate the three *absolute* attitude angles (often named *yaw*, *pitch* and *roll*), we will find out that pitch and roll can be measured absolutely—with respect to the downwards gravity acceleration—but the system is degenerate, that is, only those two tilt angles are observable. The rotational degree of freedom not observable by means of accelerometers corresponds to the yaw, that is, determining if the robot heads North, South, East, or West. It is difficult to decide if we should classify accelerometers as absolute sensors or not; if the robot is not still but moving, the situation is even more complicated and accelerometers can hardly be used to obtain any absolute reading.

Complementarily to the non-observability of the robot heading (or *yaw*) by accelerometers into a still robot, we find electronic compasses, global positioning sensors sensible to the Earth magnetic field. These devices are relatively cheap and can complement other sensors by means of determining the robot absolute orientation with respect to the surface of the Earth. Compasses have been often used in SLAM research, since reducing the uncertainty in our knowledge of the robot heading to some bounded limits alleviates the difficulty of performing loop closing (see chapter 9), one of the most difficult problems in SLAM.

Finally, we must mention the family of RFID tags, a contactless radio technology where passive or active circuits—acting like static beacons energy-fed by the receiver—can be identified by a mobile reader when passing by at a short distance. Since each tag can be uniquely identified, if they are placed at fixed known locations along the robot workspace they indeed provide a way to learn the current absolute location with an accuracy of 10-30 centimeters (Ziparo, Kleiner, Nebel, & Nardi, 2007) (see Figure 11).

13. ENERGY SUPPLY

Mobile robots intended for autonomous operation in an unknown environment must run from electric power supplied onboard, usually chemical-

Figure 11. (a) A pair of complete GPS RTK base stations, each comprising an RTK base unit, a UHF radio modem, and a portable battery. (b) A GPS RTK receiver unit, produced by JAVAD (2011). (c) A consumer grade GPS receiver, by uBlox (2011).

(a)

(b)

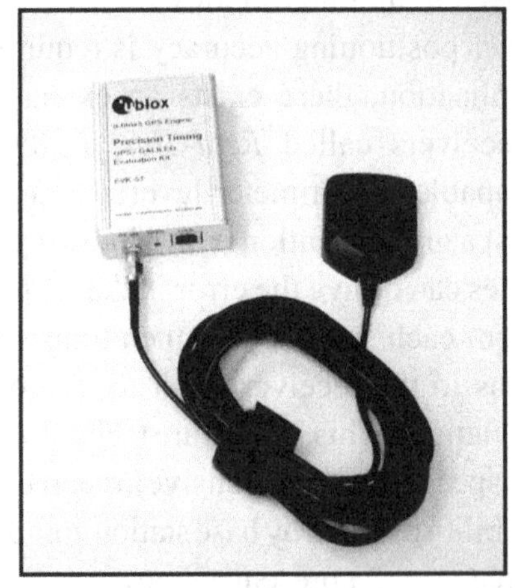

(c)

based batteries (solar driven robots are still rare). Choosing among the different battery technologies available in the market is a key step in designing a mobile robot, since their overall storage capacity will determine how long the robot will be able to operate autonomously without having to get back to the recharge station.

We could think of adding more and more batteries to a robot to fit our autonomy goals, but remember the chicken-and-egg problem explained in chapter 1: the weight of additional batteries must be carefully taken into account, since moving a heavier robot takes more energy. Therefore, as with so many other engineering decisions, the number of batteries and their types are the result of a subtle tradeoff between the batteries weight, size, maximum sustainable current output (in amperes, or A), power storage capacity (measured in amperes times hours, or Ah) and price.

An important property that characterizes each battery technology is the associated *energy density*, that is, how much electric power they can store with respect to their weight or volume. This density is measured in MJ/kg (megajoules per kilogram) or MJ/l (megajoules per liter), respectively. We next review the kinds of battery technologies more relevant to mobile robotics.

Lead-acid batteries represent a mature technology today and are one of the most widely employed solution in mid or large-size robots and in automobiles in spite of having one of the lowest energy densities, typically about 0.13 MJ/kg (McDowall, 2000).

One possible implementation, *Vented Lead-Acid* (VLA) batteries, consists of an array of individual cells of lead (Pb) and lead dioxide (PbO_2) submerged in a solution of water and sulfuric acid (H_2SO_4). These rechargeable batteries are characterized by their low cost and for being able of providing a very high output current. On the other hand, their main inconvenience is their large weight and volume, hence they must be considered as an option only for heavy mobile robots or vehicles. This is the kind of batteries employed in most non-electric automobiles where weight is not a problem due to the internal combustion motor having a much more efficient energy source at hand—the energy densities of gasoline or diesel are about 45 MJ/kg, and some terrestrial mobile robots exist that use gas motor engines (Mandow, Gómez-de-Gabriel, Martínez, Muñoz, Ollero, & García-Cerezo, 1996), but they are not common.

Mid-size robots usually employ another implementation of lead-acid batteries, known as *Valve-Regulated Lead-Acid* (VRLA) batteries. This family of batteries can be safely tilted without the risk of spilling the acid as with the VLA implementation. VRLA batteries are thus safer, and also require less maintenance and volume than VLA batteries, therefore making them the preferred choice in many commercially available robotic bases. A typical research mobile base such as the Pioneer 3-DX contains up to three VRLA batteries with capacities typically in the 4-9Ah range.

When addressing the design of small and light robots, like micro wheeled platforms or flying robots, the weight of lead-acid batteries become absolutely prohibitive: more efficient energy storage is necessary. The battery technology based on zinc (Zn) and manganese dioxide (MnO_2), commonly called *alkaline cells*, provide a better energy density (0.5 MJ/kg) than lead-acid batteries, but since most of them are not rechargeable they do not represent a cost-efficient solution and it is used just for toys and small home appliances.

A much better alternative technology that has witnessed a quick development in the last decades, particularly in the computer laptop arena, is that of *lithium-ion* (Li-ion) batteries: with an energy density of about 0.6 MJ/kg they can provide as much power as a conventional VRLA battery with only one fifth of its weight and size. A variant of this technology, *lithium-ion polymer* (Li-Po) batteries, exhibits an even higher power-to-weight ratio, hence their popularity among small flying robots. Recharging lithium batteries can be done with an efficiency of 95%, in contrast to a typical

70% for lead-acid batteries (McDowall, 2000). Without doubts, the popularity of lithium batteries will keep growing in the future due to all these advantages. The only potential problem with this technology is a security issue: if charge-discharge cycles are not observed accordingly to the specifications, there exists a certain risk of explosion or fire. Needless to say, manufacturers are constantly improving the security of commercially available cells in order to reduce this risk as much as possible, even under incorrect operation.

To put in context the energy densities of the batteries discussed above, it is illuminating to compare one of those battery technologies (Li-ion) to the energy storage we humans and many other animals use: body fat. In average, metabolism is capable of extracting a huge amount of chemical energy from fat, since its equivalent energy density is 37.7 MJ/kg (Schraer & Stoltze, 1999). This implies that 100 grams of fat stores the same energy than a 6.3kg pile of Li-ion batteries! This gap between biochemical and electric technologies explains, in part, the short energy autonomy of current robots in comparison to the humblest animal. As with many other engineering problems, evolution is millions of years ahead of us.

REFERENCES

Ambrose, R. O., Aldridge, H., Askew, R. S., Burridge, R. R., Bluethmann, W., Diftler, M., … Rehnmark, F. (2002). Robonaut: NASA's space humanoid. *IEEE Intelligent Systems and their Applications, 15*(4), 57-63.

Aoi, S., & Tsuchiya, K. (2011). Generation of bipedal walking through interactions among the robot dynamics, the oscillator dynamics, and the environment: Stability characteristics of a five-link planar biped robot. *Autonomous Robots, 30*(2), 123–141. doi:10.1007/s10514-010-9209-9

Blanco, J. L., Fernández-Madrigal, J. A., & González, J. (2008). Efficient probabilistic range-only SLAM. In *Proceedings of the IEEE/RSJ International Conference on Intelligent Robots and Systems*, (pp. 1017-1022). IEEE Press.

Borenstein, J., Everett, H. R., & Feng, L. (1996). *Where am I? Sensors and methods for mobile robot positioning. Technical Report*. Ann Arbor, MI: University of Michigan.

Boston Dynamics. (2010). *Petman - BigDog gets a big brother*. Retrieved Mar 1, 2012, from http://www.bostondynamics.com/robot_petman.html

Caballero, F., Merino, L., Ferruz, J., & Ollero, A. (2009). *Vision-based odometry and SLAM for medium and high altitude flying UAVs*. Retrieved from http://grvc.us.es/aware/papers/aware_paper_11.pdf

Ciocarlie, M., Hsiao, K., Jones, E. G., Chitta, S., Rusu, R. B., & Sucan, I. A. (2008). Towards reliable grasping and manipulation in household environments. *Robotics and Autonomous Systems, 56*(1), 54–65.

Cruz-Martín, A., Fernández-Madrigal, J. A., Galindo, C., González-Jiménez, J., Stockmans-Daou, C., & Blanco, J. L. (2012). A Lego Mindstorms NXT approach for teaching at data acquisition, control systems engineering and real-time systems undergraduate courses. *Computers & Education*. doi:10.1016/j.compedu.2012.03.026

Davison, A. J., Reid, I. D., Molton, N. D., & Stasse, O. (2007). MonoSLAM: Real-time single camera SLAM. *IEEE Transactions on Pattern Analysis and Machine Intelligence, 29*(6), 1052–1067. doi:10.1109/TPAMI.2007.1049

Desai, J. P., Ostrowski, J., & Kumar, V. (1998). Controlling formations of multiple mobile robots. In *Proceedings of the IEEE International Conference on Robotics and Automation*, (pp. 2864-2869). IEEE Press.

Doyle, J. C., Francis, B. A., & Tannenbaum, A. (1992). *Feedback control theory*. New York, NY: Dover Publications.

Endo, T., Kawasaki, H., Mouri, T., Yoshida, T., Ishigure, Y., & Shimomura, H. ... Koketsu, K. (2009). Five-fingered haptic interface robot: HIRO III. In *Proceedings of the Third Joint EuroHaptics Conference and Symposium on Haptic Interfaces for Virtual Environment and Teleoperator Systems (World Haptics Conference)*, (pp. 458-463). World Haptics Conference.

Espiau, B., Chaumette, F., & Rives, P. (1993). A new approach to visual servoing in robotics. *Geometric Reasoning for Perception and Action*, 106-136.

Gezici, S., Tian, Z., Giannakis, G. B., Kobayashi, H., Molisch, A. F., Poor, H. V., & Sahinoglu, Z. (2005). Localization via ultra-wideband radios: A look at positioning aspects for future sensor networks. *IEEE Signal Processing Magazine*, *22*(4), 70–84. doi:10.1109/MSP.2005.1458289

González, J., Blanco, J. L., Galindo, C., Ortiz-de-Galisteo, A., Fernández-Madrigal, J. A., Moreno, F. A., & Martinez, J. L. (2009). Mobile robot localization based on ultra-wide-band ranging: A particle filter approach. *Robotics and Autonomous Systems*, *57*(5), 496–507. doi:10.1016/j.robot.2008.10.022

Gozick, B., Subbu, K. P., Dantu, R., & Maeshiro, T. (2011). Magnetic maps for indoor navigation. *IEEE Transactions on Instrumentation and Measurement*, *60*(12), 3883–3891. doi:10.1109/TIM.2011.2147690

Hirose, R., & Takenaka, T. (2001). Development of the humanoid robot ASIMO. *Honda R&D Technical Review, 13*(1), 1-6.

Hokuyo. (2011). *Hokuyo Automatic Co. Ltd. corporate website*. Retrieved Mar 1, 2012, from http://www.hokuyo-aut.jp/

Huang, Q., Yokoi, K., Kajita, S., Kaneko, K., Arai, H., Koyachi, N., & Tanie, K. (2001). Planning walking patterns for a biped robot. *IEEE Transactions on Robotics and Automation*, *17*(3), 180–189.

Ishida, H., Suetsugu, K., Nakamoto, T., & Moriizumi, T. (1994). Study of autonomous mobile sensing system for localization of odor source using gas sensors and anemometric sensors. *Sensors and Actuators. A, Physical*, *45*(2), 153–157. doi:10.1016/0924-4247(94)00829-9

JAVAD. (2011). *JAVAD GNSS Inc. corporate website*. Retrieved Mar 1, 2012, from http://www.javad.com/

Kneip, L., Tache, F., Caprari, G., & Siegwart, R. (2009). Characterization of the compact Hokuyo URG-04LX 2D laser range scanner. In *Proceedings of the IEEE International Conference on Robotics and Automation*, (pp. 1447-1454). IEEE Press.

Larsen, M. B. (2000). High performance doppler-inertial navigation-experimental results. In *Proceedings of the MTS/IEEE Conference and Exhibition OCEANS 2000*, (vol 2), (pp. 1449-1456). IEEE Press.

Lee, K. H., & Ehsani, R. (2008). Comparison of two 2D laser scanners for sensing object distances, shapes, and surface patterns. *Computers and Electronics in Agriculture*, *60*(2), 250–262. doi:10.1016/j.compag.2007.08.007

Lego. (2011). *Lego mindstorms robots homepage*. Retrieved on July 12, 2011, from http://mindstorms.lego.com/en-us/Default.aspx

Lilienthal, A., & Duckett, T. (2004). Building gas concentration gridmaps with a mobile robot. *Robotics and Autonomous Systems*, *48*(1), 3–16. doi:10.1016/j.robot.2004.05.002

Mandow, A., Gómez-de-Gabriel, J. M., Martínez, J. L., Muñoz, V. F., Ollero, A., & García-Cerezo, A. (1996). The autonomous mobile robot AURORA for greenhouse operation. *IEEE Robotics & Automation Magazine*, *3*(4), 18–28. doi:10.1109/100.556479

Manoonpong, P., Geng, T., Kulvicius, T., Porr, B., & Wörgötter, F. (2007). Adaptive, fast walking in a biped robot under neuronal control and learning. *PLoS Computational Biology*, *3*(7), 134. doi:10.1371/journal.pcbi.0030134

Martínez, M. A., González, J., & Martínez, J. L. (1997). The DSP multi-frequency sonar configuration of the RAM-2 mobile robot. In *Proceedings of the Second EUROMICRO Workshop on Advanced Mobile Robots*. EUROMICRO.

McDowall, J. (2000). Conventional battery technologies-present and future. In *Proceedings of the IEEE 2000 Power Engineering Society Summer Meeting,* (vol 3), (pp. 1538-1540). IEEE Press. Mesa. (2011). *MESA imaging corporate website*. Retrieved Mar 1, 2012, from http://www.mesa-imaging.ch/

Michael, N., Mellinger, D., Lindsey, Q., & Kumar, V. (2010). The GRASP multiple micro-UAV testbed. *IEEE Robotics & Automation Magazine*, *17*(3), 56–65. doi:10.1109/MRA.2010.937855

Microsoft. (2011). *Xbox kinect web site*. Retrieved July 19, 2011, from http://www.xbox.com/en-US/kinect

Mobile Robotics. (2011). *Mobile Robotics Inc. corporate website*. Retrieved Mar 1, 2012, from http://www.mobilerobots.com/

Morris, W., Dryanovski, I., & Xiao, J. (2010). 3D indoor mapping for micro-UAVs using hybrid range finders and multi-volume occupancy grids. In *Proceedings of the RSS 2010 Workshop on RGB-D: Advanced Reasoning with Depth Cameras*. RSS.

MRPT. (2011). *The mobile robot programming toolkit website*. Retrieved Mar 1, 2012, from http://www.mrpt.org/

Newman, P., & Leonard, J. (2003). Pure range-only sub-sea SLAM. In *Proceedings of the IEEE International Conference on Robotics and Automation*, (vol 2), (pp. 1921-1926). IEEE Press.

Newman, P., Leonard, J., & Rikoski, R. J. (2005). Towards constant-time SLAM on an autonomous underwater vehicle using synthetic aperture sonar. In *Robotics Research* (pp. 409–420). Berlin, Germany: Springer.

Nistér, D., Naroditsky, O., & Bergen, J. (2006). Visual odometry for ground vehicle applications. *Journal of Field Robotics*, *23*(1), 3–20. doi:10.1002/rob.20103

Ollero, A., & Merino, L. (2004). Control and perception techniques for aerial robotics. *Annual Reviews in Control*, *28*(2), 167–178. doi:10.1016/j.arcontrol.2004.05.003

Pal Robotics. (2011). *Corporate website*. Retrieved Mar 1, 2012, from http://www.pal-robotics.com/

Parallax. (2011). *Corporate website*. Retrieved Mar 1, 2012, from http://www.parallax.com/

Quinlan, S., & Khatib, O. (1993). Elastic bands: Connecting path planning and control. In *Proceedings of the IEEE International Conference on Robotics and Automation*, (pp. 802-807). IEEE Press.

Raibert, M. H., & Tello, E. R. (1986). Legged robots that balance. *IEEE Expert*, *1*(4), 89. doi:10.1109/MEX.1986.4307016

Real World Interface. (2011). *Corporate website*. Retrieved Mar 1, 2012, from http://www.rwii.com/

Riegl. (2011). *RIEGL GmbH corporate website*. Retrieved Mar 1, 2012, from http://www.riegl.com/

Robot Electronics. (2011). *Corporate website*. Retrieved Mar 1, 2012, from http://www.robot-electronics.co.uk/

Saripalli, S., Montgomery, J. F., & Sukhatme, G. S. (2003). Visually guided landing of an unmanned aerial vehicle. *IEEE Transactions on Robotics and Automation*, *19*(3), 371–380. doi:10.1109/TRA.2003.810239

Schraer, W. D., & Stoltze, H. J. (1999). *Biology: The study of life*. Upper Saddle River, NJ: Prentice Hall.

Sharp. (2011). *Sharp microelectronics of the Americas Inc. corporate website*. Retrieved Mar 1, 2012, from http://www.sharpsma.com/

Sick. (2011). *SICK AG corporate website*. Retrieved Mar 1, 2012, from http://www.sick.com/

Sonardyne. (2011). *Sonardyne corporate website*. Retrieved Mar 1, 2012, from http://www.sonardyne.com/

Sturm, J., Magnenat, S., Engelhard, N., Pomerleau, F., Colas, F., & Burgard, W. ... Siegwart, R. (2011). Towards a benchmark for RGB-D SLAM evaluation. In *Proceedings of the RGB-D Workshop on Advanced Reasoning with Depth Cameras at Robotics: Science and Systems Conference*. RGB-D.

Tellez, R., Ferro, F., Mora, D., Pinyol, D., & Faconti, D. (2008). Autonomous humanoid navigation using laser and odometry data. In *Proceedings of the 8th IEEE-RAS International Conference on Humanoid Robots*, (pp. 500-506). IEEE Press.

Tester, J. W. (2005). *Sustainable energy: Choosing among options*. Cambridge, MA: The MIT Press.

uBlox. (2011). *u-Blox AG corporate website*. Retrieved Mar 1, 2012, from http://www.u-blox.com/

Velodyne. (2011). *Velodyne lidar corporate website*. Retrieved Mar 1, 2012, from http://velodynelidar.com/

Willow Garage. (2012). *Willow garage corporate website*. Retrieved Mar 1, 2012, from http://www.willowgarage.com/

Yuh, J. (2000). Design and control of autonomous underwater robots: A survey. *Autonomous Robots*, *8*(1), 7–24. doi:10.1023/A:1008984701078

Ziparo, V. A., Kleiner, A., Nebel, B., & Nardi, D. (2007). RFID-based exploration for large robot teams. In *Proceedings of the IEEE International Conference on Robotics and Automation*, (pp. 4606-4613). IEEE Press.

Chapter 3
Probabilistic Bases

ABSTRACT

This is the third chapter of the first section. It is a compendium of all the concepts and theorems of probability theory that are found in the problems of Bayesian estimation of a robot location and the map of its environment. It presents uncertainty as an intrinsic feature of any mobile robot that develops in a real environment. It is then discussed how uncertainty has been treated along the history of science and how probabilistic approaches have represented such a huge success in many engineering fields, including robotics. The fundamental concepts of probability theory are discussed along with some advanced topics needed in further chapters, following a learning curve as smooth and comprehensive as possible.

CHAPTER GUIDELINE

- You will learn:
  - Why dealing explicitly with uncertainty is crucial for the problems of localization and mapping.
  - A brief overview of the history of probability and statistics.
  - The foundations of representing uncertainty with probability theory: probability spaces, random variables, probability distributions, pdfs, and pmfs.
  - How uncertainty can be summarized: moments, information, and entropy.
  - How to deal with multiple random variables simultaneously.
  - The special role of the chi-squared distribution.
  - On the dependence between random variables and the concept of conditional probability.
  - How to use graphical models to represent dependences.
- Provided tools:
  - A table with the most common probability density functions in mobile robotics and their parameters.
  - Method of marginalization.
  - Methods for propagating uncertainty through arbitrary functions and some particular cases of such transformations.
  - Basic tools of conditional probability: Bayes' rule, chain rule, and law of total probability.

DOI: 10.4018/978-1-4666-2104-6.ch003

- The graphical criterion of *d*-separation to factor a distribution over a large numbers of random variables.
- The graph of correlations as a purely graphical method to reveal the sparseness of the relationships existing between the unknown random variables in any inference problem.

- Relation to other chapters:
  - This chapter provides the basic foundations of probability required for the rest of the book.
  - Transformation of r.v.s through a function (section 8) and graphical models (section 10) are pervasively employed in subsequent chapters.

1. INTRODUCTION

After studying the nuts and bolts of mobile robots in Chapter 2, we now move to a different perspective: the theoretical foundations of probabilistic mobile robot localization and mapping required for a proper and solid understanding of the methods that address these problems.

As discussed in chapter 1, one of the core issues that were preventing robotics engineers from proposing a consistent theoretical solution to practical localization and mapping before the nineties was how to deal with uncertainty. Uncertainty takes the form of noise in the data being processed, and comes from different sources exhibiting diverse behaviors. Ultimately, this reflects our lack of knowledge about the processes involved in perception and actuation of mobile robots (although quantic physics teaches us that at extremely small scales, uncertainty becomes an intrinsic feature of reality). Since we cannot build perfect models for those processes of our interest, it becomes mandatory to deal with their randomness in order to discern significant data from noise or errors.

Uncertainty begins in the sensory apparatus of the robot, that is, with noisy measurements from the physical world, and extends throughout the entire robotic system, including hardware and software, arising again at the interaction of robot actuators with the environment (see Figure 1). In general, uncertainty produces errors, which can be classified into two types: *systematic* and *random* (Scuro, 2004). Systematic errors are deviations that manifest in a persistent direction—for example, a broken mechanism often produces a systematic error in measurements, and so does a wrong model of the robot dynamics embedded into an algorithm. Random errors, on the contrary, are completely unpredictable deviations for which we can only model their long-term or expected behavior.

We can distinguish the following sources of uncertainty in the localization and mapping problems (please refer to Figure 1):

- **Environmental conditions:** There are countless parameters of the environment, many of them time-varying, that affect sensing and are uncontrollable: the color of the ambient light and its intensity, the kind of materials and surfaces of solid objects, the number of moving (unpredictable) obstacles, the temperature, etc. As explained in chapter 1, only sufficiently static parts of the environment are suitable to be modeled for localization and mapping. In addition, it is impossible to have a representation that includes *all* the aspects of all the spatial elements. Therefore, it is inevitable that the intrinsic uncertainty in the environment affects our estimates.
- **Sensor behavior in steady state:** Among the sensor limitations in this sense, no sensor is able to distinguish arbitrarily small changes in the measured signal—sensor resolution. They have limited working ranges outside of which their measurements are completely wrong, sensor linearity is not perfect, and they propagate

Figure 1. Control flow of the localization and mapping processes. Uncertainty may appear at any step (see the main text).

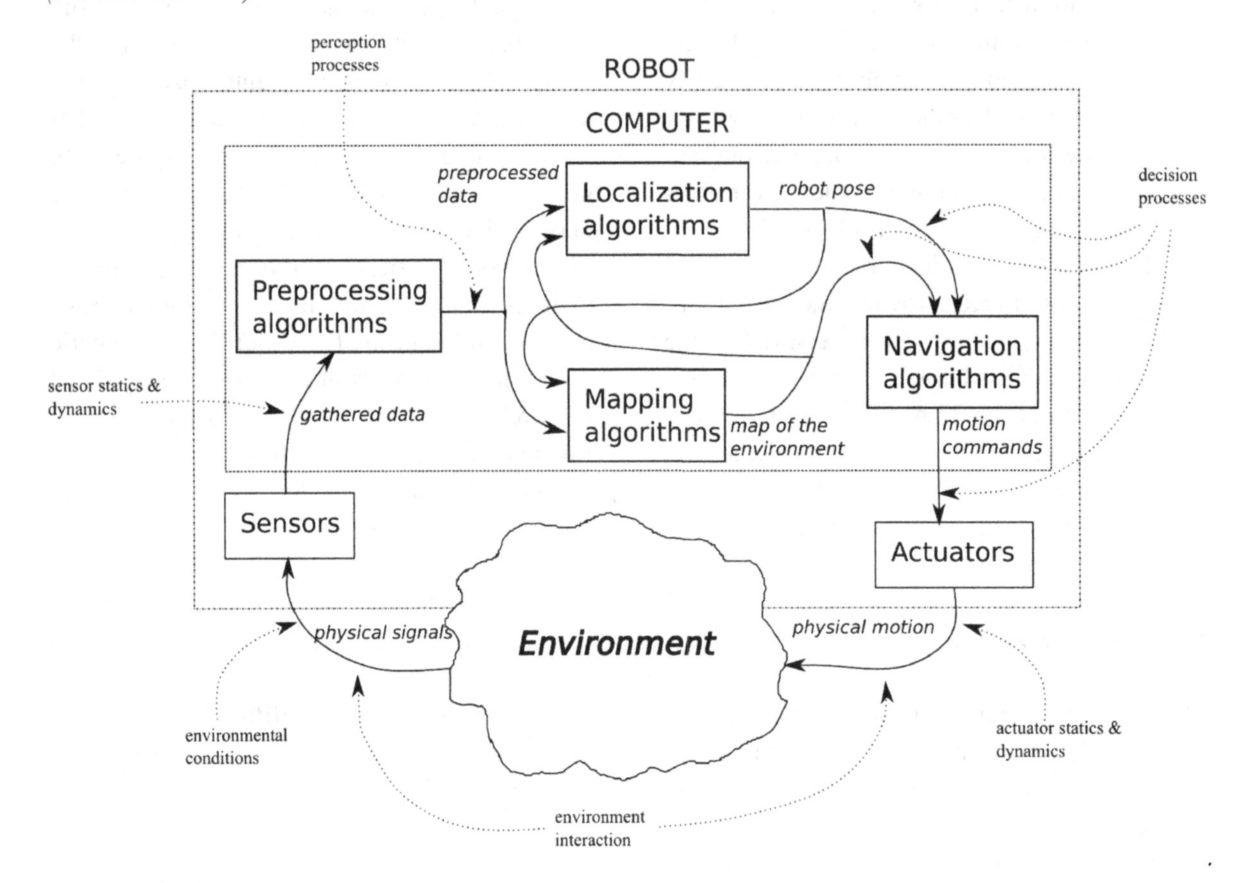

random errors produced internally to the sensory data, which leads to non-repeatability of sensors, i.e., the same measured value not repeating exactly even if we think that the environmental conditions have been kept stationary. Fortunately, many of these issues are studied in-depth by manufacturers, who provide engineers with detailed characterizations as charts and datasheets.

- **Sensor dynamics:** Apart from the random effects that sensors exhibit when they are providing a steady measurement, their dynamics—transient response—also influence the measured values (Morris, 2005). However, sensor dynamics will not be an issue for the typical sensors employed and the time scales involved in the robotic applications addressed in this text.
- **Perception processes:** Sensors often provide raw data on which further processing must be applied to extract meaningful information. For instance, a camera provides us with a matrix of intensity values from which distinctive features are to be extracted in order to perform mapping and localization. Apart from accuracy errors in the time discretization in time (frame rate), space (image resolution) and magnitude (analog to digital signal conversion), the processing algorithms in the computer introduce may new systematic or random errors, e.g. in the form of simplifications and wrong assumptions.

- **Decision processes:** A computer always makes decisions deterministically. Still, it can use random numbers as input to some numerical methods, which require stochastic inputs (the best example are particle filters, which we address in chapters 7 and 9). Strictly speaking, these would not be random but pseudo-random numbers, since a computational device has no access to an internal true random process—refer to Appendix C for more on pseudo-random numbers. The direct use of random numbers represents a clear and deliberated injection of uncertainty. There are other, less obvious sources of uncertainty that are intrinsic to computation (Lehman, 1990): computers discretize real numbers in order to store them into a finite amount of space (bits), leading to errors similar to the ones due to sensor resolution, but more complex since the encoding of numbers in floating point representations is not uniform over the real line (Morgan, 1992); computer programs cannot be completely characterized in pure mathematical form due to the Gödel theorem, thus their development depends on humans, which involves uncertainty, especially when the complexity of the algorithms is not negligible—most cases; algorithms always include assumptions and simplifications with respect to the actual processes they are modeling, mainly due to limitations of practical computers and the intractability of exact solutions to many problems of our interest, which produces unpredictable results when the program interacts with the real world.
- **Actuator steady-state characteristics:** Pretty much like sensors, actuators are devices subjected to a number of limitations in their very construction: non-linearities, resolution, repeatability, etc. In addition, they have some kind of systematic effects of their own; for example, *saturations* appear outside their working ranges (then the actuator is performing quite below what it is commanded to), and gears produce *dead-zones* (no operation in spite of non-null input commands) and *hysteresis* (no change of operation happens after a change in the direction of the command).
- **Actuator dynamics:** Unlike most common robotic sensors, the dynamics of actuators may have a strong effect on the behavior of a mobile robot. Settling times are longer than in sensors, and are quite important when considering navigation (we have to take into account the load—weight, therefore inertia—that these actuators must move). Notice that if localization and mapping processes are entangled with navigation, stability issues appear due to these delays, like in any other closed-loop control system. For the reader not familiarized with the problems of delays in a control loop, just think of the difficulties in maintaining the temperature of the water constant in your bathtub if the heater is at a certain distance.
- **Environment interaction:** Even if actuator kinematics and dynamics were perfectly modelable, their interaction with the physical world includes uncertainty. Changes produced in the world by an actuator cannot be modeled but very approximately and, what is worse, will influence the information gathered by sensors at the next iteration of the perception-action cycle, possibly increasing errors. The dynamics of environments, even the simplest ones, are at the root of the hardest problems in mobile robotics, in contrast to the case of manipulators, as it was discussed in chapter 1.

From all these arguments it becomes evident that uncertainty is pervasive in any discipline that has to deal with the physical world, but it is especially important in those that are based on perceiving that world in real-time, like mobile robotics. One cannot avoid uncertainty, so one has to handle it explicitly using some scientific theory.

The present chapter introduces the foundations of the most successful mathematical approach to modeling uncertainty: the theory of probability. It is assumed here that the reader has basic knowledge of calculus and matrix algebra; other than that, the text is self-contained. For more comprehensive introductions to the field, you can consult the uncountable books that are currently published on probability theory and statistics, for instance (Meester, 2008; Walpole, Myers, & Myers, 1997; Taylor, 1997).

In spite of the huge success of probability theory for dealing with uncertainty, that is not the only way of addressing randomness. Among the best known scientific approaches to uncertainty is also *possibility theory* (Dubois & Prade, 2001; Salicone, 2007), an extension of *fuzzy theory* (Zadeh, 1965), which in turn has been generalized recently by a General Theory of Uncertainty (GTU), being developed since 1965 by the outstanding mathematician, electrical engineer and computer scientist Lotfi Askar Zadeh (2005). In particular, fuzzy theory has been very relevant for modeling uncertainty in automatic control systems and artificial reasoning. However, it has not such presence in mobile robot localization and mapping—although it has been applied to mobile robot navigation—mainly due to the lack of precision of the results it can provide, although that is changing in the last years (Watanabe, Pathiranage, & Izumi, 2009; Havangi, Teshnehlab, Nekoui, & Taghirad, 2011; Ankishan & Efe, 2011; Chatterjee & Matsuno, 2007). In this book, we focus only on probabilistic approaches.

The rest of the chapter is structured as follows. Section 2 provides a selected historical background on probability and statistics, particularly focused on the basic concepts used later on in this and the next chapter. Sections 3 and the followings explain the basic concepts of probability, providing formalizations and, when possible, demonstrations. Each section covers an important topic out of the ones needed for the rest of the book: you will learn the basics of probability spaces, random variables, probability density and mass functions, moments, information and entropy, multivariate probability, functions of random variables, conditional probability, and graphical models.

2. HISTORICAL OVERVIEW

The history of dealing with uncertainty is the history of accepting the futility of ignoring it, as we will develop succinctly in the next paragraphs. For a more in-depth account of the historical development of probability and statistics we recommend (Tabak, 2004; Franklin, 2001), among many other books on the history of mathematics.

Before modern science, uncertainty was a religious and consequently inextricable concept: the result of the unpredictable wishes—and whims—of the gods. Therefore, it was not consistent to attempt to explain it or to examine something that was considered not to have any stable characteristics (also, it was a sacred thing that you better were not playing with). However, humanity found applications of uncertainty since prehistoric times: thousands of years before our era people threw the astragalus, or talus (Figure 2), a bone from the heels of some animals (including humans), to produce random patterns in order to make decisions under the blessing of the gods. In time this led to the invention of dice, board games—where luck was mixed with human decision making and strategy, maybe in an attempt to control the uncontrollable—and gambling. The idea that uncertainty in the physical world was a reflection of the will of the gods lasted along the centuries: actually, it can still be found as recently as two centuries before our time. Even societies with a high sense of science, such as the ancient Greeks, did not develop any scientific theory or proto-theory for the unpredictable.

Figure 2. Two astragalus bones from a lamb. They have a prismatic shape that makes them suitable as four sided dice (courtesy of Ana Martín-Padilla & Adolfo Cruz-Lobo). There exists an interesting algorithm, proposed by John Von Neumann, to obtain fair die rolls from unfair dice like these[1].

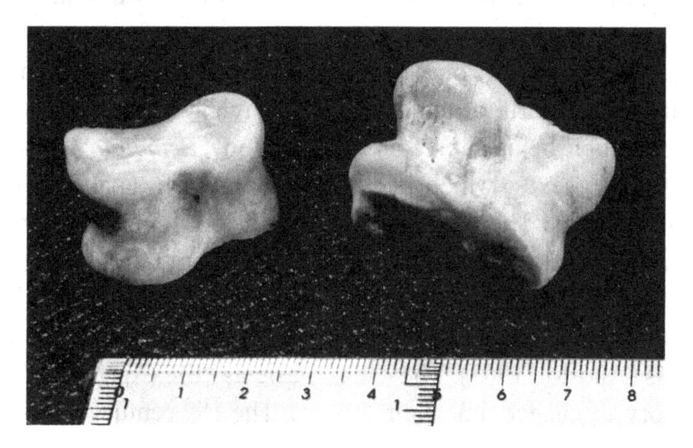

Surprisingly, in the Middle Ages, collecting data in order to extract conclusions and make decisions (the primary goal of statistics) was an idea that people did not link so strongly to mystery. We can find the first book that compiles information about properties—and their geographical location—productive land and other natural resources in the *Domesdey Book* ordered by William the Conqueror, King of England, at the end of the 11th century. It was supposed to serve as a tool for estimating the wealth of the land ruled by the king. Several centuries more were needed to develop the basic methods for analyzing such kind of data.

During the Renaissance, the apparent elusiveness of randomness started to change. The Italian mathematician Girolamo Cardano became aware that there certainly were some stable properties, independent of God's will, in the random patterns generated by dice, and tried to explain them, partly due to his love for the games of chance. His attempts were not particularly clear or successful, but planted the idea that there existed the possibility of modeling uncertainty through mathematics—without sacrificing its divine foundations. He was also the first one that started to think about randomness in terms of sequences of numbers (the outcomes of rolling dice, for example), which is the basic concept underlying modern statistics. After the death of Cardano, Galileo Galilei picked up the baton and studied again the problem of randomness from a mathematical perspective, although he had no special interest in gambling. He discarded the intervention of God in chance entirely for the first time.

It was in France in the 17th century that the history of the theory of probability, the first and most successful explicit model of uncertainty in science, began. The famous mathematicians Blaise Pascal and Pierre de Fermat tried to solve some gambling problems, for which they carried out a much more solid work than the one of their predecessors. Although they did not develop a theory, they were able to make predictions about random patterns in dice, which was a tremendous advance for the time and aroused great interest in other practical applications. This was quickly reflected in the work of other scientists, in particular in the first book on probability, written by Christian Huygens in 1657.

The history of probability is not the only one that is considered to begin in the 17th century. By the middle of that century, the shopkeeper and natural philosopher John Graunt collected the reports on deaths and christenings issued periodically by the clerks of the British parishes since around fifty years ago. Those reports, which included more than 200,000 items, also detailed the cause of each death. Graunt analyzed the data and estimated the chances of dying by each cause (he did not call those "probabilities," of course). He discovered the greater rate of birth of males in spite of the adult population being more or less equally divided into both sexes, which he attributed to war and other perils. The work of Graunt was one of the important breakthroughs in the history of statistics.

In the same century, Issac Newton and Gottfried Leibniz invented calculus (Apostol, 1967). They had no interest in a theory of chance, but their work had the utmost important implications on probability. The first in noticing that was Jacob Bernoulli, member of a family of mathematicians that made important contributions to science for generations. Jacob wrote a book, inspired by the one of Huygens, on probability and its wide applicability to human matters, where he proposed the law of large numbers, one of the links between probability theory and statistics that we will study in chapter 4.

Bernoulli's book was followed by another one by Abraham de Moivre in the 18th century, which compiled all the knowledge on probability of the time and introduced the bell-shaped or normal curve, one of the most important models for uncertainty, pervasively used in mobile robot localization and mapping, as in many other scientific fields.

By the middle of the 18th century, probability had reached the status of a new branch of mathematics, and started to be applied to other sciences and social problems that were exploding by then too. Concerning statistics, by that time British ships plying the seas and they needed to estimate their positions as accurately as possible. The most evident references when sailing the open sea were stars. Thus, tools and methods for localization by using large datasets of astronomical observations became very valuable. Edmund Halley was fascinated by both mathematics and astronomy; apart from naming the famous comet (he predicted its return using previous observations), he took new observations from hundreds of stars in order to make a map of the southern hemisphere, drew a map of the oceans which was extremely useful for navigators, revised and contributed to Isaac Newton's *Principia Mathematica*, and published a paper that, by analyzing statistically the mortality rates in the city of Breslau (Poland), set the beginning of the actuarial science: the science of assessing risks in assurance and financial businesses. Before then, concepts that look to us as basic as "life expectancy" had been unknown.

The 18th century also saw a special interest in searching for the real meaning of probability. At some point in his life, the English politician and mathematician Thomas Bayes wrote an article about probability that was published after his death. In that article, Bayes introduced a theorem which permitted the definition and calculation of the degree of "truth" of a given hypothesis, given some previous knowledge about the space of plausible hypotheses. Bayes' theorem is one of the most important concepts in mobile robot localization and mapping, as we will see in chapter 4 and beyond, and in countless problems in engineering and science. Still today, the idea of "measuring truth" (or "belief," as it is commonly stated) is controversial, due to the inclusion in the theorem of not scientifically founded prior assumptions, but Bayes had the merit of creating a new meaning for probability—that of a belief degree—in contrast with the most common one—frequency of occurrence. These paradigms of probability are now called *Bayesian* and *frequentist*, respectively.

From a wide perspective, the 18th century also saw a high degree of fragmentation in the research on probability. It was the astronomer and mathematician Pierre-Simon Laplace, who lived between the 18th and 19th centuries, the responsible for building a solid foundation for the next generations. From his astronomic studies, he concluded that the universe was deterministic, and that randomness was just the result of our lack of knowledge and errors in measurements. Thus, he proposed methods to obtain the "most probable" value of some process affected by uncertainty and also the confidence bounds for that value. He also improved Bayesian probability by reducing the bias coming from the prior assumptions, and introduced the central limit theorem, another link between probability and statistics that we will study in chapter 4.

Also in the transition between both centuries, the mathematicians Adrien-Marie Legendre and Carl Friedrich Gauss invented a new statistical method for coping with uncertainty in measurements, which was a breakthrough at the time for calculating the orbits of comets and planets. So far, errors in measurements were unmanageable: it was common to consider only a *few* measurements in order to avoid the accumulation of error, just the opposite to what nowadays we consider commonsense. These scientists were the first ones to propose taking a *large* number of observations for reducing error—variance—and thus increasing the accuracy in their estimation. Since no mathematical tool existed to cope with large datasets, they invented the method of least squares. This method is still in wide use today.

During the 19th century, the idea of a deterministic universe had begun to be questioned: some physical processes could not be predicted beyond a given point even when their mathematical models were available, as we now know. However, many of them could be still predicted *probabilistically*. Stochastic processes were studied at that time as a way of modeling the probabilistic behavior of physical phenomena that otherwise had no predictable outcome in a deterministic sense. For example, we cannot know for certain where a mobile robot exactly is (geometrically), but the stochastic process that produces its motion within the workspace, consisting of the combined behavior of computers, actuators, sensors and environment, can be used for estimating the probability of being at a number of locations, and even the most likely pose given the previous history of robot observations and actions—considering a Bayesian approach. Notice the analogy between this shift of paradigms (calculating the probabilities of the possible outcomes rather than obtaining a single outcome as the unique result of a process) and the change experimented in the eighties of the 20th century in mobile robotics, commented in chapter 1, from calculating a single pose (or map) for the robot, assuming it is the most likely truth, to estimating the likelihood of a number of them.

In 1828, the botanist Robert Brown described the first stochastic process (or, more precisely, the first random behavior produced by a stochastic process): the motion of pollen in suspension in a liquid, which is now called Brownian motion and applied to a diversity of phenomena. Almost fifty years later, the first attempt to explain mathematically a stochastic process, that is, a natural process that generates random outcomes, was developed by the most important physicist of the 19th century: James Clerk Maxwell. He, and later on the physicist Ludwig Boltzmann, developed a probabilistic theory for the overall parameters of the motion of gases, assumed to be composed of particles or *atoms*. The novelty of their theory was the possibility of obtaining accurate models for gases in spite of not knowing the exact state of motion of any individual particle, but only how they all move "as a whole."

The Maxwell-Boltzmann theory set the bases for the explanation of the Brownian motion in probabilistic terms, which was proposed by Albert Einstein and Marian Smoluchowski at the beginning of the 20th century. At that time, the mathematician Andrey Andreyevich Markov also named the dynamic probabilistic model he proposed for stochastic phenomena *Markov chains*, in the first formalization of stochastic processes. Markov chains are today a pervasive model in countless applications concerning stochastic processes, including mobile robot localization and mapping. Andrei Markov was also one of the most prominent researchers in probability theory of all times.

Modern statistics was finally born in the transition between the 19th and 20th centuries. The efforts of two great scientists is remarkable in that concern. The mathematician (among many other occupations) Karl Pearson addressed the visual representation of large datasets of statistical data and the problem of curve fitting (finding the mathematical model that fits well with those data), which he solved with the origin of our modern chi-squared test. On the other hand, the statistician and also geneticist Ronald Aymer Fisher studied how to draw conclusions from small sets of data, developed more sensitive tests of statistical significance, and studied the design of experiments.

In the early years of the 20th century, probability was already recognized as a fundamental and indisputable way of dealing with natural phenomena. There was a problem though: the mathematical power required for constructing a real probability theory was not available yet, having scientists only a bunch of techniques and methods. It was in the first two decades of the century when the mathematicians Emile Borel and Henri-Léon Lebesgue developed an extension of calculus to measure volumes or areas occupied by sets of points. Then, the mathematician Andrei Nikolayevich Kolmogorov used that measure theory for building, at last, a real probability theory: one that could use all the available human knowledge about calculus—mostly integration—to solidify and expand that of probability. For the first time, an axiomatic definition for probability was in sight. Kolmogorov is considered as the scientist that contributed the most to probability, but also outstood in his studies on information theory and complexity.

Nevertheless, Kolmogorov's approach is quite abstract, as well as the probability concept of his theory; the link to practical, real-world randomness is not always easy to justify from his perspective. This issue is related to the dispute between the Bayesian and frequentist meanings of probability: the former is more natural, easier to connect to reality (e.g., past real experiences form the prior knowledge in the approach), while the latter is more objective, since it does not depend on these assumptions, just on observed frequencies. The frequentist approach was explicitly stated as such by the work of the priest and mathematician John Venn past the middle of the 18th century and also by the mathematician Richard von Mises after the First World War. They established the meaning of probability as the frequency at which a random outcome tends over the long run, in contrast to that of a subjective degree of belief proposed by Bayesians.

Before World War 2, the frequentist paradigm was prevalent. At the beginning of the war, the multidisciplinary researcher Harold Jeffreys rescued the merits of the Bayesian approach. Today, both points of view have value. Mobile robotics is a good example of a mixture of these apparently competing approaches.

Since the middle of the 20th century, probability theory and statistics have been at the core of most modern engineering and scientific developments. One of the reasons is the development of computer science (especially for statistics). Automatic computation is of utmost importance when analyzing large amounts of data and also for approximating solutions in cases where closed mathematical forms or exact results are not available. In this regard, Monte Carlo methods are particularly relevant for modern mobile robot localization and mapping. We can trace their origin back to the work of Georges-Louis LeClerc, count of Buffon in the 18th century: he proposed a method to calculate the chances that a needle, dropped onto a floor where a number of straight parallel lines were drawn, effectively crosses one of those lines. This joined geometry and probability for the first time, but, more importantly for us, permitted us to estimate the value of π from a number of experiments with the needle, that is, Buffon used a stochastic process—i.e., thowing the needle—in order to approximate a deterministic value—the value of π. That is the core idea of Monte Carlo methods.

The first modern use of Monte Carlo methods that we know of is that by the physicist Enrico Fermi at about 1934 (Anderson, 1986). Fermi used a statistical sampling technique as a way of approximating difficult nuclear particle problems. Later on, in the early fifties, he spent a summer outside the University of Chicago at Los Alamos scientific laboratory, where the Manhattan project (responsible for creating the first atomic bombs for the Allies) had been coordinated. There he joined his friend Nick Metropolis, leader of the team that had designed and built the MANIAC, one of the first programmable computers, and started to implement his techniques as sequential programs with the collaboration of Metropolis, John von Neumann (inventor of the basic computer architecture that carries his name and many, many other things) and others. It was Metropolis who named Monte Carlo this statistical sampling method implemented on computers, and he also devised a number of techniques to improve its performance, such as importance sampling. As we will see in chapter 7, Monte Carlo methods are an essential ingredient of modern non-parametric mobile robot localization and mapping algorithms.

As it has been explained in chapter 1, the decades after World War II also saw the rise of the modern control theory, which included the automatic control of stochastic processes, mostly for aviation and military applications (also economics). This was strongly based on the Bayesian estimation of the mathematical models of dynamical stochastic processes, which is commonly called *sequential* or *recursive estimation*. The best known—and most basic—method for such estimation is the Kalman filter, developed in the early sixties by Rudolph Emil Kalman and Richard Bucy (Kalman, 1960; Kalman & Bucy, 1961). Extended Kalman filters, derived shortly after the seminal work of Kalman and Bucy (Bucy, 1965), and their variants are the counterpart of Monte Carlo methods for the case of using parametric models of uncertainty in estimation.

Both Kalman estimators and Monte Carlo-based estimators are the basis for current probabilistic localization and mapping algorithms for mobile robots. In particular, during the decades after World War 2, the statistician Calyampudi Radhakrishna Rao and the mathematician David Blackwell formulated the Rao-Blackwell-Kolmogorov theorem (Blackwell, 1947; Rao, 1965) which permits us to improve estimators in general, and also to factor some of the random variables in multivariate estimation problems. The latter led to the general theory of Rao-Blackwellization by Arnaud Doucet and others (Doucet, de Freitas, Murphy, & Russell, 2000), and to its instantiation as an effective algorithm for mobile robot simultaneous localization and mapping, which can mix parametric (Kalman-based) and non-parametric (Monte Carlo-based) filters (Montemerlo, Thrun, Koller, & Wegbreit, 2002). We have reached, finally, those years of the boom in probabilistic robotics. Most results introduced in this book arose then.

3. PROBABILITY SPACES

It seems logical to begin any attempt to introduce the modern, rigorous model of probability by defining *where* those probabilities exist and what they are exactly. We know that, under both the frequentist and Bayesian paradigms, *probabilities are numbers associated to certain phenomena in the real world*, in particular to the possible outcomes of some stochastic process. Therefore we need to formalize these "spaces of possible outcomes" to begin with.

Using the modern measure-theoretic definition of probability proposed by Kolmogorov (Shaffer & Vovk, 2005), a space which contains possible stochastic outcomes to occur is called a *probability space*, and can be modeled mathematically as a tuple of three elements (illustrated in Figure 3), denoted:

Figure 3. Illustration of the three mathematical components of a probability space, showing examples of the intersection of random events (which is another random event), a singleton event, and a non one-to-one probability function

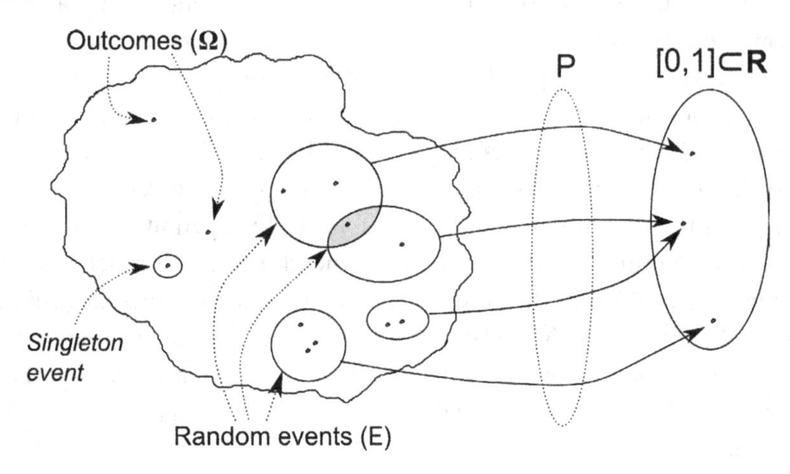

$$(\Omega, E, P) \tag{1}$$

Here, Ω represents the *sample space*: the set of all the different outcomes in which we are interested. For example, in the case of the localization of an omnidirectional mobile robot in a planar workspace, we could consider $\Omega = \mathbb{R}^2$, since the possible robot locations can be expressed as 2D geometrical coordinates[2]. However, the set of all possible outcomes is not enough to completely define the probability space.

It is common to make probabilistic reasonings not only on individual outcomes of the stochastic process (a single robot pose, for instance), but on sets of them (what is the probability that the robot is on this or that area?). That is the role of E, whose elements are sets of individual outcomes of Ω, called *random events* since they are events or situations that can occur with a certain probability. This set of events must fulfill some properties. Since we will perform certain operations on events, namely unions, intersections, complements and differences, E must be closed under them, that is, if we perform one of the mentioned operations on events of E we must obtain another event of E (in that way we can work with those operations in E without needing anything outside E). Furthermore, there are cases were we need to do these operations on more than two events at a time, even on infinite sequences of them. For coping with that, E must be closed under countable unions and intersections. The adjective *countable* indicates that the sets considered for unions and intersections could be infinite in number, but at least we should be able to enumerate them one at a time; in other words, they must be mapped to some subset of the set of natural numbers.

All these constraints are formally summarized by saying that E is a σ*–algebra* [3] of Ω. The collection of all possible subsets of Ω, those that are in E and those that are not, is called the *power set* of Ω, denoted as 2^{Ω}; therefore, $E \subseteq 2^{\Omega}$, or, in other words, not all subsets of Ω can belong to E in all cases. However, the sets that cannot be are not of interest in most probabilistic settings.

It is in order to introduce at this point the concept of *simultaneity of random events* and of outcomes, since outcomes are singleton random events. Time is not *explicitly* considered in the probability spaces formalism—it *is considered* though, as we explain now—thus simultaneity can have diverse interpretations. Typically, several events are considered simultaneous if outcomes from all of them occur within a given period of time that we cannot or do not wish to subdivide in shorter periods. Anyway, simultaneity of random events is an important concept since it has implications in many relevant aspects of the theory of probability.

A set of random events that occur simultaneously, whatever simultaneity means for us, is represented through the intersection operator, that is, the set $\{A_1, A_2, \ldots, A_n\}$ of random events occurring simultaneously is represented by the random event $\bigcap_i A_i$ just occurring. On the other hand, the union operator means the occurrence of any of the events in its arguments: $\bigcup_i A_i$ represents the situation in which some (maybe just one) outcome of the involved random events occur simultaneously, i.e., at a given time or time period. In summary: time is included in the formalism indirectly through a suitable interpretation of the union and intersection operators. Notice that individual outcomes are also random events (singleton random events), but their intersection is always an empty set unless we are dealing with only one outcome intersecting with itself: therefore, *different outcomes cannot occur simultaneously.*

The third but most relevant component of a probability space is P, the *probability function* of the probability space, a function defined on E that yields, for each random event, a value from the interval $[0,1]$ of the real numbers. Notice that this includes singleton random events, that is, individual outcomes, thus they are assigned a probability too. Actually, the function is only constrained to yield non-negative real numbers, but as a result of further restrictions imposed on it for being a measure, its resulting values lie within $[0,1]$. This can be mathematically denoted as $P : E \rightarrow [0,1]$. The value provided by P is the *probability* associated to the corresponding event, meaning either chance of occurrence of some of its outcomes—frequentist paradigm—or degree of belief about such chance—Bayesian paradigm.

In order to represent probabilities coherently, P is not free of restrictions: it must be a *measure* (Ash, 1999) that assigns the value 1 to the entire sample space. This implies satisfying the following properties, proposed by Kolmogorov as the IV-th and V-th axioms of his theory of probability:

$$P(\Omega) = 1 \tag{2}$$

$$P\left(\bigcup_i A_i\right) = \sum_i P(A_i), \quad A_i \in F,\ \forall i \neq j : A_i \parallel A_j \tag{3}$$

The purpose of constraint Equation 2 is clear: the probability of having *any* of the outcomes of the probability space must be maximum, since *some* outcome out of all the possible ones must occur.

Constraint Equation 3, also called *countable additivity*, forces the probability of occurrence of some outcome from a collection $\{A_i\}$ of events to equal the sum of the probabilities defined for each event separately, as long as all the events of that collection are pairwise *disjoint* or *mutually exclusive*. We say that two events A_i and A_j are disjoint or mutually exclusive (denoted $A_i \parallel A_j$) if and only if none of the individual outcomes of A_i can occur simultaneously to any of A_j, i.e., if and only if $A_i \cap A_j = \varnothing$—recall our previous considerations about simultaneity of events. For instance, a mobile robot that is in a corridor cannot be simultaneously outside the building, thus the probability of the robot to be at some point in the joint area formed by the corridor and the outside must equal the probability assigned of being at the corridor *plus* the probability of being outside. On the contrary, the robot being at the corridor and the robot being inside the building are not exclusive events—the corridor is inside the building—and consequently the probability of the robot being at any point inside the building, which is not only an area called "building" but also an area resulting from the union of the corridor and the building, cannot be deduced to be the sum of both.

Two remarks are in order here: firstly, any partition of Ω (a set of subsets of Ω which union is exactly Ω and which intersection is the empty set) is composed of mutually exclusive random events, as long as the partition is composed of sets of the σ-algebra of the probability space; secondly, if two events A_i and A_j are mutually exclusive, and considering that any subset of them is also a random event, all the pairs of random events that we can form by taking one subset of A_i and another one of A_j is also a pair of mutually exclusive random events. This includes pairs of outcomes, which cannot occur simultaneously ever, as explained before, it is perfectly valid mathematically (but rather useless) to talk about mutual exclusiveness of outcomes.

In addition to mutual exclusion, we could also consider another property of the probability space that also relies on simultaneity. This one is not needed for the formalization of the probability measure, but it will be used extensively in this book. It is the *mutual independence* of random events. Given a set of events $\{A_1, A_2, \ldots, A_n\}$, they are mutually independent, denoted $A_i \perp A_j, \forall i \neq j$, if and only if the following holds:

$$P\left(\bigcap_i A_i\right) = \prod_i P(A_i) \tag{4}$$

That is, the events are mutually independent if and only if the probability of them occurring simultaneously ($P\left(\bigcap_i A_i\right)$) can be calculated directly from the probabilities of each one occurring separately ($P(A_i)$), that is, no relation existing between them is reflected in that probability of simultaneous occurrence, only their individual probabilities. Mutual independence means that these events occur *independently on what the other events of the set do* (whether they occur simultaneously or not), which is different from mutual exclusion: the latter means that different events of the set *cannot occur simultaneously*. Therefore, mutual exclusion implies mutual independence, but not the other way around.

When E has a finite number of events no other constraint is needed to make P a probability measure and get a complete formalization. Otherwise, the VI-th axiom of Kolmogorov's theory is required for assuring the countable additivity pursued by constraint (3), which will not be explained in detail here:

$$\begin{aligned} &\forall A_1, A_2, A_3, \ldots \in F: \\ &\left[A_1 \subseteq A_2 \subseteq A_3 \subseteq \ldots\right] \wedge \left[\bigcap_i A_i = \varnothing\right] \\ &\Rightarrow \lim_{i \to \infty} P(A_i) = 0 \end{aligned} \tag{5}$$

For summarizing all this formulation: a probability space formally defines the set of individual outcomes of the stochastic process of interest (Ω), a useful and well-formed grouping for those outcomes into random events (E) and a probability measure over those events (P). From now on we can rely on any of the paradigms we have explained before, Bayesian or frequentist, to provide the numbers of P with human meaning.

We end this section with some interesting properties of the probability function P:

- Firstly, notice that constraints Equation 2 and Equation 3 together also imply that $P(\varnothing) = 0$ (where $\varnothing$ is the empty set), since

- $1 = P(\Omega) = P(\Omega \cup \varnothing) = P(\Omega) + P(\varnothing)$ and thus $P(\varnothing) = 1 - P(\Omega) = 0$. This can be interpreted in the same way as axiom (2), that is, as the impossibility of obtaining no outcome out of the sample space (the robot must be *somewhere*!).
- More generally, $P(\bar{A}) = 1 - P(A)$, with $\bar{A}$ denoting the complement of the set A.
- It can also be easily deduced that $P(A \cup B) = P(A) + P(B) - P(A \cap B)$, which is a more general result that when we deal with mutually exclusive events only.
- Finally, recall that P is defined on random events of E. In the case of finite sample spaces, the number of events of E is finite too, and thus we can define a probability P that is greater than zero for any outcome (singleton random event) of the sample space if we wish, and still have a probability of 1 for the entire sample space. However, in infinite (continuous) sample spaces obtaining that sum is more complicated, since the event space E can also be infinite. It can be demonstrated (Apostol, 1969; theorem 14.1) that if P must satisfy the constraints imposed on a probability measure, *we can only assign non-zero probabilities to a countable subset of events*. In continuous processes it is therefore common to assign zero probability to every singleton event, that is, events consisting of only a single outcome. For example, having a zero probability of measuring any single robot distance to target (distance that comes from a continuous sample space: a subset of the real numbers) is the common approach if we consider mobile robot navigation. This has no relevant consequence in practice. Just think of the likelihood of gathering a real number with *infinite* accuracy through any real sensor or computational process... The idea underlies the concept of *almost surely*: events, which, in spite of belonging to the sample space, have a probability of zero; we can say that those events will *almost surely* never occur. We will come back to the concept of *almost sure* events in the next chapter.

4. RANDOM VARIABLES

If all we accounted for were probability spaces, we would need a lot of different definitions for dealing with even the simplest problem, since in practice we are interested in handling several aspects of the outcome of a given stochastic process. For example, imagine an omnidirectional mobile robot moving in a planar, geometrical workspace. We can define a probability space to represent the stochastic motion of the robot, i.e., motion with uncertainty. In that example, the sample space consists, for instance, of the 2D coordinates for each possible location. Knowing the geometrical pose, i.e., a singleton random event of that space, is an important matter, but what if we also need to reason about the time-varying distance of the robot to a given target, something that is common in navigation and can be deduced from the robot pose? We have to define a different probability space that has those distances as outcomes (1D values). Undoubtedly, both probability spaces share the same underlying physical process, and our need for both does nothing but reflecting an excessive verbosity in our mathematical toolkit.

In order to define a more efficient framework, we need the concept of *random variable*. For this, knowledge of basic calculus and algebra is required from the reader again.

A random variable (r.v. for short) is just a function that translates the outcomes of a given probability space (Ω, E, P) into real numbers or vectors of real numbers. It is defined as (refer to Figure 4):

Figure 4. Example of random variable (X) defined on a probability space, along with its inverse and the corresponding cumulative distribution function

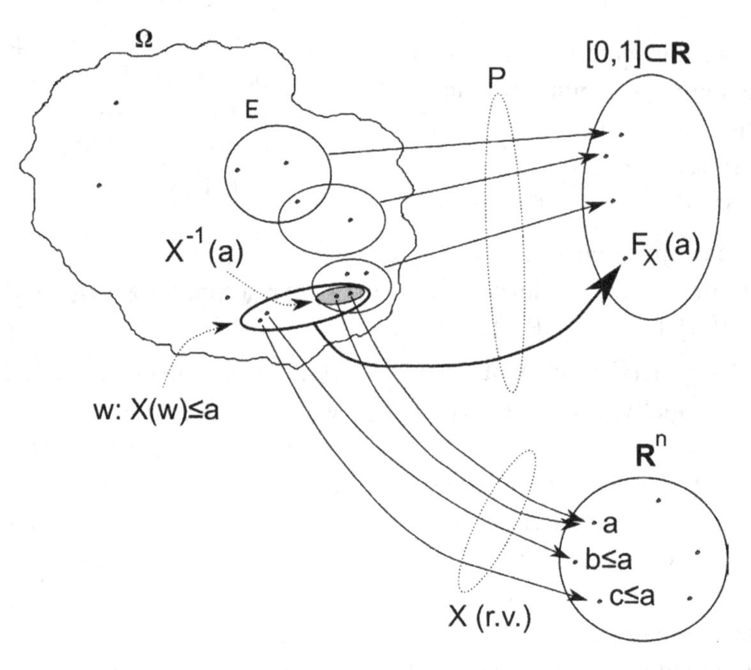

$$X:\Omega \rightarrow \mathbb{R}^n,\ n \in \mathrm{N}^+ \quad (6)$$

It is worth to recall now the aspects concerning time and simultaneity that we discussed in the previous section. The formalism of random variables does not include them: random variables are just mathematical functions from outcomes to real numbers or vectors. Since individual outcomes of the sample space are always mutually exclusive—they cannot occur simultaneously—a random variable will never map the same outcome more than once simultaneously, whichever meaning we have assigned to simultaneity. That means that the r.v. will never yield simultaneously the same value, since that means either that an outcome has occurred simultaneously more than once (Impossible! An outcome only can occur once *at a given instant of time* since otherwise we could not distinguish its several occurrences) or that two or more different outcomes has occurred simultaneously (Impossible! They are mutually exclusive by definition).

We can create as many random variables on the same probability space as we wish. Each of them represents some quantity of interest in the stochastic behavior of the process, e.g., a robot pose or a distance to a target, both deduced from the same stochastic process. The value of the r.v. varies as the current outcome of the process changes, therefore it is a stochastic value. If the outcomes of the process are real values or vectors of real values, and are themselves interesting values for us, we can also create an identity r.v., that is, $X:\Omega \rightarrow \Omega$, as a way to directly use the framework of r.v.s with the sample space.

Being a r.v. a mathematical function, we can study some of its properties. Firstly, as a function it must be defined for every element of the sample space Ω (since we have seen that some outcome must occur necessarily, i.e., $P(\Omega)=1$, this means that the r.v. will always have a value). Secondly, it may be not an *one-to-one* function (also *injective* function), that is, different outcomes taken from Ω may correspond to the same value of the r.v.—e.g., the same distance to target may be measured from different robot locations. However, notice that this might puts the distinctivity of random events at risk, thus the r.v. should not have such behavior for random events that interest us. Thirdly, it may not be an *on-to* function, that is, there may be values that lie in $\mathbb{R}^n$ that do not correspond to any outcome of the process—e.g., there may be locations in the environment where the robot cannot be located, such as inside walls.

The set of values that the r.v. can yield is its *codomain*—also called *image* or *range*—that we will denote $co(X)$. Correspondingly, the values over which the r.v. is defined, that is, Ω, is its *domain*—or *pre-image*—denoted here as $dom(X)$. Obviously, the codomain is more important than the domain, since a relevant contribution of the r.v. to our mathematical toolkit is to define a certain codomain, and not other, containing the possible values that we can observe from the stochastic process. Furthermore, within that codomain, some values will be associated a non-zero probability, in the way we will explain later. That particular subset of the codomain containing all the values that are assigned a non-zero probability is called the *support*[4] of the r.v., and will be denoted as $supp(X)$. The support of a r.v. is a pervasive concept in this book.

We can look a little closer at the codomains of random variables, which will allow us to classify them. We distinguish two important groups by doing so: *discrete* random variables and *continuous* random variables. Discrete r.v.s are functions with *countable* codomains. For example, the pose of a mobile robot in a topological map can be modeled as a discrete r.v. since the robot must be at one of a denumerable list of symbolic places. On the other hand, continuous r.v.s, such as robot metrical poses, have *uncountable* codomains. Note that finite sample spaces always lead to discrete r.v.s, and infinite sample spaces typically lead to continuous r.v.s. It is also noteworthy that this continuity feature of r.v.s is a different concept from the continuity of mathematical functions. In general, a continuous r.v. might have a mixed classification: its codomain can be split into a countable part (discrete) and an uncountable part (continuous), but such a generalization is rarely considered in probabilistic robotics.

Once we understand the basic properties and kinds of r.v.s, we need to link them to the concept of probability, that is, we need to formalize the uncertainty they capture from the underlying stochastic processes.

Since a r.v. transforms outcomes of a sample space into real numbers, and the probability function P of the corresponding probability space assigns probabilities to *random events* (sets of outcomes), we could look firstly for a way to assign a probability value to every single value yielded by the r.v. Actually, we will do that for r.v.s that have finite sample spaces (which surely are discrete r.v.s), but this is not as universally applicable as we could think: in uncountable sample spaces, the probability measure P assigns a probability of zero to singleton events, i.e., to individual outcomes; thus, assigning a probability to every single value yielded by a continuous r.v. has no practical interest: all those probabilities would be zero!

We will follow a different and more general approach, which is valid for modeling uncertainty in both discrete and continuous r.v.s.

Let us start considering the path shown in Figure 4, that goes from a r.v. to its probability space. Mathematically, this path is the inverse of the r.v.:

$$X^{-1} : \mathbb{R}^n \to 2^{\Omega} \quad (7)$$

Strictly, X^{-1} is a *partial* function in $\mathbb{R}^n$, that is, it is a function defined only on a subset of $\mathbb{R}^n$, concretely on the codomain of X (recall that 2^{Ω} denotes the set of all the subsets of Ω, that is, its powerset). On the other hand, X^{-1} will cover all the values of $dom(X) = \Omega$, i.e., X^{-1} is on-to. $X^{-1}(a)$ yields the set of outcomes w of the stochastic process that were mapped to the value a by the r.v. X, whenever $a \in co(X)$:

$$\forall a \in co(X),\ X^{-1}(a) = \{w \in \Omega : X(w) = a\} \quad (8)$$

In general, X^{-1} will yield a set that should be considered empty if X^{-1} is applied not to $co(X)$ but to the entire $\mathbb{R}^n$:

$$X^{-1} : \mathbb{R}^n \rightarrow 2^{\Omega}$$

$$\forall a \in \mathbb{R}^n,\ X^{-1}(a) = \begin{cases} \{w \in \Omega : X(w) = a\} & \text{if } a \in co(X) \\ \varnothing & \text{otherwise} \end{cases} \quad (9)$$

becoming in this last definition a total function. Thus, if $X^{-1}(a) = \varnothing$, we know that a is not in $co(X)$.

As we have said, this inverse X^{-1} maps some value of the random variable, say a, into a set of outcomes from the probability space, say $\{w_j\}$, and since the probability of such set $\{w_j\}$ is well defined as long as they constitute a random event—random events, and more concretely, the σ-algebra of the probability space, will be defined for that to occur—we can consider $P(\{w_j\})$ as the probability of a. That is the general way of assigning probabilities to values of a r.v., i.e.,

$$\forall a \in co(X), P[X = a] = P(\{w \in \Omega : X(w) = a\}).$$

However, the probability of a singleton random event is zero for continuous r.v.s, and by considering single values like a we take the risk of obtaining a singleton set $\{w_1\}$, rendering this probability mapping useless in that case. It is more reasonable to assign probabilities instead to sets of values of the r.v., say $\{a_i\}$, that necessarily come from a non-singleton random event $W = \{w_1, w_2, ...\}$ because X is a mathematical function and cannot map one value of its domain (an outcome) to several ones in its codomain. Therefore, for now we will focus on assigning probabilities not to single values of the r.v. but to sets of values, which is valid for both discrete and continuous r.v.s.

It is of particular interest to calculate the probability of a r.v. yielding a particular set of values composed of *any value that is less than or equal to a given number*[5]. This is a very important concept in probability theory, and the function that assigns a probability to such sets of values of a r.v. is called the *probability distribution function* or *cumulative distribution function* (*cdf* for short) of the r.v., denoted as F_X, with the name of the r.v. as the subscript. It is formally defined as follows:

$$F_X : \text{supp}(X) \rightarrow [0,1]$$

$$\forall a \in co(X),\ F_X(a) = P\left(X^{-1}(b) : b \leq a\right) = \\ = P\left(\{w \in \Omega : X(w) \leq a\}\right) \quad (10)$$

where P is taken from the probability space of the r.v. Equation 10 is well defined as long as the random event $\{w \in \Omega : X(w) \leq a\}$ belongs to the σ-algebra, which must be assured by choosing a suitable σ-algebra. Notice that, since $P(\varnothing)$ is always defined for any probability space, we can use here the total definition of X^{-1} given in equation (9) without problems. For the sake of brevity, it is also common to use the following notation for the probability distribution function $F_X : F_X(a) \triangleq P[X \leq a]$, which joins in the same expression both the probability space and the r.v. We use the square brackets here to highlight the fact that in this expression P is not the probability function of the space, but a more complex mathematical operator.

Figure 5. Some examples of cdfs. (a) A piecewise, discontinuous (but right-continuous) cdf corresponding to a discrete r.v. (b) A continuous cdf corresponding to a continuous r.v. (c) A cdf with only one discontinuity (a continuous r.v. with one discontinuity).

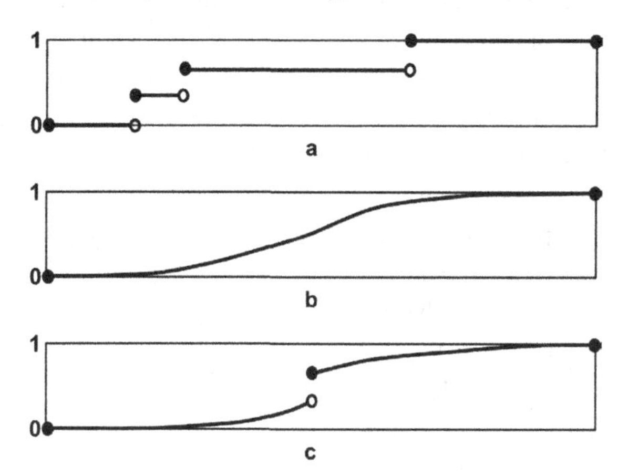

Thus, we have reached at the first general way of assigning probabilities to a r.v., or in other words, to know the uncertainty that comes with the r.v. If the r.v. yields real numbers, we can even see the graphical aspect of that uncertainty, as Figure 5 illustrates.

Let us explain some of the basic mathematical properties of a cdf F_X :

- Since F_X yields probabilities, its codomain is the same as that of P, that is, the closed interval $[0,1]$ of the real numbers.
- F_X is always *monotone non-decreasing*, that is, $F_X(a) \leq F_X(b), \forall a < b$. This makes sense, since if $a < b$ there are the same or more possible outcomes to consider in the case of b than in the case of a.
- $F_X(a)$ tends to 1 as a tends to the maximum value of the r.v., since at that place the random event to consider is the entire sample space, which has probability 1. Mathematically, $\lim_{a \to \max(\text{supp}(X))} F_X(a) = 1$. In the same fashion, $\lim_{a \to \min(\text{supp}(X))} F_X(a) = 0$. Since the maximum and minimum values of $\text{supp}(X)$ may be multidimensional, the max/min functions that appear in these limits must be understood accordingly.
- F_X might be a *continuous* function, that is, one which always exhibits small variations in its value for small variations in its independent variable, but it is only forced to be *right-continuous*[6]. This is a relevant issue since it is related to the differentiability of F_X, which we require in further steps for a complete formalization of the uncertainty of r.v.s.

Note that, if F_X is continuous (and not only right-continuous), the probability of the r.v. to yield any single value, that is, $P[X = a]$ in the previously introduced notation, is exactly zero, since

$$P[X = a] = F_X(a) - \lim_{x \to a^-} F_X(x) = F_X(a) - F_X(a) = 0.$$

This is the normal case in continuous r.v.s. On the contrary, discrete r.v.s produce only right-continuous cdfs, in the form of step-wise functions—with changes of step at each different value yielded by the r.v.—thus they can have probabilities greater than zero assigned to individual values.

5. THE SHAPE OF UNCERTAINTY

For our purposes, a r.v. is a mathematical model of an uncertain quantity of interest that is observable in a physical process. We can handle explicitly the uncertainty of such quantity by means of the probability distribution function of the r.v., or cdf, introduced in the previous section, independently on the particular class or properties of the r.v. However, the level of detail of a cdf is not as fine as we could desire; many aspects of the uncertainty cannot be observed easily due to the fact that the cdf summarizes too much: a single value of the cdf represents, for each x, all the probability information associated to an entire interval that ranges from the minimum of the support of the r.v. to x. Therefore, the set of values of the r.v. to which the cdf assigns probabilities are not a partition of its codomain (small ones are subsets of larger ones). This leads to an unsuitable level of granularity when trying to deduce properties of the underlying probability space from observation of the values yielded by the r.v., if we are restricted to use the cdf formalism.

In practice, it would be useful to have a model of the shape of uncertainty with finer granularity, one that reveals more explicitly its interesting parts in other, more flexible sets, we can define on the variable codomain; in other words, we look for a function different from the distribution function, one that concentrates less information on each value of the codomain of the r.v. In the following, we derive such functions.

The Shape of Uncertainty of Discrete R.V.s

The obvious candidate for a function that reflects the uncertainty with more detail than the cdf is one that summarizes the least: one that yields just the probability corresponding to each individual value of the codomain of the r.v. and nothing else. For continuous r.v.s this is an ill defined situation, as we have already explained, but in the discrete case it is quite the opposite.

In the discrete case, the function we are looking for is called the *probability mass function* (*pmf* for short) of the r.v. It assigns a real probability—"mass"—to each possible value of the variable. Since it is closely related to F_X, we will denote it f_X. It is formally defined as:

$$\forall a \in \mathbb{R}^n,\ f_X(a) = \begin{cases} P\left(\left\{w \in \Omega : X(w) = a\right\}\right) & \text{if } a \in co(X) \\ 0 \text{ otherwise} & \end{cases} \tag{11}$$

When the reference to the r.v. in the subscript is not needed, it is common to denote the pmf simply as P (capital letter), and write down, for example, $P(a)$ (which is analogous to the notation $P\left[X = a\right]$). We will use the subscripted version, f_X, as long as we can for the sake of uniformity, but in long formulas we will rely on $P(a)$ for simplifying the notation.

Since we are considering discrete variables, there are a countable set of values to which the pmf assigns probabilities, and any of them can be greater than zero. The sum of all the values of the pmf must equal 1, anyway, since the probability associated to the entire sample space is 1. That is:

$$\sum_{a \in \text{supp}(X)} f_X(a) = 1 \tag{12}$$

Pmfs can be plotted as bar graphs, and surely you are quite familiar with histograms, a similar representation[7].

The Shape of Uncertainty of Continuous R.V.s

For a continuous r.v. we cannot obtain the same level of detail as with discrete r.v.s, since the probability associated to a single value of the support would be zero. Thus, we are forced to summarize into a real number the probability information corresponding to more than one value of the r.v. In order to still provide more detail than the cdf, we could choose *an arbitrary region* of its support instead of regions always starting at the minimum, as the cdf does. Just to give an example of that, for an arbitrary semi-closed interval $(a,b]$ within the support of the r.v., we would have:

$$\forall a,b \in \text{supp}(X),\ a < b,$$

$$\begin{aligned}
&P\left[x \in (a,b]\right] = P\left[(x > a) \wedge (x \le b)\right] = \\
&= P\left(\left\{w \in \Omega : a < X(w) \le b\right\}\right) = \\
&= 1 - P\left(\left\{w \in \Omega : \left(X(w) \le a\right) \vee \left(X(w) > b\right)\right\}\right) = \\
&= 1 - P\left(\left\{w \in \Omega : X(w) \le a\right\} \cup \left\{w \in \Omega : X(w) > b\right\}\right) = \\
&= 1 - \begin{bmatrix} P\left(\left\{w \in \Omega : X(w) \le a\right\}\right) \\ +P\left(\left\{w \in \Omega : X(w) > b\right\}\right) \end{bmatrix} = \\
&= 1 - \begin{bmatrix} P\left(\left\{w \in \Omega : X(w) \le a\right\}\right) \\ +1 - P\left(\left\{w \in \Omega : X(w) \le b\right\}\right) \end{bmatrix} = \\
&= 1 - \left[F_X(a) + 1 - F_X(b)\right] = F_X(b) - F_X(a)
\end{aligned} \tag{13}$$

The last line of Equation 13 shows how easy is to assign a probability to this kind of regions by means of the cdf.

Now, does the last line of Equation 13 look familiar to your mathematical mind? Yes: it is the value of a definite integral of some function $\rho(x)$ calculated on the interval $(a,b]$. Thus, we can write down the following definition:

$$\begin{aligned}
&\forall (a,b] \subseteq \text{supp}(X), \\
&P\left[x \in (a,b]\right] \triangleq \int_{x \in A} \rho(x)dx = F_X(b) - F_X(a)
\end{aligned} \tag{14}$$

This relates the cdf F_X and the function ρ. In particular, Equation 14 suggests that we can force the cdf to be the indefinite integral of ρ, or, equivalently, *force* ρ to be the derivative of F_X [8]. Notice that this will be valid only as long as F_X is differentiable in the entire support of the r.v., because obviously ρ has to exist. A side effect of that approach is that since we will force F_X to be differentiable and all the differentiable functions are also continuous, the probability of any single value of the r.v. will be definitely zero, as we have assumed before.

In summary, the function ρ has been defined to assign probabilities to contiguous intervals of values of the r.v. (through its integral). The question now is: which is the meaning of the real numbers that the function ρ yields for single values a of the r.v.? It is very important to notice that these numbers *are not probabilities*; the only number that is a probability is the result of the integral in Equation 14. These number have, nevertheless, an intuitive interpretation: for each single value a of the r.v., ρ yields a number called the *probability density of* a, which due to the monotonic property of the integral, must be proportional to the likelihood of obtaining a from that r.v.

Using this important result, we can now rename $\rho \triangleq f_X$ and call f_X the *probability density function* (*pdf* for short) of the r.v. Along with pmfs, pdfs are the most common explicit representations of uncertainty in probability theory, much more than cdfs for the reasons already explained.

In contrast with pmfs, since the values of pdfs are not probabilities they can be outside the interval $[0,1]$, though they must be positive. Analogously to pmfs, the integral of a pdf over its entire support must be 1, since that integral is a probability, and the probability of observing some value of the entire support of the r.v. must be 1:

$$P\left[x \in \text{supp}(X)\right] = \int_{x \in \text{supp}(X)} f_X(x)dx = 1 \tag{15}$$

It is also very common to abbreviate f_X as p (lowercase, in contrast to the abbreviated notation used for pmfs), as long as the r.v. can be deduced from the context, writing down $p(x)$ instead of $f_x(x)$ in the integral above. However, we will maintain the f_X notation as long as possible for uniformity in notation, and shift to the abbreviation when we deal with long or complex expressions.

The Shape of Uncertainty of any R.V.: The Likelihood

A common way to denote the fact that a r.v. X has a pdf/pmf f_X is $X \sim f_X$. The mass or density function f_X clearly provides more visual information than F_X, as shown previously and illustrated in Figure 6. For instance, we can observe in it the values of the r.v. that are more likely to occur, as those that have more density or mass of probability. Those values are called *modes* of the distribution, and allow us to distinguish between *unimodal* distributions and *multimodal* distributions. This has important applications in mobile robotics. For example, when a robot tries to localize itself, the uncertainty in its position is typically multimodal for a while until it gathers information enough to discern among the potentially large number of similar places in the environment. Another value of interest that can be observed directly in a density or mass function is the *median*: the value of the support of f_X that separates it into two parts of equal area, that is, into two sets of values of the r.v. that have exactly the same probability of occurrence.

Aside from the modes of a distribution, it is noteworthy to remark that the *height* of the pmf/pdf—the number in the ordinate axis—that can be assigned to any given value of the r.v.—the number in the abscissa axis—will be called in some contexts the *likelihood* of the r.v.: it is an indication of the relative probability of that value with respect to the others, but it can also be considered just as the result of a function that coincidently has the same mathematical form of that pmf/pdf. In statistical estimation, explained in chapter 4, and in the mobile robot localization and mapping problems in particular, likelihood functions play an important role. It is also a core concept in statistical hypothesis testing (Kassam, 1988).

In section 8, we will explain how to construct new r.v.s from existing ones by transforming the r.v. values through some mathematical function, obtaining in that way a new r.v. with a shape of uncertainty that is related to the one of the original variable but which is, in general, difficult to deduce mathematically. Under that context, it is easy to see that the likelihood of a r.v., stripped off its interpretation as a probability function, can be considered as a mathematical function of the values of the variable, and thus generates a new r.v. with its own, usually complicated distribution.

6. SUMMARIZING UNCERTAINTY

As discussed in the previous section, knowing as many details as possible about the uncertainty that is present in a r.v. is of indisputable importance. However, when reaching the situation of making decisions (for example, answering the question: is the robot within one meter to some target point?) or using probabilistic information into deterministic formulas, it is useful to summarize the main properties of pdfs/pmfs in only one or a few numbers. In the following, we provide the most common approaches for that. We intentionally keep out the concept of moment generation function and its remarkable utility for deduction about random variables, especially in the fields of time series analysis and signal processing, since it would require previous knowledge about Fourier transforms that fall well outside the scope of this book—see for instance Manolakis *et al.* (2005).

Figure 6. Some examples of pdfs/pmfs. (a) A discrete r.v. yields a stepped cdf and a histogram as pmf. (b) A r.v. which yields an unimodal pdf: a chi-squared distribution with parameter k=4 (degrees of freedom), which is equivalent to a gamma distribution with parameters shape=2 and scale=2. Note how the area below the pdf, from the minimum of the support of the r.v. to a certain point, corresponds to a single value A of the cdf, and how the total area of the pdf equals 1. (c) A multimodal r.v.: a weighted sum of two Gaussians with means of 7 and 20, variances of 1 and 4, and weights of 1 and 2, respectively.

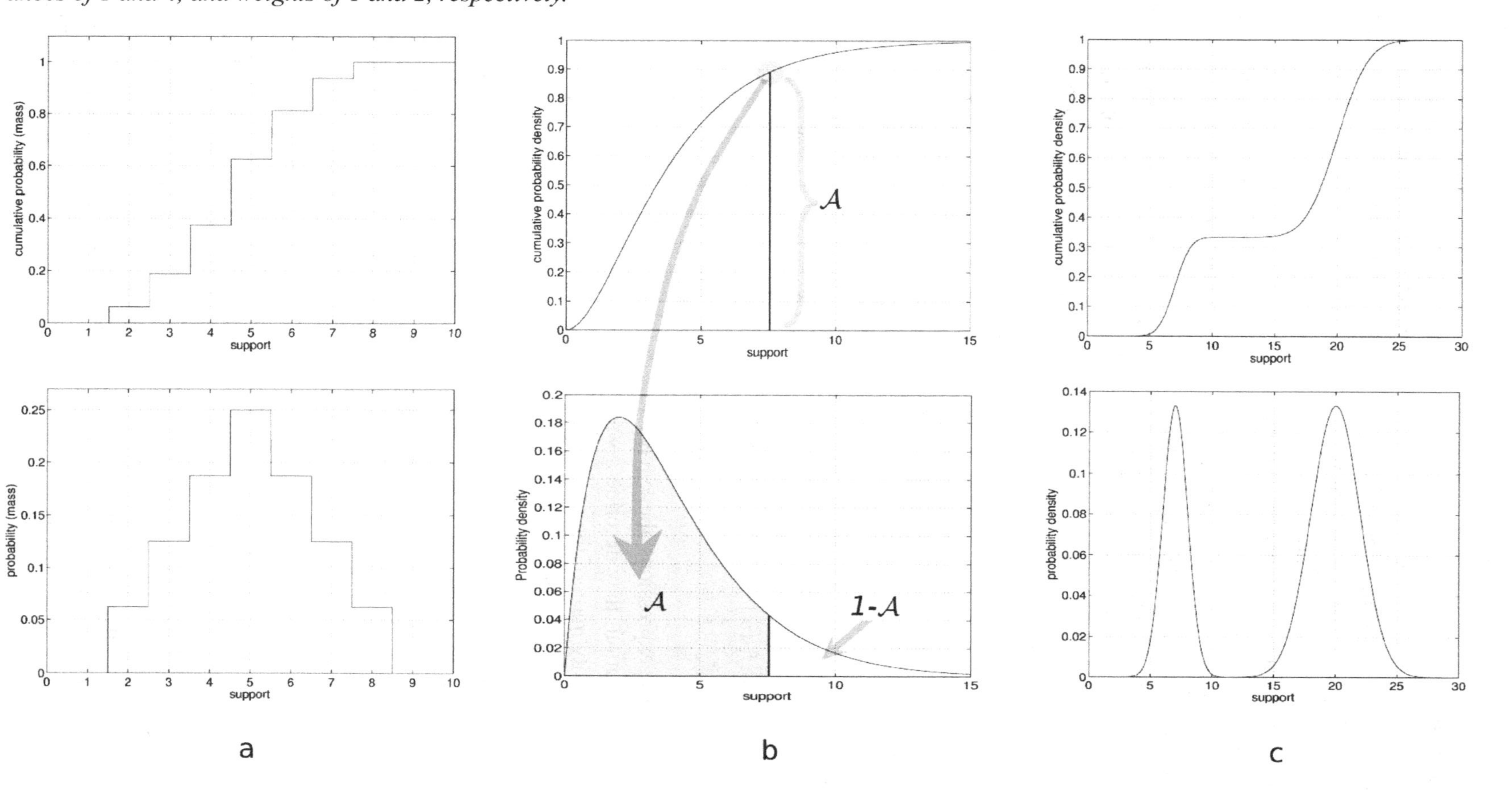

Moments of a R.V.

The most important of these numbers are the so-called *moments* of a probability distribution, which can be deduced from both pdfs and pmfs. Theoretically there can be infinitely many moments, numbered from 1 onwards, but only the first few are of interest (some distributions lack some moments, or even have none, but they are not very useful in probabilistic robotics). It can be demonstrated that, if the $n-th$ moment exists, so do all the previous moments in the sequence.

The $n-th$ moment around a given value $c \in \text{supp}(X)$ is defined as:

$$\mu_n^c \triangleq \int_{x \in \text{supp}(X)} (x-c)^n f_X(x)\,dx \tag{16}$$

for continuous r.v.s, or as:

$$\mu_n^c \triangleq \sum_{x \in \text{supp}(X)} (x-c)^n f_X(x) \tag{17}$$

for discrete ones. The expressions above have a geometrical interpretation: they describe a way of weighting the density/mass of the r.v. at each point of its support with something alike to the distance of that point to c, or, in other words, moments are a way of reflecting in different ways *how the density/mass of probability is spread over the support of the r.v.* Therefore, they become handy for summarizing the shape of uncertainty into a few real numbers. Table 1 shows some pdfs/pmfs and their first moments.

There are two common choices for the reference point c: those with $c = 0$ are called *crude* or *raw* moments; those with $c = \mu_1^0$ are called *central* moments, and they reflect how the density/mass of probability is spread around the first raw moment, which is called the mean. In addition, from the second moment on, they can be transformed by a scaling factor that cancels their units, normalizes them, and makes them invariant with respect to linear scaling, which is useful for reflecting more robustly the corresponding information about the shape. That factor is

$$\left(+\sqrt{\mu_2^{\left(\mu_1^0\right)}}\right)^n,$$

and the moments that are divided by it are called *standardized* moments.

The most important moments have special names. The first raw moment (μ_1^0) is called the *expectation* or *mean* of the r.v., and the second central moment, the *variance* of the r.v. ($\mu_2^{\mu_1^0}$). We will use them along the rest of the book intensively; they are abbreviated as $E[X]$ and $V[X]$, respectively, and when the r.v. must be made even more explicit, $E_X[X]$ and $V_X[X]$. Therefore:

$$E[X] \triangleq \int_{x \in \text{supp}(X)} x f_X(x)$$

$$V[X] \triangleq \int_{x \in \text{supp}(X)} (x - E[X])^2 f_X(x) \tag{18}$$

for continuous r.v.s, and:

$$E[X] \triangleq \sum_{x \in \text{supp}(X)} x f_X(x)$$

$$V[X] \triangleq \sum_{x \in \text{supp}(X)} (x - E[X])^2 f_X(x) \tag{19}$$

for discrete variables. In many occasions, they are also denoted μ_X and σ_X^2 respectively, and when the r.v. can be deduced from the context, μ and σ^2. The square in this symbol for the variance has its importance: the square root of the variance, although not a moment, is another value of interest. It is called the *standard deviation* of the r.v., and denoted σ_X or simply σ. Note that it has the

Table 1. Some useful shapes of unidimensional parametrical uncertainties, along with their mathematical forms and their first moments. The support of these distributions is the entire real line unless noted otherwise.

Distribution, parameters and support	E[X]	V[X]	Expression of the pmf/pdf f_X
Dirac's delta $X \sim \delta_a$ $a \in \mathbb{R}$	a	0	$\delta_a(x) = \begin{cases} \infty & \text{if } x = a \\ 0 & \text{otherwise} \end{cases}$
Bernoulli $X \sim Ber(p)$ $p \in [0,1] \subset \mathbb{R}$ supp(x)=$\{0,1\}$ (discrete)	p	$p(1-p)$	$Ber(x;p) = \begin{cases} 1-p & \text{if } x = 0 \\ p & \text{if } x = 1 \end{cases}$
Binomial $K \sim B(n,p)$ $p \in [0,1] \subset \mathbb{R}$ n $\in \mathbb{N}$ supp(x)=$\mathbb{N}$ (discrete)	np	$np(1-p)$	$B(x;n,p) = \binom{n}{x} p^x (1-p)^{n-x}$
Uniform $X \sim U(a,b)$ $a,b \in \mathbb{R}$ $a < b$	$\frac{1}{2}(a+b)$	$\frac{1}{12}(b-a)^2$	$U(x;a,b) = \begin{cases} \frac{1}{b-a} & \text{if } x \in [a,b] \\ 0 & \text{otherwise} \end{cases}$
Exponential $X \sim \exp(\lambda)$ $\lambda > 0$	$\frac{1}{\lambda}$	$\frac{1}{\lambda^2}$	$\exp(x;\lambda) = \begin{cases} \lambda e^{-\lambda x} & \text{if } x > 0 \\ 0 & \text{otherwise} \end{cases}$
Gaussian or Normal $X \sim N(\mu,\sigma^2)$ *(The especial case* $N(0,1)$ is called standard Gaussian *or* standard normal *distribution)*	μ	σ^2	$N(x;\mu,\sigma^2) = \frac{1}{\sqrt{2\pi\sigma^2}} e^{\frac{-(x-\mu)^2}{2\sigma^2}}$
Gamma $X \sim \Gamma(k,\theta)$ $k > 0$ $\theta > 0$	$k\theta$	$k\theta^2$	$\Gamma(x;k,\theta) = \frac{x^{k-1} e^{-x/\theta}}{\theta^k \underbrace{\int_{t=0}^{t=\infty} t^{k-1} e^{-t} dt}_{\Gamma(k)}}$
Chi-squared $X \sim \chi_k^2$ $k > 0$, called *degrees of freedom* *It is a especial case of the gamma distribution:* $X \sim \Gamma(k/2, 2)$	k	$2k$	$\chi_k^2(x) = \begin{cases} \frac{x^{k/2-1} e^{-x/2}}{2^{k/2} \underbrace{\int_{t=0}^{t=\infty} t^{k/2-1} e^{-t} dt}_{\Gamma(k/2)}} & \text{if } x \geq 0 \\ 0 & \text{otherwise} \end{cases}$

same units as the r.v., while the variance has those units squared.

Expectation and variance are so important that in many cases, they effectively contain *all* the information of the shape of uncertainty, that is, the shape of uncertainty—the pmf or pdf—is a mathematical function that only depends on them. In those cases, these moments are called the *parameters* of a probability distribution (pmf/pdf). For this reason, it is worth listing some of the properties of the mean and variance:

- The expectation is the value of the r.v. that one would "expect to see," which has not to be the value with the highest probability of being observed, as it happens in asymmetric or multimodal distributions. In occasions, the expected value is useful when we have to select some value of the r.v. as the "representative" of the process in order to make decisions. Note however that the expected value may be a value which is impossible to observe, since it is not necessary that it belongs to the support of the r.v.—think of a discrete distribution: the mean is not likely to be any of the values of the support.
- The variance is the expected value of the distances or divergences of all the values of the r.v. with respect to its mean, and thus an indicative of their dispersion: $V\left[X\right]= E\left[\left(X - E\left[X\right]\right)^2\right]$. This expression shows that the variance can also be seen as the expected squared error of the r.v. with respect to its own mean. A r.v. with a large variance has a flat shape, with density/mass values greater than zero in distant regions of its support, while a r.v. with a small variance often has a peaked shape. The first case is informally considered as reflecting "more uncertainty" than the latter; the probability of obtaining values of the r.v. is concentrated around few values in the latter, and thus we can be more sure of obtaining some of them in that case.
- It is not difficult to deduce algebraically that variance and expectation are also related in this way: $V\left[X\right] = E\left[X^2\right] - \left(E\left[X\right]\right)^2$, thus $E\left[X^2\right] \geq \left(E\left[X\right]\right)^2$, since $V\left[X\right] \geq 0$.
- Expectation is a linear operator, that is, $E\left[aX + bY\right] = aE\left[X\right] + bE\left[Y\right]$ for arbitrary constants a and b. We do not go into more detail with this since it involves more than one r.v., which we will address in further sections.
- In contrast, variance is not linear. Instead,

  $$V\left[aX + bY\right] = a^2V\left[X\right] + b^2V\left[Y\right] \\ +2abE\left[XY\right] - 2abE\left[X\right]E\left[Y\right].$$

  In the particular case of only one r.v., $V\left[aX\right] = a^2V\left[X\right]$.
- Finally, expectation can be extended to functions of r.v.s, that is, we can define $E\left[g(x)\right]$. This is not strictly a moment, but is equally powerful. It is defined as $E\left[g(x)\right] \triangleq \int_x g(x)f_X(x)dx$ or $E\left[g(x)\right] \triangleq \sum_x g(x)f_X(x)$ for continuous or discrete r.v.s, respectively. With this more general definition, the mean of a r.v. becomes the expectation of its identity transformation, i.e., $g(x) = x$.

The third and fourth moments of a r.v. are also important in some applications, but of little use in most robotic problems. They are the *skewness* (standardized 3rd central moment) and the *kurtosis* (standardized 4th central moment). The former reflects the symmetry of the pdf/pmf around the mean, while the latter reflects how tall and thin is the pdf/pmf with respect to a Gaussian distribution of the same variance. Gaussian distributions are the best known and most used pdfs in probability theory, as we will see.

Some Interesting Theorems about Moments

The first moments of a r.v. lead to some interesting results concerning their probability distributions. We show here two of them: the Markov's inequality and the Chebyshev's inequality. This section is a little more involved than the previous ones, and may be skipped in a first read without affecting much the understanding of the rest of the book.

Firstly, consider the *indicator random variable* I_A, defined as follows:

$$I_A(w) = \begin{cases} 1 & \text{if } w \in A \\ 0 & \text{otherwise} \end{cases} \quad (20)$$

where A is a random event from a given probability space and w is an element (outcome) of the corresponding sample space Ω. That is, I_A is a function that indicates whether an outcome of the process forms part of certain random event (=1) or not (=0). It is a r.v. since it satisfies the r.v. definition as a function of Ω. Its codomain (possible values to yield) is the set $\{0,1\}$, thus it is discrete. Its pmf is defined on the values of its support as:

$$f_{I_A}(x) = \begin{cases} P(A) & \text{if } x = 1 \\ P(\neg A) & \text{if } x = 0 \end{cases} \quad (21)$$

For clarifying the arguments of the functions $I_A(w)$ and $f_{I_A}(x)$, note that the former is a r.v., therefore a function defined on values w from the sample space Ω, while the latter is a pmf, therefore defined on the values $x \in \{0,1\}$ that its corresponding r.v. can yield.

By the definition of the expectation of discrete r.v.s (Equation 19), it is easy to see that:

$$E\left[I_A(x)\right] = 0 \cdot P(\neg A) + 1 \cdot P(A) = P(A) \quad (22)$$

That is, the expectation of the indicator r.v. is the probability of its associated random event A, according to the underlying probability space.

Now, if we have any r.v. X defined on the probability space, we can define the following new r.v. using the indicator function: $I_{|X| \geq a}$, for a given positive real number a. The random event referred to by $I_{|X| \geq a}$ is the set of all the outcomes of Ω for which the absolute value of the r.v. X (denoted $|X|$) is greater than or equal to a.

With the previous definitions, it turns out that the following holds (we use here the notation of the r.v. as a function):

$$aI_{|X| \geq a}(w) \leq |X(w)|$$

$$\forall a > 0,\ a \in \mathbb{R},\ w \in \Omega \quad (23)$$

The demonstration of Equation 23 is not difficult. If we have $I_{|X| \geq a}(b) = 0$ Equation 23 becomes $0 \leq |X(b)|$, which is trivially true. On the other hand, when $I_{|X| \geq a}(b) = 1$, Equation 23 becomes $a \leq |X(b)|$, which again is trivially true since if $I_{|X| \geq a}(b) = 1$ it must be the case that $b \in \{w \in \Omega : |X(w)| \geq a\}$, and, consequently, $|X(b)| \geq a$.

Since the expectation (either on discrete or continuous r.v.s) is a monotonic operation—it is the integral/sum of a non-negative function. It preserves the monotonicity of other previous operations on the r.v.s. Therefore, we can deduce from Equation 23 the following:

$$E\left[aI_{|X| \geq a}\right] \leq E\left[|X|\right] \quad (24)$$

And using linearity of expectation and Equation 22 on the left side of Equation 24 we obtain the Markov's inequality:

$$aE\left[I_{|X|\geq a}\right] \leq E\left[|X|\right] \quad \leftrightarrow$$

$$aP\left(|X| \geq a\right) \leq E\left[|X|\right] \quad \leftrightarrow$$

$$P\left(|X| \geq a\right) \leq \frac{E\left[|X|\right]}{a} \tag{25}$$

which gives an upper bound for the probability that the absolute value of any r.v. is greater than or equal to some positive real number. In practice, this result can be useful to estimate the probability that a r.v. yields a value outside a given band of width a centered at 0. As the band grows ($a \to \infty$) the probability of being outside it, obviously, tends to 0, as the inequality shows.

We can particularize the Markov's inequality for the case of a r.v. Y that is defined on another r.v. X in this way: $Y = \left(X - E\left[X\right]\right)^2$. By choosing a real number $c = a^2$ and provided that $a, c > 0$ we reach an interesting conclusion about X:

$$P\left(|Y| \geq c\right) \leq \frac{E\left[|Y|\right]}{c} \quad \leftrightarrow$$

$$P\left(\left|\left(X - E\left[X\right]\right)^2\right| \geq a^2\right) \leq \frac{E\left[\left|\left(X - E\left[X\right]\right)^2\right|\right]}{a^2} \quad \leftrightarrow$$

$$P\left(\left(X - E\left[X\right]\right)^2 \geq a^2\right) \leq \frac{E\left[\left(X - E\left[X\right]\right)^2\right]}{a^2} \quad \leftrightarrow$$

(by definition of variance)

$$P\left(\left(X - E\left[X\right]\right)^2 \geq a^2\right) \leq \frac{V\left[X\right]}{a^2} \quad \leftrightarrow$$

(since $a > 0$ and $z^2 \geq k^2 \leftrightarrow |z| \geq |k|$)

$$P\left(\left|X - E\left[X\right]\right| \geq a\right) \leq \frac{V\left[X\right]}{a^2} \tag{26}$$

which is the Chebyshev's inequality. It serves to bound the probability of a given value of a r.v. to deviate from the expectation of the r.v. more than a given distance a.

Information and Entropy of a R.V.

There are other ways of summarizing uncertainty that are not based on moments. Two that have a relevant role in probabilistic robotics are *information* and its counterpart, *entropy* (Cover & Thomas, 2006).

Information or, more properly, *self-information*, reflects the amount of information (sorry for the redundancy) gained by observing a certain value of the r.v. It is also called *surprisal*. The amount of information provided by a value that is very likely to appear is small, and vice versa. Self-information makes sense only for discrete r.v.s, since in the continuous case, $P\left[X = a\right] = 0$ for all the values of the support. It is formally defined as:

$$\forall a \in \text{supp}(X),\ I\left(a\right) \triangleq \log\frac{1}{P\left[X = a\right]} =$$

$$= -\log P\left[X = a\right] = -\log f_X\left(a\right) \tag{27}$$

and, if the logarithm is taken in base 2, it is measured in bits. The minus sign is responsible for obtaining smaller information for more likely values of the r.v. Notice that self-information is a mathematical function of a r.v., and as we have already seen in the Chebyshev's inequality and will study deeper later on, any mathematical function of a r.v. is another r.v. Therefore, it also contains uncertainty!

On the other hand, the *entropy* of a r.v. represents the "amount of uncertainty" contained in the whole r.v. (Klir, 2006; chapter 3), or, equivalently, the amount of information that *we expect* to obtain after observing the r.v. the next time. That expected amount of information is maximum for the case that the variable is uniformly distributed.

Entropy can be defined for both discrete and continuous r.v.s. In the latter case it is also called *differential entropy,* but it does not maintain all the properties of its discrete counterpart, thus other measures of the amount of uncertainty are proposed:

$$H(X) \triangleq E[I(X)] = -\sum_{a \in \text{support}(X)} P[X = a] \log P[X = a]$$

(discrete case)

$$H(X) \triangleq E[I(X)] = -\int_{y \in \text{support}(X)} f_X(y) \log f_X(y)\, dy$$

(continuous case) (28)

An alternative to the differential entropy is the Kullback-Leibler divergence, a measure of the relative entropy or "distance" of f_X w.r.t. g_Y, both defined on the same support:

$$D_{KL}(f_X, g_Y) \triangleq \int_{z \in \text{support}(X)} f_X(z) \log \frac{f_X(z)}{g_Y(z)}\, dz \quad (29)$$

Apart from its generic usage to compare pairs of pdfs, this measure has been applied in mobile robotics to automatically adjust localization methods to the varying uncertainty in the robot pose (Fox, Burgard, Dellaert, & Thrun, 1999).

7. MULTIVARIATE PROBABILITY

As we already reasoned in section 4, we often cannot capture all the uncertainty existing in a problem by using just one r.v. For instance, the robot location must be considered in many applications jointly with the map of the environment. However, when we work with more than one r.v. we must take into account the relations and interdependences existing among them. In this section, we cope with the complications introduces by these relations.

Joint Probability and Marginalization

To begin with, we have to analyze under which circumstances the behaviour of different r.v.s can provide us with valuable information about each other. The broadest mathematical representation of the relations existing among a number of r.v.s (either defined on the same probability space or not) is their *joint probability distribution*:

$$F_{X_1, X_2, X_3, \ldots, Xm}(a_1, a_2, \ldots, a_m) \triangleq P[X_1 \le a_1, X_2 \le a_2, X_3 \le a_3, \ldots, X_m \le a_m] \quad (30)$$

where

$$\{X_1 : \Omega_{X_1} \to \mathbb{R}^{n_1}, X_2 : \Omega_{X_2} \to \mathbb{R}^{n_2}, \ldots X_m : \Omega_{X_m} \to \mathbb{R}^{n_m}\}$$

are the r.v.s and the comma that separates the terms in equation (30) is an abbreviation for "and": the joint distribution provides information about the probability that *all* the r.v.s have values less than or equal to the given limits *simultaneously*—recall that simultaneity must be defined by the observer; it does not mean values occurring exactly at the same time unless that is what we need to model.

The joint probability distribution, as well as the single-variable probability distribution we introduced in section 4, yields probability values. Therefore, the joint distribution of Equation 30 has the same constraints and properties as the cdf of the single-variable case, and obeys all the rules of any probability measure.

Since we are just joining probabilities, without setting especial requirements to do so, we are allowed to define the corresponding joint probability mass function or joint probability density functions. In the case that all the r.v.s are discrete or continuous, these are denoted $f_{X_1 X_2 X_3 \ldots X_m}$, and, as any other probability mass/density function, they must satisfy:

$$\sum_{X_1}\sum_{X_2}\cdots\sum_{X_m} f_{X_1,X_2,X_3,\ldots,X_m} = 1 \text{ (discrete case)}$$

$$\int_{X_1}\int_{X_2}\cdots\int_{X_m} f_{X_1,X_2,X_3,\ldots,X_m}\, dx_1 dx_2 dx_3 \ldots dx_m = 1$$

(continuous case) (31)

Here we have written X_i in the indexes of the sums/integrals as an abbreviation of $\mathrm{supp}(X_i)$. You can see an example of a bivariate density function in Figure 7.

We can also define moments for multivariate distributions. For example, the first moment is:

$$E\left[X_1, X_2, \ldots, X_m\right] \triangleq \sum_{X_1}\sum_{X_2}\cdots \sum_{X_m} x_1, x_2, \ldots, x_v f_{X_1,X_2,X_3,\ldots,X_m} \text{ (discrete)}$$

$$E\left[X_1, X_2, \ldots, X_m\right] \triangleq \int_{X_1}\int_{X_2}\cdots\int_{X_m} x_1, x_2, \ldots, x_m f_{X_1,X_2,X_3,\ldots,X_m}\, dx_1 dx_2 dx_3 \ldots dx_m$$

(continuous) (32)

This can also be generalized to the expectation of functions of these r.v.s, which is defined as

$$E\left[g(x_1, x_2, \ldots, x_m)\right] \triangleq \sum_{X_1}\sum_{X_2}\cdots\sum_{X_m} g(x_1, x_2, \ldots, x_m) f_{X_1,X_2,X_3,\ldots,X_m}$$

or

$$E\left[g(x_1, x_2, \ldots, x_m)\right] \triangleq \int_{X_1}\int_{X_2}\cdots\int_{X_m} g(x_1, x_2, \ldots, x_m) f_{X_1,X_2,X_3,\ldots,X_m}\, dx_1 dx_2 dx_3 \ldots dx_m,$$

respectively.

Joint distributions can serve to deduce information about one of the r.v.s from the joint knowledge. That is precisely the basis of probabilistic inference and reasoning (Bishop, 2006; Bessière, Lugier, & Siegwart, 2008). A particularly simple operation is recovering the probability mass/density function of a single r.v. out from the joint distribution where it appears. This can be accomplished by summing

Figure 7. An example of the pdf of a bivariate Gaussian with mean at (0,0) and covariance matrix [0.25, 0.3 ; 0.3, 1]

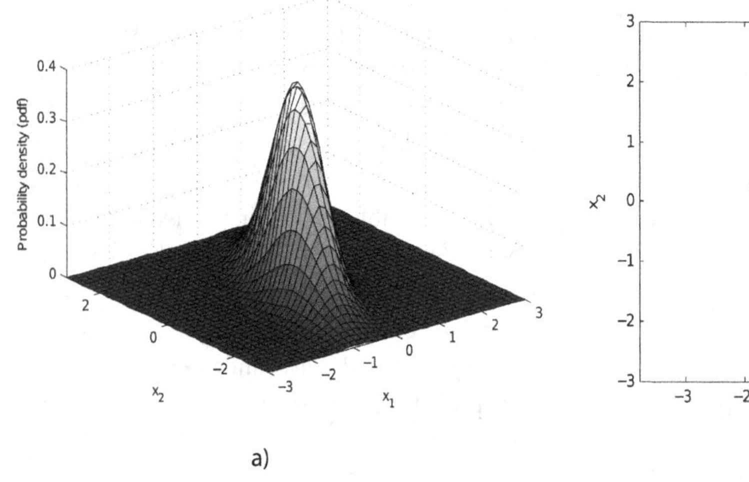

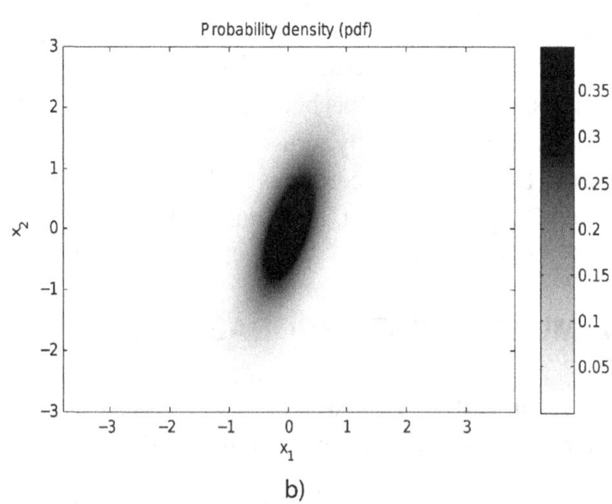

(*marginalizing out*) over the entire domain of all other variables:

$$f_{X_i} = \sum_{X_1}\sum_{X_2}\cdots\underbrace{\sum_{X_{i-1}}\sum_{X_{i+1}}}_{X_i \text{ is skipped}}\sum_{X_m} \underbrace{f_{X_1,X_2,\ldots,X_{i-1},X_i,X_{i+1},\ldots,X_m}}_{\text{all the r.v.s}}$$

(discrete)

$$f_{X_i} = \int_{X_1}\int_{X_2}\cdots\underbrace{\int_{X_{i-1}}\int_{X_{i+1}}}_{X_i \text{ is skipped}}\cdots\int_{X_m}$$

$$\underbrace{f_{X_1,X_2,\ldots X_{i-1},X_i,X_{i+1},\ldots,X_m}}_{\text{all the r.v.s}} dx_1 dx_2 \ldots \underbrace{dx_{i-1}dx_{i+1}}_{dx_i \text{ is skipped}} \ldots dx_m$$

(continuous) (33)

This procedure is called *marginalization* and the resulting functions are the *marginal distributions* for the corresponding r.v.s.

Mutual Independence and Covariances

When considering more than one r.v., the issue of the simultaneous occurrence of some of their values is of the utmost importance. We can analyze this deeper by studying the mutual exclusion and mutual independence of the r.v.s, as we did with events of probability spaces in section 3. Working with more than one r.v. at once requires matrices thus, for the next derivations we are forced to assume that the reader has some knowledge of basic matrix algebra.

We can define mutual independence of r.v.s analogously to mutual independence of random events:

$$X_1 \perp X_2 \leftrightarrow f_{X_1,X_2} = f_{X_1} f_{X_2} \quad (34)$$

When a set of r.v.s are mutual independent, the probability of occurrence of a value of any of them does not have any influence on the probability associated to the values of the others (they still could occur simultaneously if they are not mutually exclusive). For example, if we consider the two metrical coordinates of a robot pose in a plane as two different r.v.s, it is typically assumed that both are mutual independent: the probability of observing a given x does not changes when observing, simultaneously, some value of y, assuming no prohibited areas and an infinite floor. This would not occur if the floor plant is a triangle, however: the probability of having a given x varies with the current value of y.

An interesting consequence of independence is that if two r.v.s are independent, the expectation of their product is exactly the product of their expectations:

$$E_{X,Y}\left[X \cdot Y\right] = \int_{X,Y} xy f_{X,Y}(x,y) dx dy =$$

(using independence)

$$= \int_{X,Y} xy f_X(x) f_y(y) dx dy = \int_X x\left(\int_Y y dy\right) dx =$$

$$= \int_X x E_Y\left[Y\right] dx = E_Y\left[Y\right]\left(\int_X x dx\right) = E_X\left[X\right] E_Y\left[Y\right] \quad (35)$$

Now recall that we have defined the variance (the second moment of a r.v.) to reflect the divergence or distance of the values of a r.v. with respect to its mean. More concretely, it was defined as the expectation of the squared value of the divergence of the values of the r.v. with respect to its mean, that is, $V[X] = E\left[(X-\mu_X)(X-\mu_X)\right]$. This provides us with a framework to generalize the variance concept in order to reflect the relation existing between divergences of two r.v.s. We thus can define the *covariance* of two r.v.s, as long as both have a finite, non-zero variance, as:

$$Cov\left[X,Y\right] \triangleq E_{X,Y}\left[\left(X-\mu_X\right)\left(Y-\mu_Y\right)\right] =$$

$$= E_{X,Y}\left[xy - x\mu_Y - y\mu_X + \mu_X\mu_Y\right] =$$

(by linearity of expectation and constantness of means)

$$= E_{X,Y}\left[xy\right] - \mu_Y E_X\left[x\right] - \mu_X E_Y\left[y\right] + \mu_X \mu_Y =$$

$$= E_{X,Y}\left[xy\right] - \mu_Y \mu_X - \mu_X \mu_Y + \mu_X \mu_Y =$$

$$= E_{X,Y}\left[xy\right] - \mu_X \mu_Y =$$

$$= E_{X,Y}\left[X \cdot Y\right] - E_X\left[X\right] E_Y\left[Y\right] \quad (36)$$

Note that the covariance of a r.v. with respect to itself is its variance. Also, $Cov\left[X,Y\right] = 0$ when $X \perp Y$, due to Equation 35. Observe that, however, the inverse must not be always true.

It can be demonstrated that if both variances are finite and nonzero, the absolute value of the covariance is bounded by the product of the standard deviations:

$$\left|Cov\left[X,Y\right]\right| \leq \sqrt{\sigma_X^2 \sigma_Y^2} = \sigma_X \sigma_Y \quad (37)$$

This allows us to define a value that is both normalized between -1 and 1 (dimensionless) and perfectly mapped to the covariance, which is called the *linear correlation coefficient*:

$$\rho_{X,Y} = \frac{Cov\left[X,Y\right]}{\sigma_X \sigma_Y} \in \left[-1,1\right] \quad (38)$$

The correlation coefficient indicates how linear is the relation between a given pair of variables. It is an indicator widely used in probability and statistics—we will see it again in map matching in chapter 6.

Finally, we can extend the useful concept of covariance to any multivariate r.v.s, say $\mathbf{X} = \left\{X_1, X_2, X_3, ..., X_m\right\}$ and $\mathbf{Y} = \left\{Y_1, Y_2, Y_3, ..., Y_n\right\}$, by defining their *cross-covariance matrix* $\mathbf{\Sigma}_{\mathbf{X},\mathbf{Y}}$, a $m \times n$ matrix defined as:

$$\left[\mathbf{\Sigma}_{\mathbf{X},\mathbf{Y}}\right]_{i,j} \triangleq Cov\left[X_i, Y_j\right]$$

or

$$\mathbf{\Sigma}_{\mathbf{X},\mathbf{Y}} \triangleq Cov\left[\mathbf{X},\mathbf{Y}\right] = E_{\mathbf{X},\mathbf{Y}}\left[\left(\mathbf{X} - E\left[\mathbf{X}\right]\right)\left(\mathbf{Y} - E\left[\mathbf{Y}\right]\right)^T\right] \quad (39)$$

where we have used the notation $\left[\mathbf{A}\right]_{i,j}$ to represent the entry at the $i-th$ row and $j-th$ column in the matrix $\mathbf{A}$. It can be demonstrated[9] that $Cov\left[\mathbf{AX}, \mathbf{B}^T\mathbf{Y}\right] = \mathbf{A}\, Cov\left[\mathbf{X},\mathbf{Y}\right]\mathbf{B}$ for any matrices $\mathbf{A}$ and $\mathbf{B}$ that do not depend on $\mathbf{X}$ or $\mathbf{Y}$, a result which we will have to invoke throughout this book.

When the cross-covariance is defined with respect to the same variable, say $\mathbf{X} = \left\{X_1, X_2, X_3, ..., X_m\right\}$, we have the *covariance matrix* $\mathbf{\Sigma}_{\mathbf{X}} \triangleq Cov\left[\mathbf{X},\mathbf{X}\right]$. When there is no ambiguity, the covariance matrix will be denoted simply as $\mathbf{\Sigma}$. Paralleling the derivation in Equation 36 for this multivariate case, and using Equation 39, we find the following alternative expression for the covariance matrix which is sometimes more convenient:

$$\begin{aligned}
\mathbf{\Sigma}_{\mathbf{X}} &\triangleq Cov\left[\mathbf{X},\mathbf{X}\right] = E_{\mathbf{X}}\left[\left(\mathbf{X} - E_{\mathbf{X}}\left[\mathbf{X}\right]\right)\left(\mathbf{X} - E_{\mathbf{X}}\left[\mathbf{X}\right]\right)^T\right] = \\
&= E_{\mathbf{X}}\left[\left(\mathbf{X} - E_{\mathbf{X}}\left[\mathbf{X}\right]\right)\left(\mathbf{X}^T - E_{\mathbf{X}}\left[\mathbf{X}\right]^T\right)\right] = \\
&= E_{\mathbf{X}}\left[\begin{matrix}\mathbf{XX}^T - \mathbf{X}E_{\mathbf{X}}\left[\mathbf{X}\right]^T \\ -E_{\mathbf{X}}\left[\mathbf{X}\right]\mathbf{X}^T + E_{\mathbf{X}}\left[\mathbf{X}\right]E_{\mathbf{X}}\left[\mathbf{X}\right]^T\end{matrix}\right] = \\
&= E_{\mathbf{X}}\left[\mathbf{XX}^T\right] - E_{\mathbf{X}}\left[\mathbf{X}\right]E_{\mathbf{X}}\left[\mathbf{X}\right]^T \\
&\quad - E_{\mathbf{X}}\left[\mathbf{X}\right]E_{\mathbf{X}}\left[\mathbf{X}\right]^T + E_{\mathbf{X}}\left[\mathbf{X}\right]E_{\mathbf{X}}\left[\mathbf{X}\right]^T = \\
&= E_{\mathbf{X}}\left[\mathbf{XX}^T\right] - E_{\mathbf{X}}\left[\mathbf{X}\right]E_{\mathbf{X}}\left[\mathbf{X}\right]^T
\end{aligned} \quad (40)$$

A covariance matrix always has the variances of each element of the multivariate r.v. in its diagonal. It is also *positive-semidefinite*, and therefore symmetric[10] ($\Sigma_{i,j} = \Sigma_{j,i}$). A positive-semidefinite matrix $\mathbf{A}$ with real values is a symmetric matrix that, for any non-zero vector $\mathbf{x} \in \mathbb{R}^n$ satisfies $\mathbf{x}^T \mathbf{A} \mathbf{x} \geq 0$. A covariance matrix is a *positive-definite* matrix when it satisfies the more restrictive $\mathbf{x}^T \mathbf{A} \mathbf{x} > 0$ instead. A crucial property of positive-definite matrices, which positive-semidefinite ones lack, is that the former always have an inverse.

The covariance matrix contains all the information about the mutual divergences existing between pairs of elements of the r.v., and, due to matrix algebra, it can replace the variance in multivariate expressions. In addition, the expectation of a matrix of r.v.s is the matrix of the expectations of each r.v., therefore, for example, the standard expression for a normal or Gaussian pdf in the case of one and many (n) r.v.s becomes, respectively (see Figure 7):

$$f_X(x;\mu,\sigma) = N\left(x;\mu,\sigma\right) = \frac{1}{\sqrt{2\pi\sigma^2}} \exp\left(-\frac{1}{2}\frac{\left(x-\mu\right)^2}{\sigma^2}\right)$$

$$f_{\mathbf{X}}(\mathbf{x};\boldsymbol{\mu},\boldsymbol{\Sigma}) = N(\mathbf{x};\boldsymbol{\mu},\boldsymbol{\Sigma}) = \frac{1}{\sqrt{\left(2\pi\right)^n \det(\boldsymbol{\Sigma})}} \exp\left(-\frac{1}{2}\left(\mathbf{x}-\boldsymbol{\mu}\right)^T \boldsymbol{\Sigma}^{-1}\left(\mathbf{x}-\boldsymbol{\mu}\right)\right) \quad (41)$$

Notice the bold notation for $\mathbf{x}$ and $\boldsymbol{\mu}$ in the second expression, which reflects the multivariate variables and means.

8. TRANSFORMING RANDOM VARIABLES

Throughout this book, we will find numerous situations where we will need to define new random variables as transformations of others. In general, any function mapping the values of a r.v. into real numbers or vectors of real numbers is a r.v. too: since a r.v. is a function from the sample space, a function defined on its codomain that also has real numbers (or vectors) as images can be considered as another r.v. according to its mathematical composition with the former. Somehow, the probabilistic properties of the original r.v. are inherited by the new one.

Unfortunately, the closed-form formulation of the probability function for the new variable is often far from straightforward. In this section, we will focus in the simplest (but quite common) cases: those of one-dimensional continuous r.v.s (yielding real numbers) and one-to-one, monotone increasing, continuously differentiable functions of them. Even then, the following formulations may seem cumbersome, but we encourage the reader trying to follow the derivations, since this section is of the outmost importance for several developments in further chapters.

Preliminaries

Although the transformed r.v. that we are going to study is unidimensional, we need the multivariate formulation developed in the section 7 to do the derivations.

Let $\{X_1, X_2, X_3, \ldots, X_m\}$ be a set of continuous r.v.s, and $\mathbf{g} : \mathbb{R}^m \rightarrow \mathbb{R}^m$ a one-to-one, monotone increasing, continuously differentiable function, defined on the Cartesian product of the codomains of each r.v., that is, $\mathbf{g}(\mathbf{x}) = \mathbf{y}$. The bold notation is used, as before, for indicating vectors of real numbers. We are then interested in the probability distribution function of the r.v. defined by $\mathbf{g}$. Since the transformation is one-to-one (injective), it can be inverted: $\mathbf{g}^{-1} : \mathbb{R}^m \rightarrow \mathbb{R}^m$, where $\mathbf{g}^{-1}(\mathbf{y}) = \mathbf{x}$, and thus (remember that in multidimensional settings, comparison operations are defined as extending over all the dimensions of the operands):

$$F_{\mathbf{Y}}(\mathbf{a}) = P\left[\mathbf{y} \leq \mathbf{a}\right] = P\left[\mathbf{g}(\mathbf{x}) \leq \mathbf{a}\right] =$$

(using monotonicity and injectiveness of g)

$$= P\left[\mathbf{x} \leq \mathbf{g}^{-1}(\mathbf{a})\right] = \int_{\mathbf{x}=(-\infty)^m}^{\mathbf{x}=\mathbf{g}^{-1}(\mathbf{a})} f_{\mathbf{X}}(\mathbf{x}) d\mathbf{x} \tag{42}$$

We can now make the following change of variables in equation (42): $\mathbf{u} = \mathbf{g}(\mathbf{x})$, or, equivalently, $\mathbf{x} = \mathbf{g}^{-1}(\mathbf{u})$. It can be demonstrated that this leads to the following substitution for $d\mathbf{x}$:

$$d\mathbf{x} = d\mathbf{g}^{-1}(\mathbf{u}) = \left|\det\left(\mathbf{J}_{\mathbf{g}^{-1}}(\mathbf{u})\right)\right| d\mathbf{u} \tag{43}$$

where $\mathbf{J}_{\mathbf{g}^{-1}}(\mathbf{u}) \triangleq \nabla_{\mathbf{u}} \mathbf{g}^{-1}(\mathbf{u})$ is the *Jacobian* of $\mathbf{g}^{-1}(\cdot)$, evaluated at the point $\mathbf{u}$ —refer to Appendix E for a formal definition of Jacobian matrices. When $m = n$ the matrix is square and it makes sense to calculate its determinant, denoted $\det\left(\mathbf{J}_{\mathbf{g}^{-1}}\right)$.

Using the substitutions $\mathbf{x} = \mathbf{g}^{-1}(\mathbf{u})$ and $d\mathbf{x} = \left|\det\left(\mathbf{J}_{\mathbf{g}^{-1}}(\mathbf{u})\right)\right| d\mathbf{u}$ in equation (42), we find:

$$F_{\mathbf{Y}}(\mathbf{a}) = \int_{\mathbf{x}=(-\infty)^m}^{\mathbf{x}=\mathbf{g}^{-1}(\mathbf{a})} f_{\mathbf{X}}(\mathbf{x}) d\mathbf{x}$$

$$= \int_{\mathbf{g}^{-1}(\mathbf{u})=(-\infty)^m}^{\mathbf{g}^{-1}(\mathbf{u})=\mathbf{g}^{-1}(\mathbf{a})} f_{\mathbf{X}}\left(\mathbf{g}^{-1}(\mathbf{u})\right)\left|\det\left(\mathbf{J}_{\mathbf{g}^{-1}}(\mathbf{u})\right)\right| d\mathbf{u} =$$

(using again monotonicity and injectiveness of $\mathbf{g}$)

$$= \int_{\mathbf{u}=(-\infty)^m}^{\mathbf{u}=\mathbf{a}} \underbrace{f_{\mathbf{X}}\left(\mathbf{g}^{-1}(\mathbf{u})\right)\left|\det\left(\mathbf{J}_{\mathbf{g}^{-1}}(\mathbf{u})\right)\right|}_{\text{pdf of } y} d\mathbf{u} \tag{44}$$

From visual inspection of equation (44). and the definition of pdf, we can deduce that the probability density function for the multivariate $\mathbf{Y}$ must be the term that is enclosed in the integral (the one underbracketed), that is:

$$f_{\mathbf{Y}}(\mathbf{u}) = f_{\mathbf{X}}\left(\mathbf{g}^{-1}(\mathbf{u})\right)\left|\det\left(\mathbf{J}_{\mathbf{g}^{-1}}(\mathbf{u})\right)\right| \tag{45}$$

which gives us a way of obtaining the joint probability density function of the transformed random variable by knowing the one of the original random variables and the transformation from the latter into the former—refer to Figure 8 for a geometrical interpretation of this equation for the particular case of a scalar function. If, in addition, the r.v.s $\{X_1, X_2, X_3, ..., X_m\}$ are mutually independent, we can deduce the probability density function for any component of $\mathbf{Y}$ by marginalization in equation (45):

$$f_{Y_j} = \int_{Y_1}\int_{Y_2} \cdots \int_{Y_{j-1}}\int_{Y_{j+1}} \cdots \int_{Y_m} f_{\mathbf{X}}\left(\mathbf{g}^{-1}(\mathbf{u})\right)$$
$$\left|\det\left(\mathbf{J}_{\mathbf{g}^{-1}}(\mathbf{u})\right)\right| dY_1 dY_2 ... dY_{j-1} dY_{j+1} ... dY_m =$$

$$= \int_{Y_1}\int_{Y_2} \cdots \int_{Y_{j-1}}\int_{Y_{j+1}} \cdots \int_{Y_m} \left(\prod_{i\in[1,m]} f_{X_i}\left(g_i^{-1}(\mathbf{u})\right)\right)$$
$$\left|\det\left(\mathbf{J}_{\mathbf{g}^{-1}}(\mathbf{u})\right)\right| dY_1 dY_2 ... dY_{j-1} dY_{j+1} ... dY_m \tag{46}$$

Now we are prepared for using equation (45) and equation (46) in order to obtain the resulting pdf for different transformations of r.v.s, as we show in the following.

Sum of Two Continuous, Independent R.V.s

Making $\mathbf{y} = \mathbf{g}(x_1, x_2) = (x_1 + x_2, x_2)$, and thus $\mathbf{g}^{-1}(y_1, y_2) = (y_1 - y_2, y_2)$, and using Equation 45, we get $f_{\mathbf{Y}}(\mathbf{y}) = f_{X_1,X_2}(y_1 - y_2, y_2)$, or $f_{\mathbf{Y}}(\mathbf{y}) = f_{X_1}(y_1 - y_2) f_{X_2}(y_2)$. By marginalization on Y_1 we obtain:

Figure 8. Illustration of a unidimensional r.v. transformation, revealing the geometrical interpretation of the scaling by the absolute value of the determinant of the Jacobian $\mathbf{J}_{\mathbf{g}^{-1}}$. In this case, the Jacobian reduces to a single univariate derivative.

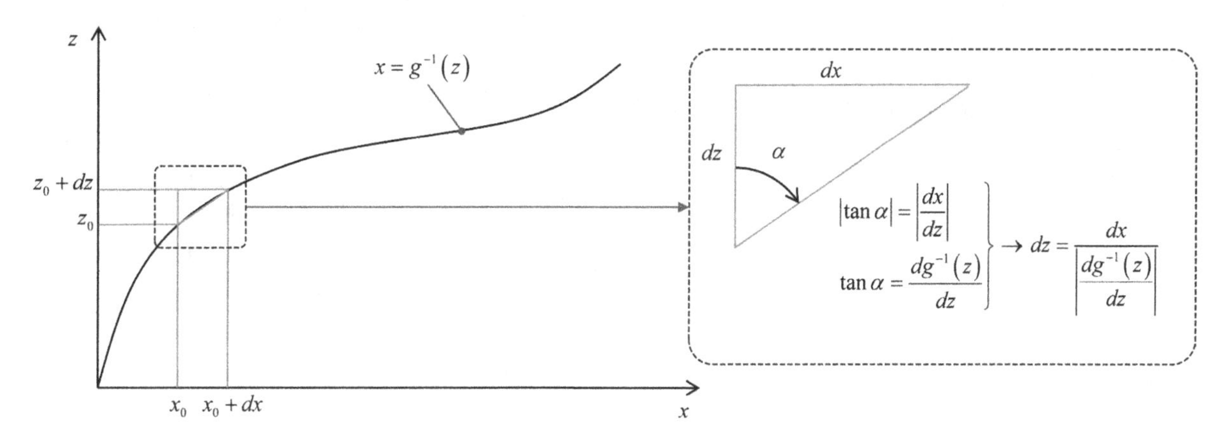

$$f_{X_1+X_2}(x_1+x_2) = f_{Y_1}(y_1) = \int_{y_2=-\infty}^{y_2=\infty} f_{\mathbf{Y}}(y_1, y_2)\, dy_2$$

$$= \int_{y_2=-\infty}^{y_2=\infty} f_{X_1}(y_1 - y_2) f_{X_2}(y_2)\, dy_2 \tag{47}$$

which is the *convolution* of the original pdfs f_{X_1} and f_{X_2}, that we will denote $f_{X_1} * f_{X_2} = f_{X_2} * f_{X_1}$ (the convolution operation is commutative). This can be extended by induction to any number of terms in the sum.

In the same way, making

$$\mathbf{y} = \mathbf{g}(x_1, x_2) = (x_1 - x_2, x_2)$$

and $\mathbf{g}^{-1}(y_1, y_2) = (y_1 + y_2, y_2)$ we have:

$$f_{X_1-X_2}(x_1 - x_2) = f_{Y_1}(y_1) = \int_{y_2=-\infty}^{y_2=\infty} f_{X_1}(y_1 + y_2) f_{X_2}(y_2)\, dy_2 \tag{48}$$

The operation of this integral is called the *cross-correlation* of f_{X_1} and f_{X_2}, that we will denote $f_{X_2} \star f_{X_1}$. Such an operation has applications to comparing the similarity between geometrical maps, as we will see in chapter 6.

Linear Combination of Continuous, Independent R.V.s

Let $\mathbf{y} = \mathbf{g}(x_1, x_2) = (ax_1 + bx_2, bx_2)$ and $\mathbf{g}^{-1}(y_1, y_2) = ((y_1 - y_2)/a, y_2/b)$, $a, b \neq 0$. This leads through Equation 45 to

$$f_{\mathbf{Y}}(\mathbf{y}) = f_{X_1,X_2}((y_1 - y_2)/a, y_2/b)\frac{1}{|ab|}, \text{ or}$$

$$f_{\mathbf{Y}}(\mathbf{y}) = f_{X_1}((y_1 - y_2)/a) f_{X_2}(y_2/b)\frac{1}{|ab|}$$

if we take the independence of the r.v.s into account. Marginalizing:

$$f_{aX_1+bX_2}(ax_1 + bx_2) = f_{Y_1}(y_1)$$

$$= \int_{y_2=-\infty}^{y_2=\infty} f_{X_1}((y_1 - y_2)/a) f_{X_2}(y_2/b)\frac{1}{|ab|}\, dy_2 \tag{49}$$

Product and Division of Continuous, Independent R.V.s

We can take for this case $\mathbf{g}(x_1, x_2) = (x_1 x_2, x_2)$ (that is, $\mathbf{g}^{-1}(y_1, y_2) = (y_1/y_2, y_2)$), which using Equation 45 leads to:

$$f_{\mathbf{Y}}(\mathbf{y}) = f_{X_1,X_2}\left(y_1 / y_2, y_2\right)\frac{1}{\left|y_2\right|} \tag{50}$$

The singularity at $y_2 = x_2 = 0$ is well treated by the Lebesgue integral and, anyway, that point does not belong to the support of $\mathbf{g}$. Marginalization yields:

$$f_{X_1 \cdot X_2}(x_1 x_2) = f_{Y_1}(y_1) = \int_{y_2=-\infty}^{y_2=\infty} f_{X_1}(y_1 / y_2) f_{X_2}(y_2)\frac{1}{\left|y_2\right|} dy_2 \tag{51}$$

Analogously, if we start with
$\mathbf{g}(x_1, x_2) = (x_1 / x_2, x_2)$ and
$\mathbf{g}^{-1}(y_1, y_2) = (y_1 y_2, y_2)$ we reach at:

$$f_{X_1/X_2}(x_1 / x_2) = f_{Y_1}(y_1) = \int_{y_2=-\infty}^{y_2=\infty} f_{X_1}(y_1 y_2) f_{X_2}(y_2)\left|y_2\right| dy_2 \tag{52}$$

Linear Transformation of Continuous, Unidimensional R.V.s

We make now $y = g(x) = ax + b$, that is, $g^{-1}(y) = (y - b) / a$ for $a \neq 0$, such that:

$$f_{aX+b}(y) = \frac{1}{|a|} f_X\left(\frac{y-b}{a}\right) \tag{53}$$

Therefore, the offset becomes a shift in the variable: $f_{X+b}(y) = f_X(y - b)$, while the scale is a scale of both the variable support and its pdf: $f_{aX}(y) = \frac{1}{|a|} f_X(y / a)$. This last result is what one could expect, since the new r.v. must be a r.v., that is, it must integrate up to the unity in its support and, therefore, the scaling in its support must be accompanied by a scaling in its shape (height) in order to keep the total area equal to the unity.

Linear Transformation of Multivariate R.V.s

By letting $\mathbf{x}$ be a vector of arbitrary length n, we can consider a more general linear transformation $\mathbf{y} = \mathbf{g}(\mathbf{x}) = \mathbf{A}\mathbf{x} + \mathbf{b}$, with $\mathbf{A}$ an $m \times n$ transformation matrix and $\mathbf{b}$ a constant (offset) vector of length m. In general, the transformed variable $\mathbf{y}$ has a dimensionality m which must not be equal to n, the dimensionality of the original variable $\mathbf{x}$, but the inverse function $\mathbf{g}^{-1}(\mathbf{y}) = \mathbf{A}^{-1}(\mathbf{y} - \mathbf{b})$ will be well-defined only when both $m = n$ (i.e. non-squared matrices do not have an inverse) and $\mathbf{A}$ is really invertible (i.e. its determinant is nonzero). Thus, only when those two conditions hold we can unambiguously determine the pdf of the transformed variable as:

$$\begin{aligned} f_{\mathbf{Y}}(\mathbf{y}) &= f_{\mathbf{X}}\left(\mathbf{g}^{-1}(\mathbf{u})\right)\left|\det\left(\mathbf{J}_{\mathbf{g}^{-1}}(\mathbf{u})\right)\right| = \\ &= f_{\mathbf{X}}\left(\mathbf{A}^{-1}(\mathbf{y} - \mathbf{b})\right)\left|\det\left(\mathbf{A}^{-1}\right)\right| \end{aligned} \tag{54}$$

Moreover, the linearity property of the expectation operator $E[\cdot]$ allows us to find out very useful information about the transformed variable $\mathbf{y} = \mathbf{A}\mathbf{x} + \mathbf{b}$ in all cases, including non-invertible $\mathbf{A}$ matrices or even when $m \neq n$. In particular, it can be demonstrated[11] that the mean and covariance of the transformed variable are given by:

$$\begin{aligned} E[\mathbf{y}] &= \mathbf{A}\, E[\mathbf{x}] + \mathbf{b} \\ Cov[\mathbf{y}] &= \mathbf{A}\, Cov[\mathbf{x}]\ \mathbf{A}^T \end{aligned} \tag{55}$$

which holds without approximations in this general case of linear transformation, disregarding the particular pdfs of the original and transformed variables—i.e., they do not have to be multivariate Gaussians.

The Special Case of the Chi-Squared Distribution

The chi-squared distribution is a especial case of the previous deductions, of great interest in statistical inference. It will appear several times along this book. Assume we have a number $k > 0$ of r.v.s, $\{X_1, X_2, \ldots, X_k\}$, all of them independent and with identical distributions, being those distributions standard Gaussian, that is, $N(x;0,1)$. Then, the distribution of the variable resulting from the sum $\sum_{i=1}^{k} X_i^2$ is said to follow a chi-squared distribution with k degrees of freedom. For instance, for only one r.v. ($k = 1$), we would have $X^2 \sim \chi_1^2$ as long as $X \sim N(0,1)$.

More generally, if each of the r.v. is distributed with a different, non-standard Gaussian, i.e., $X_i \sim N(\mu_i, \sigma_i^2)$, then the r.v. $\sum_{i=1}^{k} \left(\frac{X_i - \mu_i}{\sigma_i} \right)^2$ is distributed according to χ_k^2. For example, $\frac{(X-\mu)^2}{\sigma^2} \sim \chi_1^2$ as long as $X \sim N(\mu, \sigma^2)$ and, particularly, $\frac{X^2}{\sigma^2} \sim \chi_1^2$ if $X \sim N(0, \sigma^2)$.

Furthermore, if $\{X_1, X_2, \ldots, X_k\}$ are independent chi-squared distributed r.v.s with degrees of freedom $\{v_1, v_2, \ldots, v_k\}$, then the r.v. $\sum_{i=1}^{k} X_i$ has a chi-squared pdf χ_w^2 with $w = \sum_{i=1}^{k} v_i$ degrees of freedom.

Approximating Arbitrary Transformations

In most occasions it is very complicated, if possible at all, to obtain a closed form for the distribution of the r.v. resulting from a transformation (function) $\mathbf{g}$ of another r.v., especially when $\mathbf{g}$ is not linear. Then, we must often content ourselves with approximate solutions. In laboratory procedures, where the amount of uncertainty in the measurements has to be estimated frequently, traditional techniques exist to provide such a value, and even more modern tools such as interval analysis can be used (Rodwell & Cloud, 2012). However, we would be interested here in obtaining the whole shape of the transformed uncertainty.

One of simplest solutions we could think of is to build a new transformation $\mathbf{h}$ by linearizing $\mathbf{g}$. This can be done as a special case of a Taylor series expansion, a method to approximate functions—refer to Appendix E. A first-order Taylor approximation consists of a linear function that approximates $\mathbf{g}$, i.e.,

$$\mathbf{g}(\mathbf{x}) \approx \mathbf{h}(\mathbf{x}) = \mathbf{A}\mathbf{x} + \mathbf{b} \tag{56}$$

with $\mathbf{x}$ being the original r.v., $\mathbf{A}$ an $m \times n$ matrix and $\mathbf{b}$ an $m \times 1$ vector. Both $\mathbf{A}$ and $\mathbf{b}$ are chosen to approximate the function around a so-called "linearization point" $\mathbf{x}_0$. In particular, the matrix $\mathbf{A}$ is the Jacobian of $\mathbf{g}$ with respect to $\mathbf{x}$, evaluated at the linearization point, that is:

$$\mathbf{h}(\mathbf{x}) = \nabla_{\mathbf{x}} \mathbf{g}(\mathbf{x}_0) + \mathbf{g}(\mathbf{x}_0) \tag{57}$$

Thus, the closer $\mathbf{x}$ is to $\mathbf{x}_0$, the more similar becomes the Taylor approximation to the original function $\mathbf{g}$. In fact, at the linearization point both functions are equivalent up to their first derivatives. Far from there, or if $\mathbf{g}$ changes very quickly with small changes of $\mathbf{x}$, the linearized version $\mathbf{h}$ can become a really bad approximation.

Having a linear version of the original $\mathbf{g}$ allows us to obtain an approximated distribution for the transformed variable $\mathbf{y} = \mathbf{g}(\mathbf{x})$ just by using $\mathbf{y} \approx \mathbf{h}(\mathbf{x})$ and the linear transformation cases explained in the subsections above. It is noteworthy that this linearization approximation only guarantees to preserve similarity in the *supports* of both variables, not in the shape of the final distribution—we will see in chapter 7 an approximation that does take into account the distribution of the original variable, called the *Unscented Transformation*. It is difficult, if possible at all, to give a general assessment of the degree of similarity that will exist between the distribution obtained through this linear propagation and the actual one.

In most circumstances, one further step towards a practical solution is to assume that the target r.v. follows a Gaussian distribution, and then calculate the pdf parameters of that Gaussian. This has two main advantages: (1) forcing a particular pdf simplifies a lot of calculations and will permit us later on to apply well known results in estimation theory, such as Kalman filters, which are based on Gaussians, and (2) using a parameterized distribution such as the Gaussian reduces the information that we have to process to its first two moments, which become in fact the only data to be propagated through the transformation, thus simplifying calculation and reducing the computational requirements. Because of these, this approach is very commonly used in probabilistic mobile robot localization and mapping. It can yield very good results in many practical situations; hence, we will apply this method in several different chapters.

This method of propagating only the first moments of a r.v. through a non-linear transformation begins by approximating the expectation of the transformed r.v. $\mathbf{y}$. A way to do that is by using $\mathbf{g}$ to propagate the expectation of $\mathbf{x}$ into the support of the new variable —notice that a Taylor approximation of $\mathbf{g}$ that would use $E[\mathbf{x}]$ as the linearization point would do the same, since at that point both the linear approximation and the original function are exactly equal. Then, it is natural to assume that this propagated expectation *is* the actual expectation of our final distribution, that is, $E[\mathbf{y}] = E[\mathbf{g}(\mathbf{x})] \approx \mathbf{g}(E[\mathbf{x}])$, but always remember that, though this is acceptable in many cases, it is simply not true: it would be true if and only if $\mathbf{g}$ was linear itself.

Having an approximate expectation of the final variable $\mathbf{y}$, we need to approximate its second moment by propagating the covariance matrix $\mathbf{\Sigma}_{\mathbf{X}}$ of the original variable. The Taylor approximation of the transformation $\mathbf{g}$, around the linearization point $E[\mathbf{x}]$, becomes handy at this point. Employing the result for linear transformations in Equation 55, we see that the sought covariance can be approximated as $\mathbf{\Sigma}_{\mathbf{Y}} \approx \mathbf{J}_{\mathbf{g}} \mathbf{\Sigma}_{\mathbf{X}} \mathbf{J}_{\mathbf{g}}^T$, where $\mathbf{J}_{\mathbf{g}}$ is the Jacobian matrix of the function evaluated at the linearization point.

Note that when the transformation $\mathbf{g}$ is linear, we can still use this approach of propagating moments and assuming a parameterized form for the final distribution, typically Gaussian. In that case we use directly Equation 55. as an identity and not only as an approximation.

9. CONDITIONAL PROBABILITY

So far, we have used probabilities defined on one or several r.v.s, in a way that fits well to the frequentist (or objectivist) paradigm. However, the Bayesian (or subjectivist) perspective requires the formalization of the probability of an event *provided that (or conditioned on the fact that) other events have already occurred*. Such a probability could be translated directly to the Bayesian concept of the degree of belief in an event provided a past experience of other events. As we have explained in this chapter, this Bayesian paradigm is at the core of probabilistic robotics, and, in general, of many fields of science and engineering these days, particularly those requiring estimation of the outcome of dynamic processes.

For coping appropriately with probabilities that only have proper meaning in the context of past outcomes, we define the *conditional probability* of a set of random events $\{A_1, A_2, \ldots, A_n\}$, given another set of events $\{B_1, B_2, \ldots, B_m\}$, as:

$$P\left[A_1, A_2, \ldots, A_n \mid B_1, B_2, \ldots, B_m\right] = \frac{P\left(\left(\bigcap_{i \in [1,n]} A_i\right) \cap \left(\bigcap_{j \in [1,m]} B_j\right)\right)}{P\left(\bigcap_{j \in [1,m]} B_j\right)} \quad (58)$$

as long as $P\left(\bigcap_{j \in [1,m]} B_j\right) > 0$. It is not difficult to show that this definition satisfies all the properties of a probability measure (i.e., the same properties as the function P of the probability space), and thus, *it is* a probability. An important consequence is that *any theorem on probabilities remains valid when all its probabilities are conditioned on a common set of events.*

Interpreting Equation 58. under the subjectivist perspective, the previous experienced events are the B_j (they have already occurred), while the A_i are the events in which we are interested, but are unknown.

We enumerate next some important rules that follow from the definition of conditional probability (we maintain here the distinction in notation between the probability function $P(\cdot)$ and the conditional probability $P[\cdot|\cdot]$, but that will be relaxed later on by using parentheses for both):

The conditional probability *chain rule*, which allows us to relate the joint probability of events to their conditional probabilities, serving thus as a sort of link between the objectivist and subjectivist paradigms:

$$P\left(\bigcap_{i=1}^{n} A_i\right) = \prod_{i=n}^{1} P\left[A_i \middle| \bigcap_{j>i} A_j\right] \quad (59)$$

The *law* or *theorem of total probability*: If we have a set of mutually exclusive events $\{B_1, B_2, \ldots, B_m\}$ from a probability space (Ω, F, P) such that $\bigcup_{j \in [1,m]} B_j = \Omega$, and another event A, then:

$$P(A) = \sum_{j \in [1,m]} P\left[A \mid B_j\right] P\left(B_j\right) \quad (60)$$

The *Bayes' Rule or Theorem*, which permits obtaining a conditional probability from its inverse:

$$P\left[B_j \mid A\right] = \frac{P\left[A \mid B_j\right] P\left(B_j\right)}{\sum_{j=1}^{m} P\left[A \mid B_j\right] P\left(B_j\right)} = \frac{P\left[A \mid B_j\right] P\left(B_j\right)}{P(A)} \quad (61)$$

It is valid as long as $P(A) > 0$. More importantly, the Bayes' rule is the basis for deciding which hypothesis of a set $\{B_j\}$ is the most likely after making observations (e.g. of a random event A), having some previous information about each individual hypothesis, i.e., knowing $P(B_j)$. Notice that such previous information comes from outside the problem: it is introduced by the researcher in the expression and its shape is in principle, arbitrary. That has been the main argumentation against the Bayesian paradigm by frequentists: in a frequentist probabilistic setting, there is no room for such a subjective piece of information, and everything is the objective result of applying the scientific method to experimental data. Bayes' rule is the most basic form used today in the probabilistic estimation of the state of systems that change sequentially over time, like the pose of a mobile robot. As we will see in further chapters, it is common to call $P(B_j)$ the *prior* (knowledge before gathering new information), $P[B_j|A]$ the *posterior* (knowledge after gathering new information—occurrence of new events), and $P[A|B_j]$ the *likelihood* of having observed the occurrence of A in the case that B_j would have already occurred. $P(A)$ is often considered a normalizing factor in the expression since it is independent on any particular value that B_j could take, that is, it remains constant under the perspective of B_j, the r.v. of the posterior.

Analogously to random events, we can define conditional probabilities for r.v.s:

$f_{X|Y} = \frac{f_{XY}}{f_Y}$, for any value of Y such that $f_Y > 0$ (62)

and Bayes' rule applies here in the same form:

$$f_{X|Y} = \frac{f_{Y|X} f_X}{f_Y} \tag{63}$$

Also, we can define the *conditional independence* of a r.v. X with respect to another r.v. Y, provided that we already know the value of a third r.v. Z. This will be denoted $X \perp Y \big| Z$. Conditional independence is analogous to the concept of (unconditional) independence of r.v.s, $X \perp Y \big| Z \leftrightarrow f_{XY|Z} = f_{X|Z} f_{Y|Z}$, which is a clear application of the fact that any probability theorem remains valid when all its terms are conditioned on the same set of r.v.s. Conditional independence allows us to simplify some probabilistic reasonings that involve the chain rule explained before. For example, since $X \perp Y \big| Z \rightarrow f_{XY|Z} = f_{X|Z} f_{Y|Z}$, if we have $X \perp Y \big| Z$:

$$f_{XY|Z} = f_{X|Z} f_{Y|Z}$$

(by definition of conditional probability)

$$\frac{f_{XYZ}}{f_Z} = \frac{f_{XZ}}{f_Z} \frac{f_{YZ}}{f_Z}$$

(by algebraic manipulation, and as long as $f_{YZ} > 0$)

$$\frac{f_{XYZ}}{f_{YZ}} = \frac{f_{XZ}}{f_Z}$$

(by definition of conditional probability)

$$f_{X|YZ} = f_{X|Z} \tag{64}$$

Therefore we have deduced that

$$X \perp Y \big| Z \rightarrow f_{X|YZ} = f_{X|Z},$$

which serves to reduce terms in the chain.

When we have a set of r.v.s and some conditional independence assumptions between them we can also depict those relations graphically in the form of a *graphical model*, a powerful paradigm to which we will devote the following section.

The last concept that is useful for us in the conditional probability framework is *conditional expectation*, a particularization of expectation for conditional contexts. The conditional expectation of a r.v. X with respect to another r.v. Y is denoted as $E\left[X \big| Y\right]$ and defined as:

$$E\left[X \big| Y\right] = \sum_{x \in \text{supp}(X)} x f_{X|Y}(x) \ (X \text{ discrete})$$

$$E\left[X \big| Y\right] = \int_{x \in \text{supp}(X)} x f_{X|Y}(x) dx \ (X \text{ continuous}) \tag{65}$$

As exposed by the expressions, the conditional expectation is in general a function of the r.v. that is conditioning the equations (Y in these examples), and not a single value, as happened with expectation. Since it is a function of a r.v., it is itself a r.v., as explained in section 8. In other words, the conditional expectation itself *contains uncertainty*. This can be easily understood by realizing that we are in a subjectivist context, where conditional expectation is only valid under the assumption that Y already has an observed value, which might have been different for another observation.

A last remark on conditional expectation: for any function $r(y)$, being y the value taken by the r.v. Y:

$$E\left[r(y)E\left[X\,|Y\right]\right] = E\left[r(y)\int_x xf_{X|Y}(x,y)dx\right] =$$

(by definition of expectation)

$$= \int_y r(y)\left(\int_x xf_{X|Y}(x,y)dx\right)f_Y(y)dy$$

$$= \int_y\int_x r(y)\cdot x\cdot f_{X|Y}(x,y)f_Y(y)dxdy =$$

(by definition of conditional probability)

$$= \int_y\int_x r(y)\cdot x\cdot f_{X,Y}(x,y)dxdy =$$

(by definition of expectation, being $g(x,y) = r(y)\cdot x$)

$$= E_{X,Y}\left[g(x,y)\right] = E_{X,Y}\left[r(y)\cdot x\right] \tag{66}$$

Doing $r(y) = 1$, we get an interesting result for statistical estimation: $E\left[E\left[X\,|Y\right]\right] = E\left[X\right]$, an expression which we will need in chapter 4.

10. GRAPHICAL MODELS

The formalism of graphical models allows us to represent a set of r.v.s and their conditional independence assumptions. It was born as a fusion of probability theory and graph theory, and plays a central role in many machine learning techniques (Ghahramani, 2001; Bishop, 2006).

Definitions and Taxonomy

In a graphical model, the r.v.s of a given problem are modeled as a mathematical *graph* (a set of nodes and arcs), whose nodes stand for individual variables or, quite often, for a set of them. Arcs exist between nodes to denote that they are somehow "related," but as we will see shortly, the lack of arcs is what carries the most relevant information.

In order to be more specific about the meaning of arcs, we need to make the first broad classification of graphical models. On the one hand, if arcs are *undirected* we have a graphical model called *Markov network*, or *Markov Random Field* (MRF). In this model, a r.v. is conditionally independent of the rest of r.v.s in the graph given knowledge of its immediate neighbors, that is, of the r.v.s that are connected to it by arcs. This is the Markov property that will be mentioned in chapter 4 in the context of *temporal series of random variables*: sequences of r.v.s indexed by time (e.g., $X_1, X_2, \ldots, X_k$, being k a discrete time index) where the variable X_i *is conditionally independent of* $\left\{X_j : j < i-1\right\}$ given X_{i-1}.

On the other hand, we have graphical models with *directed* arcs, which are called *Bayesian Networks* (BNs) or, when we consider time series of variables as nodes, *Dynamic Bayesian Networks* (DBNs). The defining property of directed graphical models is that the direction of arcs carries some meaning of *causality*. However, BNs have the limitation that they must not contain loops (they are acyclic graphs), which is not a problem for MRFs. Examples of directed and undirected graphical models can be seen in Figure 9. As a convention to be held throughout this book, we will represent *observed* variables (known) as unshaded nodes, with shaded nodes being *hidden* variables (unknown).

Figure 9. Examples of a Markov random field (MRF) and a Bayesian network (BN). (a) This specific MRF is called an Ising model, and was proposed by physicists in the beginning of the 20th century to model interatomic interactions in ferromagnetic phenomena (Kindermann & Snell, 1980). (b) This BN shows a simplistic model for performing predictions and inference about the state of a vehicle. (c) The dynamic BN of a Markov process clearly reveals the conditional independence of each system state (x_k) with respect to all old states excepting the last one (x_{k-1}).

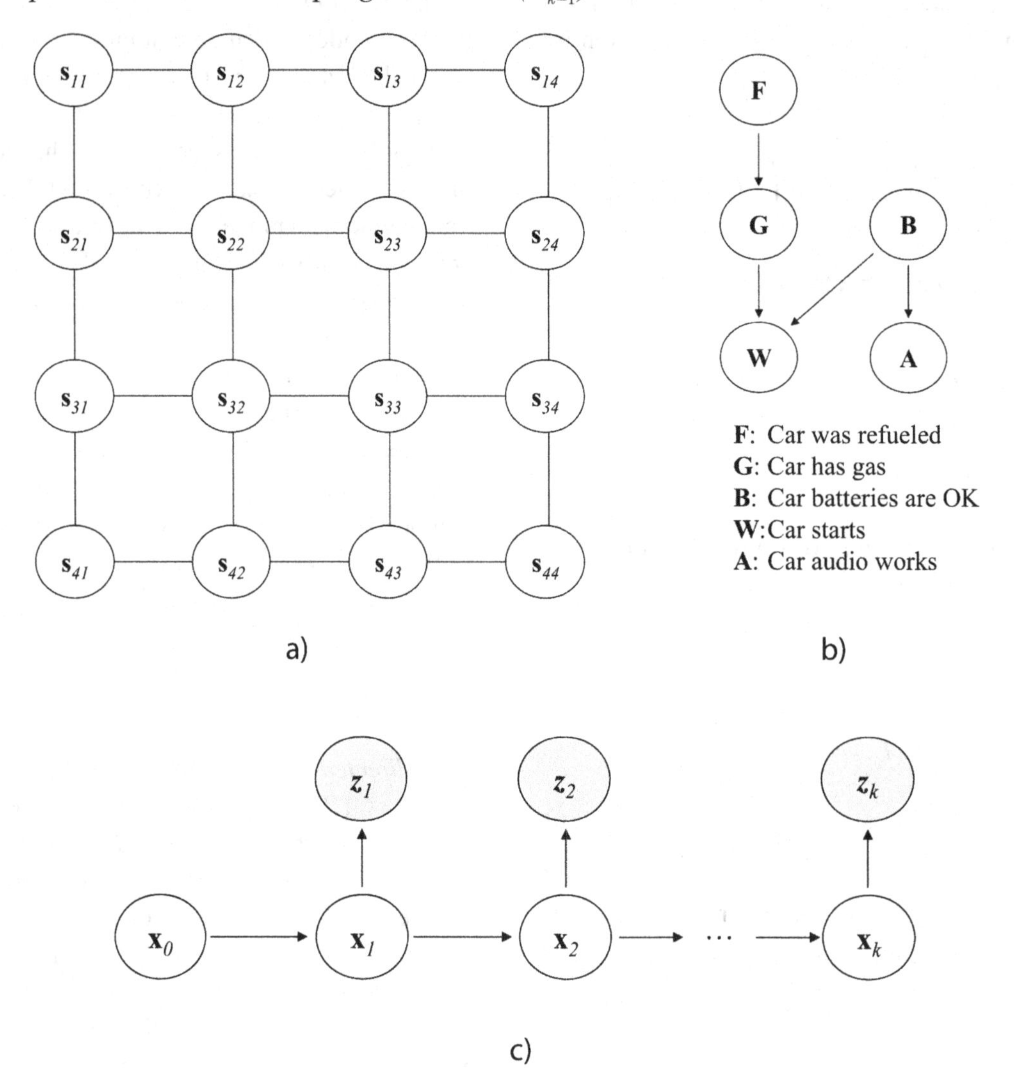

Factorizations

The first immediate application of representing a set of r.v.s $\mathbf{X} = \{X_1, X_2, ..., X_N\}$ as a graphical model is the possibility of factorizing the joint pdf of $\mathbf{X}$ by analyzing the graph structure. In the case of MRFs (undirected graphs), the joint can be expressed in the form:

$$f_{\mathbf{X}} = \frac{1}{Z} \prod_{c \in C} \phi_c(\mathbf{X}_c) \qquad (67)$$

where C is the set of all maximal cliques (and c each one of them), the $\mathbf{X}_c$ are the set of r.v.s within the specified clique, and the $\phi_c(\cdot)$ represent *potential functions*, which must not be probability densities but whose product gives us the joint pdf (Bishop, 2006). Notice that the constant Z, called the partition function, is introduced to normalize the product of all the potential functions such that the final function sums up to the unity in its entire domain, thus being a valid pdf. A *clique* in a graph is a subgraph (subset of nodes of the graph) which is a complete graph, i.e., every pair of nodes is connected by an arc. A maximal clique is one that cannot be expanded by incorporating a new node and still be a complete graph. We will come back to MRFs when studying least-squares approaches to SLAM in chapter 10.

When working with directed BNs, we can always apply the following factoring, which is a simplified version of the chain rule explained in the previous section:

$$f_{\mathbf{X}} = \prod_{i=1\ldots N} f_{X_i|pa_i} \tag{68}$$

where pa_i is the set of all the parents of the node X_i, that is, all the nodes that have an outgoing arc ending at X_i. For instance, the joint pdf of the variables in the BN of Figure 8b can be factored, using the clearer "p" notation for continuous r.v.s, as:

$$p(A,B,F,G,W) = p(F)p(B)p(G|F)p(A|B)p(W|B,G) \tag{69}$$

It is worth expending a few moments carefully analyzing the relationship between the figure and each product in this factoring.

When we construct a graphical model for a given problem three different elements must be provided: (1) the structure of the graph itself (which nodes and arcs exist), (2) all the conditional distributions existing by the factorization of the joint pdf expressed in Equation 58, and (3) the prior distributions for the nodes without parents. Given these three items, a random process is completely characterized and we can perform inference on it.

Learning and Inference with Bayesian Networks

Countless engineering problems can be addressed by means of BNs, and thus we can find quite different situations: in some applications, the conditional distributions are known but the structure of the graph is not, thus the aim is determining the minimum number of nodes or arcs that best explain the observed data; other possibilities are not having the conditional distributions for a known graph structure or, even in the hardest situation, knowing neither the structure nor the conditional distributions. In any case, all those are *learning* problems with Bayesian Networks. None of them will be needed to solve the robotic problems studied in this book.

Rather, we will focus on the *inference* problem with Bayesian Networks, which consists of, knowing the BN structure, estimating the pdf of some variables from the known values of a subset of others, which are named *observed variables* and represented as unshaded nodes in the graph. The variables in whose pdf we are interested are the unknowns, which will be named *hidden variables* and represented as shaded nodes—refer to the example in Figure 9c. Therefore, in an inference problem we intend to estimate $p(H|O)$, where we denote the observed and hidden continuous variables as O and H, respectively. For example, performing inference in the BN of Figure 9c means computing:

$$p\left(x_{0:k} \middle| z_{1:k}\right) = \prod_{i=0...k} p\left(x_i \middle| pa_{x_i}, z_{1:k}\right)$$
$$= p\left(x_0\right) \prod_{i=1...k} p\left(x_i \middle| x_{i-1}, z_{1:k}\right) \quad (70)$$
$$= p\left(x_0\right) \prod_{i=1...k} p\left(x_i \middle| x_{i-1}, z_i\right)$$

where we have abbreviated as $x_{0:k}$ the temporal sequence of r.v.s x_i, $i \in [0,k]$. Notice how the product of $k+1$ complicated distributions involving several r.v.s ($x_{0:k}$, the joint probability of the left-hand side term) has been simplified into the combination of a prior for x_0, the $p\left(x_0\right)$, and k pdfs of only one r.v. each. This has been achieved by dropping as conditioning variables those that are conditionally independent in each case. Next we provide the systematic rules to determine whether such a property holds for any set of variables in a BN.

Conditional Independence in BNs and D-Separation

As illustrated in the previous paragraph, we will frequently be interested in establishing whether a set of nodes (r.v.s) $\mathbf{x} = \left\{x_i\right\}$ are conditionally independent of another set $\mathbf{y} = \left\{y_i\right\}$ provided (conditioned on) a third set $\mathbf{z} = \left\{z_i\right\}$. That is, we want to test whether $\mathbf{x} \perp \mathbf{y} \middle| \mathbf{z}$. In other words, we are asking whether the following factoring holds for a given BN:

$$p\left(\mathbf{x},\mathbf{y} \middle| \mathbf{z}\right) \stackrel{?}{=} p\left(\mathbf{x} \middle| \mathbf{z}\right) p\left(\mathbf{y} \middle| \mathbf{z}\right)$$

(which in turn would imply the interesting simplification:)

$$p\left(\mathbf{x} \middle| \mathbf{y},\mathbf{z}\right) = p\left(\mathbf{x} \middle| \mathbf{z}\right) \quad (71)$$

For determining whether conditioning on $\mathbf{z}$ makes $\mathbf{x}$ independent on $\mathbf{y}$ in the case of a BN, we will rely upon the concept of *d-separation* (*d* stands for "dependence" in some works and for "directional" in others), which in turn is based on the concept of *path* in a graph: a path between the set of nodes $\mathbf{x}$ and the set of nodes $\mathbf{y}$ is a sequence of adjacent arcs connecting a node of the former to a node of the latter, or vice versa, having these edges any direction. $\mathbf{z}$ is said to *d-separate* $\mathbf{x}$ from $\mathbf{y}$ if *all the paths* between $\mathbf{x}$ and $\mathbf{y}$ are d-separated by $\mathbf{z}$ (as explained later on). When this occurs, $\mathbf{x}$ becomes independent on $\mathbf{y}$ if conditioning both to $\mathbf{z}$, that is, if we know $\mathbf{z}$, which contains all the "independence information" that is transmitted between $\mathbf{x}$ and $\mathbf{y}$. Unfortunately, the directness of arcs in a BN complicates testing whether a given combination of arc directions is considered to "transmit" dependence or not. We provide next two different mnemonic rules for that. Obviously, both lead to identical results for any BN.

According to the first rule, $\mathbf{z}$ d-separates a given path between $\mathbf{x}$ and $\mathbf{y}$ if and only if either: (1) the path contains a node of $\mathbf{z}$ with an incoming arc from another node of the network and an outgoing arc to another node of the network (such a configuration is called a *chain* and its existence assures that the path effectively transmits dependence information from one side to the other), (2) the path contains a node of $\mathbf{z}$ with two outgoing arcs to other nodes of the network (configuration called *fork*, whose existence confirms that there are nodes in $\mathbf{z}$ producing dependences on both $\mathbf{x}$ and $\mathbf{y}$), or (3) the path contains a node with two ingoing arcs from other nodes of the path (configuration called *collider*) and neither that node nor its descendants are in $\mathbf{z}$.

An alternative, procedural way—in contrast with the descriptive way of the first rule—for implementing the same criterion is the so-called "*Bayes ball*" algorithm (Shachter, 1998). The idea in this case is that $\mathbf{z}$ d-separates $\mathbf{x}$ and $\mathbf{y}$ if and only if an imaginary ball cannot pass from $\mathbf{x}$ to

Figure 10. A summary of the allowed and the forbidden paths in the "Bayes ball" algorithm. Notice that the "observed" variables refer in this figure to the set of variables we are conditioning on ($\mathbf{z}$), while the "hidden" ones are those whose distribution is being conditioned on the formers ($\mathbf{x}$ and $\mathbf{y}$).

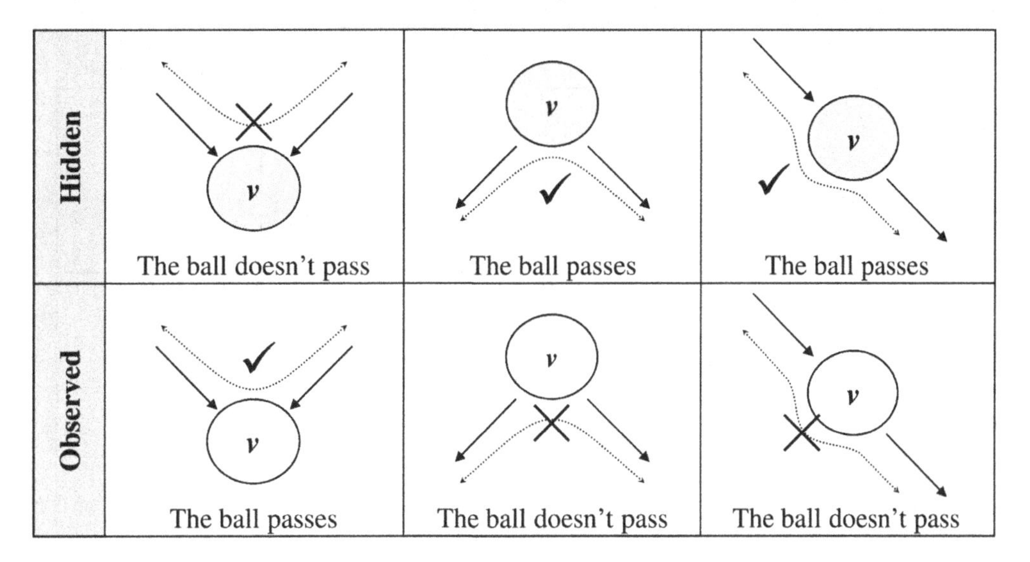

$\mathbf{y}$ given the permitted and forbidden moves reflected in Figure 10.

Marginal Distributions

Sometimes we may be interested in working with a reduced version of a BN, where some of the r.v.s (and therefore their corresponding nodes) are not of our interest and thus would be convenient to remove. As explained in section 7, this operation is called marginalization, and allows us to discard a set of r.v.s from a joint probability distribution, concentrating the information in only one or a few.

Let $\mathbf{x}$ denote the set of all the r.v.s in a given BN, and $\mathbf{y} \subset \mathbf{x}$ the subset of r.v.s which we want to left out of the BN model by concentrating their stochastic information on the rest. It is said that the $\mathbf{y}$ variables are *marginalized out*. By doing so, we must obviously modify the BN removing the nodes of $\mathbf{y}$, thus simplifying the graph regarding the node count. In contrast, new arcs may need to be inserted in order to model the dependences existing between variables in which the removed nodes mediated.

In order to determine whether the removal of a single node Y introduces an arc between any pair of nodes adjacent to it, say X_1 and X_2, we must observe the configuration of the arcs from those two nodes when meeting at Y: only if they meet head to head—the collider configuration mentioned before—Y can be removed without introducing any additional arc. If the arcs meet in any other configuration (i.e. tail to head or tail to tail), Y does indeed "transmit" dependence information between X_1 and X_2, and hence an arc between those two nodes must be added when removing Y in order to preserve that transmission. In the special case of "leaf" nodes (i.e. those with only one parent and no children), it can be shown that they can be safely marginalized out without any further modification.

We must remark that, in a general case, the pros and cons of marginalizing out r.v.s from a BN should be carefully pondered. Depending on the intended use of the BN it could be preferable to work on a graph with more nodes but sparsely connected, than on a densely-connected graph with only a few nodes. This dilemma is at the very core

Figure 11. Some common structures of Bayesian sub-networks and their corresponding graphs of correlations. Remember that shaded and unshaded nodes represent hidden ($\mathbf{h}$) and observed ($\mathbf{o}$) variables, respectively.

Bayesian network	Joint factoring $p(\mathbf{h},\mathbf{o})$	Bayes posterior $p(\mathbf{h}\mid\mathbf{o})=p(\mathbf{h},\mathbf{o})/p(\mathbf{o})$	Resulting correlation graph
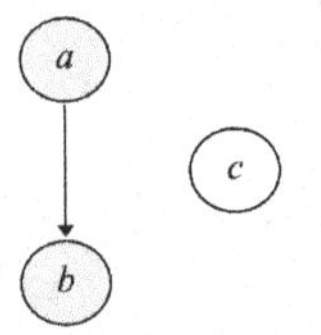 a, b, c	$=p(a)p(b\mid a)p(c)$	$=\frac{p(a)p(b\mid a)p(c)}{p(c)}=p(a)p(b\mid a)$ $=p(a,b)\neq p(a\mid\bar{\mathbf{o}})p(b\mid\bar{\mathbf{o}}) \quad ,\bar{\mathbf{o}}\subseteq\{c\}$	 a, b
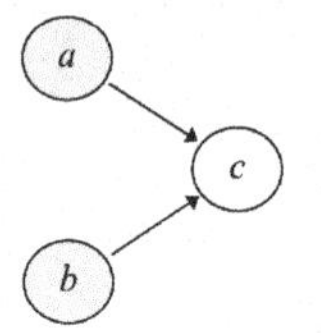 a, c, b	$=p(a)p(b)p(c\mid a,b)$	$=\frac{p(a)p(b)p(c\mid a,b)}{p(c)}$ $\neq p(a\mid\bar{\mathbf{o}})p(b\mid\bar{\mathbf{o}}) \quad ,\bar{\mathbf{o}}\subseteq\{c\}$	 a, b
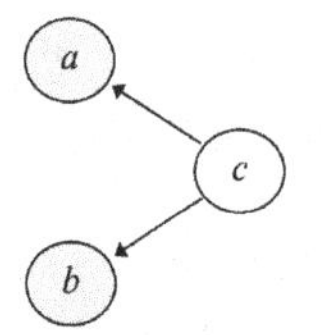 a, c, b	$=p(a\mid c)p(b\mid c)p(c)$	$=\frac{p(a\mid c)p(b\mid c)p(c)}{p(c)}=p(a\mid c)p(b\mid c)$	a, b
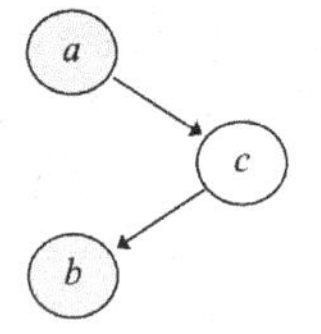 a, c, b	$=p(a)p(b\mid c)p(c\mid a)$	$=\frac{p(a)p(c\mid a)}{p(c)}p(b\mid c)\overset{\text{Bayes}}{=}p(a\mid c)p(b\mid c)$	a, b
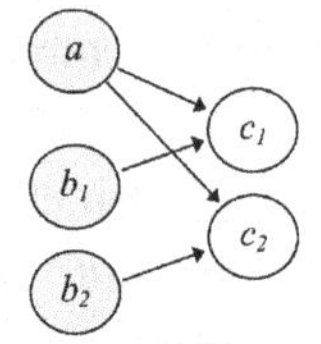 a, b_1, b_2, c_1, c_2	$=p(a)p(b_1)p(b_2)\cdot$ $\cdot p(c_1\mid a,b_1)p(c_2\mid a,b_2)$	$=\frac{p(a)p(b_1)p(b_2)p(c_1\mid a,b_1)p(c_2\mid a,b_2)}{p(c_1,c_2)}$ $\neq p(a\mid\bar{\mathbf{o}})p(b_1\mid\bar{\mathbf{o}})p(b_2\mid\bar{\mathbf{o}}) \quad ,\bar{\mathbf{o}}\subseteq\{c_1,c_2\}$	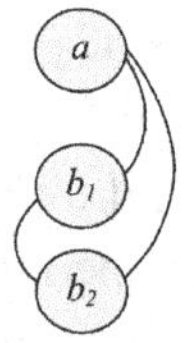 a, b_1, b_2

of the division between the two families of SLAM approaches that will be presented in chapter 9.

The Graph of Correlations

In previous section, we saw that the variance provides an indication of the amount of uncertainty existing in a given r.v., especially when that variable can be completely specified with its first two moments—as it is the case in Gaussian distributions. The covariance plays the same role in the case of two r.v.s, and the covariance matrix extends the uncertainty representation to the case of multiple r.v.s. In these last paragraphs, we are interested in some characteristics that can be deduced from the network structure about the covariance matrix of a set of r.v.s of a BN.

Consider a simple BN comprising only three nodes, a, b and c, where the first two are parents of the latter (refer to the second row in Figure 11). The variables a and b are not directly connected, thus they follow independent prior distributions such that the joint of the three variables reads $p(a,b,c)=p(a)p(b)p(c\mid a,b)$. We can reflect on what happens with the relationship between these variables after performing Bayesian inference in a BN, in this particular example when c is observable and the rest are hidden variables; or, put mathematically, when estimating the conditional distribution $p\left(a,b\middle|c\right)$.

It is easy to see that we cannot find a factorization of the distribution $p\left(a,b\middle|c\right)$ where a and b appear separately: a and b are not conditionally independent given c (c is a collider and conditioning paths cannot contain it in order to d-separate a and b), therefore we cannot separate them. It follows that performing Bayesian inference needs to consider in that case the correlation existing between their estimations. This is a well-known result and is often named in the literature as the "explaining away" effect (Bishop, 2006): the two parents of an observed node c *compete* with each other in explaining the readings of that node.

We propose here a graphical representation that generalizes this well-known effect to arbitrary graphs. Given a BN whose nodes $\{\mathbf{h},\mathbf{o}\}$ are divided into observed ($\mathbf{o}$) and hidden ($\mathbf{h}$) r.v.s, we define the *graph of correlations* associated to the inference problem of determining $p\left(\mathbf{h}\middle|\mathbf{o}\right)$ as a new undirected graph deduced from the BN: (1) with one node for each hidden variable in $\mathbf{h}$; and (2) with an arc between each pair of variables $\{h_1,h_2\}\in\mathbf{h}$ which are not conditionally independent given the observed variables, that is, with $h_1 \not\perp h_2|\mathbf{o}$.

Figure 11 shows some common structures of Bayesian sub-networks and the deduction of their graphs of correlations. As can be verified in the figure, a correlation arc has to be introduced between hidden variables that appear together (i.e. can have a nonzero correlation) in the factorization of the Bayes posterior. In general it is easier to check this condition of conditional independence by one of the abovementioned graphical methods rather than analytically.

It is clear that the graph of correlations provides us with a purely graphical method to reveal the sparseness of the relationships existing between the unknown r.v.s in any inference problem. For instance, if the intention is to model a set of hidden variables as a multivariate Gaussian, the number of arcs in the graph of correlations settles the number of nonzero entries in a sparse representation of the corresponding covariance matrix; hence, it can be used to predict and contrast the storage requirements of different BNs. In subsequent chapters, we will see such a practical application when discussing alternative solutions to SLAM.

REFERENCES

Anderson, H. L. (1986). Metropolis, Monte Carlo and the MANIAC. *Los Alamos Science*, *14*, 96–108.

Ankishan, H., & Efe, M. (2011). Adaptive neuro fuzzy supported Kalman filter approach for simultaneous localization and mapping. In *Proceedings of the IEEE 19th Conference on Signal Processing and Communications Applications (SIU 2011)*, (pp. 266-270). IEEE Press.

Apostol, T. M. (1967). *Calculus* (2nd ed., *Vol. 1*). New York, NY: John Wiley & Sons.

Apostol, T. M. (1969). *Calculus* (2nd ed., *Vol. 2*). New York, NY: John Wiley & Sons.

Arras, K. O. (1998). *An introduction to error propagation: Derivation, meaning and examples of equation* $\mathbf{C}_\mathrm{Y} = \mathbf{F}_\mathrm{X}\mathbf{C}_\mathrm{X}\mathbf{F}_\mathrm{X}^T$. Technical report no. EPFL-ASL-TR-98-01 R3. Geneva, Switzerland: Swiss Federal Institute of Technology.

Ash, R. B. (1999). *Probability and measure theory*. New York, NY: Academic Press.

Bessière, P., Laugier, C., & Siegwart, R. (Eds.). (2008). *Probabilistic reasoning and decision making in sensory-motor systems*. Berlin, Germany: Springer. doi:10.1007/978-3-540-79007-5

Bishop, C. M. (2006). *Pattern recognition and machine learning*. New York, NY: Springer.

Blackwell, D. (1947). Conditional expectation and unbiased sequential estimation. *Annals of Mathematical Statistics, 18*(1), 105–110. doi:10.1214/aoms/1177730497

Bucy, R. S. (1965). Nonlinear filtering theory. *IEEE Transactions on Automatic Control, 10*(2), 198. doi:10.1109/TAC.1965.1098109

Chatterjee, A., & Matsuno, F. (2007). A neuro-fuzzy assisted extended Kalman filter-based approach for simultaneous localization and mapping (SLAM) problems. *IEEE Transactions on Fuzzy Systems, 15*(5), 984–997. doi:10.1109/TFUZZ.2007.894972

Cover, T. A., & Thomas, J. A. (2006). *Elements of information theory*. New York, NY: Wiley-Interscience.

Doucet, A., de Freitas, N., Murphy, K., & Russell, S. (2000). Rao-Blackwellised particle filtering for dynamic Bayesian networks. In *Proceedings of the Conference on Uncertainty in Artificial Intelligence (UAI)*. UAI.

Dubois, D., & Prade, H. (2001). Possibility theory, probability theory and multiple-valued logics: A clarification. *Annals of Mathematics and Artificial Intelligence, 32*, 35–66. doi:10.1023/A:1016740830286

Fox, D., Burgard, W., Dellaert, F., & Thrun, S. (1999). Monte Carlo localization: Efficient position estimation for mobile robots. In *Proceedings of the National Conference on Artificial Intelligence (AAAI)*. AAAI.

Franklin, J. (2001). *The science of conjecture: Evidence and probability before Pascal*. Baltimore, MD: The Johns Hopkins University Press. doi:10.1007/BF02985402

Ghahramani, Z. (2001). An introduction to hidden Markov models and Bayesian networks. *International Journal of Pattern Recognition and Artificial Intelligence, 15*(1), 9–42. doi:10.1142/S0218001401000836

Havangi, R., Teshnehlab, M., Nekoui, M. A., & Taghirad, H. (2011). An adaptive neuro-fuzzy Rao-Blackwellized particle filter for SLAM. In *Proceedings of the IEEE International Conference on Mechatronics (ICM2011)*, (pp. 487-492). IEEE Press.

Kalman, R. E. (1960). A new approach to linear filtering and prediction problems. *Transactions of the ASME Journal of Basic Engineering, 82*(D), 35-45.

Kalman, R. E., & Bucy, R. S. (1961). New results in linear filtering and prediction theory. *Transactions of the ASME Journal of Basic Engineering, 83*(D), 95-108.

Kassam, S. A. (1988). *Signal detection in non-Gaussian noise*. Berlin, Germany: Springer-Verlag.

Kindermann, R., & Snell, J. L. (1980). *Markov random fields and their applications*. Providence, RI: American Mathematical Society.

Klir, G. J. (2006). *Uncertainty and information: Foundations of generalized information theory*. New York, NY: John Wiley & Sons.

Lehman, M. M. (1990). Uncertainty in computer application. *Communications of the ACM, 33*(5).

Manolakis, D. G., Ingle, V. K., & Kogon, S. M. (2005). *Statistical and adaptive signal processing*. London, UK: Artech House.

Meester, R. (2008). *A natural introduction to probability theory* (2nd ed.). Berlin, Germany: Birkhäuser Verlag.

Montemerlo, M., Thrun, S., Koller, D., & Wegbreit, B. (2002). FastSLAM: A factored solution to the simultaneous localization and mapping problem. In *Proceedings of the AAAI National Conference on Artificial Intelligence*. AAAI.

Morgan, D. (1992). *Numerical methods: Real-time and embedded systems programming*. Evansville, IN: M&T Publishing.

Morris, A. S. (2005). *Measurement and instrumentation principles*. London, UK: Elsevier.

Rao, C. R. (1965). *Linear statistical inference and its applications* (2nd ed.). New York, NY: Wiley-Interscience.

Rodwell, E. J., & Cloud, M. J. (2012). Automatic error analysis using intervals. *IEEE Transactions on Education*, *55*(1), 9–15. doi:10.1109/TE.2011.2109722

Salicone, S. (2007). *Measurement uncertainty. An approach via the mathematical theory of evidence*. Berlin, Germany: Springer.

Scuro, S. R. (2004). *Introduction to error theory. Technical Report*. College Station, TX: Texas A&M University.

Shachter, R. D. (1998). Bayes-ball: The rational pastime (for determining irrelevance and requisite information in belief networks and influence diagrams). In *Proceedings of the Fourteenth Conference in Uncertainty in Artificial Intelligence*, (pp. 480-487). IEEE.

Shaffer, G., & Vovk, V. (2005). *The origins and legacy of Kolmogorov's Grundbegriffe*. Working Paper. Retrieved February 2, 2010 from http://www.probabilityandfinance.com/articles/04.pdf

Tabak, J. (2004). *Probability and statistics: The science of uncertainty*. New York, NY: Facts on File.

Taylor, J. R. (1997). *An introduction to error analysis* (2nd ed.). New York, NY: University Science Books.

Walpole, R. E., Myers, R. L., & Myers, S. H. (1997). *Probability and statistics for engineers and scientists*. Upper Saddle River, NJ: Prentice Hall. doi:10.2307/2530629

Watanabe, K., Pathiranage, C. D., & Izumi, K. (2009). T-S fuzzy model adopted SLAM algorithm with linear programming based data association for mobile robots. In *Proceedings of the IEEE International Symposium on Industrial Electronics (ISIE 2009)*, (pp. 244-249). IEEE Press.

Zadeh, L. A. (1965). Fuzzy sets. *Control*, *8*, 338–353. doi:10.1016/S0019-9958(65)90241-X

Zadeh, L. A. (2005). Toward a generalized theory of uncertainty (GTU)—An outline. *Information Sciences*, *172*, 1–40. doi:10.1016/j.ins.2005.01.017

ENDNOTES

1. The basic Von Neumann algorithm to obtain a fair result from a loaded dice is as follows: consider a loaded dice that can show 0 or 1 in each roll (it is easy to extend this algorithm for the case of more possible outcomes); take the resulting rolls in pairs of consecutive values; if a pair is repetitive (00 or 11) discard it; if the pair is not repetitive, take the first bit of the pair as the resulting value. The basic drawback of this algorithm is that you need more outcomes from the original process than what you obtain from the resulting one. Demonstrating the validity of the algorithm is not difficult: if the probability of obtaining 1 in the loaded dice is p and the one of obtaining 0 is $1-p$, then the probability of reading 01 is (assuming independent rolls) $(1-p)p$, which is equal to the probability of reading 10; since we take a different outcome in each of these pairs, each outcome of the resulting process has the same probability and therefore the modified process is fair (Neumann, 1951).
2. A formal definition of the state spaces of 2D and 3D robot poses will be delayed until chapter 10 section 3.

3 A σ-algebra is a non-empty collection of sets (in our case, events) closed under countable union ($\bigcup$), countable intersection ($\bigcup$), relative complement or difference ($B \backslash A$) and complement ($\bar{A}$). It is not empty, therefore we can take one of its elements, say A, and the difference $A \backslash A = \varnothing$ must belong to it since it is closed under that operation. Analogously, $\bar{\varnothing} = \Omega$ belongs to it. Therefore, the σ-algebra always contains the empty ($\varnothing$) and the universal (Ω) set. Usually, F is the smallest σ-algebra that can be defined on Ω, and thus, it is unique.

4 More formally, the support of a probability measure P is the set of outcomes of the sample space Ω that do not belong to any event that has probability zero. Correspondingly, the support of a r.v. X contains all the values of $co(X)$ (values that the r.v. can produce) that have been mapped from outcomes of the support of P, that is, that correspond to events that have non-zero probability. In other words: in the support of a r.v. we have discarded those values associated to the (possibly countable infinite) outcomes of the stochastic process for which the probability measure is zero.

5 Here, "less than or equal to" must be understood in a possibly multidimensional real space, e.g., we should define some geometrical distance or order between points in 3D space if we are using a multidimensional r.v. For the sake of brevity, this will not be explicitly indicated in the rest of the book unless strongly needed.

6 Formally, F_X is continuous if

$$\forall a \in \text{supp}(X), \left[\lim_{x \to a^+} F_X(x) = F_X(a)\right] \wedge \left[\lim_{x \to a^-} F_X(x) = F_X(a)\right],$$

that is, if the limit of the function from the right toward the point and the limit of the function from the left toward the point coincide and yield the same value that the one of the function at the point (for every point in the domain of the function). A function is right-continuous if only the first squared-bracket term holds.

7 In the context of r.v.s, a histogram is a graph that shows the number of occurrences collected over a given period of time for each value of the codomain of a r.v. Therefore, the graph resulting from dividing the histogram by the total number of occurrences of all the values is an approximation to the graph of the pmf (it is not exact since the number of occurrences of a value times the total number of occurrences is not exactly the probability of the value unless we have infinite occurrences, as we will see further on in the book).

8 We do not provide here formal definitions for integrals or derivatives, but we must note that the integral in Equation 16 is actually a Lebesgue integral. Lebesgue integrals work as a classical integral in most cases, except when ρ is not continuous everywhere but *almost everywhere*. Explaining this formally is also out of the scope of this book, but you can think intuitively of any function ρ that has none or only sporadic point discontinuities; such function would be Lebesgue-integrable and thus acceptable in probability theory.

9 By basic matrix algebra,

$$Cov\left[\mathbf{AX}, \mathbf{B}^T\mathbf{Y}\right] =$$

$$E\left[\left(\mathbf{AX} - E\left[\mathbf{AX}\right]\right)\left(\mathbf{B}^T\mathbf{Y} - E\left[\mathbf{B}^T\mathbf{Y}\right]\right)^T\right];$$

since $\mathbf{A}$ and $\mathbf{B}$ do not depend on $\mathbf{X}$ or $\mathbf{Y}$, this is equivalent to

$$E\left[\left(\mathbf{AX} - \mathbf{A}E\left[\mathbf{X}\right]\right)\left(\mathbf{B}^T\mathbf{Y} - \mathbf{B}^T E\left[\mathbf{Y}\right]\right)^T\right];$$

factoring out the matrices, we have

$$E\left[\mathbf{A}\left(\mathbf{X} - E\left[\mathbf{X}\right]\right)\left(\mathbf{B}^T\left(\mathbf{Y} - E\left[\mathbf{Y}\right]\right)\right)^T\right];$$

doing the transpose this becomes

$$E\left[\mathbf{A}\left(\mathbf{X}-E\left[\mathbf{X}\right]\right)\left(\mathbf{Y}-E\left[\mathbf{Y}\right]\right)^T\mathbf{B}\right];$$

and finally, since the matrices are constant,

$$Cov\left[\mathbf{AX},\mathbf{B}^T\mathbf{Y}\right]=\mathbf{A}E\left[\left(\mathbf{X}-E\left[\mathbf{X}\right]\right)\left(\mathbf{Y}-E\left[\mathbf{Y}\right]\right)^T\right]\mathbf{B}$$
$$=\mathbf{A}\,Cov\left[\mathbf{X},\mathbf{Y}\right]\mathbf{B}.$$

10 For the case of matrices with complex numbers, they must be *Hermitian* matrices rather than symmetric (it is the analogous to symmetry, substituting transposes by conjugate transposes, i.e., the transformation of the entries of the transpose into their corresponding complex conjugate numbers), and the transposes in the two conditions for positive-semidefinite and positive-definite are then conjugate transposes.

11 We have a linear transformation $\mathbf{y}=\mathbf{Ax}+\mathbf{b}$. Due to the linear properties of the expectation, it directly follows that the mean of the transformed variable is

$$E\left[\mathbf{y}\right]=E\left[\mathbf{Ax}+\mathbf{b}\right]$$
$$=E\left[\mathbf{Ax}\right]+E\left[\mathbf{b}\right]$$
$$=\mathbf{A}E\left[\mathbf{x}\right]+\mathbf{b}.$$

Regarding its covariance, we firstly need to know that the covariance matrix of a r.v. is not affected by a shift or offset in its value; that is, is we expand $Cov\left[\mathbf{x}+\mathbf{b}\right]$ as

$$E\left[\left(\mathbf{x}+\mathbf{b}\right)\left(\mathbf{x}+\mathbf{b}\right)^T\right]-E\left[\left(\mathbf{x}+\mathbf{b}\right)\right]E\left[\left(\mathbf{x}+\mathbf{b}\right)\right]^T$$

we see that all the terms containing $\mathbf{b}$ cancel out, leading to

$$Cov\left[\mathbf{x}+\mathbf{b}\right]=E\left[\mathbf{xx}^T\right]-E\left[\mathbf{x}\right]E\left[\mathbf{x}\right]^T=Cov\left[\mathbf{x}\right].$$

Now, we particularize the result $Cov\left[\mathbf{Cw},\mathbf{D}^T\mathbf{z}\right]=\mathbf{C}\,Cov\left[\mathbf{w},\mathbf{z}\right]\mathbf{D}$, derived in a previous note, for the case of $\mathbf{Cw}=\mathbf{D}^T\mathbf{z}=\mathbf{Ax}$ (that is, $\mathbf{C}=\mathbf{A}$ and $\mathbf{D}=\mathbf{A}^T$), which gives us $Cov\left[\mathbf{Ax}\right]\triangleq Cov\left[\mathbf{Ax},\mathbf{Ax}\right]=\mathbf{A}\,Cov\left[\mathbf{x}\right]\mathbf{A}^T$. And since a shift in the variable does not alter the covariance, we finally arrive at $Cov\left[\mathbf{Ax}+\mathbf{b}\right]=\mathbf{A}\,Cov\left[\mathbf{x}\right]\mathbf{A}^T$. Please refer to Arras (1998) for further examples of linear transformations like this one.

Chapter 4
Statistical Bases

ABSTRACT

This is the fourth and last chapter of the first section. As chapter 3 introduced the mathematical tools of probability theory needed to understand all the concepts in the book, chapter 4 does the same concerning statistics. It fills the gap between probability theory and real data coming from stochastic processes, highlighting the great amount of potential applications of the different fields of statistics—particularly estimation theory—in state-of-the-art science and engineering. Topics covered in this chapter include the fundamental tools needed in probabilistic robotics: probabilistic convergence, theory of estimators, hypothesis tests, etc. Special stress is on recursive Bayesian estimators, due to their central role in the problems of probabilistic robot localization and mapping.

CHAPTER GUIDELINE

- You will learn:
  - In which ways the uncertainty in a set of observations may converge to a given probabilistic model as the number of observations increases.
  - What are the sample mean and sample variance, and their difference with the theoretical mean and variance of a distribution.
  - Why having many observations is better than having few.
  - Why the normal distribution has such a prevalent role in science and in nature.
  - What is an estimator and which of its properties shall interest us.
  - How to construct classical and Bayesian estimators.
  - The bases on estimating dynamic processes.
- Provided tools:
  - The central limit theorem and the law of large numbers.
  - The construction of maximum likelihood estimators.
  - The method of moments for constructing estimators.
  - The basic Bayesian estimators: MMSE, MAP, and MED.
  - The general Bayesian framework for estimating dynamic Markov processes.

DOI: 10.4018/978-1-4666-2104-6.ch004

- Relation to other chapters:
    - Bayesian estimators introduced in this chapter are applied to practical robotic problems in chapters 7, 8, and 9.
    - A central part of chapter 10 deals with the concept of maximum likelihood estimation and its role in the recent approaches to robot map building in especially difficult scenarios.

1. INTRODUCTION

Statistics is the branch of mathematics dealing with the collection, analysis, interpretation and presentation of numerical data (Merrian-Webster, 2011). It is, therefore, the science of making effective use of data acquired from stochastic processes in order to describe them, infer new information, or making predictions. Statistics is based on probability theory (the subject of chapter 3), but it is a different science, more practical and, sometimes, less abstract. Its historical development was intertwined with that of probability theory, as depicted in the previous chapter.

Statistics has its own diverse branches, each focused on the different aspects of the analysis of data. *Descriptive statistics* has to do with synthesizing the main characteristics of collected data, while *statistical inference* draws conclusions from these data in spite of their randomness. In mobile robot localization and mapping, the most relevant area is that of *estimation*: inferring the characteristics of the current uncertainty exposed by r.v.s from the observation of the values they yield. Other branches, such as *prediction* (estimation of the information that the r.v.s will expose in the future) have currently a marginal presence in mobile robotics, mostly in task or motion planning, which are outside the scope of this book.

In statistics, all we have are data gathered from the real world. Each datum will be called an *observation*—in particular, a real number or a vector of real numbers, thus they can be considered as the value yielded by a r.v.—while a set of gathered data will be called a *sample*. Other possibilities different from numerical observations exist, such as categorical observations, but they are rare in robotics. In the case that all the observations in a sample are considered to be gathered at the same instant of time, they can be thought of as the multidimensional value yielded by a single random variable, which is usually unknown since all we know about the process is the data. On the other hand, if we consider that our observations are gathered over time from a single stochastic process that does not change its parameters over time (dynamic processes will be explained in section 9), we can take them as the values yielded by only one r.v. X in sequential, non-simultaneous evaluations of its value. The latter will be the common scenario in the rest of this book.

In addition to observations and samples, it is also common to make plausible assumptions about the underlying stochastic process that has produced them. However, we rarely know the underlying probability space, in particular its probability function P, or the distributions (pdfs/pmfs) of the r.v.s. If needed, the most likely and/or simple forms are assumed for these unknowns, or a *ground-truth* (an estimate obtained by alternative measurements that are more precise) is used.

Statistics works very well when we have large datasets, that is, massive amounts of data. This follows from the law of large numbers, explained further on. Having important computational resources to process those datasets, typically processing them *off-line* (after all the data have been collected), often helps in obtaining better results. Unfortunately, most mobile robot applications pose important difficulties to exploit these advantages: such a robot needs to know where it is and how is its environment *at the same time* that it is collecting information from its sensors, since the results are typically required to carry on with navigation. Therefore, *on-line* estimation is the usual arrangement in robotic Simultaneous Localization And Mapping (SLAM), along with the assumption that the quality of the obtained information may require collecting data during a long-enough operation time. We will see methods that adapt well to both off-line and on-line estimation throughout this book.

Mobile robots also impose important constraints to the computational power available to statistical computations: remember the chicken-and-egg problem of the power supply vs. computation capabilities mentioned in chapter 1. This reduces the number of observations that we can work with, and also the rate of convergence to the final result (even preventing from that convergence in the worst cases!). Some of the methods that we present in further chapters have been precisely conceived for adapting to those situations. In general, computational efficiency in statistical methods is of a great concern to robotics researchers today, and the reason of an important number of simplifications introduced to the original statistical tools.

In this chapter, we still do not focus on mobile robots. Instead, we complement the previous chapter by providing an introduction to the main statistical methods and concepts that will be needed in practical probabilistic robotics. Actually, the following should be read after chapter 3, since we use here many results from there. Broadly, it copes with probabilistic convergence (section 2) and estimation (the rest). There is abundant scientific literature on probability and statistics that can be consulted for a more comprehensive study of the concepts presented here, for instance the introductory textbook (Walpole, Myers, & Myers, 1997). There also exist diverse on-line references for finding clear and thorough explanations of concrete concepts (Hazewinkel, 2002; AI Access, 2011), even including complete demonstrations (Weisstein, 2011).

2. IN BETWEEN PROBABILITY AND STATISTICS

The links between probability and statistics are, obviously, quite strong. We will select the most evident for the transition: how to model probabilistically temporal sequences of observations and some of the properties that those sequences can have.

The most general interpretation for a set of numerical, unidimensional observations $\{z_1, z_2, \ldots, z_k\}$ coming from a given stochastic process in a certain temporal order is for them to be the values taken non-simultaneously by an underlying, unknown r.v. denoted as X (i.e., $z_i \sim X$). We consider the unidimensional case here for simplicity, but generalizing to multidimensional settings should be straightforward.

Since in the real world we will have only the observations $\{z_1, z_2, \ldots, z_k\}$, we will be interested in constructing some r.v. that approaches the (unknown) mathematical definition of X with the data available. Furthermore, we will improve such a constructed $\hat{X}$ at each time index i, obtaining a sequence of r.v.s denoted $\hat{X}_i$, that are constructed from some deductive process on the available sample, which in turn includes from the first to the i-th observation. Since the observations are all identically distributed and the process to construct each $\hat{X}_i$ is the same at all time indexes, all the r.v.s $\hat{X}_i$ will be identically distributed too. Also, since each $\hat{X}_i$ yields values independently on the values provided by others $\hat{X}_i$, they will be considered independent of each other. Any set of r.v.s that are both independent and identically distributed is said to be composed of *iid* random variables.

In this scenario we can be interested in how our guess of X improves as i grows, or, mathematically, in how the sequence of iid r.v.s $\{\hat{X}_1, \hat{X}_2, \ldots, \hat{X}_k\}$ tends to exhibit the stochastic properties of the real but unknown X better as we gather more and more observations. In mathematics, this refers to the study of the *convergence* of a sequence of random variables, denoted as $\{\hat{X}_1, \hat{X}_2, ..., \hat{X}_k\} \to X$.

A more common and slightly different use of this formalism is to guess not an entire r.v., let us say Y, but some of its properties (for instance, some of its moments), for instance a value a in a unidimensional case. In that case, we will construct the $\hat{X}_i$ mentioned before to provide guesses of a at each time step, not to be similar to the r.v. Y that is producing the observations, and we will interested in how the $\hat{X}_i$ converge to the actual value of a as more and more observations arrive. Under this point of view, we can consider the X of the previous paragraphs not as the r.v. that is producing the observations, but as an unknown r.v. that has a delta pdf (δ_a)—in the continuous, unidimensional case—and therefore a deterministic behavior: it always yields a. Then the previous formalism for studying the convergence of guesses can be used without modifications, since it also reads here as $\{\hat{X}_1, \hat{X}_2, ..., \hat{X}_k\} \to X$.

Using then this general formalism, we can define three basic levels of convergence for sequences of iid r.v.s to a common r.v. We explain them in the following, from the most stringent to the least, and after that, we make use of these definitions for introducing some of the most important results in statistics.

Almost Sure Convergence or Convergence with Probability One

Formally, the sequence $\{\hat{X}_1, \hat{X}_2, ..., \hat{X}_k\}$ is said to converge *almost surely* or *with probability one* to a common r.v. X if and only if it is for certain that the difference between the value yielded by the last variable of the sequence and the values yielded by the common r.v. X tends to be as small as we want as the sequence grows and grows. Mathematically:

$$\forall \varepsilon > 0,\ P\left[\lim_{k\to\infty} \left|\hat{X}_k - X\right| < \varepsilon\right] = 1$$

This kind of convergence can be denoted either as "$\hat{X}_k \to X$ a.s. as $k \to \infty$" or as "$\hat{X}_k \to X$ w.p. 1."

This expression assures that, as time goes, the probability of obtaining a perceptible difference between the observed value of the last r.v. of the sequence and the one of X is 0, or slightly more precisely, that it is more and more difficult to see such a difference as new observations are gathered from the process. Therefore, it is assured that the stochastic process reaches a steady state. Mathematically:

$$P\left[\lim_{k\to\infty} \left|\hat{X}_k - X\right| < \varepsilon\right] = 1 \quad \leftrightarrow$$

$$1 - P\left[\lim_{k\to\infty} \left|\hat{X}_k - X\right| < \varepsilon\right] = 0 \quad \leftrightarrow$$

$$P\left[\overline{\left(\lim_{k\to\infty} \left|\hat{X}_k - X\right| < \varepsilon\right)}\right] = 0 \quad \leftrightarrow$$

$$P\left[\lim_{k\to\infty} \left|\hat{X}_k - X\right| \geq \varepsilon\right] = 0 \tag{1}$$

Convergence in Probability or in Measure

A sequence of r.v.s $\{\hat{X}_1, \hat{X}_2, ..., \hat{X}_k\}$ is said to *converge in probability* or *in measure* to a common X if and only if

$$\forall \varepsilon > 0,\ \lim_{k\to\infty} \left(P\left[\left|\hat{X}_k - X\right| < \varepsilon\right]\right) = 1.$$

This is usually denoted as "$\hat{X}_k \xrightarrow{p} X$ as $k \to \infty$." Notice the subtle difference between this expression and the one for almost sure convergence. Convergence in probability is weaker than almost sure convergence, since it *does not assures anything*: the equality to one in this expression is not established with respect to the probability, *but to the limit*. We can turn around the inequality in this expression to explain that better:

$$\lim_{k\to\infty}\left(P\left[\left|\hat{X}_k - X\right| < \varepsilon\right]\right) = 1 \quad \leftrightarrow$$

$$\lim_{k\to\infty}\left(1 - P\left[\overline{\left(\left|\hat{X}_k - X\right| < \varepsilon\right)}\right]\right) = 1 \quad \leftrightarrow$$

$$1 - \lim_{k\to\infty}\left(P\left[\left|\hat{X}_k - X\right| \geq \varepsilon\right]\right) = 1 \quad \leftrightarrow$$

$$\lim_{k\to\infty}\left(P\left[\left|\hat{X}_k - X\right| \geq \varepsilon\right]\right) = 0 \qquad (2)$$

which can be read as follows: "as time goes and the sequence is longer, the probability of observing a perceptible difference between the observation corresponding to the last r.v. of the sequence and the one coming from X *tends* to be zero, although it will not be zero for certain." In other words, regardless of how many observations we have gathered and how quickly or slowly the sequence seems to converge to X, we can always have the possibility of finding an arbitrarily large difference between the last observation and any other obtained from X, although the probability of observing that situation becomes smaller and smaller. Obviously, the sequence is tending to X, as in the previous kind of convergence.

Convergence in Distribution

A sequence of r.v.s $\left\{\hat{X}_1, \hat{X}_2, ..., \hat{X}_k\right\}$ is said to *converge in distribution* to X, denoted as "$\hat{X}_k \xrightarrow{d} X$ as $k \to \infty$," if and only if $\lim_{k\to\infty} F_{\hat{X}_k} = F_X$, that is, if and only if the distribution function of the last r.v. of the sequence tends to be the same distribution function as that of X as time goes. This assertion is weaker than the previous ones, since it does not speak about the values observed in the sequence or in X, only about the general shape of their uncertainty. For showing more formally the important implications that such weakness can have, please recall the definition of r.v. in chapter 3 section 4 and note that $\hat{X}_k = X \to F_{\hat{X}_k} = F_X$, but not vice versa: both $\hat{X}_k$ and X are mappings from outcomes of the process to real values; if both mappings coincide, so do their induced distribution functions, but two different mappings (r.v.s), possibly having different domains (outcomes of the stochastic process), can have equal distribution functions, assuming they have equal codomains. In short: convergence in distribution does not mean necessarily that the sequence of r.v.s are actually the same r.v.; only that they have the same probability distribution.

Convergence in Norm or in Q-Norm

Finally, the least stringent type of convergence of a sequence of r.v.s $\left\{\hat{X}_1, \hat{X}_2, ..., \hat{X}_k\right\}$ to a common r.v. X is the *convergence in norm* or in *q-norm*. This happens if and only if for

$$q \in \mathbb{N}^+,\ \lim_{k\to\infty}\left(E\left[\left|\hat{X}_k - X\right|^q\right]\right) = 0.$$

Typically, $q \in \{1, 2\}$. This refers to the trend of the expected value of the difference between values of the sequence and values of the common r.v., which is a more relaxed condition than using the complete shape of uncertainty like in convergence in distribution or the other kinds of convergence.

Obviously, this expression does not *assure* anything, since the value of 0 applies to the limit, not to the expectation; therefore, we could always find an expected value of the q-norm arbitrarily large, even when the sequence is very long.

Probabilistic Convergence and the Limit Laws

When we gather a series of observations (a sample) from the same stochastic process it is obvious that we know that the corresponding sequence of r.v.s $\{\hat{X}_1, \hat{X}_2, \ldots, \hat{X}_k\}$ constructed from them should converge to the ones of the underlying X that governs the process as more and more elements are added to the sequence (more available information). The problem is that X is unknown; thus, our goal is to deduce its stochastic properties only from the sample. The definitions in the previous subsections can help in that purpose, but we need more.

Before we take the step of deducing the whole shape of the uncertainty of X, we can explore the possibility of summarizing it, e.g., to deduce only its moments from the sample, which often contain less information and perhaps lead to more general and stronger results (certainly, it should be easier). The first natural summary of uncertainty that we can think of is the first moment, the expectation.

The value $E[X]$ of the expectation of the common r.v. X is unknown too, of course, but we can propose the following mathematical function to deduce the expectation of X from the available sample at time index k (in fact, this is the most basic *estimator* of the actual expectation, and in the case of Gaussian distributions, it is the best estimator in many senses):

$$\frac{\overbrace{\sum_k z_k}^{S_k}}{k} \triangleq \frac{S_k}{k} \tag{3}$$

being z_k the observation gathered at time k. This quotient is called the *sample mean* of the available observations, and is itself a r.v., since since it is a function of the observations and the observations are in turn yielded by a r.v. X. Therefore, the previous definitions of convergence are applicable to the sample mean, considering it as a sequence of iid r.v.s indexed by the time of each observation. In particular, we can study the kind of convergence of a sequence of sample means to the actual, unknown expectation $E[X]$, consequently we are using the previously explained formalism for the convergence of r.v.s to a single numerical value (you can consider that value as the one yielded by a r.v. with a delta distribution $\delta_{E[X]}$).

Precisely one of the most important results in statistics has to do with the probabilistic convergence of the sample mean to the expectation. This result says that, if $E[X]$ is finite, then:

$$\frac{S_k}{k} \xrightarrow{p} E[X] \text{ as } k \to \infty \tag{4}$$

This result was called the *weak law of large numbers* by Poisson, but it is also known as the Bernoulli's theorem after Jacob Bernoulli, who developed the first mathematical demonstration for binary r.v.s. It assures that, in spite of the actual shape of uncertainty of the sequence, its sample mean converges in probability, as we gather more and more observations, to a r.v. that always yields the actual expectation. Notice that the sample mean is very similar to the definition of expectation of a discrete r.v., but without weights. To understand this point, notice that the expectation weights the *different values* of a variable according to their corresponding probabilities, while the sample mean weights the different values appearing in the sequence of observations according to their frequency of occurrence, which is asymptotically the same as the size of the sample tends to infinity.

The demonstration of Equation 4 is easy if we assume that the variance $V[X]$ is a finite number σ^2 and use the Chebyshev's inequality that was introduced in chapter 3 section 6. Firstly, we have that:

$$V\left[\frac{S_k}{k}\right] = V\left[\frac{1}{k}\left(z_1 + z_2 + \ldots + z_k\right)\right] =$$

(by the properties of variance introduced in chapter 3 section 6, and since $z_i \sim X$)

$$= \frac{1}{k^2} V\left[\underbrace{X + X + \ldots + X}_{k \text{ times}}\right] =$$

(using again the properties of variance)

$$= \frac{1}{k^2}\sum_{i=1}^{k} V[X] = \frac{kV[X]}{k^2} = \frac{V[X]}{k} \qquad (5)$$

and also:

$$E\left[\frac{S_k}{k}\right] = \frac{1}{k}E\left[\sum_{i=1}^{k} z_i\right] = \frac{1}{k}\sum_{i=1}^{k} E[X_i] = \frac{1}{k}\sum_{i=1}^{k} E[X] = \frac{kE[X]}{k} = E[X] \qquad (6)$$

which is the property of *unbiaseness* of the sample mean, a concept that will be explained later on but that can informally expressed as follows: an estimator is unbiased iff its expected value coincides with the actual value it is intended to estimate.

Now, using the Chebyshev's inequality on the r.v. corresponding to the sample mean (S_k / k) with a given positive number $a > 0$:

$$P\left(\left|\frac{S_k}{k} - E\left[\frac{S_k}{k}\right]\right| \geq a\right) \leq \frac{V\left[\frac{S_k}{k}\right]}{a^2} \quad \leftrightarrow$$

$$P\left(\left|\frac{S_k}{k} - E[X]\right| \geq a\right) \leq \frac{V[X]}{a^2 k} \quad \leftrightarrow$$

$$-P\left(\left|\frac{S_k}{k} - E[X]\right| \geq a\right) \geq -\frac{V[X]}{a^2 k} \quad \leftrightarrow$$

$$1 - P\left(\left|\frac{S_k}{k} - E[X]\right| \geq a\right) \geq 1 - \frac{V[X]}{a^2 k} \quad \leftrightarrow$$

$$P\left(\left|\frac{S_k}{k} - E[X]\right| < a\right) \geq 1 - \frac{V[X]}{a^2 k}$$

(applying limits to both sides)

$$\lim_{k\to\infty} P\left(\left|\frac{S_k}{k} - E[X]\right| < a\right) \geq \lim_{k\to\infty}\left(1 - \frac{V[X]}{a^2 k}\right) \quad \leftrightarrow$$

$$\lim_{k\to\infty} P\left(\left|\frac{S_k}{k} - E[X]\right| < a\right) \geq 1$$

(and since a probability cannot be ever greater than 1)

$$\lim_{k\to\infty} P\left(\left|\frac{S_k}{k} - E[X]\right| < a\right) = 1 \qquad (7)$$

which is precisely the definition of convergence in probability of the sequence of r.v.s S_k / k to $E[X]$.

Since this convergence is in probability, the possibility of obtaining a sample mean arbitrarily far from the actual expectation always exist, no matter how many observations we have taken already. In general, however, such possibility diminishes over time and the estimated value will approach appropriately the actual one if a large enough dataset is available. Being as important as it is, this result is strongly conditioned by the term "large enough": what is a sufficient number of observations in some situations may not be in others. It is usually difficult or even impossible to know in advance how many observations to take in order to calculate a good estimation of the expectation.

Fortunately, the law of large numbers also holds in a stronger form if the variance of X is finite:

$$\frac{S_k}{k} \rightarrow E\left[X\right] \text{ a.s. as } k \rightarrow \infty \quad (8)$$

Note that we are guaranteeing almost sure convergence here. Therefore, when both the actual expectation and variance of the r.v.s are finite, the sample mean will converge to a steady state in which its value is almost equal to the actual expectation of the underlying r.v., without the possibility of arbitrary departures from it (we however need a large enough dataset; again the "large enough" limitation). This result is called the *strong law of large numbers*. For example, Figure 1 shows a pseudorandom sample gathered sequentially by a computer from a continuous real-valued r.v. that has a gamma density distribution with $\mu = 200$ and $\sigma^2 = 4000$. As you can see, the sample mean in this case clearly converges to the actual expectation as the size of the sample grows, and enters an steady state (defined as a $\pm 2\%$ range within the actual value) just with 150 observations approximately.

The r.v. called sample mean also conveys another important result for statistics, that we will present without demonstration since the mathematical concepts required have little use in the rest of this book —you can consult, for example Kallenberg (1998). Let us propose the same situation as in the strong law of large numbers, although the iid condition can be relaxed now to only independent r.v.s if either one of two particular conditions hold, found by Lyapunov and Lindenberg, respectively (Ash, 1999). In that case, the following can be shown to hold too:

Figure 1. A numerical simulation of the values observed from a stochastic process with a distribution modeled by a gamma with the parameters explained in the text. The sample mean converges to the true expectation of the distribution as more observations are taken, according to the law of large numbers. In this figure, you can see the strong version of that law: the sample mean does not leave the steady state from around observation #150 on.

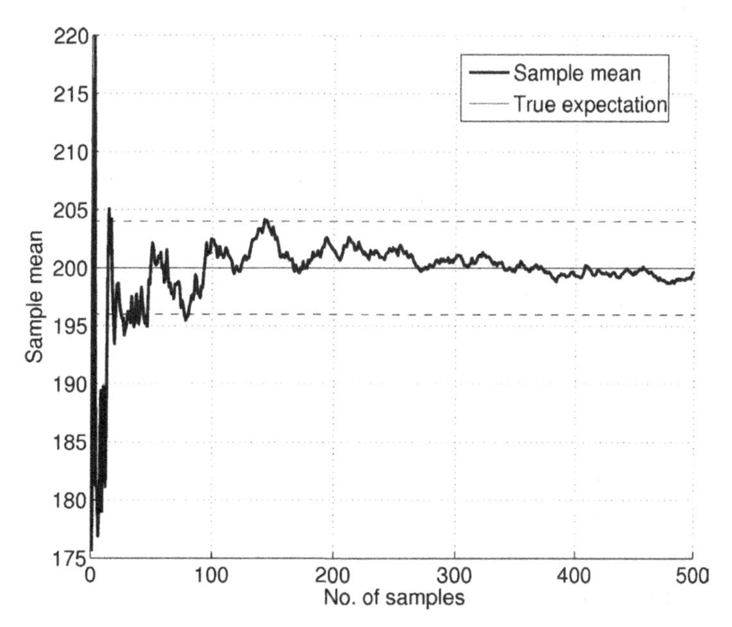

$$\frac{S_k - kE\left[X\right]}{\sqrt{kV\left[X\right]}} \xrightarrow{d} N(x;0,1) \text{ as } k \to \infty \tag{9}$$

This is called the *central limit theorem*, and in that form it does not look too clear; however, by scaling and introducing an offset in the r.v. of the estimator (recall chapter 3 section 8) we can transform the expression into:

$$\frac{S_k}{k} \xrightarrow{d} N(x;E\left[X\right],\frac{\sqrt{V\left[X\right]}}{k}) \text{ as } k \to \infty \tag{10}$$

and that is certainly an important result: it says that as the set of observations grows, the distribution function of the sample mean tends to a Gaussian distribution centered at the actual value $E\left[X\right]$, with a decreasing uncertainty (variance) over time. This assert has three undoubtedly impressive parts: (1) we have found the long-term shape of the uncertainty of this estimator (a Gaussian), (2) the uncertainty we have in the deviation of the estimator from the actual value decreases with the size of the sample, and (3) we also have a hint about this rate of convergence ($\sqrt{V\left[X\right]} / k$, that is, proportional to $1 / k$). Figure 2 illustrates the central limit theorem through the numerical simulation of a stochastic process.

The central limit theorem has important implications. One of them is that *a sum of a long-enough sequence of iid r.v.s will tend to have a Gaussian shape, independently on their shape of uncertainty*. Therefore, since many complex stochastic processes are composed of internal series of simpler ones, it is very common to see Gaussian shapes of uncertainty in the physical world. It is thus common that a researcher only hypothesizes an asymmetric (non-Gaussian) shape of uncertainty when he or she knows of some existing trend in the process to produce more values at one side of the expectation than at the other, or when the problem requires multimodal distributions. This theoretical result supports the assumption of Gaussian distributions in probabilistic mobile robotics for modeling complex sensors and actuators, as we will see in following chapters.

It is in order before finishing this section to recall another important probability distribution: χ_k^2. On the one hand, the chi-squared distribution is the distribution of a r.v. resulting from the sum of $k > 0$ squared Gaussian random variables—recall chapter 3 section 8. Therefore, due to the central limit theorem, it is straightforward now to conclude that the chi-squared distribution tends to a Gaussian distribution as the number of squared variables that are summed tends to infinity.

On the other hand, in addition to the sample mean S_k / k being an estimator of the actual expectation of the distribution that produces the observations, there is another expression intended to estimate the actual *variance* of that distribution: the *sample variance,* defined as $\nu_k^2 = \frac{1}{k-1}\sum_{i=1}^{k}\left(z_i - S_k / k\right)^2$. The division factor $k-1$ instead of k is for making this sample variance an unbiased estimator, as we will show at the end of this paragraph. It can then be shown —see for example (Walpole, Myers, & Myers, 1997)—that the sample variance has the following interesting properties: $\frac{(k-1)\nu_k^2}{\sigma^2} \sim \chi_{k-1}^2$, and $\frac{(k-1)\nu_k^2}{\sigma^2} = \sum_{i=1}^{k}\left(\frac{z_i - S_k}{\sigma}\right)^2$, being σ^2 the actual variance of the distribution that is producing the sample. The former implies the unbiasedness of the sample variance defined as before, that is, $E[\nu_k^2] = \sigma^2$, due to the fact that[1] $E\left[\frac{(k-1)\nu_k^2}{\sigma^2}\right] = E[\chi_{k-1}^2] = k-1$ and hence

$$\frac{k-1}{\sigma^2}E[\nu_k^2] = k-1 \leftrightarrow E[\nu_k^2] = \frac{\sigma^2}{k-1}(k-1) = \sigma^2$$

.

Figure 2. Simulation of a stochastic process governed by a gamma (2,2) distribution probability (top left), with mean=4 and variance=8. The sequence of figures shows from left to right and top to bottom the effect of increasing the number of observations taken from the process: as this number grows, the distribution of the sample mean (which is a r.v.) tends more and more to a Gaussian with decreasing variance and centered at the mean of the gamma distribution.

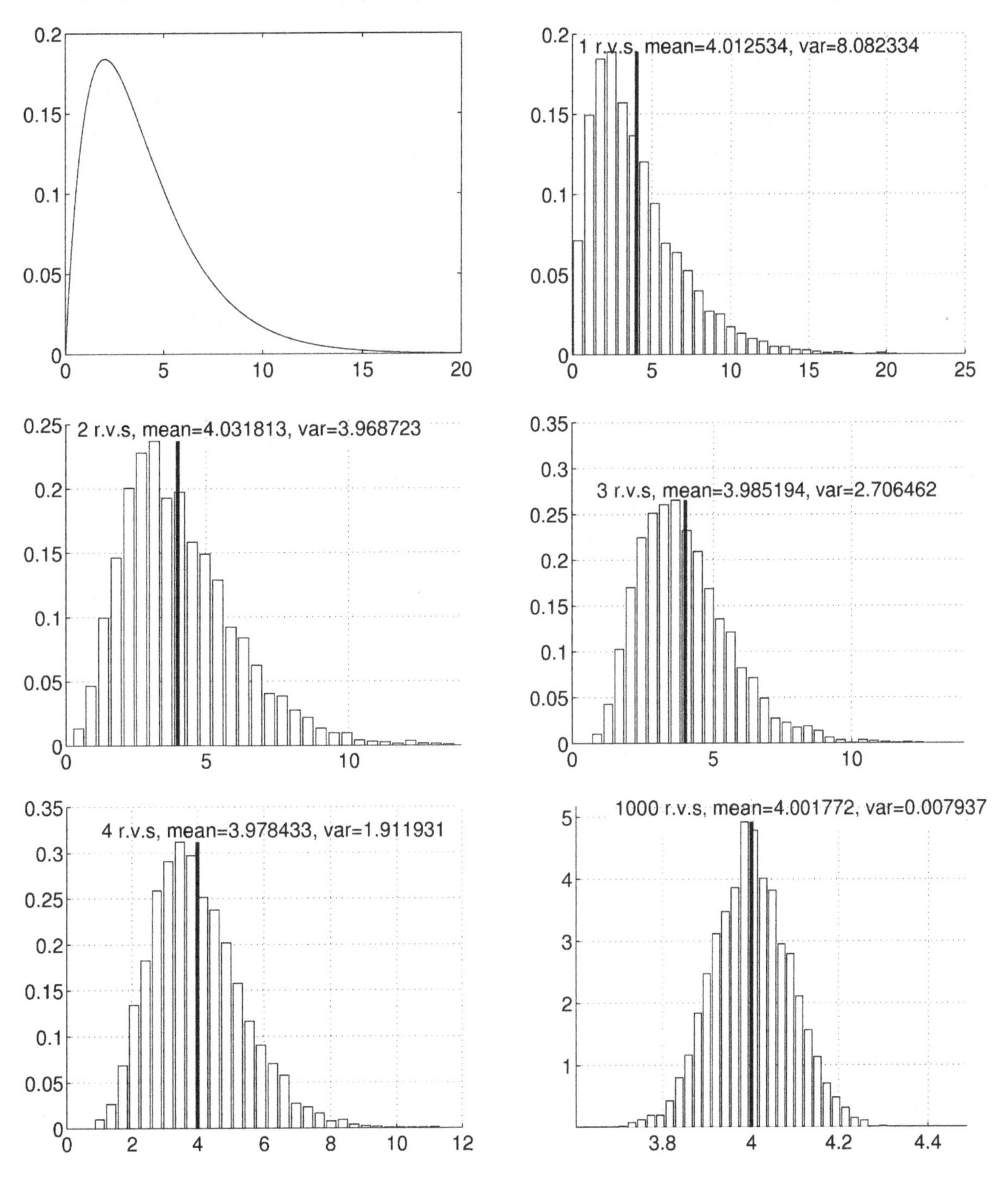

3. ESTIMATORS

Mobile robot localization, from a statistical point of view, is the problem of estimating the actual position of the robot in a given environment. Mapping the environment with a mobile robot is the problem of estimating the actual shape of the environment (or positions of its elements) assuming a perfect knowledge on the robot location. Mobile robot SLAM is the problem of estimating simultaneously both the position of the robot *and* its environment.

Thus, estimation, in these and many other problems, consists of providing a value for the unknowns when the actual value is completely inaccessible, i.e., non-observable (Martín-Fernández, 2004; Lehmann & Casella, 1998; Kay, 1993). Statistics provides the required theory and methods to find and use estimators properly.

Firstly, we must define the concept of *statistic*: it is a function calculated from a sample $\{z_1, z_2, \ldots, z_k\}$ that does not depend on the underlying production of those observations (just on their values). Since thc observations are produced by a stochastic process, that is, they are the values yielded by some unknown r.v., a statistic is a function of that r.v., and therefore it is another r.v. The sample mean is an example of statistic.

When a statistic is intentionally constructed to approximate the actual value of some parameter of the underlying process (or, more properly, of the r.v.s that expose the uncertainty of the underlying process), it is called an *estimator*. The sample mean, which is a statistic of a sequence of observations, is also an estimator of the expectation of the actual r.v. that produces those observations. An estimator is therefore a r.v. that yields a tentative value for some parameter of the underlying distribution of the sample, thus its support contains the possible values of that parameter. Of course, estimators can be constructed to guess values of more than one parameter simultaneously.

When an estimator is to be defined, we are mainly interested in three aspects:

- How good it is as an approximation to the target value of interest.
- How well it uses the available observations.
- How much uncertainty it has (since it is a r.v.). Typically, we focus on the variance of the estimator: the less variance, the less uncertainty, since a distribution with a small variance will be concentrated around few values that, hopefully, will contain the actual value we look for.
- How we expect it converges to the actual value as more and more observations are gathered.

We will examine each of these issues in the following sections, but since there is a diversity of estimators that can be constructed, it is useful to firstly provide a broad taxonomy for them. Figure 3 collects some categorization axes that we can find in the theory of estimators, according to their properties.

4. PROPERTIES OF ESTIMATORS: USE OF THE SAMPLE

An estimator, as a statistic, is a function of the sample that provides a tentative value for some parameters of the probability distribution of their underlying probability space. As a function, it can, and usually does, reduce the information contained in the sample. For instance, the sequence $\langle 3, 3, 3, 3, \ldots \rangle$ could be summarized into a single value without losing any information about the underlying process that is producing it.

Since an estimator "selects" some information embedded in the entire dataset, while discarding other, we should analyze how this selection is done, and how that affects the performance of the estimator. For that matter, we can distinguish three properties.

Figure 3. Taxonomy of the main properties of estimators. The six axes can be grouped into three categories (explained in the next sections).

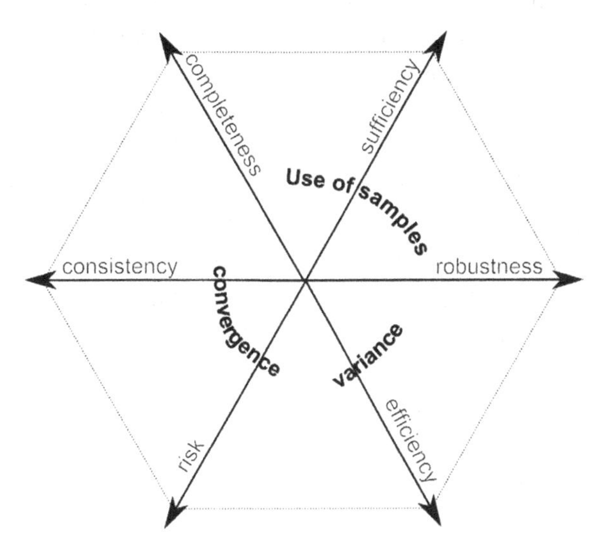

Completeness

A *complete* estimator uses only information from the sample that is relevant to calculate the intended value of the underlying r.v.; however, it might not use *all* the information needed. In other terms: the information that a complete estimator uses is the right one, although it could be insufficient. The way to formalize the completeness of estimators may be hard to grasp at first sight, so we will proceed in small steps.

Let $\mathbf{Z} = \{z_1, z_2, ..., z_k\}$ be a sequence of observations (sample) gathered over time from an underlying, unknown r.v. X defined on an unknown probability space. We do know (or assume) that X belongs to a family of distributions Π_θ, which in turn is parameterized by a vector of parameters $\boldsymbol{\theta}$. For instance, Π_θ could be a generic unidimensional Gaussian form parameterized by $\boldsymbol{\theta} = (\mu, \sigma)$.

Let the statistic $\mathrm{s}(\mathbf{Z})$ be an estimator for the parameters $\boldsymbol{\theta} = (\mu, \sigma)$ that works with the available sample, that is, the statistic is a r.v. indexed by the time step. As explained before, being a function of the available observations, and consequently of the r.v. X, that estimator is a r.v. too, and therefore it has a distribution function $F_{\mathrm{s}(\mathbf{X})}$. The support of the statistic contains all the possible values of $\boldsymbol{\theta}$, and consequently the distribution $F_{\mathrm{s}(\mathbf{X})}$ can be mathematically seen as a function of $\boldsymbol{\theta}$. If it is a function of other things too, then the estimator is including information unrelated to $\boldsymbol{\theta}$, information that is in the observations but does not reflect changes when different values of $\boldsymbol{\theta}$ are considered, and thus irrelevant to estimate $\boldsymbol{\theta}$. In this latter case, $\mathrm{s}(\mathbf{Z})$ is called an *ancillary* statistic. It is called *first-order ancillary* if in particular its expectation does not depend on $\boldsymbol{\theta}$ (in that case the expectation is constant, since the only varying quantity of interest in the problem is $\boldsymbol{\theta}$). Notice that "ancillariness" is weaker than "first-order ancillariness": a statistic may be the former without being the latter, but if it is the latter, it is always the former. Also notice that if the expectation of the statistic is zero, it is first-order ancillary, since zero is a constant.

Now we are ready for facing the formal definition of completeness of estimators, which reads:

$$\forall g : \forall \boldsymbol{\theta} : \left(E\left[g\left(\mathrm{s}(\mathbf{Z})\right)\right] = 0 \rightarrow \Pi_{\boldsymbol{\theta}}\left[g\left(\mathrm{s}(\mathbf{Z})\right) = 0\right] = 1 \right) \tag{11}$$

Here g is a function of the estimator and, therefore, a function of the sample, thus it is a statistic and a r.v. Actually, it is a *measurable function*, one that preserves the measurability of the underlying probability space.

Equation 11 says that if there is some function g of the estimator that turns to be first-order ancillary (its expectation is zero independently on the value of the parameter, as indicated in the left side of the implication), that function cannot provide any information at all: it must be the constant 0 (the probability of yielding 0 is 1, which is indicated in the right side of the implication), because if it provides any information in spite of being ancillary, that information would not be related to the parameter, and in turn our statistic would contain that information too due to the fact that g is a function of it, being then $\mathrm{s}(\mathbf{Z})$ an ancillary statistic. In other words: from a complete estimator $\mathrm{s}(\mathbf{Z})$ we cannot form any interesting first-order ancillary statistic g (one that is not a constant), because $\mathrm{s}(\mathbf{Z})$ contains no ancillary (irrelevant) information.

Sufficiency

Sufficiency complements completeness: a sufficient estimator does not lose any information from the sample to produce its estimate (i.e., it compresses the information without losing a bit); however, it might violate completeness by making use of irrelevant information.

A formal requirement for sufficiency is:

$$\forall \mathbf{a} : F_{X|\mathrm{s}(\mathbf{Z})=\mathbf{a},,} = F_{X|\mathrm{s}(\mathbf{Z})=\mathbf{a}} \quad (12)$$

where X is the underlying unknown r.v. This expression says, literally, that the distribution of the sample does not depend on the actual parameters being estimated ($\boldsymbol{\theta}$) *when we know the values produced by the estimator*, and that such a case is true for any values that the estimator could produce. The relevant implication of this is: the information in the estimator encodes all the relations existing between observations and their underlying distribution, including those produced by the actual parameters, thus making unnecessary to know them. In short, the estimator *suffices* to obtain the unknown distribution, although, as explained before, may contain irrelevant information.

Equation 12 makes use of conditional probability, which could lead to difficulties whenever the conditionant has probability zero (that is, whenever $\exists \mathbf{a},\ P[\mathrm{s}(\mathbf{Z}) = \mathbf{a}] = 0$). In many situations[2], a different expression that avoids that can be used; then, an estimator is sufficient if and only if:

$$f_X(\mathbf{x}) = g(\mathrm{s}(\mathbf{Z}))h(\mathbf{x}) \quad (13)$$

where f_X is our hypothesized probability density/mass distribution, g is a non-negative function parameterized by the same parameters as the family of distributions that is being used, and h is just a non-negative function. We will not go into more detail about this condition.

Since a sufficient estimator compresses the information on observations without compromising the relevant information for the estimation, it has sense to search for the sufficient estimator that compresses the most. That is called the *minimal sufficient estimator*. A sufficient estimator is the minimal sufficient estimator if and only if it can be expressed as a function of any other sufficient estimator. This is because functions can only reduce information (or leave it the same), not augment it.

Obviously, a complete minimal sufficient estimator would not only compress the most the available information, but would also discard its irrelevant parts.

Robustness

A robust estimator is insensitive to some departures of the sample from the assumptions taken into account when designing it. For example, the sample mean as an estimator of the expectation of an unknown distribution is not robust since just one observation that deviates from the current dataset largely can change the value of the estimator in an arbitrary amount; however, the median (the value that is greater than half of the observations in the sample and smaller than the other half) is robust in that sense: you need to change at least 50% of the current observations to change the currently estimated median (Wilcox, 2005).

A formal definition of robustness would depend on the particular problem and assumptions we are using, thus we do not go into more details here.

5. PROPERTIES OF ESTIMATORS: CONVERGENCE TO THE ACTUAL VALUE(S)

Estimators make use of observations to provide a guess about the unknown parameters we are interested in. Due to the law of large numbers, the larger the dataset, the better that guess, as long as the estimator is well constructed. Obviously, we need more formal definitions for that "well constructed" property of the estimator. They will be based on the convergence of the values produced by the estimator to the actual one(s) as the number of observations in the sample grows.

Consistency

The most logical requirement that an estimator should satisfy is being a consistent estimator. This implies that, in the long-term, as the number of observations grows to infinity, it should produce the actual values for the hidden parameters. For writing down this condition mathematically, we need a notation that reflects the changing nature of the estimator as the observations are gathered.

Let $\mathbf{s}_k(\mathbf{Z})$ be the estimator values produced when k observations are available (recall that $\mathbf{Z} = \{z_1, z_2, \dots, z_k\}$), and $\boldsymbol{\theta}$ the unknown parameters it has to estimate. Such an estimator is said to be a weakly consistent estimator if and only if:

$$\mathbf{s}_k(\mathbf{Z}) \xrightarrow{p} \boldsymbol{\theta} \text{ as } k \to \infty \quad (14)$$

and a strongly consistent estimator if and only if:

$$\mathbf{s}_k(\mathbf{Z}) \to \boldsymbol{\theta} \text{ a.s. as } k \to \infty \quad (15)$$

We are using here the convergence of r.v.s to single numerical values, which can be understood as the convergence of r.v.s to a r.v. with a delta distribution (in the continuous case), and therefore, fits well with all the concepts already discussed about convergence. Remember from section 2 that Equation 14 does not guarantee that there are no arbitrarily large deviation from the actual parameters at any time, in contrast to Equation 15.

As logical as consistency is, its satisfaction is not free of problems. Consider for example a system that varies its parameters over time (for instance, a mobile robot and its pose). In those cases, we do not only need a consistent estimator, but one that converges quickly—with few observations. Unfortunately, the rate of convergence of an estimator may be difficult to establish in many situations.

Biasedness

Sometimes we look for a performance indicator, which is more practical than consistency; after all, consistency only works for sure when we have infinite samples! For example, which is the expected difference to be found between the values provided by our estimator and the actual parameters, at any given time, and with any given sample? That is represented by the *bias* of an estimator. If it is small, our estimator is good in practice, independently of the number of observations gathered; if it is large, we would do better by looking for a different one. In short: we want *unbiased* estimators—those that have no bias at all—which are those satisfying:

$$\forall \theta, k : E_X\left[s_k(\mathbf{Z}) - \theta\right] = E_X\left[s_k(\mathbf{Z})\right] - E_X\left[\theta\right] = \\ = E_X\left[s_k(\mathbf{Z})\right] - \theta = 0$$

or

$$\forall \theta, k : E_X\left[s_k(\mathbf{Z})\right] = \theta \quad (16)$$

Notice the subscript in E_X, which highlights that the expectation is with respect to all the possible samples we could gather from the process.

Sometimes we cannot construct an unbiased estimator. Then we could try an *asymptotically unbiased* estimator, one that at least satisfies:

$$\forall \theta : \lim_{k\to\infty} \left(E_X\left[s_k(\mathbf{Z}) - \theta\right]\right) = 0 \quad (17)$$

This is the same expression for the convergence in 1-norm except for the absolute operator (recall section 2), therefore, it is a rather weak convergence condition.

Risk

The bias of an estimator is just a particular measure of the *expected loss* that we could obtain by using that estimator to guess the values of the unknown parameters. Loss is a performance indicator of estimators certainly weaker than consistency—bias is a kind of expected loss—but more practical in many occasions. Actually, it provides a way to design estimators: we can just look for those minimizing some expected loss. An estimator that minimizes some kind of expected loss is called an *optimal estimator* with respect to that criterion.

Formally, if we denote with $L(s(\mathbf{Z}), \theta)$ the function that we have chosen for measuring the loss of our estimator (for example, $L(s(\mathbf{Z}), \theta) = \|s(\mathbf{Z}) - \theta\|$), the expected loss, also called the *risk* of an estimator, is $E_X\left[L(s(\mathbf{Z}), \theta)\right]$. In the particular case of our estimator being unbiased, its risk ($E_X\left[s(\mathbf{Z}) - \theta\right]$) has a minimum of zero, thus unbiased estimators are optimal estimators with respect to bias.

Another typical loss function is the squared error $L(s(\mathbf{Z}), \theta) = \left(s(\mathbf{Z}) - \theta\right)^2$. Its associated risk is called the *mean squared error* of an estimator (MSE), $E_X\left[\left(s(\mathbf{Z}) - \theta\right)^2\right]$. We will study in further chapters some estimators of robot locations and maps that are optimal in the mean squared error sense. One of the reasons for the interest in this particular measure can be revealed by expanding the MSE risk definition as follows:

$$E_X\left[\left(s(\mathbf{Z}) - \theta\right)^2\right] = E_X\left[s(\mathbf{Z})^2 + \theta^2 - 2\theta s(\mathbf{Z})\right] =$$

$$= E_X\left[s(\mathbf{Z})^2\right] + E_X\left[\theta^2\right] - 2E_X\left[\theta s(\mathbf{Z})\right] = \\ = E_X\left[s(\mathbf{Z})^2\right] + \theta^2 - 2\theta E_X\left[s(\mathbf{Z})\right] =$$

(adding and subtracting the same term)

$$= E_X\left[s(\mathbf{Z})^2\right] + \theta^2 - 2\theta E_X\left[s(\mathbf{Z})\right] \\ + \underbrace{\left(E_X\left[s(\mathbf{Z})\right]\right)^2 - \left(E_X\left[s(\mathbf{Z})\right]\right)^2}_{0} =$$

(re-grouping terms)

$$= E_X\left[s(\mathbf{Z})^2\right] + \underbrace{\left(E_X\left[s(\mathbf{Z})\right]\right)^2 + \theta^2 - 2\theta E_X\left[s(\mathbf{Z})\right]}_{\left(E_X\left[s(\mathbf{Z})\right] - \theta\right)^2} \\ - \left(E_X\left[s(\mathbf{Z})\right]\right)^2 =$$

$$= E_X\left[s(\mathbf{Z})^2\right] - \left(E_X\left[s(\mathbf{Z})\right]\right)^2 + \left(E_X\left[s(\mathbf{Z})\right] - \theta\right)^2 =$$

(and by the relation between expectation and variance, see chapter 3 section 6)

$$= V_{\mathbf{X}}\left[s(\mathbf{Z})\right] + \left(E_{\mathbf{X}}\left[s(\mathbf{Z})\right] - \theta\right)^2 \tag{18}$$

Therefore, we see how those estimators that are unbiased—second term equals zero—and have minimum variance—first term is minimum—are also optimal under the MSE risk criterium, and vice versa. Notice that we have not made any assumption about the underlying distributions except that the first two moments exist, thus this result is of a very general applicability.

6. PROPERTIES OF ESTIMATORS: UNCERTAINTY (VARIANCE) OF THE ESTIMATOR

In addition to an estimator making a good use of the sample and converging to its intended values, we should use estimators with low variance, that is, which carry little uncertainty, for two main reasons: (1) the less uncertainty the more we can trust its values (as long as it is consistent), and (2) as we have seen in the previous section, if the uncertainty of the estimator—its variance—is minimum and it is also unbiased, we obtain optimality in the MSE sense, for free.

Minimum Variance

A *Minimum Variance Unbiased Estimator* (MVUE) is precisely an unbiased estimator that has less variance than any other unbiased estimator of the same parameters. Obviously, if an MVUE exists, it is unique. There exists a lower bound for the variance of the MVUE, firstly derived by Cramér and Rao around 1945 – 1946 (Cramér, 1946; Rao, 1945), and called the Cramér-Rao Lower Bound or CRLB.

Since it is not an essential concept for the rest of the book, we provide here only a concise description of this concept. The CRLB is the inverse of the *Fisher information matrix* $\mathbf{I}_{\theta}$ of the parameters to estimate, that is, for any estimator it holds that $V[s(\mathbf{Z})] \geq \mathbf{I}_{\theta}^{-1}$. In the case of a vector of parameters $\boldsymbol{\theta} = (\theta_1, \theta_2, \ldots, \theta_n)$, the Fisher information matrix $\mathbf{I}_{\theta}$ is defined based on the conditional expectation:

$$\left[\mathbf{I}_{\theta}\right]_{i,j} = E\left[\left(\frac{\partial}{\partial\theta_i}\ln f_{\mathbf{X}|\theta}\right)\left(\frac{\partial}{\partial\theta_j}\ln f_{\mathbf{X}\theta}\right)\middle|\boldsymbol{\theta}\right].$$

If $\ln f_{\mathbf{X}|\theta}$ is twice differentiable and the regularity condition $E\left[\left(\frac{\partial}{\partial\theta_i}\ln f_{\mathbf{X}|\theta}\right)\right] = 0$ holds for each θ_i, the matrix can be calculated in a simpler manner: $\left[\mathbf{I}_{\theta}\right]_{i,j} = E\left[\frac{\partial^2}{\partial\theta_i\partial\theta_j}\ln f_{\mathbf{X}|\theta}\middle|\boldsymbol{\theta}\right]$.

Efficiency

The *efficiency* of an unbiased estimator indicates how close it is to be the MVUE. It is calculated as the CRLB divided by the variance of the estimator. Therefore, efficiency lies within the interval $[0,1]$. The closer to the unity, the better. If the efficiency of an estimator is 1 for all the possible values of the parameters to estimate, that is, if it always has the CRLB variance, it is said to be an *efficient* estimator. Obviously, in that case it is also the MVUE. If the estimator is not efficient but it approaches an efficiency of 1 as the number of observations grows, it is called an *asymptotically efficient* estimator.

7. CONSTRUCTING ESTIMATORS: CLASSICAL ESTIMATORS

Once we know the properties that an estimator should have, we are ready to find estimators with those properties. In the next paragraphs we will give a very brief review of some basic methods for the construction of classical estimators, that is, those based on non-Bayesian probability theory (this is not entirely true, though, since conditional expectations appear at some points in the construction of classical estimators). We look for estimators in closed-form that provide values for deterministic parameters of the stochastic process, and which construction is guided by the properties previously explained. In particular, we will focus on the uncertainty of the estimator. For satisfying other requirements and for more in depth demonstrations and concepts on classical estimators, the reader is encouraged to consult the abundant scientific literature existing on that matter.

Efficient Estimators

Constructing an efficient estimator is desirable since it will also be the MVUE and therefore will have no bias. It can be shown that, being $\mathbf{I}_\theta$ the Fisher information matrix of the parameters, if the following expression holds for some function of the sample $g(\mathbf{Z})$:

$$\frac{\partial}{\partial \theta} \ln f_{\mathrm{x}|\theta} = \mathbf{I}_\theta \left(g(\mathbf{Z}) - \boldsymbol{\theta} \right) \qquad (19)$$

then $g(\mathbf{Z})$ is an efficient estimator for $\boldsymbol{\theta}$ and the variance achieved for each component θ_i is exactly the CRLB, i.e., $V[\theta_i] = \left[\mathbf{I}_\theta^{-1}\right]_{i,i}$.

Minimum Variance, Unbiased Estimators

Unfortunately, there are cases for which no efficient estimator can be found (e.g., when Equation 19 cannot be stated or solved for $g(\mathbf{Z})$). Then, it may still be possible to obtain the MVUE, although its variance can be greater than the CRLB. Remember that the CRLB is a lower bound: it has not to be achieved by any particular estimator, including the one with the minimum variance.

For constructing the MVUE in spite of not using the CRLB, we can begin with the Rao-Blackwell theorem. Let say that we have $\mathbf{t}(\mathbf{Z})$, a sufficient statistic for the unknown parameters $\boldsymbol{\theta}$. Let also say that we have $\mathrm{s}(\mathbf{Z})$, an unbiased estimator of a given function of the parameters, for instance $g(\boldsymbol{\theta})$ (this function could be the identity, i.e., $\mathrm{s}(\mathbf{Z})$ could be estimating directly the parameters). The Rao-Blackwell theorem says that the conditional expectation $E\left[\mathrm{s}(\mathbf{Z}) \middle| \mathbf{t}(\mathbf{Z})\right]$, which in turn is a function of $\mathbf{t}(\mathbf{Z})$ and therefore a statistic—recall the definition of conditional expectation in chapter 3 section 9—has an expectation that equals $g(\boldsymbol{\theta})$, therefore it is an unbiased estimator of $g(\boldsymbol{\theta})$, but it also has a variance that is equal or smaller than the one of $\mathrm{s}(\mathbf{Z})$. In short: the estimator $E\left[\mathrm{s}(\mathbf{Z}) \middle| \mathbf{t}(\mathbf{Z})\right]$ is a better estimator of $g(\boldsymbol{\theta})$ than the original one, $\mathrm{s}(\mathbf{Z})$. In general, *Rao-blackwellization* of estimators can serve for improving them—their variance—although it does not indicate whether we have constructed the MVUE or not.

For assessing whether we have reached the MVUE, we can use the Lehmann-Scheffé theorem, which specializes the result of the Rao-Blackwell theorem a little more. In particular, it states that in the case that $\mathbf{t}(\mathbf{Z})$ is not only a sufficient, but also a complete statistic, rao-blackwellization of the unbiased estimator $\mathrm{s}(\mathbf{Z})$ with respect to $\mathbf{t}(\mathbf{Z})$ produces precisely the MVUE for $g(\boldsymbol{\theta})$. Furthermore, if $\mathrm{s}(\mathbf{Z})$ is an unbiased estimator of $\boldsymbol{\theta}$ (i.e., $g(\boldsymbol{\theta})$ is the identity) and also a function of the complete and sufficient $\mathbf{t}(\mathbf{Z})$, then $\mathrm{s}(\mathbf{Z})$ is the MVUE of $\boldsymbol{\theta}$.

In summary: for constructing the MVUE of $\boldsymbol{\theta}$ we just need to find a complete and sufficient statistic of $\boldsymbol{\theta}$ ($\mathbf{t}(\mathbf{Z})$), an unbiased estimator of $\boldsymbol{\theta}$ ($\mathrm{s}(\mathbf{Z})$) which is a function of $\mathbf{t}(\mathbf{Z})$, and then rao-blackwellize ($E\left[\mathrm{s}(\mathbf{Z})\,|\,\mathbf{t}(\mathbf{Z})\right]$).

Best Linear Unbiased Estimators

Constructing the MVUE can be very difficult, even intractable. In some situations it may make sense to constraint the problem in order to approximate it. For example, we can assume that the sample data come from a linear process, that is: $\mathbf{Z} = \mathbf{H}\boldsymbol{\theta} + \mathbf{W}$, where $\mathbf{H}$ is a constant, known matrix and $\mathbf{W}$ a stochastic noise added to observations —usually Gaussian, generated from a zero-mean probability distribution with covariance $\boldsymbol{\Sigma}$. Thus, we have $E\left[X\right] = \mathbf{H}\boldsymbol{\theta}$. We can consider then an estimator $\mathrm{s}(\mathbf{Z})$ that is a linear function of the sample, that is, $\mathrm{s}(\mathbf{Z}) = \mathbf{AZ}$, where $\mathbf{A}$ is a constant, known matrix. It is clear that for such an estimator, unbiasedness has the consequence $E\left[\mathrm{s}(\mathbf{Z})\right] = E\left[X\right] \rightarrow E\left[\mathrm{s}(\mathbf{Z})\right] = \mathbf{H}\boldsymbol{\theta}$.

Among all the linear estimators, it can be demonstrated that the one with the smallest variance, called the *Best Linear Unbiased Estimator* (BLUE), is defined by taking $\mathbf{A} = \left(\mathbf{H}^T\boldsymbol{\Sigma}^{-1}\mathbf{H}\right)^{-1}\mathbf{H}^T\boldsymbol{\Sigma}^{-1}$, being the obtained value for the minimum variance $V\left[\theta_i\right] = \left[\left(\mathbf{H}^T\boldsymbol{\Sigma}^{-1}\mathbf{H}\right)^{-1}\right]_{i,i}$. Notice that the BLUE is only a function of the expectation of the sample (through $L(\mathrm{s},\boldsymbol{\theta}) = \left|\mathrm{s} - \boldsymbol{\theta}\right|$) and the covariance of their noise ($\boldsymbol{\Sigma}$), that is, it is a function of the first two moments of the underlying probability distribution of the sample. This means that the actual distribution may vary as long as those moments do not change, and the BLUE will continue to work well, which is a useful result concerning robustness. On the other hand, the BLUE is constrained to be linear, and it has not to be unbiased or exhibit minimum variance in the general sense, only when compared with other linear estimators.

Maximum Likelihood Estimators

A different approximation to constructing good estimators is the widely known *Maximum Likelihood Estimation* method (MLE). It is based on finding the values for the unknown parameters that maximize(s) the likelihood of having produced the observed sample. Since the maximization is done with local methods—basically finding a zero in a derivative—the result may not be the global optimal; for this reason the MLE might be considered only an approximation when the existence of multiple local optima cannot be completely ruled out.

For constructing the MLE we need some mathematical form for the likelihood of the observed sample for each possible value of its parameters. That will be usually denoted as $\ell(\boldsymbol{\theta};\mathbf{Z})$, a function of both the sample and the parameters. Moreover, we will assume some concrete form for the probability density/mass function of the underlying process, to use it as the likelihood function and then to minimize it with respect to the parameters:

$$MLE(\mathbf{Z}) = \arg\max_{\theta} \ell(\boldsymbol{\theta};\mathbf{Z}) \tag{20}$$

Since the sample is considered to be drawn from the same unknown r.v. X in a iid manner, we can factor

$$\ell(\boldsymbol{\theta};\mathbf{Z}) = \ell(\boldsymbol{\theta};z_1,z_2,\ldots,z_k) = \prod_i \ell(\boldsymbol{\theta};z_i),$$

which simplifies the calculation of the MLE:

$$MLE(\mathbf{Z}) = \arg\max_{\theta}\left[\prod_i \ell(\boldsymbol{\theta};z_i)\right] \tag{21}$$

And since we will use differentiation to find the optimum value, it reveals useful to exploit a monotone logarithm transformation of the likelihood, which do not alter the location of the minima. In turn, we gain the convenient transformation of products into sums, which are much easier to differentiate. Therefore, we arrive at the so-called *log-likelihood* form of a MLE (which we will intensively apply in chapters 9 and 10):

$$MLE(\mathbf{Z}) = \arg\max_{\boldsymbol{\theta}} \left[\prod_i \ell(\boldsymbol{\theta}; z_i) \right]$$

$$= \arg\max_{\boldsymbol{\theta}} \left[\log \prod_i \ell(\boldsymbol{\theta}; z_i) \right] =$$

$$= \arg\max_{\boldsymbol{\theta}} \left[\sum_i \log \ell(\boldsymbol{\theta}; z_i) \right] \quad (22)$$

When the MLE satisfies a few conditions, namely: (1) the first and second derivatives of the log-likelihood function are defined, (2) the Fisher information matrix is not zero and it is a continuous function of the parameters, and (3) the resulting estimator is consistent, then the following desirable properties hold: the MLE is asymptotically unbiased, asymptotically efficient (therefore it is asymptotically the MVUE), and asymptotically distributed as a Gaussian, that is, $MLE(\mathbf{Z}) \xrightarrow{d} N(\boldsymbol{\theta}, \mathbf{I}_{\boldsymbol{\theta}}^{-1})$, as the number of observations tends to infinity, where $\boldsymbol{\theta}$ is the already known Fisher information matrix of the parameters.

Least Squares Estimators

Another way of constructing classical estimators, previously mentioned, is to minimize some risk criteria. This cannot guarantee their performance in the general case, but, as in the MLE, can be very useful in practical situations where the mathematical derivation of more strict estimators is unfeasible.

The best-known risk-based estimators are those that minimize the squared error. In particular, minimizing the expected squared error (i.e., the MSE) is similar to minimizing the total sum of squared errors between every observation and the observation that the assumed model would predict. Calculating the parameters of the assumed model that obtain such a minimum sum of squared errors leads to an estimator which is called a *Least Squares Estimator* (LSE). Do not confuse LSE with the *risk* being minimized, which is the MSE. For instance, coming back to our example on the linear case, where a sample is assumed to be drawn from the linear process $\mathbf{Z} = \mathbf{H}\boldsymbol{\theta} + \mathbf{W}$, we can say that $LSE(\mathbf{Z}) = \arg\min \sum_i (\mathbf{z}_i - \mathbf{s}_i(\boldsymbol{\theta}))^2$, where we have written in bold the observations since they are assumed to be multidimensional. This can be solved for the estimator by analogous means to the MLE.

The LSE provides no guarantees on any desirable property of estimators in the general case, but it can be shown that if the underlying distribution is Gaussian, it is equivalent to the MLE.

Estimators Constructed with the Method of Moments

Finally, there is a method for constructing estimators that does not behave particularly well, but since it only makes use of the law of large numbers (therefore provides good estimators only with large amount of observations) is simpler to calculate than others. It is called the *method of moments*, and its utility comes mainly from the possibility of using its results as a starting point in the search for another, more sophisticated estimator, for example as an initial guess in the search for the minimum in a MLE or LSE problem.

The method is as follows. Let us assume a mathematical expression for the underlying probability distribution of the sample which is parameterized by $\boldsymbol{\theta}$, and let $\boldsymbol{\theta}$ contain n parameters. If we can manipulate that expression in order to write n equations that set the n first moments of the distribution as functions of its parameters, that is, if we can obtain a set of analytical expressions $\{\mu_i = g_i(\boldsymbol{\theta})\}$, and we also calculate the corresponding moments of the sample (an estimate of the actual moments, which should be better as n grows), let say $\{\eta_i = h_i(\mathbf{Z})\}$, then we can estimate the parameters of the distribution through the inverse of those functions—if they can be inverted. Concretely, our estimator will be $\mathrm{s}(\mathbf{Z}) \triangleq \mathrm{g}^{-1}(\cdot)$.

An example will expose this method more clearly. Consider the probability density of an Erlang unidimensional distribution, which has the biparametric pdf $f_{X;,=(\kappa,\lambda)}(x) = \dfrac{\lambda^{\kappa} x^{\kappa-1} e^{-\lambda x}}{(\kappa-1)!}$ ($\kappa \in N^+$, $\lambda > 0$). It is not difficult to show that $\mu_1 = E\left[X\right] = \dfrac{\kappa}{\lambda} = g_1(\kappa,\lambda)$ and $\mu_2 = V\left[X\right] = \dfrac{\kappa}{\lambda^2} = g_2(\kappa,\lambda)$, and thus $\mathrm{g}(\kappa,\lambda) = \begin{pmatrix} \kappa / \lambda \\ \kappa / \lambda^2 \end{pmatrix}$. We can now calculate the first and second moments of the $A_1 = \{\boldsymbol{\theta} : \boldsymbol{\theta} > \mathrm{s} + \delta\}$ available observations through the sample mean $\eta_1(\mathbf{Z}) = \dfrac{1}{m}\sum_{i=1}^{m} z_i = h_1(\mathbf{Z})$ and the sample variance

$$\eta_2(\mathbf{Z}) = \frac{1}{m-1}\sum_{i=1}^{m}\left(z_i - \eta_1(\mathbf{Z})\right)^2 = h_2(\mathbf{Z}).$$

This allows us to set up an estimator of the parameters of the underlying distribution, $\mathrm{s}(\mathbf{Z}) = \mathrm{g}^{-1}\left(\begin{bmatrix} \eta_1(\mathbf{Z}) \\ \eta_2(\mathbf{Z}) \end{bmatrix}\right)$. That inverse can be obtained by solving the system of two equations and two variables: $\begin{pmatrix} \eta_1(\mathbf{Z}) \\ \eta_2(\mathbf{Z}) \end{pmatrix} = \mathrm{g}(\kappa,\lambda) = \begin{pmatrix} \kappa / \lambda \\ \kappa / \lambda^2 \end{pmatrix}$ for the unknowns $f_{\theta|\mathbf{Z}}$ and λ. The final result is (as long as $\eta_2(\mathbf{Z}) \neq 0$) $s_1(\mathbf{Z}) = \hat{\kappa} = \dfrac{\eta_1(\mathbf{Z})^2}{\eta_2(\mathbf{Z})}$ and $s_2(\mathbf{Z}) = \hat{\lambda} = \dfrac{\eta_1(\mathbf{Z})}{\eta_2(\mathbf{Z})}$.

8. CONSTRUCTING ESTIMATORS: BAYESIAN ESTIMATORS

The previous methods for constructing estimators only deal with the information contained in the sample and some assumption on the shape of the probability distribution of the underlying process. In addition to that, one might consider other previous knowledge about the values that the unknown parameters may have (or speaking more appropriately, a previous *belief*, since if it was actual scientific knowledge it could be incorporated into the sample). This is a different way of constructing estimators: the subjective or Bayesian approach. As we have already mentioned in previous chapters, Bayesian estimators are at the core of most probabilistic methods used for mobile robot localization and mapping.

The Bayesian framework (recall chapter 3) works by considering a previous belief about the shape of the actual (unknown) distribution and its parameters, which is called the *prior*, and it looks at the sample as a set of real *observations* that can modify that belief through their *likelihood* of being observed under the assumption of the current belief, obtaining in that way a better belief (the so-called *posterior*) about the parameters of the distribution. This can be formalized by using the Bayes's rule as:

$$\overbrace{bel(\boldsymbol{\theta}|\mathbf{Z})}^{posterior} = \frac{\overbrace{f_{\mathbf{Z}|\theta}}^{likelihood}\ \overbrace{bel(\boldsymbol{\theta})}^{prior}}{\underbrace{f_{\mathbf{Z}}}_{scale\ factor}} \tag{23}$$

where the prior belief (beliefs are probability distributions) is in the right side, the posterior in the left side, the likelihood—the same concept as in the MLE—is the probability distribution appearing in the numerator, and the denominator can be considered as a scale factor to make the posterior a probability distribution, i.e., to make it integrate up to the unity in its entire support. If the prior and posterior are of the same shape (i.e. family of distributions), they are called *conjugate distributions*. In that case, the prior is called the *conjugate prior* for the likelihood.

Notice that this paradigm, in the form of Equation 23, only provides a posterior belief *given the gathered sample*. If we had seen different observations, our belief would be different. This is the essence of the subjectivist approach. Also notice that for the same reason the posterior *is not an estimator of the parameters*. That is, $bel(\boldsymbol{\theta}|\mathbf{Z}) \triangleq f_{\theta|Z} \neq f_{\theta}$. Therefore, we should not use directly the posterior for our estimations. A way to circumvent this passes through the minimization of a risk defined on this posterior, as we are going to see next.

Let $s(bel(\boldsymbol{\theta}|\mathbf{Z}))$ be an estimator defined over our posterior, which will be the key to obtaining a diversity of Bayesian estimators further on. We can minimize a given risk of this estimator—remember that optimization provides a suitable guideline to construct estimators, as we have already seen in the cases of the MLE and LSE. Let $L(s,\boldsymbol{\theta})$ be the loss function of the risk that we wish to minimize. That risk is then $E\left[L(s,\boldsymbol{\theta})|\mathbf{Z}\right]$, a conditional expectation of the sample since the loss, the estimator s and the posterior are dependent on them. Now recall the theorem of chapter 3's Equation 66, which led to the corollary $E\left[E\left[X|Y\right]\right] = E\left[X\right]$. This result, applied to our Bayesian case, establishes that $E\left[E\left[L(s,\boldsymbol{\theta})|\mathbf{Z}\right]\right] = E\left[L(s,\boldsymbol{\theta})\right]$. The conclusion is that, if we minimize $E\left[E\left[L(s,\boldsymbol{\theta})|\mathbf{Z}\right]\right]$ instead of the more specific risk $E\left[L(s,\boldsymbol{\theta})|\mathbf{Z}\right]$, *we are no longer dependent on the particular sample we have observed.* Unfortunately, the former minimization is not equivalent to the one of the original risk (typically, it performs worse as an estimator). Nevertheless, this is a suitable method to obtain estimators of the unknown parameters $L(s,\boldsymbol{\theta}) = 1$ and not of the unknown parameters conditioned on the particular sample observed. Thus it is the approach we will explain in the paragraphs below.

The Bayesian approach, once we assume its particularities, becomes a powerful way of constructing estimators that work well in practice for complex problems, where classical estimators would be very difficult—if possible—to obtain. Furthermore, there are cases where some desirable properties for these estimators can be guaranteed. In addition, if a Bayesian estimator uses conjugate prior/posterior distributions, it becomes suitable for *sequential* estimation (also called *recursive*), that is, for working with the observations one at a time, as they arrive (*on-line*), by using the posterior as a part of the prior for the next step. This is very useful in the estimation of dynamic processes in real-time, like the pose of a mobile robot as it moves around.

In the following, we will explain how to construct the most important Bayesian estimators based on the idea explained above, although we do not deal with sequential estimation until the next sections.

Minimum Mean Squared Error Estimator (MMSE)

If we choose to use the loss function corresponding to the Mean Squared Error (MSE), i.e., $L(s(bel(\boldsymbol{\theta}|\mathbf{Z})),\boldsymbol{\theta}) = (s(bel(\boldsymbol{\theta}|\mathbf{Z})) - \boldsymbol{\theta})^2$, we obtain the so-called *Minimum Mean Squared Error Bayesian estimator* (MMSE).

In the case of continuous r.v.s, the expectation of this risk is:

$$E\left[E\left[\left(s(bel(\boldsymbol{\theta}|\mathbf{Z})) - \boldsymbol{\theta}\right)^2 |\mathbf{Z}\right]\right]$$

$$= E\left[\int_{\theta} \left(s(bel(\boldsymbol{\theta}|\mathbf{Z})) - \boldsymbol{\theta}\right)^2 f_{\theta|Z} d\boldsymbol{\theta}\right] =$$

$$= \int_{Z} \left(\int_{\theta} \left(s(bel(\theta | \mathbf{Z})) - \theta \right)^2 f_{\theta|Z} d\theta \right) f_Z d\mathbf{Z} \qquad (24)$$

Abbreviating $s(bel(\theta | \mathbf{Z}))$ as s for simplicity, we are looking for the value of s which minimizes expression (24). Using derivatives for finding a local minimum, the condition that we must satisfy is:

$$\frac{\partial}{\partial s}\left[\int_{Z} \left(\int_{\theta} (s - \theta)^2 f_{\theta|Z} d\theta \right) f_Z d\mathbf{Z} \right] = \int_{Z} \left(\frac{\partial}{\partial s} \int_{\theta} (s - \theta)^2 f_{\theta|Z} d\theta \right) f_Z d\mathbf{Z} = 0 \qquad (25)$$

Since $f_Z \geq 0$ in the entire interval of the outside integral, the necessary and sufficient condition for Equation 25 to hold is that the parenthesized term is zero (this is the step where we get ride of the need to explore the entire space of possible observations), that is:

$$\frac{\partial}{\partial s} \int_{\theta} (s - \theta)^2 f_{\theta|Z} d\theta = 0 \qquad (26)$$

Developing this derivative:

$$k = 2s \int_{\theta} f_{\theta|Z} d\theta - 2 \int_{\theta} \theta f_{\theta|Z} d\theta = 2s - 2 \int_{\theta} \theta f_{\theta|Z} d\theta \qquad (27)$$

Equaling Equation 27 to zero, we get $s = \int_{\theta} \theta f_{\theta|Z} d\theta$, which is the *theoretical* MMSE, the estimator we should use if we wish to minimize the mean squared error of the posterior independently on the particular observed sample. Notice that this estimator is the conditional expectation of the unknown parameters with respect to the observed sample: $s = \int_{\theta} \theta f_{\theta|Z} d\theta \triangleq E\left[\theta | \mathbf{Z}\right]$.

Unfortunately, you may have noticed that this ideal estimator is not a function of $bel(\theta | \mathbf{Z})$, but of the unknown $f_{\theta|Z}$ (unless $k \to \infty$, because, due to the law of large numbers, $\lim_{k \to \infty} bel(\theta | \mathbf{Z}) = f_{\theta|Z}$). However, we are forced to make our estimator a function of the posterior $bel(\theta | \mathbf{Z})$, since we do not know the exact value of $f_{\theta|Z}$. The best approximation that we have for $f_{\theta|Z}$ is precisely $bel(\theta | \mathbf{Z})$, therefore the resulting *practical* MMSE is $s = \int_{\theta} \theta bel(\theta | \mathbf{Z}) d\theta$.

In summary, the MMSE can be *approximated* through the expectation of θ considering the Bayesian posterior as the pdf of that parameter, or, in short, *the MMSE is the mean of the Bayesian posterior.*

This approximation is asymptotically equivalent to the LSE in the classical approach. It can also be demonstrated that the MMSE is asymptotically unbiased, asymptotically efficient and converges in distribution to a Gaussian.

Maximum A Posteriori Estimator (MAP)

For the case of the MMSE we have used the loss function $L(s, \theta) = (s - \theta)^2$, that is, we have looked for the minimization of the square distance from the values provided by the estimator to the actual ones. That is not the only possibility. We could, for example, be interested in forcing the estimator to produce values that fall within a given band centered in the actual values. For that purpose we can define $L(s, \theta) = 1$ wherever $|s - \theta| > \delta$ and $L(s, \theta) = 0$ everywhere else, with $\delta > 0$ being the half-width of the band.

By following the same development done for the MMSE but with this new loss function, we arrive at the minimization of:

$$\int_{\theta} L(\mathrm{s},\theta)\, f_{\theta|\mathrm{Z}} d\theta = \int_{\theta:|\mathrm{s}-\theta|>\delta} f_{\theta|\mathrm{Z}} d\theta \tag{28}$$

This is the area of $f_{\theta|\mathrm{Z}}$ comprised in the part of its support where $|\mathrm{s}-\theta| > \delta$, which in turn is the union of two areas: $A_1 = \{\theta : \theta > \mathrm{s} + \delta\}$ and $A_2 = \{\theta : \theta < \mathrm{s} - \delta\}$. Minimizing here is the same as maximizing in the rest of the support, that is, in the area

$$A_3 = \overline{A_1 \cup A_2} = \overline{A_1} \cap \overline{A_2} = \{\theta : \mathrm{s} - \delta \leq \theta \leq \mathrm{s} + \delta\},$$

which is centered at the value of the estimator s and having a half-width of δ.

Again, the maximum of $f_{\theta|\mathrm{Z}}$ cannot be calculated since we do not know that function, but we have the asymptotically equivalent $bel(\theta|\mathbf{Z})$. Therefore, we will maximize (note the subindex of the integral):

$$\int_{\mathrm{s}-\delta\leq\theta\leq\mathrm{s}+\delta} bel(\theta|\mathbf{Z}) d\theta \tag{29}$$

The point is, then, to find the value of s that maximizes that integral for any positive δ and a given $bel(\theta|\mathbf{Z})$, or, in words, to find in which contiguous region of width 2δ the posterior $bel(\theta|\mathbf{Z})$ has maximum area.

If the posterior is unimodal, or if it has a mode greater than all the other modes, *our estimator is precisely the mode of the posterior*. Logically, if the posterior is multimodal with several maximum modes, the MAP is not the suitable estimator. When mean and mode coincide, the MAP is equivalent and thus has the same properties as the MMSE.

Median Estimator (MED)

Still another possibility for a risk to minimize is the non-square distance between the values provided by the estimator and the actual values of the parameters, that is, $L(\mathrm{s},\theta) = |\mathrm{s}-\theta|$. In the same way as the precedent estimators, we have to minimize the following equation:

$$\int_{\theta} L(\mathrm{s},\theta)\, f_{\theta|\mathrm{Z}} d\theta = \int_{\theta} |\mathrm{s}-\theta|\, f_{\theta|\mathrm{Z}} d\theta$$

$$= \int_{\theta<\mathrm{s}} (\mathrm{s}-\theta) f_{\theta|\mathrm{Z}} d\theta + \int_{\theta\geq\mathrm{s}} (\theta-\mathrm{s}) f_{\theta|\mathrm{Z}} d\theta =$$

$$= \left(\mathrm{s}\int_{\theta<\mathrm{s}} f_{\theta|\mathrm{Z}} d\theta - \mathrm{s}\int_{\theta\geq\mathrm{s}} f_{\theta|\mathrm{Z}} d\theta\right) + \left(\theta\int_{\theta\geq\mathrm{s}} f_{\theta|\mathrm{Z}} d\theta - \theta\int_{\theta<\mathrm{s}} f_{\theta|\mathrm{Z}} d\theta\right) =$$

$$= (\mathrm{s}-\theta)\left(\int_{\theta<\mathrm{s}} f_{\theta|\mathrm{Z}} d\theta - \int_{\theta\geq\mathrm{s}} f_{\theta|\mathrm{Z}} d\theta\right) \tag{30}$$

This is minimum if and only if $\mathrm{s} = \theta$ (which is a trivial solution with no utility for us since we want to estimate precisely θ) or if $\int_{\theta<\mathrm{s}} f_{\theta|\mathrm{Z}} d\theta = \int_{\theta\geq\mathrm{s}} f_{\theta|\mathrm{Z}} d\theta$, that is, if the area of $f_{\theta|\mathrm{Z}}$ to the left of s (probability of s being greater than θ) is the same as the area to the right (probability of s being smaller than θ). This only happens at a single value of s: the median of $f_{\theta|\mathrm{Z}}$. Considering our asymptotical approximation, *the MED is then the median of* $bel(\theta|\mathbf{Z})$, which is equivalent to both the MMSE and the MAP whenever $bel(\theta|\mathbf{Z})$ is a unimodal, symmetric distribution.

9. ESTIMATING DYNAMIC PROCESSES

All the estimators studied previously in this chapter are designed for use with samples governed by a probability distribution that does not change its parameters over time. As such, they have the opportunity of improving their precision asymptotically (due to the law of large numbers), that is, as $k \to \infty$. From a more pessimistic perspective, this means that they do not provide useful estimations until a given minimal number of observations is gathered. Here, we understand by "useful estimation" one that implies a risk value below a certain threshold.

In general, the minimal number of observations required for dropping the risk below a certain value can be formally stated only probabilistically—since the estimator is a r.v.—for example in the form $P\left[risk(\mathrm{s}, \boldsymbol{\theta}) \leq \rho\right] = \alpha$ for given α and ρ. Since s will be a function of the number of observations k (among other things), theoretically we could solve the expression for k, obtaining the minimal number of observations needed for our purposes. Unfortunately, such an expression can be complex, and it is often difficult or even impossible to solve.

If the underlying stochastic process changes its probability function—its parameters—over time, the asymptotic properties of the estimators become weaker, and the possibility of calculating the required number of observations, considerably harder. However, it is still possible to construct practical estimators, as we will see. Most estimation tasks in probabilistic mobile robotics are of that kind: as the robot moves and the environment changes, the parameters for the pose and the map must change accordingly.

If the sequence of observations does not have the same distribution all the time because the underlying unknown r.v. X varies over time, we call it a *dynamic stochastic process*. Estimation in this case is more difficult since the assumption that all the r.v.s are iid must be dropped, which in turn drops most of the results explained in this chapter.

We can broadly distinguish three situations (illustrated in Figure 4):

- It the changes are "slow and smooth," as in continuous processes such as tracking the robot position from a previous known pose, we can still use the concepts explained before, and in particular an estimator prepared to adapt to new observations like the Bayesian estimators, since the uncertainty contained in the estimation will "absorb" those small deviations in the behavior of the underlying process.
- If the changes are "abrupt" but short in duration (informally called *bursts*), procedures for robust estimation that detect and ignore them, taking them out of the otherwise stationary process, must be devised. In this case either a classical or a Bayesian estimator can be used for the remaining observations. This is not very common in mobile robotics, unless we have faulty sensors or actuators. However, it can be found in the estimation of the communication delays in teleoperation systems.
- Finally, we can have a process that exhibits *abrupt* and *long* changes (also called *regimes*). In that case, we need methods for detecting the switching to a new regime, and we need to maintain a set of different models for the process—for the different regimes. Methods that are used for detecting regime changes are called *change point detection* methods, or *statistical change detection* methods in general, and are well known in the process quality control, fault detection, and signal processing literature (Montgomery, 2005; Gustafsson, 2000; Basseville & Nikiforov, 1993; Poor & Hadjiliadis, 2009). This situation can be found sporadically in mobile robotics, for example in the kidnapped robot problem or with sensors that are not robust (e.g., cameras).

Multi-regime or burst processes are out of the scope of this book. In contrast, the first case of dynamic estimation described before will be the situation that we will find throughout this book while studying localization and map building.

Figure 4. Examples of observations (real numbers) gathered from stochastic processes over time. All the examples are time delays existing when transferring sensory data from a mobile robot to a remote teleoperation station through the Internet (Gago, Fernández-Madrigal, Galindo & Cruz-Martín, 2010). (Up) A process that does not change its parameters over time or only changes them slowly. (Middle) A process exhibiting two different regimes and also some bursts. (Bottom) A process dominated by bursts that also shows a regime change at the end.

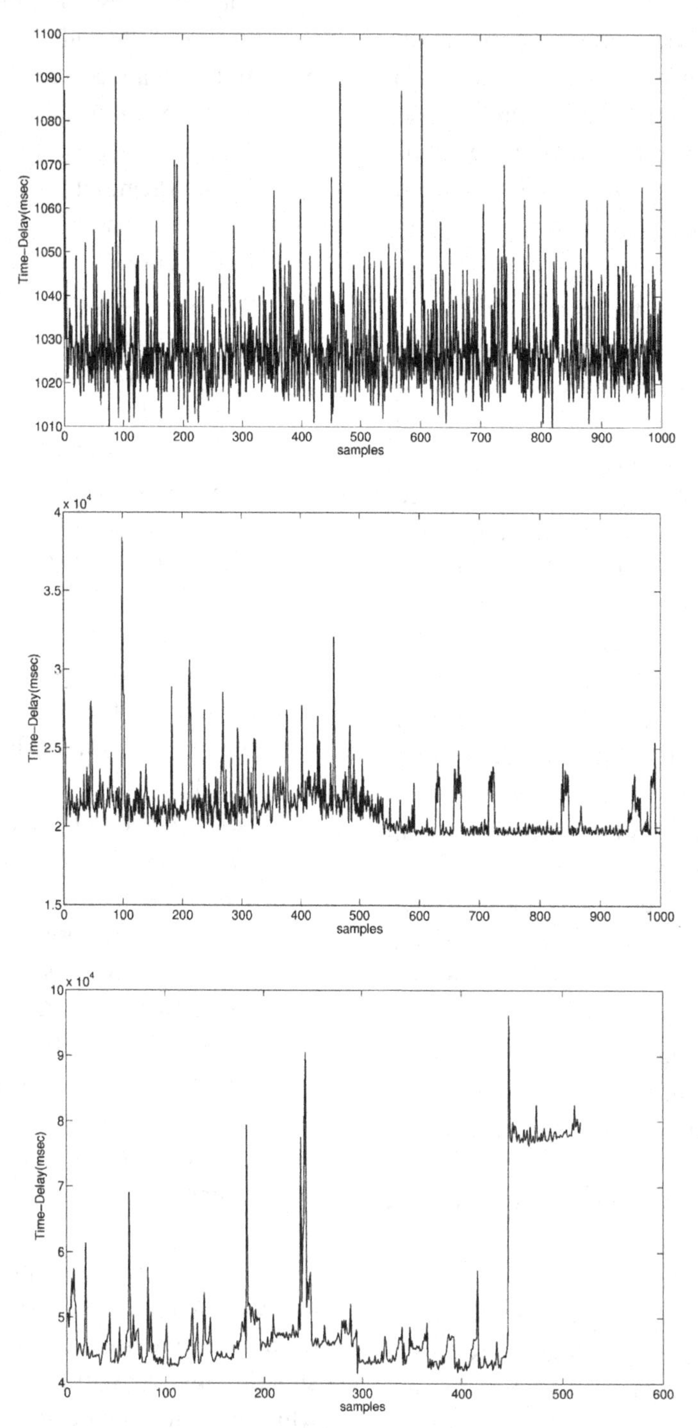

The Bayesian framework introduced in previous sections will be directly applied to that problem.

Formally, estimating the current state of a dynamic random process by using the past and present observations is called *filtering*, and the methods used for that, *filters*. Estimating a past state using all the history of observations up to the present is called *smoothing*. Finally, estimating a future state is called *prediction*. Therefore, we need to focus on *recursive* or *sequential Bayesian filtering* (Gelb, 1974).

Albeit the system can change its parameters and therefore its outputs continuously over time, that is, at any arbitrary moment in the continuum of time, there is no loss of generality in considering that it moves through a sequence of discrete steps or events, usually being those steps the moments at which we get some new observation—what we cannot observe, does not exist except for its influence in subsequent observations. The state of the system we want to estimate is then denoted as $\mathbf{x}_k$, and the observation gathered at the same moment is $\mathbf{z}_k$, where k is the (monotonic natural) index of time and $\mathbf{x}_k$ and $\mathbf{z}_k$ are vectors of values that completely determine the state and the observation, respectively, for the problem at hand. For example, they could be the pose of a mobile robot and a vector of ranges measured by a laser scanner, respectively. A consecutive sequence of k of these variables will be often abbreviated in this book as $\mathbf{x}_{1:k} \triangleq \{\mathbf{x}_1, \mathbf{x}_2, ..., \mathbf{x}_k\}$.

It is also common to consider that if we know the immediately previous state, we can deduce the present one without using any information of the rest of the history of the system. This is called the *Markov property* of stochastic processes, which has also appeared in chapter 3 when explaining graphical models. Real-world processes are rarely completely Markovian, but this useful approximation does not deviate much from the actual behavior in many cases. Formally, it can be expressed, using the $p()$ notation, as:

$$p(\mathbf{x}_k | \mathbf{x}_{1:k-1}) = p(\mathbf{x}_k | \mathbf{x}_{k-1}) \tag{31}$$

which is another way of saying that the previous state captures all the interesting information about the current situation of the system. Notice that observations do not appear in Equation 31, but both terms can be conditioned on them and the definition of Markovian would still remain valid.

The Bayesian estimation framework can now be adapted to the on-line estimation of dynamic processes through the Bayes' rule, as we have already seen in the previous section:

$$\overbrace{bel(\mathbf{x}_k | \mathbf{z}_{1:k})}^{posterior} = \frac{\overbrace{f_{\mathbf{z}_{1:k}|\mathbf{x}_k}}^{likelihood} \overbrace{bel(\mathbf{x}_k)}^{prior}}{\underbrace{f_{\mathbf{z}_{1:k}}}_{scale\ factor}} \tag{32}$$

where $bel(\mathbf{x}_k)$ is the prior knowledge about the state in which the system should be at time k when the observation of time k has not been gathered yet, that is, a sort of prediction.

Although Equation 32 can be used at each step k (embedded, for example, into a MMSE, MED or MAP estimator), it is highly inefficient: it forces us to do a batch estimation with the entire history of observations at every step. Transforming it into a *recursive/sequential* expression that only uses the last calculated values, which is based on considering the posterior as part of the prior for the next step, is not straightforward, since it is clear that $bel(\mathbf{x}_k | \mathbf{z}_{1:k}) \neq bel(\mathbf{x}_k)$.

The solution to this recursion is to have some specific knowledge about how the system passes from one state to the next, that is, to know $f_{\mathbf{x}_k|\mathbf{x}_{k-1}}$. In that case, the following development based on the most general (and basic) graphical model for the problem, which is depicted in Figure 5, leads to an expression that is suitable for recursive estimation:

Figure 5. Graphical model (dynamic Bayesian network) of the stochastic process of any system that passes through a sequence of states from which some observations can be gathered

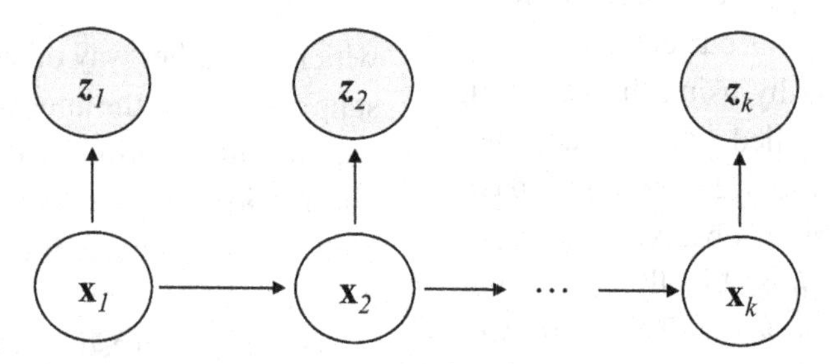

$$bel(\mathbf{x}_k|\mathbf{z}_{1:k}) = bel(\mathbf{x}_k|\mathbf{z}_k,\mathbf{z}_{1:k-1}) = \frac{f_{\mathbf{z}_{1:k}|\mathbf{x}_k} bel(\mathbf{x}_k)}{f_{\mathbf{z}_{1:k}}}$$

(conditioning all the formula to the history of past observations $\mathbf{z}_{1:k-1}$, and since $P[a|c,b,b] = P[a|c,b]$)

$$\overbrace{bel(\mathbf{x}_k|\mathbf{z}_k,\mathbf{z}_{1:k-1})}^{\text{posterior at step }k} = \frac{f_{\mathbf{z}_{1:k}|\mathbf{x}_k,\mathbf{z}_{1:k-1}} bel(\mathbf{x}_k|\mathbf{z}_{1:k-1})}{f_{\mathbf{z}_{1:k}|\mathbf{z}_{1:k-1}}} =$$

$$= \frac{f_{\mathbf{z}_k,\mathbf{z}_{1:k-1}|\mathbf{x}_k,\mathbf{z}_{1:k-1}} bel(\mathbf{x}_k|\mathbf{z}_{1:k-1})}{f_{\mathbf{z}_k,\mathbf{z}_{1:k-1}|\mathbf{z}_{1:k-1}}} =$$

(provided that $P[a,b|b,c] = \frac{P[a,b,c]}{P[b,c]} = P[a|b,c]$)

$$= \frac{f_{\mathbf{z}_k|\mathbf{x}_k,\mathbf{z}_{1:k-1}} bel(\mathbf{x}_k|\mathbf{z}_{1:k-1})}{f_{\mathbf{z}_k|\mathbf{z}_{1:k-1}}} =$$

(since $\mathbf{z}_k \perp \mathbf{z}_{1:k-1}|\mathbf{x}_k$, as can be deduced from the DBN)

$$= \frac{f_{\mathbf{z}_k|\mathbf{x}_k} bel(\mathbf{x}_k|\mathbf{z}_{1:k-1})}{f_{\mathbf{z}_k|\mathbf{z}_{1:k-1}}} =$$

(considering $bel(\mathbf{x}_k|\mathbf{z}_{1:k-1})$ as the marginalization of the multivariate

$$bel(\mathbf{x}_k,\mathbf{x}_{k-1}|\mathbf{z}_{1:k-1}) = bel(\mathbf{x}_k|\mathbf{x}_{k-1},\mathbf{z}_{1:k-1}) bel(\mathbf{x}_{k-1}|\mathbf{z}_{1:k-1})$$

,

and assuming that we know the term $bel(\mathbf{x}_k|\mathbf{x}_{k-1},\mathbf{z}_{1:k-1}) = f_{\mathbf{x}_k|\mathbf{x}_{k-1},\mathbf{z}_{1:k-1}}$)

$$= \frac{f_{\mathbf{z}_k|\mathbf{x}_k} \int_{\mathbf{x}_{k-1}=-\infty}^{+\infty} f_{\mathbf{x}_k|\mathbf{x}_{k-1},\mathbf{z}_{1:k-1}} bel(\mathbf{x}_{k-1}|\mathbf{z}_{1:k-1}) d\mathbf{x}_{k-1}}{f_{\mathbf{z}_k|\mathbf{z}_{1:k-1}}} =$$

(finally, provided that $\mathbf{x}_k \perp \mathbf{z}_{1:k-1}|\mathbf{x}_{k-1}$)

$$= \frac{\overbrace{f_{\mathbf{z}_k|\mathbf{x}_k}}^{\text{likelihood}} \overbrace{\int_{\mathbf{x}_{k-1}=-\infty}^{+\infty} \overbrace{f_{\mathbf{x}_k|\mathbf{x}_{k-1}}}^{\text{transition}} \overbrace{bel(\mathbf{x}_{k-1}|\mathbf{z}_{1:k-1})}^{\text{posterior at step }k-1} d\mathbf{x}_{k-1}}^{\text{prior at step }k}}{\underbrace{f_{\mathbf{z}_k|\mathbf{z}_{1:k-1}}}_{\text{scaling}}} \quad (33)$$

This is the fundamental framework for recursive Bayesian estimation, which can be applied to many mobile robotics problems. Subsequent chapters will work on the last equation above by looking for particular solutions depending on the form of its terms.

REFERENCES

Access, A. I. (2011). *Glossary of data modelling*. Retrieved March 15, 2011, from http://www.aiaccess.net/e_gm.htm

Ash, R. B. (1999). *Probability and measure theory*. New York, NY: Academic Press.

Basseville, M., & Nikiforov, I. V. (1993). *Detection of abrupt changes: Theory and application*. Upper Saddle River, NJ: Prentice-Hall.

Cramér, H. (1946). *Mathematical methods of statistics*. Princeton, NJ: Princeton University Press.

Gago, A., Fernández-Madrigal, J. A., Galindo, C., & Cruz-Martín, A. (2010). *Statistical characterization of the time-delay for Web-based networked telerobots*. Paper presented at the 5th International Workshop in Applied Probability (IWAP 2010). Madrid, Spain.

Gelb, A. (Ed.). (1974). *Applied optimal estimation*. Cambridge, MA: The MIT Press.

Ghahramani, Z. (2001). An introduction to hidden Markov models and Bayesian networks. *International Journal of Pattern Recognition and Artificial Intelligence*, *15*(1), 9–42. doi:10.1142/S0218001401000836

Gustafsson, F. (2000). *Adaptive filtering and change detection*. New York, NY: John Wiley & Sons.

Hazewinkel, M. (2002). *Encyclopaedia of mathematics*. Berlin, Germany: Springer-Verlag. Retrieved Mar 1, 2012, from http://www.encyclopediaofmath.org/

Kallenberg, O. (1997). *Foundations of modern probability*. New York, NY: Springer-Verlag.

Kay, S. M. (1993). Fundamentals of statistical signal processing: *Vol. 1. Estimation theory*. Upper Saddle River, NJ: Prentice-Hall.

Lehmann, E. L., & Casella, G. (1998). *Theory of point estimation* (2nd ed.). Berlin, Germany: Springer.

Martín-Fernández, M. (2004). Fundamentals on estimation theory. Boston, MA: Harvard Medical School. Retrieved March 10, 2011, from http://lmi.bwh.harvard.edu/papers/pdfs/2004/martin-fernandezCOURSE04b.pdf

Merrian Webster Inc. (2011). *Merrian-Webster on-line dictionary and thesaurus*. Retrieved Mar 1, 2012, from http://www.merriam-webster.com/

Montgomery, D. C. (2005). *Introduction to statistical quality control*. New York, NY: John Wiley & Sons.

Poor, H. V., & Hadjiliadis, O. (2009). *Quickest detection*. Cambridge, UK: Cambridge University Press.

Rao, C. R. (1945). Information and the accuracy attainable in the estimation of statistical parameters. *Bulletin of the Calcutta Mathematical Society*, *37*, 81–89.

Walpole, R. E., Myers, R. L., & Myers, S. H. (1997). *Probability and statistics for engineers and scientists*. Upper Saddle River, NJ: Prentice Hall. doi:10.2307/2530629

Weisstein, E. (2011). *Wolfram mathworld: On-line mathematical encyclopedia*. Retrieved Mar 1, 2012, from http://mathworld.wolfram.com

Wilcox, R. R. (2005). *Introduction to robust estimation and hypothesis testing* (2nd ed.). London, UK: Elsevier.

ENDNOTES

1. Avoid the temptation to follow the more direct but false path that starts by deducing

that $\frac{(k-1)\nu_k^2}{\sigma^2} \sim \chi_{k-1}^2 \to \nu_k^2 \sim \frac{\sigma^2}{k-1}\chi_{k-1}^2$; the "being drawn from"/"is distributed according to" symbol, $\sim$, does not obeys the same rules as equality.

2 Specifically, when the family P of distributions that we are proposing as models are dominated by a σ-finite measure μ, or, equivalently, are absolutely continuous with respect to μ. A distribution $Q \in P$ defined on a probability space is absolutely continuous with respect to a measure μ also defined on that space if whenever $\mu(A) = 0$, $Q(A) = 0$. Note: any measure that only yields finite values is σ-finite.

Section 2
Mobile Robot Localization

Chapter 5
Robot Motion Models

ABSTRACT

This is the first chapter of the second section, a section devoted to mobile robot localization. Before presenting the general Bayesian framework for that problem at chapter 7, it is first required to study the different probabilistic models of robot motion. This chapter explores some of the reasons why any real robot cannot move as perfectly as planned, thus demanding a probabilistic model of the robot actions—mainly, its movements. Special emphasis is put on the most common ground wheeled robots, although other configurations (including non-robotic ones) with more degrees of freedom, such as arbitrarily-moving hand-held sensors or aerial vehicles, are also mentioned. The best-known approximate probabilistic models for robot motion are provided and justified.

CHAPTER GUIDELINE

- You will learn:
  - The particular place of odometry and proprioceptive sensors in the Bayesian filtering framework.
  - Six different kinematic models and their corresponding motion models with uncertainty, including closed-form formulas for the case of all random variables being Gaussians.
- Provided tools:
  - A table which relates kinematic models with physical locomotion implementations.
  - A systematic method to obtain a probabilistic counterpart of any ideal kinematic model under the assumptions of Gaussian distributions.
  - Practical formulas to settle hard-to-tune parameters in the constant velocity and the no-motion models.
- Relation to other chapters:
  - Kinematic model equations are directly applicable to the particle filter methods introduced in chapters 7 and 9 for localization and map building, respectively.
  - Probabilistic motion models for Gaussian models find their utility for localization and mapping when using parametric distributions in chapters 7, 9, and 10.

DOI: 10.4018/978-1-4666-2104-6.ch005

1. INTRODUCTION

In the previous chapters, we have studied the mathematical foundations needed for addressing the problems of mobile localization and mapping. Now it is time to turn again to our robotic realm and begin to set the particular bases for solving robotics problems.

As explained in chapter 1, as a mobile robot moves throughout its environment, it constantly needs to update estimation about its instantaneous pose (i.e. the problem of localization) and, if the environment itself is unknown, a model of the world around (i.e. the problem of mapping; tackling both problems together is SLAM). In both cases the robot must rely on the readings from its sensors, both *exteroceptive* and *proprioceptive*, to perform these updates.

Both localization and SLAM consist of dynamic processes, thus we can apply recursive Bayesian estimation, introduced in chapter 4 section 9, when dealing with them. Recall that recursive here means iteratively computing the belief about some unknowns $\mathbf{x}_k$ from the belief for the previous time step, that is, $\mathbf{x}_{k-1}$. Most probabilistic models of robot motion can be directly applied to either localization or SLAM; the only difference will be in the vector of unknowns $\mathbf{x}_k$, which in localization comprises the pose of the robot while in SLAM additionally includes a parameterized model of the world. In any case, we showed in chapter 4's Equation 29 that the update equation for this recursive Bayesian estimator can be written as:

$$\overbrace{bel(\mathbf{x}_k | \mathbf{z}_{1:k})}^{posterior\ at\ step\ k} \propto \propto \overbrace{f_{\mathbf{z}_k|\mathbf{x}_k}}^{likelihood} \int_{\mathbf{x}_{k-1}=-\infty}^{+\infty} \overbrace{f_{\mathbf{x}_k|\mathbf{x}_{k-1}}}^{transition} \overbrace{bel(\mathbf{x}_{k-1} | \mathbf{z}_{1:k-1})}^{posterior\ at\ step\ k-1} d\mathbf{x}_{k-1} \quad (1)$$

where the transition distribution is the part of interest in this chapter.

The transition distribution is conditioned in Equation 1 to the previous state of the system. Thus, $f_{\mathbf{x}_k|\mathbf{x}_{k-1}}$ is a probabilistic function of that previous state that *predicts* the new state. The natural way of providing $f_{\mathbf{x}_k|\mathbf{x}_{k-1}}$ with a mathematical form is to use some physical model of the motion of the vehicle, and the most basic of such models is the vehicle kinematics, augmented with the appropriate uncertainty. This chapter is devoted to providing some useful probabilistic models based on kinematics.

Equation 1 stands for a generic Bayesian estimation problem where the only information is the sequence of observations $\mathbf{z}_{1:k}$. As it will be detailed in chapters 7 and 9, in localization and SLAM we will also have knowledge about another sequence of variables: the robot *actions*, denoted as $\mathbf{u}_{1:k}$—we already briefly discussed robot actions in chapter 2. It will be shown there that recursive Bayesian estimation in those problems leads to transition models conditioned not only on the previous pose, but on the latest action as well. That is, we have to define $f_{\mathbf{x}_k|\mathbf{x}_{k-1},\mathbf{u}_k}$ instead of $f_{\mathbf{x}_k|\mathbf{x}_{k-1}}$. As we will see next, the actual meaning of these action vectors $\mathbf{u}_{1:k}$ depends on the kinematics of the robot and the particular instrumentation of its motion. Typically, if the robot is equipped with some sort of odometry, it simply consists of its readings.

Notice that odometry is provided by sensors, which implies that the transition must include sensory information, in addition to the sensory information included in the observation likelihood part of Equation 1. At first sight, there exist no evident reasons for treating each kind of sensor differently in the recursive update equation. However, under a Bayesian approach such as the one at hand some differences start to emerge. We will distinguish two types of data inputs to the formulation: on the one hand, the transition model will be devised from the kinematics of the robot being particularized with data coming from proprioceptive sensors (e.g. odometry); on the other hand, readings from exteroceptive sensors will constitute the input for observation models, i.e., used in the observation likelihood of the formulation—chapter 6 is devoted to them.

It may seem an unnecessary complexity to introduce a probabilistic treatment for the transition model from a previous robot pose $\mathbf{x}_{k-1}$ to the next one $\mathbf{x}_k$ if the robot already provides odometry measurements of such pose increments. After all, robots may be equipped with wheel encoders with sub-millimeter resolution, so, why should not we just trust these accurate measurements? Truth is that there is only an indirect relationship between the readings of odometry sensors and the actual displacement of the robot, due to the existence of:

- Inaccuracies of the kinematic models. Each model makes a set of assumptions, which may hold, or not. The most common hypothesis is that the robot moves following a fixed-shaped path between consecutive time intervals. The smaller those time steps, the more exact the approximation, but in any case it will never exactly match the actual robot trajectory.
- Parameter imprecision. Motion models also rely on system parameters that reflect some physical properties of the robot (e.g. the distance between wheels) and which are assumed to be static and perfectly known. In practice, however, one always has a certain error determining those parameters and they may even be dynamic. For instance, consider the effective diameter of pneumatic tires, which can change with temperature or slowly drift over long periods of time. This kind of inaccuracies leads to systematic errors, since they introduce a non-stochastic bias in the motion model.
- Small errors and perturbations. Think of the particular case of a wheeled mobile robot, which can find small irregularities on an otherwise even floor, thus introducing deviations in the odometry-induced pose estimation. In addition, the mechanical links of motor wheels could induce small deviations in the wheel orientation as it moves. Due to the unpredictability of these and other similar error sources, and invoking the central limit theorem (see chapter 4 section 2), it seems plausible to model them as Gaussian distributions. These errors are non-systematic, in contradistinction to the systematic errors mentioned above.

At this point we have to clearly state what we define as the kinematic model of a robot and what as a probabilistic motion model (the $f_{\mathbf{x}_k|\mathbf{x}_{k-1},\mathbf{u}_k}$), as well as the relation between them. Kinematic models are functions that map the previous robot pose $\mathbf{x}_{k-1}$ and an associated action $\mathbf{u}_k$ into the predicted new pose $\mathbf{x}_k$ as accurately as possible:

$$\mathbf{x}_k \leftarrow \underbrace{\mathbf{g}\left(\mathbf{x}_{k-1}, \mathbf{u}_k\right)}_{\text{Kinematic model}} \tag{2}$$

A robot has, by design, only one kinematic model. But in some cases, alternative odometry instrumentations are possible for the same physical configuration, as we briefly mentioned in chapter 2 section 2. In those cases we will find different kinematic model equations for each odometry arrangement.

Since most kinematic models involve non-linear transformations, when we aim at a probabilistic treatment of the kinematics we find that even assuming normally-distributed input variables ($\mathbf{x}_{k-1}$ and $\mathbf{u}_k$) the resulting distribution of $\mathbf{x}_k$ will not have, in general, any closed-form expression. However, following the linearization method explained in chapter 3 section 8 we will be able to approximate the actual distribution $f_{\mathbf{x}_k|\mathbf{x}_{k-1},\mathbf{u}_k}$ to obtain practical probabilistic motion models. When dealing with non-parameterized estimation algorithms, as particle filters introduced in chapter 7 section 4, there will be no need for such simplification and the exact kinematic equations could be employed then.

In the following sections, we describe the most common kinematic models and their associated probabilistic—approximate—motion models for mobile robots. To clarify the correspondence of these models to the different physical implementations of a robot mobility and the possible odometry instrumentations (in the cases where more than one possibility exists), we have summarized all of them in Table 1.

2. CONSTANT VELOCITY MODEL

Kinematic Equations

This simple model can be always applied even in the lack of any odometry sensor on the robot. Hence it is often the model of choice when performing localization or SLAM with hand-held cameras or small flying robots.

The constant velocity model requires the state vector $\mathbf{x}_k$ to include not only the pose of the robot, but its velocity vector as well. Then, the kinematic model function $\mathbf{g}\left(\mathbf{x}_{k-1},\mathbf{u}_k\right)$ assumes that the velocity has not changed since the last time step and approximates the robot trajectory with segments of constant linear and angular velocities. In other words: in the lack of any odometry information, this model simply assumes that if the robot was standing still, it will still be there for the next time step, or that if it was moving forward with a certain speed, its future motion will be a straight line in that direction. For this motion model, the action vector $\mathbf{u}_k$ only comprises one value: the length of the time lapse between two consecutive time steps. When this time interval is fixed and assumed to be known by the reader, some authors simply drop it from the formulation, rewriting the model function $\mathbf{g}\left(\mathbf{x}_{k-1},\mathbf{u}_k\right)$ as $\mathbf{g}\left(\mathbf{x}_{k-1}\right)$.

Table 1. Summary of the different kinematic models presented in this chapter along with their corresponding physical realizations

Kinematic model \ Physical locomotion type	Hand-held, submarine or flying devices with no odometry at all	Synchro-drive, omniwheels, mecanum wheels	Two or four-wheel differential drive	Tricycle-like wheeled robot	Ackerman steered wheeled robot
Constant velocity model (section 5.2)	✓	✓	✓	✓	✓
Holonomic model (section 5.3)		✓			
Non-holonomic model: two increments (section 5.4)			✓	✓	✓
Non-holonomic model: one increment, one angle (section 5.5)				✓	✓
Black-box model (section 5.6)		✓	✓	✓	✓
No-motion motion model (section 5.7)	✓	✓	✓	✓	✓

As simple as the approximation implied by this model may be, its power comes from the lack of any assumption on the physical implementation of the robot mobility, thus it enjoys universal applicability. In practice, however, its effectiveness in performing accurate predictions depends on a number of factors:

- The smoothness of the actual motion: a clear weakness of this model is that abrupt changes in the motion have a catastrophic impact on the quality of predictions. The larger the modulus of the acceleration vector (including both translation and rotation), the worse the correctness of this model which assumes null accelerations.
- The update rate: given a fixed trajectory, the accuracy of this model can be arbitrarily improved by updating the system state at a faster and faster rate. That is, as the interval between consecutive time steps becomes smaller, the constant-velocity approximation becomes more realistic. Obviously, this introduces a computational burden, thus in practical systems some sort of compromise between computational power and accepted errors must be achieved. Notice that this and the previous points are different aspects of the same thing, since the smoothness of a path is actually linked to the rate at which it is sampled.
- The correct modeling of accelerations: A probabilistic approach to the constant velocity model gives us an estimation of the uncertainty in the pose of the robot and also in its velocities. It will be shown below that a designer must supply parameters related to the range of typical accelerations attained by the robot. Those parameters must be carefully tuned since an underestimation of the accelerations may cause a localization or SLAM algorithm to be overconfident on the robot being at a wrong position, which is a serious situation from which it is difficult to recover.

Next, we describe the implementation of this kinematic model and its associated probabilistic motion model for robots operating on planar surfaces, and later on, we extend it for those operating in three-dimensional workspaces.

Let the vector $\mathbf{x}_k$ denote the robot pose (that is, a 2D position and a heading) and its estimated velocity, that is:

$$\mathbf{x}_k = \begin{bmatrix} \mathbf{p}_k \\ \mathbf{v}_k \end{bmatrix} = \begin{bmatrix} x_k \\ y_k \\ \theta_k \\ v_{x_k} \\ v_{y_k} \\ v_{\theta_k} \end{bmatrix} \begin{matrix} \left.\begin{matrix} \\ \\ \\ \end{matrix}\right\} \text{Robot pose} \\ \left.\begin{matrix} \\ \\ \\ \end{matrix}\right\} \text{Velocities} \end{matrix} \tag{3}$$

where the robot pose $\mathbf{p}_k$ is given with respect to an arbitrary global coordinates frame. The vector of velocities $\mathbf{v}_k$ is often more conveniently modeled as relative to the local robot coordinates for the last time step, as sketched in Figure 1. Regarding the action vector $\mathbf{u}_k$, it only contains the time increment between time steps k and $k+1$ which we will denote as Δt_k. The k subscript can be dropped if all time increments are exactly identical.

Under such parameterization, the kinematic model reads:

$$\mathbf{x}_k = \mathbf{g}\left(\mathbf{x}_{k-1}, \mathbf{u}_k\right)$$

(splitting the vectors into their two parts:)

$$\begin{bmatrix} \mathbf{p}_k \\ \mathbf{v}_k \end{bmatrix} = \begin{bmatrix} \mathbf{p}_{k-1} \oplus \left(\mathbf{v}_{k-1} \Delta t_k\right) \\ \mathbf{v}_{k-1} \end{bmatrix} \tag{4}$$

Figure 1. Representation of all the coordinate frames involved in the constant velocity model

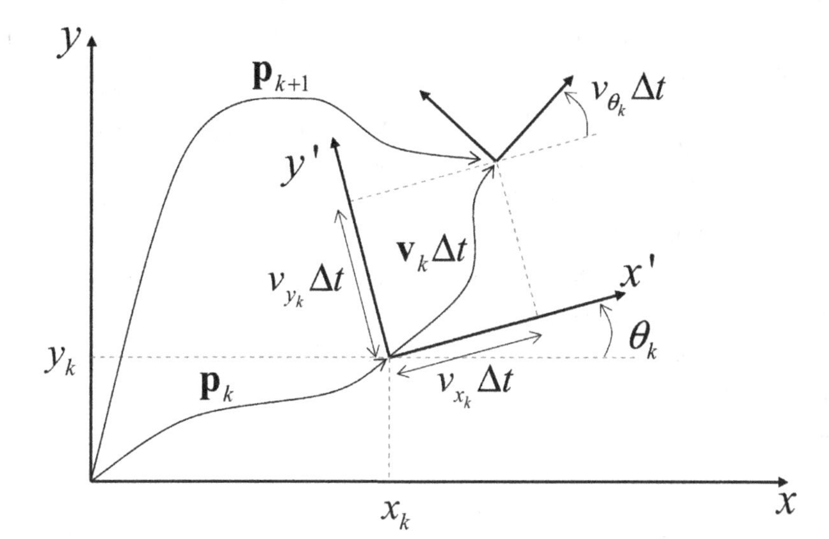

This equation clearly reveals the constant-velocity assumption made by this motion model, since $\mathbf{v}_k = \mathbf{v}_{k-1}$. Correspondingly, the estimation of the trajectory described by the robot during the time step becomes $\mathbf{v}_{k-1}\Delta t_k$ (the velocity times the time interval), which is in local coordinates with respect to the last pose $\mathbf{p}_k$. Both poses are concatenated (or *composed*) via the pose composition operator $\oplus$ (refer to the Appendix A). For clarifying the operation of $\oplus$, we illustrate the result for the 2D case (notice we do not write in bold the components of $\mathbf{x}_k$ since they are unidimensional r.v.s):

$$\begin{bmatrix} x_k \\ y_k \\ \theta_k \end{bmatrix} = \mathbf{p}_k = \mathbf{p}_{k-1} \oplus \left(\mathbf{v}_{k-1}\Delta t_k\right) = \begin{bmatrix} \underbrace{x_{k-1}}_{\in \mathbf{p}_{k-1}} + v_{x_{k-1}}\Delta t_k \cos\theta_{k-1} - v_{y_{k-1}}\Delta t_k \sin\theta_{k-1} \\ \underbrace{y_{k-1}}_{\in \mathbf{p}_{k-1}} + v_{x_{k-1}}\Delta t_k \sin\theta_{k-1} + v_{y_{k-1}}\Delta t_k \cos\theta_{k-1} \\ \underbrace{\theta_{k-1}}_{\in \mathbf{p}_{k-1}} + v_{\theta_{k-1}}\Delta t_k \end{bmatrix} \quad (5)$$

Probabilistic Motion Model

The probabilistic motion model corresponding to this kinematic model consists of a pdf modeling the random variable $\mathbf{x}_k$ provided known distributions for the other random variables, $\mathbf{x}_{k-1}$ and $\mathbf{u}_k$. In this particular motion model it is common practice to ignore the uncertainty in the action $\mathbf{u}_k$ (remember, this variable only contains the time lapse Δt_k between time steps), although we will not discard it in the following to provide a more generic solution.

In order to account for the inaccuracy of the constant-velocity assumption, the kinematic Equation 4 can be expanded by introducing a new variable $\mathbf{r}_k$, a zero-mean additive Gaussian noise that models the effects of the actual (unknown) accelerations:

$$\begin{bmatrix} \mathbf{p}_k \\ \mathbf{v}_k \end{bmatrix} = \mathbf{x}_k = \mathbf{g}'\left(\mathbf{x}_{k-1}, \mathbf{u}_k, \mathbf{r}_k\right) = \begin{bmatrix} \mathbf{p}_{k-1} \oplus \left(\mathbf{v}_{k-1}\Delta t_k\right) \\ \mathbf{v}_{k-1} + \mathbf{r}_k\sqrt{\Delta t_k} \end{bmatrix} \quad (6)$$

where the square root of time increments was taken to assure that the uncertainty in velocity grows linearly with time and becomes independent of the time step resolution[1]. Thus, the motion model consists of transforming the multidimensional random variables $\left(\mathbf{x}_{k-1}, \mathbf{u}_k, \mathbf{r}_k\right)$ into the variable $\mathbf{x}_k$ using the non-linear transformation $\mathbf{g}'$. Since this variable has no closed-form probability distribution, we need to approximate it as it was shown in chapter 3 section 8. For small displacements, as those typically found in motion models, multivariate Gaussians are a quite good approximation for the involved uncertainties, thus the common approach is based on assuming Gaussian distributions for every r.v. We can parameterize the four random variables involved in this model as follows—means are denoted with bars over the variables (see Table 2).

By means of a first-order Taylor series expansion—refer to chapter 3 section 8—we can approximate the mean (the mathematical expectation) of the resulting variable $\mathbf{x}_k$, transformed through $\mathbf{g}'$, with the transformation of the means of the original r.v.s., that is:

$$\overline{\mathbf{x}}_k \approx \mathbf{g}'\left(\overline{\mathbf{x}}_{k-1}, \overline{u}_k, 0\right) = \mathbf{g}\left(\overline{\mathbf{x}}_{k-1}, \overline{u}_k\right) \tag{7}$$

Assuming independence between the uncertainty in the time increment (the action $\mathbf{u}_k$), the acceleration noise $\mathbf{r}_k$ and the previous robot state $\mathbf{x}_{k-1}$, the covariance matrix for the new state can be approximated using the Jacobian of $\mathbf{g}'$ in the Taylor linearization point $\left(\overline{\mathbf{x}}_{k-1}, \overline{u}_k, 0\right)$. We will write here the Jacobian of $\mathbf{g}'$ as $\frac{\partial \mathbf{g}'}{\partial\left\{\mathbf{x}_{k-1}, u_k, \mathbf{r}_k\right\}}$ instead of the less explicit notation $\mathbf{J}_{\mathbf{g}'}$ used in previous chapters:

$$\begin{aligned}
\boldsymbol{\Sigma}_{\mathbf{x}_k} &= \frac{d\mathbf{g}'}{d\left\{\mathbf{x}_{k-1}, u_k, \mathbf{r}_k\right\}} \begin{pmatrix} \boldsymbol{\Sigma}_{\mathbf{x}_{k-1}} & \mathbf{0}_{6\times1} & \mathbf{0}_{6\times3} \\ \mathbf{0}_{1\times6} & \sigma^2_{u_k} & \mathbf{0}_{1\times3} \\ \mathbf{0}_{3\times6} & \mathbf{0}_{3\times1} & \boldsymbol{\Sigma}_r \end{pmatrix} \frac{d\mathbf{g}'}{d\left\{\mathbf{x}_{k-1}, u_k, \mathbf{r}_k\right\}}^T \\
&= \begin{pmatrix} \frac{\partial \mathbf{g}'}{\partial \mathbf{x}_{k-1}} & \frac{\partial \mathbf{g}'}{\partial u_k} & \frac{\partial \mathbf{g}'}{\partial \mathbf{r}_k} \end{pmatrix} \begin{pmatrix} \boldsymbol{\Sigma}_{\mathbf{x}_{k-1}} & \mathbf{0}_{6\times1} & \mathbf{0}_{6\times3} \\ \mathbf{0}_{1\times6} & \sigma^2_{u_k} & \mathbf{0}_{1\times3} \\ \mathbf{0}_{3\times6} & \mathbf{0}_{3\times1} & \boldsymbol{\Sigma}_r \end{pmatrix} \begin{pmatrix} \frac{\partial \mathbf{g}'}{\partial \mathbf{x}_{k-1}}^T \\ \frac{\partial \mathbf{g}'}{\partial u_k}^T \\ \frac{\partial \mathbf{g}'}{\partial \mathbf{r}_k}^T \end{pmatrix} \\
&= \frac{\partial \mathbf{g}'}{\partial \mathbf{x}_{k-1}} \boldsymbol{\Sigma}_{\mathbf{x}_{k-1}} \frac{\partial \mathbf{g}'}{\partial \mathbf{x}_{k-1}}^T + \frac{\partial \mathbf{g}'}{\partial u_k} \sigma^2_{u_k} \frac{\partial \mathbf{g}'}{\partial u_k}^T + \frac{\partial \mathbf{g}'}{\partial \mathbf{r}_k} \boldsymbol{\Sigma}_r \frac{\partial \mathbf{g}'}{\partial \mathbf{r}_k}^T
\end{aligned} \tag{8}$$

These Jacobians, which must be evaluated at the mean value of the original random variables, can be easily derived from Equation 4, Equa-

Table 2. Parameterization of the four random variables

Random variable	Description	Dimensionality
$\mathbf{x}_k \sim N\left(\overline{\mathbf{x}}_k, \boldsymbol{\Sigma}_{x_k}\right)$	State vector for the current time step.	6
$\mathbf{x}_{k-1} \sim N\left(\overline{\mathbf{x}}_{k-1}, \boldsymbol{\Sigma}_{x_{k-1}}\right)$	State vector for the previous time step.	6
$u_k \sim N\left(\overline{u}_k, \sigma^2_{u_k}\right)$	Action vector (a time increment).	1
$\mathbf{r}_k \sim N\left(0, \boldsymbol{\Sigma}_{r_k}\right)$	Noise vector (variation in velocity per time square root).	3

Box 1.

$$\frac{\partial \mathbf{g}'}{\partial \mathbf{x}_{k-1}} = \left(\begin{array}{ccc|ccc} 1 & 0 & -v_{x_{k-1}} \Delta t_k \sin\theta_{k-1} - v_{y_{k-1}} \Delta t_k \cos\theta_{k-1} & \Delta t_k \cos\theta_{k-1} & -\Delta t_k \sin\theta_{k-1} & 0 \\ 0 & 1 & v_{x_{k-1}} \Delta t_k \cos\theta_{k-1} - v_{y_{k-1}} \Delta t_k \sin\theta_{k-1} & \Delta t_k \sin\theta_{k-1} & \Delta t_k \cos\theta_{k-1} & 0 \\ 0 & 0 & 1 & 0 & 0 & \Delta t_k \\ \hline & \mathbf{0}_{3\times 3} & & & \mathbf{I}_3 & \end{array}\right) \tag{9}$$

tion 5, and Equation 6 (see Box 1 and Equations (10)-(11)).

$$\frac{\partial \mathbf{g}'}{\partial u_k} = \left(\begin{array}{c} v_{x_{k-1}} \cos\theta_{k-1} - v_{y_{k-1}} \sin\theta_{k-1} \\ v_{x_{k-1}} \sin\theta_{k-1} + v_{y_{k-1}} \cos\theta_{k-1} \\ v_{\theta_{k-1}} \\ \hline \mathbf{0}_{3\times 1} \end{array}\right) \tag{10}$$

$$\frac{\partial \mathbf{g}'}{\partial \mathbf{r}_k} = \left(\begin{array}{c} \mathbf{0}_{3\times 3} \\ \sqrt{\Delta t_k}\, \mathbf{I}_3 \end{array}\right) \tag{11}$$

A few words are in order to clarify the role and physical meaning of each individual covariance within the block-diagonal covariance matrix of Equation 8. Firstly, the matrix $\Sigma_{\mathbf{x}_{k-1}}$ is simply the covariance of the robot state vector for the previous time step. Since in this motion model the action vector $\mathbf{u}_k$ only comprises the time interval Δt_k, the second diagonal component $\sigma^2_{u_k}$ is a scalar value: the variance of each such time interval. It is common to consider this variance as zero for systems with sensors that provide measures at highly accurate frequencies, such as, for example, cameras. Finally, the most difficult parameter to tune is the third matrix in the diagonal, namely Σ_r, which models the uncertainty in the zero-mean accelerations that we assumed the system is undergoing. Making the reasonable assumption of independence between these stochastic accelerations in each degree of freedom, we can model this covariance as a diagonal matrix, that is:

$$\Sigma_r = \begin{pmatrix} r^2_{v_x} & 0 & 0 \\ 0 & r^2_{v_y} & 0 \\ 0 & 0 & r^2_{v_\theta} \end{pmatrix} \tag{12}$$

Understanding the physical meaning of the variances on the diagonal is essential since these three values are parameters that should be adapted for each specific problem. We will find next how to set these parameters such that they fulfill the following condition: provided a maximum acceleration $a_i^{\max}$ for the $i-th$ component of the robot pose, we want to impose that, after a time increment of Δt seconds, the probability of any velocity component to be out of the range of the attainable velocities for that maximum acceleration must be below a certain limit $1-c$. Typically, the confidence interval c is in the range of 95% to 99%. Regarding the maximum velocity attainable in Δt seconds, it is easily shown to be $v_i^{\max} = \Delta t \; a_i^{\max}$, assuming without loss of generality an initial velocity of zero. On the other hand, from Equation 8 and Equation 11, we can analyze how the uncertainty (variance) of each velocity evolves with time. Under the Gaussian assumption that has guided us to here, the marginal probability distribution of v_i (the $i-th$ component of the velocity vector denoted above as $\mathbf{v}_k$) turns out to be a one-dimensional Gaussian with a zero mean and a variance of $\sigma^2_{v_i} = \Delta t \; r^2_{v_i}$

In chapter 3 section 8 we arrived at an interesting result linking the chi-squared distribution to the Gaussian distribution: $X^2 / \sigma^2 \sim \chi_1^2$ as long as $X \sim N(0,\sigma^2)$. Here, the r.v. of interest (X) is v_i, which follows the Gaussian distribution $N(0,\sigma_{v_i}^2)$ with $\sigma_{v_i}^2 = \Delta t^2\, r_{v_i}^2$, hence we know that $v_i^2 / \sigma_{v_i}^2$ must follow a χ_1^2 distribution. We are interested in the probability of the expression $v_i^2 / \sigma_{v_i}^2$ being below a certain value. Such probability is given by the cdf of χ_1^2, that we can denote here as φ_1^2, that is, $P\left[v_i^2 / \sigma_{v_i}^2 < z\right] = \varphi_1^2(z)$, being z any point in the support of the variable. If we take that probability to be c, i.e., $\varphi_1^2\left(z\right) = c$, we have $z = \left[\varphi_1^2\right]^{-1}\left(c\right)$. This use of the inverse of the chi-squared cdf (which is a function that is guaranteed to exist) is often abbreviated as $\chi_{1,c}^2 \triangleq \left[\varphi_1^2\right]^{-1}\left(c\right)$. Using this notation we have $z = \chi_{1,c}^2$ and $\varphi_1^2\left(z\right) = c$, which leads to the following statement free of z and particularized for c:

$$P\left[v_i^2 / \sigma_{v_i}^2 < \chi_{1,c}^2\right] = c \leftrightarrow$$

(since the variance is positive)

$$P\left[v_i^2 < \chi_{1,c}^2 \sigma_{v_i}^2\right] = c \rightarrow$$

(since the square root is a monotonic function, taking the square root at both sides of the inequality does not alter it)

$$P\left[\left|v_i\right| < \sqrt{\chi_{1,c}^2}\, \sigma_{v_i}\right] = c \tag{13}$$

Our goal is to assure that:

$$P\left[\left|v_i\right| > \Delta t\ a_i^{\max}\right] = 1 - c$$

(or, equivalently:)

$$P\left[\left|v_i\right| < \Delta t\ a_i^{\max}\right] = c \tag{14}$$

It follows then from the last two equations that:

$$\left.\begin{array}{l} P\left[\left|v_i\right| < \sqrt{\chi_{1,c}^2}\, \sigma_{v_i}\right] = c \\ P\left[\left|v_i\right| < \Delta t\ a_i^{\max}\right] = c \end{array}\right\} \quad \leftrightarrow \quad \Delta t\ a_i^{\max} = \sqrt{\chi_{1,c}^2}\, \sigma_{v_i} \tag{15}$$

Notice that the right-hand equality can be established only because we assumed that the velocity follows a normal distribution, which assures $p\left(\left|v_i\right|\right) \neq 0$ for any v_i and therefore there exists only one solution to the pair of equations at the left hand. Moreover, since we know that $\sigma_{v_i}^2 = \Delta t\ r_{v_i}^2$, that is, $\sigma_{v_i} = \sqrt{\Delta t}\ \ r_{v_i}$, the condition above can be also written down as:

$$\begin{array}{l} \Delta t\ a_i^{\max} = \sqrt{\chi_{1,c}^2}\ \sqrt{\Delta t}\ r_{v_i} \quad \rightarrow \\ r_{v_i} = a_i^{\max} \sqrt{\dfrac{\Delta t}{\chi_{1,c}^2}} \end{array} \tag{16}$$

which is an expression that links the acceleration parameters r_{v_i} needed to define the matrix $\mathbf{\Sigma}_r$ in Equation 12, to a physical magnitude such as the maximum acceleration the system is expected to experience. Therefore, this equation allows us to tune the uncertainty in the velocities from any reasonably accurate dynamic analysis of the system. Or seen the other way around: given some arbitrary values for the elements of matrix $\mathbf{\Sigma}_r$, it imposes the maximum admissible acceleration which should be observed during operation.

Application Example

In order to best illustrate the accuracy we can expect from the linearized approximation employed in this motion model, it is instructive to discuss on some numerical results obtained from simulations. First, consider the Figure 2a, which displays a Monte Carlo simulation[2] of the presented constant-velocity model. At the beginning, the robot is at the origin of coordinates. Then, we simulate four time steps of the motion model with some arbitrary values of translational and rotational velocities. A small acceleration uncertainty (Σ_r) has been employed, simulating the conditions of a smooth path without abrupt movements. Each pose sampled within the clouds that appear in the figure is depicted as a thick dot with a small line indicating its heading. As with any other motion model, which in itself does not use any external observation to correct its errors, uncertainty can only increase at each time step, which is reflected in the larger dispersion of the pose observations over time. By computing the Gaussian that best fit all the samples from the Monte Carlo simulation, we obtain an approximate[3] ground truth for the distribution of the pose at each time step, depicted as dashed ellipses in Figure 2b. It can be seen that the linearized approximation, whose Gaussian is represented by the solid ellipse in the same figure, is quite good for this example even after the four time steps. The bottom row of Figure 2 illustrates the same experiment but for a larger acceleration uncertainty, where this model clearly achieves a poorer accuracy, that is, there exists a larger mismatch between the ground truth and the Gaussian from the linearized model.

We must remark that this is a general result that can be extrapolated to all the linearized approximations to probabilistic motion models. As uncertainty grows (especially in the dimensions related to robot orientation) so does the degree of non-linearity of the involved equations, thus turning the first-order Taylor approximation involved in the propagation of covariances less accurate. Though, this should not be a problem when motion models are employed within localization or SLAM frameworks, since they include corrections through sensory observations, providing uncertainty reduction that compensates the increase from the one of the motion models alone.

Extension for the 3D Case

It is interesting to make the constant-velocity model, illustrated up to now for planar mobile robots, applicable to three-dimensional motions since this model is especially interesting for handheld cameras or other systems that move freely.

To achieve this, the first hurdle to be addressed concerns the several ways that poses in 3D can be parameterized. Such poses have six degrees of freedom, three translations plus three independent rotations, thus six is the minimum dimensionality for any parameterization. The translational part is always represented as an $\begin{bmatrix} x & y & z \end{bmatrix}^T$ vector, but modeling the rotation is more complex that may seem at first sight. A common approach is to directly employ a set of three angles, each representing the rotation around a predefined axis. Depending on the axis of reference, this method leads to the so-called Euler's angles, for which several conventions exist. For one particular setting (see Appendix A), those angles are known in robotics as *yaw*, *pitch* and *roll*. In spite of their minimum-size parameterization, employing those three-angle parameterizations cause some problems. Hence, alternative representations such as quaternions or linearized increments in the manifold $\mathbf{SE}(3)$ are becoming increasingly popular. Further discussion on basic 3D geometry is provided in Appendix A, while more advanced topics regarding manifold increments will be postponed until chapter 10.

In the following, we will assume a unit quaternion representation of rotations, which should be

Figure 2. A simulation of the probabilistic motion model for the constant-velocity kinematic model. Both top and bottom graphs represent the evolution of the robot pose after four consecutive time steps, but the uncertainty in the angular velocity is slightly larger in the simulation at the bottom. A Monte-Carlo simulation is represented in (a) and (c), whose Gaussian fit is shown in (b) and (d) as the dashed ellipse, while the solid ellipses correspond to the Taylor first-order approximation proposed in the text.

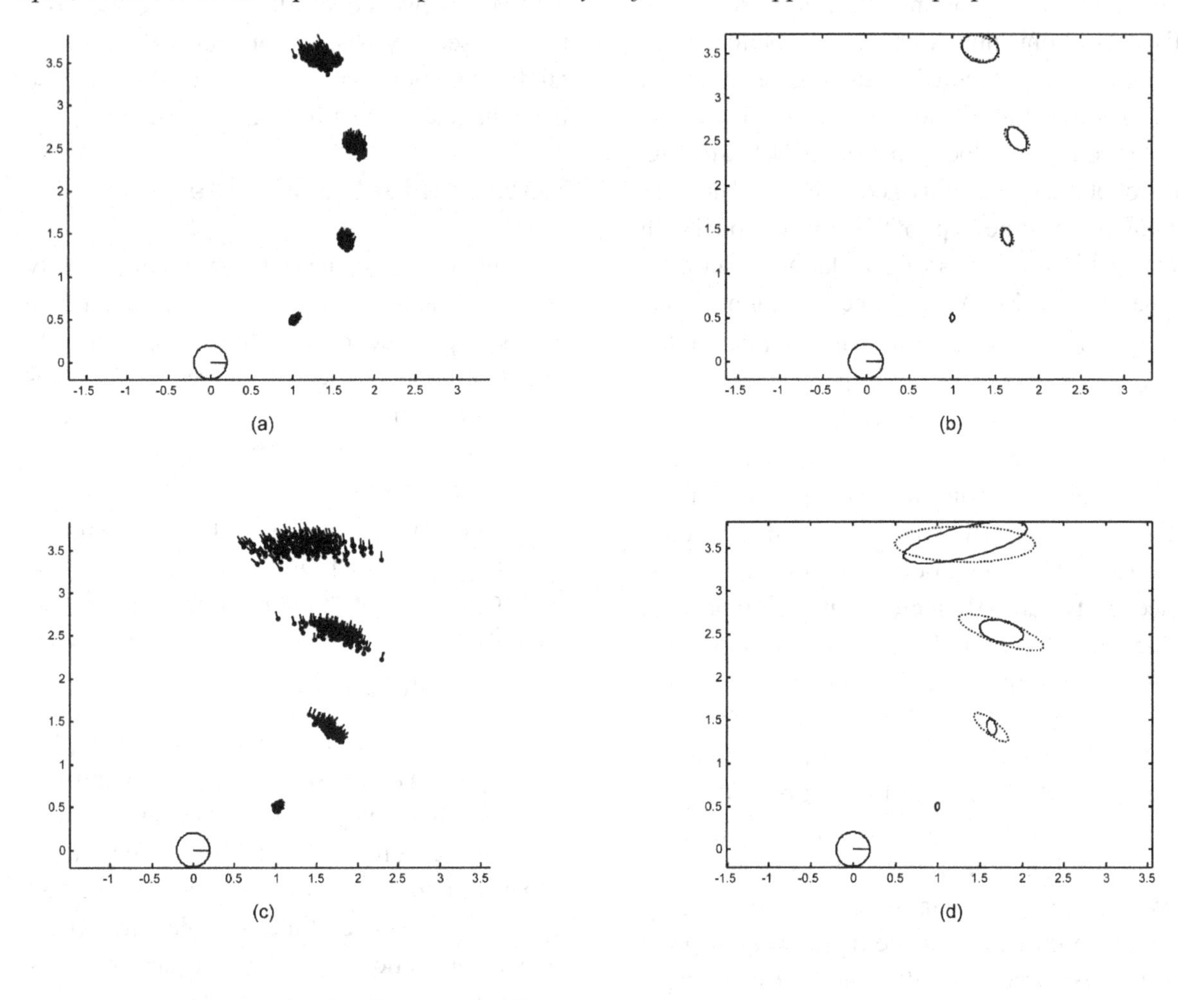

a convenient parameterization for many applications of the constant-velocity model. As explained in Appendix A, a unit quaternion is a four-element entity which describes a rotation in 3D and which (interpreted as a 4-vector) has a unit norm. All in all, the state vector of our system now comprises the following fourteen values (seven for each of the relevant parts of the state; quaternions indicated with q's):

$$\mathbf{x}_k = \begin{bmatrix} \mathbf{p}_k \\ \mathbf{v}_k \end{bmatrix} = \begin{bmatrix} \overbrace{x_k \quad y_k \quad z_k \quad q_{rk} \quad q_{xk} \quad q_{yk} \quad q_{zk}}^{\text{Robot pose } \mathbf{p}_k} \quad \overbrace{v_{x_k} \quad v_{y_k} \quad v_{z_k} \quad v_{q_{rk}} \quad v_{q_{xk}} \quad v_{q_{yk}} \quad v_{q_{zk}}}^{\text{Velocity vector } \mathbf{v}_k} \end{bmatrix}^T \tag{17}$$

The kinematic equation, in Equation 4, still has the same form as in the 2D case (repeated here for convenience):

$$\mathbf{x}_k = \mathbf{g}\left(\mathbf{x}_{k-1}, \mathbf{u}_k\right)$$

(splitting the vectors into its two parts):

$$\begin{bmatrix} \mathbf{p}_k \\ \mathbf{v}_k \end{bmatrix} = \begin{bmatrix} \mathbf{p}_{k-1} \oplus \left(\mathbf{v}_{k-1}\Delta t_k\right) \\ \mathbf{v}_{k-1} \end{bmatrix} \tag{18}$$

The main difference now is the pose composition operator $\oplus$ being a bit more complex in the 3D quaternion-based case:

$$\mathbf{p}_k = \mathbf{p}_{k-1} \oplus \overbrace{\left(\mathbf{v}_{k-1}\Delta t_k\right)}^{\mathbf{p}_{"}}$$

$$\begin{bmatrix} x_k \\ y_k \\ z_k \\ q_{rk} \\ q_{xk} \\ q_{yk} \\ q_{zk} \end{bmatrix} = \begin{bmatrix} x_{k-1} + x_\Delta + 2\begin{pmatrix} -\left(q_{y_{k-1}}^2 + q_{z_{k-1}}^2\right) x_\Delta \\ +\left(q_{x_{k-1}} q_{y_{k-1}} - q_{r_{k-1}} q_{z_{k-1}}\right) y_\Delta \\ +\left(q_{r_{k-1}} q_{y_{k-1}} + q_{x_{k-1}} q_{z_{k-1}}\right) z_\Delta \end{pmatrix} \\ y_{k-1} + y_\Delta + 2\begin{pmatrix} \left(q_{r_{k-1}} q_{z_{k-1}} + q_{x_{k-1}} q_{y_{k-1}}\right) x_\Delta \\ -\left(q_{x_{k-1}}^2 + q_{z_{k-1}}^2\right) y_\Delta \\ +\left(q_{y_{k-1}} q_{z_{k-1}} - q_{r_{k-1}} q_{x_{k-1}}\right) z_\Delta \end{pmatrix} \\ z_{k-1} + z_\Delta + 2\begin{pmatrix} \left(q_{x_{k-1}} q_{z_{k-1}} - q_{r_{k-1}} q_{y_{k-1}}\right) x_\Delta \\ +\left(q_{r_{k-1}} q_{x_{k-1}} + q_{y_{k-1}} q_{z_{k-1}}\right) y_\Delta \\ -\left(q_{x_{k-1}}^2 + q_{y_{k-1}}^2\right) z_\Delta \end{pmatrix} \\ q_{r_{k-1}} q_{r\Delta} - q_{x_{k-1}} q_{x\Delta} - q_{y_{k-1}} q_{y\Delta} - q_{z_{k-1}} q_{z\Delta} \\ q_{r_{k-1}} q_{x\Delta} + q_{x_{k-1}} q_{r\Delta} + q_{y_{k-1}} q_{z\Delta} - q_{z_{k-1}} q_{y\Delta} \\ q_{r_{k-1}} q_{y\Delta} + q_{y_{k-1}} q_{r\Delta} + q_{z_{k-1}} q_{x\Delta} - q_{x_{k-1}} q_{z\Delta} \\ q_{r_{k-1}} q_{z\Delta} + q_{z_{k-1}} q_{r\Delta} + q_{x_{k-1}} q_{y\Delta} - q_{y_{k-1}} q_{x\Delta} \end{bmatrix} \tag{19}$$

where the following auxiliary pose was introduced for the sake of notation:

$$\begin{aligned} \mathbf{p}_{"} &= \mathbf{v}_{k-1}\Delta t_k \\ &= \begin{bmatrix} x_\Delta & y_\Delta & z_\Delta & q_{r\Delta} & q_{x\Delta} & q_{y\Delta} & q_{z\Delta} \end{bmatrix}^T \end{aligned} \tag{20}$$

Regarding the probabilistic motion model associated to this 3D kinematic model, we can undertake the same approach of approximating the function $\mathrm{g}(\cdot)$ via first-order Taylor expansion, which would take us to the very same expression obtained in Equation 8. However, the expressions for the Jacobians are quite more involved in this case —they are provided in Equation 28 of Appendix A.

3. HOLONOMIC MODEL WITH A DIRECTION AND A DISTANCE

Kinematic Equations

The present kinematic model assumes a *holonomic* mobile robot moving in a planar surface. A robot, or in general any mechanical system, is said to be holonomic if none of its constraints involve the velocity vectors. For our purposes, this means that the velocity vector of a holonomic robot can point into any arbitrary direction at any time and, therefore, move into any direction without any internal restriction from its mechanics[4]. Motion models for non-holonomic robots are explored in subsequent sections, thus we will not further discuss them at this point.

Holonomic mobile robots are also called omnidirectional robots, due to their abovementioned ability of moving in any direction at any instant, disregarding its actual heading direction. Such agility can be accomplished by a few different means. A synchro-drive mechanism, where all wheels rotate at unison, was already described in chapter 2 section 2. Other alternatives are the usage of omnidirectional wheels or mecanum wheels (dubbed "Swedish wheels"). All those three designs have been embedded into research and commercial robotic vehicles, although their presence is rather marginal.

Ignoring its actual physical implementation, we will assume in the following that the movement

Figure 3. A motion model for omnidirectional robots. Specifically, the figure represents a synchro-drive mobile base. Each discrete step into which we decompose the robot movements comprises two stages: first, the robot stays at a pose $\mathbf{x}_k$ with all their wheels aligned into the direction $\theta_k + \beta$ (with respect to a global frame of reference) and describes a straight path in that direction; second, the entire robot rotates an angle of $\Delta\theta$ on itself, without further displacements.

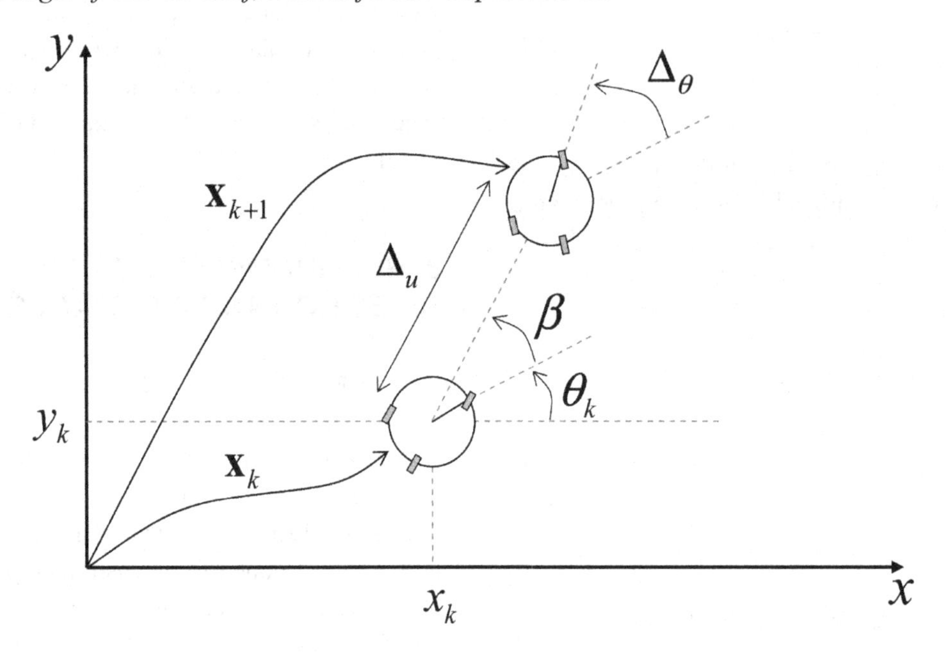

of any omnidirectional robot can be approximated as a straight motion in the desired direction followed by an in-place rotation of the entire robot. One must keep in mind that for these robots, the robot heading and the direction of motion are independent variables. The Figure 3 schematically represents this model along with all the involved variables, introduced in the next paragraphs. More complex motion models have been proposed to compensate the most important systematic errors that appear in these robots (Doh, Choset, & Chung, 2003), but they will be not discussed in this introductory text.

We assume the robot is instrumented with the appropriate odometry such that its straight movements can be measured by means of two variables: the direction angle β (relative to the robot platform heading) and the traveled distance Δu. For example, in a synchro-drive robot those variables could be directly mapped to a pair of rotary encoders attached to the direction mechanism and to the wheel axes, respectively. A third encoder must be present to measure the rotation $\Delta\theta$ of the robot platform on its own center at the end of the movement (refer to Figure 3).

We will define the action vector $\mathbf{u}_k$ of this kinematic model as being composed of those three odometric readings, that is:

$$\mathbf{u}_k = \begin{bmatrix} \Delta u \\ \beta \\ \Delta\theta \end{bmatrix} \tag{21}$$

Unlike the previous motion model, the state vector for the robot pose in this case only contains the planar position and the robot heading:

$$\mathbf{x}_k = \begin{bmatrix} x_k \\ y_k \\ \theta_k \end{bmatrix} \quad (22)$$

Notice that the direction of motion, the β angle measured by the robot odometry, cannot be an absolute direction with respect to any global frame of reference. Instead, we insist on it being relative to the robot, such that the direction $\beta = 0$ corresponds to an arbitrary axis fixed to the robot body.

After these definitions, the holonomic kinematic model can be simply described as:

$$\mathbf{x}_k = \mathbf{g}\left(\mathbf{x}_{k-1}, \mathbf{u}_k\right) =$$

(seeing the new pose as an increment relative to the previous robot pose)

$$= \mathbf{x}_{k-1} \oplus \begin{bmatrix} \Delta u \cos \beta \\ \Delta u \sin \beta \\ \Delta \theta \end{bmatrix} =$$

(substituting the expression for the pose composition operator)

$$= \begin{bmatrix} x_{k-1} + \cos \theta_{k-1} \Delta u \cos \beta - \sin \theta_{k-1} \Delta u \sin \beta \\ y_{k-1} + \sin \theta_{k-1} \Delta u \cos \beta + \cos \theta_{k-1} \Delta u \sin \beta \\ \theta_{k-1} + \Delta \theta \end{bmatrix} =$$

(using the trigonometric identities for the sine and cosine of the sum of two angles)

$$= \begin{bmatrix} x_{k-1} + \Delta u \cos\left(\theta_{k-1} + \beta\right) \\ y_{k-1} + \Delta u \sin\left(\theta_{k-1} + \beta\right) \\ \theta_{k-1} + \Delta \theta \end{bmatrix} =$$

(simplifying via the vector notation)

$$= \mathbf{x}_{k-1} + \begin{bmatrix} \Delta u \cos\left(\beta + \theta_{k-1}\right) \\ \Delta u \sin\left(\beta + \theta_{k-1}\right) \\ \Delta \theta \end{bmatrix} \quad (23)$$

Probabilistic Motion Model

Correspondingly to these equations for updating a robot pose from an odometry reading $\mathbf{u}_k$, we define a probabilistic version by now interpreting all the variables as random variables and introducing an additive Gaussian error $\mathbf{r}_k = \begin{bmatrix} r_{\Delta u} & r_\beta & r_{\Delta\theta} \end{bmatrix}^T$ that affects odometry, such as:

$$\mathbf{x}_k = \mathbf{g}'\left(\mathbf{x}_{k-1}, \mathbf{u}_k, \mathbf{r}_k\right) =$$

(by definition, $\mathbf{r}_k$ is an additive noise of the odometry $\mathbf{u}_k$)

$$= \mathbf{g}\left(\mathbf{x}_{k-1}, \mathbf{u}_k + \mathbf{r}_k\right) =$$

(from Equation 23):

$$= \mathbf{x}_{k-1} + \begin{bmatrix} \left(\Delta u + r_{\Delta u}\right) \cos\left(\theta_{k-1} + \beta + r_\theta\right) \\ \left(\Delta u + r_{\Delta u}\right) \sin\left(\theta_{k-1} + \beta + r_\theta\right) \\ \theta_{k-1} + \Delta \theta + r_{\Delta\theta} \end{bmatrix} \quad (24)$$

As in other motion models, one should have clear that the intention of introducing this random error $\mathbf{r}_k$ is not to account for accuracy errors in the odometry encoders (errors which is most cases will be negligible), but for the imperfections in the wheel-ground interactions and in the wheels themselves: even if an omnidirectional robot accurately aligns all its wheels in exactly one direction and make them rotate at unison exactly the same number of revolutions, it will not follow an

Table 3. Means and covariances

Random variable	Description	Dimensionality
$\mathbf{x}_k \sim N\left(\overline{\mathbf{x}}_k, \boldsymbol{\Sigma}_{x_k}\right)$	Robot pose for the current time step.	3
$\mathbf{x}_{k-1} \sim N\left(\overline{\mathbf{x}}_{k-1}, \boldsymbol{\Sigma}_{x_{k-1}}\right)$	Robot pose for the previous time step.	3
$\mathbf{u}_k \sim N\left(\overline{\mathbf{u}}_k, \boldsymbol{\Sigma}_{u_k}\right)$	Action vector (odometry).	3

exact straight path for a variety of reasons (Borenstein, 1996).

Also notice at this point how assuming an additive Gaussian noise in the odometry can be directly modeled by considering a normally-distributed odometry $\mathbf{u}_k$ whose covariance matrix already incorporates the noise $\mathbf{r}_k$ plus, optionally, any other modeled inaccuracies due to imperfections in the traction system.

Therefore, we assume that all random variables in Equation 24 are normally distributed with the following means and covariances (see Table 3).

Then, the Gaussian parameters for $\mathbf{x}_k$ can be estimated in the same way as the previous model. Thus, the mean is simply the propagation of the mean of the input variables through the motion model function:

$$\overline{\mathbf{x}}_k = \mathbf{g}'\left(\overline{\mathbf{x}}_{k-1}, \overline{\mathbf{u}}_k, 0\right) = \mathbf{g}\left(\overline{\mathbf{x}}_{k-1}, \overline{\mathbf{u}}_k\right) \tag{25}$$

while the covariance can be obtained from the uncertainty in each of the random variables (previous robot pose $\mathbf{x}_k$, odometry $\mathbf{u}_k$ and noise $\mathbf{r}_k$). A very good approximation in this case is to assume statistical independence between those three variables, leading to a simpler expression for the covariance as the sum of three independent components:

$$\begin{aligned}
\boldsymbol{\Sigma}_{x_{k-1}} &= \frac{d\mathbf{g}'}{d\left\{\mathbf{x}_{k-1}, \mathbf{u}_k\right\}} \begin{pmatrix} \boldsymbol{\Sigma}_{\mathbf{x}_{k-1}} & \mathbf{0}_{3\times3} \\ \mathbf{0}_{3\times3} & \boldsymbol{\Sigma}_{\mathbf{u}_k} \end{pmatrix} \frac{d\mathbf{g}'}{d\left\{\mathbf{x}_{k-1}, \mathbf{u}_k\right\}}^T \\
&= \begin{pmatrix} \dfrac{\partial \mathbf{g}'}{\partial \mathbf{x}_{k-1}} & \dfrac{\partial \mathbf{g}'}{\partial \mathbf{u}_k} \end{pmatrix} \begin{pmatrix} \boldsymbol{\Sigma}_{\mathbf{x}_{k-1}} & \mathbf{0}_{3\times3} \\ \mathbf{0}_{3\times3} & \boldsymbol{\Sigma}_{\mathbf{u}_k} \end{pmatrix} \begin{pmatrix} \dfrac{\partial \mathbf{g}'}{\partial \mathbf{x}_{k-1}}^T \\ \dfrac{\partial \mathbf{g}'}{\partial \mathbf{u}_k}^T \end{pmatrix}^T \\
&= \frac{\partial \mathbf{g}'}{\partial \mathbf{x}_{k-1}} \boldsymbol{\Sigma}_{\mathbf{x}_{k-1}} \frac{\partial \mathbf{g}'}{\partial \mathbf{x}_{k-1}}^T + \frac{\partial \mathbf{g}'}{\partial \mathbf{u}_k} \boldsymbol{\Sigma}_{\mathbf{u}_k} \frac{\partial \mathbf{g}'}{\partial \mathbf{u}_k}^T
\end{aligned} \tag{26}$$

From Equation 24 we can easily derive the two Jacobian matrices which, as follows from the application of first-order Taylor expansion, must be evaluated at the mean value of all the input variables:

$$\frac{\partial \mathbf{g}'}{\partial \mathbf{x}_{k-1}} = \begin{pmatrix} 1 & 0 & -\Delta u \sin\left(\theta_{k-1} + \beta\right) \\ 0 & 1 & \Delta u \cos\left(\theta_{k-1} + \beta\right) \\ 0 & 0 & 1 \end{pmatrix} \tag{27}$$

$$\frac{\partial \mathbf{g}'}{\partial \mathbf{u}_{k-1}} = \begin{pmatrix} \cos\left(\theta_{k-1} + \beta\right) & -\Delta u \sin\left(\theta_{k-1} + \beta\right) & 0 \\ \sin\left(\theta_{k-1} + \beta\right) & \Delta u \cos\left(\theta_{k-1} + \beta\right) & 0 \\ 0 & 0 & 1 \end{pmatrix} \tag{28}$$

Regarding the covariance matrix of the odometry $\boldsymbol{\Sigma}_{\mathbf{u}_k}$, it must be determined empirically for each specific robot from heuristic criteria. A trivial approach would consist of always employing a constant matrix, that is, assume that the uncertainty in each odometry measurement (distance Δu, rotation $\Delta\theta$ and direction β) is constant no matter how much or how little the robot moves in a given time step. Obviously, doing so would underestimate the uncertainty when the robot is moving fast and overestimating it when the robot is still. Both situations are undesirable since may lead to a divergence of the estimation in the localization problem or, in the best case, to an unnecessary computational cost in some implementations as particle filters (a technique to be discussed in chapter 7).

Therefore, we derive next another heuristic matrix which, still being quite simple, provides us a good representation of typical odometry errors. The first point is to make an assumption on what those errors should look like. Focusing on the first component of the odometry, the distance Δu, it is plausible to assume that its uncertainty grows as the traveled distance, and that it is largely independent of the direction of motion β. Regarding the robot rotation $\Delta\theta$, its value should not affect the uncertainty in Δu if the robot really moved first in straight line and then rotating, but in practice both movements are hardly decoupled, thus we will also assume that larger rotations should imply larger uncertainties. A similar reasoning can be followed regarding the other two odometry components, and thus we conclude that the uncertainty in the all three components must grow with the distance and rotation, but do not change with the direction angle.

Next, it must be determined in which way the uncertainty grows: the relationship could be linear, quadratic or anything else. The issue can be resolved by imposing a simple requisite: the uncertainty of any odometry increment should be independent of a given robot motion being divided into one or several consecutive odometry increments. In other words, the uncertainty due to odometry should be equal after moving three meters forward than after moving exactly the same distance but considering two time steps instead of just one. It can be easily shown that to fulfill this condition, the *variance* of an odometry component must grow *linearly to the absolute value* of the parameter it depends on[5].

Putting the two previous paragraphs together, we arrive at the following plausible model for a covariance matrix that models the uncertainty for the present motion model:

$$\mathbf{\Sigma}_{\mathbf{u}_k} = \begin{pmatrix} \sigma^2_{\Delta_u} & 0 & 0 \\ 0 & \sigma^2_{\beta} & 0 \\ 0 & 0 & \sigma^2_{\Delta\theta} \end{pmatrix} \qquad (29)$$

where each diagonal element describes the variance of one odometry component, given by:

$$\begin{aligned} \sigma^2_{\Delta_u} &= \varsigma^2_{\Delta_u} + \alpha_1 \left|\Delta u\right| + \alpha_2 \left|\Delta\theta\right| \\ \sigma^2_{\beta} &= \varsigma^2_{\beta} + \alpha_3 \left|\Delta u\right| + \alpha_4 \left|\Delta\theta\right| \\ \sigma^2_{\Delta\theta} &= \varsigma^2_{\Delta\theta} + \alpha_5 \left|\Delta u\right| + \alpha_6 \left|\Delta\theta\right| \end{aligned} \qquad (30)$$

Notice that the α_i coefficients are the proportionality constants that force each variance to increase linearly with the absolute value of the traveled distance or rotation. The optional terms ς^2_i, which can be set to zero, allow us fixing a minimum variance in each component, which may become necessary in sample-based probabilistic algorithms for localization to avoid the problem of degeneracy (see chapter 7 section 4). Regarding the proportionality constants, these are the physical meanings and units associated to each of them:

- α_1 $(\mathrm{m}^2/\mathrm{m})$: Uncertainty in the traveled distance as the robot moves farther away in a straight line. This term accounts for calibration inaccuracies of the encoder-ticks-to-distance ratio, the radius of the wheels and minor slippage effects.

- α_2 $\left({}^{\mathrm{m}^2}\!/_{\mathrm{rad}}\right)$ and α_5 $\left({}^{\mathrm{rad}^2}\!/_{\mathrm{m}}\right)$: Uncertainty in traveled distance as the robot rotates and vice versa, respectively. If we do not need to consider the case that straight motion and rotation overlap, that is, the robot only executes rotations when standing still, then these two parameters can be set to zero. Otherwise, their values should be larger as both movements overlap in time more often.
- α_3 $\left({}^{\mathrm{rad}^2}\!/_{\mathrm{m}}\right)$ and α_4 $\left({}^{\mathrm{rad}^2}\!/_{\mathrm{rad}}\right)$: Uncertainty in the direction of motion as the robot moves farther away or rotates larger angles, respectively. In this case the parameters model potential deviations in the straight path caused by even floors, different slippage in each wheel, etc.
- α_6 $\left({}^{\mathrm{rad}^2}\!/_{\mathrm{rad}}\right)$: Uncertainty in the rotation angle as it becomes larger. This last parameter accounts for possible defects in the encoder that measures the robot driving direction, the effects of slack gearings in that driving mechanism, etc.

Recall that all these parameters must be determined experimentally for a given robotic platform, ideally running some benchmarking tests to measure each of them independently. Once measured, it is convenient to slightly overestimate the uncertainty in a robot odometry to avoid the algorithm used for localization to become overconfident, which may lead to catastrophic results. This overestimation can be seen as a *security margin* for the odometry uncertainty, preparing the model to account for more adverse situations than those analyzed in the laboratory benchmarks. As a rule of thumb, an overestimation of 10% – 20% in all the parameters above is a good practice in general.

4. NON-HOLONOMIC MODEL WITH TWO WHEEL ENCODERS

Kinematic Equations

We address now the motion model with the most widespread applicability in mobile robotics at present. As summarized in Table 1, this model is applicable to robots with tricycle-like or differential drive locomotion system, the latter in its two variants: two and four wheel-drive. All what is needed to apply this model is the existence of two odometry encoders to measure the revolution of the left and right wheels. Such locomotion model is often employed in hobbyist robots but also in a variety of commercial mobile robots, such as in the Pioneer, Powerbot, and Peoplebot families from MobileRobots, in the B21 robots from Real World Interface (RWI) or in popular configurations of the recent Lego Mindstorm NXT robotic kits.

Since this model accounts for planar movements only, the state vector describing the robot pose at time step k comprises its position and heading:

$$\mathbf{x}_k = \begin{bmatrix} x_k \\ y_k \\ \theta_k \end{bmatrix} \tag{31}$$

We will assume that the encoder measurements are readily available as the distance traveled by each wheel. Such values can be obtained from the tick counts of each encoder by means of the appropriate conversion parameters estimated from a calibration process. As a result of that calibration, the distance between both wheels (denoted as D) must be also available as a known parameter. In this kinematic model, therefore, the action vector $\mathbf{u}_k$ is built up from those distances:

Figure 4. (a) The trajectory of a non-holonomic robot instrumented with two odometers can be described as composed of sequences of circular arcs. (b) Equivalent representation of the arc segment trajectory as seen in a frame of coordinates attached to the original vehicle pose.

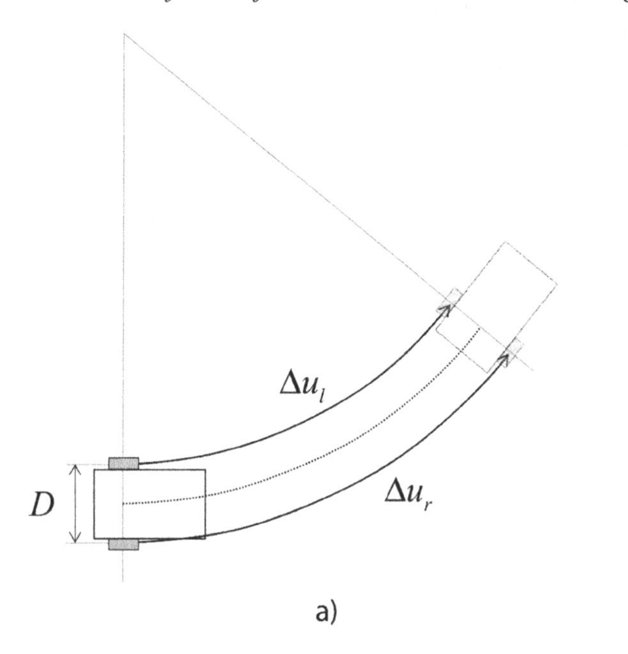

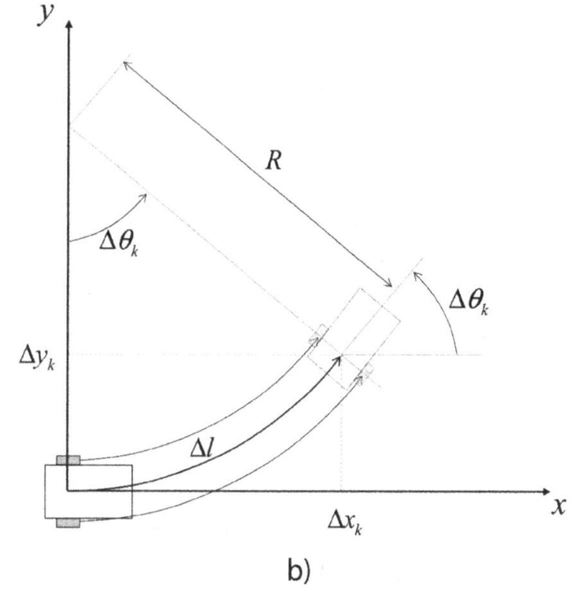

$$\mathbf{u}_k = \begin{bmatrix} \Delta u_l \\ \Delta u_r \end{bmatrix} \tag{32}$$

where the physical meaning of these values are represented in Figure 4a.

At this point, it is mandatory to introduce the concept of *non-holonomic restrictions* in any physical model—recall that a definition of what a holonomic robot is was already given in the previous section. Mathematically, non-holonomic constraints are equations that involve not only the robot state vector, but at least one of its time derivatives, i.e., velocities. In the present kinematic model, there exists such a restriction: the instantaneous robot velocity vector (which is a time derivative of the robot position in global coordinates) must be always aligned to its instantaneous heading. Just like an automobile, the robot cannot move sideways. Therefore, we are facing a non-holonomic kinematic model, and the relevance of this classification comes from the fact that in such systems, the robot pose depends not only on the total action $\mathbf{u}_k$ applied during a certain period of time, but also on how those actions have varied throughout that period.

Consequently, some assumption must be made about how actions $\mathbf{u}_k$ are applied over time. The most practical approach is to assume that both the left and the right wheel velocities have remained constant within each time step, an approximation which in fact is very accurate when odometry is sampled at a high frequency (e.g. at 10Hz or higher). In that case, the robot describes a path with the exact shape of a circular arc, as already illustrated in Figure 4a. Also, the actual velocities will be revealed as irrelevant in the formulation, with all the information that is needed being the overall distance measured by each encoder in the time step, as already stated in our definition of the action vector in Equation 32. Note as well that when both wheels spin at exactly the same speed the path degenerates into a straight line, which can be seen as a special case of a circular arc with an infinite radius.

Once stated the assumptions taken by this model, we aim to derive a closed-form expression for this kinematic model. Firstly, we decompose the problem in two parts for clarity:

$$\begin{aligned}\mathbf{x}_k &= \mathbf{g}\left(\mathbf{x}_{k-1}, \mathbf{u}_k\right) \\ &= \mathbf{x}_{k-1} \oplus \mathbf{f}\left(\mathbf{u}_k\right)\end{aligned} \tag{33}$$

That is, the updated robot pose is simply the concatenation of the pose increment associated to the latest odometry $\mathbf{u}_k$, composed from the coordinates reference of the last robot pose $\mathbf{x}_{k-1}$. We denote that pose increment as $\mathbf{f}\left(\mathbf{u}_k\right)$, whose components $\mathbf{f}\left(\mathbf{u}_k\right) = \left[\Delta x_k \quad \Delta y_k \quad \Delta\theta_k\right]^T$ are derived next.

The diagram in Figure 4b represents the pose increment $\mathbf{f}\left(\mathbf{u}_k\right)$ for an arbitrary odometry reading $\mathbf{u}_k = \left[\Delta u_l \quad \Delta u_r\right]^T$. Note that the coordinate frame is attached to the previous robot pose, that is, the robot path always starts at the origin of coordinates. We also introduced in the figure other two auxiliary variables with important physical meanings: the distance traveled by the robot reference point (the point that lays exactly between the pair of instrumented wheels) and the radius of curvature of the path, denoted as Δl and R, respectively.

From the geometry of the problem we can state the following two identities regarding the distances Δu_l and Δu_r:

$$\begin{aligned}\Delta u_l &= \left(R - \frac{D}{2}\right)\Delta\theta_k \\ \Delta u_r &= \left(R + \frac{D}{2}\right)\Delta\theta_k\end{aligned} \tag{34}$$

Notice that these equations are not specific to left-hand arcs as the one shown in the figure, but also hold for right-hand arcs by just employing negative radius values. In a similar way, we can write the equation for the path length of the robot central point:

$$\Delta l = R\Delta\theta_k \tag{35}$$

or, equivalently:

$$\Delta\theta_k = \frac{\Delta l}{R} \tag{36}$$

Then, if we sum the two identities from Equation 34 and replace the value of the robot angle increment from Equation 36, we obtain:

$$\Delta u_r + \Delta u_l = \left(R + \frac{D}{2} + R - \frac{D}{2}\right)\Delta\theta_k = 2R\Delta\theta_k =$$

(using Equation 36):

$$= 2\not{R}\frac{\Delta l}{\not{R}} =$$

$$= 2\Delta l \rightarrow$$

$$\Delta l = \frac{\Delta u_r + \Delta u_l}{2} \tag{37}$$

while if we subtract the same two identities from Equation 34 we arrive at:

$$\begin{aligned}\Delta u_r - \Delta u_l &= \left(\not{R} + \frac{D}{2} - \not{R} + \frac{D}{2}\right)\Delta\theta_k \\ &= D\Delta\theta_k \qquad \rightarrow \\ \Delta\theta_k &= \frac{\Delta u_r - \Delta u_l}{D}\end{aligned} \tag{38}$$

Once we have derived the values of Δl and $\Delta\theta_k$ we also know the radius of curvature from Equation 35, that is:

$$R = \frac{\Delta l}{\Delta \theta_k} = \frac{\frac{\Delta u_r + \Delta u_l}{2}}{\frac{\Delta u_r - \Delta u_l}{D}} = \frac{D}{2}\frac{\Delta u_r + \Delta u_l}{\Delta u_r - \Delta u_l} \quad (39)$$

which is a valid expression for all cases except when the robot moves in an exact straight line $(\Delta\theta_k = 0)$ where this radius becomes infinity.

At this point the expression for the coordinates of the pose increment $\mathbf{f}(\mathbf{u}_k)$ can be written by means of straightforward trigonometric relationships:

$$\mathbf{f}(\mathbf{u}_k) = \begin{bmatrix} \Delta x_k \\ \Delta y_k \\ \Delta\theta_k \end{bmatrix} = \begin{bmatrix} R\sin\Delta\theta_k \\ R(1-\cos\Delta\theta_k) \\ \Delta\theta_k \end{bmatrix} = \begin{cases} \begin{bmatrix} \frac{D}{2}\frac{\Delta u_r + \Delta u_l}{\Delta u_r - \Delta u_l}\sin\frac{\Delta u_r - \Delta u_l}{D} \\ \frac{D}{2}\frac{\Delta u_r + \Delta u_l}{\Delta u_r - \Delta u_l}\left(1-\cos\frac{\Delta u_r - \Delta u_l}{D}\right) \\ \frac{\Delta u_r - \Delta u_l}{D} \end{bmatrix} & \text{, if } \Delta u_r \neq \Delta u_l \\ \begin{bmatrix} \frac{\Delta u_r + \Delta u_l}{2} \\ 0 \\ 0 \end{bmatrix} & \text{, otherwise} \end{cases} \quad (40)$$

Therefore, by putting together Equations 33 and 40 we have this kinematic model completely defined by means of closed-form expressions. Our interest is next on deriving the probabilistic version of those equations.

Probabilistic Motion Model

As in previous sections, we rely on the approximation of assuming that all the involved variables can be modeled as Gaussians (see Table 4).and we will approximate the parameters of $\mathbf{x}_k$ by a first-order approximation of the kinematic equation in Equation 33. Consequently, the mean of the new robot pose simply becomes:

$$\bar{\mathbf{x}}_k = \mathbf{g}(\bar{\mathbf{x}}_{k-1}, \bar{\mathbf{u}}_k) \quad (41)$$

while its covariance matrix is approximated as:

$$\begin{aligned} \mathbf{\Sigma}_{x_{k-1}} &= \frac{dg}{d\{\mathbf{x}_{k-1}, \mathbf{u}_k\}} \begin{pmatrix} \mathbf{\Sigma}_{\mathbf{x}_{k-1}} & \mathbf{0}_{3\times2} \\ \mathbf{0}_{2\times3} & \mathbf{\Sigma}_{\mathbf{u}_k} \end{pmatrix} \frac{dg}{d\{\mathbf{x}_{k-1}, \mathbf{u}_k\}}^T \\ &= \begin{pmatrix} \frac{\partial \mathbf{g}}{\partial \mathbf{x}_{k-1}} & \frac{\partial \mathbf{g}}{\partial \mathbf{u}_k} \end{pmatrix} \begin{pmatrix} \mathbf{\Sigma}_{\mathbf{x}_{k-1}} & \mathbf{0}_{3\times2} \\ \mathbf{0}_{2\times3} & \mathbf{\Sigma}_{\mathbf{u}_k} \end{pmatrix} \begin{pmatrix} \frac{\partial \mathbf{g}}{\partial \mathbf{x}_{k-1}}^T \\ \frac{\partial \mathbf{g}}{\partial \mathbf{u}_k}^T \end{pmatrix} \\ &= \frac{\partial \mathbf{g}}{\partial \mathbf{x}_{k-1}} \mathbf{\Sigma}_{\mathbf{x}_{k-1}} \frac{\partial \mathbf{g}}{\partial \mathbf{x}_{k-1}}^T + \frac{\partial \mathbf{g}}{\partial \mathbf{u}_k} \mathbf{\Sigma}_{\mathbf{u}_k} \frac{\partial \mathbf{g}}{\partial \mathbf{u}_k}^T \end{aligned} \quad (42)$$

Table 4. Involved variables modeled as Gaussians

Random variable	Description	Dimensionality
$\mathbf{x}_k \sim N(\bar{\mathbf{x}}_k, \mathbf{\Sigma}_{x_k})$	Robot pose for the current time step.	3
$\mathbf{x}_{k-1} \sim N(\bar{\mathbf{x}}_{k-1}, \mathbf{\Sigma}_{x_{k-1}})$	Robot pose for the previous time step.	3
$\mathbf{u}_k \sim N(\bar{\mathbf{u}}_k, \mathbf{\Sigma}_{u_k})$	Action vector (odometry).	2

The first Jacobian can be easily obtained from Equation 33 by replacing the pose composition operator $\oplus$ by its analytical expression, finally leading to:

$$\frac{\partial \mathbf{g}}{\partial \mathbf{x}_{k-1}} = \begin{pmatrix} 1 & 0 & R\left(-s_{k-1}s_{\Delta} - c_{k-1}\left(1-c_{\Delta}\right)\right) \\ 0 & 1 & R\left(c_{k-1}s_{\Delta} - s_{k-1}\left(1-c_{\Delta}\right)\right) \\ 0 & 0 & 1 \end{pmatrix} \tag{43}$$

where we have used the substitutions $c_{k-1} = \cos\theta_{k-1}$ $s_{k-1} = \sin\theta_{k-1}$, $c_{\Delta} = \cos\Delta\theta_k$ and $s_{\Delta} = \sin\Delta\theta_k$ for the sake of notation.

About the second Jacobian, the straightforward approach of substituting Equation 40 into Equation 33 and taking derivatives leads to excessively lengthy expressions. Instead, we will take the alternative way of applying the derivative chain rule, such that the Jacobian is obtained as the multiplication of two simpler matrices:

$$\frac{\partial \mathbf{g}}{\partial \mathbf{u}_k} = \frac{\partial \mathbf{g}}{\partial\left\{R, \Delta\theta_k\right\}} \frac{\partial\left\{R, \Delta\theta_k\right\}}{\partial \mathbf{u}_k} = \begin{pmatrix} c_{k-1}s_{\Delta} - s_{k-1}\left(1-c_{\Delta}\right) & R\left(c_{k-1}c_{\Delta} - s_{k-1}s_{\Delta}\right) \\ s_{k-1}s_{\Delta} + c_{k-1}\left(1-c_{\Delta}\right) & R\left(s_{k-1}c_{\Delta} + c_{k-1}s_{\Delta}\right) \\ 0 & 1 \end{pmatrix} \cdot \begin{pmatrix} D\frac{\Delta u_r}{\left(\Delta u_r - \Delta u_l\right)^2} & -D\frac{\Delta u_l}{\left(\Delta u_r - \Delta u_l\right)^2} \\ -\frac{1}{D} & \frac{1}{D} \end{pmatrix} \tag{44}$$

Before ending the description of this motion model, we must mention that the covariance of the odometry, that is:

$$\mathbf{\Sigma}_{\mathbf{u}_k} = \begin{pmatrix} \sigma^2_{\Delta u_l} & \sigma_{\Delta u_l \Delta u_r} \\ \sigma_{\Delta u_l \Delta u_r} & \sigma^2_{\Delta u_r} \end{pmatrix} \tag{45}$$

has to be provided by means of any empirical model calibrated for the specific robot. In the absence of any customized model, the following generic approximation could be employed:

$$\mathbf{\Sigma}_{\mathbf{u}_k} = \begin{pmatrix} \alpha_1\left|\Delta u_l\right| & 0 \\ 0 & \alpha_2\left|\Delta u_r\right| \end{pmatrix} \tag{46}$$

Here, the errors in each wheel encoder are assumed to be independent (i.e. the cross covariance terms are zero), and the uncertainty in the variance of each reading grows linearly with the absolute value of that reading value. As mentioned earlier in this chapter, this relationship leads to uncertainty values with the desirable property of remaining fairly independent of the sample rate at which time steps are processed. About the proportionality constants α_1 and α_2, given in units of m^2/m, their values must be estimated empirically for each robotic platform.

5. NON-HOLONOMIC MODEL WITH ONE ANGULAR AND ONE WHEEL ENCODER

Kinematic Equations

This section addresses the kinematic equations of a vehicle with tricycle-like or Ackermann steering and whose steering wheels have been instrumented with odometry.

Firstly, we must review the spatial structure of these two steering schemes, briefly mentioned in chapter 2. As it can be seen in Figure 5a, a tricycle-like vehicle comprises two rear wheels and only one central, front steering wheel. The radius of curvature of the path described by such a vehicle is controlled by rotating the front steering wheel. In fact, that radius is easily determined from the geometric constraint that requires the center of the arcs to lie on the intersection of the

Figure 5. Geometry of the (a) tricycle-like and (b) Ackermann steering

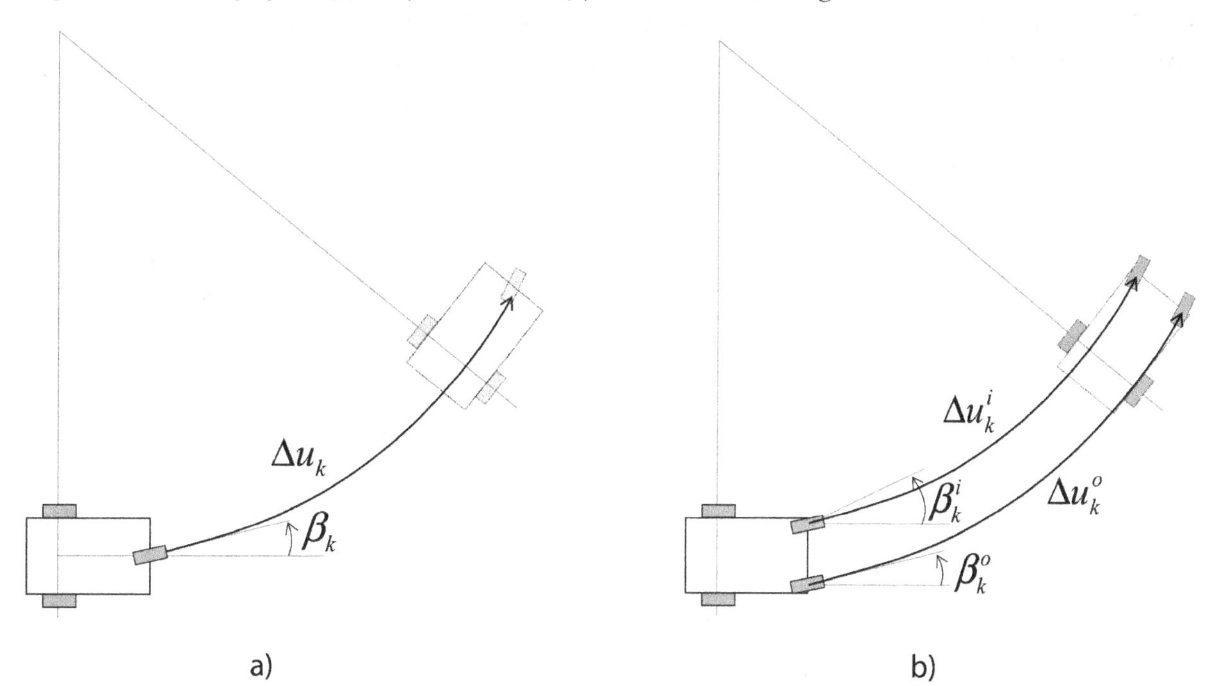

extended axes of all the vehicle wheels, as shown in the figure. Regarding the Ackermann steering, illustrated in Figure 5b, it is widely known due to its usage in virtually all automobiles. The difference with the tricycle model is the existence of two front steering wheels which, in order to fulfill the abovementioned geometric constraint, must be turned in slightly different angles to avoid unnecessary frictions with the road.

In both the tricycle and the Ackermann models, the torque from the motor can be transmitted to the front wheels, but more often, the rear wheels are the driving wheels. Under such design, commercial automobiles employ a mechanism called *differential* to transmit the torque to both wheels by means of two separate shafts, which can rotate at different velocities. Smaller vehicles, such as those used in robotics, often rely on having only one single axle for both driving wheels, a more inexpensive solution that in turn increases the slippage in curves.

In any case, the point of interest for us about these locomotion systems is how they are instrumented with odometry sensors. As it was reflected in Table 1, one particular vehicle can be instrumented in different ways and each scheme would imply the application of different equations. For tricycle-like and car-like vehicles, the two choices for odometry instrumentation are either two wheel encoders in each rear wheel or one wheel encoder and one angular encoder in each steering wheel. In the former case, the differential drive (described in chapter 5 section 4) model must be employed, while the latter is the scheme addressed here.

In the following, we will focus on a tricycle-like vehicle instrumented with one wheel and one angular encoder, both attached to its front steering wheel as shown in Figure 5a. These sensors provide us with the length of the path described by that wheel and its orientation with respect to a fixed reference on the vehicle, respectively. For a time step k we will denote those odometry readings as Δu_k and β_k, respectively.

The Ackermann model will be not explicitly addressed below since its odometry readings can be always transformed into an equivalent tricycle

Figure 6. (a) A vehicle with Ackermann steering can be instrumented by measuring the angle of the steering wheel and the distance it traverses, which can then be transformed into (b) an equivalent tricycle motion model measurement.

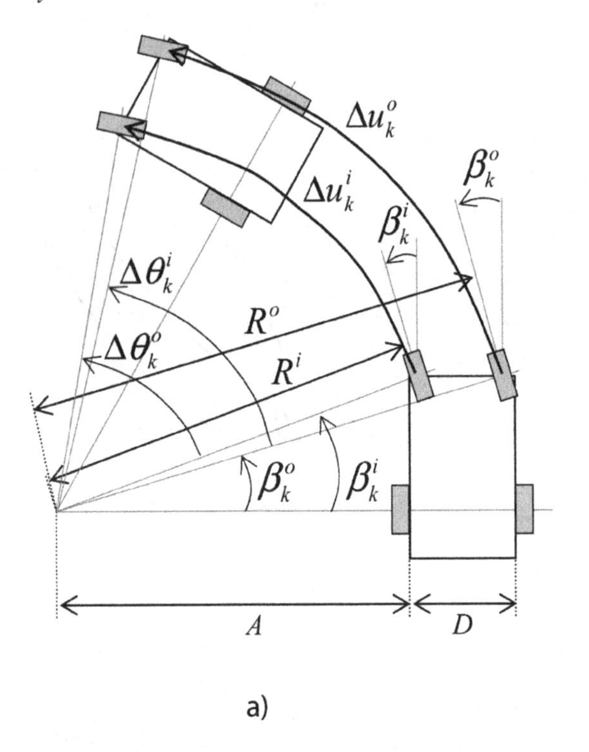

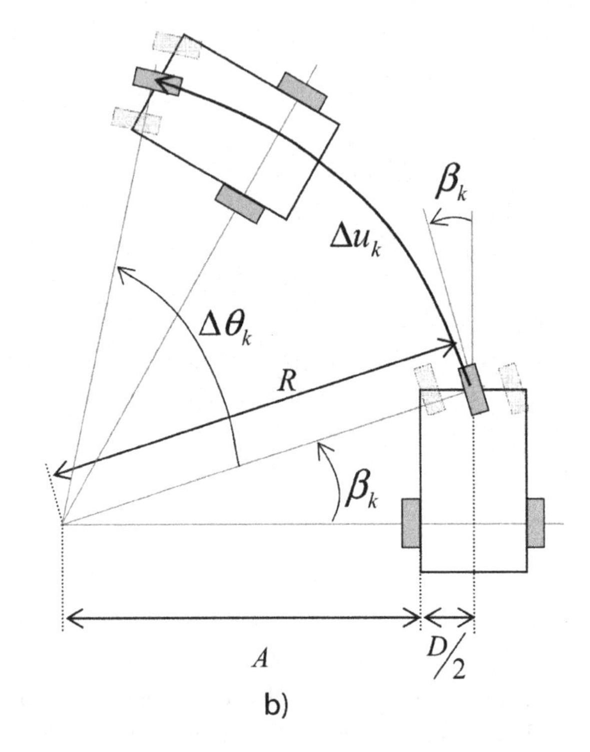

model. Let β_k^i and β_k^o denote the steering angles of the inner and outer wheel, respectively, and Δu_k^i and Δu_k^o the corresponding distance readings from each of the two wheels, as shown in Figure 5b. Our aim is to find an equivalent angle β_k and distance Δu_k such as the one-wheel tricycle model describes exactly the same path.

To begin with, we derive the Ackerman equation, which establishes the relationship between the angles of the two steering wheels. From the geometry of the problem, represented in Figure 6a, we can write:

$$\begin{aligned}\tan\beta_k^o &= \frac{L}{A+D} \\ \tan\beta_k^i &= \frac{L}{A}\end{aligned} \tag{47}$$

where L and D stand for the vehicle length and width, respectively, and A is the lateral distance from the inner wheel to the concurring center of all the circular arcs described by the vehicle wheels. Subtracting the inverse of both identities above gives us:

$$\begin{aligned}\frac{1}{\tan\beta_k^o} - \frac{1}{\tan\beta_k^i} &= \frac{A+D}{L} - \frac{A}{L} \\ \cot\beta_k^o - \cot\beta_k^i &= \frac{D}{L}\end{aligned} \tag{48}$$

which is the desired Ackerman equation. Following now a similar approach for the equivalent angle β_k, represented in Figure 6b, it turns out that:

$$\tan \beta_k = \frac{L}{A + \frac{D}{2}} \quad \rightarrow \quad \cot \beta_k = \frac{A + \frac{D}{2}}{L}$$

(subtracting the value of $\cot \beta_k^i$ from Equation 48)

$$\begin{aligned} \cot \beta_k - \cot \beta_k^i &= \frac{A + \frac{D}{2}}{L} - \frac{A}{L} = \frac{D}{2L} && \rightarrow \\ \cot \beta_k &= \frac{D}{2L} + \cot \beta_k^i && \rightarrow \\ \beta_k &= \text{arc cot}\left(\frac{D}{2L} + \cot \beta_k^i\right) \end{aligned} \tag{49}$$

Regarding the equivalent distance Δu_k, we can only obtain its value after establishing the value of the three different radius of curvature in the problem (one for each real wheel plus that for the "virtual" central wheel). Analyzing the geometry in Figures 6a, b, we can state that:

$$R^i = \frac{L}{\sin \beta_k^i} \quad , \quad R^o = \frac{L}{\sin \beta_k^o} \quad , \quad R = \frac{L}{\sin \beta_k} \tag{50}$$

where we must remind that all three angles are known: the two first are readings from the odometry, and an expression for the last one was just derived above. Denoting the rigid rotation of the vehicle with respect to its initial position as $\Delta\theta_k$, it is clear that:

$$\Delta\theta_k = \frac{\Delta u_k^i}{R^i} \quad , \quad \Delta\theta_k = \frac{\Delta u_k}{R}$$

(replacing the values of Equation 50):

$$\begin{aligned} \Delta\theta_k &= \frac{\Delta u_k^i}{\frac{L}{\sin \beta_k^i}} = \Delta u_k^i \frac{\sin \beta_k^i}{L}, \\ \Delta\theta_k &= \frac{\Delta u_k}{\frac{L}{\sin \beta_k}} = \Delta u_k \frac{\sin \beta_k}{L} \end{aligned}$$

(joining both identities)

$$\begin{aligned} \Delta u_k^i \frac{\sin \beta_k^i}{\not{L}} &= \Delta u_k \frac{\sin \beta_k}{\not{L}} \quad \rightarrow \\ \Delta u_k &= \Delta u_k^i \frac{\sin \beta_k^i}{\sin \beta_k} \end{aligned} \tag{51}$$

That is, by using Equations 49 and 51 we will be always able to convert odometry readings from the two front wheels of an Ackerman-like vehicle into equivalent one-wheel tricycle odometry readings. Hence, we can focus only on the kinematic equations of the latter without loss of generality.

We will define an action vector comprising only one pair of odometry readings: a distance and an angle:

$$\mathbf{u}_k = \begin{bmatrix} \Delta u_k \\ \beta_k \end{bmatrix} \tag{52}$$

As in previous models, the vehicle pose at time step k is represented as a 2D position plus a heading angle:

$$\mathbf{x}_k = \begin{bmatrix} x_k \\ y_k \\ \theta_k \end{bmatrix} \tag{53}$$

and yet again, it becomes convenient to decompose the kinematic function $\mathbf{g}(\cdot)$ into the geometric pose composition of the circular arc described by the odometry increment $\mathbf{u}_k$ concatenated to the

Figure 7. Detailed look at the geometry of an odometry increment for the tricycle kinematic model

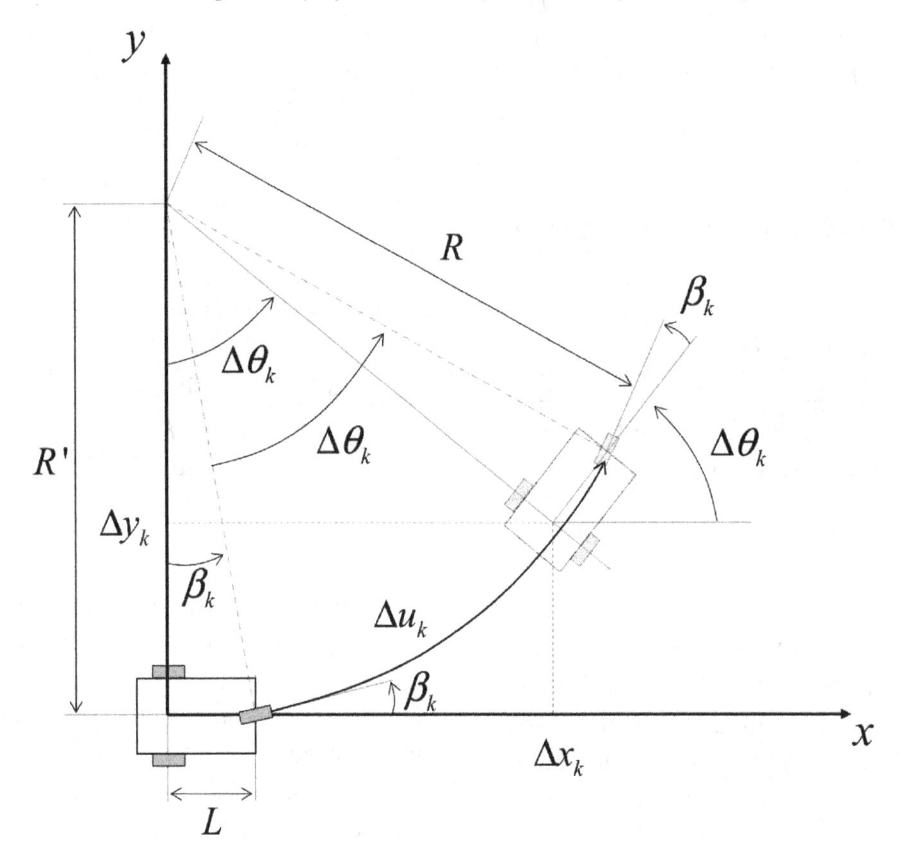

previous robot pose. Denoting the coordinates of the updated robot pose with respect to the previous pose as $\mathbf{f}\left(\mathbf{u}_k\right) = \left[\Delta x_k \quad \Delta y_k \quad \Delta \theta_k\right]^T$, we can write:

$$\begin{aligned} \mathbf{x}_k &= \mathbf{g}\left(\mathbf{x}_{k-1}, \mathbf{u}_k\right) \\ &= \mathbf{x}_{k-1} \oplus \mathbf{f}\left(\mathbf{u}_k\right) \end{aligned} \tag{54}$$

As shown in Figure 7, the robot pose $\mathbf{x}_k$ actually defines the location and orientation of one particular reference point on the vehicle, the midpoint between the rear wheels, which is consistent with the convention employed in previous sections. From the geometry of the problem, it can be seen that the radius of curvature for the arc described by the front wheel, R in the figure, is related to the wheel angle β_k by means of:

$$\sin \beta_k = \frac{L}{R} \quad \rightarrow \quad R = \frac{L}{\sin \beta_k} \tag{55}$$

Now, given that we know the length of the circular arc described by that point (it is measured by the odometry as Δu_k), we can obtain the rigid rotation of the vehicle as:

$$R\,\Delta\theta_k = \Delta u_k \quad \rightarrow \quad \Delta\theta_k = \frac{\Delta u_k}{R} = \frac{\Delta u_k \sin \beta_k}{L} \tag{56}$$

In order to compute the coordinates $\left(\Delta x_k, \Delta y_k\right)$ of the reference point of the vehicle after the odometry increment, it must be observed that the radius of curvature described by that point is slightly different than the radius described by the front wheel. Both radii are denoted in the Figure

7 as R' and R, respectively. The former is easily computed from the following trigonometric identity:

$$\tan\beta_k = \frac{L}{R'} \quad \rightarrow \quad R' = \frac{L}{\tan\beta_k} \tag{57}$$

and, referring once more to the problem geometry in the figure, it follows that:

$$\sin\Delta\theta_k = \frac{\Delta x_k}{R'}$$
$$\rightarrow \quad \Delta x_k = R'\sin\Delta\theta_k = L\frac{\sin\Delta\theta_k}{\tan\beta_k}$$

$$\cos\Delta\theta_k = \frac{R' - \Delta y_k}{R'}$$
$$\rightarrow \quad \Delta y_k = R'\left(1-\cos\Delta\theta_k\right) = L\frac{1-\cos\Delta\theta_k}{\tan\beta_k} \tag{58}$$

Notice that this kinematic model is affected by a singularity (i.e. a denominator that becomes zero) whenever the vehicle moves exactly along a straight line, that is, when the angle of the steering wheel β_k is zero. In that case, it becomes straightforward to prove that the pose increment we are looking for is simply $\left(\Delta x_k, \Delta y_k, \Delta\theta_k\right) = \left(\Delta u_k, 0, 0\right)$.

Probabilistic Motion Model

Next, we analyze the corresponding probabilistic version of this motion model, assuming as usual that all the variables can be properly modeled as Gaussians (see Table 5).

The parameters of $\mathbf{x}_k$ can be approximated, like in previous models, by first-order linearization of the kinematic equations in Equation 54, giving us a mean value of:

$$\bar{\mathbf{x}}_k = \mathbf{g}\left(\bar{\mathbf{x}}_{k-1}, \bar{\mathbf{u}}_k\right) \tag{59}$$

and the following approximated covariance matrix:

$$\begin{aligned}
\boldsymbol{\Sigma}_{x_{k-1}} &= \frac{d\mathbf{g}}{d\left\{\mathbf{x}_{k-1}, \mathbf{u}_k\right\}} \begin{pmatrix} \boldsymbol{\Sigma}_{\mathbf{x}_{k-1}} & \mathbf{0}_{3\times2} \\ \mathbf{0}_{2\times3} & \boldsymbol{\Sigma}_{\mathbf{u}_k} \end{pmatrix} \frac{d\mathbf{g}}{d\left\{\mathbf{x}_{k-1}, \mathbf{u}_k\right\}}^T \\
&= \begin{pmatrix} \frac{\partial\mathbf{g}}{\partial\mathbf{x}_{k-1}} & \frac{\partial\mathbf{g}}{\partial\mathbf{u}_k} \end{pmatrix} \begin{pmatrix} \boldsymbol{\Sigma}_{\mathbf{x}_{k-1}} & \mathbf{0}_{3\times2} \\ \mathbf{0}_{2\times3} & \boldsymbol{\Sigma}_{\mathbf{u}_k} \end{pmatrix} \begin{pmatrix} \frac{\partial\mathbf{g}}{\partial\mathbf{x}_{k-1}}^T \\ \frac{\partial\mathbf{g}}{\partial\mathbf{u}_k}^T \end{pmatrix} \\
&= \frac{\partial\mathbf{g}}{\partial\mathbf{x}_{k-1}} \boldsymbol{\Sigma}_{\mathbf{x}_{k-1}} \frac{\partial\mathbf{g}}{\partial\mathbf{x}_{k-1}}^T + \frac{\partial\mathbf{g}}{\partial\mathbf{u}_k} \boldsymbol{\Sigma}_{\mathbf{u}_k} \frac{\partial\mathbf{g}}{\partial\mathbf{u}_k}^T
\end{aligned} \tag{60}$$

The first Jacobian above can be easily obtained by expanding the expression for the composition operator $\oplus$, giving us:

Table 5. Corresponding probabilistic version of motion model

Random variable	Description	Dimensionality
$\mathbf{x}_k \sim N\left(\bar{\mathbf{x}}_k, \boldsymbol{\Sigma}_{x_k}\right)$	Robot pose for the current time step.	3
$\mathbf{x}_{k-1} \sim N\left(\bar{\mathbf{x}}_{k-1}, \boldsymbol{\Sigma}_{x_{k-1}}\right)$	Robot pose for the previous time step.	3
$\mathbf{u}_k \sim N\left(\bar{\mathbf{u}}_k, \boldsymbol{\Sigma}_{u_k}\right)$	Action vector (odometry).	2

$$\frac{\partial \mathbf{g}}{\partial \mathbf{x}_{k-1}} = \begin{pmatrix} 1 & 0 & R'\left(-s_{k-1}s_{\Delta} - c_{k-1}\left(1-c_{\Delta}\right)\right) \\ 0 & 1 & R'\left(c_{k-1}s_{\Delta} - s_{k-1}\left(1-c_{\Delta}\right)\right) \\ 0 & 0 & 1 \end{pmatrix} \quad (61)$$

where the following replacements have been employed for clarity: $c_{k-1} = \cos\theta_{k-1}$, $s_{k-1} = \sin\theta_{k-1}$, $c_{\Delta} = \cos\Delta\theta_k$, and $s_{\Delta} = \sin\Delta\theta_k$

In order to obtain a simple expression for the second Jacobian, we will repeat the approach already employed in the previous section: to decompose it into the product of two matrices by means of the chain rule for Jacobians. Instead of taking derivatives with respect to the odometry readings directly, an intermediary step will be considered instead which takes the derivatives with respect to the radius R' and the pose increment $\Delta\theta_k$. Proceeding in that way, we have:

$$\frac{\partial \mathbf{g}}{\partial \mathbf{u}_{k-1}} = \frac{\partial \mathbf{g}}{\partial\left\{R', \Delta\theta_k\right\}} \frac{\partial\left\{R', \Delta\theta_k\right\}}{\partial \mathbf{u}_{k-1}} = \begin{pmatrix} c_{k-1}s_{\Delta} - s_{k-1}\left(1-c_{\Delta}\right) & R'\left(c_{k-1}c_{\Delta} - s_{k-1}s_{\Delta}\right) \\ s_{k-1}s_{\Delta} + c_{k-1}\left(1-c_{\Delta}\right) & R'\left(s_{k-1}c_{\Delta} + c_{k-1}s_{\Delta}\right) \\ 0 & 1 \end{pmatrix} \cdot \begin{pmatrix} 0 & -L\dfrac{\cot\beta_k}{\sin\beta_k} \\ \dfrac{\sin\beta_k}{L} & \dfrac{\Delta u_k}{L}\cos\beta_k \end{pmatrix} =$$

(using the trigonometric identities for the sine and cosine of the sum of two angles and the fact that $\theta_{k-1} + \Delta\theta_k$ is θ_k)

$$= \begin{pmatrix} \sin\theta_k - s_{k-1} & R'\cos\theta_k \\ \cos\theta_k + c_{k-1} & R'\sin\theta_k \\ 0 & 1 \end{pmatrix} \begin{pmatrix} 0 & -L\dfrac{\cot\beta_k}{\sin\beta_k} \\ \dfrac{\sin\beta_k}{L} & \dfrac{\Delta u_k}{L}\cos\beta_k \end{pmatrix} \quad (62)$$

Finally, regarding the modeling of the odometry uncertainty $\Sigma_{\mathbf{u}_k}$, a simple approach would be to consider the following approximation:

$$\boldsymbol{\Sigma}_{\mathbf{u}_k} = \begin{pmatrix} \sigma_{\Delta u}^2 & 0 \\ 0 & \sigma_{\beta}^2 \end{pmatrix} \quad (63)$$

that is, to assume independence between the errors in the distance and the angle measured by the two wheel encoders. A plausible way to establish both variances above is outlined below:

$$\begin{aligned} \sigma_{\Delta u}^2 &= \alpha_1 \left|\Delta u_k\right| \\ \sigma_{\beta}^2 &= \varsigma_{\beta}^2 + \alpha_2 \left|\Delta u_k\right| \end{aligned} \quad (64)$$

where:

- ς_{β}^2 $\left(\text{rad}^2\right)$: A term to introduce an optional minimum uncertainty in the steering wheel direction.
- α_1 $\left(\text{m}^2/\text{m}\right)$: This term establishes how the uncertainty of the traveled distance Δu_k grows as that distance increases.
- α_2 $\left(\text{rad}^2/\text{m}\right)$: This value determines how much does the distance traveled affect the uncertainty in the angle measured by encoders in the steering wheel.

Heuristic values must be determined for all these parameters in order to adapt them to any specific vehicle.

6. A BLACK BOX UNCERTAINTY MODEL FOR COMMERCIAL ROBOTS

Kinematic Equations

Most models previously analyzed in this chapter assume that we have access to the raw readings from odometry sensors (i.e. the linear or angular encoders on each wheel). In some commercial mobile bases, however, this information may not be available. Instead, the robot firmware may work on those readings and output already processed estimations of odometry as pose increments, according to some internal model. Moreover, some platforms allow users to access both kinds of data: raw encoder readings and the processed odometry estimation. In those situations, the user must decide which one is more advantageous, since both choices have pros and cons.

Reconstructing the robot motion directly from the encoder readings forces us to employ the actual robot kinematic equations (those introduced in previous sections), hence the uncertainty in the robot pose would be accurate since it was obtained from a realistic physical model. On the other hand, all kinematic models assume constant velocities between consecutive time steps, thus processing the raw encoder data requires doing so at a high rate in order to reduce the errors implied by this approximation. Relying on platform-specific firmware to process the encoder data at high rates (typically, at least 100Hz) relaxes the real-time runtime requirements of any localization of SLAM algorithm, which can safely operates at a much lower rate (typically, at 1—10Hz). Obviously, if a robotic application ignores the underlying kinematics and just relies on the odometry as reconstructed by the robotic base firmware, there is a loss of information about the exact path described in between two time steps and, therefore, the modeling of the robot pose uncertainty must be based on a very generic model without any connection to the robot kinematics. This is exactly the aim of the present model.

Assume a planar robot state vector identical to that employed in previous sections, that is, for a time step k, the robot pose is:

$$\mathbf{x}_k = \begin{bmatrix} x_k \\ y_k \\ \theta_k \end{bmatrix} \tag{65}$$

By hypothesis, we assume the robot equipped with the appropriate sensors and software such that it provides us with its latest dead-reckoning odometry pose, which will be denoted as $\mathbf{o}_k$, comprising these three components:

$$\mathbf{o}_k = \begin{bmatrix} o_{x_k} \\ o_{y_k} \\ o_{\theta_k} \end{bmatrix} \tag{66}$$

Note that these coordinates are typically output by the robot in the arbitrary frame of reference, which has its origin at the robot location after powering up. Since the localization or SLAM algorithms that will run on the robot must be independent of this arbitrary coordinate frame, it is convenient to work on incremental odometry readings only. That is, the odometry reading for time step k consists of the robot pose according to $\mathbf{o}_k$, taken in coordinates relative to the last odometry reading $\mathbf{o}_{k-1}$. This operation, which must not be confused with a simple component-wise subtraction, is denoted as $\ominus$ (see Appendix A). Therefore, the action vector $\mathbf{u}_k$ is defined in this model simply as:

$$\mathbf{u}_k = \begin{bmatrix} \Delta x_k \\ \Delta y_k \\ \Delta \theta_k \end{bmatrix} = \mathbf{o}_k \ominus \mathbf{o}_{k-1} =$$

(replacing the analytical expression for the $\ominus$ operator)

Table 6. Next probabilistic version of this kinematic model

Random variable	Description	Dimensionality
$\mathbf{x}_k \sim N\left(\overline{\mathbf{x}}_k, \boldsymbol{\Sigma}_{x_k}\right)$	Robot pose for the current time step.	3
$\mathbf{x}_{k-1} \sim N\left(\overline{\mathbf{x}}_{k-1}, \boldsymbol{\Sigma}_{x_{k-1}}\right)$	Robot pose for the previous time step.	3
$\mathbf{u}_k \sim N\left(\overline{\mathbf{u}}_k, \boldsymbol{\Sigma}_{u_k}\right)$	Action vector (odometry-based pose increment).	3

$$= \begin{bmatrix} \left(o_{x_k} - o_{x_k-1}\right)\cos o_{\theta_k-1} + \left(o_{y_k} - o_{y_{k-1}}\right)\sin o_{\theta_k-1} \\ -\left(o_{x_k} - o_{x_k-1}\right)\sin o_{\theta_k-1} + \left(o_{y_k} - o_{y_{k-1}}\right)\cos o_{\theta_k-1} \\ o_{\theta_k} - o_{\theta_k-1} \end{bmatrix} \quad (67)$$

that is, the odometry estimation of the pose increment since the last time step. Then, the kinematic equation $\mathbf{g}(\cdot)$ is simply the concatenation of the last robot pose and this latest increment:

$$\mathbf{x}_k = \mathbf{g}\left(\mathbf{x}_{k-1}, \mathbf{u}_k\right) = \mathbf{x}_{k-1} \oplus \mathbf{u}_k \quad (68)$$

Probabilistic Motion Model

As with all previous motion models, we examine next the probabilistic version of this kinematic model. Assuming that all the variables follow Gaussian distributions with parameters (See Table 6).we can approximate the distribution of the new pose $\mathbf{x}_k$ by first-order Taylor series expansion of Equation 68, as usual, leading to a mean value of:

$$\overline{\mathbf{x}}_k = \mathbf{g}\left(\overline{\mathbf{x}}_{k-1}, \overline{\mathbf{u}}_k\right) \quad (69)$$

and to the approximated covariance matrix:

$$\begin{aligned} \boldsymbol{\Sigma}_{x_{k-1}} &= \frac{dg}{d\left\{\mathbf{x}_{k-1}, \mathbf{u}_k\right\}} \begin{pmatrix} \boldsymbol{\Sigma}_{\mathbf{x}_{k-1}} & \mathbf{0}_{3\times2} \\ \mathbf{0}_{2\times3} & \boldsymbol{\Sigma}_{\mathbf{u}_k} \end{pmatrix} \frac{dg}{d\left\{\mathbf{x}_{k-1}, \mathbf{u}_k\right\}}^T \\ &= \begin{pmatrix} \frac{\partial \mathbf{g}}{\partial \mathbf{x}_{k-1}} & \frac{\partial \mathbf{g}}{\partial \mathbf{u}_k} \end{pmatrix} \begin{pmatrix} \boldsymbol{\Sigma}_{\mathbf{x}_{k-1}} & \mathbf{0}_{3\times2} \\ \mathbf{0}_{2\times3} & \boldsymbol{\Sigma}_{\mathbf{u}_k} \end{pmatrix} \begin{pmatrix} \frac{\partial \mathbf{g}}{\partial \mathbf{x}_{k-1}}^T \\ \frac{\partial \mathbf{g}}{\partial \mathbf{u}_k}^T \end{pmatrix} \\ &= \frac{\partial \mathbf{g}}{\partial \mathbf{x}_{k-1}} \boldsymbol{\Sigma}_{\mathbf{x}_{k-1}} \frac{\partial \mathbf{g}}{\partial \mathbf{x}_{k-1}}^T + \frac{\partial \mathbf{g}}{\partial \mathbf{u}_k} \boldsymbol{\Sigma}_{\mathbf{u}_k} \frac{\partial \mathbf{g}}{\partial \mathbf{u}_k}^T \end{aligned} \quad (70)$$

In this expression, both Jacobians can be easily obtained from Equation 68 by replacing the expression of the composition operator $\oplus$, leading to:

$$\frac{\partial \mathbf{g}}{\partial \mathbf{x}_{k-1}} = \begin{pmatrix} 1 & 0 & -\Delta x_k \sin\theta_{k-1} - \Delta y_k \cos\theta_{k-1} \\ 0 & 1 & \Delta x_k \cos\theta_{k-1} - \Delta y_k \sin\theta_{k-1} \\ 0 & 0 & 1 \end{pmatrix} \quad (71)$$

and:

$$\frac{\partial \mathbf{g}}{\partial \mathbf{u}_k} = \begin{pmatrix} \cos\theta_{k-1} & -\sin\theta_{k-1} & 0 \\ \sin\theta_{k-1} & \cos\theta_{k-1} & 0 \\ 0 & 0 & 1 \end{pmatrix} \quad (72)$$

Similarly to all other motion models, we must supply a heuristic for the uncertainty in the odometry readings to account for slippage, imperfections in the wheels, etc. A possible model for the covariance matrix of odometry increments is to assume statistical independence between each term:

$$\Sigma_{\mathbf{u}_k} = \begin{pmatrix} \sigma^2_{\Delta x} & 0 & 0 \\ 0 & \sigma^2_{\Delta y} & 0 \\ 0 & 0 & \sigma^2_{\Delta\theta} \end{pmatrix} \quad (73)$$

and to assume that each variance grows with the traversed distance and the increment in rotation:

$$\begin{aligned} \sigma^2_{\Delta x} &= \varsigma^2_{\Delta xy} + \alpha_1 \sqrt{\Delta x_k^2 + \Delta y_k^2} + \alpha_2 \left|\Delta\theta_k\right| \\ \sigma^2_{\Delta y} &= \sigma^2_{\Delta x} \\ \sigma^2_{\Delta\theta} &= \varsigma^2_{\Delta\theta} + \alpha_3 \sqrt{\Delta x_k^2 + \Delta y_k^2} + \alpha_4 \left|\Delta\theta_k\right| \end{aligned} \quad (74)$$

where the parameters have the following meaning:

- $\varsigma^2_{\Delta xy}\ \left(\mathrm{m}^2\right)$ and $\varsigma^2_{\theta}\ \left(\mathrm{rad}^2\right)$: These optional terms introduce a minimum uncertainty for the case of the robot standing still, since a covariance matrix of exactly zero may cause problems in the particular case of using particle filters as estimation algorithm (they will be introduced in chapter 7 section 4).
- $\alpha_1\ \left(\mathrm{m}^2/\mathrm{m}\right)$ and $\alpha_3\ \left(\mathrm{rad}^2/\mathrm{m}\right)$: Establish how much the uncertainty in the 2D position increments and the rotation grows, respectively, with the length of the pose increment.
- $\alpha_2\ \left(\mathrm{m}^2/\mathrm{rad}\right)$ and $\alpha_4\ \left(\mathrm{rad}^2/\mathrm{rad}\right)$: The equivalent to the pair of parameters above, but establishing how the uncertainty is affected by robot rotations.

Calibration experiments should help determining heuristic values for all these parameters for each particular robot.

7. AN ALTERNATIVE MODEL: THE NO-MOTION MOTION MODEL

Kinematic Equations

We began this chapter pointing out that the need for a probabilistic motion model is related to the need to provide a transition model in the framework of Bayesian sequential estimation. In the case of robot localization or SLAM, such a transition model is embodied with the physical meaning of a probabilistic motion model for the vehicle.

Most approaches described in previous sections for providing us with such a model have something in common: they rely on one or another kind of proprioceptive sensors (i.e. odometry sensors) for predicting the most up-to-date pose of the vehicle. However, in certain situations a robot can be designed without any of those sensors, thus exclusively relying on exteroceptive sensors. The main condition for allowing a robot designer to save equipping the platform with odometry sensors is that exteroceptive sensors must be accurate and provide rich and redundant data as to keep the robot well localized with that information alone. Moreover, there will be practical situations where having an appropriate odometry sensor is not as easy as in the common case of ground wheeled vehicles, for example, when working with handheld cameras or with flying or undersea robots.

No matter the reason a robot is not equipped with odometry, we must still supply a motion model for the use of sequential Bayesian estimators. We have already analyzed a possible model for these situations in chapter 5 section 2, where a constant-velocity model was proposed to incorporate an estimation of the instantaneous velocity in the system state vector. But an even simpler alternative exists: we call it the *no-motion* motion model. It can be seen as a further simplification of the constant velocity model, where velocities are not part of the state vector and where the kinematic equation consists in just propagating the pose from the last time step without changes:

Table 7. Parameters

Random variable	Description	Dimensionality
$\mathbf{x}_k \sim N\left(\overline{\mathbf{x}}_k, \mathbf{\Sigma}_{x_k}\right)$	Robot pose for the current time step.	3
$\mathbf{x}_{k-1} \sim N\left(\overline{\mathbf{x}}_{k-1}, \mathbf{\Sigma}_{x_{k-1}}\right)$	Robot pose for the previous time step.	3
$\mathbf{r}_k \sim N\left(0, \mathbf{R}\right)$	Zero-mean velocity vector (an unknown)	3

$$\mathbf{x}_k = \mathbf{g}\left(\mathbf{x}_{k-1}, \mathbf{u}_k\right) = \mathbf{x}_{k-1} \tag{75}$$

About the action vector $\mathbf{u}_k$, it has exactly the same meaning as in the constant-velocity model, that is, it comprises the time length Δt between consecutive time steps. Although this interval does not appear in the kinematic equation above, we will see how it must be accounted for in the corresponding probabilistic motion model while updating the uncertainty of the vehicle pose.

Probabilistic Motion Model

Basically, the idea is to model the pdf of the robot pose as a distribution centered at the latest known pose but to grow its uncertainty into all directions. The underlying assumption is that the velocity $\mathbf{r}_k$ (an unknown) remains constant between time steps, but its values follow a zero-mean Gaussian distribution, a reasonable choice in the absence of odometry. Notice that this model implies that most of the time the vehicle remains still or moves slowly, since higher likelihoods are assigned to smaller velocities. Put explicitly:

$$\mathbf{x}_k = \mathbf{g}'\left(\mathbf{x}_{k-1}, \mathbf{u}_k, \mathbf{r}_k\right) = \mathbf{x}_{k-1} + \Delta t\ \mathbf{r}_k \tag{76}$$

where the uncertainty in the time increment Δt will be ignored for the sake of simplicity, but all the other variables are modeled as Gaussian distributions with Table 7's parameters.

It follows from the linear relationship in Equation 68 that the mean of the robot pose for the latest time step is simply:

$$\overline{\mathbf{x}}_k = \mathbf{g}'\left(\overline{\mathbf{x}}_{k-1}, \mathbf{u}_k, \overline{\mathbf{r}}_k\right) = \overline{\mathbf{x}}_{k-1} + \Delta t\ 0 = \overline{\mathbf{x}}_{k-1} \tag{77}$$

Regarding its covariance, the increase in uncertainty is straightforward to obtain:

$$\mathbf{\Sigma}_{x_{k-1}} = \mathbf{\Sigma}_{\mathbf{x}_{k-1}} + \Delta t^2 \mathbf{R} \tag{78}$$

with the unknown velocity modeled as a diagonal covariance:

$$\mathbf{R} = \begin{pmatrix} r_x^2 & 0 & 0 \\ 0 & r_y^2 & 0 \\ 0 & 0 & r_\theta^2 \end{pmatrix} \tag{79}$$

Paralleling our derivation in chapter 5 section 2, we can also employ here the chi-squared inverse cdf to devise a relationship between the diagonal elements of $\mathbf{R}$ and a physical quantity, namely, the expected maximum velocities of the vehicle in each dimension $\left(v_x^{\max}, v_y^{\max}, v_\theta^{\max}\right)$. Provided a desired confidence interval c (typically in the range 95%—99%), it can be shown that the above equations model a velocity vector whose $i-th$ component is below the limit $v_i^{\max}$ with a probability of c as long as:

$$r_i = \frac{v_i^{\max}}{\sqrt{\chi_{1,c}^2}} \tag{80}$$

Therefore, given an expected vehicle dynamics one can easily determine the values for the $\mathbf{R}$ matrix with the equation above.

8. IMPROVEMENTS OF THE BASIC KINEMATIC MODELS

All motion models described in this chapter are more than suitable for usage as transition models within sequential Bayesian frameworks for either localization or SLAM. In many situations, more than one model could be employed, thus it is important to note that some may be more advantageous than others. The ideal would be to use models that account for the *actual* uncertainty of the robot pose for each new time step, avoiding both optimism (overconfidence) and pessimism (an excessively large uncertainty). The former is more dangerous than the latter, but an excess in uncertainty with respect to the *actual* errors in the kinematic system may also degrade the performance of Bayesian estimators. Thus, it comes at no surprise that enhanced techniques exist to achieve such a tradeoff by reducing uncertainty as much as possible.

Firstly, any of the motion models suitable for robot moving in planar scenarios can be modified to account for the distribution of free space in the environment. Obviously, this is only possible in the case we have some kind of representation of the obstacles, i.e., a *map*, which might be known in advance (the localization problem) or estimated on-line (SLAM). In any case, the idea is to modify the predicted density distribution of the robot pose $f_{\mathbf{x}_k|\mathbf{u}_k,\mathbf{x}_{k-1}}$ by imposing the constraint that the robot cannot physically go through obstacles, obtaining the improved distribution (Thrun, Burgard and Fox, 2005):

$$f'_{\mathbf{x}_k|\mathbf{u}_k,\mathbf{x}_{k-1}} \propto f_{\mathbf{x}_k|\mathbf{m}} f_{\mathbf{x}_k|\mathbf{u}_k,\mathbf{x}_{k-1}} \tag{81}$$

with the density $f_{\mathbf{x}_k|\mathbf{m}}$ being uniform at all places but at those robot poses that are forbidden for obstacle collisions, where it has a value of zero. Please note that the equation above represents a heuristic approach to a modified kinematic model, thus it does not arise from any rule of statistics. Also, note that the map is considered static, i.e. it does not change over time.

Another technique, this one specific for robots equipped with odometry, is to exploit *sensor fusion*. Typically, odometry sensors are reasonably accurate in measuring the length of straight displacements, but their performance degrades when the robot rotates. In order to compensate this deficiency, one can fuse the information of different proprioceptive sensors to obtain an enhanced odometry. A common situation is the usage of gyroscopes or electronic compasses to reduce orientation errors coming from odometry. One possibility is to employ an Extended Kalman filter (a probabilistic technique to be introduced in chapter 7 for the problem of mobile robot localization) in order to fuse the sensors, taking into account the different levels of uncertainty that each one provides for each component of the robot pose.

Finally, regarding the constant-velocity motion model, we could enhance its predictions if velocity profiles or velocity histograms are known in advance for some given vehicle or application. For instance, in a work on visual-based SLAM where the sensor was known to be a hand-held camera

(Klein & Murray, 2007), the authors proposed a "decaying velocity model" where the magnitude of all velocities decays with time. The advantages of such an approach are clear in the context of hand-held cameras if displacements tend to be short and cameras move very slowly most of the time. In other application fields, different heuristics may have to be devised.

REFERENCES

Borenstein, J., Everett, H. R., & Feng, L. (1996). *Where am I? Sensors and methods for mobile robot positioning. Technical Report*. Ann Arbor, MI: University of Michigan.

Doh, N., Choset, H., & Chung, W. K. (2003). Accurate relative localization using odometry. In *Proceedings of the IEEE International Conference on Robotics and Automation*, (vol 2), (pp. 1606-1612). IEEE Press.

Klein, G., & Murray, D. (2007). Parallel tracking and mapping for small AR workspaces. In *Proceedings of the 2007 6th IEEE and ACM International Symposium on Mixed and Augmented Reality*. IEEE Press.

Thrun, S., Burgard, W., & Fox, D. (2005). *Probabilistic robotics*. Cambridge, MA: The MIT Press.

ENDNOTES

1 It is instructive to prove that Equation 6 provides a consistent prediction of uncertainty for the velocities after any fixed period of time, independently of the number of discrete time steps into which we divide that period. That is, the covariance after N time steps of value Δt each, should match that after one single step of length $N\Delta t$. The random noise $\mathbf{r}_k$ models an unknown acceleration, which we will assume to be constant over short periods Δt_k and with values following the Gaussian distribution $N\left(0, \boldsymbol{\Sigma}_{r_k}\right)$. The equation for a velocity update with constant acceleration is simply $\mathbf{v}_k = \mathbf{f}\left(\mathbf{v}_{k-1}, \Delta t_k, \mathbf{r}_k\right) = \mathbf{v}_{k-1} + \mathbf{r}_k\sqrt{\Delta t_k}$. Since this is a linear equation, if we assume that $\mathbf{v}_{k-1}$ follows the Gaussian distribution $N\left(\overline{\mathbf{v}}_{k-1}, \boldsymbol{\Sigma}_{\mathbf{v}_{k-1}}\right)$ it turns out that the updated $\mathbf{v}_k$ also follows a Gaussian distribution, whose covariance matrix happens to be

$$\boldsymbol{\Sigma}_{\mathbf{v}_k} = \frac{\partial \mathbf{f}}{\partial v_{k-1}} \boldsymbol{\Sigma}_{\mathbf{v}_{k-1}} \frac{\partial \mathbf{f}}{\partial v_{k-1}}^T + \frac{\partial \mathbf{f}}{\partial r_k} \boldsymbol{\Sigma}_{\mathbf{r}_k} \frac{\partial \mathbf{f}}{\partial r_k}^T = \boldsymbol{\Sigma}_{\mathbf{v}_{k-1}} + \sqrt{\Delta t_k}\, \boldsymbol{\Sigma}_{\mathbf{r}_k} \sqrt{\Delta t_k} = \boldsymbol{\Sigma}_{\mathbf{v}_{k-1}} + \boldsymbol{\Sigma}_{\mathbf{r}_k} \Delta t_k,$$

an expression that indeed leads to the same values for any final Δt disregarding the number of intermediary time steps.

2 In a Monte Carlo simulation, the pdf of the state vector (the robot pose $\mathbf{x}_k$ in this case) is represented non-parametrically by means of a large sample, or set of hypotheses, each one being transformed through the system model, the Equation 6, with different values of the random noises. Given an enough number of elements in the sample, the obtained distribution closely resembles the actual pdf without the need for it to be modeled as a Gaussian or any other parametric distribution. More on Monte Carlo methods will be discussed later on in the context of Bayesian estimation in chapter 7.

3 The only approximations of this ground truth are: (1) the non-infinite number of elements in the sample and (2) the assumption of that sample being distributed following a Gauss-

ian, which given the non-linear nature of the kinematic equations will never be an exact assumption. However, since our goal here is to compare this pdf to another Gaussian (that one obtained from the linearized motion model) the small errors of assuming a Gaussian distribution can be ignored.

4 Dynamic constraints do not account for the effects of this definition. That is, if a robot needs a few instants to orientate its drive wheels but afterwards it can move into any direction, it is still holonomic for our purposes.

5 Or, put otherwise, the variance of the odometry increments should have a standard deviation proportional to the *square root* of the increments themselves. This principle was already employed in Equation 6 while discussing the constant velocity model.

Chapter 6
Sensor Models

ABSTRACT

This is the second chapter of the second section. Analogously to chapter 5, here the authors study probabilistic models of sensors, which is the second fundamental component of the general Bayesian framework for localization. In this chapter, they explain common mathematical models of sensors, stressing their differences and effects in further estimation techniques, in particular whether they are parametrical or not. The chapter also points out the existence of the association problem between observations and known elements of maps for some kinds of sensors, and presents solutions to that problem. Finally, some methods for matching local maps provided by particular kinds of sensors are also included.

CHAPTER GUIDELINE

- You will learn:
  - The general role of sensors in the Bayesian estimation framework for localization and mapping.
  - The basic probabilistic models for the most important kinds of robotic sensors.
  - The data association problem and its best known solutions.
  - A glimpse on the different types of maps that a mobile robot can use.
- Provided tools:
  - The NN and JCBB methods for solving the data association problem.
  - Basic methods for matching pairs of maps.
- Relation to other chapters:
  - The proposed probabilistic models are one piece within the proposed solutions to localization and mapping presented in chapters 7, 9, and 10.
  - Chapter 8 provides a comprehensive overview of the different kinds of maps for mobile robotics.

DOI: 10.4018/978-1-4666-2104-6.ch006

1. INTRODUCTION

In addition to modeling the uncertainty of robot actions, all Bayesian estimators employed for localization or SLAM require some mathematical model for the uncertainty introduced by sensors. In particular, before addressing mobile robot localization in chapter 7 and mapping and SLAM in chapters 8-10, we will need to provide explicit forms for $f_{\mathbf{z}_k|\mathbf{x}_k,\mathbf{m}}$, the likelihood of gathering a given observation $\mathbf{z}_k$ at the end of time step k if the robot moves within a static, known or hypothetical environment $\mathbf{m}$, assuming an also known or hypothetical robot pose $\mathbf{x}_k$. Since current sensors do not provide topological information directly, we will assume in the following that both the robot pose and measurements are continuous (metrical) random variables.

The distribution $f_{\mathbf{z}_k|\mathbf{x}_k,\mathbf{m}}$, considered as a function, is commonly called the *observation likelihood* in the context of sequential Bayesian estimators for robotics. This likelihood is not to be necessarily understood as a probabilistic model applicable when the robot gathers one single measurement; quite commonly, a mobile robot will be equipped with several sensors on board or with sensors that produce several simultaneous measurements each. In general, the richer the information we can retrieve at each time step, the better the result of the estimation processes due to the law of large numbers and the asymptotic convergence of sequential estimators (recall chapter 4). Including more than one observation at each step in a sequential Bayesian estimation framework is, in principle, straightforward: if we have multiple observations $\left\{\mathbf{z}_{k,i}\right\}$, we can just stack all those variables into a single vector $\mathbf{z}_k = \left\{\mathbf{z}_{k,1}, \mathbf{z}_{k,2}, \ldots, \mathbf{z}_{k,n}\right\}$, thus the observation likelihood becomes a multivariate probability density of a higher dimensionality. However, having such a set of multiple simultaneous observations presents two difficulties: (1) it is usually more difficult to provide a form for a joint density function than for a single-observation one (although some interesting approaches exist, such as Plagemann, Kersting, Pfaff and Burgard [2007]), and (2) the computational cost of its evaluation is also higher (we will introduce some heuristics for that in section 4). Therefore, it is more practical to assume statistical independence between observations—actually, conditional independence, given the hypothetical robot pose and the known map—in order to reduce complexity. This leads to factoring the likelihood as $f_{\mathbf{z}_k|\mathbf{x}_k,\mathbf{m}} = \prod_i f_{\mathbf{z}_{k,i}|\mathbf{x}_k,\mathbf{m}}$.

In general, this independence assumption does not produce relevant complications, although it is unrealistic as it completely ignores the influences of other simultaneous observations. For example, if observation $\mathbf{z}_{k,i}$ indicates the presence of some part of a wall nearby, it is more likely to observe another part of the same wall just next to it (in $\mathbf{z}_{k,i+1}$) than if we had seen open space in the first place. Thus, the independence assumption leads to a gross approximation of the actual probability distribution, but still work reasonably well in many practical settings. For coping with these deviations from reality, common approaches include over-estimating uncertainty (often called *inflating* the uncertainty in the literature) and employing robust statistical techniques when considering more than one simultaneous observations, for instance removing observations that are likely produced by dynamic obstacles or using some sort of consensus to fuse all the observations (Blanco, González, & Fernández-Madrigal, 2007). Overestimating uncertainty—being pessimistic—may lead to poor convergence in sequential estimators, while underestimating uncertainty—being optimistic—may lead to losing the consistency of the estimator, which is much worse. This is the reason why inflating uncertainty is a common practice.

Providing an explicit form for the observation likelihood of different kinds of sensory data is the goal of this chapter. It depends, obviously, on the particular sensor we use. In theory, for modeling this likelihood for a particular sensor we should take large amounts of real measurements and build $f_{\mathbf{z}_k|\mathbf{x}_k,\mathbf{m}}$ on them—that is another estimation problem! However, as we will see, it is possible to propose quite generic models that avoid such thorough work.

The rest of the chapter is structured as follows. Section 2 presents a popular probabilistic model for range-bearing sensors: the *beam* model. Sections 3 and 4 study sensors that provide features and the particular problems that arise from having to distinguish these features from each other. Finally, section 5 touches the issue of comparing small maps typically built from several measurements of a sensor with a global known map, which is also applicable for comparing different portions of the same map.

2. THE BEAM MODEL AND RAY-CASTING

The most popular sensors for mobile robots during the 1990s were those providing distances and orientations to obstacles. At that time, their cost had dropped and their accuracy raised enough to make affordable their use for precise localization and mapping, while the limits existing in computational power still prevented to include more complex devices such as video cameras for achieving the same goals. Today, computer vision and 3D sensors are clearly replacing this central role in SLAM, but many mobile robots are still equipped with range-bearing sensors.

A range-bearing sensor gathers information about the distance (*range*) to the closest obstacle in the direction (*bearing*) at which it points, being this bearing selectable from a set of values in some kind of sensors—radial scanners—fixed at only one value for the simplest ones—e.g. sonars—and even governed by an actuator if a single-beam sensor is attached to a controllable motor. In the following we will assume that the bearing is implicit (not measured by the sensor but known), while the range is the data actually measured. We introduce a popular model for the uncertainty in such range measurements, the so-called *beam model* (Thrun, Burgard, & Fox, 2005), which is practical for a variety of range and range-bearing sensors.

In the mobile robot localization problem we know the map of the environment, which is considered static, and the observation (range) obtained by the sensor at some instant of time, and we want to model the uncertainty existing in that observation given a possible pose of the robot where it gathered that measurement, that is, we look for the value of $f_{z_k|\mathbf{x}_k,\mathbf{m}}$ for that observation, map, and hypothetical pose. Since we are considering range measurements, z_k is here a real positive number. If the sensor provides more than one range (corresponding to different bearings, such as radial laser scanners or rings of ultrasound sensors) they can be treated as independent measures as explained in the introduction to this chapter, or they can form a partial map of the environment and then we could apply the methods explained in section 4.

In the *beam model*, the support of the density $f_{z_k|\mathbf{x}_k,\mathbf{m}}$ is $[0, z_{\max}]$, being $z_{\max} > 0$ the maximum distance that can be provided by the sensor in any circumstance. We also should take into account that sensors often return $z_{\max}$ when no obstacle is detected and that most of them cannot work properly in short distances, imposing an actual minimum greater than zero.

The shape of uncertainty in the beam model is assumed not to depend on the absolute position of the sensor ($\mathbf{x}_k$). This is true in most range sensors, but there are exceptions: for instance, RFID sensors, when used for ranging, have a stochastic behavior that strongly depends on the absolute position of the sensor and the object to detect, being completely different when both of them change, even if they keep their relative distance (this is due to reflections of the signal in the environment, mostly); UWB emitters and receivers used for ranging also may exhibit a similar behavior (González-Jiménez, Blanco, Galindo, Ortiz-de-Galisteo, Fernández-Madrigal, Moreno, & Martínez, 2009). In the following, we only deal with the original, absolute-position-independent beam model, which is as follows (see Figure 1).

Firstly, it makes sense to consider that the attainable measurements will tend to concentrate around the actual distance to the obstacle, say d_k, although there will be other possible sources of uncertainty in addition to that. Therefore, the first step in order to instantiate the beam model is to calculate the value for d_k, the distance that the sensor would have measured from the robot hypothetical pose if we had a perfect, ideal sensor. For that hypothetical pose of the robot—and therefore a pose of the sensor[1], say $\mathbf{y}_k$—and a known or estimated map of the environment, let d_k be the distance measured along a straight line from $\mathbf{y}_k$ to the closest obstacle in the environment (i.e., along the line of sight of the sensor, as shown in Figure 1). Tracing that line and calculating its intersection with the map of the environment is called *ray-casting* (Roth, 1982), and can be done with standard geometrical methods for any robot pose. Obviously, its calculation depends on the particular representation of the environment we are using. For some specific types of maps, it even can be pre-calculated for all possible robot poses before the robot is operating, improving run-time efficiency. For example, in the case of feature maps constituted by straight segments, ray-casting consists of solving for the intersection between the sensor line of sight and some of these segments, then choosing the one that produces the smallest distance to the sensor; in the case of regular tessellations of space (grids), it can be implemented as a algorithm for detecting the closest occupied cell in the line of sight of the sensor. As we will see in chapter 8, occupancy

Figure 1. The components of the uncertainty modeled in the general beam model when applied to a range sensor that is looking towards a certain point of the environment. Please refer to the main text for the rationale behind each one.

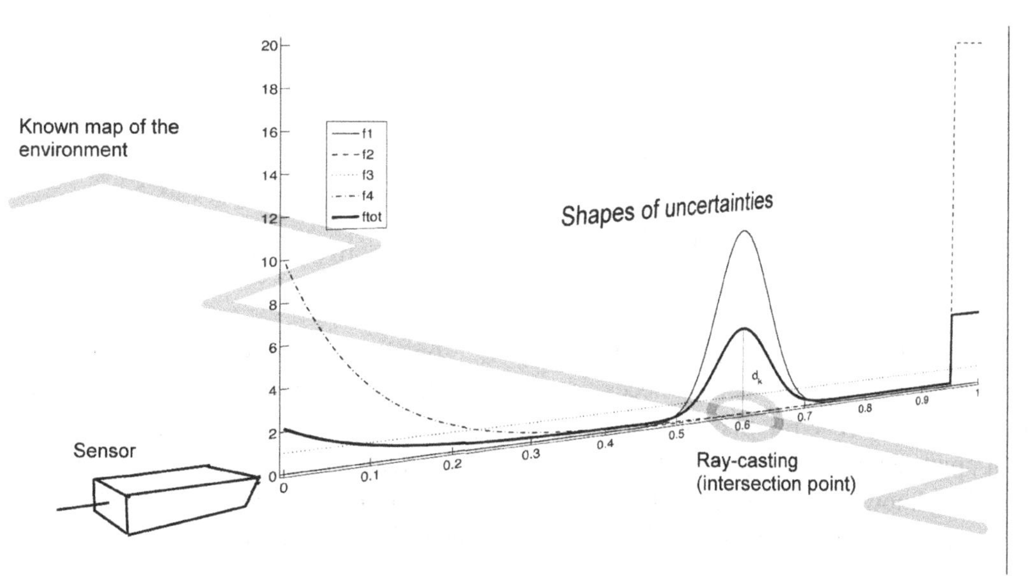

grids actually keep the *probability* of each part of the space to be occupied, therefore, in order to decide whether a ray-traced beam collides or not with an obstacle, we must define a threshold for the probability of a cell to be occupied. This is a clear indication that maps, in spite of being known, can contain uncertainty. In the general case, such uncertainty should be propagated to the likelihood appropriately; however, for the beam model, it can be ignored without affecting much the resulting practical behavior of the localization or mapping filter if we inflate the one of the beam model, which is commonly done. A different case occurs in other models for the likelihood, such as the ones based on features explained in section 3. In those cases the uncertainty that the map may carry should definitely propagated into the likelihood, as we will explain there.

The beam model assumes that the uncertainty in the sensor readings has, roughly, a Gaussian shape (pdf) centered at d_k with variance σ_1^2, that is:

$$f_{1;z_k,\mathbf{x}_k,\mathbf{m}} = \begin{cases} \frac{1}{\eta_1} N(z_k; d_k, \sigma_1^2) & \text{if } z_k \in [0, z_{\max}] \\ 0 & \text{otherwise} \end{cases} \tag{1}$$

where $\eta_1 > 0$ is a normalizing factor for the integral of $f_{1;z_k,\mathbf{x}_k}$ to be 1 in its support.

This principal uncertainty accounts for the imprecision of the sensor. Including in the beam model additional uncertainty due to the real (non-zero) width of the beam is not easy, although there exist approaches in that sense (Plagemann, Kersting, Pfaff, & Burgard, 2007); it is therefore more common to work with the simple Gaussian of Equation 1 and inflate its uncertainty as necessary. An efficient alternative to ray-tracing is precomputing an approximation to Equation 1 for each location of the known map independently on the position from where the robot would be sensing the obstacle. This approach, called *likelihood field* (Thrun, 2001), works by computing the distance r (the clearance) from each possible location to its closest obstacle, then assigning a likelihood of $f_{1,LF;z_k,\mathbf{x}_k,\mathbf{m}} = N(r_k; 0, \sigma_{LF}^2)$ to each beam whose end point, i.e., the point detected as an obstacle at the end of the beam, has a clearance of r_k (notice that r_k is a function of $\mathbf{x}_k$ and the map). This heuristic model often provides a robust alternative to Equation 1.

Depending on the kind of range sensor we have, other sources of uncertainty can be considered that do not complicate the model much. For example, we could take into account the possibility of obtaining the maximum distance in situations where the sensor beam is lost (due, for example, to reflections, absorptions, etc.). This uncertainty can be modeled by a narrow uniform distribution at the maximum value, with some small width h

$$f_{2;z_k,\mathbf{x}_k,\mathbf{m}} = \begin{cases} \frac{1}{h} & \text{if } z_k \in [z_{\max} - h, z_{\max}] \\ 0 & \text{otherwise} \end{cases} \tag{2}$$

It is also possible to consider unexplainable measurements due to bounces, interferences with other sensors, etc. In these cases the sensor can produce nearly any value, thus a new uniform uncertainty that covers the entire support can be added to the model:

$$f_{3;z_k,\mathbf{x}_k,\mathbf{m}} = \begin{cases} \frac{1}{z_{\max}} & \text{if } z_k \in [0, z_{\max}] \\ 0 & \text{otherwise} \end{cases} \tag{3}$$

Finally, the model can be augmented when we know that there are mobile unpredictable (and temporary) obstacles in the line of sight, not belonging to the static map. These obstacles can appear at any distance from the sensor, but the ones closest to it may occlude farther ones.

Therefore, it is more likely to have a short distance measured due to this effect than a long one. For modeling that gradual fading of that density with the distance to the sensor pose it makes sense to use an exponential pdf:

$$f_{4;z_k,\mathbf{x}_k,\mathbf{m}} = \begin{cases} \frac{1}{\eta_4}\lambda_4 e^{-\lambda_4 z_k} & \text{if } z_k \in [0, z_{\max}] \\ 0 & \text{otherwise} \end{cases} \quad (4)$$

where $\eta_4 > 0$ is another normalizing factor for this to integrate up to one in its entire support and where λ_4 is the mean of the exponential. Notice that the inclusion of this uncertainty does not help if the environment is cluttered: in that case, especial methods to distinguish the dynamic obstacles and reject them from the sensor measurements may be required.

The general form for the observation likelihood of the beam model is therefore the result of summing up Equations 1, 2, 3, and 4, weighting each term by an appropriate importance factor (notice that we are adding uncertainties, i.e., we are considering that all of them are independent and thus it is correct to sum them for representing their "or-ing"):

$$f_{z_k,\mathbf{x}_k,\mathbf{m}} = w_1 f_{1;z_k,\mathbf{x}_k,\mathbf{m}} + w_2 f_{2;z_k,\mathbf{x}_k,\mathbf{m}} + \\ + w_3 f_{3;z_k,\mathbf{x}_k,\mathbf{m}} + w_4 f_{4;z_k,\mathbf{x}_k,\mathbf{m}} \quad (5)$$

where $\sum_i w_i = 1$ to assure that the resulting function is a pdf. Figure 1 depicts as a thick solid line the resulting density when all the terms have weights greater than zero.

Notice that in the general beam model we have to provide six parameters, namely $\{w_1, w_2, w_3, w_4, \sigma_1^2, \lambda_4\}$. This is easier than estimating the whole shape for the real uncertainty from scratch, but a bad adjustment of these can even lead to the divergence of the estimation results; therefore, their choice is not straightforward.

If we have at least suitable values for the weights of the pdfs, we can obtain the best values for σ_1^2 and λ_4 through maximum likelihood estimation (recall chapter 4), provided a large enough dataset of measurements taken by the sensor in different controlled situations. This leads to closed-form equations for both parameters. Unfortunately, if we do not know the weights, there is no closed-form for calculating anything. In that case, we still can obtain an estimation for them using recursive local optimization methods, such as gradient descent (Snyman, 2005) or expectation-maximization (Dempster, Laird, & Rubin, 1977). Any local optimization algorithm (such as the Gauss-Newton method introduced in chapter 10) will work well as long as the actual pdf is smooth and has not many local optima.

3. FEATURE SENSORS: PROBABILISTIC MODELS

Sensors more sophisticated than range-bearing ones do not provide simple scalar measures such as distances. For example, cameras provide bidimensional arrays of pixels that must usually undergo several processing and detection stages before obtaining anything directly usable for metrical localization and mapping.

In general, we will call feature sensors to all those physical sensors, perhaps including post-processing software stages, which are capable of detecting interesting or especially distinctive elements in what they perceive, e.g. colored regions, corners, vertical lines, purposely placed distinctive marks, etc. We will refer to all these elements as *features* or *landmarks*. Sensors that detect features are usually more powerful than range-only or range-bearing sensors, but in turn, they introduce the new problem of data association, as we will see in the next section.

When dealing with metrical maps, features must be assigned a location in space. For now, let us assume it is a 2D position with respect to the sensor. A common mathematical form for the observed $i-th$ feature is then $\mathbf{z}_i = (r_i, b_i, \phi_i)$, where r_i and b_i are the range and bearing of the feature with respect to the sensor—you can think of them as polar coordinates of the feature w.r.t. the sensor device—and ϕ_i its unique identifier, an information that may or may not be available depending on the type of sensor and the processing that is done on the raw data.

The kind of map $\mathbf{m}$ required for feature sensors should comprise a set of L features and their location in space, that is, $\mathbf{m} = \{\mathbf{m}_i\}$, where $\mathbf{m}_i = (x_i, y_i)$—this is only one possibility; other parameterizations of feature positions will be seen in chapter 8. At this point, we will assume that we know such a map, in contrast to the SLAM problem where it will be also estimated from observations. However, since the location of landmarks can hardly be determined without some degree of uncertainty even for completely known maps, we will assume that the map $\mathbf{m}$ contains uncertainty, that is, it is represented as the following multivariate Gaussian:

$$\mathbf{m} \sim N(\mu, \Sigma)$$
$$\mu = \begin{bmatrix} \mu_1 \\ \mu_1 \\ \vdots \\ \mu_L \end{bmatrix}, \text{with } \mu_i = \begin{bmatrix} \overline{x}_i \\ \overline{y}_i \end{bmatrix} \text{ the expected positions of the landmarks in the map}$$
$$\Sigma = \begin{pmatrix} \Sigma_{\mathbf{m}_1} & \cdots & \Sigma_{\mathbf{m}_1\mathbf{m}_L} \\ \vdots & \ddots & \vdots \\ \Sigma_{\mathbf{m}_L\mathbf{m}_1} & \cdots & \Sigma_{\mathbf{m}_L} \end{pmatrix}, \text{with } \Sigma_{\mathbf{m}_1} = \begin{pmatrix} \sigma^2_{x_i} & \sigma_{x_i,y_i} \\ \sigma_{y_i,x_i} & \sigma^2_{y_i} \end{pmatrix} \text{ their uncertainties} \quad (6)$$

In the case the feature locations are known by manual measurement in a controlled environment, the coordinates of the $i-th$ feature could be directly assigned to the corresponding mean μ_i. The physical interpretation of the diagonal block matrices $\Sigma_{\mathbf{m}_i}$ is the uncertainty in the location of each landmark. Off-diagonal entries $\Sigma_{\mathbf{m}_i,\mathbf{m}_j}$ represent the cross-correlation between features, which will be automatically calculated if the map has been constructed using SLAM techniques (refer to chapters 9 and 10). If the map is provided by hand, it makes sense to set all cross covariance blocks to zero and only populate the diagonal entries with reasonable values depending on the measurement accuracy—for instance, to set $\sigma^2_{x_i} = \sigma^2_{y_i} = 0.01^2$ if the accuracy is in the order of 1cm, assuming that coordinates are represented in meters.

Our goal is therefore finding a model for the observation likelihood that represents the probability of gathering a given observation $\mathbf{z}_k$ if the robot is at some hypothetical pose $\mathbf{x}_k$ and given the map $\mathbf{m}$. Assuming that we are able to detect features *without misclassifying them* is a rather strong assumptions in many cases, as we will see further on, but we will take that for granted by now. Then the observation likelihood has to account only for the uncertainty in the continuous variables (i.e., the observed landmark position) but not in the discrete aspect of identifying each observed landmark with its correspondence in the map.

We start by defining a function $\mathbf{h}(\mathbf{x}_k, \mathbf{m})$ that predicts the value of observations given the landmarks in the map $(\mathbf{m})$ and the hypothetical pose of the *sensor* ($\mathbf{x}_k$ from now on, please take care that it is not necessary the same as the pose of the *robot*):

$$\mathbf{z}_k = \mathbf{h}(\mathbf{x}_k, \mathbf{m}) + \delta_k \quad (7)$$

where $\delta_k \sim N(0, \mathbf{Q}_k)$ stands for an additive Gaussian noise that models imperfections in the sensor measurements themselves. What we look for at the end is a mathematical form for $f_{z_k, \mathbf{x}_k, \mathbf{m}}$, which will be a particular distribution centered at the value of $\mathbf{z}_k$ provided by Equation 7. It must be stressed that the observation function $\mathbf{h}(\mathbf{x}_k, \mathbf{m})$ will usually predict the gathering of several simultaneous observations, that is, $\mathbf{z}_k = \{\mathbf{z}_{k,1}, \mathbf{z}_{k,2}, \ldots, \mathbf{z}_{k,M}\}$.

Under the plausible assumption of statistical independence between the additive noises of each individual observation, it follows that the covariance $\mathbf{Q}_k$ has a block diagonal structure:

$$\mathbf{Q}_k = \begin{pmatrix} \mathbf{Q}_{k,1} & 0 & \cdots & 0 \\ 0 & \mathbf{Q}_{k,2} & & 0 \\ \vdots & & \ddots & \vdots \\ 0 & 0 & \cdots & \mathbf{Q}_{k,M} \end{pmatrix} = \begin{pmatrix} \blacksquare & & & \\ & \blacksquare & & \\ & & \ddots & \\ & & & \blacksquare \end{pmatrix} \quad (8)$$

where for clarity we have represented all non-zero submatrices as solid blocks ($\blacksquare$), a convention which will be repeated throughout following chapters. It is also common to simplify and assume that all observations are corrupted by iid Gaussian noises, that is, that all the $\mathbf{Q}_{k,i}$ become equal to some given $\mathbf{Q}$. In the case that the sensor provides range-bearing information on the observations (the (r_i, b_i) pairs mentioned before), we have:

$$\mathbf{Q} = \begin{pmatrix} \sigma_r^2 & 0 \\ 0 & \sigma_b^2 \end{pmatrix} \quad (9)$$

where we find the sensor noise (variances) in range (σ_r^2) and in bearing (σ_b^2). These values should be set such that the standard deviations σ_r and σ_b are in the order of magnitude of the sensor typical noise in range and bearing, respectively.

However, the sensor noise is not the only source of uncertainty in our prediction in Equation 7, since the map $\mathbf{m} \sim N(\mu, \Sigma)$ could carry its own uncertainty, since it has been possibly constructed from noisy sensor readings. This uncertainty is going to be propagated through the function $\mathbf{h}(\mathbf{x}_k, \mathbf{m})$ into the observation space—the space of distances or ranges measured from the sensor. Let us assume a multivariate Gaussian distribution for $f_{z_k, \mathbf{x}_k, \mathbf{m}}$, i.e., the uncertainty in the output of the function ($\mathbf{z}_k$), such that:

$$\begin{aligned} \mathbf{z}_k &= \mathbf{h}'(\mathbf{x}_k, \mathbf{m}, \delta_k) = \mathbf{h}(\mathbf{x}_k, \mathbf{m}) + \delta_k \\ \mathbf{z}_k &\sim N(\overline{\mathbf{z}}_k, \mathbf{S}_k) = f_{z_k, \mathbf{x}_k, \mathbf{m}} \end{aligned} \quad (10)$$

Now, we proceed as usual for propagating the uncertainty (recall chapter 3 section 8) by applying a first-order Taylor series expansion to approximate the mean:

$$\overline{\mathbf{z}}_k = \mathbf{h}'(\overline{\mathbf{x}}_k, \mu, \mathbf{0}) = \mathbf{h}(\overline{\mathbf{x}}_k, \mu) \quad (11)$$

and then propagating through the linearized function the covariance matrix of the variables $\{\mathbf{m}, \delta_k\}$, with uncertainties Σ and $\mathbf{Q}_k$, respectively (recall that we are assuming a completely deterministic pose, without uncertainty since it is a hypothetical robot location):

$$\mathbf{S}_k = \mathbf{J}_{\mathbf{h}'} \begin{pmatrix} \Sigma & \mathbf{0} \\ \mathbf{0} & \mathbf{Q}_k \end{pmatrix} \mathbf{J}_{\mathbf{h}'}^{T} =$$

(decomposing the Jacobian into two parts)

$$= \begin{pmatrix} \frac{\partial \mathbf{h}'}{\partial \mathbf{m}} & \frac{\partial \mathbf{h}'}{\partial \delta_k} \end{pmatrix} \begin{pmatrix} \Sigma & \mathbf{0} \\ \mathbf{0} & \mathbf{Q}_k \end{pmatrix} \begin{pmatrix} \frac{\partial \mathbf{h}'}{\partial \mathbf{m}}^T \\ \frac{\partial \mathbf{h}'}{\partial \delta_k}^T \end{pmatrix} =$$

(and since $\frac{\partial \mathbf{h}'}{\partial \delta_k}$ is the identity and $\frac{\partial \mathbf{h}'}{\partial \mathbf{m}}$ is $\frac{\partial \mathbf{h}}{\partial \mathbf{m}}$, as follows from Equation 10)

$$= \frac{\partial \mathbf{h}}{\partial \mathbf{m}} \Sigma \frac{\partial \mathbf{h}}{\partial \mathbf{m}}^T + \mathbf{Q}_k \tag{12}$$

where all the Jacobians are evaluated at the mean of each random variable, although we did not reflect this in the equations to simplify the notation.

An interesting issue is determining the structure of the observation uncertainty $\mathbf{S}_k$, that is, which parts of the matrix will be zeros, if any at all. In general, the submatrices of $\mathbf{S}_k$ can be split as:

$$\mathbf{S}_k = \begin{pmatrix} \mathbf{S}_1 & \mathbf{S}_{1,2} & \cdots & \mathbf{S}_{1,M} \\ \mathbf{S}_{2,1} & \mathbf{S}_2 & & \mathbf{S}_{2,M} \\ \vdots & & \ddots & \vdots \\ \mathbf{S}_{M,1} & \mathbf{S}_{M,2} & \cdots & \mathbf{S}_M \end{pmatrix} \tag{13}$$

The relevance of studying the structure of this matrix is that simultaneously observing several features gives us observations that may be correlated to each other (i.e., the cross correlation term between the $i-th$ and the $j-th$ observed landmarks, $\mathbf{S}_{i,j}$, may be not all zeros), an information which must be taken into account while establishing correspondences between the real observations and predicted ones from the map—this is the data association problem addressed later on. In those cases when observations are uncorrelated, we could exploit this knowledge to simplify the operations required for evaluating data association.

The diagonal elements of the $\mathbf{S}_k$ matrix have a very clear geometrical meaning: each $\mathbf{S}_i$ represents the uncertainty of the $i-th$ observed landmark in the observation space, which includes[2] the map uncertainty (Σ) and the noise introduced by the sensor itself $(\mathbf{Q})$. In a sensor that provides pairs (r_i, b_i) as described before, the first and second diagonal entries of $\mathbf{S}_i$ stand for the variances of the ranges and bearing of the feature, respectively.

To illustrate this concept with an example, please refer to Figure 2, where three landmarks with uncertainty in their locations in the map are projected into a camera sensor, whose pose is perfectly known. If we plot the confidence intervals for the three diagonal blocks $\mathbf{S}_1$, $\mathbf{S}_2$ and $\mathbf{S}_3$ in the observation space (in the image provided by the camera, with coordinates of pixels), then we would obtain something like Figure 2b. This idea has been often employed in computer vision to constrain the search for specific visual features only on those parts of an image where we predict they could appear.

Regarding the possible structures of $\mathbf{S}_k$ in Equation 12 we can distinguish three different situations:

1. The features in the map are uncorrelated to each other, thus Σ is block diagonal. This will happen when the map has been created by hand or when the map was built using a specific family of mapping algorithms known as Rao-Blackwellized Particle Filters (which we will address in chapter 9). In this case, $\mathbf{S}_k$ will be also block diagonal, thus all observations are uncorrelated. This can be easily verified by using the solid blocks notation (■) and developing the matrix products (as seen in Box 1, there are four observation predictions from a map with six landmarks):

Figure 2. An illustration of uncertainty propagation through a camera-like (bearing-only) sensor model. (a) Three landmarks with position uncertainty in a map, and the sensor pose from which to obtain an observation prediction. (b) The prediction for each landmark, in the observation space, i.e. the figure represents the camera image plane. All ellipsoids and ellipses represent 3σ confidence intervals of the corresponding covariance matrices.

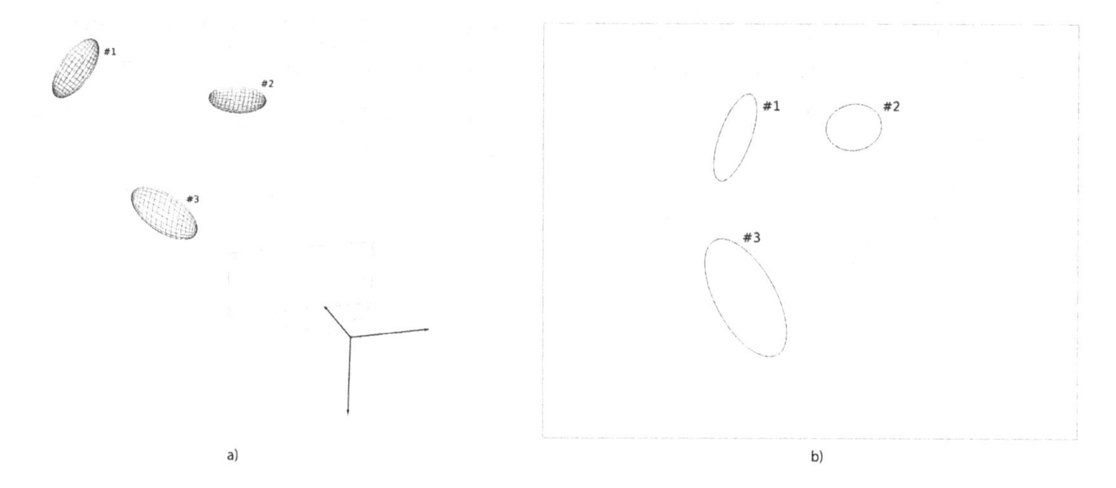

Box 1.

$$\mathbf{S}_k = \frac{\partial \mathbf{h}}{\partial \mathbf{m}} \; \Sigma \; \frac{\partial \mathbf{h}}{\partial \mathbf{m}}^T + \mathbf{Q}_k$$

$$= \begin{pmatrix} \frac{\partial \mathbf{h}_1}{\partial \mathbf{m}_1} & \mathbf{0} & \mathbf{0} & \mathbf{0} & \mathbf{0} & \mathbf{0} \\ \mathbf{0} & \frac{\partial \mathbf{h}_2}{\partial \mathbf{m}_2} & \mathbf{0} & \mathbf{0} & \mathbf{0} & \mathbf{0} \\ \mathbf{0} & \mathbf{0} & \mathbf{0} & \frac{\partial \mathbf{h}_3}{\partial \mathbf{m}_4} & \mathbf{0} & \mathbf{0} \\ \mathbf{0} & \mathbf{0} & \mathbf{0} & \mathbf{0} & \mathbf{0} & \frac{\partial \mathbf{h}_4}{\partial \mathbf{m}_6} \end{pmatrix} \begin{pmatrix} \Sigma_1 & & 0 \\ & \ddots & \\ 0 & & \Sigma_6 \end{pmatrix} \begin{pmatrix} \frac{\partial \mathbf{h}_1}{\partial \mathbf{m}_1}^T & \mathbf{0} & \mathbf{0} & \mathbf{0} \\ \mathbf{0} & \frac{\partial \mathbf{h}_2}{\partial \mathbf{m}_2}^T & \mathbf{0} & \mathbf{0} \\ \mathbf{0} & \mathbf{0} & \mathbf{0} & \mathbf{0} \\ \mathbf{0} & \mathbf{0} & \frac{\partial \mathbf{h}_3}{\partial \mathbf{m}_4}^T & \mathbf{0} \\ \mathbf{0} & \mathbf{0} & \mathbf{0} & \mathbf{0} \\ \mathbf{0} & \mathbf{0} & \mathbf{0} & \frac{\partial \mathbf{h}_4}{\partial \mathbf{m}_6}^T \end{pmatrix} + \begin{pmatrix} \mathbf{Q} & \mathbf{0} & \mathbf{0} & \mathbf{0} \\ \mathbf{0} & \mathbf{Q} & \mathbf{0} & \mathbf{0} \\ \mathbf{0} & \mathbf{0} & \mathbf{Q} & \mathbf{0} \\ \mathbf{0} & \mathbf{0} & \mathbf{0} & \mathbf{Q} \end{pmatrix}$$

$$= \begin{pmatrix} \blacksquare & & & & & \\ & \blacksquare & & & & \\ & & & \blacksquare & & \\ & & & & & \blacksquare \end{pmatrix} \begin{pmatrix} \blacksquare & & \\ & \ddots & \\ & & \blacksquare \end{pmatrix} \begin{pmatrix} \blacksquare & & & \\ & \blacksquare & & \\ & & & \\ & & \blacksquare & \\ & & & \\ & & & \blacksquare \end{pmatrix} + \begin{pmatrix} \blacksquare & & & \\ & \blacksquare & & \\ & & \blacksquare & \\ & & & \blacksquare \end{pmatrix}$$

$$= \begin{pmatrix} \blacksquare & & & \\ & \blacksquare & & \\ & & \blacksquare & \\ & & & \blacksquare \end{pmatrix} \qquad (14)$$

2. The features in the map are correlated to each other, that is, Σ is a dense matrix without (many) zeros. In this case the $\mathbf{S}_k$ will be always dense, as seen in Box 2.
3. Finally, we could have also accounted for the uncertainty in the robot pose instead of assuming it was a perfectly known, deterministic value. Since this situation will only be found in the context of simultaneous localization and mapping, we delay explaining this case until chapter 9 section 2. Anyway, we can advance here that the uncertainty in the robot pose introduces correlations among all observation predictions, no matter whether landmarks were already correlated in the map or not.

Going back to our case of the feature sensor, all we need to complete the description of its probabilistic model is providing the corresponding function $\mathbf{h}(\mathbf{x}_k, \mathbf{m})$ with a particular form, recall Equation 10. If we denote the coordinates and orientation of the planar sensor pose as $\mathbf{x}_k = (x_s, y_s, \theta_s)^T$, then we can write:

Box 2.

$$
\begin{aligned}
\mathbf{S}_k &= \frac{\partial \mathbf{h}}{\partial \mathbf{m}} \quad \Sigma \quad \frac{\partial \mathbf{h}}{\partial \mathbf{m}}^T + \mathbf{Q}_k \\
&= \begin{pmatrix} \frac{\partial \mathbf{h}_1}{\partial \mathbf{m}_1} & 0 & 0 & 0 & 0 & 0 \\ 0 & \frac{\partial \mathbf{h}_2}{\partial \mathbf{m}_2} & 0 & 0 & 0 & 0 \\ 0 & 0 & 0 & \frac{\partial \mathbf{h}_3}{\partial \mathbf{m}_4} & 0 & 0 \\ 0 & 0 & 0 & 0 & 0 & \frac{\partial \mathbf{h}_4}{\partial \mathbf{m}_6} \end{pmatrix}
\begin{pmatrix} \Sigma_1 & \cdots & \Sigma_{1,6} \\ \vdots & \ddots & \vdots \\ \Sigma_{1,6} & \cdots & \Sigma_6 \end{pmatrix}
\begin{pmatrix} \frac{\partial \mathbf{h}_1}{\partial \mathbf{m}_1}^T & 0 & 0 & 0 \\ 0 & \frac{\partial \mathbf{h}_2}{\partial \mathbf{m}_2}^T & 0 & 0 \\ 0 & 0 & 0 & 0 \\ 0 & 0 & \frac{\partial \mathbf{h}_3}{\partial \mathbf{m}_4}^T & 0 \\ 0 & 0 & 0 & 0 \\ 0 & 0 & 0 & \frac{\partial \mathbf{h}_4}{\partial \mathbf{m}_6}^T \end{pmatrix}
+ \begin{pmatrix} \mathbf{Q} & 0 & 0 & 0 \\ 0 & \mathbf{Q} & 0 & 0 \\ 0 & 0 & \mathbf{Q} & 0 \\ 0 & 0 & 0 & \mathbf{Q} \end{pmatrix} \\
&= \begin{pmatrix} \blacksquare & & & \\ & \blacksquare & & \\ & & \blacksquare & \\ & & & \blacksquare \end{pmatrix}
\begin{pmatrix} \blacksquare & \cdots & \blacksquare \\ \vdots & \ddots & \vdots \\ \blacksquare & \cdots & \blacksquare \end{pmatrix}
\begin{pmatrix} \blacksquare & & & \\ & \blacksquare & & \\ & & \blacksquare & \\ & & & \blacksquare \end{pmatrix}
+ \begin{pmatrix} \blacksquare & & & \\ & \blacksquare & & \\ & & \blacksquare & \\ & & & \blacksquare \end{pmatrix} \\
&= \begin{pmatrix} \blacksquare & \blacksquare & \blacksquare & \blacksquare \\ \blacksquare & \blacksquare & \blacksquare & \blacksquare \\ \blacksquare & \blacksquare & \blacksquare & \blacksquare \\ \blacksquare & \blacksquare & \blacksquare & \blacksquare \end{pmatrix}
\end{aligned} \tag{15}
$$

$$\mathbf{z}_k = \mathbf{h}(\mathbf{x}_k, \mathbf{m})$$

$$\begin{bmatrix} \mathbf{z}_{k,1} \\ \vdots \\ \mathbf{z}_{k,M} \end{bmatrix} = \begin{bmatrix} \mathbf{h}(\mathbf{x}_k, \mathbf{m}_1) \\ \vdots \\ \mathbf{h}(\mathbf{x}_k, \mathbf{m}_M) \end{bmatrix}$$

$$\rightarrow \mathbf{z}_{k,i} = \begin{pmatrix} r_i \\ b_i \end{pmatrix} = \mathbf{h}(\mathbf{x}_k, \mathbf{m}_i)$$

with:

$$\begin{cases} r_i = \sqrt{(x_i - x_s)^2 + (y_i - y_s)^2} \\ b_i = \mathrm{atan2}(y_i - y_s, x_i - x_s) - \theta_s \end{cases} \quad (16)$$

The trigonometric function $\mathrm{atan2}(\Delta y, \Delta x)$ yields the arc tangent of $\Delta y \,/\, \Delta x$ extended to cover a whole $[-\pi, \pi]$ range, that is, it is able to identify the quadrant where the angle is.

4. FEATURE SENSORS: DATA ASSOCIATION

Up to this point, we have assumed a perfect identification between features being observed and their corresponding models in the map. However, in most practical cases this correspondence is not straightforward to establish. For instance, think of a sensor capable of detecting vertical lines in the environment and providing their range and bearing with respect to it. As long as all vertical lines look identical to the sensor, we cannot be sure on which map feature has to be associated to each given observation. This is called the *Data Association problem (DA)* (Bar-Shalom & Fortmann, 1988) and is one of the main complications that arise when employing feature sensors for mobile robot localization and mapping.

In general, if the robot gathers a number M of observations (features), the map has a number L of known features indexed from 1 to L, and we also take into account the possibility of having observed something that is not in the map (which will be represented by a "virtual" feature $\mathbf{m}_0$), there are $(L+1)^M$ possible associations from which we have to select the most likely one in order to use Equation 7. This clearly shows the intractability issue that arises in the DA problem: usually it is not possible to explore all the associations exhaustively, especially with sensors like cameras, which may detect tenths or even hundreds of features in a single image. Approximations to the problem exist, as we explain below, but at the risk of making wrong associations, which may lead to a disastrous localization of the robot. Furthermore, if such errors occur while building a map (not when only performing localization) we face one of the worst possible situations for mapping algorithms, since most of them are unable to backtrack and correct previously established wrong associations.

Formally, the DA problem consists of finding a one-to-one function $H_k : [1, M] \rightarrow [0, L]$ that maximizes the likelihood $f_{\mathbf{z}_k | \mathbf{x}_k, \mathbf{m} \cup \mathbf{m}_0}$ of the set of observations gathered at time k with respect to the given map and robot pose hypothesis. For historical reasons, the common approach in robotics is to minimize instead the Mahalanobis distance between the observations and the predicted locations of map features for the hypothetical pose of the sensor. Although this leads to good results in general, it is easily shown that the Mahalanobis distance is not the optimal statistic to determine correspondences (Blanco, González, & Fernández-Madrigal, 2012). However, we first describe its application to DA due to its wide-spread usage and later on we discuss how to replace it with the *matching likelihood*, the optimal statistic.

The Mahalanobis distance is a measure between two multidimensional points, $\mathbf{q} = (q_1\ q_2 \ldots\ q_r)^T$ and $\boldsymbol{\mu} = (\mu_1\ \mu_2 \ldots\ \mu_r)^T$, but not in Euclidean space, which would give us a distance of $\|\mathbf{q} - \boldsymbol{\mu}\|$, but in a modified one which is scaled according to some covariance matrix Σ. Mathematically, the squared Mahalanobis distance (SMD) is defined such that:

$$D^2_{M;\mu,\Sigma}\left(\mathbf{q}\right) \triangleq \left(\mathbf{q}-\mu\right)^T \Sigma^{-1}\left(\mathbf{q}-\mu\right)$$

(sometimes also denoted as):

$$D^2_{M;\mu,\Sigma}\left(\mathbf{q}\right) \triangleq \left\|\mathbf{q}-\mu\right\|^2_{\Sigma} \quad (17)$$

This can be interpreted as the distance of the point $\mathbf{q}$ to the multivariate Gaussian $N\left(\mu,\Sigma\right)$, such that distances are weighted in a way inversely proportional to the uncertainty present in each direction of space. To clarify this concept, it becomes illuminating to consider what would be the Mahalanobis distance between two 2D points according to a diagonal covariance matrix:

$$\begin{aligned} D^2_{M;\mu,\Sigma}\left(\mathbf{q}\right) &= \left(\mathbf{q}-\mu\right)^T \Sigma^{-1}\left(\mathbf{q}-\mu\right) = \\ &= \begin{bmatrix} q_x-\mu_x \\ q_y-\mu_y \end{bmatrix}^T \begin{pmatrix} \sigma_x^2 & 0 \\ 0 & \sigma_y^2 \end{pmatrix}^{-1} \begin{bmatrix} q_x-\mu_x \\ q_y-\mu_y \end{bmatrix} = \\ &= \begin{bmatrix} q_x-\mu_x & q_y-\mu_y \end{bmatrix} \begin{pmatrix} \frac{1}{\sigma_x^2} & 0 \\ 0 & \frac{1}{\sigma_y^2} \end{pmatrix}^{-1} \begin{bmatrix} q_x-\mu_x \\ q_y-\mu_y \end{bmatrix} = \\ &= \left(\frac{q_x-\mu_x}{\sigma_x}\right)^2 + \left(\frac{q_y-\mu_y}{\sigma_y}\right)^2 \end{aligned} \quad (18)$$

In comparison to the squared Euclidean distance $\left\|\mathbf{q}-\mu\right\|^2 = \left(q_x-\mu_x\right)^2 + \left(q_y-\mu_y\right)^2$ it becomes clear the role of the covariance as a scale for each dimension. The same interpretation can be applied to non-diagonal covariances, since there will always[3] exist a change of coordinates (according to the eigenvectors of the covariance, see Appendix E) under which the covariance is diagonalized.

The intuitive meaning of this distance is that the "closer" the point $\mathbf{q}$ is to the expected value μ of the distribution—taking into account its covariance Σ adequately—the more likely should be that the point is produced by that distribution. However, notice that a small SMD does not imply any particularly high *likelihood* value. Even for a distance of zero, all we know is that the point $\mathbf{q}$ is exactly at the mean of the Gaussian, but we have no information at all about how large is its covariance, thus we could have *any* likelihood between almost zero up to almost infinity. Thus, if we want to find out whether a point $\mathbf{q}$ is more likely generated by a Gaussian $N\left(\mu_1,\Sigma_1\right)$ or another one $N\left(\mu_2,\Sigma_2\right)$, the Mahalanobis distance is *not* the best statistic, against the wide spread misconception. For a more in-depth discussion of this topic, refer to (Blanco, González, & Fernández-Madrigal, 2012).

In spite of this loss of information about the scale of covariance matrices, DA is usually addressed in the literature by minimizing the SMD between observations and the predicted location of map features under the given hypothetical sensor pose. The goal is therefore searching for the mapping between observations and predicted map features that minimizes the SMD of the overall observation vector, i.e. taking into account its full covariance $\mathbf{S}_k$ in Equation 12.

In general, the search space for finding the optimal total distance has the form of a mathematical tree[4] called an *interpretation tree* (Grimson, 1991; Grimson & Lozano-Pérez, 1987; see Figure 3), that has N_k levels and a branching factor (number of children of each node) of $M+1$ A node in the tree fixes an association for all the previous nodes (observations), while a complete path in that tree, from the root node to one of the leaves, represents a complete association between all the observations and map features.

Since the exhaustive search of all the paths in the interpretation tree has a prohibitive cost, we should use some heuristic algorithm that discards parts of the tree without exploration. In the following, we explain two of the most used algorithms of that kind. We will also discuss why the maxi-

Figure 3. A simple example of an interpretation tree. Two features (e.g. "corners") have been detected in a scene (a), and the known map also has two features (b). (c) The interpretation tree, which contains all the possible associations between observations and map features in the form of paths, of which we highlight two examples: (1) the combination $\{z_1 = m_1, z_2 = m_0\}$ which seems to be the most likely in this case (i.e. observation z_1 corresponds to the known m_1 and z_2 is a new feature not mapped yet), and (2) the combination $\{z_1 = m_2, z_2 = m_2\}$, which seems very unlikely in this case.

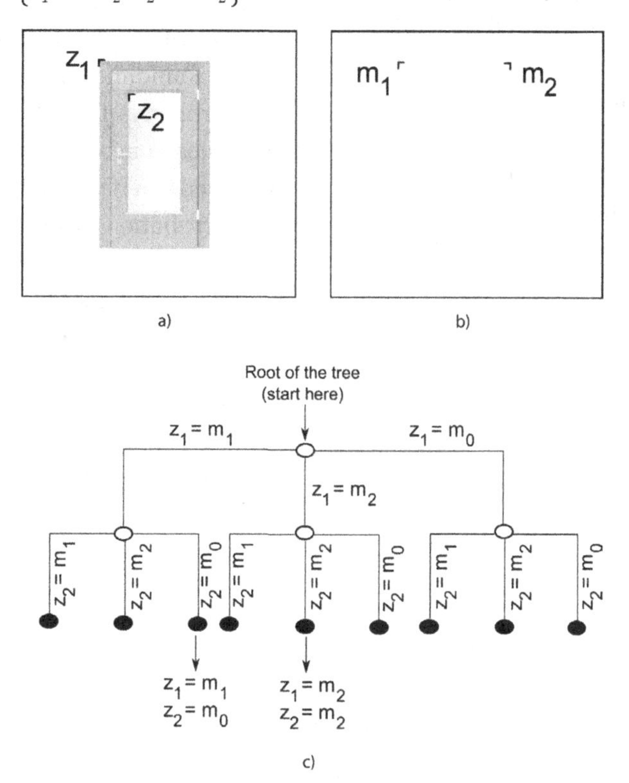

mum likelihood, instead of the Mahalanobis distance we have proposed initially, is the natural measure to be optimized in the DA problem.

The Nearest Neighbor DA Algorithm (NN)

This first DA algorithm is really simple and fast, but more sensitive to wrong associations than others. It is an example of a *greedy* algorithm, that is, one that is strongly based on local decisions—usually unchangeable afterwards—while searching for a global solution.

Broadly, the NN works as follows: it first selects an arbitrary observation feature $\mathbf{z}_i$ and calculates its SMD to the predicted observations for each feature $\mathbf{m}_j$ in the map; the distances below a given threshold are included in the set of possible candidates for association to $\mathbf{z}_i$ (if the maximum likelihood is used instead of the SMD, we would look for the maximum value of the likelihood and check if it is *above* a threshold). If the resulting set of candidates is empty, $\mathbf{z}_i$ is associated to the invalid map feature $\mathbf{m}_0$, meaning that a feature not contained into the map has been detected (this is mostly useful in the SLAM problem); otherwise, the association having the minimum distance among the candidates is selected as the correct one. Notice that, in principle, several observations could be associated to the same map feature by this method.

Now we go into the specifics. For selecting candidates for $\mathbf{z}_i$, taking into account the existing uncertainty, we need a test that is able to tell which distances to predicted map features can be considered actually greater than desirable. For modeling the uncertainty in those distances, we begin by focusing on the chi-squared probability distribution, which, as explained in chapter 3 section 8, is the distribution of a r.v. resulting from the sum of k independent, squared r.v.s with normal distributions. Just to remember its mathematical form, the chi-squared pdf in the unidimensional case is[5]:

$$f_{x;k} = \chi_k^2(x) = \begin{cases} \dfrac{1}{2^{k/2}\Gamma(k/2)} x^{x/2-1} e^{-x/2} & \text{if } x \geq 0 \\ 0 & \text{otherwise} \end{cases} \tag{19}$$

where the parameter k is called the *degrees of freedom* of the chi-squared distribution. Figure 4 shows a unidimensional example.

A property of the chi-squared distribution that is of interest in our problem is that it can be demonstrated that if we have a k-dimensional Gaussian distribution with mean μ and covariance matrix Σ, then the SMD that exists from the vectors $\mathbf{q}$ that could be produced stochastically by that Gaussian to the mean of the Gaussian, that is, $\left\|\mathbf{q}-\mu\right\|_{\Sigma}^{2}$, follows a chi-squared unidimensional distribution with k degrees of freedom[6]. Particularizing for our case, the probability that

Figure 4. The pdf of a chi-squared distribution with 4 degrees of freedom. The figure also shows the p-value of a given statistic value (a point in the support of the distribution), and also a level of significance α established arbitrarily for hypothesis testing and its corresponding $\chi^2_{k,\alpha}$ (see the text). The value of the statistic and of $\chi^2_{k,\alpha}$ are points in the support of the chi-squared distributed r.v.; the p-value and the level of significance are areas of the chi-squared pdf.

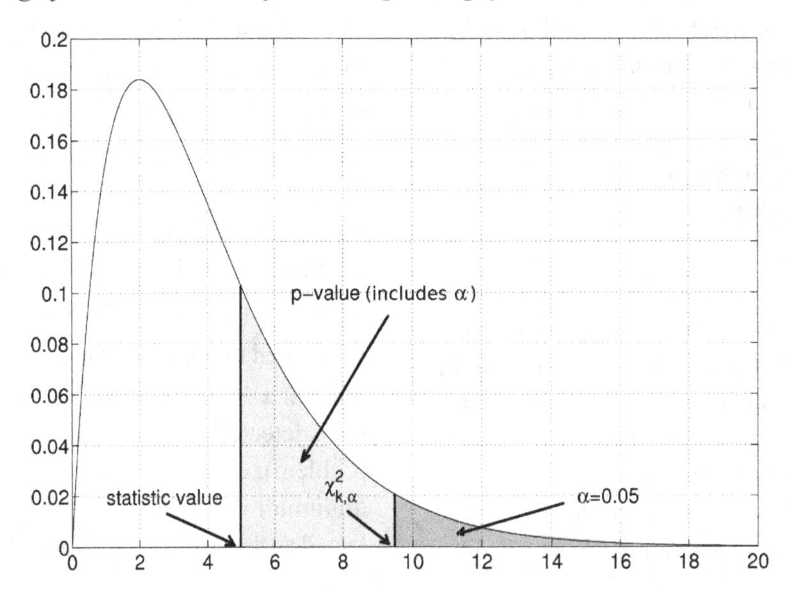

a given observed feature $\mathbf{z}_i$ corresponds to a map feature $\mathbf{m}_j$ is clearly related to the probability of occurrence of the SMD $D^2_{M;\mathbf{m}_j}(\mathbf{z}_i)$ existing from the location of $\mathbf{z}_i$ to the mean of the Gaussian centered at the theoretical location where $\mathbf{m}_j$ should be seen from the robot pose[7].

As said above, in the NN algorithm we require the probability of correspondence between $\mathbf{z}_i$ and $\mathbf{m}_j$ to be larger than a certain threshold in order to accept such correspondence as valid. Using the Mahalanobis distance and the chi-squared distribution, the probability of correspondence between $\mathbf{z}_i$ and $\mathbf{m}_j$ becomes the probability that $D^2_{M;\mathbf{m}_j}(\mathbf{z}_i)$ is less than or equal to the integral of the chi-squared distribution in a certain interval $[0, p]$; inversely, the probability that $\mathbf{z}_i$ does not correspond to $\mathbf{m}_j$ becomes the probability that $D^2_{M;\mathbf{m}_j}(\mathbf{z}_i)$ is larger than or equal to the integral of the chi-squared from p to infinity (the so-called *right-tail* of a pdf), an area of the pdf that is called the *p-value* of $D^2_{M;\mathbf{m}_j}(\mathbf{z}_i)$. As we increase p, the probability of having an observation with equal or greater distance than p vanishes. If the distance we measure is equal or greater than such a p, we cannot tell that $\mathbf{z}_i$ corresponds to $\mathbf{m}_j$: it would more likely correspond to another map feature, to a newly discovered feature or even it could be a spurious reading. Therefore, we can choose a limit value—a threshold—q and decide that if $D^2_{M;\mathbf{m}_j}(\mathbf{z}_i) \geq q$, the hypothesis that $\mathbf{z}_i$ corresponds to $\mathbf{m}_j$ is not sustainable. The p-value corresponding to such q is called the level of significance of our test, and denoted α.

In statistics, this is called a *test of significance,* and in the case of using the chi-squared distribution, a *chi-squared test.* In such a test we have a statistic ($D^2_{M;\mathbf{m}_j}(\mathbf{z}_i)$ in our case) that yields values in the support of a given r.v. that has certain distribution (chi-squared in our case), and we check whether the p-value associated to that statistic value (given by the area under the chi-squared pdf that ranges from that value to infinity) is greater or equal than an arbitrarily fixed level of significance α (usually, $\alpha = 0.05$ or 0.01). The level of significance indicates with which probability, at most, we will reject an actual, valid correspondence—a *false negative* or *error of type I* in the terminology of statistics. The value in the support of the chi-squared variable that corresponds to the minimum limit of the area α can be calculated through the inverse cdf of the chi-squared distribution, as it was explained in chapter 5, and is commonly denoted as $\chi^2_{k,\alpha}$.

Therefore, our test for including $\mathbf{m}_j$ in the set of candidates for association to $\mathbf{z}_i$, taking into account the uncertainty in the observation, will be:

$$D^2_{M;\mathbf{m}_j}(\mathbf{z}_i) < \chi^2_{k,\alpha} \tag{20}$$

where k is the dimensionality of individual feature observations, typically 2 or 3. Remember that data association takes place in the observation space, disregarding of the dimensionality of feature representations in the map.

Since the chi-squared pdf in Equation 19 has no closed solution, due to an integral that appears within the gamma function, numerical approximations are needed to obtain the values of $\chi^2_{k,\alpha}$ for any given α and degrees of freedom. Pre-calculated tables can be found in most statistics books and good closed-form approximations also exist (Press, Teukolsky, Vetterling, & Flannery, 1992), thus this can be done in constant time and is not included in the computational cost of testing for associations. The computational complexity[8] of the NN algorithm is therefore $O(N_k M)$, linear in the number of observations, instead of the $O(M^{N_k})$ of an exhaustive search.

The Joint Compatibility Branch and Bound DA Algorithm (JCBB)

Obtaining more robust results than the NN can only be done at the expense of a higher computational cost. Notice that the NN does not consider any of the existing dependencies between the simultaneous observations of different features. The way to cope with this is to take into account not only the individual compatibility between an observation and a map feature, but also joint compatibilities among sets of such pairs. Furthermore, we need an algorithm that is able to consider different hypotheses for pairings and that does not become stuck in previous (wrong) association decisions. One of the most interesting algorithms in that sense is the JCBB (Neira & Tardós, 2001), which is based on a well known heuristic search strategy used in optimization problems called *Branch and Bound (BB)*.

The BB strategy is a recursive procedure to search for the global optimal value of a numeric function (actually, of any function whose codomain can be ordered). Due to its heuristic nature, the global optimum is not guaranteed to be found, thus it is recommended in situations where an exhaustive search is prohibitive. It partitions the domain of the function into a finite number of sets which union yields the original domain (the vertical regions shown in Figure 5). That partition is the *branch* part of the strategy. Then, bounds for the minimum and maximum values of the function within each set are estimated in an efficient way; that constitutes the *bound* part. When we compare the bounds of a set in the partition to the bounds of the other sets we can deduce whether the former set can contain or not the optimum, and discard further explorations of that set in the second case. For example, if we look for the minimum of the total Mahalanobis distance between observations

Figure 5. Illustration of the branch and bound strategy for the search of the global minimum of a mathematical function. The branch part partitions the search space into three, while the bound part estimates minimum and maximum bounds for each set in the partition (shaded regions). Those sets that cannot contain the global optimum (the middle and right sets in this figure) since their bounds prevent it, are discarded from further exploration. The figure only shows the first step of the BB execution. In the next step, the left set will be the only one to be partitioned again.

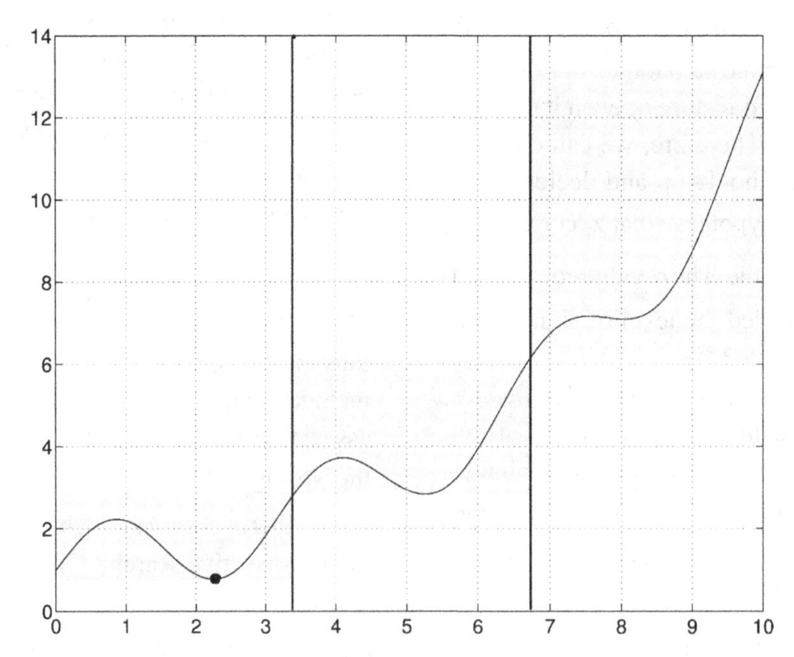

and map features, we can divide the set of all the paths in the interpretation tree into several subsets of paths. If for each subset we could estimate efficiently lower and upper bounds for the total distance of its paths, we could discard from further exploration those subsets with a lower bound greater than any upper bound of the other subsets: the former are not likely to contain the minimum total distance we are looking for. If we repeat this procedure in each of the surviving sets of this level of the branching, until reaching levels with singleton sets, we could select at the final step the surviving path with less total distance, and that would be our approximately optimal path.

The JCBB is based on the *joint compatibility* of a set of correspondences between observations and map features (i.e., a path in the interpretation tree). The joint compatibility of a given path in the interpretation tree, which does not need to reach a leaf, i.e., to be complete, is defined as the *joint* Mahalanobis distance of all the nodes in the path gathered together as a single multidimensional correspondence. That is, if each correspondence (node) is of dimension 2 in a typical 2D scenario, a sequence of k nodes in the tree can be considered as a problem of dimension $2k$ and treated as a single Mahalanobis distance in that higher dimensional space. The usage of the SMD over the *joint* distribution of all the observations at once, as opposed to the one-by-one policy seen in the NN, is the key for the robustness of JCBB.

Notice that an incomplete path in the interpretation tree (i.e., one that does not contain leaves) is actually a set of complete paths: it represents all the possible explorations of the children of its deepest node. Furthermore: an incomplete path $\wp_a$ represents a set of complete paths that has no element in common with the set of complete paths represented by a different incomplete path $\wp_b$, as long as $\wp_b \not\subset \wp_a$. That is the kind of partition of the search space that the JCBB produces.

Using this concept of joint compatibility, the JCBB searches the interpretation tree in a depth-first fashion: it first selects a child of the root node, then repeats the exploration with a child of that child, and so on until a leaf node is reached. At that point it, backtracks—goes upwards the tree—and selects for exploration one of the next deepest nodes that has not been explored yet; this procedure continues until no more nodes remain unexplored in the tree. As you easily observe, a pure depth-first search like the one described would be exhaustive, therefore the JCBB does not perform all the explorations: at some nodes it decides not to explore deeper. That is where the heuristic of the JCBB for bounding plays a determinant role.

The bound heuristic is as follows. Since the algorithm knows the number of pairings of each explored path that represent a correspondence with the invalid feature $\mathbf{m}_0$, it can stop the exploration of any incomplete path that contains more such pairings than the ones contained in some other explored (complete or incomplete) path. This is so since the discarded path cannot improve (decrease) the number of invalid pairings in further explorations, only increase them. In other words: the JCBB is designed to minimize the number of pairings with the invalid feature $\mathbf{m}_0$, which represents the association of an observation with a feature that is not in the known map.

In the above depth-first procedure, we have not specified which child of a given node v is chosen to be explored first. That is the branch heuristic. In the JCBB, since the joint compatibility of the incomplete path that reaches v has already been calculated, we can efficiently calculate the joint compatibility of the one-step extensions of that path with each of the children of v. The branch heuristic then consists of selecting, for first exploration, the child of v yielding the best value for the joint compatibility test; that is, yielding the lowest value for the squared multidimensional Mahalanobis distance. Thus, the algorithm gives more precedence to a better joint compatibility in the next step of exploration. Note that this does not prevent the exploration of paths with worse joint compatibility in the future: it only establishes an ordering for that exploration.

Unfortunately, branch and bound does not guarantee not to explore the entire tree: depending on the suitability of heuristics for the particular situation at hand, it can reduce more or less the search cost. In the case of the JCBB, the reduction is typically high, having an asymptotic computational complexity of $O(1.53^{N_k})$—measured experimentally in Neira and Tardós (2001).

Mahalanobis Distance vs. Matching Likelihood

The methods described in the previous paragraphs rely on the SMD to test whether a given feature in the observation corresponds to another feature in the global map. This is an approximation to the real problem, which is to maximize the likelihood that the observation corresponds to the feature: theoretically, minimizing the SMD is not equivalent to maximizing the likelihood of the correspondence, although they are related. Therefore, the analysis of their respective behaviors may be interesting.

Under the assumption of Gaussian uncertainty, the likelihood for a given set $\mathbf{z}_k$ of r observations taken from a hypothetical robot pose $\mathbf{x}_k$, given that the robot operates in a known map (set of features) $\mathbf{m}$, and assuming a set of pairings $\mathbf{n}_k$ between observations in $\mathbf{z}_k$ and features in $\mathbf{m}$ such that each element of the vector $\mathbf{n}_k$ is associated to an observation and contains the index of the map feature that corresponds to it (or 0 if the correspondence with no map feature is selected), is:

$$f_{\mathbf{z}_k|\mathbf{x}_k,\mathbf{m},\mathbf{n}_k} = N\left(\mathbf{z}_k;\mathbf{h}\left(\mathbf{x}_k,\mathbf{n}_k\right),\Sigma\left(\mathbf{n}_k\right)\right) \quad (21)$$

Here, $\mathbf{h}\left(\mathbf{x}_k,\mathbf{n}_k\right)$ gives us the expected location of the corresponding map feature according to the pose of the robot, and $\Sigma\left(\mathbf{n}_k\right)$ is its covariance. The unknown element here is the association vector $\mathbf{n}_k$

On the other hand, the SMD for this case is:

$$D_M^2\left(\mathbf{n}_k\right) = \left(\mathbf{z}_k - \mathbf{h}\left(\mathbf{x}_k,\mathbf{n}_k\right)\right)^T \Sigma\left(\mathbf{n}_k\right)^{-1}\left(\mathbf{z}_k - \mathbf{h}\left(\mathbf{x}_k,\mathbf{n}_k\right)\right) \quad (22)$$

We can take logarithms in Equation 21 (being monotonically increasing, they do not alter the location of the optimal values of the function):

$$\begin{aligned}
\ln f_{\mathbf{z}_k|\mathbf{x}_k,\mathbf{m},\mathbf{n}_k} &= \ln N\left(\mathbf{z}_k;\mathbf{h}\left(\mathbf{x}_k,\mathbf{n}_k\right),\Sigma\left(\mathbf{n}_k\right)\right) = \\
&= \ln\left(\frac{1}{\left(2\pi\right)^{r/2}\left|\Sigma\left(\mathbf{n}_k\right)\right|^{1/2}}\exp\left(-\frac{1}{2}\left\|\mathbf{z}_k - \mathbf{h}\left(\mathbf{x}_k,\mathbf{n}_k\right)\right\|^2_{\Sigma(\mathbf{n}_k)}\right)\right) = \\
&= -\frac{r}{2}\ln 2\pi - \frac{1}{2}\ln\left|\Sigma\left(\mathbf{n}_k\right)\right| - \frac{1}{2}\underbrace{\left\|\mathbf{z}_k - \mathbf{h}\left(\mathbf{x}_k,\mathbf{n}_k\right)\right\|^2_{\Sigma(\mathbf{n}_k)}}_{\text{The SMD } D_M^2(\mathbf{n}_k)} = \\
&= -\frac{r}{2}\ln 2\pi - \frac{1}{2}\ln\left|\Sigma\left(\mathbf{n}_k\right)\right| - \frac{1}{2}D_M^2\left(\mathbf{n}_k\right)
\end{aligned} \quad (23)$$

Maximizing this expression is equivalent (by dropping the multiplying factor and inverting all signs) to minimizing:

$$ML = r\ln 2\pi + \ln\left|\Sigma\left(\mathbf{n}_k\right)\right| + D_M^2\left(\mathbf{n}_k\right) \quad (24)$$

which clearly states why minimizing the SMD alone, $D_M^2(\mathbf{n}_k)$, is not the same as maximizing the Matching Likelihood (ML). We can take both measures of optimality as being approximately equivalent only when the determinant of the covariances, i.e., $\left|\Sigma\left(\mathbf{n}_k\right)\right|$, is nearly constant for any possible pairing given by $\mathbf{n}_k$. In practice, the latter assumption is what makes the SMD to give us fairly good results. However, the ML form should be preferred instead of the SMD, since it has been proven to reduce the number of wrong associations with respect to the SMD (Blanco, González, & Fernández-Madrigal, 2012).

5. "MAP" SENSORS

In previous models we considered each of the observations gathered by the sensor to be independent from the others, thus we assign an individual likelihood to each separate observation. Instead, in this section we will explore a completely different idea: gathering a set of simultaneous observations that can be considered as a *local map* of the environment, as seen from the current position of the robot, and then evaluating the likelihood of the two maps, the local, and the global one, to correspond (or *match*) to each other. Given a hypothetical robot pose, which defines the translation and rotation between the maps, evaluating the likelihood of the local map to coincide with another map (the *global map*) means testing how well the two maps overlap and coincide with each other. A high likelihood would indicate that the robot is well located in the global map, or, when addressing the SLAM problem, can also indicate the revisiting of an already known place. Another application of this idea is comparing local maps with the portions (or *submaps*) in which a large metric map has been split. This perspective is closer to higher-level (i.e. more abstract) approaches to SLAM, thus we postpone the discussion of this point of view until chapter 10.

In any case, if we consider that our observations consist of local maps, the sensor model (observation likelihood) becomes $f_{\mathbf{m}_{local,k}|\mathbf{x}_k,\mathbf{m}}$, where $\mathbf{m}_{local,k}$ is the observed local map at time step k. We will assume that both the local and global maps are modeled as the same kind of metric map, e.g. two grid maps. In addition, we will consider that the robot pose fixes the translation and rotation of the local map with respect to the global map.

Several methods for map matching in such a scenario can be found in the literature, since they are highly dependent of the nature of the map representations—chapter 8 is devoted to provide detailed descriptions of the most common map representations. Next, we address two useful cases: grid maps and point maps.

Grid Map Matching

Grid maps or occupancy grids are regular tessellations of space whose cells reflect the evidence of occupancy in each region. More on the probabilistic interpretation of grid maps will be addressed in chapter 8 section 3, but at this point all we need to know is that grids have a structure that resembles grayscale images. Therefore, methods from computer vision for image registration can be also applied to matching occupancy grids.

Matching two grid maps can be seen as the problem of finding the rigid transformation (a rotation and a translation) such that, applied to the local map, maximizes its *matching* with a global map (Konolige, 1999). In our case the transformation is fixed by the hypothetical pose of the robot, thus we only need to obtain a matching quality value for that transformation, avoiding the optimization step. Providing an exact form for that matching function under a given transformation is difficult since we should take into account issues such as the impossibility of having observed a cell as occupied if there is another cell in the line of sight that is closer to the sensor and is already detected as occupied. Therefore, we have to rely on approximate or heuristic methods. In the following, we explain three of them, whose results are demonstrated with a small example in Figure 6.

The first (and simplest) approximation to measuring the "amount or quality of matching" existing between two maps under a given rigid

Figure 6. Illustration of the three methods for map matching explained in the text. (a) A local map obtained through an ultrasound sensor turning within a real environment. (b) The known, global map of the environment where the local map was acquired. (c) – (e) The results of the three grid map matching methods discussed in the text as for different displacements of the local map in the range -25 to +25 cells along the horizontal axis, assuming a perfect alignment in the vertical direction and in the orientation The three of them clearly show a maximum peak at the origin, where both maps are known to actually coincide. Figure 6c is the result of Equation 26 with $\alpha = 1 / N$, where N is the number of cells of the grid. Figure 6d is the application of Equation 30. Figure 6e is the result of Equation 31.

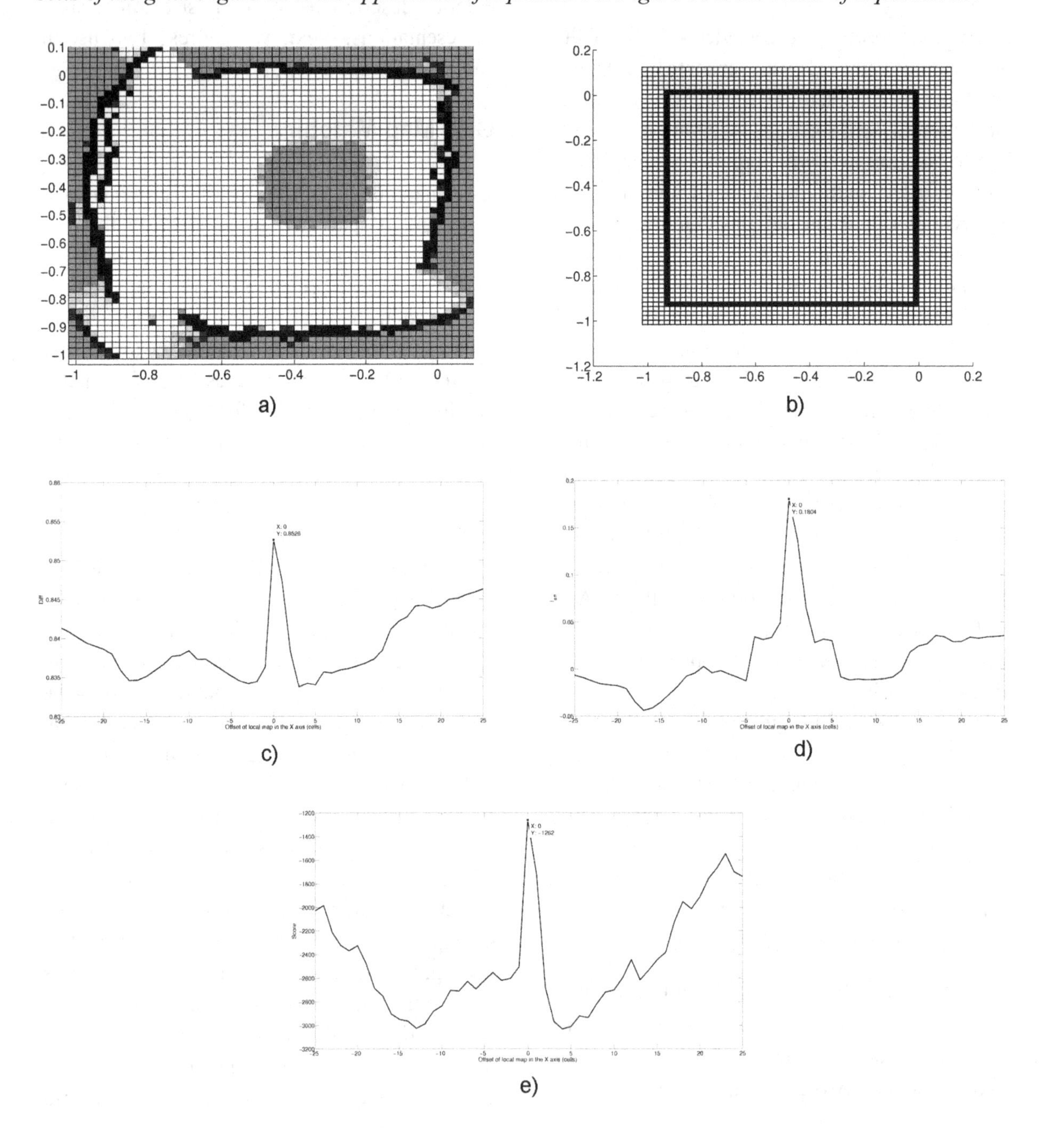

transformation consists of just accounting for their individual differences at every cell:

$$diff(\mathbf{m}_{local};\mathbf{m},\mathbf{x}_k) = \sum_{(x',y')\to(x,y)} (m_{local,x,y} - m_{x',y'})^2 \quad (25)$$

where we assume 2D grids and the existence of a relation from any cell (x',y') of the local map to a single cell (x,y) in the global map. Thus the alignment of both maps through $\mathbf{x}_k$ must be such that the areas of their cells overlap one-to-one, which happens rarely in practice (for example, such alignment does not occur whenever non-zero rotations are involved). A solution is to map the content of $\mathbf{m}_{local}$, once it is shifted and rotated by $\mathbf{x}_k$, into a new grid structure that coincides with the one of $\mathbf{m}$.

Notice that the larger the result of Equation 25, the smaller the possibility that a good match exists between both grid maps under the given transformation. In addition, the range of attainable values extends from 0 up to a number that depends on the size of the smallest of the two maps. Therefore, the use of this kind of matching into a observation likelihood requires: (1) that Equation 25 is modified in order to yield higher values for better matches, and (2) that the integral of the resulting function in its entire support equals one, i.e., that the integral of the likelihood over all the possible local maps—the likelihood variable—equals one. Issue (2) is clearly overwhelming, thus we can provide a solution to issue (1) that is proportional to the likelihood, and then consider its proportionality constant by the normalizing constant of the localization or SLAM filter, something that is explained in chapter 7. For example, we can use for the likelihood the following expression:

$$f_{\mathbf{m}_{local,k}|\mathbf{x}_k,\mathbf{m}} \propto \exp\left(-\alpha \cdot diff(\mathbf{m}_{local};\mathbf{m},\mathbf{x}_k)\right) \quad (26)$$

being α a suitable constant for avoiding large values of the exponent.

A more powerful approximation to measuring the "amount of matching" is based on the Baron's cross-correlation coefficient, also called *normalized cross-correlation* in the computer vision literature, which also requires both maps to be represented with respect to the same coordinate frame and have corresponding regions of space for each cell in the grid. According to Baron (1981), computing the cross-correlation function between two images amounts to comparing them against one another in every possible position when shifting both vertically and horizontally.

This can also be seen from the perspective of random variables if we consider the index of a cell in the final grid as a r.v., with dimension the one of the grid, typically 2, and with its support spanning the entire grid; in this case we need to assume that the local map shares exactly the same grid—structure and size—as the global map, with zeroes in those cells not originally considered in the former. Having such indexes as r.v.s, we can provide them with meaning as follows: the probability associated to such an index is the evidence of having non-empty space in the cell of that index (we need to scale appropriately the grid contents in order for these probabilities to sum up to 1). In other words, the content of the map becomes the pmf of the index of the map, when considering that index as a discrete r.v.

In our problem we have two pmfs, $f_{\mathbf{m}_{local}}(\mathbf{u})$ and $f_{\mathbf{m}}(\mathbf{w})$, that yield the probability of indexes $\mathbf{u}$ and $\mathbf{w}$, taken from the occupancy content of the corresponding grids as explained before. The local map $\mathbf{m}_{local}$ is the result of translating and rotating the data gathered by the sensor according to the hypothetical robot pose. The global map $\mathbf{m}$ is constant. We are interested in the probability of that translation and rotation of the local map to make its content similar to the one of the global map, that is, the probability that there is no difference between the contents of coincident cells after the transformation, or, mathematically, the probability that $f_{\mathbf{m}_{local}}(\mathbf{u})$ equals $f_{\mathbf{m}}(\mathbf{w})$. It is clear that if both $\mathbf{u}$ and $\mathbf{w}$ are the same r.v. (that is, $\mathbf{u}=\mathbf{w}$ or, equivalently, $\mathbf{u}-\mathbf{w}=\mathbf{0}$), then both have the same shapes of uncertainty, that is, $f_{\mathbf{m}_{local}}(\mathbf{u})=f_{\mathbf{m}}(\mathbf{w})$. Therefore we can use the probability that $\mathbf{u}-\mathbf{w}=\mathbf{0}$ for our calculations of the likelihood of the map matching: if that probability is high it is likely that both r.v.s share the same pmf, that is, that both maps coincide.

From chapter 3 section 8, we know that the pdf of the subtraction of two continuous r.v.s is the cross-correlation of both. In the case of discrete $\mathbf{u}$ and $\mathbf{w}$, we have the following analogy for pmfs:

$$f_{diff}(\mathbf{z})=f_{\mathbf{m}}\star f_{\mathbf{m}_{local}}(\mathbf{z})=\sum_{\mathbf{w}} f_{\mathbf{m}}(\mathbf{w})f_{\mathbf{m}_{local}}(\mathbf{z}+\mathbf{w}) \tag{27}$$

Here, $\mathbf{z}$ and $\mathbf{w}$ are indexes within the support defined by the size of the grids, and $\mathbf{z}$ comes from a subtraction: $\mathbf{z}=\mathbf{u}-\mathbf{w}$. We are interested in the case for $\mathbf{z}=\mathbf{0}$, thus we have $f_{diff}(\mathbf{0})=\sum_{\mathbf{w}} f_{\mathbf{m}}(\mathbf{w})f_{\mathbf{m}_{local}}(\mathbf{w})$ as our approximation to the likelihood of that transformation of the local map. We will denote it as ℓ_{diff} for simplicity. In the case of 2D grid maps, this equation becomes:

$$\ell_{diff}=\sum_{i,j} f_{m,i,j} f_{m_{local},i,j} \tag{28}$$

where $f_{m_{local},i,j}$ and $f_{m,i,j}$ denote the contents of the local and global grid maps at the cell (i,j), respectively.

We can make an alternative deduction of this matching measure from the simpler measure of map difference given in Equation 25:

$$\begin{aligned} diff(\mathbf{m}_{local};\mathbf{m},\mathbf{x}_k) &= \sum_{(x',y')\to(x,y)} (m_{local,x,y}-m_{x',y'})^2 = \\ &= \sum_{(x',y')\to(x,y)} \begin{pmatrix} m_{x',y'}^2+m_{local,x,y}^2 \\ -2m_{local,x,y}m_{x',y'} \end{pmatrix} = \\ &= \sum_{(x',y')\to(x,y)} m_{x',y'}^2 + \sum_{(x',y')\to(x,y)} m_{local,x,y}^2 \\ &\quad -2\sum_{(x',y')\to(x,y)} m_{local,x,y}m_{x',y'} \end{aligned} \tag{29}$$

It is easy to see that the first term in the last expression is constant for comparisons with any local map, the second term is very close to be also constant for a given local map as long as rotations and translations have minimal effects in the distortion of its cell shapes, and therefore a minimum distance between both grids is obtained when maximizing the absolute value of the third term, which is equivalent to maximize ℓ_{diff} in Equation 28.

Cross-correlation yields values in the interval $[0,N]$ if the cells contain values in the interval $[0,1]$, being N the number of cells of the grids ($N=width\times height$ in the common 2D case), and its values are higher when a higher likelihood exists. Thus, it would be a satisfactory candidate to form our observation likelihood, although its values widely depend on the variety of values of the cells of the grids. In order to better compare grids with different characteristics, it is common to take the Baron's cross-correlation *coefficient*, which is like the cross-correlation but with the pmfs previously centered at their means and divided by their standard deviations:

$$\begin{aligned} \overline{\ell}_{diff} &= \\ &= \frac{1}{N-1}\frac{1}{\sqrt{\sigma_m^2\sigma_{m_{local}}^2}}\sum_{i,j}\left(f_{m,i,j}-\mu_m\right)\left(f_{m_{local},i,j}-\mu_{m_{local}}\right) \end{aligned} \tag{30}$$

where, naturally, sample means and variances must be used: $\mu_m=N^{-1}\sum_{i,j} f_{m,i,j}$, $\mu_{m_{local}}=N^{-1}\sum_{i,j} f_{m_{local},i,j}$,

$\sigma^2_{m} = (N-1)^{-1}\sum_{i,j}(f_{m,i,j} - E[f_m])^2$ and

$\sigma^2_{m_{local}} = (N-1)^{-1}\sum_{i,j}(f_{m_{local},i,j} - E[f_{m_{local}}])^2$.

This coefficient is analogous to the linear correlation coefficient explained in chapter 3, which measures the "amount of linearity" existing in the relation of two r.v.s. Consider the maps indexed by a single number instead of two ("unstacking" the grids) and you will see that Equation 30 measures in fact that linear relation: when it is 1, the values of cells in the local map perfectly match the ones of corresponding cells in the global map[9]; when it is -1, the values in one of them are completely the opposite to what they should be for a perfect match.

The Baron's cross-correlation *coefficient* provides values in the interval $[-1,1]$, and closer to 1 (i.e., higher) whenever the likelihood of matching is higher. In order to embed it into the likelihood formulation of a filter, we need a shift for avoiding negative numbers (the scaling to integrate to 1 over all possible local maps is left as a normalizing constant). We can use $f_{\mathbf{m}_{local,k}|\mathbf{x}_k,\mathbf{m}} \propto \max(0,\overline{\ell}_{diff})$ or $f_{\mathbf{m}_{local,k}|\mathbf{x}_k,\mathbf{m}} \propto \overline{\ell}_{diff}+1$ for that shift.

Still a third approach to account for the similarities existing between two grid maps is the *score* (Martin & Moravec, 1996). This measure has only sense if both maps contain evidence about their cells being occupied, that is, the cells hold numbers in the interval $[0,1]$, where 0 indicates that the cell is free for sure, 1 that it is occupied for sure, and $1/2$ that we do not have evidence of it being free or occupied. Firstly, a value called the *match* of two evidence grids is defined. Since the probability that two coinciding cells at index (i,j) are both occupied is $m_{i,j}m_{local,i,j}$ and the probability that they are both free is $(1-m_{i,j})(1-m_{local,i,j}) = \overline{m}_{i,j}\overline{m}_{local,i,j}$, assuming the evidence to be mutually independent among all cells, we have that the probability of both maps containing the same evidence is $\prod_{i,j}\left(m_{i,j}m_{local,i,j} + \overline{m}_{i,j}\overline{m}_{local,i,j}\right)$. If both maps contain exactly the same evidence, that value is 1, while if both contain completely contradictory evidence in all cells (even in only one cell!), it becomes 0. Therefore, the number is quite small for practical use. For scaling it we can use its logarithm (typically in base 2), which is comprised within the interval $[-\infty,0]$; we can also add a constant for obtaining a value that equals N, the total number of cells in the grids, when both maps have a completely perfect match (the log equals 0 in that case since the probability is 1). The form for the match of two evidence grids becomes then:

$$\begin{aligned} &match(\mathbf{m},\mathbf{m}_{local}) = \\ &= N + \log_2\left(\prod_{i,j}\left(m_{i,j}m_{local,i,j} + \overline{m}_{i,j}\overline{m}_{local,i,j}\right)\right) = \\ &= N + \sum_{i,j}\log_2\left(m_{i,j}m_{local,i,j} + \overline{m}_{i,j}\overline{m}_{local,i,j}\right) = \\ &= \sum_{i,j}\left(1 + \log_2\left(m_{i,j}m_{local,i,j} + \overline{m}_{i,j}\overline{m}_{local,i,j}\right)\right) \end{aligned} \tag{31}$$

Since we have included a logarithm operation in this measure, we are working now with a *log-likelihood* (or more properly, with something that is proportional to a log-likelihood, since the likelihood must integrate to 1 and we are not including the normalization constant here). Notice that the range of values that Equation 31 can yield cannot be shifted in order to make it positive, thus a different approach must be taken for embedding this measure in an observation likelihood of a filter. Since a base two logarithm is proportional (by a constant) to a base e logarithm, we can apply natural logarithms to the rest of the components of the filter and then insert this log-likelihood (with a proportionality constant) into the filter. A similar technique will be illustrated in chapter 8 when estimating grid maps with a Bayesian approach.

When the global map is known, like in the mobile robot localization problem, it can be the case that it contains only 0, 1, or $1/2$ in its cells. In that case, the same expression of Equation 28 is called the *score* of both maps. If other values in the cells are possible, the *entropy* can be defined instead, but that is out of the scope of this section.

Point Map Matching

Grid maps are among the most popular metric representations of space in mobile robotics, mainly due to their suitability for including in the same map data gathered from different sensors and their good fitting in the probabilistic frameworks for localization and mapping. However, they can consume excessive computer memory since they include all the free space, the dominant form of space in any practical map; some operations (e.g. as rotating the map or expanding its limits) become computationally expensive too.

Another popular alternative form of metrical map is a point map. A point map only represents non-free space. In particular, it contains clouds of points that correspond to the intersection of the lines of sight of some sensor (e.g. the beams of a laser scanner) with solid parts of the environment. This leads to several computational savings, although the loss of information about empty space makes inference less robust. As in the grid map case, point maps have been traditionally also 2D, but in recent years, they are more often used to build 3D models of the world. The same kind of sensors (range and bearing) that are of common use with grid maps are also common in point maps: any sensor that can provide a set of points metrically located in space with respect to the robot may serve.

Obtaining an observation likelihood from a perceived local point map with l points $\mathbf{z}_k = \{\mathbf{r}_1, \mathbf{r}_2, \ldots, \mathbf{r}_l\}$, a global point map with g points $\mathbf{m} = \{\mathbf{q}_1, \mathbf{q}_2, \ldots, \mathbf{q}_g\}$ and a hypothetical robot pose $\mathbf{x}_k$ from which the local map has been observed can be achieved with the following three steps.

Firstly, the set of points in the local map $\mathbf{z}_k$ is transformed (rotated and translated) according to the hypothetical pose, becoming $\mathbf{m}_{local} = \{\mathbf{p}_1, \mathbf{p}_2, \ldots, \mathbf{p}_l\}$ with each point transformed as $\mathbf{p}_i = \mathbf{x}_k \oplus \mathbf{r}_i$ —refer to Appendix A for the expressions of the pose-point $\oplus$ composition operator.

Secondly, we need to establish *correspondences* between the points $\mathbf{m}_{local}$ and those in the global or "reference" map $\mathbf{m}$. As a result, we will have P pairings $\Omega = \{(l_k, g_k)\}_{k=1\ldots P}$, such that the local point $\mathbf{p}_{l_k}$ is associated to the global point $\mathbf{q}_{g_k}$ for each $k = 1\ldots P$. There exists several heuristics for establishing these pairings, but the simplest one is searching for the *closest point* in $\mathbf{m}$ for each local point in $\mathbf{m}_{local}$.

Searching for candidate pairings between point clouds is a well studied topic since it is, by far, the most computationally expensive stage of most Iterative Closest Point (ICP) algorithms, which aim at *optimizing* the matching likelihood in whose evaluation we are interested—that is, ICP can be seen as a sort of MLE of the optimal robot pose with respect to the optimum alignment of two point clouds. More elaborated alternatives for searching pairings have been proposed in the literature, such that searching for the closest distance not to points in a global map $\mathbf{m}$, but to some continuous interpolation of surfaces built from those global points (Besl & McKay, 1992; Lu & Milios, 1997; Censi, 2008). Another common optimization, which we strongly recommend, is to employ KD-trees or Approximate Nearest Neighbors (ANN) techniques to improve the search of nearby points (Rusinkiewicz & Levoy, 2001; Qiu, May, & Nüchter, 2009).

In addition to these steps, it is recommended to detect and discard bad pairings: those where we can hardly consider a real correspondence. The simplest criterion is discarding those pairings whose Euclidean distance is larger than some heuristically determined threshold. Other alternatives have been proposed allowing larger distances to more distant points, which makes sense in we realize that any small uncertainty in the robot orientation (or heading) becomes a large error when projecting a sensed point from the sensor frame of reference into the local map (Minguez, Montesano, & Lamiraux, 2006).

Finally, the third stage consists in the evaluation of the likelihood given the final set of correspondences Ω. We can consider that the distance between each pair of corresponding points follows a Gaussian distribution with some heuristic variance σ^2, whose square root—the standard deviation—must reflect a corresponding distance which could be considered acceptable. Notice that this model parallels the *likelihood field* approach (Thrun, 2001) for grid maps mentioned in section 2. Then, assuming statistical independence between all the correspondence distances, the likelihood can be written down as:

$$f_{\mathbf{m}_{local,k}|\mathbf{x}_k,\mathbf{m}} \propto \prod_{(l_k,g_k)\in\Omega} \exp\left(-\frac{1}{2}\frac{\left(\mathbf{p}_{l_k}-\mathbf{q}_{g_k}\right)^2}{\sigma^2}\right) \quad (32)$$

However, in practice this number will be so small that we can find numerical problems when implementing this equation in a computer program (e.g. underflows). Thus, it becomes advisable here to take logarithms and evaluate the log-likelihood instead:

$$\ln\left(f_{\mathbf{m}_{local,k}|\mathbf{x}_k,\mathbf{m}}\right) = \underbrace{\eta}_{\substack{\text{Arbitrary}\\ \text{constant}\\ \text{(to ignore)}}} - \frac{1}{2\sigma^2}\sum_{(l_k,g_k)\in\Omega}\left(\mathbf{p}_{l_k}-\mathbf{q}_{g_k}\right)^2 \quad (33)$$

Note that different $\mathbf{x}_k$ transformations produce different values of the total distance (remember that $\mathbf{p}_i = \mathbf{x}_k \oplus \mathbf{r}_i$) and thus, different values of log-likelihood.

REFERENCES

Aho, A. V., Ullman, J. D., & Hopcroft, J. E. (1983). *Data structures and algorithms*. Reading, MA: Addison-Wesley.

Bar-Shalom, Y., & Fortmann, T. E. (1988). *Tracking and data association*. San Diego, CA: Academic Press Professional, Inc.

Baron, R. J. (1981). Mechanisms of human facial recognition. *International Journal of Man-Machine Studies*, *15*(2), 137–178. doi:10.1016/S0020-7373(81)80001-6

Besl, P. J., & McKay, N. D. (1992). A method for registration of 3D shapes. *IEEE Transactions on Pattern Analysis and Machine Intelligence*, *14*(2), 239–256. doi:10.1109/34.121791

Blanco, J. L., González, J., & Fernández-Madrigal, J. A. (2007). A consensus-based approach for estimating the observation likelihood of accurate range sensors. In *Proceedings of the IEEE International Conference on Robotics and Automation*. IEEE Press.

Blanco, J. L., González, J., & Fernández-Madrigal, J. A. (2012). An alternative to the Mahalanobis distance for determining optimal correspondences in data association. *IEEE Transactions on Robotics*, *28*(4), 980-986.

Censi, A. (2008). An ICP variant using a point-to-line metric. In *Proceedings of the IEEE International Conference on Robotics and Automation (ICRA)*, (pp. 19-25). IEEE Press.

Craig, J. J. (1989). *Introduction to robotics mechanics and control* (2nd ed.). Reading, MA: Addison-Wesley.

Dempster, A. P., Laird, N. M., & Rubin, D. B. (1977). Maximum likelihood from incomplete data via the EM algorithm. *Journal of the Royal Statistical Society. Series B. Methodological, 39*(1), 1–38.

Garey, M. R., & Johnson, D. S. (1979). *Computers and intractability: A guide to the theory of NP-completeness*. New York, NY: W. H. Freeman.

González-Jiménez, J., Blanco, J. L., Galindo, C., Ortiz-de-Galisteo, A., Fernández-Madrigal, J. A., Moreno, F., & Martínez, J. L. (2009). Mobile robot localization based on ultra-wide-band ranging: A particle filter approach. *Robotics and Autonomous Systems, 57*(5), 496–507. doi:10.1016/j.robot.2008.10.022

Grimson, W. E. L. (1991). *Object recognition by computer: The role of geometric constraints*. Cambridge, MA: The MIT Press.

Grimson, W. E. L., & Lozano-Pérez, T. (1987). Localizing overlapping parts by searching the interpretation tree. *IEEE Transactions on Pattern Analysis and Machine Intelligence, 9*(4), 469–482. doi:10.1109/TPAMI.1987.4767935

Konolige, K. (1999). Markov localization using correlation. In *Proceedings of the International Joint Conference on Artificial Intelligence*. IEEE.

Lu, F., & Milios, E. (1997). Robot pose estimation in unknown environments by matching 2d range scans. *Journal of Intelligent & Robotic Systems, 18*, 249–275. doi:10.1023/A:1007957421070

Martin, M. C., & Moravec, H. P. (1996). *Robot evidence grids*. CMU Technical Report CMU-RI-TR-96-06. Pittsburgh, PA: Carnegie Mellon University.

Minguez, J., Montesano, L., & Lamiraux, F. (2006). Metric-based iterative closest point scan matching for sensor displacement estimation. *IEEE Transactions on Robotics, 22*(5), 1047–1054. doi:10.1109/TRO.2006.878961

Neira, J., & Tardós, J. D. (2001). Data association in stochastic mapping using the joint compatibility test. *IEEE Transactions on Robotics and Automation, 17*(6), 890–897. doi:10.1109/70.976019

Plagemann, C., Kersting, K., Pfaff, P., & Burgard, W. (2007). Gaussian beam processes: A nonparametric Bayesian measurement model for range finders. In *Proceedings of Robotics: Science and Systems Conference*. IEEE. Press, W. H., Teukolsky, S. A., Vetterling, W. T., & Flannery, B. P. (1992). *Numerical recipes in C: The art of scientific programming* (2nd ed). Cambridge, UK: Cambridge University Press.

Qiu, D., May, S., & Nüchter, A. (2009). GPU-accelerated nearest neighbor search for 3D registration. In *Proceedings of the 7th International Conference on Computer Vision Systems: Computer Vision Systems*, (pp. 194-203). IEEE.

Roth, S. D. (1982). Ray casting for modeling solids. *Computer Graphics and Image Processing, 18*, 109–144. doi:10.1016/0146-664X(82)90169-1

Snyman, J. A. (2005). *Practical mathematical optimization: An introduction to basic optimization theory and classical and new gradient-based algorithms*. Berlin, Germany: Springer Publishing.

Thrun, S. (2001). A probabilistic on-line mapping algorithm for teams of mobile robots. *The International Journal of Robotics Research, 20*(5), 335. doi:10.1177/02783640122067435

Thrun, S., Burgard, W., & Fox, D. (2005). *Probabilistic robotics*. Cambridge, MA: The MIT Press.

ENDNOTES

1 If the robot is at a hypothetical pose $\mathbf{x}_k$ with respect to a global reference frame and the sensor is at a pose $\tilde{\mathbf{y}}_k$ with respect to an egocentric reference frame attached to the robot, the pose of the sensor in the global reference frame is then given by $\mathbf{y}_k = \mathbf{x}_k \oplus \tilde{\mathbf{y}}_k$, where the expressions for the pose composition operator $\oplus$ can be found in Appendix A.

2 The covariance matrix of the observation prediction will also contains the uncertainty of the robot pose when it is evaluated in the context of simultaneous localization and mapping, as we discuss in chapter 9.

3 Even for semidefinite-positive covariance matrices (see Appendix E for the definition of this concept) we could find such a diagonalized covariance, but in that case the Mahalanobis distance is not well defined, since the matrix cannot be inverted.

4 A mathematical tree is a very common data structure in computer science (Aho, Ullman, & Hopcroft, 1983); mathematically it is a type of connected, acyclic directed graph, that is, a set of nodes connected by directed arcs. Arcs go from "parent" to "child" nodes. Each node has only one parent (except the "root" node, which has no parent) and any number of children (those nodes without children are called "leafs"). The children of a node cannot contain any ancestor of that node or the node itself (for avoiding cycles and loops), and have an order.

5 $\Gamma(z)$ is the gamma function, a generalization of the factorial $z!$ They are equivalent if z is an integer; more generally, $\Gamma(z) = \int_0^\infty u^{z-1} e^{-u} du$ for any complex number z. The gamma function has no closed solution in the general case.

6 Since $\mathbf{q} - \boldsymbol{\mu}$ has zero mean and, when divided by the variance of the Gaussian, unit variance, the resulting distribution is the product of squared zero mean, unit variance Gaussians.

7 That Gaussian should include both the uncertainty in the robot pose and in the observation. We will see in further chapters how to estimate those uncertainties in the case of having Gaussian form.

8 The computational complexity of an algorithm indicates its cost in time or in consumed memory (Garey & Johnson, 1979). The most common choice is the cost in time, and the notation for complexity is the "big-O" notation: an algorithm with an input of length n is said to has time complexity $O(g(n))$ if, when $n \to \infty$, it runs for less than or exactly $\alpha g(n)$ units of time, being $\alpha > 0$ a real constant. Notice that we can choose whatever unit of time we want (more precisely, unit of computational time), since that will only influence the value of the constant, which does not appear in the big-O expression. In other words: the power of the particular computer we use is not relevant as long as it is equivalent to a deterministic Turing machine. Depending on the time-complexity, the most important classes of algorithms, from the most tractable to the least, are *constant* ($O(1)$), *linear* ($O(n)$), *log-linear* or *logarithmic* ($O(n \log n)$), *polynomial* ($O(n^s)$, with a constant $s > 1$) and *exponential* ($O(s^n)$). Algorithms of exponential complexity or worse are considered intractable, since even for inputs of small size they spend impractical amounts

of time in execution. For example, the exhaustive search in an interpretation tree for the DA problem is intractable even considering a map of constant size (number of features).

9 Another way of deducing that is the following. If both grid maps have a perfect match, we can write $f_{m,i,j} = f_{m_{local},i,j} = c_{i,j}$, $E[f_m] = E[f_{m_{local}}] = \mu$ and $V[f_m] = V[f_{m_{local}}] = \sigma^2$ Substituting these in Equation 15 and developing the sum it can be easily seen that the Baron's cross-correlation coefficient is 1.

Chapter 7
Mobile Robot Localization with Recursive Bayesian Filters

ABSTRACT

In this last chapter of the second section, the authors present probabilistic solutions to mobile robot localization that bring together the recursive filters introduced in chapter 4 and all the components and models already discussed in the preceding chapters. It presents the general, Bayesian framework for a probabilistic solution to localization and mapping. The problem is formally described as a graphical model (in particular a dynamic Bayesian network), and the characteristics that can be exploited to approach it efficiently are elaborated. Among parametric Bayesian estimators, the family of the Kalman filters is introduced with examples and practical applications. Then, the more modern non-parametric filters, mainly particle filters, are explained. Due to the diversity of filters available for localization, comparative tables are included.

CHAPTER GUIDELINE

- You will learn:
  - The general structure of the recursive estimation problem underlying mobile robot localization.
  - The different approaches to instantiate that structure: parametric and non-parametric filters.
  - The rationale behind the different factors and parameters of a filter for localization.
  - How to cope with non-linear systems and non-Gaussian random variables.
  - The pros and cons of each localization filter: how to choose the fittest for your application.
- Provided tools:
  - Complete formulation and justification of each localization filter.
  - Tables of computational costs for the presented filtering algorithms.
- Relation to other chapters:
  - This chapter closes the first part of the book (localization) and sets in motion the basic tools needed for solving the mapping problem.
  - Modified versions of the filters introduced in this chapter are used for constructing maps in chapter 8 and for SLAM in chapter 9.

DOI: 10.4018/978-1-4666-2104-6.ch007

1. INTRODUCTION

Previous chapters have already introduced all the elements required to build a general probabilistic framework for mobile robot localization. Now it is time to glue all the components together to arrive at complete and practical solutions.

Firstly, a word is in order about notation. Up to this point, we have used the $f_{\mathbf{x}}$ notation in order to highlight the similarities between discrete and continuous distributions (pmfs and pdfs). In contrast, from this chapter on we will switch to the simpler notation $p(\mathbf{x})$ (a continuous $f_{\mathbf{x}}$) and $P(\mathbf{x})$ (a discrete $f_{\mathbf{x}}$), since otherwise the most interesting expressions would always appear as subindices.

The starting point for deriving solutions to localization is the Bayesian inference framework we introduced in chapter 4 for estimating the state of a dynamic system. For convenience, we repeat it here using the $p()$ notation:

$$\overbrace{bel(\mathbf{x}_k | \mathbf{z}_{1:k})}^{posterior} = \frac{\overbrace{p(\mathbf{z}_{1:k} | \mathbf{x}_k)}^{likelihood} \overbrace{bel(\mathbf{x}_k)}^{prior}}{\underbrace{p(\mathbf{z}_{1:k})}_{scale\ factor}} \tag{1}$$

As demonstrated in chapter 4, when the system is Markovian and we know the transition function for the system state, this expression can be turned into an incremental form, becoming the base of a sequential or recursive Bayesian filter (MMSE, MED, MAP, etc.):

$$\overbrace{bel(\mathbf{x}_k | \mathbf{z}_{1:k})}^{posterior\ at\ step\ k} = \frac{\overbrace{p(\mathbf{z}_k | \mathbf{x}_k)}^{likelihood} \overbrace{\int_{\mathbf{x}_{k-1}} \overbrace{p(\mathbf{x}_k | \mathbf{x}_{k-1})}^{transition} \overbrace{bel(\mathbf{x}_{k-1} | \mathbf{z}_{1:k-1})}^{posterior\ at\ step\ k-1} d\mathbf{x}_{k-1}}^{prior\ at\ step\ k}}{\underbrace{p(\mathbf{z}_k | \mathbf{z}_{1:k-1})}_{scaling}} \tag{2}$$

Here we can see the role played by the motion (transition) and sensor (likelihood) models discussed in chapters 5 and 6, respectively. Equation 2 is general enough to be applicable to any dynamic Markovian estimation problem, and thus it needs to be instantiated for the particular problems we encounter in mobile robotics. This is the goal of this and the following chapters.

Figure 1 shows the Dynamic Bayesian Network (DBN) corresponding to the first of this problems that we will address: mobile robot localization. It clearly reveals the Markovian assumption:

Figure 1. The DBN that represents the mobile robot localization problem. The figure illustrates the different random variables involved (poses, actions, observations, and the map) and their dependences. As usual, we represent hidden and observed variables as shaded and unshaded nodes, respectively.

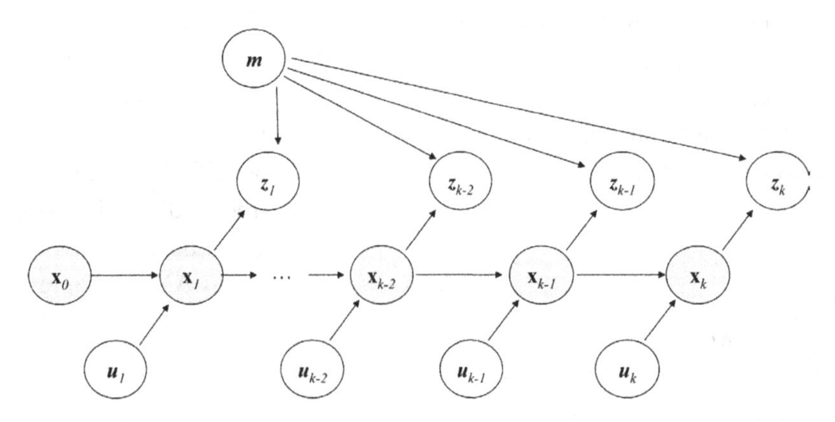

given a previous state (pose) of the robot, say $\mathbf{x}_{k-1}$, and the action executed to go from $\mathbf{x}_{k-1}$ to $\mathbf{x}_k$, namely $\mathbf{u}_k$, the current pose $\mathbf{x}_k$ does not depend on the past history of poses. On the other hand, given the current state $\mathbf{x}_k$ and the known map $\mathbf{m}$ (which appears without any time step index since it is considered static), the current observation $\mathbf{z}_k$ does not depend on any other variable.

This DBN exhibits one difference with respect to the general formulation of Equation 2, and that is the explicit consideration of actions. In mobile robots this is mandatory, since the transition between consecutive poses is clearly conditioned on the *actions* that the robot performs; if we ignored the action, the transition model would be so general that it would render almost useless. Recall from chapter 5 that actions (the $\mathbf{u}_k$) where introduced while studying transition models that represent different robot motion models. Another particularity of the DBN is the inclusion of a "prior" pose of the robot, $\mathbf{x}_0$, which represents the initial location of the robot before start moving and gathering observations. We must always provide the pdf of this initial pose, which does not imply that we must know the exact initial pose of the robot: that distribution might be a Gaussian centered around an approximate starting location or a uniform distribution that spans over the entire environment if we have no information at all (i.e., the robot "global localization" problem).

We can particularize the Bayesian framework for robot localization by conditioning all parts of Equation 2 on the history of actions of the robot (remember that conditioning all the parts does not alter the validity of the expression) and then making use of the Markovian assumptions and the conditional independence structure of the DBN of Figure 1:

$$bel(\mathbf{x}_k | \mathbf{z}_{1:k}, \mathbf{u}_{1:k}) = \\ = \frac{p(\mathbf{z}_k | \mathbf{x}_k, \mathbf{u}_{1:k}) \int_{\mathbf{x}_{k-1}} p(\mathbf{x}_k | \mathbf{x}_{k-1}, \mathbf{u}_{1:k}) bel(\mathbf{x}_{k-1} | \mathbf{z}_{1:k-1}, \mathbf{u}_{1:k}) d\mathbf{x}_{k-1}}{p(\mathbf{z}_k | \mathbf{z}_{1:k-1}, \mathbf{u}_{1:k})} =$$

(by the conditional independences $\mathbf{z}_k \perp \mathbf{u}_{1:k} | \mathbf{x}_k$, $\mathbf{x}_k \perp \mathbf{u}_{1:k-1} | \mathbf{x}_{k-1}$ and $\mathbf{x}_{k-1} \perp \mathbf{u}_k$)

$$= \frac{p(\mathbf{z}_k | \mathbf{x}_k) \int_{\mathbf{x}_{k-1}} p(\mathbf{x}_k | \mathbf{x}_{k-1}, \mathbf{u}_k) bel(\mathbf{x}_{k-1} | \mathbf{z}_{1:k-1}, \mathbf{u}_{1:k-1}) d\mathbf{x}_{k-1}}{p(\mathbf{z}_k | \mathbf{z}_{1:k-1}, \mathbf{u}_{1:k})} \tag{3}$$

We should also include the map, in the same way, since in the localization problem it is known and there are variables that depend on it:

$$bel(\mathbf{x}_k | \mathbf{z}_{1:k}, \mathbf{u}_{1:k}, \mathbf{m}) = \\ = \frac{p(\mathbf{z}_k | \mathbf{x}_k, \mathbf{m}) \int_{\mathbf{x}_{k-1}} p(\mathbf{x}_k | \mathbf{x}_{k-1}, \mathbf{u}_k, \mathbf{m}) bel(\mathbf{x}_{k-1} | \mathbf{z}_{1:k-1}, \mathbf{u}_{1:k-1}, \mathbf{m}) d\mathbf{x}_{k-1}}{p(\mathbf{z}_k | \mathbf{z}_{1:k-1}, \mathbf{u}_{1:k}, \mathbf{m})} =$$

(by the independence $\mathbf{x}_k \perp \mathbf{m}$)

$$= \frac{p(\mathbf{z}_k | \mathbf{x}_k, \mathbf{m}) \int_{\mathbf{x}_{k-1}} p(\mathbf{x}_k | \mathbf{x}_{k-1}, \mathbf{u}_k) bel(\mathbf{x}_{k-1} | \mathbf{z}_{1:k-1}, \mathbf{u}_{1:k-1}, \mathbf{m}) d\mathbf{x}_{k-1}}{p(\mathbf{z}_k | \mathbf{z}_{1:k-1}, \mathbf{u}_{1:k}, \mathbf{m})}$$

(Since the denominator $p(\mathbf{z}_k | \mathbf{z}_{1:k-1}, \mathbf{u}_{1:k}, \mathbf{m})$ is not a function of the independent variable of the posterior, $\mathbf{x}_k$, that is, it is constant for any value that $\mathbf{x}_k$ can take when a given set of observations have been gathered)

$$bel(\mathbf{x}_k | \mathbf{z}_{1:k}, \mathbf{u}_{1:k}, \mathbf{m}) \propto \\ \propto \underbrace{p(\mathbf{z}_k | \mathbf{x}_k, \mathbf{m}) \underbrace{\int_{\mathbf{x}_{k-1}} p(\mathbf{x}_k | \mathbf{x}_{k-1}, \mathbf{u}_k) bel(\mathbf{x}_{k-1} | \mathbf{z}_{1:k-1}, \mathbf{u}_{1:k-1}, \mathbf{m}) d\mathbf{x}_{k-1}}_{\text{prediction stage}}}_{\text{update stage}} \tag{4}$$

This is the general structure for the recursive Bayesian filtering in mobile robot localization.

We have also highlighted the two main stages involved in the calculation of the posterior:

1. The integral, which permits us to propagate our prior belief (the previous posterior) to a new pose according to the measured action (transition), that is, to *predict* our new pose from our motion using any of the models discussed in chapter 5.
2. The observation is taken into account for *updating* that prediction. Among other possibilities, here one can apply the sensor models introduced in chapter 6.

In Figure 2 you can see an informal illustration of this general procedure, which applies to all the filters studied in this chapter.

Unfortunately, the integral above has no closed-form solution in the general case, which is the main reason why there exist diverse filters for solving the same problem. The particularizations of Equation 4can be categorized at a first glance into two broad classes:

- **Parametric filters.** They are based on the assumption of a known distribution shape, in the form of a mathematical function with some arguments, for the probabilistic components of the formulation. Formally, a pdf $p(\mathbf{x}|\theta)$ that is defined on a set of parameters $\Theta \subseteq \mathbb{R}^p$, being p a finite natural value, it is said to be a parametric distribution (Kassam, 1988). Filters based on the election of particular parametric distributions do not directly estimate the posterior distribution itself, but this finite number of parameters. These filters provide optimal or near-optimal results if their assumptions are correct, although in practice they suffer inaccuracies due to deviations from these idealized assumptions. Furthermore, few distribution shapes (pre-eminently, Gaussians) lead to filters in closed form. Multimodal problems (i.e., multiple hypotheses for the robot location) are also difficult to address and require special treatment.

Figure 2. An informal illustration of the two main stages in the estimation of the posterior pose for the mobile robot localization problem. (1) The robot moves. (2) – (4) Prediction stage. (5) – (6) Update stage.

1) **Prior** to any movement toward that door, I **believe** that I'm at *a*, but I'm not certain: better to believe that I'm **likely to be around** *a*...

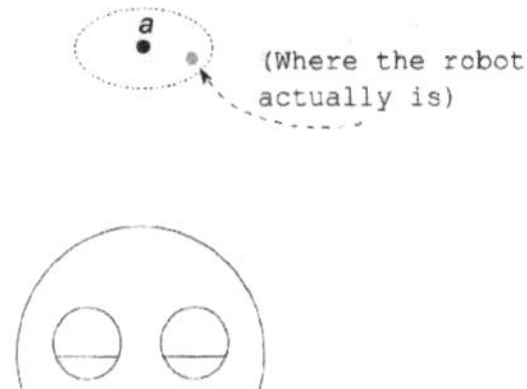

2) Now I move toward the door, paying attention only to my **odometry**. After moving, and taking into account my **kinematics**, I am expected to be at *b*...

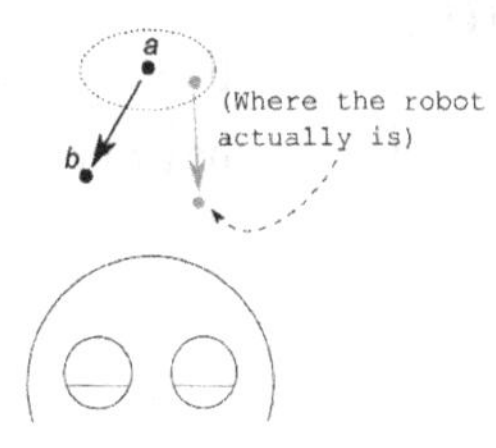

3) However, since I have **my eyes closed**, my mechanics **is not deterministic**, and I had **uncertainty** about my prior pose, I can only **predict** that I am **around** *b*.

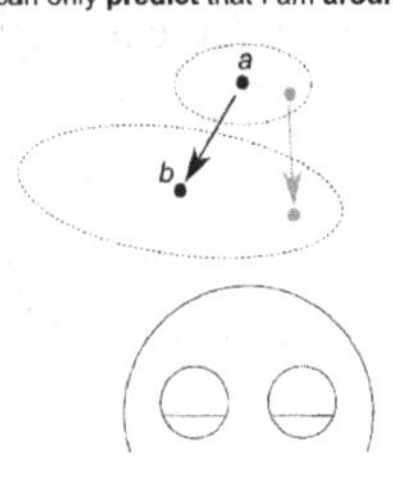

4) I will **open my eyes** to see where I am with more certainty...

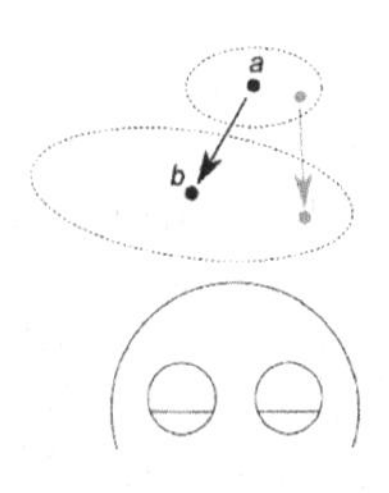

5) Oops! I do not **see** the door in front of me, but to the right... I must have drifted to the left!

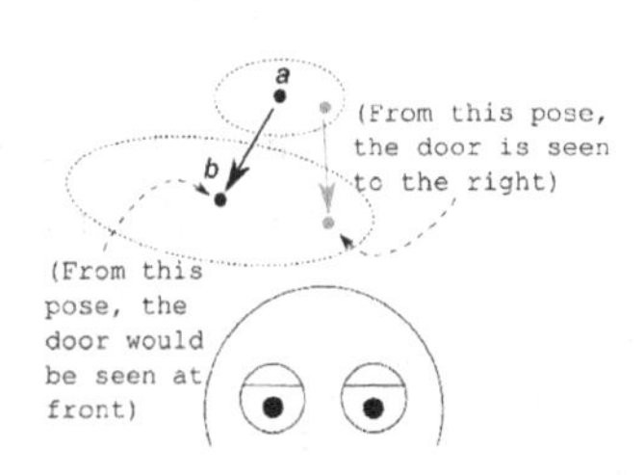

6) I must **update** my prediction with that observation, but I also must take into account the **uncertainty** in the observation itself, since my sensors are not perfect.

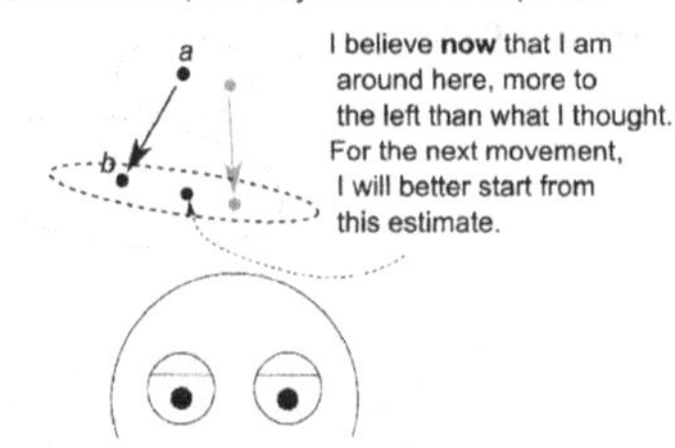

- **Non-parametric filters.** They do not assume any particular distribution shape: their results are approximations of the actual shapes of the involved distributions. Their estimations are often sub-optimal, but they can better handle multimodal and non-Gaussian uncertainties. Normally, these methods present a tradeoff between computational cost and accuracy in the results, thus we can adjust them to the computational power available on a mobile robot.

In summary: parametric filters are more exact and efficient due to their nice mathematical properties, but at the cost of sometimes making too idealized assumptions. In turn, non-parametric filters are more flexible and thus much better suited to a complex reality, which must be paid with a higher computational cost and being only asymptotically exact.

It is noteworthy to mention that probabilistic localization filters are not limited to *metrical* localization: they have been also reported in works on topological and hybrid metric-topological localization and mapping in the scientific literature. However, these works are still scarce, since they present a clear advantage only in special situations, such as in very large-scale scenarios.

In the following, we focus on the best known and most common filters for mobile robot localization. Sections 2 and 3 deal with parametric and non-parametric filters, respectively. For a deeper study of filters for the estimation of dynamic systems the reader can consult (Gelb, 1974; Jazwinsky, 1970), among many other texts on this topic.

2. PARAMETRIC FILTERS FOR LOCALIZATION

The fundamental formulation of Equation 4 for recursive Bayesian estimation has to be instantiated with concrete probability distributions in order to obtain a parametric filter. If we wish to do that, we must choose the same mathematical form for the prior and posterior distribution, since obviously only the parameters of the distribution will change with each filter iteration. We have already seen in chapter 4 that priors and posteriors that have such a property in a Bayesian filter are called *conjugate*. The prior is then called the *conjugate prior* of the likelihood.

There are a number of conjugate distributions both in the discrete and continuous cases (Raiffa & Schlaifer, 2000; Whetherill, 1961), but few of them are of real utility for modeling the kind of uncertainty found in mobile robotics. The one with the most applications, due to the central limit theorem, is the Gaussian, which is conjugate prior with itself and allows us to represent uncertainties coming from complex stochastic sources as long as they are symmetric. Few other distributions remain practical; the gamma distribution, for example, would allow for asymmetries, but is much less tractable mathematically due to the need to evaluate the gamma function that it includes.

Therefore it is no wonder that most parametric filters rely on an assumption of Gaussianity for the uncertainty in the state of the system (prior and posterior), for the system transition model[1] and for the observation likelihood—recall how Gaussian uncertainties were assumed in most motion models of chapter 5 and observation models of chapter 6. In this section we firstly study the most classical of those filters: the original Kalman Filter (KF); it is not directly usable in practical robotics due to its strong assumptions, but with a small modification leads to the widely used Extended Kalman Filter (EKF) and others, which will be described next. The reader can find diverse tutorials on these filters elsewhere, such as Welch and Bishop (2001).

The Kalman Filter (KF)

The Kalman filter (Kalman, 1960) was developed in the early 1960s for estimating trajectories in the Apollo space program. It was designed with robustness and efficiency in mind, since it was to be run on the navigation computer of lunar modules. The KF assumes the following three system characteristics, called a State Space Model or SSM of a system (Li, 2003):

- **Normality of Uncertainties:** All the involved uncertainties are Gaussian, typically multivariate. These include the uncertainties in the motion and observations models (which are assumed to be independent), and also in the prior (including the first prior, $bel(\mathbf{x}_0)$) and posterior states of the system, that is:

$$bel\left(\mathbf{x}_k \middle| \mathbf{z}_{1:k}, \mathbf{u}_{1:k}, \mathbf{m}\right) = N\left(\mathbf{x}_k; \mu_k, \Sigma_k\right) \propto \exp\left(-\frac{1}{2}\left(\mathbf{x}_k - \mu_k\right)^T \Sigma_k^{-1} \left(\mathbf{x}_k - \mu_k\right)\right) \quad (5)$$

- **Linearity of the Transition:** The transition model is assumed to be perfectly linear, that is:

$$\mathbf{x}_k = \mathbf{A}_k \mathbf{x}_{k-1} + \mathbf{B}_k \mathbf{u}_k + \mu_k \quad (6)$$

where $\mathbf{A}_k$ and $\mathbf{B}_k$ are matrices not depending on the state or actions of the system and μ_k is an additive Gaussian noise modeling the uncertainty in the transition, i.e., $\mu_k \sim \mathrm{N}(\mathbf{x}; 0, \mathbf{R}_k)$ where $\mathbf{R}_k$ its covariance matrix. Notice that all the matrices of this linear model can be time-variant, that is, their content can vary over different time steps. Assuming a perfectly known robot pose for a previous time step, $\mathbf{x}_{k-1}$, we can formulate the pdf of the robot motion model as the Gaussian:

$$p(\mathbf{x}_k \middle| \mathbf{x}_{k-1}, \mathbf{u}_k) = N(\mathbf{x}_k; \mathbf{A}_k \mathbf{x}_{k-1} + \mathbf{B}_k \mathbf{u}_k, \mathbf{R}_k) \quad (7)$$

- **Linearity of the Observation:** The observation model is also assumed to be perfectly linear on the state of the system (pose of the robot): $\mathbf{z}_k = \mathbf{C}_k \mathbf{x}_k + \delta_k$, where $\mathbf{C}_k$ is a matrix depending on the map and $\delta_k \sim \mathrm{N}(\mathbf{x}; 0, \mathbf{Q}_k)$ is additive Gaussian noise in the observation model, which is (plausibly) assumed to be independent of the motion noise. Therefore, all this can be written down as $p(\mathbf{z}_k \middle| \mathbf{x}_k, \mathbf{m}) = N(\mathbf{z}_k; \mathbf{C}_k \mathbf{x}_k, \mathbf{Q}_k)$.

Other sequential filters can be devised that are not restricted by the SSM assumption, such as the *Exponentially Weighted Recursive Least Squares* (EW-RLS), a particular form of *Infinite Impulse Response filter* (IIR), that are very useful, for instance, in signal processing and econometrics, but they are suboptimal due to their relaxed assumptions (Li, 2003), thus we will focus here on filters that assume the SSM in the parametric case. Taking the SSM assumptions into the recursive Bayesian filter of Equation 4 we arrive at:

$$bel(\mathbf{x}_k \middle| \mathbf{z}_{1:k}, \mathbf{u}_{1:k}, \mathbf{m}) =$$

$$= N(\mathbf{x}_k; \mu_k, \Sigma_k) \propto p(\mathbf{z}_k \middle| \mathbf{x}_k, \mathbf{m}) \int_{\mathbf{x}_{k-1}} p(\mathbf{x}_k \middle| \mathbf{x}_{k-1}, \mathbf{u}_k) bel(\mathbf{x}_{k-1} \middle| \mathbf{z}_{1:k-1}, \mathbf{u}_{1:k-1}, \mathbf{m}) d\mathbf{x}_{k-1} =$$
$$= N(\mathbf{z}_k; \mathbf{C}_k \mathbf{x}_k, \mathbf{Q}_k) \int_{\mathbf{x}_{k-1}} N(\mathbf{x}_k; \mathbf{A}_k \mathbf{x}_{k-1} + \mathbf{B}_k \mathbf{u}_k, \mathbf{R}_k) N(\mathbf{x}_{k-1}; \mu_{k-1}, \Sigma_{k-1}) d\mathbf{x}_{k-1} =$$

(sub-indexing the exponential terms for an easier identification in the following)

$$= \psi_z \exp_z(\mathbf{z}_k, \mathbf{x}_k) \int_{\mathbf{x}_{k-1}} \psi_t \exp_t(\mathbf{x}_k, \mathbf{x}_{k-1}) \psi_i \exp_i(\mathbf{x}_{k-1}) d\mathbf{x}_{k-1}$$
$$\propto \exp_z(\mathbf{z}_k, \mathbf{x}_k) \int_{\mathbf{x}_{k-1}} \exp_t(\mathbf{x}_k, \mathbf{x}_{k-1}) \exp_i(\mathbf{x}_{k-1}) d\mathbf{x}_{k-1} \quad (8)$$

which will serve us as a roadmap for the derivation of the Kalman filter. Notice that the terms inside this last integral must be rearranged if we want to solve it, since both depends on the variable of the integral and it would be more convenient to have only one exponent as the integrand.

The forthcoming derivation of the Kalman filter is not the shortest one, but uses only basic algebra (refer to Appendix E if needed) and, in that sense, is minimalistic. We will use the canonical form extensively in the deduction, in particular as a method for simplifying certain operations (mainly sums) within the Gaussian exponents. However, if the Gaussian uncertainties involved in the filter are in canonical form from the beginning, the entire filter could be deduced in that form, only needing to pass to standard form exponents in one of the steps. That kind of filter is called the *Information Filter* (IF) and also suffers from the same limitations than the KF (mainly, its linearity assumption), which however can be addressed in similar ways (see the Extended Kalman filter further on; accordingly, there also exists the Extended Information filter). Both Kalman and Information filters are considered dual to each other, although some versions of the latter (the Square Root Information Filter, or SRIF, and the Square Root Information Smoother, or SRIS) have been shown to be more stable numerically than their covariance counterparts (Bierman, 1977; Dellaert & Kaess, 2006). However, due to the most intuitive meaning of covariance matrices and mean vectors, the KF and derivatives are still widely in use, hence that we devote this chapter to them. We will not mention information filters again until chapter 10, in relation with the latest approaches to SLAM.

Our derivation of the KF consists of two main stages: one concerns the integral of Equation 4, and is called the *prediction* stage of the KF, while the other solves the rest and is called the *update* stage of the KF. The details of both of them are illustrated in Figures 3 and 4, which complement the next derivations. The first stage focuses only on the prior of the filter at step k:

$$\overbrace{\int_{\mathbf{x}_{k-1}} \exp_t(\mathbf{x}_k, \mathbf{x}_{k-1}) \exp_i(\mathbf{x}_{k-1}) d\mathbf{x}_{k-1}}^{\textit{proportional to the prior of step k}}$$

where, using the Kalman assumptions for the transition and previous posterior,

$$\begin{aligned}
&\exp_t(\mathbf{x}_k, \mathbf{x}_{k-1}) = \\
&= \exp\begin{pmatrix} -\frac{1}{2}\left(\mathbf{x}_k - \mathbf{A}_k\mathbf{x}_{k-1} - \mathbf{B}_k\mathbf{u}_k\right)^T \\ \mathbf{R}_k^{-1}\left(\mathbf{x}_k - \mathbf{A}_k\mathbf{x}_{k-1} - \mathbf{B}_k\mathbf{u}_k\right) \end{pmatrix} \\
&\exp_i(\mathbf{x}_{k-1}) = \\
&= \exp\left(-\frac{1}{2}\left(\mathbf{x}_{k-1} - \mu_{k-1}\right)^T \Sigma_{k-1}^{-1}\left(\mathbf{x}_{k-1} - \mu_{k-1}\right)\right)
\end{aligned} \tag{9}$$

A first goal is to change the form of the integrand $\exp_t(\mathbf{x}_k, \mathbf{x}_{k-1}) \exp_i(\mathbf{x}_{k-1})$ to the product of two different exponentials: one corresponding to a Gaussian pdf in the variable $\mathbf{x}_{k-1}$, which will be integrated and therefore converted into a constant and discarded from the whole derivation, and another one that does not depend on $\mathbf{x}_{k-1}$, which will be taken out of the integral and used in further steps. That is, we seek for the following changes in the roadmap for the prediction stage of the derivation (notice the different subindices in the exponentials):

$$\begin{aligned}
&\int_{\mathbf{x}_{k-1}} \exp_t(\mathbf{x}_k, \mathbf{x}_{k-1}) \exp_i(\mathbf{x}_{k-1}) d\mathbf{x}_{k-1} \propto \\
&\propto \int_{\mathbf{x}_{k-1}} \exp_g(\mathbf{x}_{k-1}) \exp_c(\mathbf{x}_k) d\mathbf{x}_{k-1} = \\
&= \exp_c(\mathbf{x}_k) \int_{\mathbf{x}_{k-1}} \exp_g(\mathbf{x}_{k-1}) d\mathbf{x}_{k-1} \propto \\
&\propto \exp_c(\mathbf{x}_k)
\end{aligned} \tag{10}$$

Figure 3. Illustration of the roadmap we follow for the derivation of the prediction stage of the Kalman filter (see the text)

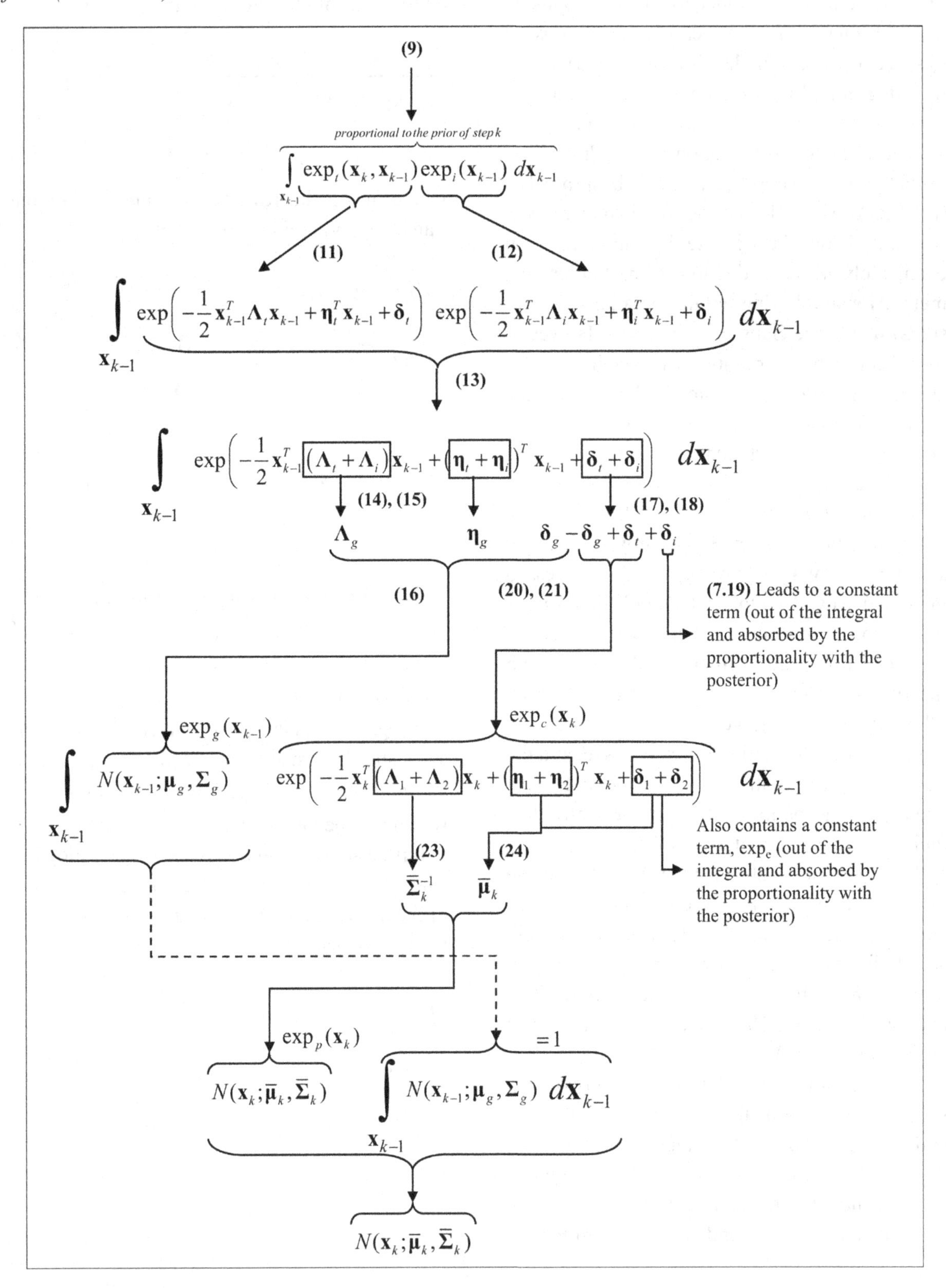

Figure 4. Illustration of our derivation of the update stage of the Kalman filter (see the text)

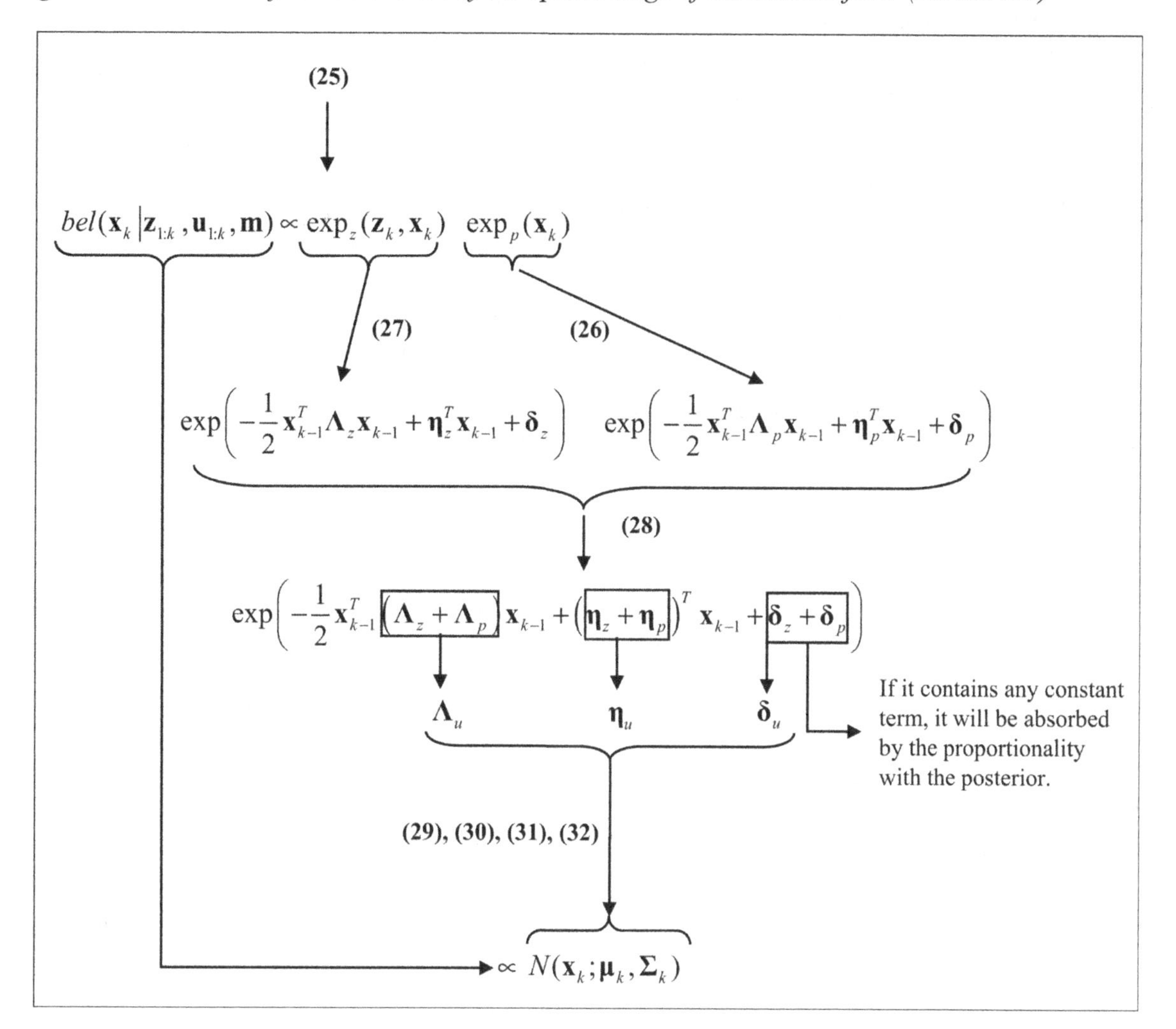

For achieving that, we will employ the canonical forms of both $\exp_t(\mathbf{x}_k, \mathbf{x}_{k-1})$ and $\exp_i(\mathbf{x}_{k-1})$ and will consider that the former is centered (has its mean) at the variable $\mathbf{x}_{k-1}$, that is, we rearrange the exponent of $\exp_t(\mathbf{x}_k, \mathbf{x}_{k-1})$ in this way:

$$\exp_t(\mathbf{x}_k, \mathbf{x}_{k-1}) = = -\frac{1}{2}\left(\mathbf{x}_k - \mathbf{A}_k\mathbf{x}_{k-1} - \mathbf{B}_k\mathbf{u}_k\right)^T \mathbf{R}_k^{-1}\left(\mathbf{x}_k - \mathbf{A}_k\mathbf{x}_{k-1} - \mathbf{B}_k\mathbf{u}_k\right) = = -\frac{1}{2}\left(\underbrace{-\mathbf{A}_k}_{\mathbf{E}_t}\mathbf{x}_{k-1} - \underbrace{\left(\mathbf{B}_k\mathbf{u}_k - \mathbf{x}_k\right)}_{\mu_t}\right)^T \underbrace{\mathbf{R}_k^{-1}}_{\Sigma_t^{-1}}\left(-\mathbf{A}_k\mathbf{x}_{k-1} - \left(\mathbf{B}_k\mathbf{u}_k - \mathbf{x}_k\right)\right) =$$

(changing to the canonical form—refer to Appendix E)

$$= -\frac{1}{2}\mathbf{x}_{k-1}^T \Lambda_t \mathbf{x}_{k-1} + \eta_t^T \mathbf{x}_{k-1} + \delta_t$$

where

$$\begin{aligned} \Lambda_t &= \mathbf{A}_k \mathbf{R}_k^{-1} \mathbf{A}_k^T \\ \eta_t &= \mathbf{A}_k^T \mathbf{R}_k^{-1} \left(\mathbf{x}_k - \mathbf{B}_k \mathbf{u}_k\right) \\ \delta_t &= -\frac{1}{2}\left(\mathbf{B}_k \mathbf{u}_k - \mathbf{x}_k\right)^T \mathbf{R}_k^{-1} \left(\mathbf{B}_k \mathbf{u}_k - \mathbf{x}_k\right) \end{aligned} \tag{11}$$

Likewise for the exponent of the prior exponential function:

$$\begin{aligned} \exp_i(\mathbf{x}_{k-1}) &= -\frac{1}{2}\left(\mathbf{x}_{k-1} - \mu_{k-1}\right)^T \Sigma_{k-1}^{-1} \left(\mathbf{x}_{k-1} - \mu_{k-1}\right) \\ &= -\frac{1}{2}\left(\underbrace{\mathbf{I}}_{\mathbf{E}_i} \mathbf{x}_{k-1} - \underbrace{\mu_{k-1}}_{\mu_i}\right)^T \underbrace{\Sigma_{k-1}^{-1}}_{\Sigma_i^{-1}} \left(\mathbf{I}\mathbf{x}_{k-1} - \mu_{k-1}\right) \\ &= -\frac{1}{2}\mathbf{x}_{k-1}^T \Lambda_i \mathbf{x}_{k-1} + \eta_i^T \mathbf{x}_{k-1} + \delta_i \end{aligned}$$

where

$$\begin{aligned} \Lambda_i &= \Sigma_{k-1}^{-1} \\ \eta_i &= \Sigma_{k-1}^{-1} \mu_{k-1} \\ \delta_i &= -\frac{1}{2}\mu_{k-1}^T \Sigma_{k-1}^{-1} \mu_{k-1} \end{aligned} \tag{12}$$

Now the integrand of the prior of the filter has been transformed into the product of two exponentials in canonical form centered at the same variable, $\mathbf{x}_{k-1}$. The original product $\exp_t(\mathbf{x}_k, \mathbf{x}_{k-1}) \exp_i(\mathbf{x}_{k-1})$ becomes then a single exponential with the sum of the two exponents developed in Equations 11 and 12. Since both are in canonical form, the resulting exponent is straightforward to compute:

$$\begin{aligned} &\exp_t(\mathbf{x}_k, \mathbf{x}_{k-1}) \exp_i(\mathbf{x}_{k-1}) = \\ &= -\frac{1}{2}\mathbf{x}_{k-1}^T \Lambda_t \mathbf{x}_{k-1} + \eta_t^T \mathbf{x}_{k-1} \\ &+\delta_t - \frac{1}{2}\mathbf{x}_{k-1}^T \Lambda_i \mathbf{x}_{k-1} + \eta_i^T \mathbf{x}_{k-1} + \delta_i = \\ &= -\frac{1}{2}\mathbf{x}_{k-1}^T \left(\Lambda_t + \Lambda_i\right) \mathbf{x}_{k-1} \\ &+ \left(\eta_t^T + \eta_i^T\right) \mathbf{x}_{k-1} + \delta_t + \delta_i = \\ &= -\frac{1}{2}\mathbf{x}_{k-1}^T \left(\Lambda_t + \Lambda_i\right) \mathbf{x}_{k-1} \\ &+ \left(\eta_t + \eta_i\right)^T \mathbf{x}_{k-1} + \delta_t + \delta_i \end{aligned}$$

where

$$\begin{aligned} \Lambda_t + \Lambda_i &= \mathbf{A}_k \mathbf{R}_k^{-1} \mathbf{A}_k^T + \Sigma_{k-1}^{-1} \\ \eta_t + \eta_i &= \mathbf{A}_k^T \mathbf{R}_k^{-1} \left(\mathbf{x}_k - \mathbf{B}_k \mathbf{u}_k\right) + \Sigma_{k-1}^{-1} \mu_{k-1} \\ \delta_t + \delta_i &= -\frac{1}{2}\Big(\left(\mathbf{B}_k \mathbf{u}_k - \mathbf{x}_k\right)^T \mathbf{R}_k^{-1} \left(\mathbf{B}_k \mathbf{u}_k - \mathbf{x}_k\right) + \\ &\qquad + \mu_{k-1}^T \Sigma_{k-1}^{-1} \mu_{k-1}\Big) \end{aligned} \tag{13}$$

As we explained before (Equation 10) we look for this exponential to be the product of two: one corresponding to a Gaussian on variable $\mathbf{x}_{k-1}$, that we denoted as $\exp_g(\mathbf{x}_{k-1})$, and one free from that variable, $\exp_c(\mathbf{x}_k)$. We need to distinguish the sum of both exponents in Equation 13.

The exponent of $\exp_g(\mathbf{x}_{k-1})$ in standard form would be:

$$\exp_g(\mathbf{x}_{k-1}) = -\frac{1}{2}\left(\mathbf{x}_{k-1} - \mu_g\right)^T \Sigma_g^{-1} \left(\mathbf{x}_{k-1} - \mu_g\right) \tag{14}$$

that must correspond to the canonical form:

$$\exp_g(\mathbf{x}_{k-1}) = -\frac{1}{2}\mathbf{x}_{k-1}^T \Lambda_g \mathbf{x}_{k-1} + \eta_g^T \mathbf{x}_{k-1} + \delta_g$$

where

$$\begin{aligned}\Lambda_g &= \Sigma_g^{-1} \quad &&\text{, therefore, } \Sigma_g = \Lambda_g^{-1}\\ \eta_g &= \Sigma_g^{-1}\mu_g &&\text{, therefore, } \mu_g = \Sigma_g\eta_g\\ \delta_g &= -\frac{1}{2}\mu_g^T\Sigma_g^{-1}\mu_g\end{aligned} \tag{15}$$

In Equation 13 we can distinguish part of this already: we can do $\Lambda_t + \Lambda_i = \Lambda_g$ and $\eta_t + \eta_i = \eta_g$. These are sufficient to deduce the parameters of the standard form Gaussian of Equation 14, as follows:

$$\begin{aligned}&\Sigma_g = \Lambda_g^{-1} = \left(\Lambda_t + \Lambda_i\right)^{-1} = \\ &= \left(\mathbf{A}_k\mathbf{R}_k^{-1}\mathbf{A}_k^T + \Sigma_{k-1}^{-1}\right)^{-1}\\ &\mu_g = \Sigma_g\eta_g = \Sigma_g\left(\eta_t + \eta_i\right) = \\ &= \left(\mathbf{A}_k\mathbf{R}_k^{-1}\mathbf{A}_k^T + \Sigma_{k-1}^{-1}\right)^{-1}\left(\mathbf{A}_k^T\mathbf{R}_k^{-1}\left(\mathbf{x}_k - \mathbf{B}_k\mathbf{u}_k\right) + \right.\\ &\qquad\left. + \Sigma_{k-1}^{-1}\mu_{k-1}\right)\end{aligned} \tag{16}$$

According to these standard form parameters, we can turn back to the canonical form of the exponent of $\exp_g(\mathbf{x}_{k-1})$ and look for its third canonical parameter, that is, δ_g, which must be:

$$\begin{aligned}&\delta_g = -\frac{1}{2}\mu_g^T\Sigma_g^{-1}\mu_g = \\ &= -\frac{1}{2}\left[\left(\mathbf{A}_k\mathbf{R}_k^{-1}\mathbf{A}_k^T + \Sigma_{k-1}^{-1}\right)^{-1}\left(\mathbf{A}_k^T\mathbf{R}_k^{-1}\left(\mathbf{x}_k - \mathbf{B}_k\mathbf{u}_k\right) + \Sigma_{k-1}^{-1}\mu_{k-1}\right)\right]^T\\ &\left(\mathbf{A}_k\mathbf{R}_k^{-1}\mathbf{A}_k^T + \Sigma_{k-1}^{-1}\right)\left[\left(\mathbf{A}_k\mathbf{R}_k^{-1}\mathbf{A}_k^T + \Sigma_{k-1}^{-1}\right)^{-1}\left(\mathbf{A}_k^T\mathbf{R}_k^{-1}\left(\mathbf{x}_k - \mathbf{B}_k\mathbf{u}_k\right) + \Sigma_{k-1}^{-1}\mu_{k-1}\right)\right] =\end{aligned}$$

(applying the first transpose)

$$\begin{aligned}&= -\frac{1}{2}\left(\mathbf{A}_k^T\mathbf{R}_k^{-1}\left(\mathbf{x}_k - \mathbf{B}_k\mathbf{u}_k\right) + \Sigma_{k-1}^{-1}\mu_{k-1}\right)^T\\ &\underbrace{\left[\left(\mathbf{A}_k\mathbf{R}_k^{-1}\mathbf{A}_k^T + \Sigma_{k-1}^{-1}\right)^{-1}\right]^T\left(\mathbf{A}_k\mathbf{R}_k^{-1}\mathbf{A}_k^T + \Sigma_{k-1}^{-1}\right)\left(\mathbf{A}_k\mathbf{R}_k^{-1}\mathbf{A}_k^T + \Sigma_{k-1}^{-1}\right)^{-1}}_{\mathbf{I}}\\ &\left(\mathbf{A}_k^T\mathbf{R}_k^{-1}\left(\mathbf{x}_k - \mathbf{B}_k\mathbf{u}_k\right) + \Sigma_{k-1}^{-1}\mu_{k-1}\right) =\end{aligned}$$

(by exchangeability of transpose and inverse in the underlined factor)

$$\begin{aligned}&= -\frac{1}{2}\left(\mathbf{A}_k^T\mathbf{R}_k^{-1}\left(\mathbf{x}_k - \mathbf{B}_k\mathbf{u}_k\right) + \Sigma_{k-1}^{-1}\mu_{k-1}\right)^T\\ &\left[\left(\mathbf{A}_k\mathbf{R}_k^{-1}\mathbf{A}_k^T + \Sigma_{k-1}^{-1}\right)^T\right]^{-1}\\ &\left(\mathbf{A}_k^T\mathbf{R}_k^{-1}\left(\mathbf{x}_k - \mathbf{B}_k\mathbf{u}_k\right) + \Sigma_{k-1}^{-1}\mu_{k-1}\right) =\end{aligned}$$

(and since $\left(\mathbf{A}_k\mathbf{R}_k^{-1}\mathbf{A}_k^T + \Sigma_{k-1}^{-1}\right)^T = \mathbf{A}_k\mathbf{R}_k^{-1}\mathbf{A}_k^T + \Sigma_{k-1}^{-1}$)

$$\begin{aligned}&= -\frac{1}{2}\left(\mathbf{A}_k^T\mathbf{R}_k^{-1}\left(\mathbf{x}_k - \mathbf{B}_k\mathbf{u}_k\right) + \Sigma_{k-1}^{-1}\mu_{k-1}\right)^T\\ &\left(\mathbf{A}_k\mathbf{R}_k^{-1}\mathbf{A}_k^T + \Sigma_{k-1}^{-1}\right)^{-1}\\ &\left(\mathbf{A}_k^T\mathbf{R}_k^{-1}\left(\mathbf{x}_k - \mathbf{B}_k\mathbf{u}_k\right) + \Sigma_{k-1}^{-1}\mu_{k-1}\right)\end{aligned} \tag{17}$$

Notice that this δ_g does not coincide with the third parameter of the canonical form that appears in the integrand we had ($\delta_t + \delta_i$ in Equation 13) which means that that integrand effectively contains two parts: the Gaussian we are looking for, $\exp_g(\mathbf{x}_{k-1})$, and a free term $\exp_c(\mathbf{x}_k)$, as we expected. The way to retrieve the latter is to rewrite the exponent of Equation 13 in this way:

Box 1.

$$\begin{aligned}
-\delta_g + \delta_t + \delta_i &= \\
&= \frac{1}{2}\left(\mathbf{A}_k^T\mathbf{R}_k^{-1}\left(\mathbf{x}_k - \mathbf{B}_k\mathbf{u}_k\right) + \boldsymbol{\Sigma}_{k-1}^{-1}\boldsymbol{\mu}_{k-1}\right)^T\left(\mathbf{A}_k\mathbf{R}_k^{-1}\mathbf{A}_k^T + \boldsymbol{\Sigma}_{k-1}^{-1}\right)^{-1} \\
&\qquad \left(\mathbf{A}_k^T\mathbf{R}_k^{-1}\left(\mathbf{x}_k - \mathbf{B}_k\mathbf{u}_k\right) + \boldsymbol{\Sigma}_{k-1}^{-1}\boldsymbol{\mu}_{k-1}\right) + \\
&\quad + \underbrace{\left[-\frac{1}{2}\left(\mathbf{B}_k\mathbf{u}_k - \mathbf{x}_k\right)^T\mathbf{R}_k^{-1}\left(\mathbf{B}_k\mathbf{u}_k - \mathbf{x}_k\right)\right] + \left[-\frac{1}{2}\right]\left(\boldsymbol{\mu}_{k-1}^T\boldsymbol{\Sigma}_{k-1}^{-1}\boldsymbol{\mu}_{k-1}\right)}_{\delta_t + \delta_i} \\
&= -\frac{1}{2}\left(\mathbf{A}_k^T\mathbf{R}_k^{-1}\left(\mathbf{x}_k - \mathbf{B}_k\mathbf{u}_k\right) + \boldsymbol{\Sigma}_{k-1}^{-1}\boldsymbol{\mu}_{k-1}\right)^T\left[-\left(\mathbf{A}_k\mathbf{R}_k^{-1}\mathbf{A}_k^T + \boldsymbol{\Sigma}_{k-1}^{-1}\right)^{-1}\right] \\
&\qquad \left(\mathbf{A}_k^T\mathbf{R}_k^{-1}\left(\mathbf{x}_k - \mathbf{B}_k\mathbf{u}_k\right) + \boldsymbol{\Sigma}_{k-1}^{-1}\boldsymbol{\mu}_{k-1}\right) + \\
&\quad + \underbrace{\left[-\frac{1}{2}\left(\mathbf{B}_k\mathbf{u}_k - \mathbf{x}_k\right)^T\mathbf{R}_k^{-1}\left(\mathbf{B}_k\mathbf{u}_k - \mathbf{x}_k\right)\right]}_{\delta_t} + \\
&\quad + \underbrace{\left[-\frac{1}{2}\right]\left(\boldsymbol{\mu}_{k-1}^T\boldsymbol{\Sigma}_{k-1}^{-1}\boldsymbol{\mu}_{k-1}\right)}_{\delta_i}
\end{aligned} \tag{19}$$

Box 2.

$$\begin{aligned}
-\delta_g + \delta_t &= \\
&= -\frac{1}{2}\left(\mathbf{A}_k^T\mathbf{R}_k^{-1}\left(\mathbf{x}_k - \mathbf{B}_k\mathbf{u}_k\right) + \boldsymbol{\Sigma}_{k-1}^{-1}\boldsymbol{\mu}_{k-1}\right)^T\left[-\left(\mathbf{A}_k\mathbf{R}_k^{-1}\mathbf{A}_k^T + \boldsymbol{\Sigma}_{k-1}^{-1}\right)^{-1}\right] \\
&\qquad \left(\mathbf{A}_k^T\mathbf{R}_k^{-1}\left(\mathbf{x}_k - \mathbf{B}_k\mathbf{u}_k\right) + \boldsymbol{\Sigma}_{k-1}^{-1}\boldsymbol{\mu}_{k-1}\right) + \\
&\quad \left[-\frac{1}{2}\left(\mathbf{B}_k\mathbf{u}_k - \mathbf{x}_k\right)^T\mathbf{R}_k^{-1}\left(\mathbf{B}_k\mathbf{u}_k - \mathbf{x}_k\right)\right] \\
&= -\frac{1}{2}\left(\underbrace{\mathbf{A}_k^T\mathbf{R}_k^{-1}}_{\mathbf{E}_1}\mathbf{x}_k - \underbrace{\left(\mathbf{A}_k^T\mathbf{R}_k^{-1}\mathbf{B}_k\mathbf{u}_k - \boldsymbol{\Sigma}_{k-1}^{-1}\boldsymbol{\mu}_{k-1}\right)}_{\boldsymbol{\mu}_1}\right)^T\underbrace{\left[-\left(\mathbf{A}_k\mathbf{R}_k^{-1}\mathbf{A}_k^T + \boldsymbol{\Sigma}_{k-1}^{-1}\right)^{-1}\right]}_{\boldsymbol{\Sigma}_1^{-1}} \\
&\qquad \left(\mathbf{A}_k^T\mathbf{R}_k^{-1}\mathbf{x}_k - \left(\mathbf{A}_k^T\mathbf{R}_k^{-1}\mathbf{B}_k\mathbf{u}_k - \boldsymbol{\Sigma}_{k-1}^{-1}\boldsymbol{\mu}_{k-1}\right)\right) + \\
&\quad + \left(-\frac{1}{2}\left(\underbrace{-\mathbf{I}}_{\mathbf{E}_2}\mathbf{x}_k - \underbrace{\left(-\mathbf{B}_k\mathbf{u}_k\right)}_{\boldsymbol{\mu}_2}\right)^T\underbrace{\mathbf{R}_k^{-1}}_{\boldsymbol{\Sigma}_2^{-1}}\left(-\mathbf{I}\mathbf{x}_k - \left(-\mathbf{B}_k\mathbf{u}_k\right)\right)\right)
\end{aligned} \tag{20}$$

$$
\begin{aligned}
&-\frac{1}{2}\mathbf{x}_{k-1}^T\left(\Lambda_t+\Lambda_i\right)\mathbf{x}_{k-1}\\
&\quad+\left(\eta_t+\eta_i\right)^T\mathbf{x}_{k-1}+\delta_t+\delta_i=\\
&=-\frac{1}{2}\mathbf{x}_{k-1}^T\left(\Lambda_t+\Lambda_i\right)\mathbf{x}_{k-1}\\
&\quad+\left(\eta_t+\eta_i\right)^T\mathbf{x}_{k-1}+\underbrace{\delta_g-\delta_g}_{=0}+\delta_t+\delta_i=\\
&=\underbrace{-\frac{1}{2}\mathbf{x}_{k-1}^T\Lambda_g\mathbf{x}_{k-1}+\eta_g^T\mathbf{x}_{k-1}+\delta_g}_{\text{canonical form of the exponent of a Gaussian on variable }\mathbf{x}_{k-1}}\\
&\quad\underbrace{-\delta_g+\delta_t+\delta_i}_{\text{term free of variable }\mathbf{x}_{k-1}}
\end{aligned}
\tag{18}
$$

The exponent of the free part $\exp_c(\mathbf{x}_k)$ is then shown in Box 1.

We can observe two relevant facts here: the terms corresponding to $-\delta_g$ and δ_t (the first two summands in the previous equation) could be rearranged as two standard form exponents of two Gaussians on the variable $\mathbf{x}_k$. Thus, they must correspond to the exponential $\exp_c(\mathbf{x}_k)$ in our roadmap. On the other hand, the term for δ_i makes up a constant exponent and therefore a constant value for its exponential, which is taken out of our whole derivation (it is considered another proportional factor of the posterior).

Let us focus then on the sum of the two Gaussian pdf exponents that lead to $\exp_c(\mathbf{x}_k)$, δ_g and δ_t. We will first rearrange them to state clearly their role as Gaussian exponents on the variable $\mathbf{x}_k$, as we have said, shown in Box 2.

The same expression in canonical form is:

$$
\begin{aligned}
&-\frac{1}{2}\left(\mathbf{E}_1\mathbf{x}_k+\boldsymbol{\mu}_1\right)^T\boldsymbol{\Sigma}_1^{-1}\left(\mathbf{E}_1\mathbf{x}_k+\boldsymbol{\mu}_1\right)\\
&-\frac{1}{2}\left(\mathbf{E}_2\mathbf{x}_k+\boldsymbol{\mu}_2\right)^T\boldsymbol{\Sigma}_2^{-1}\left(\mathbf{E}_2\mathbf{x}_k+\boldsymbol{\mu}_2\right)=\\
&=-\frac{1}{2}\mathbf{x}_k^T\boldsymbol{\Lambda}_1\mathbf{x}_k+\boldsymbol{\eta}_1^T\mathbf{x}_k+\boldsymbol{\delta}_1-\frac{1}{2}\mathbf{x}_k^T\boldsymbol{\Lambda}_2\mathbf{x}_k+\boldsymbol{\eta}_2^T\mathbf{x}_k+\boldsymbol{\delta}_2=\\
&=-\frac{1}{2}\mathbf{x}_k^T\left(\boldsymbol{\Lambda}_1+\boldsymbol{\Lambda}_2\right)\mathbf{x}_k+\left(\boldsymbol{\eta}_1^T+\boldsymbol{\eta}_2^T\right)\mathbf{x}_k+\boldsymbol{\delta}_1+\boldsymbol{\delta}_2=\\
&=-\frac{1}{2}\mathbf{x}_k^T\left(\boldsymbol{\Lambda}_1+\boldsymbol{\Lambda}_2\right)\mathbf{x}_k+\left(\boldsymbol{\eta}_1+\boldsymbol{\eta}_2\right)^T\mathbf{x}_k+\boldsymbol{\delta}_1+\boldsymbol{\delta}_2
\end{aligned}
$$

where:

$$
\begin{aligned}
\boldsymbol{\Lambda}_1&=\mathbf{E}_1^T\boldsymbol{\Sigma}_1^{-1}\mathbf{E}_1=\\
&=\mathbf{R}_k^{-1}\mathbf{A}_k\left(-\left(\mathbf{A}_k^T\mathbf{R}_k^{-1}\mathbf{A}_k+\boldsymbol{\Sigma}_{k-1}^{-1}\right)^{-1}\right)\mathbf{A}_k^T\mathbf{R}_k^{-1}\\
\boldsymbol{\eta}_1&=\mathbf{E}_1^T\boldsymbol{\Sigma}_1^{-1}\boldsymbol{\mu}_1=\\
&=\mathbf{R}_k^{-1}\mathbf{A}_k\left(-\left(\mathbf{A}_k^T\mathbf{R}_k^{-1}\mathbf{A}_k+\boldsymbol{\Sigma}_{k-1}^{-1}\right)^{-1}\right)\\
&\quad\left(\mathbf{A}_k^T\mathbf{R}_k^{-1}\mathbf{B}_k\mathbf{u}_k-\boldsymbol{\Sigma}_{k-1}^{-1}\boldsymbol{\mu}_{k-1}\right)\\
\boldsymbol{\Lambda}_2&=\mathbf{E}_2^T\boldsymbol{\Sigma}_2^{-1}\mathbf{E}_2=\mathbf{R}_k^{-1}\\
\boldsymbol{\eta}_2&=\mathbf{E}_2^T\boldsymbol{\Sigma}_2^{-1}\boldsymbol{\mu}_2=\mathbf{R}_k^{-1}\mathbf{B}_k\mathbf{u}_k
\end{aligned}
\tag{21}
$$

Now we will see that the resulting exponent in Equation 21 contains the canonical form of the exponent of the Gaussian $\exp_c(\mathbf{x}_k)$, but also a term free from the variable $\mathbf{x}_k$ that will be taken out and considered as another proportionality constant, that we will denote $\exp_e$.

We are about to end the first stage of the filter, where we pursued a closed form for the prior of the Kalman filter at step k, thus it is time to remember and update our roadmap for the prediction stage of the KF:

$$
\begin{aligned}
&\int_{\mathbf{x}_{k-1}}\exp_t(\mathbf{x}_k,\mathbf{x}_{k-1})\exp_i(\mathbf{x}_{k-1})d\mathbf{x}_{k-1}\propto\\
&\propto\int_{\mathbf{x}_{k-1}}\exp_g(\mathbf{x}_{k-1})\exp_c(\mathbf{x}_k)d\mathbf{x}_{k-1}=\\
&=\exp_c(\mathbf{x}_k)\int_{\mathbf{x}_{k-1}}\exp_g(\mathbf{x}_{k-1})d\mathbf{x}_{k-1}\propto\\
&\propto\exp_c(\mathbf{x}_k)=\\
&=\exp_p(\mathbf{x}_k)\exp_e\propto\\
&\propto\exp_p(\mathbf{x}_k)
\end{aligned}
\tag{22}
$$

We have reached at $\exp_c(\mathbf{x}_k)$ in Equation 21. It is clear that for finishing this first stage of the filter we are interested in the final Gaussian $\exp_p(\mathbf{x}_k)$, whose exponent in standard form must be of the form $-(1/2)\left(\mathbf{x}_k-\bar{\boldsymbol{\mu}}_k\right)^T\bar{\boldsymbol{\Sigma}}_k^{-1}\left(\mathbf{x}_k-\bar{\boldsymbol{\mu}}_k\right)$.

Box 3.

$$\bar{\mu}_k = \bar{\Sigma}_k\left(\eta_1 + \eta_2\right) =$$

$$= \left(\mathbf{R}_k + \mathbf{A}_k\Sigma_{k-1}\mathbf{A}_k^T\right)\begin{pmatrix}\mathbf{R}_k^{-1}\mathbf{A}_k\left[-\left(\mathbf{A}_k^T\mathbf{R}_k^{-1}\mathbf{A}_k + \Sigma_{k-1}^{-1}\right)^{-1}\right]\left(\mathbf{A}_k^T\mathbf{R}_k^{-1}\mathbf{B}_k\mathbf{u}_k - \Sigma_{k-1}^{-1}\mu_{k-1}\right) + \\ + \mathbf{R}_k^{-1}\mathbf{B}_k\mathbf{u}_k\end{pmatrix} =$$

$$= \underbrace{\mathbf{R}_k\mathbf{R}_k^{-1}}_{\mathbf{I}}\mathbf{A}_k\left[-\left(\mathbf{A}_k^T\mathbf{R}_k^{-1}\mathbf{A}_k + \Sigma_{k-1}^{-1}\right)^{-1}\right]\left(\mathbf{A}_k^T\mathbf{R}_k^{-1}\mathbf{B}_k\mathbf{u}_k - \Sigma_{k-1}^{-1}\mu_{k-1}\right) + \underbrace{\mathbf{R}_k\mathbf{R}_k^{-1}}_{\mathbf{I}}\mathbf{B}_k\mathbf{u}_k +$$
$$+ \mathbf{A}_k\Sigma_{k-1}\mathbf{A}_k^T\mathbf{R}_k^{-1}\mathbf{A}_k\left[-\left(\mathbf{A}_k^T\mathbf{R}_k^{-1}\mathbf{A}_k + \Sigma_{k-1}^{-1}\right)^{-1}\right]\left(\mathbf{A}_k^T\mathbf{R}_k^{-1}\mathbf{B}_k\mathbf{u}_k - \Sigma_{k-1}^{-1}\mu_{k-1}\right) +$$
$$+ \mathbf{A}_k\Sigma_{k-1}\mathbf{A}_k^T\mathbf{R}_k^{-1}\mathbf{B}_k\mathbf{u}_k =$$

$$= \mathbf{A}_k\underline{\left[-\left(\mathbf{A}_k^T\mathbf{R}_k^{-1}\mathbf{A}_k + \Sigma_{k-1}^{-1}\right)^{-1}\right]}\left(\mathbf{A}_k^T\mathbf{R}_k^{-1}\mathbf{B}_k\mathbf{u}_k - \Sigma_{k-1}^{-1}\mu_{k-1}\right) + \mathbf{B}_k\mathbf{u}_k +$$
$$+ \mathbf{A}_k\Sigma_{k-1}\mathbf{A}_k^T\mathbf{R}_k^{-1}\mathbf{A}_k\underline{\left[-\left(\mathbf{A}_k^T\mathbf{R}_k^{-1}\mathbf{A}_k + \Sigma_{k-1}^{-1}\right)^{-1}\right]}\left(\mathbf{A}_k^T\mathbf{R}_k^{-1}\mathbf{B}_k\mathbf{u}_k - \Sigma_{k-1}^{-1}\mu_{k-1}\right) +$$
$$+ \mathbf{A}_k\Sigma_{k-1}\mathbf{A}_k^T\mathbf{R}_k^{-1}\mathbf{B}_k\mathbf{u}_k =$$

$$= \left(\mathbf{A}_k + \mathbf{A}_k\Sigma_{k-1}\mathbf{A}_k^T\mathbf{R}_k^{-1}\mathbf{A}_k\right)\left[-\left(\mathbf{A}_k^T\mathbf{R}_k^{-1}\mathbf{A}_k + \Sigma_{k-1}^{-1}\right)^{-1}\right]\left(\mathbf{A}_k^T\mathbf{R}_k^{-1}\mathbf{B}_k\mathbf{u}_k - \Sigma_{k-1}^{-1}\mu_{k-1}\right) +$$
$$+ \mathbf{B}_k\mathbf{u}_k + \mathbf{A}_k\Sigma_{k-1}\mathbf{A}_k^T\mathbf{R}_k^{-1}\mathbf{B}_k\mathbf{u}_k =$$

$$= -\mathbf{A}_k\left(\mathbf{I} + \Sigma_{k-1}\mathbf{A}_k^T\mathbf{R}_k^{-1}\mathbf{A}_k\right)\left(\mathbf{A}_k^T\mathbf{R}_k^{-1}\mathbf{A}_k + \Sigma_{k-1}^{-1}\right)^{-1}\left(\mathbf{A}_k^T\mathbf{R}_k^{-1}\mathbf{B}_k\mathbf{u}_k - \Sigma_{k-1}^{-1}\mu_{k-1}\right) +$$
$$+ \mathbf{B}_k\mathbf{u}_k + \mathbf{A}_k\Sigma_{k-1}\mathbf{A}_k^T\mathbf{R}_k^{-1}\mathbf{B}_k\mathbf{u}_k =$$

$$= -\mathbf{A}_k\underbrace{\Sigma_k\Sigma_k^{-1}}_{\mathbf{I}}\left(\mathbf{I} + \Sigma_{k-1}\mathbf{A}_k^T\mathbf{R}_k^{-1}\mathbf{A}_k\right)\left(\mathbf{A}_k^T\mathbf{R}_k^{-1}\mathbf{A}_k + \Sigma_{k-1}^{-1}\right)^{-1}\left(\mathbf{A}_k^T\mathbf{R}_k^{-1}\mathbf{B}_k\mathbf{u}_k - \Sigma_{k-1}^{-1}\mu_{k-1}\right) +$$
$$+ \mathbf{B}_k\mathbf{u}_k + \mathbf{A}_k\Sigma_{k-1}\mathbf{A}_k^T\mathbf{R}_k^{-1}\mathbf{B}_k\mathbf{u}_k =$$

$$= -\mathbf{A}_k\Sigma_{k-1}\underbrace{\left(\Sigma_{k-1}^{-1} + \mathbf{A}_k^T\mathbf{R}_k^{-1}\mathbf{A}_k\right)\left(\mathbf{A}_k^T\mathbf{R}_k^{-1}\mathbf{A}_k + \Sigma_{k-1}^{-1}\right)^{-1}}_{\mathbf{I}}\left(\mathbf{A}_k^T\mathbf{R}_k^{-1}\mathbf{B}_k\mathbf{u}_k - \Sigma_{k-1}^{-1}\mu_{k-1}\right) +$$
$$+ \mathbf{B}_k\mathbf{u}_k + \mathbf{A}_k\Sigma_{k-1}\mathbf{A}_k^T\mathbf{R}_k^{-1}\mathbf{B}_k\mathbf{u}_k =$$

$$= -\mathbf{A}_k\Sigma_{k-1}\left(\mathbf{A}_k^T\mathbf{R}_k^{-1}\mathbf{B}_k\mathbf{u}_k - \Sigma_{k-1}^{-1}\mu_{k-1}\right) + \mathbf{B}_k\mathbf{u}_k + \mathbf{A}_k\Sigma_{k-1}\mathbf{A}_k^T\mathbf{R}_k^{-1}\mathbf{B}_k\mathbf{u}_k =$$

$$= \underline{-\mathbf{A}_k\Sigma_{k-1}\mathbf{A}_k^T\mathbf{R}_k^{-1}\mathbf{B}_k\mathbf{u}_k} + \mathbf{A}_k\mu_{k-1} + \mathbf{B}_k\mathbf{u}_k + \underline{\mathbf{A}_k\Sigma_{k-1}\mathbf{A}_k^T\mathbf{R}_k^{-1}\mathbf{B}_k\mathbf{u}_k} =$$

$$= \mathbf{A}_k\mu_{k-1} + \mathbf{B}_k\mathbf{u}_k \qquad (24)$$

At this point we can already deduce the values for these parameters from the exponent of $\exp_c(\mathbf{x}_k)$. Firstly, the covariance of the Gaussian must be:

$$\bar{\boldsymbol{\Sigma}}_k^{-1} = \boldsymbol{\Lambda}_1 + \boldsymbol{\Lambda}_2 =$$
$$= \mathbf{R}_k^{-1}\mathbf{A}_k\left(-\left(\mathbf{A}_k^T\mathbf{R}_k^{-1}\mathbf{A}_k + \boldsymbol{\Sigma}_{k-1}^{-1}\right)^{-1}\right)\mathbf{A}_k^T\mathbf{R}_k^{-1} + \mathbf{R}_k^{-1} =$$

(changing the order of the two terms in the sum)

$$= \mathbf{R}_k^{-1} - \mathbf{R}_k^{-1}\mathbf{A}_k\left(\mathbf{A}_k^T\mathbf{R}_k^{-1}\mathbf{A}_k + \boldsymbol{\Sigma}_{k-1}^{-1}\right)^{-1}\mathbf{A}_k^T\mathbf{R}_k^{-1} =$$

(applying the matrix inversion lemma—refer to Appendix E)

$$= \left(\mathbf{R}_k + \mathbf{A}_k\boldsymbol{\Sigma}_{k-1}\mathbf{A}_k^T\right)^{-1} \quad (23)$$

Therefore, the covariance of the Gaussian $\exp_p(\mathbf{x}_k)$ resulting from the first stage of the filter is $\bar{\boldsymbol{\Sigma}}_k = \mathbf{R}_k + \mathbf{A}_k\boldsymbol{\Sigma}_{k-1}\mathbf{A}_k^T$, which is the covariance $\mathbf{R}_k$ of the motion model plus the covariance $\boldsymbol{\Sigma}_{k-1}$ of the previous pose of the robot propagated by the lineal motion $\mathbf{A}_k$, as explained in chapter 3 section 8 in the paragraphs about propagation of covariances. However, here it is not an approximation but an exact result, since we are working with linear functions.

On the other hand, the mean of the Gaussian must be (do not be overwhelmed by the length of this derivation shown in Box 3: it is just basic matrix manipulation, and the result is pretty short!).

Thus, the mean of the Gaussian $\exp_p(\mathbf{x}_k)$, which summarizes in closed form the first stage of the filter, is $\bar{\boldsymbol{\mu}}_k = \mathbf{A}_k\boldsymbol{\mu}_{k-1} + \mathbf{B}_k\mathbf{u}_k$, which is the propagation of the mean $\boldsymbol{\mu}_{k-1}$ of the previous posterior through the lineal motion model.

So far, we have derived the integral part of the Kalman filter: the one corresponding to the prior at step k, and we have seen that in this prediction stage we obtain a Gaussian distribution centered at $\bar{\boldsymbol{\mu}}_k = \mathbf{A}_k\boldsymbol{\mu}_{k-1} + \mathbf{B}_k\mathbf{u}_k$ and with covariance matrix $\bar{\boldsymbol{\Sigma}}_k = \mathbf{R}_k + \mathbf{A}_k\boldsymbol{\Sigma}_{k-1}\mathbf{A}_k^T$. In spite of their lengthy derivation, these values are quite easy to interpret and understand: before we consider the effect of the observation likelihood in the second stage of the filter, we can *predict* that the robot is expected to be at the expected pose it was at step $k-1$ transformed by the linear motion model (i.e, $\bar{\boldsymbol{\mu}}_k = \mathbf{A}_k\boldsymbol{\mu}_{k-1} + \mathbf{B}_k\mathbf{u}_k$); in addition, and maybe more importantly, the uncertainty that we have about that prediction is the same we already had at step $k-1$ once it is propagated by the transition function that corresponds to the linear transformation described by $\mathbf{A}_k$ (giving a covariance of $\mathbf{A}_k\boldsymbol{\Sigma}_{k-1}\mathbf{A}_k^T$) and augmented with the uncertainty from the additive Gaussian noise (with covariance $\mathbf{R}_k$).

In other words, *before it takes into account new observations, the robot increments its uncertainty according to its motion model.* That is the reason why a mobile robot that only uses odometry (i.e., the action term $\mathbf{u}_k$) and no exteroceptive sensor (no second stage in the filter) only can see its pose uncertainty to increase over time.

Let us go for the second stage of the filter, called the *update stage* of the KF, where the observation is incorporated into the Bayes filter to reduce the estimation uncertainty as much as possible provided the observation model and the uncertainty of observations. We first recall the entire roadmap of the derivation of the KF, now including the observation likelihood:

$$\begin{aligned}
&bel(\mathbf{x}_k \mid \mathbf{z}_{1:k}, \mathbf{u}_{1:k}, \mathbf{m}) = N(\mathbf{x}_k; \boldsymbol{\mu}_k, \boldsymbol{\Sigma}_k) \propto \\
&\propto \exp_z(\mathbf{z}_k, \mathbf{x}_k) \int_{\mathbf{x}_{k-1}} \exp_t(\mathbf{x}_k, \mathbf{x}_{k-1}) \exp_i(\mathbf{x}_{k-1}) d\mathbf{x}_{k-1} \propto \\
&\propto \exp_z(\mathbf{z}_k, \mathbf{x}_k) \int_{\mathbf{x}_{k-1}} \exp_g(\mathbf{x}_{k-1}) \exp_c(\mathbf{x}_k) d\mathbf{x}_{k-1} = \\
&= \exp_z(\mathbf{z}_k, \mathbf{x}_k) \exp_c(\mathbf{x}_k) \int_{\mathbf{x}_{k-1}} \exp_g(\mathbf{x}_{k-1}) d\mathbf{x}_{k-1} \propto \\
&\propto \exp_z(\mathbf{z}_k, \mathbf{x}_k) \exp_c(\mathbf{x}_k) = \\
&= \exp_z(\mathbf{z}_k, \mathbf{x}_k) \exp_p(\mathbf{x}_k) \exp_e \propto \\
&\propto \exp_z(\mathbf{z}_k, \mathbf{x}_k) \exp_p(\mathbf{x}_k)
\end{aligned} \tag{25}$$

We have here the product of two Gaussian exponentials: a first one, $\exp_z(\mathbf{z}_k, \mathbf{x}_k)$, on the variable $\mathbf{z}_k$ (but also depending on $\mathbf{x}_k$) and another one, $\exp_p(\mathbf{x}_k)$ resulting from the prediction stage and only depending on $\mathbf{x}_k$. We can rearrange the former to have the form of a Gaussian on the variable $\mathbf{x}_k$, and thus the result will be an exponential with exponent the sum of two exponents. In that final exponent there will be a part depending only on $\mathbf{x}_k$ (that forms the exponent of the Gaussian posterior, which we are looking for from the beginning) and another free of that variable (which will disappear within a proportionality constant).

We develop the update stage making use again of the canonical forms:

$$\begin{aligned}
\exp_p(\mathbf{x}_k) &= \exp\left(-\frac{1}{2}\left(\mathbf{x}_k - \bar{\boldsymbol{\mu}}_k\right)^T \bar{\boldsymbol{\Sigma}}_k^{-1}\left(\mathbf{x}_k - \bar{\boldsymbol{\mu}}_k\right)\right) \\
&= \exp\left(-\frac{1}{2}\left(\underbrace{\mathbf{I}}_{\mathbf{E}_p}\mathbf{x}_k - \underbrace{\bar{\boldsymbol{\mu}}_k}_{\boldsymbol{\mu}_p}\right)^T \underbrace{\bar{\boldsymbol{\Sigma}}_k^{-1}}_{\boldsymbol{\Sigma}_p^{-1}}\left(\mathbf{I}\mathbf{x}_k - \bar{\boldsymbol{\mu}}_k\right)\right) \\
&= \exp\left(-\frac{1}{2}\mathbf{x}_k^T \boldsymbol{\Lambda}_p \mathbf{x}_k + \boldsymbol{\eta}_p^T \mathbf{x}_k + \boldsymbol{\delta}_p\right)
\end{aligned}$$

where

$$\begin{aligned}
\boldsymbol{\Lambda}_p &= \bar{\boldsymbol{\Sigma}}_k^{-1} \\
\boldsymbol{\eta}_p &= \bar{\boldsymbol{\Sigma}}_k^{-1}\bar{\boldsymbol{\mu}}_k \\
\boldsymbol{\delta}_p &= -\frac{1}{2}\bar{\boldsymbol{\mu}}_k^T \bar{\boldsymbol{\Sigma}}_k^{-1}\bar{\boldsymbol{\mu}}_k
\end{aligned} \tag{26}$$

and

$$\exp_z(\mathbf{z}_k, \mathbf{x}_k) = \exp\left(-\frac{1}{2}\left(\mathbf{z}_k - \mathbf{C}_k\mathbf{x}_k\right)^T \mathbf{Q}_k^{-1}\left(\mathbf{z}_k - \mathbf{C}_k\mathbf{x}_k\right)\right) =$$

(centering the Gaussian at $\mathbf{x}_k$)

$$\begin{aligned}
&= \exp\left(-\frac{1}{2}\left(\underbrace{-\mathbf{C}_k}_{\mathbf{E}_z}\mathbf{x}_k - \underbrace{(-\mathbf{z}_k)}_{\boldsymbol{\mu}_z}\right)^T \underbrace{\mathbf{Q}_k^{-1}}_{\boldsymbol{\Sigma}_z^{-1}}\left(-\mathbf{C}_k\mathbf{x}_k - (-\mathbf{z}_k)\right)\right) = \\
&= \exp\left(-\frac{1}{2}\mathbf{x}_k^T \boldsymbol{\Lambda}_z \mathbf{x}_k + \boldsymbol{\eta}_z^T \mathbf{x}_k + \boldsymbol{\delta}_z\right)
\end{aligned}$$

where

$$\begin{aligned}
\boldsymbol{\Lambda}_z &= \mathbf{C}_k^T \mathbf{Q}_k^{-1}\mathbf{C}_k \\
\boldsymbol{\eta}_z &= \mathbf{C}_k^T \mathbf{Q}_k^{-1}\mathbf{z}_k \\
\boldsymbol{\delta}_z &= -\frac{1}{2}\mathbf{z}_k^T \mathbf{Q}_k^{-1}\mathbf{z}_k
\end{aligned} \tag{27}$$

The exponential from the second stage of the filter, Equation 25, has the following exponent in canonical form:

$$-\frac{1}{2}\mathbf{x}_k^T \underbrace{(\boldsymbol{\Lambda}_z + \boldsymbol{\Lambda}_p)}_{\boldsymbol{\Lambda}_u}\mathbf{x}_k + \underbrace{(\boldsymbol{\eta}_z + \boldsymbol{\eta}_p)^T}_{\boldsymbol{\eta}_u^T}\mathbf{x}_k + \underbrace{\boldsymbol{\delta}_z + \boldsymbol{\delta}_p}_{\boldsymbol{\delta}_u}$$

where

$$\begin{aligned}
\boldsymbol{\Lambda}_u &= \mathbf{C}_k^T \mathbf{Q}_k^{-1}\mathbf{C}_k + \bar{\boldsymbol{\Sigma}}_k^{-1} \\
\boldsymbol{\eta}_u &= \mathbf{C}_k^T \mathbf{Q}_k^{-1}\mathbf{z}_k + \bar{\boldsymbol{\Sigma}}_k^{-1}\bar{\boldsymbol{\mu}}_k \\
\boldsymbol{\delta}_u &= -\frac{1}{2}\left(\mathbf{z}_k^T \mathbf{Q}_k^{-1}\mathbf{z}_k + \bar{\boldsymbol{\mu}}_k^T \bar{\boldsymbol{\Sigma}}_k^{-1}\bar{\boldsymbol{\mu}}_k\right)
\end{aligned} \tag{28}$$

With these parameters, we have all the information that is needed to deduce the mean and covariance of the posterior of the filter, by transformation to the corresponding parameters of the standard form. The covariance is then:

$$\mathbf{\Sigma}_k = \mathbf{\Lambda}_u^{-1} = \left(\mathbf{C}_k^T \mathbf{Q}_k^{-1} \mathbf{C}_k + \bar{\mathbf{\Sigma}}_k^{-1}\right)^{-1} =$$

(by the matrix inversion lemma)

$$= \underbrace{\bar{\mathbf{\Sigma}}_k - \bar{\mathbf{\Sigma}}_k \mathbf{C}_k^T \left(\underbrace{\mathbf{Q}_k + \mathbf{C}_k \bar{\mathbf{\Sigma}}_k \mathbf{C}_k^T}_{\text{Innovation Covariance } = \mathbf{S}_k} \right)^{-1}}_{\text{Kalman Gain } = \mathbf{K}_k = \bar{\mathbf{\Sigma}}_k \mathbf{C}_k^T \mathbf{S}_k^{-1}} \mathbf{C}_k \bar{\mathbf{\Sigma}}_k =$$

$$= \begin{cases} \bar{\mathbf{\Sigma}}_k - \mathbf{K}_k \mathbf{C}_k \bar{\mathbf{\Sigma}}_k & \text{(first form)} \\ \left(\mathbf{I} - \mathbf{K}_k \mathbf{C}_k\right) \bar{\mathbf{\Sigma}}_k & \text{(second form)} \\ \bar{\mathbf{\Sigma}}_k - \mathbf{K}_k \left(\bar{\mathbf{\Sigma}}_k \mathbf{C}_k^T\right)^T & \text{(third form)} \end{cases} \quad (29)$$

where the factor $\mathbf{K}_k$ is called the *Kalman gain* and $\mathbf{S}_k$ the *innovation covariance* of the filter, being both related by $\mathbf{K}_k = \bar{\mathbf{\Sigma}}_k \mathbf{C}_k^T \mathbf{S}_k^{-1}$. Notice that this innovation covariance has the same meaning that that we saw in chapter 6 section 3 in the context of studying sensor models, although in this case we assume uncertainty exists only in the robot pose and in the sensor noise—we will come back to this innovation covariance below.

We have developed three forms of the covariance of the posterior $\left(\mathbf{\Sigma}_k\right)$ in order to highlight diverse things: the first form is the easiest to interpret, since it says that the covariance (uncertainty) of the posterior is the one predicted by the motion $\left(\bar{\mathbf{\Sigma}}_k\right)$ minus a term $\mathbf{K}_k \mathbf{C}_k \bar{\mathbf{\Sigma}}_k$ that clearly depends on the observation, that is, it says that *making observations can decrease the uncertainty produced by motion* (we will be more specific below); the second form is the most concise and may be more convenient for implementation; the third form includes a special factor, the $\bar{\mathbf{\Sigma}}_k \mathbf{C}_k^T$, on which a few words are in order.

We will deduce now that the factor $\bar{\mathbf{\Sigma}}_k \mathbf{C}_k^T$ is a cross-covariance matrix representing the uncertainty existing in the relation between two different r.v.s. Obviously, $\bar{\mathbf{\Sigma}}_k$ is a covariance matrix: $\bar{\mathbf{\Sigma}}_k = Cov\left[\bar{\mathbf{x}}_k, \bar{\mathbf{x}}_k\right]$, being $\bar{\mathbf{x}}_k$ a r.v. representing the predicted pose of the robot in the first stage of the filter, with mean $\bar{\mu}_k$. On the other hand, $\mathbf{C}_k$ is the matrix that relates robot poses with observations, and does not depend on $\bar{\mathbf{x}}_k$. Recalling the theorem that we proved in chapter 3 section 7 about covariance matrices, i.e., that $\mathbf{A}\, Cov\left[\mathbf{X}, \mathbf{Y}\right] \mathbf{B} = Cov\left[\mathbf{A}\mathbf{X}, \mathbf{B}^T \mathbf{Y}\right]$, it is straightforward to deduce, by taking $\mathbf{A} = \mathbf{I}$ and $\mathbf{B} = \mathbf{C}_k^T$ that:

$$\bar{\mathbf{\Sigma}}_k \mathbf{C}_k^T = Cov\left[\bar{\mathbf{x}}_k, \bar{\mathbf{x}}_k\right] \mathbf{C}_k^T = Cov\left[\bar{\mathbf{x}}_k, \mathbf{C}_k \bar{\mathbf{x}}_k\right] \quad (30)$$

or put in words, that the $\bar{\mathbf{\Sigma}}_k \mathbf{C}_k^T$ factor is the cross-covariance between $\bar{\mathbf{x}}_k$ and $\mathbf{C}_k \bar{\mathbf{x}}_k$.

Let us further explore what this cross-covariance represents. If $\bar{\mathbf{x}}_k$ is a r.v. with mean $\bar{\mu}_k$, then $\mathbf{C}_k \bar{\mathbf{x}}_k$ has a mean of $\mathbf{C}_k \bar{\mu}_k$. We can then abbreviate the r.v. $\mathbf{C}_k \bar{\mathbf{x}}_k$ as $\bar{\mathbf{z}}_k$, and its mean as $\bar{\mu}_{z_k} = \mathbf{C}_k \bar{\mu}_k$. In words: $\bar{\mathbf{z}}_k$ is the r.v. representing the value that the observation would have if the robot pose was exactly $\bar{\mathbf{x}}_k$, and $\bar{\mu}_{z_k}$ is its mean, that is, *the expected value of the observation assuming that the predicted pose is true*. What is the expected value of the observation before we gather it? The answer is $\bar{\mu}_{z_k}$. What is the uncertainty that we have in that expected value? The answer comes from the innovation covariance $\mathbf{S}_k = \mathbf{Q}_k + \mathbf{C}_k \bar{\mathbf{\Sigma}}_k \mathbf{C}_k^T$, which is the uncertainty resulting from the addition of the observation uncertainty ($\mathbf{Q}_k$) to the uncertainty in the pose predicted by the robot motion ($\bar{\mathbf{\Sigma}}_k$) once it is propagated by the linear observation model

($\mathbf{C}_k$) to obtain the predicted observation—remember the propagation of covariances through linear models, introduced in chapter 3 section 8, and already used in the Kalman filter. Notice that the present formulation assumes a perfectly known map, although it would be easy to consider uncertainty in the map as well[2]. All in all, $\mathbf{S}_k$ *reflects the uncertainty we have about the expected predicted observation* $\bar{\boldsymbol{\mu}}_{z_k}$.

We know already the meaning of the two r.v.s $\bar{\mathbf{x}}_k$ and $\bar{\mathbf{z}}_k$ involved in the factor $\bar{\boldsymbol{\Sigma}}_k \mathbf{C}_k^T$. Thus, $\bar{\boldsymbol{\Sigma}}_k \mathbf{C}_k^T$ is $Cov\left[\bar{\mathbf{x}}_k, \bar{\mathbf{z}}_k\right]$, which makes clear that *it is the cross-covariance between the predicted robot pose and the predicted observation.* For highlighting its role, we will denote it as $\bar{\boldsymbol{\Sigma}}_{po,k}$, and write $\mathbf{K}_k = \bar{\boldsymbol{\Sigma}}_{po,k} \mathbf{S}_k^{-1}$ from now on. Note that this cross-covariance appears in different parts of the covariance of the Kalman filter posterior, namely in the third form of that posterior $\boldsymbol{\Sigma}_k = \bar{\boldsymbol{\Sigma}}_k - \mathbf{K}_k \left(\bar{\boldsymbol{\Sigma}}_k \mathbf{C}_k^T\right)^T = \bar{\boldsymbol{\Sigma}}_k - \mathbf{K}_k \bar{\boldsymbol{\Sigma}}_{po,k}^T$ in the definition of Kalman gain $\mathbf{K}_k = \bar{\boldsymbol{\Sigma}}_k \mathbf{C}_k^T \mathbf{I}_k^{-1} = \bar{\boldsymbol{\Sigma}}_{po,k} \mathbf{S}_k^{-1}$, and in the definition of innovation covariance

$$\mathbf{S}_k = \mathbf{Q}_k + \mathbf{C}_k \bar{\boldsymbol{\Sigma}}_k \mathbf{C}_k^T = \mathbf{Q}_k + \mathbf{C}_k \bar{\boldsymbol{\Sigma}}_{po,k}.$$

By abbreviating this factor as $\bar{\boldsymbol{\Sigma}}_{po,k}$, we can now be more specific about our previous statement that "making observations can decrease the uncertainty of motion." The Kalman gain can thus be written as follows (where matrix division A/B denotes AB^{-1}):

$$\mathbf{K}_k = \frac{\bar{\boldsymbol{\Sigma}}_{po,k}}{\mathbf{Q}_k + \mathbf{C}_k \bar{\boldsymbol{\Sigma}}_k \mathbf{C}_k^T} = \frac{\bar{\boldsymbol{\Sigma}}_{po,k}}{\mathbf{Q}_k + \mathbf{C}_k \bar{\boldsymbol{\Sigma}}_{po,k}} \tag{31}$$

From this expression it is clear that when the uncertainty in the observation is small ($\mathbf{Q}_k \to \mathbf{0}$), then $\mathbf{K}_k \to \mathbf{C}_k^{-1}$, and thus the uncertainty in the filter $\boldsymbol{\Sigma}_k = \left(\mathbf{I} - \mathbf{K}_k \mathbf{C}_k\right) \bar{\boldsymbol{\Sigma}}_k \to \mathbf{0}$. The same occurs in this last expression if the uncertainty of the robot pose predicted by the motion is small ($\bar{\boldsymbol{\Sigma}}_k \to \mathbf{0}$). In words*: if we are highly certain about either the observation or the pose predicted by the motion, the resulting posterior uncertainty will be small.* Notice that this argument only makes sense for square, and invertible, $\mathbf{C}_k$ matrices. However, the same principle holds in general, excepting degenerate situations. An example of such a problematic case would be when no observation depends on one or more of the robot pose coordinates (there is a rank deficiency), in which case it does not matter how precise the observation is, we will not able to estimate all the components of a robot pose. A real-world scenario where this could happen is when observing one single landmark with a range-bearing sensor. The reader is encouraged to draw the geometry of this problem and to reflect on how many degrees of freedom will remain completely undetermined in this case for both, a robot pose in 2D (3 DOFs) and in 3D (with 6 DOFs).

We have analyzed thoroughly enough the covariance—uncertainty—of the posterior of the Kalman filter. Let us see, finally, which is the mean of that posterior, that is, what is our belief in the pose of the robot at the end of the update stage of step k:

$$\begin{aligned}
\boldsymbol{\mu}_u = \boldsymbol{\Lambda}_u^{-1} \boldsymbol{\eta}_u &= \underbrace{\left(\mathbf{C}_k^T \mathbf{Q}_k^{-1} \mathbf{C}_k + \bar{\boldsymbol{\Sigma}}_k^{-1}\right)^{-1}}_{\boldsymbol{\Sigma}_k} \left(\mathbf{C}_k^T \mathbf{Q}_k^{-1} \mathbf{z}_k + \bar{\boldsymbol{\Sigma}}_k^{-1} \bar{\boldsymbol{\mu}}_k\right) = \\
&= \left(\mathbf{I} - \mathbf{K}_k \mathbf{C}_k\right) \bar{\boldsymbol{\Sigma}}_k \left(\mathbf{C}_k^T \mathbf{Q}_k^{-1} \mathbf{z}_k + \bar{\boldsymbol{\Sigma}}_k^{-1} \bar{\boldsymbol{\mu}}_k\right) = \\
&= \left(\mathbf{I} - \mathbf{K}_k \mathbf{C}_k\right) \left(\bar{\boldsymbol{\Sigma}}_k \mathbf{C}_k^T \mathbf{Q}_k^{-1} \mathbf{z}_k + \bar{\boldsymbol{\mu}}_k\right) = \\
&= \bar{\boldsymbol{\Sigma}}_k \mathbf{C}_k^T \mathbf{Q}_k^{-1} \mathbf{z}_k + \bar{\boldsymbol{\mu}}_k - \mathbf{K}_k \mathbf{C}_k \bar{\boldsymbol{\Sigma}}_k \mathbf{C}_k^T \mathbf{Q}_k^{-1} \mathbf{z}_k - \mathbf{K}_k \mathbf{C}_k \bar{\boldsymbol{\mu}}_k = \\
&= \bar{\boldsymbol{\mu}}_k - \mathbf{K}_k \mathbf{C}_k \bar{\boldsymbol{\mu}}_k + \underbrace{(\bar{\boldsymbol{\Sigma}}_k \mathbf{C}_k^T \mathbf{Q}_k^{-1} - \mathbf{K}_k \mathbf{C}_k \bar{\boldsymbol{\Sigma}}_k \mathbf{C}_k^T \mathbf{Q}_k^{-1})}_{\mathbf{K}_k} \mathbf{z}_k
\end{aligned} \tag{32}$$

The identity indicated in the underbracket of the last line can be demonstrated as follows:

$$\begin{aligned}
&\bar{\Sigma}_k C_k^T Q_k^{-1} = K_k (I + C_k \bar{\Sigma}_k C_k^T Q_k^{-1}) \quad \leftrightarrow \\
&\bar{\Sigma}_k C_k^T Q_k^{-1} (I + C_k \bar{\Sigma}_k C_k^T Q_k^{-1})^{-1} = K_k \quad \leftrightarrow \\
&\bar{\Sigma}_k C_k^T Q_k^{-1} \left(\left(Q_k + C_k \bar{\Sigma}_k C_k^T \right) Q_k^{-1} \right)^{-1} = K_k \quad \leftrightarrow \\
&\bar{\Sigma}_k C_k^T Q_k^{-1} \left[Q_k \left(Q_k + C_k \bar{\Sigma}_k C_k^T \right)^{-1} \right] = K_k \quad \leftrightarrow \\
&\bar{\Sigma}_k C_k^T \left(\underbrace{Q_k + C_k \bar{\Sigma}_k C_k^T}_{S_k} \right)^{-1} = K_k \quad \leftrightarrow \\
&\bar{\Sigma}_k C_k^T S_k^{-1} = K_k
\end{aligned} \tag{33}$$

which is true by the definition of the Kalman gain. Therefore, we have:

$$\begin{aligned}
\mu_u &= \bar{\mu}_k - K_k C_k \bar{\mu}_k + K_k z_k \\
&= \bar{\mu}_k + K_k \left(z_k - C_k \bar{\mu}_k \right) \\
&= \bar{\mu}_k + K_k \left(z_k - \bar{\mu}_{z_k} \right)
\end{aligned} \tag{34}$$

This expression for the mean of the posterior is also worth to interpret. It says that *the expected*

Table 1. Closed-form equations of the KF in a suitable sequence for their implementation. The last column indicates the computational complexity of each step.

Prediction stage	Prediction stage - pose	$\bar{\mu}_k = A_k \mu_{k-1} + B_k u_k$ *(mean of the predicted pose)*	$O(d^2)$
		$\bar{\Sigma}_k = R_k + A_k \Sigma_{k-1} A_k^T$ *(covariance of the predicted pose)*	$O(d^3)$ *(see note[3])*
	Prediction stage - observation	$\bar{\mu}_{z_k} = C_k \bar{\mu}_k$ *(mean of the predicted observation)*	$O(cd)$
		$S_k = Q_k + C_k \bar{\Sigma}_k C_k^T$ *(covariance of the predicted observation)*	$O(cd^2 + c^2 d)$
Update stage	Update stage - intermediate matrices	$\bar{\Sigma}_{po,k} = \bar{\Sigma}_k C_k^T$ *(cross-covariance between predicted pose and predicted observation)*	$O(cd^2)$
		$K_k = \bar{\Sigma}_{po,k} S_k^{-1}$ *(KALMAN gain)*	$O(c^2 d + c^3)$ *(see note[4])*
	Update stage - posterior	$\mu_k = \bar{\mu}_k + K_k \left(z_k - \bar{\mu}_{z_k} \right)$ *(mean of the posterior)*	$O(dc)$
		$\Sigma_k = \left(I - K_k C_k \right) \bar{\Sigma}_k$ *(covariance of the posterior)*	$O(cd^2 + d^3)$ *(see note[5])*

posterior pose of the robot $(\mu_k = \mu_u)$ is the expected predicted one $(\bar{\mu}_k)$ corrected through the Kalman gain by the difference existing between the gathered observation $(\mathbf{z}_k)$ and the expected predicted observation $(\bar{\mu}_{z_k})$. That is, the *residual error* of the KF, existing between the observed and expected data, i.e., $\mathbf{z}_k - \bar{\mu}_{z_k}$, serves to correct the prediction of the pose, to a greater extent if the Kalman gain is large, that is, if the uncertainty in the observation and/or the uncertainty in the predicted pose are small. On the contrary, if the gathered observation agrees with the expected one or if the uncertainties in the observation and/or predicted pose are large, then *the posterior pose follows closely the pose predicted by the motion only, and the observation has little influence in revising our belief.*

Now the derivation of the filter is complete. The Kalman filter can be summarized in a few closed-form equations, shown in Table 1. This table can also serve as a guide to implement the KF algorithm.

Table 1 also shows the computational complexity of the KF for the case that the system state is a $d \times 1$ vector and the observation is a $c \times 1$ vector. The total complexity is then $O(cd^2 + c^2d + c^3 + d^3)$, which is typically dominated by $O(c^3)$ since the number of observations is typically larger than the dimensionality of the state vector. Except for very large-scale problems, this computational complexity is more than reasonable: the Kalman filter and most of its derivatives are considered, in general, quite efficient as long as their computational complexity does not grow excessively over time. In robot localization, as long as the number of observations remains bounded (i.e. there exists a maximum) it will be as well the time taken by the KF —as we will see in chapter 9, this is not the case for the problem of SLAM. Also, notice that the complexity analyses we will perform in this chapter do not take into account solving the problem of data association, which should be addressed after obtaining the prediction-stage mean and covariance ($\bar{\mu}_k$ and $\bar{\Sigma}_k$, respectively) and before the update stage.

The KF is optimal in the least-squares sense. Actually, it is often derived not from the Bayesian framework, as we have done here, but from the problem of minimizing the mean squared error to the actual pose of the robot. Therefore, the KF is the MMSE explained in chapter 4 and, as discussed there, it is also asymptotically equivalent to the classical—non-Bayesian—LSE (Least-Squares Estimator), asymptotically unbiased and asymptotically efficient, that is, i.e., the Minimum Variance Unbiased Estimator (MVUE).

These nice properties, however, clash with reality: we would only achieve optimality if the real world was populated by perfectly linear processes with Gaussian uncertainties. The robot awakening problem (global localization), for instance, is difficult to address with the KF, since the initial belief is not Gaussian (and even less it is the observation model under such circumstances!). Another issue with the KF is its incapability of coping with a number of different hypotheses for the robot position at once. We are also in trouble with the robot kidnapping problem: if the initial belief was centered at a position very different from the one the robot has been moved to, it would be very difficult for the filter to engage to the actual pose, that is, the filter will diverge. The problem in this case would be actually due to incorrect data associations, but the ultimate cause behind that lies at the impossibility of keeping multimodal distributions with a standard KF.

The only practical use in mobile robot localization of the KF as is would be *tracking* the robot pose, that is, performing continuous localization, but only when the uncertainty is limited to a small (and unimodal) region of the state space. The KF could be directly applied to that problem by defining the initial belief to be a Gaussian centered at the starting, known robot pose. Unfortunately, there exist two important limitations that hamper the application of KF to pose tracking: (1) the linearity assumption (as seen in previous chapters, the motion and observation models are far from being linear) and (2) the fact that unmodeled obstacles often lead to pose belief distributions far from being unimodal and Gaussian. In short: the KF as such is not used in mobile robot localization and needs to be extended to cope with, at least, non-linear models.

The Extended Kalman Filter (EKF)

The simplest and also most used variation of the KF in mobile robotics is the extended Kalman filter. In mobile robot localization, the EKF appeared for the first time in (Smith & Cheeseman, 1986).

The idea behind the EKF is quite simple: to approximate the non-linear motion and observation models by linear functions that are close to the actual ones around the current estimated location of the robot. The assumption of Gaussianity for the filter prior and posterior distributions remains without changes.

Note that by linearizing the models, the filter gets apart from the perfect Bayesian solution to localization, since we are no longer estimating *that* problem, but a similar (linearized) one. In other words, the EKF estimates the wrong parameters in an optimal way (Schön, 2006). However, when the filter is updated with a high enough frequency, so that the linearizations are kept close to the real non-linear models continuously, its results are quite accurate, while not losing much of the computational efficiency of the original KF.

The EKF estimates the state of a Gaussian, non-linear system defined with the following SSM:

$$\mathbf{x}_k = \mathbf{g}(\mathbf{x}_{k-1}, \mathbf{u}_k) + \varepsilon_k$$
$$\mathbf{z}_k = \mathbf{h}(\mathbf{x}_k) + \delta_k$$

where

$$\mathbf{g} : \mathbb{R}^{d+u} \rightarrow \mathbb{R}^d, \text{ non-linear function}$$

$$\mathbf{h} : \mathbb{R}^d \rightarrow \mathbb{R}^c, \text{ non-linear function}$$

$$\varepsilon_k \sim N(\mathbf{x}; 0, \mathbf{R}_k)$$

$$\delta_k \sim N(\mathbf{x}; 0, \mathbf{Q}_k)$$

$$\varepsilon_k \perp \delta_k \tag{35}$$

The linearization in the EKF is based on the two first terms of the Taylor expansion of the non-linear functions g and h—refer to Appendix E. For approximating the system of Equations 35 the EKF linearizes the non-linear part of the motion model, that is, the function $\mathbf{g}(\mathbf{x}_{k-1}, \mathbf{u}_k)$—among all the possible models described in chapter 5—around the most likely point of its domain: the expected pose from the last step, that is, $\mathbf{x}_{k-1} = \mu_{k-1}$, using whatever value we have measured for the action $\mathbf{u}_k$, that becomes here a fixed parameter for the function:

$$\begin{aligned}\mathbf{g}(\mathbf{x}_{k-1}, \mathbf{u}_k) &\approx \underbrace{\mathbf{g}(\mathbf{x}_{k-1}, \mathbf{u}_k)\Big|_{\mathbf{x}_{k-1}=\mu_{k-1}}}_{\mathbf{g}_k \text{ (a constant during step } k)} \\ &\quad + \underbrace{\mathbf{J}_{\mathbf{g},\mathbf{x}_{k-1}}(\mathbf{x}_{k-1}, \mathbf{u}_k)\Big|_{\mathbf{x}_{k-1}=\mu_{k-1}}}_{\mathbf{G}_k \text{ (a constant matrix during step } k)} (\mathbf{x}_{k-1} - \mu_{k-1}) \\ &= \mathbf{g}_k + \mathbf{G}_k(\mathbf{x}_{k-1} - \mu_{k-1})\end{aligned} \tag{36}$$

Therefore, $\mathbf{x}_k \approx \mathbf{g}_k + \mathbf{G}_k(\mathbf{x}_{k-1} - \mu_{k-1}) + \varepsilon_k$. Since the noise ε_k is symmetric around $\mathbf{0}$, this corresponds to a Gaussian centered at the rest of terms:

$$p(\mathbf{x}_k | \mathbf{x}_{k-1}, \mathbf{u}_k) = N(\mathbf{x}_k; \mathbf{g}_k + \mathbf{G}_k(\mathbf{x}_{k-1} - \mu_{k-1}), \mathbf{R}_k) \tag{37}$$

Again, we must remark that this corresponds to the transition model assuming a perfectly known pose in the last time step, $\mathbf{x}_{k-1}$. We could then derive the prediction step of the EKF from this, step by step, in the same way as we did for the KF (we leave that as an exercise for *very* patient readers). However, we can distinguish quickly two mappings from the KF to the EKF: in the EKF, $\mathbf{G}_k$ substitutes the $\mathbf{A}_k$ matrix of the KF, and the expression $\mathbf{g}_k + \mathbf{G}_k(\mathbf{x}_{k-1} - \mu_{k-1})$ plays the same role as $\mathbf{B}_k\mathbf{u}_k$ in the KF. Then, we can directly use both mappings in the final result of Equations 23 and 24 for the prediction stage of the filter, obtaining:

$$\begin{aligned}\bar{\mu}_k &= \mathbf{g}_k \\ \bar{\Sigma}_k &= \mathbf{R}_k + \mathbf{G}_k \Sigma_{k-1} \mathbf{G}_k^T\end{aligned} \tag{38}$$

In addition to the linearized transition model of Equation 36, the EKF also linearizes the non-linear part of the observation model, $\mathbf{h}(\mathbf{x}_k)$—which was described in chapter 6 section 3. The best point to linearize around is the most likely estimate that we have of $\mathbf{x}_k$. Since we have the prediction stage finished, that point is $\bar{\mu}_k$. Using Taylor series again:

$$\begin{aligned}\mathbf{h}(\mathbf{x}_k) \approx &\underbrace{\mathbf{h}(\mathbf{x}_k)\Big|_{\mathbf{x}_k=\bar{\mu}_k}}_{\mathbf{h}_k \text{(a constant during step } k)} \\ &+ \underbrace{\mathbf{J}_{\mathbf{h},\mathbf{x}_k}(\mathbf{x}_k)\Big|_{\mathbf{x}_k=\bar{\mu}_k}}_{\mathbf{H}_k \text{(a constant matrix during step } k)} (\mathbf{x}_k - \bar{\mu}_k) \\ &= \mathbf{h}_k + \mathbf{H}_k(\mathbf{x}_k - \bar{\mu}_k)\end{aligned} \tag{39}$$

The dependence on the map is encoded into the calculation of the value of $\mathbf{h}_k$. This leads to the linearized observation uncertainty:

$$p(\mathbf{z}_k | \mathbf{x}_k, \mathbf{m}) = N(\mathbf{z}_k; \mathbf{h}_k + \mathbf{H}_k(\mathbf{x}_k - \bar{\mu}_k), \mathbf{Q}_k) \tag{40}$$

We can use again the same trick as in the prediction stage. The update stage differs in the EKF with respect to the KF in two terms: in the first one, the matrix $\mathbf{C}_k$ is substituted by $\mathbf{H}_k$, and $\mathbf{z}_k$ becomes $\mathbf{h}_k - \mathbf{H}_k\bar{\mu}_k - \mathbf{z}_k$. Since $\mathbf{C}_k$ and $\mathbf{z}_k$ are indivisible terms in the KF derivation, we can translate these substitutions directly into the final result of the update stage Equations 32 and 34, obtaining:

$$\begin{aligned}\Sigma_k &= \bar{\Sigma}_k - \underbrace{\underbrace{\bar{\Sigma}_k \mathbf{H}_k^T}_{\bar{\Sigma}_{po,k}} \Big(\underbrace{\mathbf{Q}_k + \mathbf{H}_k \bar{\Sigma}_k \mathbf{H}_k^T}_{\text{Innovation Covariance } = \mathbf{S}_k} \Big)^{-1}}_{\text{Kalman Gain } = \mathbf{K}_k} \mathbf{H}_k \bar{\Sigma}_k \\ &= (\mathbf{I} - \mathbf{K}_k \mathbf{H}_k) \bar{\Sigma}_k\end{aligned}$$

$$\mu_k = \bar{\mu}_k + \mathbf{K}_k(\mathbf{z}_k - \mathbf{h}_k) \tag{41}$$

As you can see in these equations, the EKF is very similar in its implementation to the original KF. For example, the new factor $\bar{\Sigma}_k \mathbf{H}_k^T$ in the Kalman gain is the old cross-covariance between the predicted robot pose and the predicted observation, $\bar{\Sigma}_{po,k}$ (which has now an expected value of $\mathbf{h}_k$). The sizes of the involved matrices are shown in Table 2, and the computational cost of the EKF in Table 3, which only differs from the KF in the computation of the two linearized functions. As commented above, the EKF is one of the most popular estimations algorithms in probabilistic robotics and is considered as a standard solution for localization, especially with featured-based maps. Its advantages are its low computational complexity (except for high-dimensional spaces) and its good behavior: it is close to be optimal when the motion and observation models are "smooth" enough and the iteration period of the filter is short. These limits can be somewhat relaxed by repeating the update stage more than once in the same iteration of the filter

Table 2. Sizes of all the matrices involved in the closed-form of the EKF

Matrix	Size
$\boldsymbol{\mu}_{k-1}$, $\bar{\boldsymbol{\mu}}_k$, $\boldsymbol{\mu}_k$, $\mathbf{g}_k$	$d \times 1$ (d being the dimension of the state vector or pose)
$\boldsymbol{\Sigma}_{k-1}$, $\bar{\boldsymbol{\Sigma}}_k$, $\boldsymbol{\Sigma}_k$, $\mathbf{R}_k$, $\mathbf{G}_k$	$d \times d$
$\mathbf{z}_k$, $\bar{\boldsymbol{\mu}}_{z_k}$, $\mathbf{h}_k$	$c \times 1$ (c being the dimension of the observation)
$\mathbf{S}_k$, $\mathbf{Q}_k$	$c \times c$
$\bar{\boldsymbol{\Sigma}}_{po,k}$, $\mathbf{K}_k$	$d \times c$
$\mathbf{H}_k$	$c \times d$

Table 3. Computational complexity of the different steps of the EKF

Algorithm step	Complexity
$\bar{\boldsymbol{\mu}}_k = \mathbf{g}_k$	The one of computing the function, say $O(g)$.
$\bar{\boldsymbol{\Sigma}}_k = \mathbf{R}_k + \mathbf{G}_k \boldsymbol{\Sigma}_{k-1} \mathbf{G}_k^T$	$O(d^3)$ (the same notes as in the KF apply; the cost of computing the Jacobian is d^2 and thus it is dominated by the cubic term)
$\bar{\boldsymbol{\mu}}_{z_k} = \mathbf{h}_k$	$O(h)$ (the cost of computing $\mathbf{h}_k$)
$\mathbf{S}_k = \mathbf{Q}_k + \mathbf{H}_k \bar{\boldsymbol{\Sigma}}_k \mathbf{H}_k^T$	$O(cd^2 + c^2 d)$
$\bar{\boldsymbol{\Sigma}}_{po,k} = \bar{\boldsymbol{\Sigma}}_k \mathbf{H}_k^T$	$O(cd^2)$
$\mathbf{K}_k = \bar{\boldsymbol{\Sigma}}_{po,k} \mathbf{S}_k^{-1}$	$O(c^2 d + c^3)$
$\boldsymbol{\mu}_k = \bar{\boldsymbol{\mu}}_k + \mathbf{K}_k \left(\mathbf{z}_k - \bar{\boldsymbol{\mu}}_{z_k}\right)$	$O(cd)$
$\boldsymbol{\Sigma}_k = \left(\mathbf{I} - \mathbf{K}_k \mathbf{H}_k\right) \bar{\boldsymbol{\Sigma}}_k$	$O(cd^2 + d^3)$
Total (dominant) complexity:	$O(cd^2 + c^2 d + c^3 + d^3 + g + h)$, typically $O(c^3)$.

(what is called *Iterated Extended Kalman Filter*) without acquiring any new observation, or by increasing the number of terms in the Taylor expansions, although these approaches involve additional computational burden.

The EKF, however, has still drawbacks that can be important in practical situations: (1) its linearization approximation can make the filter to diverge from the optimal result if the Taylor expansion becomes a poor approximation for some reason (i.e. a too large covariance or the presence of very strong non-linearities); (2) we have to provide expressions for the Jacobians of two non-linear functions, which can be difficult to obtain sometimes; (3) it has to be tuned, mainly through values for the covariances $\mathbf{R}_k$ and $\mathbf{Q}_k$, which is not straightforward in the real world; and (4) it is unimodal, that is, it does not work well for the robot awakening or robot kidnapping problems (as any Gaussian filter). However, in local localization (pose tracking), it can be really powerful. There exist variants of the EKF in the scientific literature that break the unimodal limitation by considering mixtures of Gaussians, for instance the *Multi-Hypothesis or Extended Mixture Kalman Filter* (EMKF), where the uncertainties are of the form $\sum_{i=1}^{l} \omega_i N(\mathbf{x};\boldsymbol{\mu}_i,\boldsymbol{\Sigma}_i)$ for a given finite set of weights $\{\omega_i\}$ chosen appropriately for the total uncertainty to be a pdf (Chen & Liu, 2000).

Figure 5 shows an example of the execution of an EKF for a real mobile robot that travels along the sides of a rectangle within a known environment. It is a low-cost microrobot by Lego (2011). The known map of the environment is depicted in thick black lines (up). The sonar readings are the point clouds shown in two different levels of gray: lighter for where the robot believes they are if it uses only its odometry for estimating its pose, and darker for where the robot believes the observations are when it estimates its pose through the EKF (observe how the latter fits better with the real environment). The history of poses of the robot according to the EKF, along with some confidence intervals on them, is shown in the center of the figure. A close-up look at the trajectory of the robot, according to the EKF (bottom left). Detail of the trajectory where the robot is following the wall to its left (bottom right). Since the sonar is providing observations of the distance to that wall but not to the up wall, the uncertainty (ellipses in the figure) expands vertically. The observations are simply the distances measured by the sonar sensor (one at each time step), and the likelihood is implemented with a beam model and ray casting (recall chapter 6). You can observe how errors in the estimated pose increase when the robot reaches the corners of the rectangle (it does not see many things there) but decreases again progressively when it turns around and sees again some features. Remember from the derivation of the KF that acquiring observations is the way to decrease uncertainty in the pose estimation.

Another example, in this case corresponding to a simulation, can be seen in Figure 6. Here, an EKF employs a constant-velocity motion model for tracking a robot whose pose must be estimated from just one range-bearing sensor that measures the location of one fixed landmark (arbitrarily at the origin of coordinates). Interestingly, the EKF estimations of the robot position agree well with the ground truth, as can be verified by checking in Figure 6b, c how the latter always falls within the 99% confidence interval of the EKF estimates. The complete equations of this example, along with C++ source code and additional multimedia material, are all available online as part of the MRPT toolkit (EKF Web, 2012).

The Unscented Transform and the Unscented Kalman Filter (UKF)

An interesting alternative for approaching the non-linear parts of the estimation problem in a Kalman context is the *Unscented Kalman Filter* (Julier & Uhlmann, 1997; Wan & Van der Merwe, 2000),

Figure 5. Real example of localization with EKF carried out in a mobile robot endowed with an ultrasound sensor that looks towards its left, suitable for wall-following.

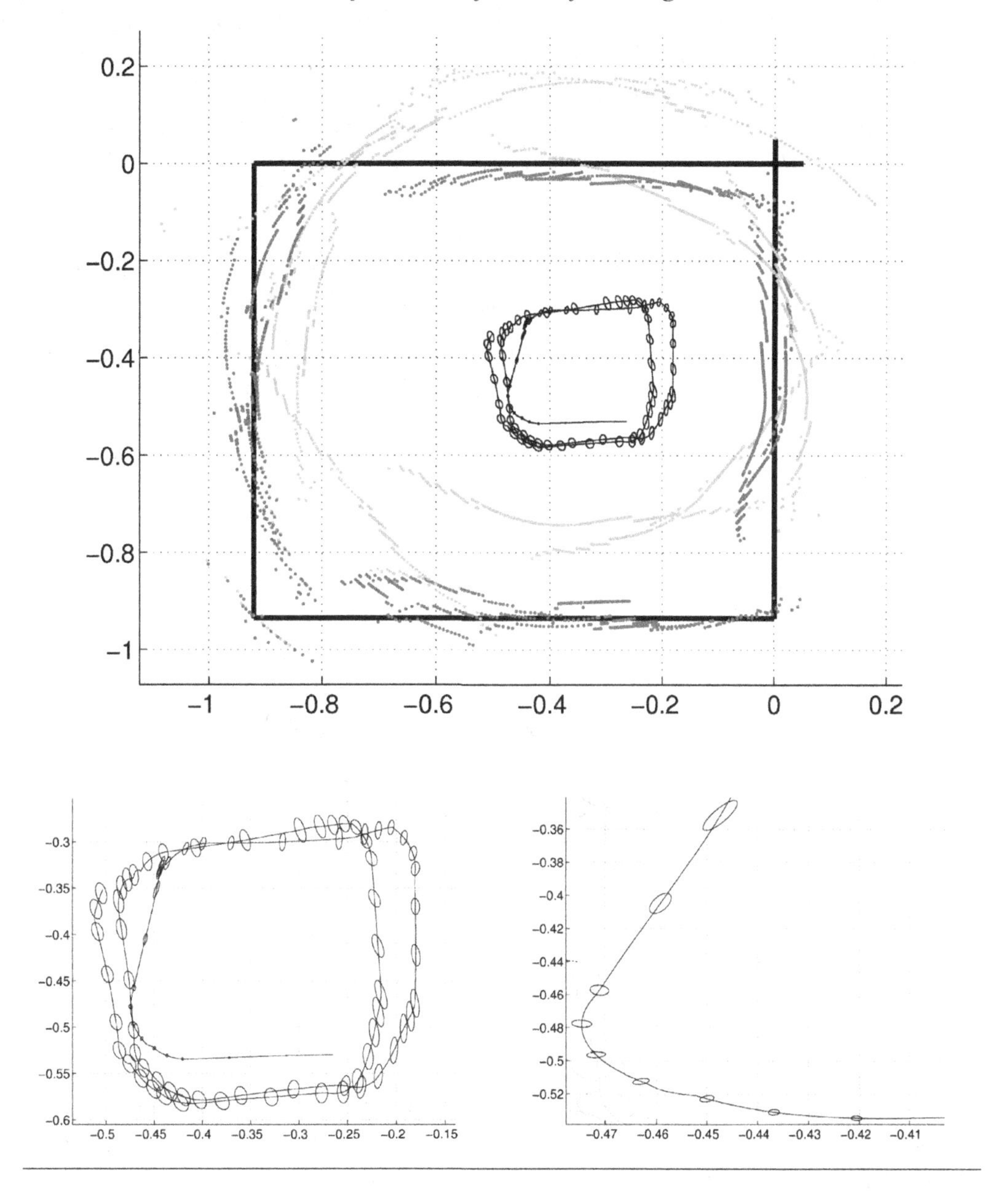

which improves the accuracy of the approximation of the EKF for non-linear models with roughly the same computational cost.

As in the EKF, we will denote the transition function of a non-linear system as $\mathbf{x}_k = \mathbf{g}(\mathbf{x}_{k-1}, \mathbf{u}_k) + \varepsilon_k$, and the observation function as $\mathbf{z}_k = \mathbf{h}(\mathbf{x}_k) + \delta_k$. Both are random (Gaussian) variables. The EKF substitutes them by their truncated first-order linearizations, which does not guarantee to keep the correct moments for the resulting Gaussian.

The UKF, on the contrary, avoids the Taylor linearization and thus also avoids the calculation of the Jacobians, which is a relevant advantage

Figure 6. A simulation illustrating a setup for mobile robot localization (pose tracking) with an EKF. (a) A snapshot of the simulator state, showing the ground truth position of the robot (the dark point), the 99% confidence interval of its position as estimated by the EKF (the ellipse), the most likely velocity of the robot according to the EKF (the thick line from the center of the ellipse) and the latest range-bearing observation of the robot position from the origin of coordinates (the thin line). (b) – (c) A comparison over time of the tracked robot position (in x and y) with the ground truth (thick line). The mean of the EKF estimation is the dashed line while its 99% confidence intervals are shown as shaded areas.

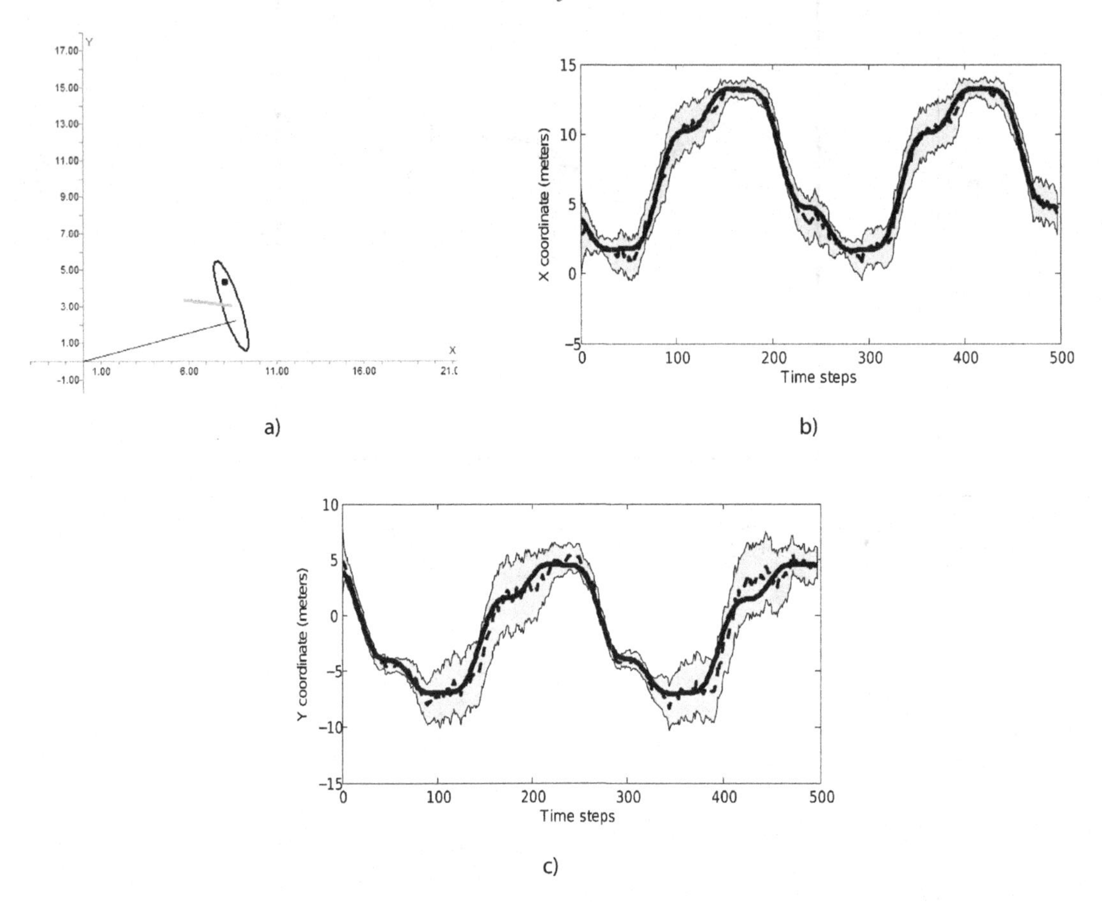

with respect to the EKF: it approximates the original distributions of the r.v.s involved in the non-linear functions by a set of values selected from their support so as to convey information about the first two moments of their distributions. This is a general form of filtering called *sigma-point filtering*, when only those points are actually transformed through the actual non-linear functions $\mathbf{g}(\mathbf{x}_{k-1}, \mathbf{u}_k)$ and $\mathbf{h}(\mathbf{x}_k)$, without any approximation. Notice that this does not involve any significant computational cost, since the functions only need to be evaluated pointwise.

Consequently, the UKF can be studied as being based on an alternative method of transforming a random variable through a non-linear function, different from those we introduced in chapter 3 section 8. This is called the *Unscented Transform* (UT), and we describe it next.

Assume we have a random variable $\mathbf{x}$ with dimensionality d and a non-linear function that maps it into another r.v., say $\mathbf{y} = \mathbf{f}(\mathbf{x})$. The first two moments of $\mathbf{x}$ are given by its mean $\mu_{\mathbf{x}}$ (a column vector) and its covariance matrix $\Sigma_{\mathbf{x}}$. The latter is typically a positive-definite matrix (only exceptionally it will be positive-semidefinite), thus we can find a Cholesky factorization of such matrix (refer to Appendix E for further details):

$$\Sigma_{\mathbf{x}} = \mathbf{L}\mathbf{L}^T \text{ (that is, } \sqrt{\Sigma_{\mathbf{x}}} = \mathbf{L}\text{)} \tag{42}$$

This means that we can "condense" all the information of $\Sigma_{\mathbf{x}}$ in the non-zero entries of the lower triangular matrix $\mathbf{L}$

Now, we are interested in finding the first and second moments of the transformed r.v. $\mathbf{y}$, that is, $\mu_{\mathbf{y}}$ and $\Sigma_{\mathbf{y}}$. What the UT does is to select the following $2d+1$ values[6] from the distribution of $\mathbf{x}$, called *sigma-points,* which preserve the mean and covariance of $\mathbf{x}$ in a statistical sense. If we denote the $k-th$ column of a matrix $\mathbf{A}$ as $\left[\mathbf{A}\right]_{*,k}$ and the entry at the $j-th$ row and $k-th$ column of $\mathbf{A}$ as $\left[\mathbf{A}\right]_{j,k}$, we can write the equation for the $2d+1$ sigma-points as:

$$\chi_i = \begin{cases} \mu_{\mathbf{x}} & \text{, for } i = 0 \\ \mu_{\mathbf{x}} + \left[\sqrt{(d+\lambda)\Sigma_{\mathbf{x}}}\right]_{*,i} & \text{, for } i \in \left[1, d\right] \\ \mu_{\mathbf{x}} - \left[\sqrt{(d+\lambda)\Sigma_{\mathbf{x}}}\right]_{*,(i-d)} & \text{, for } i \in \left[d+1, 2d\right] \end{cases} \tag{43}$$

where λ is a real-valued parameter whose purpose will be clarified later on.

Recall from chapter 4 that the sample mean of a set of observations $S = \left\{\mathbf{s}_1, \mathbf{s}_2, \ldots, \mathbf{s}_n\right\}$ is an unbiased statistical estimator of the mean of the distribution that is generating that sample, and it is calculated as $\mu_S = \frac{1}{n}\sum_{i=1}^{n}\mathbf{s}_i$. The sigma-points preserve the first moment of the original distribution, because their sample mean can be seen to match $\mu_{\mathbf{x}}$:

$$\begin{aligned} \mu_{\chi} &= \frac{1}{2d+1}\sum_{i=0}^{2d}\chi_i = \frac{1}{2d+1}\left(\mu_{\mathbf{x}} + \sum_{i=1}^{d}\chi_i + \sum_{i=d+1}^{2d}\chi_i\right) \\ &= \frac{1}{2d+1}\left(\mu_{\mathbf{x}} + \sum_{i=1}^{d}\left(\mu_{\mathbf{x}} + \left[\sqrt{(d+\lambda)\Sigma_{\mathbf{x}}}\right]_{*,i}\right) + \right. \\ &\qquad \left. + \sum_{i=1}^{d}\left(\mu_{\mathbf{x}} - \left[\sqrt{(d+\lambda)\Sigma_{\mathbf{x}}}\right]_{*,i}\right)\right) \\ &= \frac{1}{2d+1}\left(\mu_{\mathbf{x}} + d\mu_{\mathbf{x}} + \sum_{i=1}^{d}\left[\sqrt{(d+\lambda)\Sigma_{\mathbf{x}}}\right]_{*,i} + \right. \\ &\qquad \left. + d\mu_{\mathbf{x}} - \sum_{i=1}^{d}\left[\sqrt{(d+\lambda)\Sigma_{\mathbf{x}}}\right]_{*,i}\right) \\ &= \frac{1}{2d+1}\left(\mu_{\mathbf{x}} + d\mu_{\mathbf{x}} + d\mu_{\mathbf{x}}\right) \\ &= \mu_{\mathbf{x}} \end{aligned} \tag{44}$$

Similarly, we can define the sample covariance of a sample $S = \left\{\mathbf{s}_1, \mathbf{s}_2, \ldots, \mathbf{s}_n\right\}$ as $\Sigma_S = \frac{1}{n-1}\sum_{i=1}^{n}\left(\mathbf{s}_i - \mu_S\right)\left(\mathbf{s}_i - \mu_S\right)^T$, which is an unbiased statistical estimator of the covariance of the distribution that is generating that sample. The sigma-points preserve the covariance of the original variable because their sample covariance is proportional to the former:

$$\Sigma_{\chi} = \frac{d+\lambda}{d}\Sigma_{\mathbf{X}} \tag{45}$$

which can be demonstrated as follows:

Figure 7. Illustration of the UT. (Left) The pdf of a scalar Gaussian r.v. with mean 5 and variance 1. Three sigma-points are selected (with $d = 1, \lambda = 2$*) and then transformed by the non-linear function* $f(x) = 2x^3$*. (Right) The continuous, asymmetric black curve is the actual pdf of the r.v. resulting from the transformation, calculated numerically with high accuracy. .*

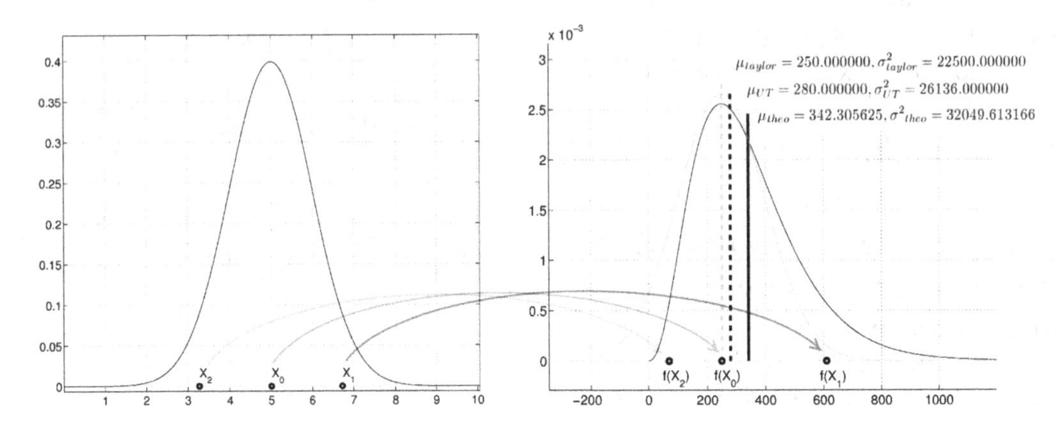

$$
\begin{aligned}
\boldsymbol{\Sigma}_{\mathbf{X}} &= \frac{d}{d+\lambda}\boldsymbol{\Sigma}_{\mathbf{x}} \\
&= \frac{d}{d+\lambda}\frac{1}{2d}\sum_{i=0}^{2d}\left(\chi_i - \mu_{\mathbf{x}}\right)\left(\chi_i - \mu_{\mathbf{x}}\right)^T \\
&= \frac{1}{2(d+\lambda)}\sum_{i=1}^{2d}\left(\chi_i - \mu_{\mathbf{x}}\right)\left(\chi_i - \mu_{\mathbf{x}}\right)^T \\
&= \frac{1}{2(d+\lambda)}\left(\sum_{i=1}^{d}\left(\chi_i - \mu_{\mathbf{x}}\right)\left(\chi_i - \mu_{\mathbf{x}}\right)^T + \right. \\
&\qquad \left. + \sum_{i=d+1}^{2d}\left(\chi_i - \mu_{\mathbf{x}}\right)\left(\chi_i - \mu_{\mathbf{x}}\right)^T\right) \\
&= \frac{1}{2(d+\lambda)}\left(\sum_{i=1}^{d}\left[\sqrt{(d+\lambda)\boldsymbol{\Sigma}_{\mathbf{x}}}\right]_{*,i}\left[\sqrt{(d+\lambda)\boldsymbol{\Sigma}_{\mathbf{x}}}\right]_{*,i}^T\right. \\
&\qquad \left. + \sum_{i=1}^{d}\left[-\sqrt{(d+\lambda)\boldsymbol{\Sigma}_{\mathbf{x}}}\right]_{*,i}\left[-\sqrt{(d+\lambda)\boldsymbol{\Sigma}_{\mathbf{x}}}\right]_{*,i}^T\right) \\
&= \frac{1}{2(d+\lambda)}\left((d+\lambda)\sum_{i=1}^{d}\left[\sqrt{\boldsymbol{\Sigma}_{\mathbf{X}}}\right]_{*,i}\left[\sqrt{\boldsymbol{\Sigma}_{\mathbf{X}}}\right]_{*,i}^T\right. \\
&\qquad \left. +(d+\lambda)\sum_{i=1}^{d}\left[\sqrt{\boldsymbol{\Sigma}_{\mathbf{X}}}\right]_{*,i}\left[\sqrt{\boldsymbol{\Sigma}_{\mathbf{X}}}\right]_{*,i}^T\right)
\end{aligned}
$$

(simplifying and using the auxiliary result of section 3 in Appendix E)

$$
\begin{aligned}
\boldsymbol{\Sigma}_{\mathbf{X}} &= \frac{2(d+\lambda)}{2(d+\lambda)}\sum_{i=1}^{d}\left[\sqrt{\boldsymbol{\Sigma}_{\mathbf{X}}}\right]_{*,i}\left[\sqrt{\boldsymbol{\Sigma}_{\mathbf{X}}}\right]_{*,i}^T \\
&= \sum_{i=1}^{d}\left[\sqrt{\boldsymbol{\Sigma}_{\mathbf{X}}}\right]_{*,i}\left[\sqrt{\boldsymbol{\Sigma}_{\mathbf{X}}}\right]_{*,i}^T \\
&= \boldsymbol{\Sigma}_{\mathbf{X}}
\end{aligned}
\tag{46}
$$

As can be deduced from this expression, the parameter λ, which has no effect in the mean, is a way of condensing or spreading the sigma-points around that mean. Actually, this parameter is a heuristic with a typical value of $\lambda = 3 - d$ which serves for adjusting the approximation errors implied by the unscented transform.

Once we have selected the sigma-points in such a way that they preserve the first two statistical moments of the original r.v., the UT continues by propagating them through the non-linear function $\mathbf{y} = \mathbf{f}(\mathbf{x})$ to obtain values of the support of the new r.v.: $\gamma_i = \mathbf{f}(\chi_i)$. An approximation to the mean and covariance of $\mathbf{y}$ can be recovered through the sample mean and sample covariance of the resulting set of values (it can also be shown that the UT can approximate higher moments of the transformed variable, being clearly superior to Taylor first-order linearization). The procedure is the following:

$$\mu_y \approx \sum_{i=0}^{2d} \nu_i \gamma_i$$
$$\Sigma_y \approx \sum_{i=0}^{2d} \nu_i \left(\gamma_i - \mu_i\right)\left(\gamma_i - \mu_i\right)^T$$

where the weights are defined as:

$$\nu_i = \begin{cases} \dfrac{\lambda}{d+\lambda} & \text{, for } i = 0 \\ \dfrac{1}{2(d+\lambda)} & \text{, for } i \in \left[1, 2d\right] \end{cases}$$

such that:

$$\sum_{i=0}^{2d} \nu_i = \frac{\lambda}{d+\lambda} + \frac{2d}{2(d+\lambda)} = 1 \qquad (47)$$

Figure 7 illustrates the use of the unscented transform to obtain the first two moments of an r.v. which is the result of transforming a Gaussian r.v. through a non-linear function. The mean and variance of that r.v. have also been calculated as a reference (the mean is plotted as a vertical, thick continuous line). The transformed sigma points are indicated in the support of the new r.v. with circles. The thick, discontinuous vertical line shows the approximation to the mean of the transformed r.v. by using the UT. The gray discontinuous vertical line shows the approximation to the same mean by using a Taylor expansion with two terms (as in the EKF). The gray continuous curve is the Gaussian pdf corresponding to the Taylor approximation of the transformed r.v

As mentioned earlier, the unscented transform can be used in the non-linear parts of the Kalman filter equation as an alternative to the linearization employed in the EKF. In particular, it can be applied to approximating the Gaussians resulting from the non-linear transformations $\mathbf{g}(\mathbf{x}_{k-1}, \mathbf{u}_k)$ and $\mathbf{h}(\mathbf{x}_k)$ of the filter, which appear in the prediction and update stages, respectively.

The prediction stage of the Kalman filter obtains a preliminary estimate of the new pose of the robot, $\mathbf{x}_k$, using only knowledge about its previous pose (estimate) $\mathbf{x}_{k-1}$ and the action performed, $\mathbf{u}_k$. In Equation 23 of the KF we saw that the predicted covariance is the sum of the one of $\mathbf{x}_{k-1}$ propagated through the linear process, plus the covariance of the motion, $\mathbf{R}_k$. In the UKF we can remake this process by substituting the entire prediction stage of the KF as follows:

Selection of sigma-points from $\mathbf{x}_{k-1}$:

$$\boldsymbol{\sigma}_{k-1} = \left\{\boldsymbol{\sigma}_{k-1}^{[i]}\right\}, \text{ with } i = 1, \ldots, 2d$$

$$\boldsymbol{\sigma}_{k-1}^{[i]} = \begin{cases} \boldsymbol{\mu}_{k-1} & \text{, for } i = 0 \\ \boldsymbol{\mu}_{k-1} + \left[\sqrt{(d+\lambda)\boldsymbol{\Sigma}_{k-1}}\right]_{\bullet,i} & \text{, for } i \in \left[1, d\right] \\ \boldsymbol{\mu}_{k-1} - \left[\sqrt{(d+\lambda)\boldsymbol{\Sigma}_{k-1}}\right]_{\bullet,(i-d)} & \text{, for } i \in \left[d+1, 2d\right] \end{cases}$$

Transformation of the sigma-points through the non-linear function(propagation of the previous pose):

$$\gamma_i = \mathbf{g}\left(\boldsymbol{\sigma}_{k-1}^{[i]}, \mathbf{u}_k\right)$$

Approximation of the resulting Gaussian (including the uncertainty of the motion, $\mathbf{R}_k$):

$$\bar{\boldsymbol{\mu}}_k \approx \sum_{i=0}^{2d} \boldsymbol{\nu}_i \gamma_i$$
$$\bar{\boldsymbol{\Sigma}}_k \approx \sum_{i=0}^{2d} \boldsymbol{\nu}_i \left(\gamma_i - \bar{\boldsymbol{\mu}}_k\right)\left(\gamma_i - \bar{\boldsymbol{\mu}}_k\right)^T + \mathbf{R}_k \qquad (48)$$

Similarly, the update stage of the Kalman filter involves obtaining, in a first step, the distribution of the observation given the predicted pose

$\bar{\mu}_k$ and its covariance $\bar{\Sigma}_k$ by means of the non-linear function $\mathbf{h}(\mathbf{x}_k)$. In Equation 29 we saw that the resulting covariance (the innovation covariance) is the sum of the predicted covariance $\bar{\Sigma}_k$ propagated by the linear observation function plus the observation covariance $\mathbf{Q}_k$. Consequently, in the UKF we can use the UT for replicating that reasoning:

Selection of sigma-points from the prediction $\bar{\mathbf{x}}_k$:

$$\bar{\sigma}_k = \left\{\bar{\sigma}_k^{[i]}\right\}, \text{ with } i = 1, \ldots, 2d$$

$$\bar{\sigma}_k^{[i]} = \begin{cases} \bar{\mu}_k & , \text{ for } i = 0 \\ \bar{\mu}_k + \left[\sqrt{(d+\lambda)\bar{\Sigma}_k}\right]_{*,i} & , \text{ for } i \in [1, d] \\ \bar{\mu}_k - \left[\sqrt{(d+\lambda)\bar{\Sigma}_k}\right]_{*,(i-d)} & , \text{ for } i \in [d+1, 2d] \end{cases}$$

Transformation of the sigma-points by means of the non-linear observation function:

$$\bar{\gamma}_i = \mathbf{h}\left(\bar{\sigma}_k^{[i]}\right)$$

Approximation of the resulting Gaussian (including the observation covariance $\mathbf{Q}_k$):

$$\bar{\mu}_{z_k} = \sum_{i=0}^{2d} \nu'_i \bar{\gamma}_i$$
$$\mathbf{S}_k = \sum_{i=0}^{2d} \nu'_i \left(\bar{\gamma}_i - \bar{\mu}_{z_k}\right)\left(\bar{\gamma}_i - \bar{\mu}_{z_k}\right)^T + \mathbf{Q}_k \tag{49}$$

Also, we saw in the Kalman filters that the Kalman gain is the cross-covariance $\bar{\Sigma}_{po,k}$ between the predicted robot pose and the predicted observation multiplied by the inverse of the innovation covariance $\mathbf{S}_k$. Since the expectations of both the predicted pose and the predicted observation are already calculated, we can do:

(approximating the cross-covariance between the predicted pose and the predicted observation)

$$\bar{\Sigma}_{po,k} = \sum_{i=0}^{2d} \nu''_i \left(\gamma_i - \bar{\mu}_k\right)\left(\bar{\gamma}_i - \bar{\mu}_{z_k}\right)^T$$

with the Kalman gain being:

$$\mathbf{K}_k = \bar{\Sigma}_{po,k} \mathbf{S}_k^{-1} \tag{50}$$

The mean of the posterior can now be calculated as in the EKF:

$$\mu_k = \bar{\mu}_k + \mathbf{K}_k \left(\mathbf{z}_k - \bar{\mu}_{z_k}\right) \tag{51}$$

However, the posterior covariance, as defined in Equation 41, cannot be used directly since it involves the Jacobian $\mathbf{H}_k$,

i.e., $\Sigma_k = \left(\mathbf{I} - \mathbf{K}_k \mathbf{H}_k\right) \bar{\Sigma}_k$.

This has a simple workaround, since by the definition of the Kalman gain we can derive the following:

$$\mathbf{K}_k = \bar{\Sigma}_{po,k} \mathbf{S}_k^{-1} \quad \leftrightarrow$$

(transposing both sides)

$$\mathbf{K}_k^T = \left(\mathbf{S}_k^{-1}\right)^T \bar{\Sigma}_{po,k}^T \quad \leftrightarrow$$

(by exchangeability of transpose and inverse)

$$\mathbf{K}_k^T = \left(\mathbf{S}_k^T\right)^{-1} \bar{\Sigma}_{po,k}^T \quad \leftrightarrow$$

Table 4. Sequence of steps that implement the UKF, with their corresponding computational costs

Algorithm step	Complexity
Selection of sigma-points from $\mathbf{x}_{k-1}$	$O(d^3)$ for the Cholesky decomposition of Σ_{k-1}. $O(d^2)$ for the calculation of the $2d+1$ sigma-points.
Propagation of those sigma-points through $\mathbf{g}$	$O(dg)$ where the computational cost of $\mathbf{g}$ is $O(g)$.
Computation of the mean of the predicted pose, $\bar{\mu}_k$	$O(d^2)$
Computation of the covariance of the predicted pose, $\bar{\Sigma}_k$	$O(d^3)$
Selection of the sigma-points from the predicted pose $\bar{\mathbf{x}}_k$	$O(d^3)$ (the same as the selection of sigma-points from $\mathbf{x}_{k-1}$)
Propagation of those sigma-points through $\mathbf{h}$	$O(dh)$ where the computational cost of $\mathbf{h}$ is $O(h)$.
Computation of the mean of the predicted observation, $\bar{\mu}_{z_k}$	$O(cd)$
Computation of the covariance of the predicted observation, $\mathbf{S}_k$	$O(c^2d)$
Computation of the cross-covariance between the predicted pose and the predicted observation, $\bar{\Sigma}_{po,k}$	$O(d^2c)$
Computation of the Kalman gain, $\mathbf{K}_k$	$O(c^2d + c^3)$
Computation of the mean of the posterior, μ_k	$O(cd)$
Computation of the covariance of the posterior, Σ_k	$O(d^2 + dc^2 + cd^2)$
Total (dominant) complexity:	$O(cd^2 + c^2d + c^3 + d^3 + dg + dh)$, typically $O(c^3)$.

(pre-multiplying both sides by $\mathbf{S}_k^T$)

$$\mathbf{S}_k^T\mathbf{K}_k^T = \bar{\mathbf{\Sigma}}_{po,k}^T \quad \leftrightarrow$$

(and since the innovation covariance is a covariance and thus a symmetric matrix)

$$\mathbf{S}_k\mathbf{K}_k^T = \bar{\mathbf{\Sigma}}_{po,k}^T \tag{52}$$

This can be used to obtain a new form of the posterior covariance that does not include the Jacobian of the observation model:

$$\begin{aligned}\mathbf{\Sigma}_k &= \bar{\mathbf{\Sigma}}_k - \underbrace{\bar{\mathbf{\Sigma}}_k\mathbf{H}_k^T\left(\mathbf{Q}_k + \mathbf{H}_k\bar{\mathbf{\Sigma}}_k\mathbf{H}_k^T\right)^{-1}}_{\text{Kalman Gain } = \mathbf{K}_k}\underbrace{\mathbf{H}_k\bar{\mathbf{\Sigma}}_k}_{\bar{\mathbf{\Sigma}}_{po,k}^T} \\ &= \bar{\mathbf{\Sigma}}_k - \underbrace{\bar{\mathbf{\Sigma}}_{po,k}\mathbf{S}_k^{-1}}_{\mathbf{K}_k}\underbrace{\bar{\mathbf{\Sigma}}_{po,k}^T}_{\substack{\mathbf{S}_k\mathbf{K}_k^T \\ \text{(equation (53))}}} \\ &= \bar{\mathbf{\Sigma}}_k - \mathbf{K}_k\mathbf{S}_k\mathbf{K}_k^T\end{aligned} \tag{53}$$

which is the one to include in the last step of the UKF.

Table 4 summarizes the computational cost of the UKF. As you can see, the differences with the EKF are only the terms $O(dg)$ and $O(dh)$, since the UKF propagates more than one point through the non-linear motion and observation models. The overall complexity is usually still dominated by $O(c^3)$ when applied to the problem of robot localization, where the problem dimensionality, the d, remains constant.

3. NON-PARAMETRIC FILTERS FOR LOCALIZATION

In the previous section, we have approached the Bayesian problem of estimating the pose of the robot by pursuing a closed-form solution for the recursive filter. That is, instead of estimating the entire shape of the pose uncertainty in a general case we pursued estimating only its parameters after assuming it has some particular mathematical form. This was possible at the expense of constraining the uncertainties to be Gaussians, since a Gaussian prior is one of the few probability distributions that are conjugate with respect to themselves as posteriors and thus fits well into a recursive framework. The main disadvantage of restricting ourselves to a particular shape, concretely an unimodal one, is the inability to cope with multi-hypothesis settings, for instance to represent the belief of being located within a large region of space (the robot awakening problem) or in a number of indistinguishable spots—something familiar to anyone that works in structured environments with highly symmetric structures, like many office buildings.

The good news is that there exists a way to cope with arbitrarily complex shapes of uncertainty in a recursive Bayesian filter, even including non-linear motion and observation models. The bad news is that those solutions come at the price of being approximations whose accuracy can be improved by increasing the computational cost. Nevertheless, due to the current power of computers and their good results in practice, they represent a compelling approach and, in fact, are widely employed nowadays in mobile robotics.

The rest of this section introduces the main representatives of non-parametric recursive Bayesian filters for solving mobile robot localization. As in section 2, we will start with a not much practical filter, which will make easier to understand the more complex ones discussed next.

The Discrete Bayes Filter (DBF)

Clearly, the main issue with the recursive Bayesian filter for mobile robot localization of Equation 4 is the integral, which does not allow us to obtain closed-form solutions for an arbitrary shape of uncertainty. That problem, however, totally vanishes when we consider discrete random variables, working, correspondingly, with the so-called *Discrete Bayes Filter* (DBF):

$$bel(\mathbf{x}_k | \mathbf{z}_{1:k}, \mathbf{u}_{1:k}, \mathbf{m}) \propto \\ \propto \underbrace{\underbrace{p(\mathbf{z}_k | \mathbf{x}_k, \mathbf{m})}_{\text{observation likelihood}} \underbrace{\sum_{\mathbf{x}_{k-1}} P(\mathbf{x}_k | \mathbf{x}_{k-1}, \mathbf{u}_k) bel(\mathbf{x}_{k-1} | \mathbf{z}_{1:k-1}, \mathbf{u}_{1:k-1}, \mathbf{m})}_{\text{prediction}}}_{\text{update}} \tag{54}$$

In this formulation all the uncertainty is modeled by pmfs, except the observation likelihood, which can be also modeled as a continuous pdf ($p(\mathbf{z}_k | \mathbf{x}_k, \mathbf{m})$ instead of $P(\mathbf{z}_k | \mathbf{x}_k, \mathbf{m})$) since it will be evaluated only pointwise for a given map, a hypothetical robot pose and an observation.

Like parametric filters, the DBF in Equation 54 can clearly be split into two stages: a prediction (the sum) and an update (its multiplication by the observation likelihood), the latter adjusting the prediction by taking into account the observed data. Unlike parametric filters, here we do not need any approximation if the transition or observation models are non-linear, which also permits any type of (discrete) probability distribution for all its terms.

The obvious drawback of this approach is its limitation to problems with a discrete domain. The only such problems that naturally arise in mobile robotics are related to topological maps: maps composed of a finite number of places (regions) connected by some kind of spatial relation, usually modeled by graphs. In these maps, a "robot pose" is just one of the places included in the map, and the transition function represents the probability of moving from a region to another, while the observation likelihood must recognize one of those regions, typically by means of sophisticated recognition algorithms that rely on rich sensory information; these are the so-called *appearance-based* approaches (Ulrich & Nourbakhsh, 2000; Cummins & Newman, 2008).

The issue with topological localization is that the most common methods for constructing and maintaining such kind of maps use metrical information to deduce the topology of the environment—a remarkable exception is the work reported by Kuipers (1977) and his collaborators. But once the robot has the metrical information, it could use any non-discrete filter to estimate its location far more accurately and with much less sensory effort than if it waits for the topology to be ready and then uses a DBF. This is the main reason why few methods proposed in the literature rely on a pure DBF approach.

The Histogram Filter (HF)

The DBF is the basis for another more successful filter: the *grid-based* or *Histogram Filter* (HF). The idea of the HF is simple: approximate a continuous state space by a tessellation of that space, which is discrete and therefore permits us to use a DBF. In that way we obtain the benefits of both the discrete formulation and of continuous sensory and motion information; furthermore, we can regulate the accuracy of the approximation by changing the dimensions of the spatial elements of the tessellation. Actually, the HF became a successful solution to practical mobile robot localization in the late 90s, under the name of *Markov localization* (Fox, Burgard, & Thrun, 1999).

Commonly, the tessellation of space used in a HF is similar to a grid map, with the particularity that all the dimensions of a robot pose must be accounted for. That is, for a robot moving in 2D we would need a three-dimensional grid, whose third dimension corresponds to all the possible robot headings. Such a grid comprises a set of cells $\{\mathbf{v}_i\}$, typically all of the same dimensions, each one representing a volume of the robot workspace $|\mathbf{v}_i|$ and having a representative metrical pose $\mathbf{c}_i$. The typical representative pose of such a cell is its *center of mass*, as defined in physics (Serway & Jewett, 2009)—in a solid, if we have uniform mass density, the center of mass is also the geometrical centroid of the shape of the solid; if we consider a d-dimensional cell as being filled with matter of constant density ρ totaling a mass of M, and denote with $\mathbf{x}$ the metrical position within the cell of an infinitesimal portion dm of that mass, that occupies an infinitesimal portion $d\mathbf{v}$ of space ($dm = \rho d\mathbf{v}$, and $d\mathbf{v} = dx_1 dx_2 ... dx_d = d\mathbf{x}$), the center of mass of the cell is defined as:

$$\mathbf{c}_i = \frac{\int_{dm \in \mathbf{v}_i} \mathbf{x}\, dm}{M} = \frac{\int_{\mathbf{v}} \mathbf{x} \rho d\mathbf{v}}{\int_{\mathbf{v}} \rho d\mathbf{v}} = \frac{\rho \int_{\mathbf{x} \in \mathbf{v}_i} \mathbf{x} d\mathbf{x}}{\rho |\mathbf{v}_i|} = \frac{1}{|\mathbf{v}_i|} \int_{\mathbf{x} \in \mathbf{v}_i} \mathbf{x} d\mathbf{x} \tag{55}$$

All the cells are connected by the natural adjacency relations arising from their metrical positions. The estimates of the filter will be more precise (albeit more computationally expensive) as the sizes of the cells get smaller. Typical values are in the range of $10cm^2$ and 5° for planar scenarios—poses being a 2D position plus a heading angle.

For setting up a DBF like the one in Equation 54 when we have a tessellation like the one described above, we need to define discrete beliefs and transition models over the cells, besides an observation likelihood that refers to an entire cell. In the following we will see how these discretizations of the continuous transition/belief/observation uncertainties lead to the HF approximation.

The robot pose is a continuous variable. Still, the belief about the pose to be considered in the filter must be uniform within each cell, since we will not distinguish any variation in the information about actuators or sensors inside a cell volume. Thus, we define a discrete r.v. i_k that yields the cell index at which we belief the robot is for a time step k. Correspondingly, a probability mass function $P_{cell}(i_k)$ is defined over all the cells for representing our entire belief on the robot pose at that step. The relation of the pmf $P_{cell}(i_k)$ with its continuous counterpart $p_{pose}(\mathbf{x}_k)$ is $P_{cell}(j) = \int_{\mathbf{x}\in\mathbf{v}_j} p_{pose}(\mathbf{x})d\mathbf{x}$, that is, the belief of being at a certain cell j is simply the infinitesimal summation of the likelihood of being anywhere within the cell. Since we lack more concrete information to localize the robot at any particular point within the cell, we should assume that $p_{pose}(\mathbf{x}_k)$ follows a *uniform* distribution within each volume $\mathbf{v}_i$. Mathematically, we can write this as:

$$p_{pose}(\mathbf{x}) = \frac{P_{cell}(i)}{\left|\mathbf{v}_i\right|} = \frac{1}{\left|\mathbf{v}_i\right|}\int_{\mathbf{x}\in\mathbf{v}_i} p_{pose}(\mathbf{x})d\mathbf{x},\ \forall\mathbf{x}\in\mathbf{v}_i \tag{56}$$

an expression which we will need later.

Regarding the transition pmf, it represents the discrete probability of the robot being at cell i_k given that it was previously at j_{k-1} and has measured a (continuous) motion $\mathbf{u}_k$. This pmf is related to the probabilistic motion model $p(\mathbf{x}_k|\mathbf{x}_{k-1},\mathbf{u}_k)$ described in chapter 5, as follows:

$$P_{trans}(i_k|j_{k-1},\mathbf{u}_k) =$$

(by definition of conditional probability, if we consider the expression to be conditioned on $\mathbf{u}_k$, that is, $P_{trans}(i_k|j_{k-1},\mathbf{u}_k) = "P_{trans}(i_k|j_{k-1}|\mathbf{u}_k)"$)

$$= \frac{P(i_k, j_{k-1}|\mathbf{u}_k)}{P(j_{k-1}|\mathbf{u}_k)} =$$

(by the relation existing between the discrete P_{cell} and the continuous p_{pose})

$$= \frac{\int_{\mathbf{x}_k\in\mathbf{v}_{i_k}}\int_{\mathbf{x}_{k-1}\in\mathbf{v}_{j_{k-1}}} p(\mathbf{x}_k,\mathbf{x}_{k-1}|\mathbf{u}_k)d\mathbf{x}_k d\mathbf{x}_{k-1}}{\int_{\mathbf{x}_{k-1}\in\mathbf{v}_{j_{k-1}}} p(\mathbf{x}_{k-1}|\mathbf{u}_k)d\mathbf{x}_{k-1}} =$$

(by the chain rule of conditional probability—conditioned on $\mathbf{u}_k$)

$$= \frac{\int_{\mathbf{x}_k\in\mathbf{v}_{i_k}}\int_{\mathbf{x}_{k-1}\in\mathbf{v}_{j_{k-1}}} p(\mathbf{x}_k|\mathbf{x}_{k-1},\mathbf{u}_k)p(\mathbf{x}_{k-1}|\mathbf{u}_k)d\mathbf{x}_k d\mathbf{x}_{k-1}}{\int_{\mathbf{x}_{k-1}\in\mathbf{v}_{j_{k-1}}} p(\mathbf{x}_{k-1}|\mathbf{u}_k)d\mathbf{x}_{k-1}} =$$

(provided that $\mathbf{x}_{k-1}\perp\mathbf{u}_k$ in the dynamic Bayesian network of the localization problem)

$$= \frac{\int_{\mathbf{x}_k\in\mathbf{v}_{i_k}}\int_{\mathbf{x}_{k-1}\in\mathbf{v}_{j_{k-1}}} p(\mathbf{x}_k|\mathbf{x}_{k-1},\mathbf{u}_k)p_{pose}(\mathbf{x}_{k-1})d\mathbf{x}_k d\mathbf{x}_{k-1}}{\int_{\mathbf{x}_{k-1}\in\mathbf{v}_{j_{k-1}}} p_{pose}(\mathbf{x}_{k-1})d\mathbf{x}_{k-1}} =$$

(by the uniform nature of $p_{pose}(\mathbf{x}_{k-1})$ within the cell where we are integrating)

$$= \frac{\int_{\mathbf{x}_k\in\mathbf{v}_{i_k}}\int_{\mathbf{x}_{k-1}\in\mathbf{v}_{j_{k-1}}} p(\mathbf{x}_k|\mathbf{x}_{k-1},\mathbf{u}_k)\frac{P_{cell}(j_{k-1})}{\left|\mathbf{v}_{j_{k-1}}\right|}d\mathbf{x}_k d\mathbf{x}_{k-1}}{\int_{\mathbf{x}_{k-1}\in\mathbf{v}_{j_{k-1}}} \frac{P_{cell}(j_{k-1})}{\left|\mathbf{v}_{j_{k-1}}\right|}d\mathbf{x}_{k-1}} =$$

$$= \frac{\int\limits_{\mathbf{x}_k \in \mathbf{v}_{i_k}} \int\limits_{\mathbf{x}_{k-1} \in \mathbf{v}_{j_{k-1}}} p(\mathbf{x}_k \left| \mathbf{x}_{k-1}, \mathbf{u}_k \right.) d\mathbf{x}_k d\mathbf{x}_{k-1}}{\int\limits_{\mathbf{x}_{k-1} \in \mathbf{v}_{j_{k-1}}} d\mathbf{x}_{k-1}} =$$

$$= \frac{1}{\left| \mathbf{v}_{j_{k-1}} \right|} \int\limits_{\mathbf{x}_k \in \mathbf{v}_{i_k}} \int\limits_{\mathbf{x}_{k-1} \in \mathbf{v}_{j_{k-1}}} p(\mathbf{x}_k \left| \mathbf{x}_{k-1}, \mathbf{u}_k \right.) d\mathbf{x}_k d\mathbf{x}_{k-1} \tag{57}$$

The term inside the double integral corresponds to the probability of being at the metrical pose $\mathbf{x}_1 \in \mathbf{v}_i$ after having been at $\mathbf{y}_1 \in \mathbf{v}_j$ *or* being at metrical pose $\mathbf{x}_1 \in \mathbf{v}_i$ after having been at $\mathbf{y}_2 \in \mathbf{v}_j$ *or*... so on (for all combinations of metrical poses in both cells). An exact expression for that is difficult to obtain, but we can consider that all these probabilities are the same and equal to that of being at the representative pose of the current cell after having been at the representative pose of the previous cell, that is, we can consider that:

$$p(\mathbf{x}_k \left| \mathbf{x}_{k-1}, \mathbf{u}_k \right.) \approx \eta p(\mathbf{c}_{i_k} \left| \mathbf{c}_{j_{k-1}}, \mathbf{u}_k \right.), \quad \forall \mathbf{x}_k \in \mathbf{v}_{i_k}, \mathbf{x}_{k-1} \in \mathbf{v}_{j_{k-1}} \tag{58}$$

with the normalization factor η needed for having a valid probability distribution. This last expression is a constant in the intervals of integration, and thus:

$$P(i_k \left| j_{k-1}, \mathbf{u}_k \right.) \approx$$
$$\approx \eta \frac{1}{\left| \mathbf{v}_{j_{k-1}} \right|} \int\limits_{\mathbf{x}_k \in \mathbf{v}_{i_k}} \int\limits_{\mathbf{x}_{k-1} \in \mathbf{v}_{j_{k-1}}} p(\mathbf{c}_{i_k} \left| \mathbf{c}_{j_{k-1}}, \mathbf{u}_k \right.) d\mathbf{x}_k d\mathbf{x}_{k-1} =$$
$$= \eta \frac{p(\mathbf{c}_{i_k} \left| \mathbf{c}_{j_{k-1}}, \mathbf{u}_k \right.)}{\left| \mathbf{v}_{j_{k-1}} \right|} \int\limits_{\mathbf{x}_k \in \mathbf{v}_{i_k}} \int\limits_{\mathbf{x}_{k-1} \in \mathbf{v}_{j_{k-1}}} d\mathbf{x}_k d\mathbf{x}_{k-1} =$$
$$= \eta \frac{p(\mathbf{c}_{i_k} \left| \mathbf{c}_{j_{k-1}}, \mathbf{u}_k \right.)}{\left| \mathbf{v}_{j_{k-1}} \right|} \left| \mathbf{v}_{i_k} \right| \left| \mathbf{v}_{j_{k-1}} \right| =$$
$$= \eta \left| \mathbf{v}_{i_k} \right| p(\mathbf{c}_{i_k} \left| \mathbf{c}_{j_{k-1}}, \mathbf{u}_k \right.) \tag{59}$$

which is our final approximation to the discrete transition probability $P_{trans}(i_k \left| j_{k-1}, \mathbf{u}_k \right.)$.

We also need a form for the observation likelihood that fits into the discretization of the grid. Given a hypothetical cell i_k where the robot may be, the map m and an actual observation $\mathbf{z}_k$, the observation likelihood must yield the density of probability that $\mathbf{z}_k$ has of having been observed while the robot was at the cell i_k. That probability is the one of having observed $\mathbf{z}_k$ while being at a metrical pose $\mathbf{x}_1 \in \mathbf{v}_i$ *or* having observed $\mathbf{z}_k$ while being at a metrical pose $\mathbf{x}_2 \in \mathbf{v}_i$ *or*... so on, for all the possible $\mathbf{x} \in \mathbf{v}_{i_k}$. A deduction similar to the one of the transition probability can be done:

(by definition of conditional probability, conditioned on $\mathbf{m}$)

$$p(\mathbf{z}_k \left| i_k, \mathbf{m} \right.) = \frac{p(\mathbf{z}_k, i_k \left| \mathbf{m} \right.)}{p(i_k \left| \mathbf{m} \right.)} = \frac{p(\mathbf{z}_k, i_k \left| \mathbf{m} \right.)}{p(\mathbf{x}_k \in \mathbf{v}_{i_k} \left| \mathbf{m} \right.)} =$$

(by definition of probability density)

Table 5. Computational cost of the different elements of the HF, with r standing for the number of cells in the filter

Algorithm step	Complexity
Prediction (sum term)	$O(rg)$ per cell of the posterior, being $O(g)$ the cost of evaluating the discretized transition.
Update	$O(h)$ per cell of the posterior: the cost of evaluating the discretized observation likelihood.
Total (dominant) complexity:	$O(r^2 g + rh)$ for the entire posterior, typically $O(r^2)$.

$$= \frac{\int_{\mathbf{x}\in\mathbf{v}_{i_k}} p(\mathbf{z}_k, \mathbf{x} | \mathbf{m}) d\mathbf{x}}{\int_{\mathbf{x}\in\mathbf{v}_{i_k}} p(\mathbf{x}|\mathbf{m}) d\mathbf{x}} =$$

(by the chain rule and the fact that $\mathbf{x} \perp \mathbf{m}$)

$$= \frac{\int_{\mathbf{x}\in\mathbf{v}_{i_k}} p(\mathbf{z}_k | \mathbf{x}, \mathbf{m}) p(\mathbf{x}|\mathbf{m}) d\mathbf{x}}{\int_{\mathbf{x}\in\mathbf{v}_{i_k}} p(\mathbf{x}|\mathbf{m}) d\mathbf{x}} = \frac{\int_{\mathbf{x}\in\mathbf{v}_{i_k}} p(\mathbf{z}_k | \mathbf{x}, \mathbf{m}) p_{pose}(\mathbf{x}) d\mathbf{x}}{\int_{\mathbf{x}\in\mathbf{v}_{i_k}} p_{pose}(\mathbf{x}) d\mathbf{x}} =$$

(by the uniformity assumption of the pose probability within a cell)

$$= \frac{\int_{\mathbf{x}\in\mathbf{v}_{i_k}} p(\mathbf{z}_k | \mathbf{x}, \mathbf{m}) \frac{P_{cell}(i_k)}{|\mathbf{v}_{i_k}|} d\mathbf{x}}{\int_{\mathbf{x}\in\mathbf{v}_{i_k}} \frac{P_{cell}(i_k)}{|\mathbf{v}_{i_k}|} d\mathbf{x}} = \frac{\int_{\mathbf{x}\in\mathbf{v}_{i_k}} p(\mathbf{z}_k | \mathbf{x}, \mathbf{m}) d\mathbf{x}}{\int_{\mathbf{x}\in\mathbf{v}_{i_k}} d\mathbf{x}} =$$
$$= \frac{1}{|\mathbf{v}_{i_k}|} \int_{\mathbf{x}\in\mathbf{v}_{i_k}} p(\mathbf{z}_k | \mathbf{x}, \mathbf{m}) d\mathbf{x} \tag{60}$$

Again, this exact expression can be difficult to calculate. For finding a suitable approximation we can consider that the probability $p(\mathbf{z}_k | \mathbf{x}, \mathbf{m})$ of having observed $\mathbf{z}_k$ when being at any position $\mathbf{x}$ within the cell i_k equals the one of having observed $\mathbf{z}_k$ when being at the representative position of the cell, $\mathbf{c}_{i_k}$, that is:

$$p(\mathbf{z}_k | \mathbf{x}, \mathbf{m}) \approx \eta' p(\mathbf{z}_k | \mathbf{c}_{i_k}, \mathbf{m}), \quad \forall \mathbf{x} \in \mathbf{v}_{i_k} \tag{61}$$

leading to:

$$\begin{aligned} p(\mathbf{z}_k | i_k, \mathbf{m}) &\approx \frac{1}{|\mathbf{v}_{i_k}|} \int_{\mathbf{x}\in\mathbf{v}_{i_k}} \eta' p(\mathbf{z}_k | \mathbf{c}_{i_k}, \mathbf{m}) d\mathbf{x} \\ &= \eta' \frac{1}{|\mathbf{v}_{i_k}|} p(\mathbf{z}_k | \mathbf{c}_{i_k}, \mathbf{m}) \int_{\mathbf{x}\in\mathbf{v}_{i_k}} d\mathbf{x} \\ &= \eta' \frac{1}{|\mathbf{v}_{i_k}|} p(\mathbf{z}_k | \mathbf{c}_{i_k}, \mathbf{m}) |\mathbf{v}_{i_k}| \\ &= \eta' p(\mathbf{z}_k | \mathbf{c}_{i_k}, \mathbf{m}) \end{aligned} \tag{62}$$

We already have the elements required for using the DBF formulation in the context of a histogram filter. Notice that the proportionality constants η and η' are absorbed by the proportionality of the filter, as well as the factor $|\mathbf{v}_{i_k}|$ if all the cells in the grid have the same volume (a regular grid); in that particular case, the HF becomes:

$$bel(i_k \left| \mathbf{z}_{1:k}, \mathbf{u}_{1:k}, \mathbf{m} \right.) \propto$$
$$\propto p(\mathbf{z}_k \left| \mathbf{c}_{i_k}, \mathbf{m} \right.) \sum_{j_{k-1}} p(\mathbf{c}_{i_k} \left| \mathbf{c}_{j_{k-1}}, \mathbf{u}_k \right.) bel(j_{k-1} \left| \mathbf{z}_{1:k-1}, \mathbf{u}_{1:k-1}, \mathbf{m} \right.) \quad (63)$$

Care must be taken when evaluating the pdfs appearing in the filter: typically, their support is infinite (e.g., Gaussians), but the grid or tessellation is not, and therefore it is not correct to use their densities as such, since part of them will fall outside the map. A solution is to clip their support to the area of the map and to scale them appropriately to integrate to one in that new support.

The computational cost of the Histogram filter is analyzed in Table 5, provided that the grid has r cells. As discussed before, it increases proportionally to the square of the size of the grid, which can easily be cumbersome in many practical applications. One solution to this is to use adaptive sizes for the grid (subdividing in more depth around the parts of interest in the environment) or switching to a particle filter, which we introduced next.

The Particle Filter (PF)

Mobile robot localization through recursive Bayesian filters witnessed a breakthrough around the late nineties with the work of Dieter Fox and colleagues, who adapted to mobile robotics an idea previously applied to several other fields: computer vision, target tracking, probabilistic inference, etc. They proposed a non-parametric estimator for general recursive non-linear estimation problems, intended as an alternative to the popular EKF algorithm (Doucet & Johansen, 2008). That filter was based, in turn, on Monte Carlo methods, which, as explained in chapter 3, can be traced back as earlier as the 18th century with the count of Buffon, but were actually developed and applied by Enrico Fermi and subsequently given its present name by Nick Metropolis in the 1930s.

A Monte Carlo method, broadly speaking, consists of estimating some parameter of a complicated system—for example, the expected value of some of its properties—using random observations of it instead of a closed-form formula. For instance, it serves for approximating numerically the value of difficult integrals by sampling random values of the integral variable and evaluating the integrand function pointwise, but it also serves for approximating some information of interest in incompletely modeled distributions of probability. Today, Monte Carlo methods are pervasive in science and engineering and can find applications in very diverse ways to a diversity of problems.

In the context of recursive Bayesian estimation, a Monte Carlo approach can be used to approximate the uncertainty of arbitrary pdfs. For example, the posterior density can be approximated with a number of observations which are assumed to be drawn from it. In this consist the approaches called *Sequential Monte Carlo estimation* or SMC—recall that recursive Bayesian filters are also called *sequential* Bayesian filters or *sequential* estimators—and, particularly, *Monte Carlo localization* in our context (Fox, Burgard, Dellaert, & Thrun, 1999).

This approach breaks the analytical barrier of the integral of the prior in the prediction stage of a continuous recursive Bayesian filter, although at the expense of the computational power needed to maintain a suitable number of observations that is typically large (one that reflects all the information we are interested in). Only with infinite observations, the resulting uncertainties would be exact and the filter is guaranteed not to diverge in *any* situation it may encounter, since Monte Carlo methods rely on the law of large numbers explained in chapter 4. However, with the power of today's computers, we can choose a number that yields both a reasonable computational cost and a high likelihood of convergence for most practical settings.

Since the observations used for Monte Carlo methods are commonly named *particles*, we can find in the literature the term *Particle Filter* (PF) referring to a general class of sequential Monte

Carlo estimators that have been proposed in diverse fields under different names, like the *bootstrap filter* (Gordon, Salmond, & Smith, 1993), the *condensation algorithm* or *the survival of the fittest* algorithm.

A PF represents a probability density distribution, for instance the posterior of the pose of the robot, $bel(\mathbf{x}_k | \mathbf{z}_{1:k}, \mathbf{u}_{1:k}, \mathbf{m})$, as follows (we use the symbol $\triangleright$ for "is represented by"):

$$bel(\mathbf{x}_k | \mathbf{z}_{1:k}, \mathbf{u}_{1:k}, \mathbf{m}) \triangleright \left\{ \left(w_k^{[i]}, \mathbf{x}_k^{[i]} \right) \right\} \text{ (indexed set)}$$

where

$\mathbf{x}_k^{[i]}$ is the i-th particle

$w_k^{[i]}$ is the importance factor of the i-th particle, provided that $\sum_i w_k^{[i]} = 1$ (64)

The importance factor indicates the relevance of the $i-th$ particle to represent information of interest from the distribution. More specifically, the idea of importance sampling is that the *density of particles* within a given volume of the state space—a given volume of the support of the r.v.—weighted by their importance factors, should be proportional to the actual value of the pdf which is being approximated.

We will demonstrate further on in a formal way how the PF is designed to represent the posterior with a suitable set of observations assumedly to be drawn from it, but first it is illustrative to provide an informal derivation from the Bayesian framework of Equation 4, in order to keep the same line of reasoning as with previous filters (Arumlampalam, Maskell, Gordon, & Clapp, 2002).

Firstly, the goal of representing a pdf by a set of particles is being able to recover information of the distribution from the particles. Typically its moments, but also the probability of any given random event. For that purpose, we can use the indicator function $I_{\mathbf{A}}(\mathbf{x})$, already explained in chapter 4, which yields 1 if $\mathbf{x} \in \mathbf{A}$ and 0 otherwise, being $\mathbf{A}$ some part of the support of the r.v. $\mathbf{x}$. For instance, if we wish to estimate $p(\mathbf{x} \in \mathbf{A})$, in the continuous case we have:

$$p(\mathbf{x} \in \mathbf{A}) = \int_{\mathbf{x} \in \mathbf{A}} p(\mathbf{x}) d\mathbf{x} = \int_{\mathbf{x}} I_{\mathbf{A}}(\mathbf{x}) p(\mathbf{x}) d\mathbf{x} = E\left[I_{\mathbf{A}}(\mathbf{x}) \right]$$

Then, it is reasonable to propose this estimator of $E\left[I_{\mathbf{A}}(\mathbf{x}) \right]$:

$$E\left[I_{\mathbf{A}}(\mathbf{x}) \right] \approx \sum_{i=1}^{N} w^{[i]} I_{\mathbf{A}}(\mathbf{x}^{[i]})$$

By the law of large numbers we have:

$$p(\mathbf{x} \in \mathbf{A}) = \lim_{N \to \infty} \sum_{i=1}^{N} w^{[i]} I_{\mathbf{A}}(\mathbf{x}^{[i]}) \quad (65)$$

However, when working with pdfs in robotic filters we need to refer to individual random values rather than random events (sets of values). For that, we will employ the Dirac's delta, denoted $\delta_{\mathbf{a}}(\mathbf{x})$, instead of the indicator function. The delta is a generalized function (Kanwal, 1998) that yields $\mathbf{0}$ for any $\mathbf{x} \neq \mathbf{a}$. It can be thought as the limiting pdf of a Gaussian with mean $\mathbf{a}$ as its variance tends to zero, or as the limit of a uniform pdf centered in $\mathbf{a}$ as its support tends to have zero width. It is not strictly a function, since any function that is zero except in one point must have an integral (over its entire domain) of zero. In contrast, the Dirac's delta has the following properties[7]:

$$\int_{\mathbf{x} \in \mathbf{A}} \delta_{\mathbf{a}}(\mathbf{x}) d\mathbf{x} = \begin{cases} 1 & \text{if } \mathbf{a} \in \mathbf{A} \\ 0 & \text{otherwise} \end{cases} \leftrightarrow$$
$$\int_{\mathbf{x} \in \mathbf{A}} \delta_{\mathbf{a}}(\mathbf{x}) d\mathbf{x} = I_{\mathbf{A}}(\mathbf{a}) \int_{\mathbf{x}} f(\mathbf{x}) \delta_{\mathbf{a}}(\mathbf{x}) d\mathbf{x} =$$
$$= f(\mathbf{a}) f(\mathbf{x}) \delta_{\mathbf{a}}(\mathbf{x}) = f(\mathbf{a}) \delta_{\mathbf{a}}(\mathbf{x}) \quad (66)$$

With the Dirac's delta, we can deduce the following, starting from the approximation of Equation 65:

$$p(\mathbf{x} \in \mathbf{A}) \approx \sum_{i=1}^{N} w^{[i]} I_{\mathbf{A}}(\mathbf{x}^{[i]}) =$$

(using the equivalence between the indicator function and the Dirac's delta)

$$= \sum_{i=1}^{N} w^{[i]} \int_{\mathbf{x}\in\mathbf{A}} \delta_{\mathbf{x}^{[i]}}(\mathbf{x}) d\mathbf{x} = \int_{\mathbf{x}\in\mathbf{A}} \underbrace{\sum_{i=1}^{N} w^{[i]} \delta_{\mathbf{x}^{[i]}}(\mathbf{x})}_{\text{must be } p(\mathbf{x})} d\mathbf{x}$$

(by the definition of probability density, the integrand must be the pdf $p(\mathbf{x})$, thus):

$$p(\mathbf{x}) \approx \sum_{i=1}^{N} w^{[i]} \delta_{\mathbf{x}^{[i]}}(\mathbf{x}) \quad (67)$$

That is, we can obtain information from a continuous distribution using only an approximate distribution $\sum_i w_k^{[i]} \delta_{\mathbf{x}_k^{[i]}}(\mathbf{x})$, which is guaranteed to be exact only if we have infinite particles. From this approximation of a pdf by a set of particles and importance weights, we can estimate moments or other information of interest of the actual distribution.

Now, if we substitute this definition of a particle-based pdf into the Bayesian filtering framework for localization, we have:

$$bel(\mathbf{x}_k | \mathbf{z}_{1:k}, \mathbf{u}_{1:k}, \mathbf{m}) \propto$$
$$\propto p(\mathbf{z}_k | \mathbf{x}_k, \mathbf{m}) \int_{\mathbf{x}_{k-1}} p(\mathbf{x}_k | \mathbf{x}_{k-1}, \mathbf{u}_k)$$
$$bel(\mathbf{x}_{k-1} | \mathbf{z}_{1:k-1}, \mathbf{u}_{1:k-1}, \mathbf{m}) d\mathbf{x}_{k-1} \approx$$

(inserting the particle representation of the posterior of step $k-1$)

$$\approx p(\mathbf{z}_k | \mathbf{x}_k, \mathbf{m}) \int_{\mathbf{x}_{k-1}} p(\mathbf{x}_k | \mathbf{x}_{k-1}, \mathbf{u}_k) \left(\sum_{i=1}^{N} w_{k-1}^{[i]} \delta_{\mathbf{x}_{k-1}^{[i]}}(\mathbf{x}) \right) d\mathbf{x}_{k-1} =$$
$$= p(\mathbf{z}_k | \mathbf{x}_k, \mathbf{m}) \int_{\mathbf{x}_{k-1}} \left(\sum_{i=1}^{N} w_{k-1}^{[i]} p(\mathbf{x}_k | \mathbf{x}_{k-1}, \mathbf{u}_k) \delta_{\mathbf{x}_{k-1}^{[i]}}(\mathbf{x}) \right) d\mathbf{x}_{k-1} =$$
$$= p(\mathbf{z}_k | \mathbf{x}_k, \mathbf{m}) \sum_{i=1}^{N} w_{k-1}^{[i]} \int_{\mathbf{x}_{k-1}} p(\mathbf{x}_k | \mathbf{x}_{k-1}, \mathbf{u}_k) \delta_{\mathbf{x}_{k-1}^{[i]}}(\mathbf{x}) d\mathbf{x}_{k-1} =$$

(using the second property of the Dirac's delta)

$$= p(\mathbf{z}_k | \mathbf{x}_k, \mathbf{m}) \sum_{i=1}^{N} w_{k-1}^{[i]} p(\mathbf{x}_k | \mathbf{x}_{k-1}^{[i]}, \mathbf{u}_k) \quad (68)$$

We find two likelihood terms where the unknown $\mathbf{x}_k$ appears: in $p(\mathbf{x}_k | \mathbf{x}_{k-1}^{[i]}, \mathbf{u}_k)$ inside the sum and in the observation likelihood $p(\mathbf{z}_k | \mathbf{x}_k, \mathbf{m})$. One of the possibilities for representing the posterior $bel(\mathbf{x}_k | \mathbf{z}_{1:k}, \mathbf{u}_{1:k}, \mathbf{m})$ as a set of particles is to approximate the pdf inside the sum by particles too (this approach is the most convenient in the context of robot localization since we will typically know how to draw a sample from the pdf inside the sum but only know how to evaluate pointwise the observation likelihood). The simplest way to perform this is to draw a sample of size N from $p(\mathbf{x}_k | \mathbf{x}_{k-1}^{[i]}, \mathbf{u}_k)$, one value for each $\mathbf{x}_{k-1}^{[i]}$ from the previous time step, and assigning them the same importance factors as the $\mathbf{x}_{k-1}^{[i]}$ from which they come, that is:

$$p(\mathbf{x}_k | \mathbf{x}_{k-1}^{[i]}, \mathbf{u}_k) \approx \sum_{j=1}^{N} \overline{w}_k^{[j]} \delta_{\overline{\mathbf{x}}_k^{[j]}}$$

where

$$\overline{\mathbf{x}}_k^{[j]} \sim p(\mathbf{x}_k | \mathbf{x}_{k-1}^{[j]}, \mathbf{u}_k)$$

$$\overline{w}_k^{[j]} = w_{k-1}^{[j]} \quad (69)$$

Table 6. Execution steps of the basic implementation of a PF as the SIS algorithm, along with their computational complexity

Algorithm step	Operations	Complexity
Prediction	*-Draw a sample of size* N, $\{\bar{\mathbf{x}}_k^{[j]}\}$, from $p(\mathbf{x}_k \mid \mathbf{x}_{k-1}, \mathbf{u}_k)$, taking for the $j-th$ element in the sample the conditioning value $\mathbf{x}_{k-1} = \mathbf{x}_{k-1}^{[j]}$ and the measured motion $\mathbf{u}_k$. *-Assign to the particle* $\bar{\mathbf{x}}_k^{[j]}$ the importance factor of the particle we used for it, i.e., $\bar{w}_k^{[j]} = w_{k-1}^{[j]}$	$O(Ng)$, *where* $O(g)$ is the cost of drawing a sample (a set of values of dimension d each) from the transition pdf. Typically dominated by $O(N)$.
Update	*-Take the particles of the posterior as the predicted particles:* $\mathbf{x}_k^{[j]} = \bar{\mathbf{x}}_k^{[j]}$. *-Calculate importance factors from the ones of the predicted particles using the observation likelihood:* $\hat{w}_k^{[j]} = p(\mathbf{z}_k \mid \bar{\mathbf{x}}_k^{[j]}, \mathbf{m})\bar{w}_k^{[j]}$. *-Scale the importance factors of the posterior to sum up 1:* $w_k^{[j]} = \frac{\hat{w}_k^{[j]}}{\sum_{s=1}^{N} \hat{w}_k^{[s]}}$.	$O(Nh)$, *being* $O(h)$ the cost of evaluating pointwise the observation likelihood (dimension c). Typically dominated by $O(N)$.
Total (dominant) complexity:		*Typically dominated by* $O(N)$.

In this way, our approximation of the posterior becomes (here η is the proportionality factor implicit in Equation 4):

$$\begin{aligned} bel(\mathbf{x}_k \mid \mathbf{z}_{1:k}, \mathbf{u}_{1:k}, \mathbf{m}) &\approx \eta p(\mathbf{z}_k \mid \mathbf{x}_k, \mathbf{m}) \sum_i w_{k-1}^{[i]} p(\mathbf{x}_k \mid \mathbf{x}_{k-1}^{[i]}, \mathbf{u}_k) \\ &\approx \eta p(\mathbf{z}_k \mid \mathbf{x}_k, \mathbf{m}) \sum_{i=1}^{N} w_{k-1}^{[i]} \sum_{j=1}^{N} \bar{w}_k^{[j]} \delta_{\bar{\mathbf{x}}_k^{[j]}}(\mathbf{x}) \\ &= \eta p(\mathbf{z}_k \mid \mathbf{x}_k, \mathbf{m}) \sum_{i=1}^{N} \sum_{j=1}^{N} \bar{w}_k^{[j]} \delta_{\bar{\mathbf{x}}_k^{[j]}}(\mathbf{x}) w_{k-1}^{[i]} \\ &= \eta p(\mathbf{z}_k \mid \mathbf{x}_k, \mathbf{m}) \sum_{j=1}^{N} \bar{w}_k^{[j]} \delta_{\bar{\mathbf{x}}_k^{[j]}}(\mathbf{x}) \underbrace{\sum_{i=1}^{N} w_{k-1}^{[i]}}_{=1} \end{aligned}$$

$$= \eta \sum_{j=1}^{N} p(\mathbf{z}_k \mid \mathbf{x}_k, \mathbf{m}) \bar{w}_k^{[j]} \delta_{\bar{\mathbf{x}}_k^{[j]}}(\mathbf{x})$$

(substituting back $\bar{w}_k^{[j]}$)

$$= \eta \sum_{j=1}^{N} p(\mathbf{z}_k \mid \mathbf{x}_k, \mathbf{m}) w_{k-1}^{[j]} \delta_{\bar{\mathbf{x}}_k^{[j]}}(\mathbf{x})$$

(applying the third property of the Dirac's delta we have)

$$bel(\mathbf{x}_k \mid \mathbf{z}_{1:k}, \mathbf{u}_{1:k}, \mathbf{m}) \approx \sum_{j=1}^{N} \underbrace{\left(\eta p(\mathbf{z}_k \mid \bar{\mathbf{x}}_k^{[j]}, \mathbf{m}) w_{k-1}^{[j]}\right)}_{\text{importance weight}} \delta_{\bar{\mathbf{x}}_k^{[j]}}(\mathbf{x}) \quad (70)$$

In the last expression we can see a particle representation of the posterior of the filter: the particles of that posterior are drawn from the transition probability when it is instantiated with the particles of the posterior of the previous step, i.e., $\bar{\mathbf{x}}_k^{[j]} \sim p(\mathbf{x}_k \mid \mathbf{x}_{k-1}^{[i]}, \mathbf{u}_k)$; their importance factors are then chosen to be proportional to $p(\mathbf{z}_k \mid \bar{\mathbf{x}}_k^{[j]}, \mathbf{m}) w_{k-1}^{[j]}$. A Particle Filter could therefore be implemented by means of the algorithm presented in Table 6. As we will see further on, this algorithm is not yet ready to be applied to mobile robot localization as is, due to some important drawbacks, which we address next.

The computational complexity of the PF algorithm presented so far is typically linear, which may be a surprise since we have commented at the beginning of this section that the advantages

Figure 8. Example of the application of importance sampling to set appropriately the weights of the particles drawn from a simple distribution (left, a bi-dimensional Gaussian $N(\mathbf{x};\mu,\Sigma) = N((x_1,x_2);(\mu_1,\mu_2),\Sigma)$ *with mean* $\mu = [5,5]$ *and covariance* $\Sigma = [1,1.69\ ;\ 1.69,4]$*) to reflect another, complex distribution (right, the target distribution is* $\propto \left|2N(\mathbf{x};\mu,\Sigma)^3 \cos(x_1)\right|$*). The darker the element of the sample drawn in the right, the greater is its importance weight.*

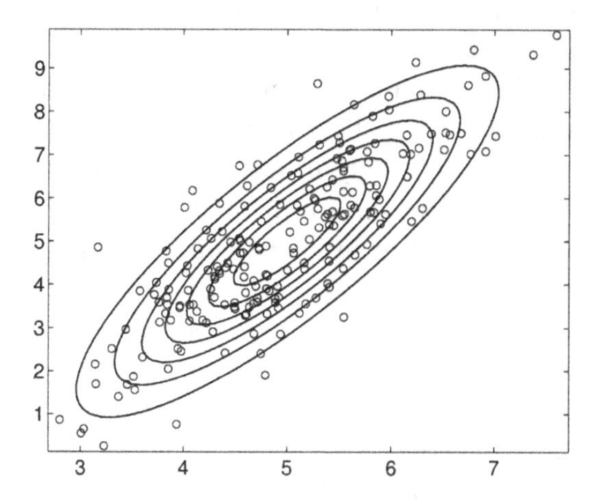

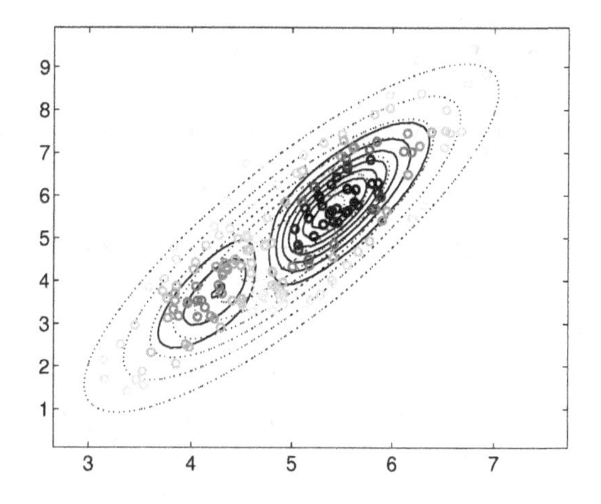

of particle filters come at the price of their computational cost. However, you should notice that the complexity is linear… *in the number of particles*. The problem is that depending on the complexity of the shape of the posterior pdf (especially when coping with multiple hypotheses), *we may need a large number of particles*. In practice, at least a few hundred particles have to be used for pose tracking in planar scenarios, that is, for 2D robot poses with a 2D position plus a heading angle. The problem gets worse in higher dimensional spaces—the so-called *curse of dimensionality*—requiring the introduction of smart tricks to make more complex problems, such as SLAM, tractable with PFs. We will defer those advanced PF algorithms until chapter 9, since simpler ones are more than suitable for robot localization.

In order to compare the complexities of PF to that of parametric filters, recall that the latter typically exhibit a cubic complexity with the *dimensionality* of the state space, which, however, remains constant for the problem of robot localization. Therefore, both PF and parametric filters have a constant time complexity, disregarding the variable size of samples required at different stages of localization with a particle filter—i.e. an initial state of high uncertainty would require a large sample of poses, which may be reduced when the estimate converges to a single most-likely spot. In any case, the time constants are such that particle filters typically end up being more computationally demanding.

Before proceeding with the enhancements needed for this filter to work in practical applications, we will provide an alternative, more rigorous derivation for it. Derived in this new way, the filter is also called the *Sequential Importance Sampling (SIS)* algorithm.

We start by stressing the goal of a PF: obtaining a particle-based representation of the posterior of the desired variables *that consists of particles that have been drawn from the true posterior*. Of course, that is not directly possible, since the true posterior is precisely what we are trying to estimate in the first place! We need some technique that permits us to sample from the posterior *indirectly*. Here comes handy *importance sampling*, a statistical technique that complements Monte Carlo methods and is used for generating samples from distributions $p(\mathbf{x})$ from which it is difficult to directly draw samples. It works by drawing samples from a different, *arbitrary* distribution $q(\mathbf{x})$ which is called the *proposal* distribution and from which we know how to easily draw samples. Then, we assign "importance weights" to the elements of the sample drawn from $q(\mathbf{x})$ such as the *weighted* density of particles within any volume of the state space is proportional to the target distribution $p(\mathbf{x})$. The good news is that the only operation required to be performed on $p(\mathbf{x})$ under this approach is its pointwise evaluation, which is much simpler than drawing samples from it.

It can be shown that if importance weights $w^{[i]}$ are assigned the following value:

Unnormalized importance weights:

$$\tilde{w}^{[i]} = \frac{p(\mathbf{x}^{[i]})}{q(\mathbf{x}^{[i]})}$$

Normalized importance weights:

$$w^{[i]} = \frac{\tilde{w}^{[i]}}{\sum_i \tilde{w}^{[i]}} \propto \frac{p(\mathbf{x}^{[i]})}{q(\mathbf{x}^{[i]})} \tag{71}$$

and, as long as $q(\mathbf{x}) > 0$ implies $p(\mathbf{x}) > 0$, this importance sampling technique produces samples that *tend* to represent the ideal $p(\mathbf{x})$ as their size *tends* to infinite. Intuitively, in those points (particles) where $q(\mathbf{x})$ is similar to $p(\mathbf{x})$, the unnormalized importance weights will be close to 1; in the points where $q(\mathbf{x})$ is clearly larger than $p(\mathbf{x})$ that is, it is little relevant for representing the shape of $p(\mathbf{x})$, importance weights will be small; and finally, when $q(\mathbf{x})$ is smaller than $p(\mathbf{x})$ we have a point that is relevant for $p(\mathbf{x})$ albeit not for $q(\mathbf{x})$, thus its unnormalized importance weight is boosted up. In the end, if we normalize the weights so they sum up the unity we obtain a particle-based representation of $p(\mathbf{x})$ that approximates the ideal case of drawing directly samples from it. An example of this process can be seen in Figure 8. In practice, the wide dynamic range of the attainable weights makes almost mandatory to implement particle filters by keeping the logarithm of weights instead of the weights themselves.

It is clear from Equation 71 that the importance weights will equal each other and therefore be constant if and only if $q(\mathbf{x}) \propto p(\mathbf{x})$. This is an ideal situation, since if that hold we would not need to calculate any importance weight at all! In fact, this case is achievable by means of the so-called *optimal proposal distribution* (Doucet, Godsill, & Andrieu, 2000; Doucet, De Freitas, Murphy, & Russell, 2000; Blanco, González, & Fernández-Madrigal, 2010). Some techniques relying on this proposal will be studied in the context of PFs for SLAM in chapter 9 section 4.

We can find a simpler and easier-to-implement proposal distribution by taking a detour. Let us assume that we are estimating not the state at time k but the full history of states, $\mathbf{x}_{0:k}$. Of course, this is unfeasible in real implementations due to the dimensionality increase over time; however, the inconvenient terms will be dropped along the reasoning.

Estimating the full history $\mathbf{x}_{0:k}$ means that we have at each time step a set of particles, each one representing *all the poses* from the beginning, that is, one such particle contains a pose hypothesis for each time step, including the one for time k, which is of interest for us. This full posterior can be developed as follows:

$$\underbrace{bel(\mathbf{x}_{0:k} | \mathbf{z}_{1:k}, \mathbf{u}_{1:k}, \mathbf{m})}_{\text{full posterior at time step } k} = bel(\mathbf{x}_{0:k} | \mathbf{z}_k, \underbrace{\mathbf{z}_{1:k-1}, \mathbf{u}_{1:k}, \mathbf{m}}_{\text{conditioning r.v.s}}) =$$

(applying Bayes' Rule)

$$= \frac{p(\mathbf{z}_k | \mathbf{x}_{0:k}, \mathbf{z}_{1:k-1}, \mathbf{u}_{1:k}, \mathbf{m}) p(\mathbf{x}_{0:k} | \mathbf{z}_{1:k-1}, \mathbf{u}_{1:k}, \mathbf{m})}{p(\mathbf{z}_k | \mathbf{z}_{1:k-1}, \mathbf{u}_{1:k}, \mathbf{m})} =$$

(since $\mathbf{z}_k \perp \mathbf{z}_{1:k-1}, \mathbf{u}_{1:k} | \mathbf{x}_k, \mathbf{m}$, according to the DBN for localization)

$$= \frac{p(\mathbf{z}_k | \mathbf{x}_k, \mathbf{m}) p(\mathbf{x}_{0:k} | \mathbf{z}_{1:k-1}, \mathbf{u}_{1:k}, \mathbf{m})}{p(\mathbf{z}_k | \mathbf{z}_{1:k-1}, \mathbf{u}_{1:k}, \mathbf{m})} \propto$$
$$\propto p(\mathbf{z}_k | \mathbf{x}_k, \mathbf{m}) p(\mathbf{x}_{0:k} | \mathbf{z}_{1:k-1}, \mathbf{u}_{1:k}, \mathbf{m}) =$$

(expanding the last term through the chain rule of conditional probability)

$$= p(\mathbf{z}_k | \mathbf{x}_k, \mathbf{m}) p(\mathbf{x}_k | \mathbf{x}_{0:k-1}, \mathbf{z}_{1:k-1}, \mathbf{u}_{1:k}, \mathbf{m})$$
$$p(\mathbf{x}_{0:k-1} | \mathbf{z}_{1:k-1}, \mathbf{u}_{1:k}, \mathbf{m}) =$$

(since $\mathbf{x}_k \perp \mathbf{x}_{0:k-2} | \mathbf{x}_{k-1}$, $\mathbf{x}_k \perp \mathbf{z}_{1:k-1}, \mathbf{m}$ and $\mathbf{x}_k \perp \mathbf{u}_{1:k-1} | \mathbf{x}_{k-1}$)

$$= p(\mathbf{z}_k | \mathbf{x}_k, \mathbf{m}) p(\mathbf{x}_k | \mathbf{x}_{k-1}, \mathbf{u}_k) p(\mathbf{x}_{0:k-1} | \mathbf{z}_{1:k-1}, \mathbf{u}_{1:k}, \mathbf{m}) =$$

(and since $\mathbf{x}_{0:k-1} \perp \mathbf{u}_k$)

$$= p(\mathbf{z}_k | \mathbf{x}_k, \mathbf{m}) p(\mathbf{x}_k | \mathbf{x}_{k-1}, \mathbf{u}_k) p(\mathbf{x}_{0:k-1} | \mathbf{z}_{1:k-1}, \mathbf{u}_{1:k-1}, \mathbf{m}) =$$
$$= \underbrace{p(\mathbf{z}_k | \mathbf{x}_k, \mathbf{m})}_{\text{observation likelihood}} \underbrace{p(\mathbf{x}_k | \mathbf{x}_{k-1}, \mathbf{u}_k)}_{\text{transition distribution}} \underbrace{bel(\mathbf{x}_{0:k-1} | \mathbf{z}_{1:k-1}, \mathbf{u}_{1:k-1}, \mathbf{m})}_{\text{full posterior at time step } k-1} \quad (72)$$

Note how this expression is recursive in spite of not having included the integral of the transition model: this is because we are using the full posterior, which encodes that information.

Now, we have in the last expression a term from which it is easy to draw samples: the transition distribution, which in localization may correspond to any of the kinematic models introduced in chapter 5. Therefore that is the next most logical candidate as a proposal distribution, $q(\mathbf{x}_k) = p(\mathbf{x}_k | \mathbf{x}_{k-1}, \mathbf{u}_k)$, which we will call the *standard proposal* distribution for PF in mobile robotics. If we use it for the SIS, the weights of the particles should be as follows for representing well the full posterior density:

$$w^{[i]} \propto$$
$$\propto \frac{\overbrace{p(\mathbf{z}_k | \mathbf{x}_k^{[i]}, \mathbf{m}) p(\mathbf{x}_k^{[i]} | \mathbf{x}_{k-1}, \mathbf{u}_k) bel(\mathbf{x}_{0:k-1} | \mathbf{z}_{1:k-1}, \mathbf{u}_{1:k-1}, \mathbf{m})}^{\text{full posterior}}}{\underbrace{p(\mathbf{x}_k^{[i]} | \mathbf{x}_{k-1}, \mathbf{u}_k)}_{\text{standard proposal}}} =$$
$$= p(\mathbf{z}_k | \mathbf{x}_k^{[i]}, \mathbf{m}) bel(\mathbf{x}_{0:k-1} | \mathbf{z}_{1:k-1}, \mathbf{u}_{1:k-1}, \mathbf{m}) \quad (73)$$

Now a gentle twist. This expression yields the importance weight of a particle that encodes the full history of poses of the robot. If we disregard the dependencies existing between the poses at different time steps of that full posterior, we could assign to each of them the same weight, in particular to the last pose (the one at time step k). In other words, such an approximation allows us to consider a particle of the full posterior as being equivalent to a set of $k+1$ "sub-particles" corresponding to each time step (including the initial $\mathbf{x}_0$), each of them having the same weight of Equation 73. This rationale is the one that supports the following version of the SIS for a sequential filter:

$$\mathbf{x}_k^{[i]} \sim p(\mathbf{x}_k | \mathbf{x}_{k-1}, \mathbf{u}_k)$$
$$w^{[i]} \propto p(\mathbf{z}_k | \mathbf{x}_k^{[i]}, \mathbf{m}) bel(\mathbf{x}_{k-1} | \mathbf{z}_{1:k-1}, \mathbf{u}_{1:k-1}, \mathbf{m})$$

(and since the posterior at time step $k-1$ is approximated as the set of particles $\left\{\left(x_{k-1}^{[j]}, w_{k-1}^{[j]}\right)\right\}$, we have:)

$$\begin{aligned} \mathbf{x}_k^{[i]} &\sim p(\mathbf{x}_k \left| \mathbf{x}_{k-1}^{[i]}, \mathbf{u}_k \right.) \\ w^{[i]} &\propto p(\mathbf{z}_k \left| \mathbf{x}_k^{[i]}, \mathbf{m} \right.) bel(\mathbf{x}_{k-1}^{[i]} \left| \mathbf{z}_{1:k-1}, \mathbf{u}_{1:k-1}, \mathbf{m} \right.) = \\ &= p(\mathbf{z}_k \left| \mathbf{x}_k^{[i]}, \mathbf{m} \right.) w_{k-1}^{[i]} \end{aligned} \tag{74}$$

which coincides with the derivation based on the Dirac's delta that we developed before.

So far we have developed the vanilla particle filter (the SIS algorithm), one that is guaranteed to work well only when an unfeasibly large number of particles is used. In short, this is because it has no provision for "driving" the suitable amount of sampled particles towards the interesting parts of the true posterior.

This leads to several important drawbacks, such as a systematic bias if the number of particles is small. To understand the origin of these defects, consider how this filter works in practice. In the prediction step, all the particles coming from the previous posterior are propagated through the transition function, becoming the particles of the new posterior after adjusting their importance weights with the observation likelihood. This includes not only those that will arrive at interesting parts of the posterior, but also the ones that will not. Notice however, that the latter will receive very small importance weights. As a result, in the long term we can expect that most particles may decrease their weights. How fast this will occur depends on the probability of arriving at wrong places of the posterior when propagated through the transition, i.e., depending on the similarity between the proposal distribution and the real posterior. This is called the *particle depletion* problem, and consists of ending up with most of the particles having small (negligible) importance weights and, therefore, with a sample-based representation that exhibits an undesirable large variance in the particle weights.

The obvious solution to the depletion problem is to replace those particles that are not relevant for the posterior by duplicating others which are relevant. This, however, is not a trivial issue since some of those "bad" particles might become "good" a few iterations later.

A proper way to address this replacement problem is through an evolution of the SIS method called the *Sampling Importance/Resampling* (SIR) algorithm (Rubin, 1987). Basically, the SIR introduces a resampling stage of N particles resulting from the last update stage, such that the probability of picking a particle is proportional to its importance weight. Then, all the importance weights are set to a constant value since the old values already have been used to determine the existence or not of elements of the sample at each point of the state space. This process has certain resemblance to Darwinian selection, which is the reason why the SIR method is also called *survival of the fittest*.

There exist a number of ways of implementing this resampling step (Douc, Capp, & Moulines, 2005). Probably the best-known method is *multinomial resampling*, which iterates N times, generating at each step a random number from the discrete distribution defined by the importance weights (therefore, the higher weight has more probability of being selected), and selecting the particle corresponding to that index. Other resampling methods and their computational complexities are described in Appendix B.

Notice that, since the resampling is done with replacement, the same particle can be selected more than once if it has a relevant importance weight. Equal particles, however, are likely to be propagated to different places at the next time step, since the transition function is stochastic.

Unfortunately, there is certainly one problem with the SIR algorithm: if we resample at every time step of the filter, it is easy to eliminate particles that have small weight at that time but could represent important regions of the posterior of future steps. The SIR, when applied every time, leads to a loss of the diversity of particles, what is known as the *particle impoverishment* problem. That is another subtle effect of maintaining a finite

Table 7. Steps in the execution of the SIR filter, along with their corresponding computational costs. For further references on the time complexities of resampling algorithms see Appendix B.

Algorithm step	Operations	Complexity
Prediction	*-Draw a sample of size* N, $\{\bar{\mathbf{x}}_k^{[j]}\}$, from $p(\mathbf{x}_k \mid \mathbf{x}_{k-1}, \mathbf{u}_k)$, taking for the $j-th$ element of the sample the conditioning value $\mathbf{x}_{k-1} = \mathbf{x}_{k-1}^{[j]}$ (and the measured motion $\mathbf{u}_k$). *-Assign to the particle* $\bar{\mathbf{x}}_k^{[j]}$ the importance factor of the particle we used for it, i.e., $\bar{w}_k^{[j]} = w_{k-1}^{[j]}$	$O(Ng)$, *where* $O(g)$ is the cost of drawing a sample (a set of values of dimension d each) from the transition pdf. Typically dominated by $O(N)$.
Update	*-Take the particles of the posterior as the predicted particles:* $\mathbf{x}_k^{[j]} = \bar{\mathbf{x}}_k^{[j]}$. *-Calculate importance factors from the ones of the predicted particles using the observation likelihood:* $\hat{w}_k^{[j]} = p(\mathbf{z}_k \mid \bar{\mathbf{x}}_k^{[j]}, \mathbf{m})\bar{w}_k^{[j]}$. *Scale the importance factors of the posterior to sum up 1:* $w_k^{[j]} = \frac{\hat{w}_k^{[j]}}{\sum_{s=1}^{N} \hat{w}_k^{[s]}}$.	$O(Nh)$, *being* $O(h)$ the cost of evaluating pointwise the observation likelihood (dimension c). Typically dominated by $O(N)$.
Resampling	*-Calculate* $N_{eff} = \frac{1}{\sum_{i=1}^{N}(w_k^{[i]})^2}$. *-If* $N_{eff} \geq \frac{N}{2}$, end step. *-Otherwise, resample* N new particles with replacement from the set $\{(x_k^{[i]}, w_k^{[i]})\}$ according to some resampling method, and then set all $w_k^{[i]} = \frac{1}{N}$.	Compute ESS: $O(N)$ Resampling (not in all iterations): $O(N)$
Total (dominant) complexity:		*Typically dominated by* $O(N)$.

size of the sample from the posterior. The result is that the filter loses its ability to cope with multi-hypotheses situations (one of the main advantages of PFs) since it keeps only the best particles. The pose estimate loses diversity.

This problem caused by resampling can be addressed in different ways, for instance by including new elements in the sample, randomly selected in the support of the posterior, or by not performing resampling at every step. The latter is quite a commonly used solution: whether the updated particles should be resampled or not at the current time step is determined by evaluating the *diversity* of the particle weights. The most common measure of this diversity is the *Effective Sample Size (ESS)*, which is defined as:

$$N_{eff} = \frac{1}{\sum_{i=1}^{N}(w^{[i]})^2} \tag{75}$$

In the worst possible situation, where only one particle has the maximum weight (1) and the rest have no weight at all (0), we have $N_{eff} = 1$. On the other hand, for the best weight diversity, all the particles would have equal weights ($1/N$), and then

$$N_{eff} = 1 / \sum_{i=1}^{N}\left(1/N^2\right) = 1/\left(N/N^2\right) = N.$$

Therefore, the ESS yields values in the range $[1, N]$, being closer to 1 when the importance

Figure 9. (Top) Real experiment we have conducted on the application of a SIR filter to the localization of the same low-cost educational Lego robot and scenario illustrated in Figure 5 (EKF). The trajectory estimated by the filter is the continuous thin curve. The point clouds correspond to the location of the sonar readings in the case that the robot pose is the one given both by odometry (light gray) and by the particle filter (dark gray), which clearly fits better the real environment, indicating a good performance of the filter (given the low-cost hardware and the fact that only one range sensor is used!). The bottom figures show selected steps of the execution of the filter, where it can be seen how the initially disperse area where we believed the robot could be shrinks as it gathers observations (the star marks each observation, located relative to the robot estimated pose).

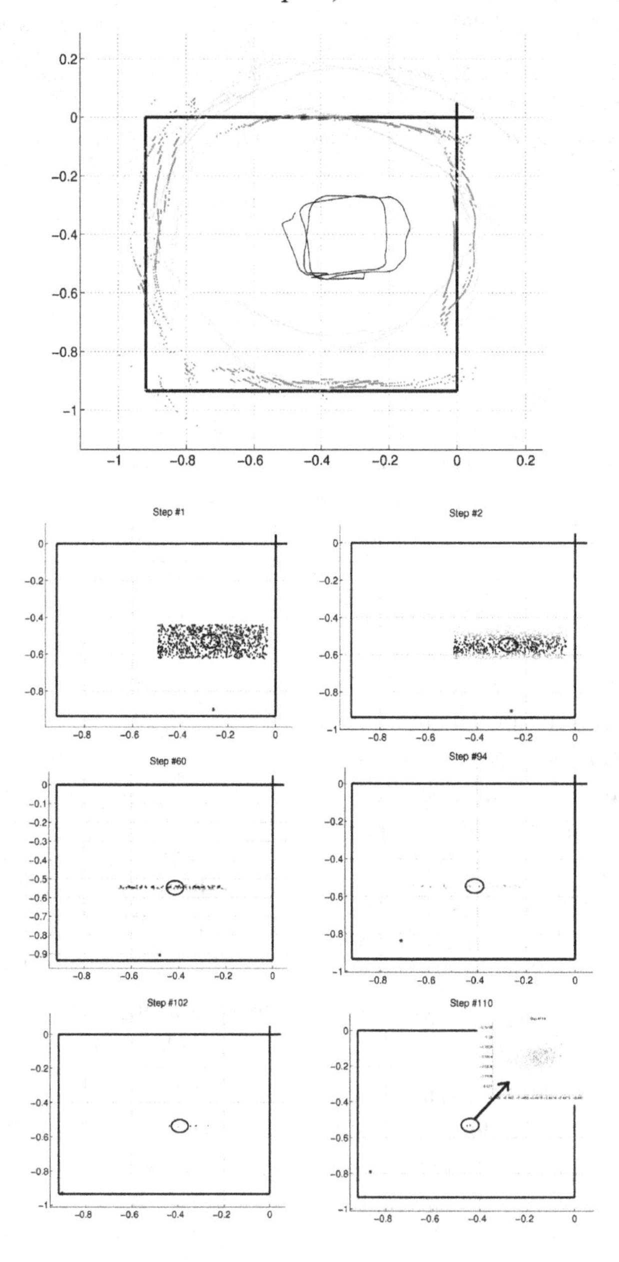

Figure 10. Snapshots of the evolution of an adaptive PF at three different instants of time: (a) initial state, (b) after moving a few meters, and (c) when the PF has almost completely converged to the actual robot pose. The map is modeled as an occupancy grid map and the sensor is a 2D laser scanner. Note that particles are represented as dots with an orientated head to reflect the heading dimension of the robot pose. In all cases, the 3σ confidence interval of the Gaussian that best fit the 2D location of all the particles is also overlaid for illustration purposes. Notice how in (b) we can deduce that most of the particles in the upper room have negligible weights since the covariance does not include them. Subsequently, all those particles would be discarded in a resampling stage, as can be verified in (c) where they have already disappeared.

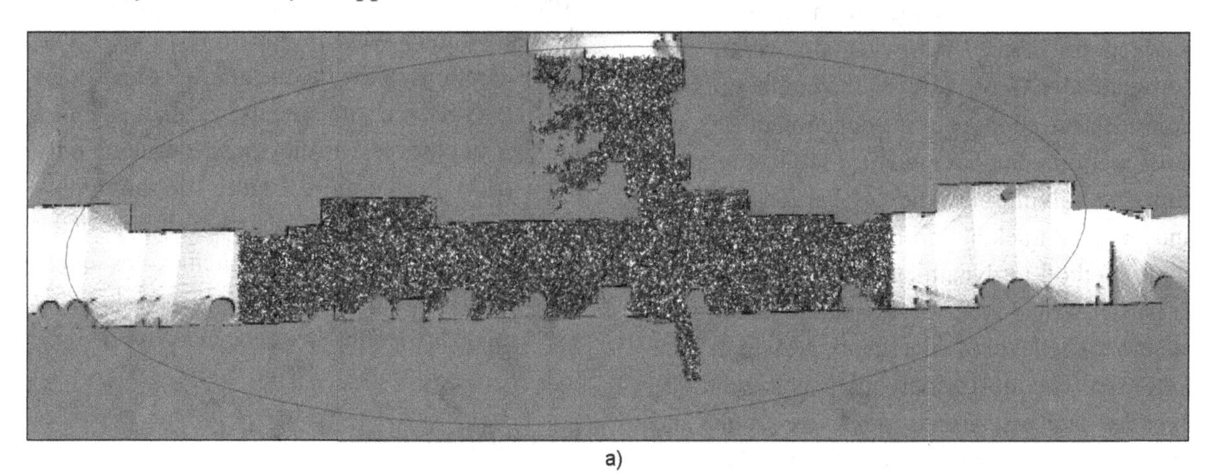

a)

b)

c)

weights are more concentrated in few particles. Precisely in those cases, there is evidence that we are maintaining many uninteresting particles, which is a motivation for resampling. In practice, a threshold in the value of the ESS (typically, $N_{eff} < N / 2$) is established for deciding whether to resample or not at each iteration.

The computational cost of the SIR particle filter algorithm and the algorithm itself is summarized in Table 7, showing an overall cost dominated by $O(N)$ with N the number of particles. Before ending our introduction of PFs, we must mention that any practical implementation aimed at solving both the "global localization" and the "pose tracking" problems should introduce some method to dynamically adapt the number of particles (and thus, its computational requirements) to the level of uncertainty present at each time step. The most widely used such technique is called and was proposed by Fox (2003) and relies on a smart approximation of the robot state space as a grid-like tessellation. If we denote the dynamic number of particles as N, and the inverse of the chi-squared cdf as $\chi^2_{1,c}$ (recall that we used it in chapter 5 section 2), it can then be shown that for:

$$N = \frac{1}{2\varepsilon}\chi^2_{k-1,1-\delta} \tag{76}$$

we can assure with a probability of $1-\delta$ that the Kullback-Leibler Divergence (KLD) between the pdf modeled in the PF and an (approximated and idealized) ground truth distribution is below a certain limit ε. The number k above corresponds to the number of bins in the tessellation of the state-space which contains at least one particle. Typical values which we use daily on our robots are $\varepsilon = 0.01$ and $\delta = 0.01$ for a grid resolution of 10cm (for both spatial dimensions) and 10 degrees (for the robot heading).

Finally, in Figure 9 you can see a real example of the execution of the SIR for a mobile robot endowed with an ultrasound range sensor, navigating along the walls of a small environment. The evolution of the particles during the robot trajectory can be clearly appreciated there. Another experiment with a real robot is shown in Figure 10. Here, a robotic wheelchair equipped with a 2D laser scanner successfully localizes itself without any prior information on its pose, i.e. it solves the "global localization" problem. Figure 10a represents the initial state of the PF, where particles spread all over the free space of the known map. After moving a few meters and a few iterations of the PF, the number of particles drastically reduces, due the usage of an adaptive number of particles, as they concentrate on only a few spots. Furthermore, Figure 10c shows how the particles have converged to the real pose of the robot. As with the EKF experiment, C++ source code based on MRPT and datasets to reproduce this experiment are available online (PF Web, 2012).

REFERENCES

Arumlampalam, M. S., Maskell, S., Gordon, N., & Clapp, T. (2002). A tutorial on particle filters for online nonlinear/non-Gaussian Bayesian tracking. *IEEE Transactions on Signal Processing, 50*(2), 174–188. doi:10.1109/78.978374

Bierman, G. J. (1977). *Factorization methods for discrete sequential estimation*. New York, NY: Academic Press.

Blanco, J. L., González, J., & Fernández-Madrigal, J. A. (2010). Optimal filtering for non-parametric observation models: Applications to localization and SLAM. *The International Journal of Robotics Research, 29*(14), 1726–1742. doi:10.1177/0278364910364165

Chen, R., & Liu, J. S. (2000). Mixture Kalman filters. *Journal of the Royal Statistical Society. Series B. Methodological, 62*(3), 493–508. doi:10.1111/1467-9868.00246

Coppersmith, D., & Winograd, S. (1990). Matrix multiplication via arithmetic progressions. *Journal of Symbolic Computation, 9*(3), 251–280. doi:10.1016/S0747-7171(08)80013-2

Cummins, M., & Newman, P. (2008). FAB-MAP: Probabilistic localization and mapping in the space of appearance. *The International Journal of Robotics Research, 27*(6), 647–665. doi:10.1177/0278364908090961

Dellaert, F., & Kaess, M. (2006). Square root SAM: Simultaneous localization and mapping via square root information smoothing. *The International Journal of Robotics Research, 25*(12), 1181–1203. doi:10.1177/0278364906072768

Douc, R., Capp, O., & Moulines, E. (2005). Comparison of resampling schemes for particle filtering. In *Proceedings of the 4th International Symposium on Image and Signal Processing and Analysis,* (pp. 64-69). IEEE.

Doucet, A., De Freitas, N., Murphy, K., & Russell, S. (2000). Rao-Blackwellised particle filtering for dynamic Bayesian networks. In *Proceedings of the Sixteenth Conference on Uncertainty in Artificial Intelligence*, (pp. 176-183). IEEE.

Doucet, A., Godsill, S., & Andrieu, C. (2000). On sequential Monte Carlo sampling methods for Bayesian filtering. *Statistics and Computing, 10*(3), 197–208. doi:10.1023/A:1008935410038

Doucet, A., & Johansen, A. M. (2008). *A tutorial on particle filters and smoothing: Fifteen years later*. British Columbia, Canada: University of British Columbia.

Fox, D. (2003). Adapting the sample size in particle filters through KLD-sampling. *The International Journal of Robotics Research, 22*(12), 985–1003. doi:10.1177/0278364903022012001

Fox, D., Burgard, W., Dellaert, F., & Thrun, S. (1999). Monte Carlo localization: Efficient position estimation for mobile robots. In *Proceedings of the National Conference on Artificial Intelligence (AAAI)*. AAAI.

Fox, D., Burgard, W., & Thrun, S. (1999). Markov localization for mobile robots in dynamic environments. *Journal of Artificial Intelligence Research, 11*, 391–427.

Gelb, A. (Ed.). (1974). *Applied optimal estimation*. Cambridge, MA: The MIT Press.

Georgakis, C. (1994, February). A note on the Gaussian integral. *Mathematics Magazine*, 47. doi:10.2307/2690556

Gordon, N. J., Salmond, D. J., & Smith, A. F. M. (1993). Novel approach to nonlinear/non-Gaussian Bayesian state estimation. *IEE Proceedings. Part F. Communications, Radar and Signal Processing, 140*(2), 107–113. doi:10.1049/ip-f-2.1993.0015

Jazwinsky, A. H. (1970). *Stochastic processes and filtering theory*. New York, NY: Academic Press.

Julier, S. J., & Uhlmann, J. K. (1997). A new extension of the Kalman filter to nonlinear systems. In *Proceedings of the International Symposium Aerospace/Defense Sensing, Simulation and Controls,* (vol 3068), (pp. 182-193). Berlin, Germany: Springer.

Kalman, R. E. (1960). A new approach to linear filtering and prediction problems. *Transactions of the ASME Journal of Basic Engineering, 82*, 35–45. doi:10.1115/1.3662552

Kanwal, R. P. (1998). *Generalized functions, theory and technique* (2nd ed.). Berlin, Germany: Birkhäuser.

Kassam, S. A. (1988). *Signal detection in non-Gaussian noise*. Berlin, Germany: Springer-Verlag.

Koller, D., & Friedman, N. (2009). *Probabilistic graphical models: Principles and techniques*. Cambridge, MA: The MIT Press.

Kuipers, B. J. (1977). *Representing knowledge of large-scale space.* (Doctoral Dissertation). MIT. Cambridge, MA.

Lego. (2011). *Lego mindstorms robots homepage*. Retrieved July 12, 2011, from http://mindstorms.lego.com/en-us/Default.aspx

Li, T. S. (2003). *On exponentially weighted recursive least squares for estimating time-varying parameters*. Hawthorne, NY: IBM. doi:10.1080/15598608.2008.10411879

Raiffa, H., & Schlaifer, R. (2000). *Applied statistical decision theory*. New York, NY: Wiley-Interscience.

Rubin, D. B. (1987). A noniterative sampling/importance resampling alternative to the data augmentation algorithm for creating a few imputations when fractions of missing information are modest: The SIR algorithm. *Journal of the American Statistical Association, 82*(398), 543–546. doi:10.2307/2289460

Schön, T. B. (2006). *Estimation of non-linear dynamic systems: Theory and applications.* (Doctoral Dissertation). Linköpings Universitet. Linköpings, Sweden.

Serway, R. A., & Jewett, J. W. (2009). *Physics for scientists and engineers*. New York, NY: Brooks Cole.

Smith, R. C., & Cheeseman, P. (1986). On the representation and estimation of spatial uncertainty. *The International Journal of Robotics Research, 5*(4), 56–68. doi:10.1177/027836498600500404

Strassen, V. (1969). Gaussian elimination is not optimal. *Numerische Mathematik, 13*(4), 354–356. doi:10.1007/BF02165411

Thomas, D. B., Leong, P. H. W., Luk, W., & Villaseñor, J. D. (2007). Gaussian random number generators. *ACM Computing Surveys, 39*(4). doi:10.1145/1287620.1287622

Ulrich, I., & Nourbakhsh, I. (2000). Appearance-based place recognition for topological localization. In *Proceedings of the IEEE International Conference on Robotics and Automation*, (pp. 1023-1029). IEEE Press.

Wan, E., & Van der Merwe, R. (2000). The unscented Kalman filter for nonlinear estimation. In *Proceedings of Symposium on Adaptive Systems for Signal Processing, Communication and Control (AS-SPCC)*. AS-SPCC.

Web, E. K. F. (2012). *Kalman filters website of the MRPT project*. Retrieved Mar 1, 2012, from http://www.mrpt.org/Kalman_Filters

Web, P. F. (2012). *Application:pf-localization website of the MRPT project*. Retrieved Mar 1, 2012, from http://www.mrpt.org/Application:pf-localization

Section 3
Mapping the Environment of Mobile Robots

Chapter 8
Maps for Mobile Robots:
Types and Construction

ABSTRACT

This is the first chapter of the third section. It describes the kinds of mathematical models usable by a mobile robot to represent its spatial reality, and the reasons by which some of them are more useful than others, depending on the task to be carried out. The most common metric, topological, and hybrid map representations are described from an introductory viewpoint, emphasizing their limitations and advantages for the localization and mapping problems. It then addresses the problem of how to update or build a map from the robot raw sensory data, assuming known robot positions, a situation that becomes an intrinsic feature of some SLAM filters. Since the process greatly depends on the kind of map and sensors, the most common combinations of both are shown.

CHAPTER GUIDELINE

- You will learn:
  - The main kinds of maps a mobile robot can use.
  - Their main characteristics and advantages/disadvantages in the context of recursive Bayesian localization and mapping.
  - The connection between purely geometrical approaches and more abstract (cognitive) ones.
  - How to estimate some kinds of maps assuming perfectly known robot localization.
- Provided tools:
  - A comprehensive discussion on the pros and cons of each kind of map, which can be used to choose the more appropriate for your application.
  - Detailed formulations for building occupancy grids and landmark maps. For the latter we include the particular problems of having range-bearing, bearing-only, and range-only sensors.

DOI: 10.4018/978-1-4666-2104-6.ch008

- Relation to other chapters:
  - The extensive discussion on the different kinds of maps in this chapter complements some concepts that appeared already in chapters 6 and 7.
  - Solving the mapping-only problem in this chapter serves as a base for addressing the more complex SLAM problem in chapters 9 and 10.

1. INTRODUCTION

Some approaches to autonomous robots intend not to use any explicit representation of the environment, even not to use *any* representation at all, aiming at employing the environment itself as its own best model and considering the robot as part of it (Brooks, 1991). However, having an internal model of space—a map—is currently the only known practical way of efficient and optimally planning actions (taking decisions) to operate in the long-term. Furthermore, maintaining such a map has not been ruled out as a real possibility by the modern embodied cognition paradigm (Anderson, 2003), which has extended the seminal work based on not employing any model at all. It has been shown that humans use a so-called *cognitive map* to plan routes and locate in our environments (Tolman, 1948); a different story is whether the map is explicitly stored in our brains—up to date this possibility seems controversial at least—or emerges from a set of complex interactions with the environment, if we choose a developmental perspective. In state-of-the-art mobile robot mapping and localization, the former is the dominant approach, and consequently that is the one followed in this book.

While discussing robot localization in previous chapters we already faced the most common metrical environment representations, namely grid maps and landmark or feature-based maps. This revealed the tight coupling existing between localization and mapping: the robot cannot localize well if it does not have a good map, but on the other hand, a map cannot be accurately reconstructed if the robot is poorly localized. In this chapter, we will widen and organize this perspective on maps by describing the best-known types of explicit spatial representations for robots. A few of them are quite common in the robotic localization and mapping literature, while others have very restricted niches of usability, i.e., they respond to very specific sensors, environments or tasks. The existing variety makes evident the diversity and complexity of the problems arising when an automatic mobile device is intended to operate autonomously in a given spatial region: no single map representation seems to be universally valid for all the tasks of a given robot; in fact, the choice of the map has important implications in these tasks (e.g. navigation, manipulation, etc.) that extend well beyond the issues of localization and mapping addressed in this book.

After describing the different map types, this chapter addresses how to build and update those maps from the experiences of a mobile robot, that is, from its raw sensory data. Although localization and mapping are tightly coupled, within the scope of this chapter we will assume a perfectly known robot pose in order to clarify the problem of map building. In general, metrical mapping-only can be performed through Bayesian filters much like the ones described in the localization section of this book, although the much higher dimensionality of the problem sometimes forces us to adopt approximations to make it tractable. Note that approaching only the mapping part of the localization+mapping problem could be seen as a dual perspective to that already studied in chapter 7 while explaining robot localization, where a perfect knowledge about the map was assumed in order to study how to estimate the robot pose alone. However, while we can easily think of situations where the robot is endowed with a map of its environment by its designers or operators and then left on its own to localize itself, to know its position without having a map is a rarer situation—though a possible one. The utility of artificially decoupling these problems also comes from the fact that to update a map assuming perfectly known robot poses has become part of one family of popular SLAM methods, namely those relying on a Rao-Blackwellized Particle Filter (RBPF) approach, which will be introduced in chapter 9. Thus, the mapping-only methods provided in this chapter are not only a suitable way of introducing the mapping topic, but essential algorithms included at the core of many state-of-the-art SLAM implementations.

In summary, the chapter is structured as follows. Section 2 describes the most commonly used types of spatial representations, ranging from purely metrical maps to the most abstract representations. Then, the next two sections are devoted to probabilistic techniques for learning grid and landmark maps, respectively, always under the perspective of recursive Bayesian estimators. Finally, section 5 covers some map building algorithms which are quite common and thus worth knowing, in spite of not relying heavily (or even at all) on probabilistic foundations.

2. EXPLICIT REPRESENTATIONS OF THE SPATIAL ENVIRONMENT OF A MOBILE ROBOT

In this section, we describe the most relevant types of maps in mobile robotics. Many of them arose in the first decades of the discipline, and some have survived almost unchanged until today. Others were so specific that their use is now rather limited or obsolete.

Robotic maps can be classified, at a first glance, into two broad classes. Although there exists no consensus about how to refer to them, we will write here about *symbolic* maps and *sub-symbolic* maps. Maps within the former category comprise discrete elements (typically, real-world objects of certain complexity) that can be distinguished from the much less informative "background" of the environment through some sort of processing of the sensory data. This does not necessarily imply non-metrical maps: maybe such a map includes topological relations between these elements or maybe only their metrical location. In general, symbolic maps minimize storage needs, but at the expense of a higher computational cost for detecting, identifying (the problem of data-association introduced in chapter 6 section 4), tracking and maintaining the distinctive elements. On the contrary, sub-symbolic maps do not deal (much) with the problem of distinguishing things: they aim at representing perceptions, either containing special objects or not. Obviously, this shifts the cost from computation to storage, following the general principle in computer science that if one wishes to reduce the computational cost of any program it is likely that he or she will have to pay an increase in its storage needs.

A problem with representing the environment through a symbolic approach (with elements that have some kind of "meaning" or "semantics," at least for the human observer) is the automatic creation of the first symbols from sub-symbolic information, which at the end is the only one available through the robot sensors: this is the linguistic and also philosophical *symbol grounding problem* (Harnad, 1990). Few practical solutions to this problem have been reported in the robotics literature, except for *anchoring* (Coradeschi & Saffiotti, 2003), a general framework that allows the robot to link perceptions of external physical objects to symbols in its internal map—the symbol grounding problem also copes, in addition, with non-physical phenomena. A comprehensive and coherent implementation of anchoring is still an open issue, but some promising research regarding the emergence of symbols from the interaction of the robot with its environment is being conducted in the areas of developmental robotics and others.

To choose the right kind of representation for a particular problem is not straightforward, and a good knowledge of the pros and cons of each one is required. Furthermore, the frontier between symbolic and sub-symbolic representations is far from clear, as we will see in the next paragraphs. The particular kind of map to use depends on: the robot tasks, the scale of the environment, whether the scenario is indoor or outdoor, the required precision and available computational resources, the need of fusing information from different sensors, the nature of these sensors, etc. In addition, the kind of map will determine which estimation procedures will be applicable. For instance, in the case of sub-symbolic, metrical maps, occupancy grids are usually linked to non-parametric filters, while landmark maps, where the uncertainty is typically assumed to be Gaussian, are more akin to parametric (EKF-based) estimators; in the case of symbolic, topological maps, Bayesian estimation has been applied only recently.

It is worth to notice from the beginning that the current most common and practical approaches to mapping lie on the sub-symbolic category: they represent the environment using just metrical data such as landmarks, distances to obstacles, free-space, etc., that are obtained from sensory data with little processing. That is, the most used maps are currently pure geometrical representations. Therefore, there exists a solid mathematical background to automatically construct metrical maps, as we will see. However, the problem of modeling under such approaches large-scale space—physical space that cannot be entirely perceived from a single vantage point (Kuipers, 1977)—does not have a well-established solution yet, mostly due to the hurdles associated to the problem of revisiting already known places, or *loop closure* in a sub-symbolic setting. Metrical simultaneous localization and mapping in large-scale spaces will be dealt with in chapter 10.

In the next paragraphs, we review the most representative kinds of explicit maps that have been in use during the history of mobile robots, ordered from the "less symbolic" to the most. We provide a few bibliographical references for each map type, such that the reader could explore deeper if interested.

Grid Maps

Possibly the best known, most popular and most basic type of map for mobile robots—in the sense of the little processing done on sensory data—is a regular tessellation of space, or grid, which cells represent the probability of the corresponding spatial region to be occupied by solid objects (see Figure 1). A grid map or occupancy grid is a *random field* (a spatial region where each point has an associated r.v.) composed of cells from which we are interested just in one property: its occupancy. Being this occupancy a discrete-valued property with only two possible real values (occupied or free), each cell can be modeled as a Bernoulli distribution. Hence, a grid map consists of a collection of such distributions, one for each cell, although that does not mean that the r.v.s are independent. The most common form of grid map is two-dimensional (where cells are square areas), although sometimes it has been employed in a three-dimensional form (with cells being small cubes, also called *voxels*). The main ideas date back to almost thirty year ago, when grid maps were proposed for mapping the environment of mobile robots by using ultrasound range sensors (Moravec & Elfes, 1985), but today they are still pervasively used. They are closely related to localization and mapping methods as remarkable as Markov Localization or Particle Filters.

Since the cells of a grid map are used for holding evidence-of-occupation values, this kind of map seamlessly fits into Bayesian recursive estimators. In particular, it naturally fits into certain family of non-parametric RBE techniques for SLAM, as will be seen in chapter 9 section 3. These maps also fit well into possibilistic representations of uncertainty, as in the case of fuzzy grid maps. In addition, most grid maps assume that each cell is conditionally independent on each other (as discussed in section 3), thus there is no a priori knowledge about the spatial structure of the environment. For example, these maps do not assume we are in an indoor or outdoor scenario.

Figure 1. Real examples of occupancy grid maps built for (a) our Málaga 2006 dataset (Blanco, Fernández-Madrigal, & González, 2008), (b) the Intel dataset (Fox, 2003), and (c) the New College dataset (Smith, Baldwin, Churchill, Paul, & Newman, 2009). All maps have been created with applications from the MRPT (2011). In section 3, we present an RBE that is able to estimate occupancy grids conditioned on the (known) poses and observations of the robot.

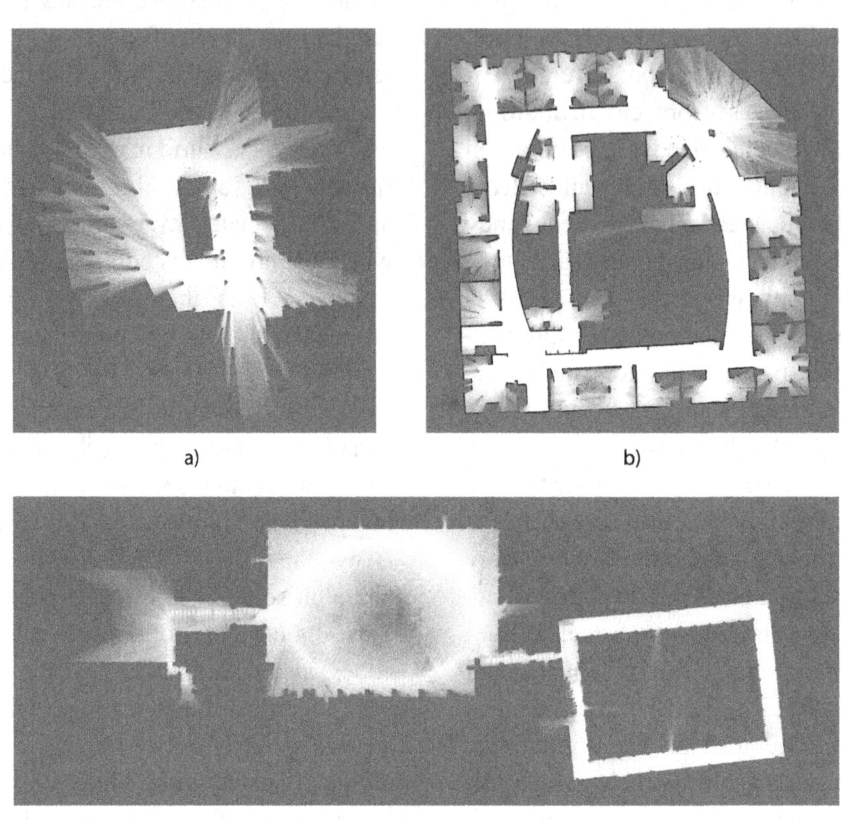

Since the only information of interest for each cell is its occupancy, grid maps are a valuable tool when merging information coming from different sensors, as long as all of them can provide evidence of occupancy. When a robot is equipped with sensors capable of detecting different kinds of obstacles (e.g. ultrasonic sensors detect glass doors but cannot perceive some textile fabrics, while a laser range scanner behaves exactly oppositely), care must be taken with sensor fusion since the evidence for occupation yielded by different sensors may be contradictory to each other, so different grid maps should be maintained in that case and fused together only through functions that mix appropriately the uncertainty. Letting this issue apart, the general adequacy for sensor fusion is a valuable characteristic of grid maps that increases the robustness of many operations: localization, navigation, exploration, etc. Still another advantage of grid maps is their close correspondence to the metrical motion of the robot: both the map and the motion refer to the same frame of reference, thus, for instance, translating motion into the map to find out the robot position within the grid is straightforward.

The most evident disadvantage of grid maps is that they impose important storage requirements, especially in their three-dimensional version, which can render them unfeasible for large-scale scenarios. For coping with that, they can be constrained to represent separate, small areas of the environment, linking several grids for modeling the whole space; this approach, however, is associated to hybrid metric-topological SLAM, which introduces its own challenges (see chapter 10). Another solution to the curse of dimensionality is adaptive tessellation, i.e., increasing the resolution of the grid around the most interesting areas, those with a higher variability in their occupancy. A recent development in this sense is the use of the abstract data type called *octree* to structure 3D grids: an octree is a mathematical tree in which each node represents the space contained in a cubic volume of space, or voxel; this volume is recursively subdivided into eight sub-volumes until a given minimum voxel size is reached, typically constrained by the sensor noise. Unknown regions or regions containing uniform information are not subdivided further. Different views of space can be obtained by selecting the appropriate amount of detail. An octree implemented for robot mapping is called an *octomap* (Wurm, Hornung, Bennewitz, Stachniss, & Burgard, 2010).

Another difficulty with grid maps is to integrate them with mobile robots that use high-level reasoning, which is mostly symbolic. Some methods exist to extract "more symbolic" information out from grids, for instance distinguishing wide open spaces and narrow passages connecting them (Fabrizi & Saffiotti, 2000), which leads to a topology of the environment very convenient for planning routes for navigation. Processing the information from a grid map to obtain higher level knowledge—i.e., a "more symbolic" map—can be done, for example, by means of image processing and computer vision techniques, considering the grid to be a gray-level image; one could identify then discrete entities such as walls by means of the Hough transform for lines (Duda & Hart, 1972), and employ those objects as landmarks within a landmark-based map.

Finally, a few techniques have been proposed in the literature for merging different grid maps (Konolige, Fox, Limketkai, Ko, & Stewart, 2003; Birk & Carpin, 2006). This problem requires identifying which parts of two grids overlap, provided that an arbitrary rotation may exist and that the overlap may typically be only partial, and it is related to the map matching methods described in chapter 6.

Point-Based Maps

An alternative to grid maps which alleviates their problems with storage is to use a discrete metrical representation that does not include empty space. For example, point maps represent solid parts of the environment that can be sampled by the robot through suitable sensors, usually range or range-bearing devices (recall chapter 2 sections 6 – 8). Due to that sample-based nature, it is common for these maps to be informally called *point clouds*, especially when representing three-dimensional objects or environments (see Figure 2).

Point maps were first devised in the early 90s for working with range scanners. They do not include any explicit model of uncertainty, which is one of their main drawbacks for our purposes. Also, this lack of probabilistic foundations complicates the fusion of point clouds produced by different kinds of sensors; unlike with occupancy grid maps, there is not a mathematically well-founded method for fusing several observations, even when they are captured by the same scanner from different positions in an environment, being the most common approach the brute-force accumulation of all the sensed points.

Point maps can also be processed to obtain "more symbolic" items, for example finding segments that represent walls or corners, a kind of processing that was common in mobile robotics in the nineties. The result can be considered as a sort of feature or landmark map (see further on). Nowadays, in part thanks to the recently renewed

Figure 2. (a) A point map of the same environment already shown in Figure 1a as a grid map. (b) 3D point cloud representing a "bunny figure" scanned with a Cyberware 3030MS optical triangulation scanner (data set courtesy of the Computer Graphics Laboratory/Stanford University).

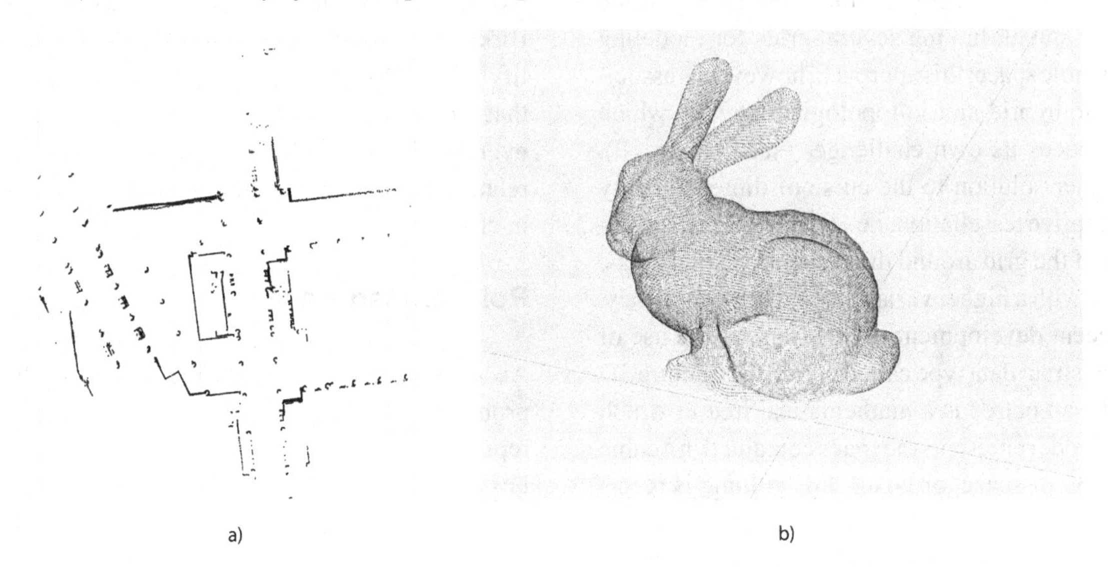

interest in point clouds especially when coming from 3D range cameras, the identification of interest points in point clouds is again an active topic in the research community (Johnson & Hebert, 1999; Rusu, Marton, Blodow, Dolha, & Beetz, 2008; Tipaldi & Arras, 2010).

Free-Space Maps

A dual alternative to point-based maps is that of representing *only free space*. Historically, this representation had an especial interest for motion planning in mobile robotics, since the robot can only move in the portion of the environment that contains no obstacles. Today, however, these maps have less interest since free space can be easily deduced from its dual information (the presence of solid objects, or obstacles), which is provided, for example, by point maps or occupancy grid maps.

Maps can represent free-space in different ways (see Figure 3): by using geometrical shapes such as trapezoids, generalized cones and others (Brooks, 1982); by partitioning space (Lozano-Perez & Wesley, 1979); or by finding regions of particular interest for some operation (typically, for collision-free navigation). For instance, Generalized Voronoi Graphs represent a convenient tool for modeling those regions of space that are equidistant to all obstacles, that is, those places where the robot minimizes its risk to collide with the environment (Rotwat, 1979; Choset & Burdick, 1995). Since in all these approaches free space must be deduced from occupied space—the only one perceived by sensors—some processing of the latter is needed (i.e., computational cost). In addition, since they do not explicitly include uncertainty, they are not a direct choice for probabilistic frameworks. On the contrary, they were suitable in classic motion planning algorithms (Latombe, 1991).

As any map resulting from processing raw data gathered by sensors, free-space maps are closer to symbolic representations than point-based or grid maps; for example, it is easy or even straightforward to obtain a topology of space from them.

Figure 3. Two ways of representing free space for the same environment: (a) by means of geometrical shapes and (b) with generalized Voronoi graphs

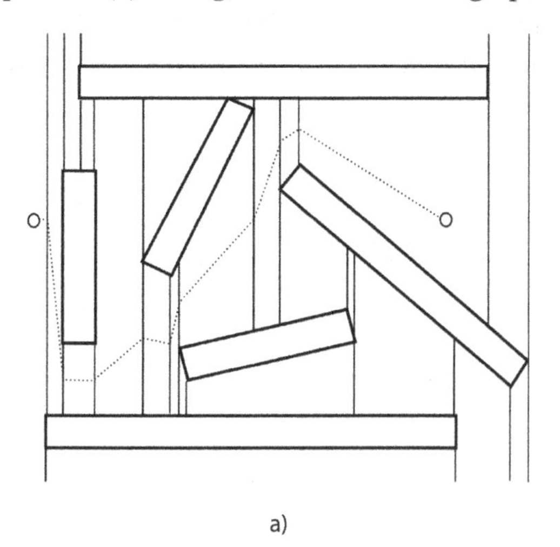

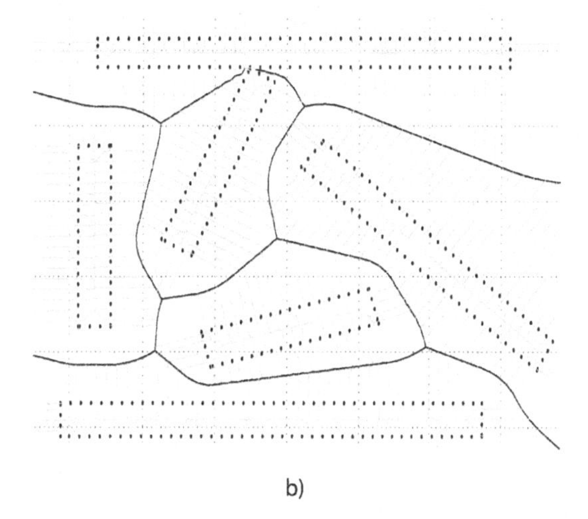

Feature or Landmark Maps

The maps described above reflect either the data coming directly from sensors (with no processing) or sensory data processed just to obtain information about regions of the environment that are of general interest. The next logical step is to process the data less generically in order to detect distinctive physical elements of the environment. These elements may have some meaning for us humans (walls, corners, doors—in indoor scenarios—trees, roads, other vehicles—in outdoor scenarios) or not (abrupt changes in perceived brightness, regions filled with some color, lines, textures, etc.). All of these lead to the category of *feature* or *landmark* maps, which are, along with grid and point maps, one of the most important kinds of maps used today in mobile robotics.

In contrast to grid maps, which do not infer any spatial structure from the data beyond the imposed, fixed tessellation of space, landmark maps do contain well-distinguished elements along with their (estimated) spatial location. Both of them are metric maps—when a topology is included in feature maps they fall into a different category—and both can include assumptions about the uncertainty of their content, therefore both can be (and are) used extensively in probabilistic frameworks, as we have already seen in the chapters on localization. In particular, it is common in feature maps to model uncertainty with Gaussians; hence, they are commonly associated to parametric EKF-like filters (Leonard & Durrant-Whyte, 1991).

Landmark maps can be built from any sensor from which salient, well-distinguished features are extracted by means of a suitable detection algorithm. Such detectors have been proposed for 3D range cameras (Johnson & Hebert, 1999; Rusu, Marton, Blodow, Dolha, & Beetz, 2008), for the more conventional 2D range scanners (Núñez, Vázquez-Martín, del Toro, Bandera, & Sandoval, 2010) and even for sequences of sonar readings (Tardós, Neira, Newman, & Leonard, 2002). However, the most active research field for landmark maps in the last years has probably been localization and SLAM with imaging sensors. Some examples of features detected in different sensory data are shown in Figure 4. One reason for this favoring of landmarks in the computer vision community is that working with a few landmarks extracted from each video frame (ranging from a dozen up to a few hundreds, depending on the approach) means an immense reduction of the

Figure 4. Examples of features detected in readings from different types of sensors. (a) A picture of the Champ de Mars and (b) the visual features detected on it by the FAST detector algorithm. (c) An original points sensed by a laser scanner, and (d) the multiresolution features identified by FLIRT. The latter two pictures were generated from data and software courtesy of Gian Diego Tipaldi (Tipaldi & Arras, 2010).

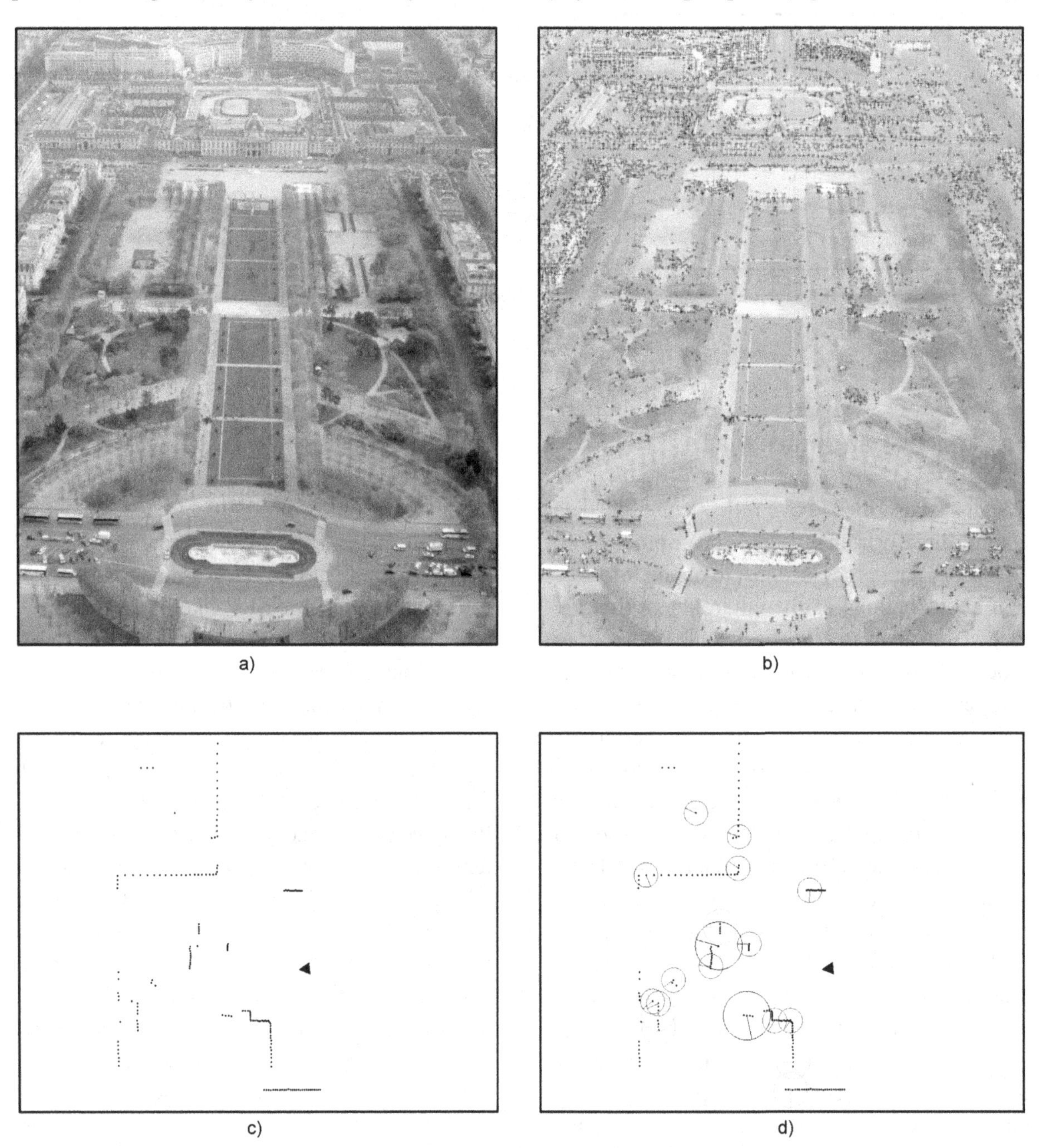

information volume provided by the sensor, that would be hard to handle in any other way if one pretends to achieve real-time performance. Additionally, the process of detecting salient features assures that the entities or objects considered in the maps are easy to redetect from nearby locations, thus facilitating the association problem. The consequence is that methods to extract features from the images provided by vision sensors can be easily found in the extensive existing literature about computer vision (Lowe, 2004; Mikolajczyk, & Schmid, 2005; Trucco & Verri, 1998; Nixon & Aguado, 2008).

We must make a warning here. Since performing vision-based localization and SLAM in static, indoor scenarios has become increasingly functional and robust during the last decade with state-of-the-art approaches, and that research progresses with steady pace, readers without a computer vision background could fall in the wrong preconception that detecting interesting objects or features in images is an uncomplicated task, due to the astonishing effortlessly way in which we humans can interpret our surroundings from a quick look around. We must remark now that even with the latest techniques from computer vision, we are still far from automatically and robustly detecting in images what we, humans fitted with a powerful and not well understood sense of sight—which does not include only our "sensors," the eyes—consider natural and distinctive characteristics. Today, computers have reached a performance level that makes most theoretical methods applicable, but not all mobile robots can carry on such a computational power, and anyway the cost grows with the complexity and number of features to detect, becoming impractical in some situations. Not to mention the lack of intelligence of computers performing this detection task (when a human "sees" a distinctive object, many cognitive processes are running to make the perception task converge).

An additional issue with landmark maps when working with cameras is that the spatial location of features is not an observable variable, or at least not all its dimensions are observable. This simply means that we cannot tell the depth of a particular pixel in the image from one single video frame. Probabilistic (and some non-probabilistic) techniques can however deal with this ambiguity by means of collecting observations of features from slightly different point of views and then fusing all the information (Klein & Murray, 2007; Civera, Davison, & Montiel, 2008). All those methods, however, must introduce a *scale factor* for the map, since the real size of objects cannot be estimated from pure mathematics out of images gathered by one single camera. When we see a picture or a movie, our brain is able to interpret the scale of the objects only because we identify them within some particular context, from which our experience—cognitive processing!—infers the correct scale. In particular, a problem still open in mapping with a single camera is to avoid the drift of this world scale factor over time, which in practice leads to important inconsistencies in the reconstructed maps (Strasdat, Montiel, & Davison, 2010). The non-observability of depth is a trouble not found when employing more than one camera (e.g. stereo camera pairs) or 3D range cameras (refer to chapter 2 section 8).

Returning to the more generic discussion on landmark maps, they have a unique facet not shared by any other map type: as long as we attempt to detect individual features from within the environment, we could wrongly detect a feature (false positive) or could miss it (false negative). Even more importantly: features must be associated to previously detected features in order to decide whether they refer to the same physical element, since they should not be introduced in the map twice as different landmarks. This is what leads to the Data Association problem (DA), already discussed in chapter 6, which is computationally intractable in its exact form and must be approximated.

On the other hand, a clear advantage of the extraction of features from sensory information is that they do not contain all the details of the underlying physical elements, that is, they are more abstract, and consequently, more robust: they vary less over time. A feature does not include all the noise conveyed by the sensor, either because we have corrected it or because we have ignored it. In contrast, a grid map, for example, is more subjected to reflect small variations in the data due to the stochastic sensor behavior, which is directly reflected in the map.

Another advantage, as we have already mentioned in the case of cameras, is that, in general, feature maps can highly reduce the storage needs with respect to grid maps, which leads to important improvements in efficiency in mobile robot operations, if we disregard the cost of feature extraction and data association.

Finally, feature maps are sub-symbolic maps but very close to be symbolic. As we have explained before, there does not exist a crisp frontier between both broad map categories. We can highlight here the fact that most types of features in the literature do not represent objects to which a human would assign any meaning, and logic relationships between them are not included in the map, but feature-based maps can be a logical and very natural basis for constructing higher-level, purely symbolic representations of the environment; therefore, they play an important role when the robot is to be enhanced with artificial reasoning capabilities.

Relational Maps and Topological Maps

A map from any of the kinds discussed above contains an intrinsic relationship between its constituents (cells, points, free-space regions, features) and the underlying space where the corresponding physical elements exist: the location of the formers onto the latter. This spatial relationship is the one that provides those maps with their metrical nature, but it is not the only one that can be included and exploited in a metrical map.

The simplest types of explicit relationship that can be added to a metrical map, apart from location, are still metrical, although they are not defined between the elements of the map and the underlying space, but among the elements themselves. For example, we can include metrical constraints that should be satisfied by the spatial locations of elements or groups of elements of the map. These so-called *constraint maps* are particularly suited for minimizing the number of hypotheses arising in a probabilistic framework and for correcting errors that are inconsistent with the imposed constraints. When the elements in the map are not world objects but robot poses, we have the popular *pose constraint map* representation (Konolige, 2005; Arras, Castellanos, Schilt, & Siegwart, 2003; Grisetti, Grzonka, Stachniss, Pfaff, & Burgard, 2007), which lies at the core of graph-SLAM approaches, described in chapter 10. To provide these maps with information, individual robot observations are stored at each pose node, leading to the so-called *view-based maps* (Konolige, Bowman, Chen, Mihelich, Calonder, Lepetit, & Fua, 2010), where an explicit representation of the environment itself does not exist. Those maps can be used to aid in splitting the environment into different areas or sub-maps, an approach that facilitates dealing with large-scale environments (Blanco, González, & Fernández-Madrigal, 2009).

Whenever some method is available to split the spatial environment into regions, these regions can be connected by *topological* relations, which are spatial properties unaffected by continuous changes of shape or size of the regions; examples of use of such relations are "element A is to the left of element B" or "region A can be reached from region B through navigation." A map that explicitly includes this type of relations is called in mobile robotics a *topological map* (Kuipers, 1978). A topological relation can also be constructed from non-metrical data, although that is rather uncommon. Topological relations and, in general, any non-metrical relation, are "elastic":

you can project a topological map onto a plane, placing its elements at arbitrary positions of the metrical space, and the map will always be valid. They have a direct use in mobile robot operations at levels higher than motion control, such as route planning (navigation) or manipulation, and can also serve for spatial reasoning. Finally, metrical and topological maps can be combined in the so-called *hybrid* metrical-topological maps or *multi-resolution metrical maps*, which are a promising approach to represent large-scale space efficiently (Blanco, Fernández-Madrigal, & González, 2008).

Most maps including explicit relationships existing among their elements can be implemented through the abstract data type and corresponding mathematical entity called mathematical *graph* (Trudeau, 1994). A graph is a set of nodes (elements) and arcs (relations). It can be directed (if arcs represent a one-way relation) or not, can contain loops (relations affecting only one element), can have some information attached to arcs (annotated graphs, also called *networks* in the mathematical literature) and can provide support to different types of relations in the same map (*multigraphs*). A huge body of literature on graphs and their uses is available, and thus we do not include here any formalization or computational specification of graphs. Basic computational operations on graphs include path searching or routing, that is, finding a sequence of adjacent arcs with certain properties, which is at the core of the route planning operation for mobile robots. Since mathematical trees are a special kind of graphs, graph theory is useful more generally in mobile robotics, and applied on diverse problems: data association, octomaps, etc. However, care must be taken when using graphs since many operations on them are computationally intractable.

An issue with maps that include explicit relationships is their fitting into probabilistic frameworks, especially when those relations are not metrical, as is the case of topological maps. In hybrid metrical-topological maps, the metrical part does not suffer from this, and thus a common solution is to handle the uncertainty of the topology separately from the metrical uncertainty, using especial approaches for that (of course, you have to take into account the fact that both uncertainties are interdependent). Recently, this has been addressed by dealing probabilistically with the space of topological maps: a probability is assigned to each possible topology (Ranganathan & Dellaert, 2011; Blanco, Fernández-Madrigal, & González, 2008). The obvious problem with that is the combinatorial size of this space, which makes intractable its exhaustive exploration and forces us to deal with approximations.

Explicit spatial relationships are, in general, well suited for representing information from the environment that does not change much over time, i.e., that is robust, as we have mentioned about feature maps. Some examples of these maps are shown in Figure 5. They are the second step towards symbolic maps.

Symbolic Maps and Semantic Maps

We must insist once more in the fact that the frontier between sub-symbolic and symbolic representations of space is quite vague. To begin with, *symbols* belong to the scientific area that studies human cognition. Although they are obviously also used in other areas, such as linguistics, it is necessary to always start from the concept of what a symbol is in our brains. It seems very likely that the human mind uses symbols in some form (and probably some other animal species), but it has not been demonstrated yet which entity within the processes of a brain could possibly be associated with symbols: neural connection patterns? neural activation patterns? the dynamics of other cells different from neurons? The hypothetical cognitive processes that produce symbols from sub-symbolic information are also unknown. There is, in fact, a philosophical problem concerning symbols, the already mentioned symbol grounding problem: since symbols, by definition, must convey some meaning, how that meaning is

Figure 5. (a) A pose-constraint map, where robot poses (or "frames") and constraints are represented as corners and edges among them, respectively. The representation has been built with applications from our software library MRPT (2011), using a two-dimensional dataset published by (Grisetti, Stachniss, Grzonka, & Burgard, 2007). (b) Example of how an environment can be assigned a set of "distinctive" places. (c) An example of a topology for a map.

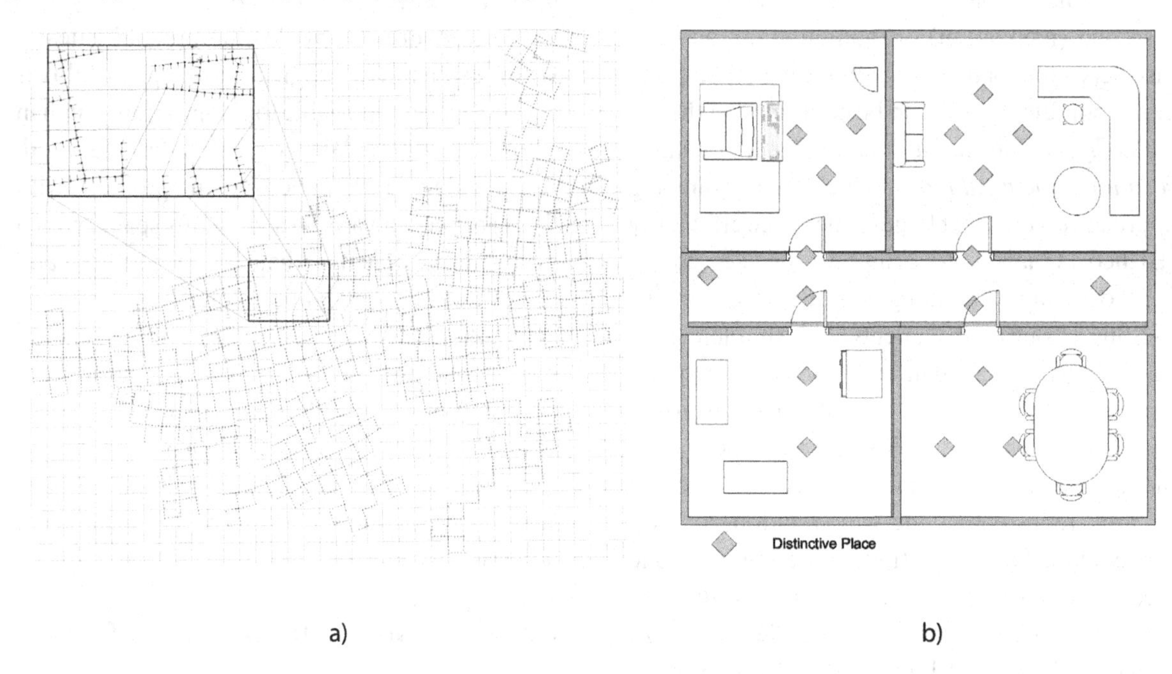

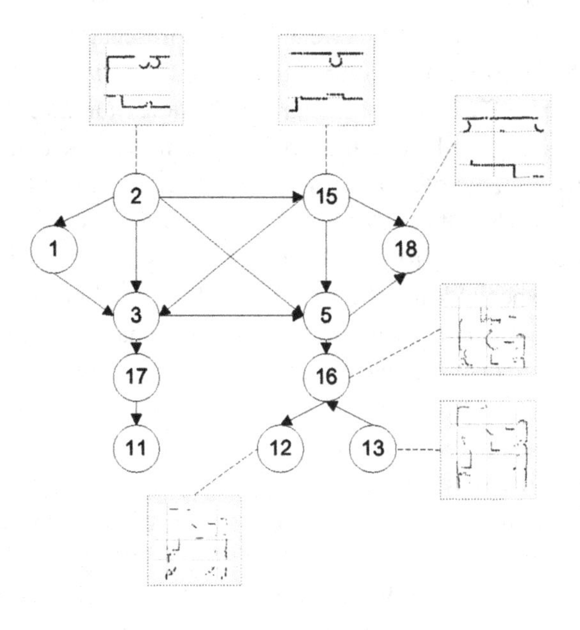

associated to them from within, not from the perspective of an external observer, if the robot—or the human—only has non-symbolic information coming in from sensors? (Harnad, 1990). Actually, the problem is more general, including not only the emergence, but the maintenance of the "links" between symbols and the non-symbolic reality on which they must be grounded. A way of reducing this complexity is to consider only symbols that refer to physical elements of the world, obtaining the so-called anchoring problem, mentioned before too, but anchoring is currently far from being figured out.

In spite of this, we have to provide some concrete definition for *symbol* and for *symbolic map* in order to give form to maps of higher levels of abstraction than the ones introduced before, assuming that other definitions, even contradictory with ours, may also exist. In this chapter—and in most robotics literature—we will call *symbol* an explicit computational representation, (i.e., one with storage needs in a computer and associated algorithms that use it), of some physical, distinctive and stable element of the robot environment that can be used explicitly in high level, usually called *cognitive*, processes.

We should explain with some more detail the adjectives used in this definition to bind its vagueness a little. With "physical," we refer to representations of parts of the environment that can be perceived by the robot sensors. All the maps previously described contain information about physical elements of the environment, but we need to explicitly include that word in our definition in order to reduce the intractable (up to date) symbol grounding problem to the intuitively more practical anchoring problem. With "distinctive," we stress the fact that if a part of the environment is to be represented by a symbol, it must be distinguished from the background. This rules out both points and grid cells: they are clearly not considered symbolic maps. Finally, with "stable" we do not mean static or unchanging, but the quality of a physical element to exist during a certain period of time that is enough to perceive its distinctive constituents and also to process them.

We also mentioned in our definition that a symbol "can be used explicitly in high level processes." By "high level processes," we refer to those that resemble human cognition: planning, reasoning, decision making, communicating with other cognitive agents, etc. Without this constraint in our definition, free-space, and feature maps could enter the symbolic class of maps: the highest level operation that can be carried out by a mobile robot with a free-space map is to control its motion to navigate (and not collide with obstacles); the one with a feature map is to locate itself in the environment. It is for certain that both kinds of maps can be used for more sophisticated operations... if they are enriched with explicit relationships or semantic knowledge, or their elements are further processed to obtain more complex ones. That is the reason why we do not consider them as symbolic.

Thus, we reach to relational maps. For a map to contain explicit relationships, it must certainly contain elements to relate (features, free-space regions, etc.), and these elements could be interpreted as symbols: they correspond to physical, distinctive and stable parts of the environment—in the sense explained before—and they *can* be used by high level processes. The emphasis in "can" is the reason why we have left relational maps as a separate class of maps and not consider them as symbolic: because they can be thought of as symbolic maps or not, depending on the situation. It is true that any symbolic map should be relational: cognitive processes require the existence of a diversity of relationships between the elements of discourse. But the reverse may not always hold: one can have relationships in a map and use them for finding routes in the environment, an operation (typical for topological maps) that is just in the edge of what can be considered a cognitive process, much more when taking into account that finding a path in a graph is a quite simple computational operation.

Therefore, our proposed rule is, if the map contains more than one type of explicit relation, and its elements have enough complexity in their defining characteristics, true high-level processes are possible and thus we can confidently classify the map as symbolic. We insist in that this is our way of classifying maps in this book, and others, perfectly valid, may exist.

The first kind of symbolic map that we can implement in a mobile robot comes from the inclusion of explicit *hierarchical* relations on a topological map: groups of distinctive places in the environment can be abstracted to a single symbol that represents, for instance, the room that they define from the perspective of the sensorimotor apparatus of the robot. This abstraction process can go on recursively, building up a hierarchy of abstraction that ends with a single symbol that represents the whole environment. Furthermore, the abstraction operation can be defined in different ways for the same base data, leading to multiple hierarchies of abstraction on a single map. These kinds of maps have demonstrated to improve efficiency of some mobile robot operations, to be able to optimize these operations in the long term (adapting to the particular environment where they are used) and also to make easier the communication with humans (Fernández-Madrigal & González, 2002; Galindo, Fernández-Madrigal, & González, 2007). You can see an example in Figure 6a.

Another kind of symbolic map comes from the enrichment of the attributes of symbols such that they can be classified into "semantic categories." If explicit relationships are added to these categories, especially the "is-a" relation, you obtain the so-called *semantic maps* (Galindo, Fernández-Madrigal, González, & Saffiotti, 2008). A semantic map is a symbolic map that allows the robot to deduce new knowledge from the general properties (semantics) of the categories of objects in the world. This can be exploited by the robot to perform better in some particular task. For example, while planning complex tasks involving moving, manipulating, communicating results, etc., semantic inference can extend the scope of the planner by providing the robot with the possibility of reasoning about elements of the environment that it has not perceived yet; the semantic structure of the map can also be used to plan at a more abstract level than with a topological representation, thus reducing computa-

Figure 6. (a) Example of a multi-hierarchy in a topological map and (b) a semantic map used by our robots (Images courtesy of Cipriano Galindo-Andrades, University of Málaga)

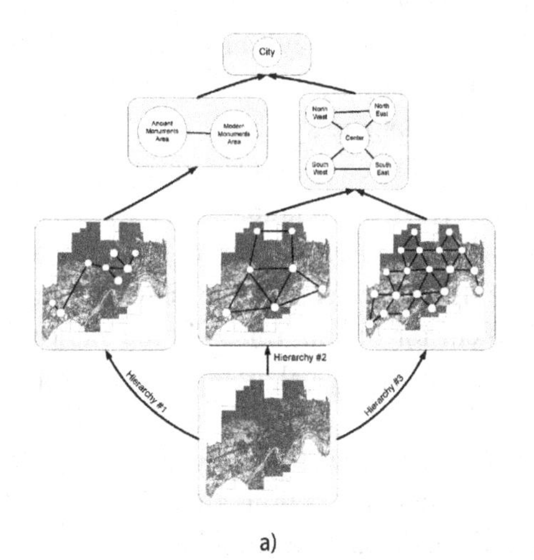

a)

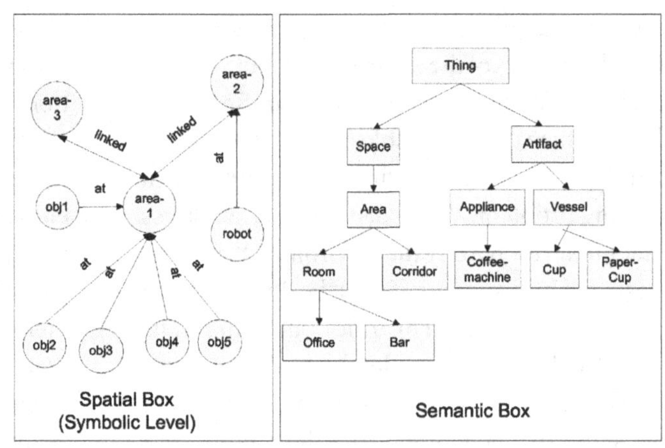

b)

tional cost. An example of a semantic map is shown in Figure 5b.

Still a different approach to symbolic maps is to create relations and symbols from non-metrical data. The work of the psychologist Jean Piaget (1948) showed that children acquire *first* relational information from the environment and *then* metrical one. This idea led to Benjamin J. Kuipers and his colleagues to propose a computational implementation of the human cognitive map with several ontological levels, the *Spatial Semantic Hierarchy* (Kuipers, 2000), in which firstly symbols represent stable operation points of basic motion behaviors (e.g., distinctive junctions or spots in free-space), then causal relations are added representing the fact of getting at that distinctive place if executing that operation from that origin place, then a topology is deduced from the causal map, and finally metrical information is added to the topological relations. A hierarchy of abstraction similar to the one described in previous paragraphs can also fit at the highest ontologies of this model (Remolina, Fernández-Madrigal, Kuipers, & González-Jiménez, 1999).

In general, symbolic maps are out of the scope of the problems of this book, thus they are entirely confined in this section. Their use for localization and mapping is an interesting way to explore in the future, though. The inclusion of uncertainty in this kind of maps is in general an open issue.

3. BAYESIAN ESTIMATION OF GRID MAPS

Now that we have a general and broad vision on the kinds of explicit representations of space that a robot can have, and assuming that we know the pose of the robot, we can consider the problem of estimating a map. As we have seen along this book, a rigorous mathematical way of taking into account the uncertainty in measurements and motion is probability theory and statistics and, in particular, Bayesian recursive estimation (in the sense of *sequential* with time) is especially well suited for on-line estimation of the dynamics of continuous systems that do not exhibit abrupt changes. When estimating maps, they are usually

Figure 7. The DBN for the problem of grid map building. Shaded nodes stand for the hidden variables that will be estimated. Notice how the map variable $\mathbf{m}$ *can be considered as a vector of scalar variables* m_i*, one for each cell in the grid map. The indices of map variables have been numbered as for a grid map with* c *columns and* $r-1$ *rows.*

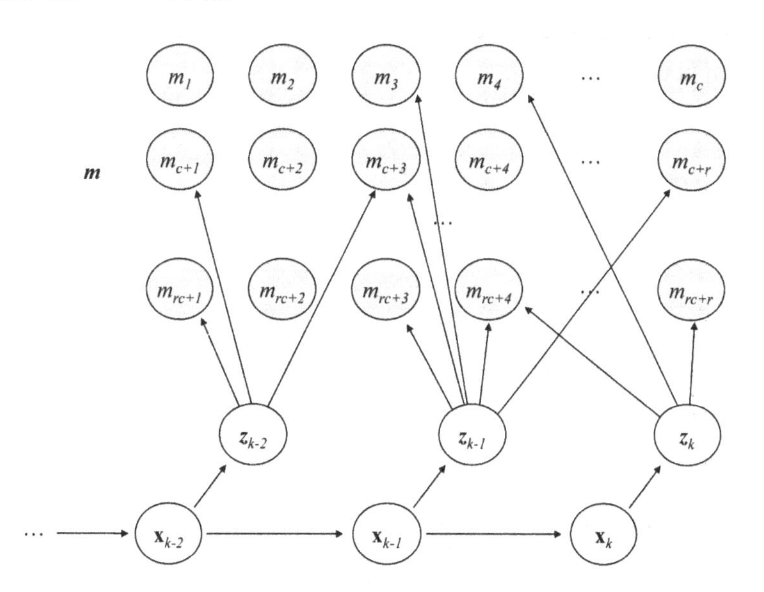

considered static, thus the reason to use a sequential approach is not for coping with changes in the map, but rather for allowing us to learn them incrementally, as new observations are gathered while the robot explores its environment.

One of most common types of map that can be estimated under a probabilistic framework is the occupancy grid map. The Dynamic Bayesian Network (DBN) for the case of estimating a grid map from robot observations, considering the map a static random variable, is shown in Figure 7. We can pose the problem mathematically as finding out the following pmf:

$$P(\mathbf{m}|\mathbf{z}_{1:t},\mathbf{x}_{1:t}) \tag{1}$$

A problematic aspect of this representation is the extremely high dimensionality of that pmf. Notice that in a grid map, the probability distribution should include the interdependences (covariances) between any pair of map cells. If we know that a cell shows evidence of containing an obstacle, the probability of nearby cells containing an obstacle too would not be the same as if we learned that the first one was free. Estimating the full probability distributions in such a way soon becomes intractable even for small maps. Therefore, the first simplification we will make is to consider the individual elements of the map conditionally independent on each other, with which our estimation problem becomes:

$$P(\mathbf{m}|\mathbf{z}_{1:t},\mathbf{x}_{1:t}) = P(\{m_i\}_{i=1}^{N}|\mathbf{z}_{1:t},\mathbf{x}_{1:t}) \approx$$

(assuming $m_i \perp m_j | \mathbf{z}_{1:t},\mathbf{x}_{1:t}$ for any $i \neq j$)

$$\approx \prod_{i=1}^{N} P(m_i|\mathbf{z}_{1:t},\mathbf{x}_{1:t}) \tag{2}$$

that is, the factoring of the full joint distribution into the product of the individual pmf for each one of the N map cells m_i. Hence, in the following we will focus on estimating those individual distributions $P(m_i|\mathbf{z}_{1:t},\mathbf{x}_{1:t})$ instead of the joint. Notice that this simplification is not realistic at all, since each observation z_i (e.g. a laser range scan or a sonar range) is actually affected by several grid cells, thus, statistically speaking, estimating all those cells from the observation should introduce a strong correlation between them. However, the assumption of conditional independence is almost universal in the literature because it leads to efficient and convenient update equations whose results are, in practice, quite satisfactory.

Recall that each variable m_i can yield only two values: occupied or free, representing the state of that portion of space. That is, all m_i are binary random variables. Actually, we could also write down the problem in this alternative form:

$$P(\neg m_i|\mathbf{z}_{1:t},\mathbf{x}_{1:t}) \tag{3}$$

being $\neg m_i$ the r.v. that represents the $i-th$ cell being free (and not occupied). Of course, $P(\neg m_i|\mathbf{z}_{1:t},\mathbf{x}_{1:t}) = 1 - P(m_i|\mathbf{z}_{1:t},\mathbf{x}_{1:t})$.

In order to find out the RBE for this setting, we can apply the Bayes' rule to the distribution of the $i-th$ cell at the last line of Equation 2:

$$P(m_i|\mathbf{z}_{1:t},\mathbf{x}_{1:t}) = P(m_i|\mathbf{z}_t, \underbrace{\mathbf{z}_{1:t-1},\mathbf{x}_{1:t}}_{\substack{\text{this conditions}\\ \text{the entire}\\ \text{expression}}}) =$$

$$= \frac{p(\mathbf{z}_t|m_i,\mathbf{z}_{1:t-1},\mathbf{x}_{1:t})P(m_i|\mathbf{z}_{1:t-1},\mathbf{x}_{1:t})}{} =$$

(by the conditional independence $\mathbf{z}_t \perp \mathbf{z}_{1:t-1}|\mathbf{x}_t, m_i$ that can be seen in the DBN)

$$= \frac{p(\mathbf{z}_t | m_i, \mathbf{x}_{1:t}) P(m_i | \mathbf{z}_{1:t-1}, \mathbf{x}_{1:t})}{p(\mathbf{z}_t | \mathbf{z}_{1:t-1}, \mathbf{x}_{1:t})} =$$

(using again the Bayes' rule in the first term of the numerator, considering that $\mathbf{x}_{1:t}$ is conditioning every factor of the decomposition)

$$= \frac{\dfrac{P(m_i | \mathbf{z}_t, \mathbf{x}_{1:t}) p(\mathbf{z}_t | \mathbf{x}_{1:t})}{P(m_i | \mathbf{x}_{1:t})} P(m_i | \mathbf{z}_{1:t-1}, \mathbf{x}_{1:t})}{p(\mathbf{z}_t | \mathbf{z}_{1:t-1}, \mathbf{x}_{1:t})} =$$

(applying $m_i \perp \mathbf{x}_{1:t}$ in some places)

$$= \frac{\dfrac{P(m_i | \mathbf{z}_t, \mathbf{x}_t) p(\mathbf{z}_t | \mathbf{x}_{1:t})}{P(m_i)} P(m_i | \mathbf{z}_{1:t-1}, \mathbf{x}_{1:t-1})}{p(\mathbf{z}_t | \mathbf{z}_{1:t-1}, \mathbf{x}_{1:t})} =$$

$$= \frac{P(m_i | \mathbf{z}_t, \mathbf{x}_t) p(\mathbf{z}_t | \mathbf{x}_{1:t}) P(m_i | \mathbf{z}_{1:t-1}, \mathbf{x}_{1:t-1})}{P(m_i) p(\mathbf{z}_t | \mathbf{z}_{1:t-1}, \mathbf{x}_{1:t})} \quad (4)$$

This is the farthest we can reach by pursuing terms with known expressions. However, there still appear probabilities that are difficult to calculate, such as $p(\mathbf{z}_t | \mathbf{x}_{1:t})$ or $p(\mathbf{z}_t | \mathbf{z}_{1:t-1}, \mathbf{x}_{1:t})$. Since these terms do not depend on the r.v. we are estimating (m_i), we can get rid of them by firstly using the fact that the same deduction shown in Equation 4 applies to the dual problem of estimating the "freeness" of the cell (Equation 3):

$$P(\neg m_i | \mathbf{z}_{1:t}, \mathbf{x}_{1:t}) =$$
$$= \frac{P(\neg m_i | \mathbf{z}_t, \mathbf{x}_t) p(\mathbf{z}_t | \mathbf{x}_{1:t}) P(\neg m_i | \mathbf{z}_{1:t-1}, \mathbf{x}_{1:t-1})}{P(\neg m_i) p(\mathbf{z}_t | \mathbf{z}_{1:t-1}, \mathbf{x}_{1:t})} \quad (5)$$

Then, as long as $P(\neg m_i | \mathbf{z}_{1:t}, \mathbf{x}_{1:t}) > 0$, we can divide Equation 4 by Equation 5, obtaining what is called the *odds* of m_i:

$$\frac{P(m_i | \mathbf{z}_{1:t}, \mathbf{x}_{1:t})}{P(\neg m_i | \mathbf{z}_{1:t}, \mathbf{x}_{1:t})} =$$
$$= \frac{P(m_i | \mathbf{z}_t, \mathbf{x}_t) P(m_i | \mathbf{z}_{1:t-1}, \mathbf{x}_{1:t-1}) P(\neg m_i)}{P(\neg m_i | \mathbf{z}_t, \mathbf{x}_t) P(\neg m_i | \mathbf{z}_{1:t-1}, \mathbf{x}_{1:t-1}) P(m_i)}$$

and applying logarithms to both sides in order to simplify calculations, we get the so-called *log-odds*:

$$\begin{aligned}
\ln \frac{P(m_i | \mathbf{z}_{1:t}, \mathbf{x}_{1:t})}{P(\neg m_i | \mathbf{z}_{1:t}, \mathbf{x}_{1:t})} &= \ln \frac{P(m_i | \mathbf{z}_t, \mathbf{x}_t)}{P(\neg m_i | \mathbf{z}_t, \mathbf{x}_t)} + \\
+ \ln \frac{P(m_i | \mathbf{z}_{1:t-1}, \mathbf{x}_{1:t-1})}{P(\neg m_i | \mathbf{z}_{1:t-1}, \mathbf{x}_{1:t-1})} &+ \ln \frac{P(\neg m_i)}{P(m_i)} = \\
= \ln \frac{P(m_i | \mathbf{z}_t, \mathbf{x}_t)}{P(\neg m_i | \mathbf{z}_t, \mathbf{x}_t)} &+ \ln \frac{P(m_i | \mathbf{z}_{1:t-1}, \mathbf{x}_{1:t-1})}{P(\neg m_i | \mathbf{z}_{1:t-1}, \mathbf{x}_{1:t-1})} - \\
- \ln \frac{P(m_i)}{P(\neg m_i)}
\end{aligned} \quad (6)$$

This expression involves terms that either can be calculated or are recursive functions of other terms. We name each term for convenience:

$$\underbrace{\ln \frac{P(m_i | \mathbf{z}_{1:t}, \mathbf{x}_{1:t})}{P(\neg m_i | \mathbf{z}_{1:t}, \mathbf{x}_{1:t})}}_{\ell_t(m_i)} = \underbrace{\ln \frac{P(m_i | \mathbf{z}_t, \mathbf{x}_t)}{P(\neg m_i | \mathbf{z}_t, \mathbf{x}_t)}}_{\tau_t(m_i)} +$$
$$+ \underbrace{\ln \frac{P(m_i | \mathbf{z}_{1:t-1}, \mathbf{x}_{1:t-1})}{P(\neg m_i | \mathbf{z}_{1:t-1}, \mathbf{x}_{1:t-1})}}_{\ell_{t-1}(m_i)} - \underbrace{\ln \frac{P(m_i)}{P(\neg m_i)}}_{\ell_0(m_i)}$$
$$\ell_t(m_i) = \tau_t(m_i) - \ell_0(m_i) + \ell_{t-1}(m_i) \quad (7)$$

This is the final *log-odds* recursive formulation for the on-line estimation of a binary r.v. m_i. It is not equivalent to estimating the posterior occupancy *probability* of m_i, but the latter can be deduced from it, since:

$$\ell_t(m_i) = \ln \frac{P(m_i | \mathbf{z}_{1:t}, \mathbf{x}_{1:t})}{P(\neg m_i | \mathbf{z}_{1:t}, \mathbf{x}_{1:t})} =$$

$$= \ln \frac{P(m_i | \mathbf{z}_{1:t}, \mathbf{x}_{1:t})}{1 - P(m_i | \mathbf{z}_{1:t}, \mathbf{x}_{1:t})} \quad \leftrightarrow$$

$$e^{\ell_t(m_i)} = \frac{P(m_i | \mathbf{z}_{1:t}, \mathbf{x}_{1:t})}{1 - P(m_i | \mathbf{z}_{1:t}, \mathbf{x}_{1:t})} \quad \leftrightarrow$$

$$e^{\ell_t(m_i)} - e^{\ell_t(m_i)} P(m_i | \mathbf{z}_{1:t}, \mathbf{x}_{1:t}) = P(m_i | \mathbf{z}_{1:t}, \mathbf{x}_{1:t}) \quad \leftrightarrow$$

$$e^{\ell_t(m_i)} = P(m_i | \mathbf{z}_{1:t}, \mathbf{x}_{1:t}) \left(e^{\ell_t(m_i)} + 1 \right) \quad \leftrightarrow$$

($e^{\ell_t(m_i)} \neq -1$, always)

$$P(m_i | \mathbf{z}_{1:t}, \mathbf{x}_{1:t}) = \frac{e^{\ell_t(m_i)}}{1 + e^{\ell_t(m_i)}} = 1 - \frac{1}{1 + e^{\ell_t(m_i)}} \tag{8}$$

Therefore, we could store the log-odds values (in the range $\ell_t(m_i) \in (-\infty, +\infty)$) for the grid cells at each time step, and then retrieve their posterior occupancy estimation (in the range $P(m_i | \mathbf{z}_{1:t}, \mathbf{x}_{1:t}) \in (0,1)$) just when needed by using Equation 8.

The log-odds representation stands as the most convenient implementation form for Bayesian grid maps in mobile robotics. In practice, updating a cell amounts to adding a positive or negative value to its current log-odds, according to the sensor readings (Equation 7). Furthermore, it is advisable for the sake of efficiency to always store log-odds as *integer* values to avoid the more costly operation of floating point addition, as long as saturation arithmetic is observed while adding the integers (i.e., taking care of avoiding overflow and underflow conditions). As demonstrated in grid map implementations such as the one of the second author within the MRPT (2011), representing the log-odds value of each cell as one 8-bit signed integer provides an excellent computational performance and negligible rounding errors for virtually any practical mapping application.

Returning to the log-odds formulation of Equation 7, we need to provide values for $\tau_t(m_i)$ and $\ell_0(m_i)$ at each step (the latter is a constant), which requires to provide values for $P(m_i | \mathbf{z}_t, \mathbf{x}_t)$ and $P(m_i)$. The former is called the *inverse sensor model*, while the latter is the a priori information that we have about the map occupancy. If we do not know anything in particular about the occupancy of each cell at the first step, we could set $P(m_i) = P_{undefined}(m_i) = 0.5$ as a matter of convenience, since then

$$\ell_0(m_i) = \ln(P(m_i) / P(\neg m_i)) = \ln(0.5 / 0.5) = 0$$

and we would save one sum in the update of each cell.

Obtaining an expression for the inverse sensor model is more complicated, and different approaches exist. We already studied the *forward* sensor model in chapter 6, used there as the likelihood of the sensor, which in the case of the complete map would be $p(\mathbf{z}_t | \mathbf{m}, \mathbf{x}_t)$. If this is known, using Bayes' rule we can deduce the inverse model:

$$P(\mathbf{m} | \mathbf{z}_t, \mathbf{x}_t) = \frac{p(\mathbf{z}_t | \mathbf{m}, \mathbf{x}_t) P(\mathbf{m} | \mathbf{x}_t)}{p(\mathbf{z}_t | \mathbf{x}_t)} =$$

(since $\mathbf{m} \perp \mathbf{x}_t$)

$$= \frac{p(\mathbf{z}_t | \mathbf{m}, \mathbf{x}_t) P(\mathbf{m})}{p(\mathbf{z}_t | \mathbf{x}_t)} \propto$$
$$\propto p(\mathbf{z}_t | \mathbf{m}, \mathbf{x}_t) P(\mathbf{m}) \tag{9}$$

In the case of grid maps, we are interested in $P(m_i = a | \mathbf{z}_t, \mathbf{x}_t)$ for every cell m_i, which can be retrieved from Equation 9 by considering all the possible maps that have the value a at that cell:

$$\begin{aligned} P\left(m_i = a \middle| \mathbf{z}_t, \mathbf{x}_t\right) &= \sum_{\mathbf{m}:m_i=a} P(\mathbf{m} | \mathbf{z}_t, \mathbf{x}_t) \\ &= \sum_{\mathbf{m}:m_i=a} \frac{p(\mathbf{z}_t | \mathbf{m}, \mathbf{x}_t) P(\mathbf{m})}{p(\mathbf{z}_t | \mathbf{x}_t)} \\ &\propto \sum_{\mathbf{m}:m_i=a} p(\mathbf{z}_t | \mathbf{m}, \mathbf{x}_t) P(\mathbf{m}) \end{aligned} \quad (10)$$

Obviously, this sum is intractable: the number of potential maps that have a certain value in a cell is overwhelming. See Figure 8. It could be approximated though through advanced techniques such as neural networks that fall out of the scope of this book.

A more practical, albeit not rigorous, approximation to the inverse sensor model is as follows (Thrun, Burgard, & Fox, 2005). Consider a general beam model for the sensor consisting of a cone defined by an angle of aperture β —see Figure 7a. We have to provide a value $P(m_i | \mathbf{z}_t, \mathbf{x}_t)$ of the inverse sensor model for every cell in the grid after acquiring observation $\mathbf{z}_t$ from location $\mathbf{x}_t$, in order to update the log-odds of the map. The focus of the cone, therefore, will be placed at $\mathbf{x}_t$, and its bisector line will point along the orientation of the line of sight of the sensor. In such a situation we can distinguish three zones in the grid—commonly, the region a cell of the grid lie in is decided by considering its center of mass or geometrical center.

Firstly, it is reasonable to assume that all the cells that fall outside the cone do not obtain any new evidence about their occupancy after that observation, thus they can keep the old log-odds previously stored in the grid. The same is valid for those that lie within the area of the cone but beyond the observed distance $\mathbf{z}_t$, since the observation of $\mathbf{z}_t$ indicates the presence of something solid at that distance (with which the beam has hit) that is occluding the sight beyond. In order

Figure 8. (a) Grid mapping beam model. (b) An example of a grid map built from an ultrasonic sensor for the same navigation experiment we have conducted for the EKF and the PF localization methods in chapter 7, with the previously introduced educational robot Lego Mindstorms NXT. The robot had to move within a small square box of 92 cm x 92 cm.

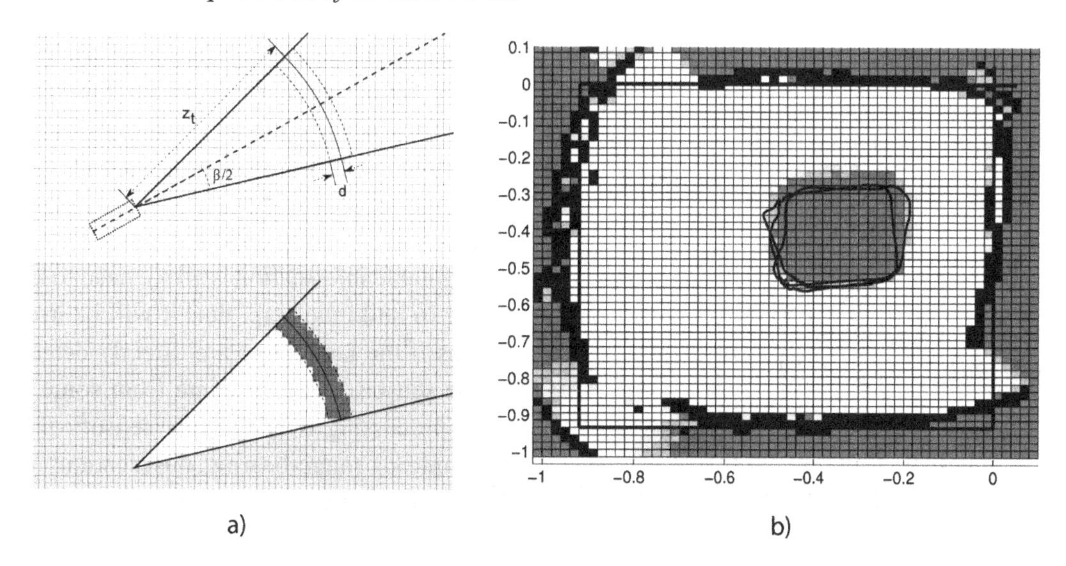

to include a margin of error in this measurement, we can take $\mathbf{z}_t + d$ instead, with some small $d > 0$. Thus, all the cells outside the cone remain unchanged. We will denote this first region of the heuristic inverse model as R_u.

Secondly, those cells lying within the cone of the beam and closer than the observation $\mathbf{z}_t$ have evidence of being free—otherwise the beam would have not reached as far as $\mathbf{z}_t$. For including a security margin in this area, we can take $\mathbf{z}_t - d$ instead. This region will be denoted as R_f.

Finally, all the cells lying in a region R_o defined within the distances $[\mathbf{z}_t - d, \mathbf{z}_t + d]$ from the location of the sensor and within angles $[-\beta / 2, \beta / 2]$ from the bisector of the beam have evidence of being occupied since we do not know at which point (or points) in that region did the beam bounce exactly.

In summary, we only need to update, according to Equation 7, the log-odds of the cells belonging to areas R_f and R_o. The value that we need for that is $P(m_i | \mathbf{z}_t, \mathbf{x}_t)$. We can use any value for this inverse sensor model distribution, for example

$$P(m_i | \mathbf{z}_t, \mathbf{x}_t, m_i \in R_f) = P(\neg m_i | \mathbf{z}_t, \mathbf{x}_t) = 0.25$$

and

$$\begin{aligned} &P(m_i | \mathbf{z}_t, \mathbf{x}_t, m_i \in R_o) = \\ &= P(m_i | \mathbf{z}_t, \mathbf{x}_t) = 1 - P(\neg m_i | \mathbf{z}_t, \mathbf{x}_t) = 0.75, \end{aligned}$$

as long as

$$P(\neg m_i | \mathbf{z}_t, \mathbf{x}_t) < P_{undefined}(m_i) < P(m_i | \mathbf{z}_t, \mathbf{x}_t),$$

which is satisfied with the proposed values: $0.25 < 0.5 < 0.75$. In practice, the values of $P(\neg m_i | \mathbf{z}_t, \mathbf{x}_t)$ and $P(m_i | \mathbf{z}_t, \mathbf{x}_t)$ may be chosen heuristically or from trial and error: values too close to 0.5 will require many repeated observations of the same cells for their probability to noticeably change from their default initial value, while, on the other hand, too extreme values (close to 0 and 1, respectively) may give too much weight to spurious or noisy readings. In any case, the exact values of zero and one should never be employed as the probabilities of the sensor model update, since they lead to inconsistencies, which reflect as infinities in the log-odds formulation.

Finally, you can see in Figure 7b the result of the application of this inverse sensor model to the mapping of an environment by a simple mobile robot endowed with an ultrasound sensor; the figure displays the occupancy probabilities, recovered from the log-odds through Equation 8. Notice the near-circular area inside the map that shows an undefined occupancy: it is bounded by the circumference along which the sensor has rotated.

4. BAYESIAN ESTIMATION OF LANDMARK MAPS: GENERAL APPROACH

We now address the estimation of a map of landmarks (or features) assuming a perfect knowledge of the robot poses from which the landmark observations took place. As it will be shown, the approach may vary depending on the kind of information provided by the observations: range and bearing, bearing only or range only. The kind of available information is all we need to know in order to build the map, thus it will be irrelevant for us if the landmarks are directly detected by some sensor or if they come from some sort of post-processing over the raw sensory data. An example of range-bearing observations are the results of applying salient feature detectors to

2D laser range scans, such as the detection of tree trunks in an outdoor environment (Guivant & Nebot, 2001) or the detection of corners in indoors (Tipaldi & Arras, 2010). The most common case of bearing-only observations is, by far, image features detected in video frames from a camera: given a feature in an image, all we can say about the three-dimensional spatial location of the corresponding landmark is that it falls somewhere along a semi-infinite line emerging from the camera focus along a direction specified by the pixel coordinates (Davison, Reid, Molton, & Stasse, 2007). Finally, sensors providing range-only observations are more uncommon but include interesting families of devices such as radio or ultrasonic beacons—recall chapter 2 section 9.

In all cases, however, our aim is the same: obtaining the spatial location of each observed landmark within an arbitrary global frame of reference along with an estimation of the uncertainty of its position. We could state the problem mathematically as the estimation of the following pdf:

$$p\left(\mathbf{m}\left|\mathbf{z}_{1:t},\mathbf{x}_{1:t}\right.\right) = p\left(\left\{\mathbf{m}_i\right\}_{i=1}^{N}\left|\mathbf{z}_{1:t},\mathbf{x}_{1:t}\right.\right) \tag{11}$$

where the map $\mathbf{m}$ comprises in this case a set of N variables $\mathbf{m}_i$, each one representing the spatial location of the $i-th$ landmark in the map. Notice that since these quantities are continuous values, the distribution of interest is a pdf instead of a pmf as it was the case with grid mapping—compare Equation 11 to Equation 1.

The DBN for this estimation problem, shown in Figure 9, is also independent of the actual kind of observations (range-bearing, bearing-only, or range-only). Observe how we have expanded in the DBN the map variable $\mathbf{m}$ into the corresponding sequence of individual map elements: the spatial location of each landmark. Unlike in grid mapping, where a single observation was affected by several grid cells, observations in landmark mapping always consist of a sequence of *individual* landmark measurements. For instance, a range-bearing observation $\mathbf{z}_k$ taken at some time step i may contain the range and bearing of the first and the second landmarks stored in the map; we would denote such observation as $\mathbf{z}_k = \left\{\mathbf{z}_{k,1},\mathbf{z}_{k,2}\right\}$, with the second subscript index denoting the index in the map of the observed

Figure 9. The DBN for landmark map estimation from a sequence of known robot poses. The map and observation variables have been split into its elemental constituents. As usual, shaded nodes represent the hidden variables to be estimated.

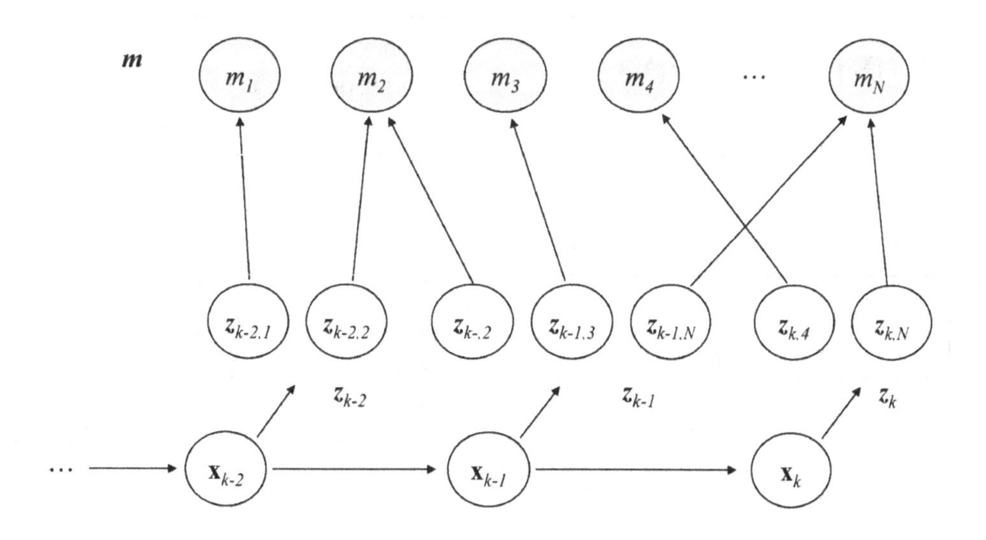

landmark. Obviously, finding out those indices may be not straightforward since it implies solving the data association problem, already explained in chapter 6 section 4.

Observing this graphical model, we arrive at two important realizations: (1) conditioned on the whole robot path and all observations, map landmarks are conditionally independent of each other; and (2) each landmark is conditionally independent of all the robot poses and observations that do not directly observe it, given all the poses and observations that do really observe it directly. Such conclusions easily emerge by applying the concept of d-separation to the graph, as we discussed in chapter 3 section 10. Then, denoting the set of time steps from which the $i-th$ landmark was observed as Ω_i, we can factor and simplify the target pdf in Equation 11 as follows:

$$p\left(\left\{\mathbf{m}_i\right\}_{i=1}^{N}\middle|\mathbf{z}_{1:t},\mathbf{x}_{1:t}\right)=$$

(factoring due to $\mathbf{m}_i \perp \mathbf{m}_j \middle| \mathbf{z}_{1:t}, \mathbf{x}_{1:t}$)

$$=\prod_{i=1}^{N} p\left(\mathbf{m}_i\middle|\mathbf{z}_{1:t},\mathbf{x}_{1:t}\right)=$$

(and since

$$\mathbf{m}_i \perp \left\{\mathbf{z}_{k,j}\right\}_{\substack{k\notin\Omega_i\\ j\neq i}},\left\{\mathbf{x}_k\right\}_{k\notin\Omega_i}\middle|\left\{\mathbf{z}_{k,i}\right\}_{k\in\Omega_i},\left\{\mathbf{x}_k\right\}_{k\in\Omega_i})$$

$$=\prod_{i=1}^{N} p\left(\mathbf{m}_i\middle|\left\{\mathbf{z}_{k,i}\right\}_{k\in\Omega_i},\left\{\mathbf{x}_k\right\}_{k\in\Omega_i}\right) \tag{12}$$

It is important not to get lost into the formulation details but keep clear the real significance of the expression at which we have arrived. The r.v. of each landmark position in the world can be estimated independently, just from the observations directly associated with it. There exists no cross-covariance terms linking different landmarks, but not due to any approximation or simplification, only because we assumed a perfect knowledge of the robot pose at each time step (that cross-covariances do arise in the SLAM problem). In this mapping-only scenario, the only possible source of cross-covariance between landmarks would be in the sensor noise, which in virtually all sensors would never be a real possibility.

To realize why this independence between landmarks is so important, assume we are estimating a map with N three-dimensional landmarks. According to Equation 12, all we need is to estimate N separate pdfs of dimensionality 3. In turn, directly estimating the joint pdf $p\left(\mathbf{m}\middle|\mathbf{z}_{1:t},\mathbf{x}_{1:t}\right)$ implies estimating a pdf of dimensionality $3N$. Since it is common to implement the RBE with Kalman-like filters, which typically exhibit a cubic computational complexity with the problem dimensionality (recall chapter 7 section 3), the factoring of the pdf means to pass from a complexity $O\left(\left(3N\right)^3\right)\equiv O\left(N^3\right)$ down to a much simpler $O\left(N\ 3^3\right)\equiv O\left(N\right)$. More on performance will be discussed in chapter 9 section 3 when dealing with RBPF-based SLAM.

An additional advantage of this pdf factoring, not always exploited in the literature, is the possibility of modeling each landmark with a different pdf parameterization (Blanco, González, & Fernández-Madrigal, 2008b), an issue touched later on.

So far our aim is to address the Bayesian estimation of the $p\left(\mathbf{m}_i\middle|\left\{\mathbf{z}_{k,i}\right\}_{k\in\Omega_i},\left\{\mathbf{x}_k\right\}_{k\in\Omega_i}\right)$ distributions in Equation 12, which for the sake of clarity we will refer to simply as $p\left(\mathbf{m}_i\middle|\mathbf{z}_{1:t},\mathbf{x}_{1:t}\right)$ in the following, advising the reader to keep in mind that the robot poses and the observations appearing in this expression are only those directly related to the $i-th$ landmark. The first step is to apply Bayes' rule conditioned on the latest observation and some conditional independences that follow from the DBN:

$$\underbrace{p\left(\mathbf{m}_i \left| \mathbf{z}_{1:t}, \mathbf{x}_{1:t} \right.\right)}_{\text{Posterior pdf for } t} = p\left(\mathbf{m}_i \left| \mathbf{z}_t, \mathbf{z}_{1:t-1}, \mathbf{x}_{1:t} \right.\right) =$$

(Bayes' rule applied on $\mathbf{z}_t$)

$$= \frac{p\left(\mathbf{m}_i \left| \mathbf{z}_{1:t-1}, \mathbf{x}_{1:t} \right.\right) p\left(\mathbf{z}_t \left| \mathbf{m}_i, \mathbf{z}_{1:t-1}, \mathbf{x}_{1:t} \right.\right)}{\underbrace{p\left(\mathbf{z}_t \left| \mathbf{m}_i, \mathbf{z}_{1:t-1}, \mathbf{x}_{1:t} \right.\right)}_{\text{This term is a constant w.r.t. the posterior}}} \propto$$

$$\propto p\left(\mathbf{m}_i \left| \mathbf{z}_{1:t-1}, \mathbf{x}_{1:t} \right.\right) p\left(\mathbf{z}_t \left| \mathbf{m}_i, \mathbf{z}_{1:t-1}, \mathbf{x}_{1:t} \right.\right) =$$

(and since $\mathbf{m}_i \perp \mathbf{x}_t \left| \mathbf{z}_{1:t-1}, \mathbf{x}_{1:t-1} \right.$)

$$= p\left(\mathbf{m}_i \left| \mathbf{z}_{1:t-1}, \mathbf{x}_{1:t-1} \right.\right) p\left(\mathbf{z}_t \left| \mathbf{m}_i, \mathbf{z}_{1:t-1}, \mathbf{x}_{1:t} \right.\right) \quad (13)$$

Looking at the second term in the last product, one can see that the landmark location $\mathbf{m}_i$ appears as a conditioning variable, while in fact we do not know its real value (that is exactly the problem we are trying to solve!). The solution is to sum all the contributions to this density conditioned on the likelihood of each possible value of the landmark location; since $\mathbf{m}_i$ belongs to a continuous domain, the sum becomes an integration. Doing so and using further simplifications leads to:

$$p\left(\mathbf{m}_i \left| \mathbf{z}_{1:t-1}, \mathbf{x}_{1:t-1} \right.\right) p\left(\mathbf{z}_t \left| \mathbf{m}_i, \mathbf{z}_{1:t-1}, \mathbf{x}_{1:t} \right.\right) =$$

(integrating over all the possible values of $\mathbf{m}_i$)

$$= p\left(\mathbf{m}_i \left| \mathbf{z}_{1:t-1}, \mathbf{x}_{1:t-1} \right.\right) \int_{-\infty}^{\infty} \underbrace{p\left(\mathbf{z}_t \left| \tilde{\mathbf{m}}_i, \mathbf{z}_{1:t-1}, \mathbf{x}_{1:t} \right.\right)}_{\text{can be simplified}} p\left(\tilde{\mathbf{m}}_i \left| \mathbf{z}_{1:t-1}, \mathbf{x}_{1:t} \right.\right) d\tilde{\mathbf{m}}_i =$$

(from the DBN we have $\mathbf{z}_t \perp \mathbf{z}_{1:t-1}, \mathbf{x}_{1:t-1} \left| \tilde{\mathbf{m}}_i, \mathbf{x}_t \right.$)

$$= p\left(\mathbf{m}_i \left| \mathbf{z}_{1:t-1}, \mathbf{x}_{1:t-1} \right.\right) \int_{-\infty}^{\infty} p\left(\mathbf{z}_t \left| \tilde{\mathbf{m}}_i, \mathbf{x}_t \right.\right) \underbrace{p\left(\tilde{\mathbf{m}}_i \left| \mathbf{z}_{1:t-1}, \mathbf{x}_{1:t} \right.\right)}_{\text{can be simplified}} d\tilde{\mathbf{m}} =$$

(and applying again that $\mathbf{m}_i \perp \mathbf{x}_t \left| \mathbf{z}_{1:t-1}, \mathbf{x}_{1:t-1} \right.$)

$$= \underbrace{p\left(\mathbf{m}_i \left| \mathbf{z}_{1:t-1}, \mathbf{x}_{1:t-1} \right.\right)}_{\text{Posterior pdf at } t-1} \int_{-\infty}^{\infty} p\left(\mathbf{z}_t \left| \tilde{\mathbf{m}}_i, \mathbf{x}_t \right.\right) \underbrace{p\left(\tilde{\mathbf{m}}_i \left| \mathbf{z}_{1:t-1}, \mathbf{x}_{1:t-t} \right.\right)}_{\text{Posterior pdf at } t-1} d\tilde{\mathbf{m}}_i \quad (14)$$

Since we aim at estimating the individual pdf for the $i-th$ landmark, we must assume that the correspondence between observations and the $i-th$ map landmarks has been already done following any of the methods explained in chapter 6 section 4 for data association. As an outcome of such methods we could get two possible results: either the observed landmark (1) corresponds to any of the existing ones or (2) it is a new one not mapped yet (and the index i simply stands for any unoccupied landmark index). The former means that we need to fuse the new information with the latest filtered posterior, which must be done in a way that depends on the kind of available observations (this will be addressed below). In the latter case, if we interpret what occurs to Equations 13 and 14, we obtain:

$$\underbrace{p\left(\mathbf{m}_i \middle| \mathbf{z}_{1:t}, \mathbf{x}_{1:t}\right)}_{\text{New posterior pdf}} \propto \underbrace{p\left(\mathbf{m}_i \middle| \mathbf{z}_{1:t-1}, \mathbf{x}_{1:t-1}\right)}_{\text{Previous posterior pdf}}$$
$$\int_{-\infty}^{\infty} p\left(\mathbf{z}_t \middle| \tilde{\mathbf{m}}_i, \mathbf{x}_t\right) \underbrace{p\left(\tilde{\mathbf{m}}_i \middle| \mathbf{z}_{1:t-1}, \mathbf{x}_{1:t-1}\right)}_{\text{Previous posterior pdf}} d\tilde{\mathbf{m}}_i$$

(instantiating for the first observation of a landmark, that is, $t = 1$)

$$p\left(\mathbf{m}_i \middle| \mathbf{z}_1, \mathbf{x}_1\right) \propto \underbrace{p\left(\mathbf{m}_i\right)}_{\text{A priori pdf}} \int_{-\infty}^{\infty} p\left(\mathbf{z}_t \middle| \tilde{\mathbf{m}}_i, \mathbf{x}_t\right) \underbrace{p\left(\tilde{\mathbf{m}}_i\right)}_{\text{A priori pdf}} d\tilde{\mathbf{m}}_i \tag{15}$$

where it can be seen that, as could be expected, the first time a landmark is observed some sort of *a priori* distribution will be needed for its spatial location. The most generic attitude we could take here is to assume that no information is available apart from the observations themselves, thus the a priori distributions become uniform pdfs:

$$\begin{aligned} p\left(\mathbf{m}_i \middle| \mathbf{z}_1, \mathbf{x}_1\right) &\propto \underbrace{p\left(\mathbf{m}_i\right)}_{\text{Constant}} \int_{-\infty}^{\infty} p\left(\mathbf{z}_t \middle| \tilde{\mathbf{m}}_i, \mathbf{x}_t\right) \underbrace{p\left(\tilde{\mathbf{m}}_i\right)}_{\text{Constant}} d\tilde{\mathbf{m}}_i \\ &\propto p\left(\mathbf{z}_t \middle| \mathbf{m}_i, \mathbf{x}_t\right) \end{aligned} \tag{16}$$

that is, the pdf of the landmark after its first observation coincides with the *inverse sensor model*, which is the name of the distribution $p\left(\mathbf{z}_t \middle| \mathbf{m}_i, \mathbf{x}_t\right)$ when all the terms are known values except the map. Recall that the same distribution was named *sensor observation model* in chapter 6 when the unknown term was the observation itself.

To sum up, we have learned that updating a landmark map requires an inverse sensor model for the first time a landmark is detected, and a generic Bayesian filtering algorithm for solving Equation 14 in subsequent observations. The next sections expose some solutions for those two situations for the three different kinds of observations that we can obtain from robotic sensors.

5. BAYESIAN ESTIMATION OF LANDMARK MAPS: RANGE-BEARING SENSORS

For simplicity in the exposition, we will assume a robot moving on a planar surface, and all landmarks contained in a single plane. In this set up, we can model each map landmark as a r.v. comprising its two coordinates, that is, $\mathbf{m}_i = \left(m_{x_i} \quad m_{y_i}\right)^T$. As already discussed in chapter 6 section 3, a range-bearing sensor provides us with observations $\mathbf{z}_k = \left(r_k \quad b_k \quad \varphi_k\right)^T$, having one range r_k and one bearing angle b_k that describe the landmark position as detected from the instantaneous pose of the sensor $\mathbf{s}_k = \left(s_{x_k} \quad s_{y_k} \quad s_{\theta_k}\right)^T$. The value φ_k represents the identification of the sensed landmark, and will be present only if the sensor is able to uniquely identify it in the environment. Since it was assumed above that data association was already solved at this point, we will go on with $\mathbf{z}_k = \left(r_k \quad b_k\right)^T$. Regarding the sensor pose, it is straightforwardly computed given the robot pose $\mathbf{x}_k = \left(x_k \quad y_k \quad \theta_k\right)^T$ but for clarity we will assume that both coincide, i.e., the sensor is exactly at the origin of the robocentric coordinate reference.

The Inverse Sensor Model

Once we stated the parameterization of the problem variables, we aim at providing the inverse sensor model, for which we have to start from the sensor observation model (or direct model):

$$\mathbf{z}_k = \begin{bmatrix} r_i \\ b_i \end{bmatrix} = \mathbf{h}_i\left(\mathbf{x}_k, \mathbf{m}_i\right) + \mathbf{n}_k \tag{17}$$

Here, $\mathbf{n}_k \sim N(\mathbf{0}, \mathbf{R})$ is an additive zero-mean Gaussian noise and the observation function is the one already presented in chapter 6's Equation 6, repeated here for convenience:

$$\mathbf{h}_i(\mathbf{x}_t, \mathbf{m}_i) = \begin{pmatrix} \sqrt{(m_{x_i} - x_t)^2 + (m_{y_i} - y_t)^2} \\ \operatorname{atan2}(m_{y_i} - y_t, m_{x_i} - x_t) - \phi_t \end{pmatrix} \quad (18)$$

Obtaining an inverse sensor model implies having a probabilistic relationship such that, given a sensor reading from a known robot pose, yields a pdf for the location of a landmark $\mathbf{m}_i$ that explains the known data. By finding the value of $\mathbf{m}_i$ in Equation 17 we have:

$$\begin{aligned} \mathbf{z}_k &= \mathbf{h}_i(\mathbf{x}_k, \mathbf{m}_i) + \mathbf{n}_k && \rightarrow \\ \mathbf{z}_k - \mathbf{n}_k &= \mathbf{h}_i(\mathbf{x}_k, \mathbf{m}_i) && \rightarrow \\ \mathbf{m}_i &= \mathbf{h}_i^{-1}(\mathbf{x}_k, \mathbf{z}_k - \mathbf{n}_t) \end{aligned} \quad (19)$$

where the inverse observation function[1] takes one robot pose $\mathbf{x}_k$ and an observation $\mathbf{z}_k$ and returns the projected location of the landmark. In this case, it can be shown that this function is:

$$\begin{pmatrix} m_{x_i} \\ m_{y_i} \end{pmatrix} = \mathbf{m}_i = \mathbf{h}_i^{-1}(\mathbf{x}_k, \mathbf{z}_k) = \begin{pmatrix} x_k + r_i \cos(\phi_k + b_i) \\ y_k + r_i \sin(\phi_k + b_i) \end{pmatrix} \quad (20)$$

Up to this point, all the derivation was based on statistical bases and was totally generic, but now we need to decide what specific distribution will be used to model the uncertainty in the map landmarks. For range-bearing sensors, it turns out that approximating that uncertainty as Gaussian is quite reasonable. Thus, the inverse sensor model $p(\mathbf{z}_k | \mathbf{m}_i, \mathbf{x}_k)$ in this case equals the distribution $N(\mathbf{m}_i; \bar{\mathbf{m}}_i, \Sigma_{\mathbf{m}_i})$, the parameters of which are derived next.

Notice from Equation 19 that the noise $\mathbf{n}_k$ is the unique input to the inverse function $\mathbf{h}_i^{-1}(\cdot)$ that is not perfectly known, but a pdf: it is the sensor noise uncertainty, the only one that leads to uncertainty in the landmark position. As usual when faced with uncertainty transformations, an appealing solution is to apply linearization, arriving at an expected value of:

$$\bar{\mathbf{m}}_i = \mathbf{h}_i^{-1}(\mathrm{E}[\mathbf{x}_k], \mathrm{E}[\mathbf{z}_k - \mathbf{n}_k]) \quad \rightarrow$$

(since $\mathbf{x}_k$ and $\mathbf{n}_k$ are known values and the mean of $\mathbf{n}_k$ is zero)

$$\bar{\mathbf{m}}_i = \mathbf{h}_i^{-1}(\mathbf{x}_k, \mathbf{z}_k) \quad (21)$$

and a covariance of:

$$\Sigma_{\mathbf{m}_i} = \frac{\partial \mathbf{h}_i^{-1}(\cdot)}{\partial \mathbf{z}_k} \mathbf{R} \frac{\partial \mathbf{h}_i^{-1}(\cdot)}{\partial \mathbf{z}_k}^T$$

With the Jacobian, evaluated at the mean of $\mathbf{n}_k$, given by:

$$\frac{\partial \mathbf{h}_i^{-1}(\cdot)}{\partial \mathbf{z}_k} = \begin{pmatrix} \cos(\phi_k + b_i) & -r_i \sin(\phi_k + b_i) \\ \sin(\phi_k + b_i) & r_i \cos(\phi_k + b_i) \end{pmatrix} \quad (22)$$

Notice that all of this is possible only because landmarks are completely *observable* with range-bearing sensors, as was already mentioned in chapter 2 section 3. That means that there exists a well-defined inverse sensor function $\mathbf{h}_i^{-1}(\cdot)$, something that not always occur with other sensors.

Recursive Bayesian Estimation

We now address the update of a map with successive observations of a landmark, which was already mapped upon its first detection. This implies providing a concrete implementation for Equation 15. Recall that we already decided to represent uncertainty in landmarks as Gaussian distributions and that the perfect knowledge of the robot path allows us to update the pdf of each landmark independently. The Gaussianity assumption and the non-linearity of the observation function $\mathbf{h}_i(\cdot)$ defined in Equation 18 makes the Extended Kalman Filter (EKF) a good choice as the algorithm to estimate the pdf for each individual landmark. In this particular problem the EKF equations are simpler since there is no prediction stage (the transition or motion model) due to the assumption of a static map. Recall that each landmark pdf is parameterized as $\mathbf{m}_i \sim N\left(\bar{\mathbf{m}}_i, \boldsymbol{\Sigma}_{\mathbf{m}_i}\right)$, thus the EKF equations must provide the updated mean and covariance matrix from the previous ones and the new robot pose and observation. From chapter 7's Equation 44, the equations turn out to be:

$$\bar{\mathbf{m}}_i \leftarrow \bar{\mathbf{m}}_i + \mathbf{K}_{k,i}\left(\mathbf{z}_k - \mathbf{h}_i\right)$$

$$\boldsymbol{\Sigma}_{\mathbf{m}_i} \leftarrow \left(\mathbf{I} - \mathbf{K}_{k,i}\mathbf{H}_{k,i}\right)\boldsymbol{\Sigma}_{\mathbf{m}_i}$$

with the Kalman gain matrix being:

$$\mathbf{K}_{k,i} = \boldsymbol{\Sigma}_{\mathbf{m}_i}\mathbf{H}_{k,i}^T\left(\mathbf{R} + \mathbf{H}_{k,i}\boldsymbol{\Sigma}_{\mathbf{m}_i}\mathbf{H}_{k,i}^T\right)^{-1} \quad (23)$$

$\mathbf{H}_{k,i}$ is the Jacobian of $\mathbf{h}_i(\cdot)$ with respect to the landmark coordinates, which easily follows from Equation 18 to be:

$$\mathbf{H}_{k,i} = \frac{\partial \mathbf{h}_i\left(\mathbf{x}_k, \mathbf{m}_i\right)}{\partial \mathbf{m}_i} = \begin{pmatrix} \frac{m_{x_i} - x_k}{r_i} & \frac{m_{y_i} - y_k}{r_i} \\ -\frac{m_{y_i} - y_k}{r_i^2} & \frac{m_{x_i} - x_k}{r_i^2} \end{pmatrix} \quad (24)$$

In the case that the uncertainties for a particular experimental set up are so large that the linearization in EKF represents a poor approximation to the actual pdfs, the UKF algorithm could be used instead. The reader can review all the properties and equations of both filters in chapter 7 section 3.

6. BAYESIAN ESTIMATION OF LANDMARK MAPS: BEARING-ONLY SENSORS

This kind of landmark observations is of the greatest interest since they are the ones obtained from image features while performing vision-based mapping or SLAM. It is possible to address two-dimensional mapping with these observations, for example, by detecting vertical features in images of the environment (like doorframes or corridor corners). Although ideas like this were proposed during the last two decades, the truth is that they never became really popular, probably because of the much richer information that one can obtain by considering the full 3D information embedded in the images instead.

Therefore, we will focus here on a full 3D approach, which complicates the formulation and introduces new challenges with respect to previously seen methods. When working on a planar environment, a robot pose simply consists of a pair of coordinates that define the translation and a third parameter for the rotation angle; it will be always like that and there is no room for further complications. In contrast, the first hurdle when dealing with three-dimensional localization, mapping or SLAM is to choose a parameterization for the spatial pose of the camera—quite often visual mapping is performed with hand-held cameras, thus we will refer to the camera pose instead of the robot pose all along this section. Employing unit *quaternions* to represent 3D rotations is the preferred approach when those rotations are to be directly estimated within a Bayesian filter, as is our case—an alternative using more mathematically advanced techniques is briefly discussed in chapter 10. This approach, common in the literature (Civera, Davison, & Montiel, 2008; Davison, Reid, Molton, & Stasse, 2007), consists in describing the camera pose by means of its three spatial coordinates (the translational part) plus other four coordinates interpreted as a unit quaternion (rotational part), that is:

$$\mathbf{x}_k = \begin{pmatrix} x_k & y_k & z_k & q_{k,r} & q_{k,x} & q_{k,y} & q_{k,z} \end{pmatrix}^T \tag{25}$$

In order to grasp the geometrical meaning of the four quaternion coordinates, one can visualize them as a rotation of a magnitude proportional to $q_{k,r}$ around the spatial direction defined by the vector $\begin{pmatrix} q_{k,x} & q_{k,y} & q_{k,z} \end{pmatrix}^T$—refer to Appendix A.

The next important complication that arises with visual-based mapping or SLAM is the need to perform data association from visual and geometrical information. Typically, feature points are selected with salient keypoint detectors, and then are assigned some sort of feature descriptor, which can be as simple as a patch of the surrounding image or quite involved high-dimensional descriptors (Lowe, 2004). The challenge is to achieve a combination of detectors, descriptors and matching algorithms for pairing features between different frames such that it is fast enough to be executed in real-time (at the camera frame rate) and while keeping low the ratio of false positives and false negatives. In any case, all those operations are far beyond the scope of this text, thus we will assume next that data association is already solved, just like in the previous case for range-bearing observations.

Thus, we will define a bearing-only observation as $\mathbf{z}_{k,i} = \begin{pmatrix} \alpha_{k,i} & \beta_{k,i} \end{pmatrix}^T$, comprising an azimuth (*yaw*) and elevation (*pitch*) angles, respectively. The relationship of these angles with the location of the landmark can be easily established from geometry to be:

$$\begin{cases} \alpha_{k,i} = \operatorname{atan2}\left(l_{i,z}, l_{i,x}\right) \quad , \text{ or } \tan^{-1}\left(\dfrac{l_{i,z}}{l_{i,x}}\right) \\ \beta_{k,i} = \sin^{-1}\left(\dfrac{l_{i,y}}{\sqrt{l_{i,x}^2 + l_{i,y}^2 + l_{i,z}^2}}\right) \end{cases} \tag{26}$$

Box 1.

$$\mathbf{M}'_{k,i}\left(\mathbf{M}_i, \mathbf{x}_k\right) = \mathbf{M}_i \ominus \mathbf{x}_k = \begin{pmatrix} l_{i,x} \\ l_{i,y} \\ l_{i,z} \end{pmatrix} =$$

$$= \begin{pmatrix} g_{i,x} - x_k \\ g_{i,y} - y_k \\ g_{i,z} - z_k \end{pmatrix} + 2 \begin{pmatrix} -q_{k,y}^2 - q_{k,z}^2 & q_{k,x}q_{k,y} - q_{k,r}q_{k,z} & q_{k,r}q_{k,y} + q_{k,x}q_{k,z} \\ q_{k,r}q_{k,z} + q_{k,x}q_{k,y} & -q_{k,x}^2 - q_{k,z}^2 & q_{k,y}q_{k,z} - q_{k,r}q_{k,x} \\ q_{k,x}q_{k,z} - q_{k,r}q_{k,y} & q_{k,r}q_{k,x} + q_{k,y}q_{k,z} & -q_{k,x}^2 - q_{k,y}^2 \end{pmatrix} \begin{pmatrix} g_{i,x} - x_k \\ g_{i,y} - y_k \\ g_{i,z} - z_k \end{pmatrix} \tag{27}$$

Figure 10. The real uncertainty of an inverse sensor model for bearing-only observations are cone-shaped, as shown in the four individual observations shown in (a) and (c). In each situation, the estimation of the potential landmark location from two observations can be obtained by fusing both cone-like shapes, leading to the pdfs of (b) and (d). A larger parallax, as in (a) – (b), leads to a pdf closer to a Gaussian, while a reduced parallax, as in (c)—(d), makes the Gaussian a poorer approximation and favors the inverse depth parameterization. (e) Confidence intervals for an inverse-depth parameterization, compared to the actual cone-like pdf which is being approximated. (f) The convention used in the text regarding the axes of local coordinates with respect to the camera.

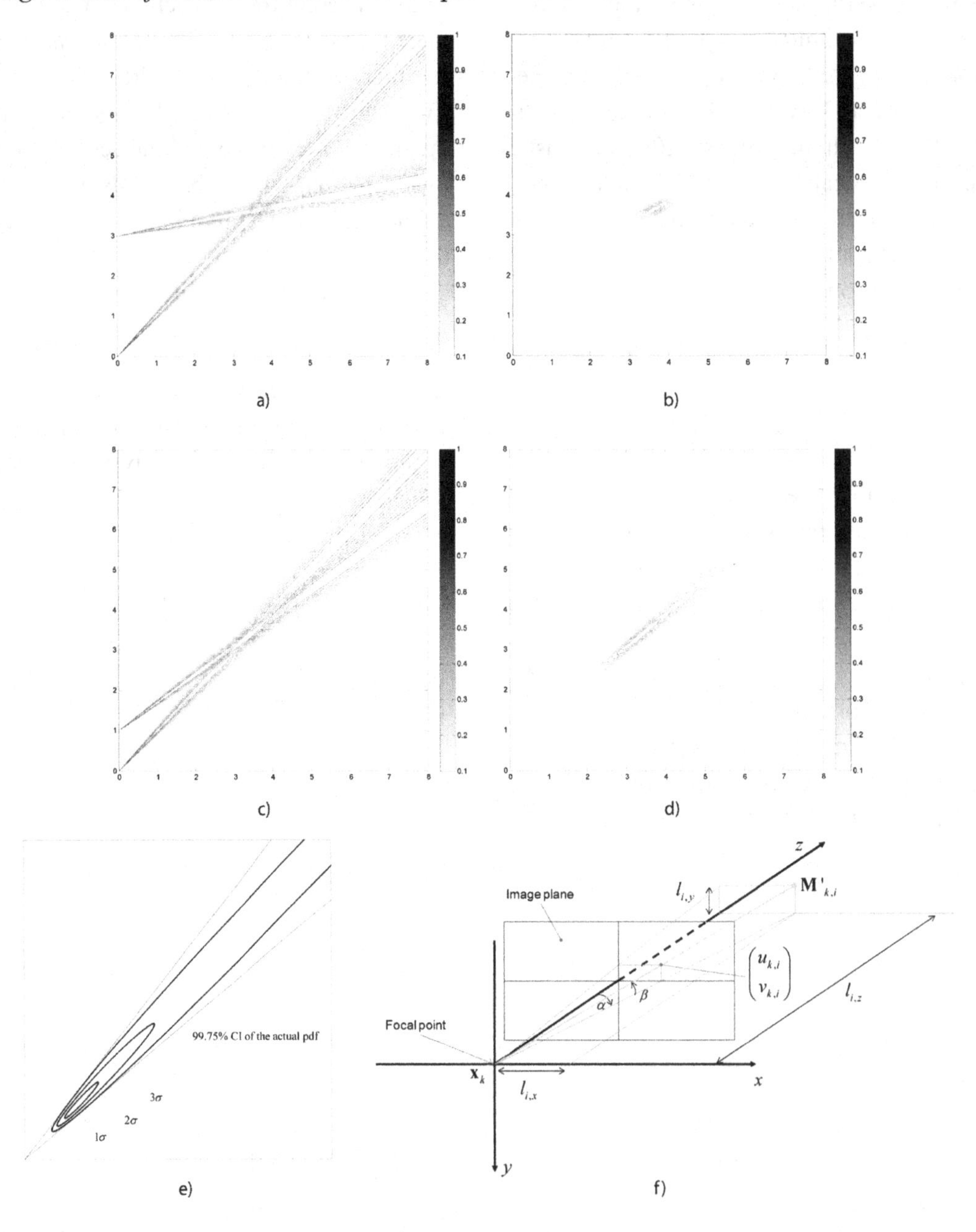

when given in relative (local) coordinates $\mathbf{M}'_{k,i} = \begin{pmatrix} l_{i,x} & l_{i,y} & l_{i,z} \end{pmatrix}^T$ with respect to the camera location $\mathbf{x}_k$—please refer to Figure 10f for the axes convention. These local coordinates can be computed from the landmark global coordinates $\mathbf{M}_i = \begin{pmatrix} g_{i,x} & g_{i,y} & g_{i,z} \end{pmatrix}^T$ and the quaternion-based representation of the camera pose as shown in Box 1.

The Inverse Sensor Model

Let us clearly define first the sensor observation model for this kind of observations. It consists of an observation function $\mathbf{h}(\cdot)$ which takes as input the relative position of the landmark ($\mathbf{M}_{k,i}'$) and, via Equation 26 gives us the pair of observed angles:

$$\mathbf{z}_{k,i} = \begin{pmatrix} \alpha_{k,i} \\ \beta_{k,i} \end{pmatrix} = \mathbf{h}\left(\mathbf{M}_{k,i}'\right) + \mathbf{n}_{k,i} \quad (28)$$

As usual, we assume an additive zero-mean Gaussian noise $\mathbf{n}_{k,i} \sim N\left(\mathbf{0}, \mathbf{R}\right)$ on this observation, which in this case models the uncertainty in the image feature detectors due to the discrete nature of pixels, possibly blurred images, etc. As a matter of fact, given the wide-spread application of bearing-only observations in computer vision, the observation model can be (and usually is) directly formulated in terms of pixel coordinates, which are much closer to the vision-based front-end algorithms than the pair of angles in the equations above (Nüchter, 2009). In this form, the two components of the observation are the pixel coordinates $\begin{pmatrix} u_{k,i} & v_{k,i} \end{pmatrix}^T$, computed as:

$$\mathbf{z}_{k,i} = \begin{pmatrix} u_{k,i} \\ v_{k,i} \end{pmatrix} = \mathbf{h}\left(\mathbf{M}'_{k,i}\right) + \mathbf{n}_{k,i}$$

with:

$$\mathbf{h}\left(\mathbf{M}'_{k,i}\right) = \begin{pmatrix} c_x + f_x \dfrac{l_{i,x}}{l_{i,z}} \\ c_y + f_y \dfrac{l_{i,y}}{l_{i,z}} \end{pmatrix} \quad (29)$$

where c_x and c_y and the pixel coordinates of the camera optical center and f_x and f_y are both the focal distance, measured in units of horizontal and vertical pixels, respectively. This equation above is called the camera *pinhole projective model*, and can be further improved to reflect real cameras by the inclusion of distortion parameters. For the introductory nature of this book, no more details will be given here on camera projective geometry, thus we recommend the interested reader to consult the rich existing literature (Mikhail, Bethel, & McGlone, 2001; Hartley & Zisserman, 2003).

What is really significant for the present discussion is the realization that bearing-only observations reduce the dimensionality of the observed landmarks from three (i.e. its spatial location with respect to the observer) to only two (i.e. the two pixel coordinates). Therefore, there is one degree of freedom, which is irremediably lost: the depth of the landmark. This has one practical consequence of paramount importance: if we were to try finding out the inverse sensor model just as with range-bearing sensors, we would not find any such function $\mathbf{h}^{-1}(\cdot)$ that maps pairs of pixel coordinates into three dimensional coordinates, simply because such a bijective relationship does not exist.

Still, under a probabilistic viewpoint this is not a limitation. A probabilistic inverse sensor model can assign a uniform distribution to the unknown depth, starting at the camera location and extending up to some arbitrarily large maximum distance. Combining that depth uncertainty with the uncertainty of the two angles (or equivalently, of the pixel coordinates) we end up with a pdf for the potential location of a landmark upon its first observation which is cone-like shaped, as illustrated in Figure 10.

The fundamental problem with initializing a landmark from a bearing-only observation relies on that uncommon, cone-like shape of the uncertainty, which should be modeled. A Gaussian in three-dimensional space whose variables are the three Cartesian coordinates $\left(x_i, y_i, z_i\right)$ makes quite a poor job in approximating the real shape of the inverse sensor model. As a workaround, tricks have been proposed in the literature such as delaying the insertion of landmarks in the Bayesian filter until their location uncertainty has been reduced, with other auxiliary Bayesian filters, and can be appropriately modeled as Gaussian (Davison, 2003). That approach discards valuable information until each landmark converges, which could help localizing the camera when these maps are used within a SLAM framework; also, it discards distant landmarks ("features at the infinity") which are known to help estimating the orientation of the camera.

In order to avoid all those disadvantages, a different and smart solution was proposed in the literature (Civera, Davison, & Montiel, 2008): instead of parameterizing landmarks with their three Cartesian coordinates $\mathbf{M}_i = \left(g_{i,x} \; g_{i,y} \; g_{i,z}\right)^T$, one can explicitly store the first pose from which they were first observed $\left(o_{x_i}, o_{y_i}, o_{z_i}\right)$, the associated observation direction (two angles θ_i and ϕ_i) and the inverse of the depth (ρ_i) from the observing point. These parameters are related to the Cartesian coordinates by means of:

$$\mathbf{M}_i\left(\mathbf{m}_i\right) = \begin{pmatrix} o_{x_i} \\ o_{y_i} \\ o_{z_i} \end{pmatrix} + \frac{1}{\rho_i} \begin{pmatrix} \cos\phi_i \sin\theta_i \\ -\sin\phi_i \\ \cos\phi_i \cos\theta_i \end{pmatrix} \tag{30}$$

At the cost of a clear over-parameterization, it has been demonstrated that a Gaussian distribution over these six parameters resembles the cone-like shape of the actual uncertainty—refer to the examples in Figure 10. Historically, this was the first parameterization that succeeded in unifying the representation of close and distant features in such a way that both kinds could be used simultaneously for localization, mapping, and SLAM.

Therefore, we can summarize the probabilistic inverse sensor model based on the inverse-depth parameterization as follows. Each landmark will be represented as a vector:

$$\mathbf{m}_i = \begin{pmatrix} o_{x_i} & o_{y_i} & o_{z_i} & \theta_i & \phi_i & \rho_i \end{pmatrix}^T \tag{31}$$

over which we define a Gaussian distribution such that $\mathbf{m}_i \sim N\left(\bar{\mathbf{m}}_i, \mathbf{\Sigma}_{\mathbf{m}_i}\right)$, with:

$$\underbrace{p(\mathbf{z}_k | \mathbf{m}_i, \mathbf{x}_k)}_{\text{Inverse sensor model}} = N\left(\mathbf{m}_i; \bar{\mathbf{m}}_i, \mathbf{\Sigma}_{\mathbf{m}_i}\right) \rightarrow \begin{cases} \bar{\mathbf{m}}_i = \begin{pmatrix} \bar{o}_{x_i} & \bar{o}_{y_i} & \bar{o}_{z_i} & \bar{\theta}_i & \bar{\phi}_i & \bar{\rho}_i \end{pmatrix}^T \\ \mathbf{\Sigma}_{\mathbf{m}_i} = \left(\begin{array}{c|ccc} \mathbf{\Sigma}_{\mathbf{x}} & & \mathbf{0}_{3\times3} & \\ \hline & \sigma_\theta^2 & 0 & 0 \\ \mathbf{0}_{3\times3} & 0 & \sigma_\phi^2 & 0 \\ & 0 & 0 & \sigma_\rho^2 \end{array}\right) \end{cases} \tag{32}$$

The first five values in the mean vector $\bar{\mathbf{m}}_i$ directly correspond to the camera position $\begin{pmatrix} o_{x_i} & o_{y_i} & o_{z_i} \end{pmatrix}^T$ and to the direction in which the landmark was observed, with respect to the global frame of reference. Within the covariance matrix, the first 3×3 diagonal block represents the uncertainty in the camera location. Since in this chapter we are assuming it is perfectly known, all the matrix entries become zeros. However, when we will revisit bearing-only observations in the context of SLAM in chapter 9, this matrix shall be set to the actual uncertainty of the camera pose. Notice how the covariance of the camera location and the three other parameters (the pair of 3×3 off-diagonal symmetric blocks) is exactly zero. This follows from the realistic assumption of statistical independence of the sensor noises (θ_i, ϕ_i) and the depth (ρ_i) with respect to the uncertainty of the camera pose.

The sixth value in the mean vector (i.e., the inverse depth $\bar{\rho}_i$) and the three standard deviations σ_θ, σ_ϕ and σ_ρ are free parameters to be settled heuristically. Firstly, σ_θ and σ_ϕ can be adjusted to account for the inaccuracies in the bearing angles, typically, due to errors in the interest keypoint detector on the images. And secondly, $\bar{\rho}_i$ and its associated standard deviation σ_ρ can be determined by defining some arbitrary minimum and maximum depths (d_0 and d_1, respectively) as the desired limits of the (for example) 99.7% confidence interval of the modeled uncertainty. Then, $\bar{\rho}_i$ and σ_ρ can be found from:(using the equivalence of ±3 sigmas for a 99.7% confidence interval of a one-dimensional Gaussian distribution)

$$\begin{cases} d_0 = \dfrac{1}{\bar{\rho}_i + 3\sigma_\rho} \\ d_1 = \dfrac{1}{\bar{\rho}_i - 3\sigma_\rho} \end{cases}$$

(and solving for the desired parameters)

$$\rightarrow \begin{cases} \bar{\rho} = \dfrac{1}{2}\left(\dfrac{1}{d_0} + \dfrac{1}{d_1}\right) \\ \sigma_\rho = \dfrac{1}{6}\left(\dfrac{1}{d_0} - \dfrac{1}{d_1}\right) \end{cases} \tag{33}$$

Recursive Bayesian Estimation

Once the three-dimensional landmark has been initialized in the map with an inverse-depth parameterization, information from subsequent observations must be fused to reduce its uncertainty. For this kind of observations, as soon as the landmark is observed from a slightly different direction, which in computer vision is called the observation *parallax*, the uncertainty in its depth will be drastically reduced.

Since we employ one Gaussian distribution for each landmark, the particular implementation of Equation 15 of choice for this case is also an Extended Kalman filter. Due to the assumption of a static map (i.e., landmarks do not move on their own), the recursive form of the EKF in chapter 7's Equation 44 simplifies in our case to:

$$\bar{\mathbf{m}}_i \leftarrow \bar{\mathbf{m}}_i + \mathbf{K}_{k,i}\left(\mathbf{z}_k - \mathbf{h}_{k,i}\right)$$

$$\mathbf{\Sigma}_{\mathbf{m}_i} \leftarrow \left(\mathbf{I} - \mathbf{K}_{k,i}\mathbf{H}_{k,i}\right)\mathbf{\Sigma}_{\mathbf{m}_i}$$

with the Kalman gain matrix being

$$\mathbf{K}_{k,i} = \mathbf{\Sigma}_{\mathbf{m}_i}\mathbf{H}_{k,i}^T\left(\mathbf{R} + \mathbf{H}_{k,i}\mathbf{\Sigma}_{\mathbf{m}_i}\mathbf{H}_{k,i}^T\right)^{-1} \tag{34}$$

The Jacobian matrix $\mathbf{H}_{k,i}$ contains the derivatives of the function $\mathbf{h}(\cdot)$ defined in Equations 27 and 28 respect to the six parameters of the landmark. Due to the complexity of the transformation involved, here is more convenient to apply the chain rule in order to evaluate this Jacobian. Recalling that we denote the parameterization of a landmark as $\mathbf{m}_i$, its Cartesian coordinates in the global frame of reference as $\mathbf{M}_i$ and its local coordinates with respect to the camera as $\mathbf{M}'_{k,i}$, we can write:

$$\mathbf{H}_{i,t} = \frac{\partial \mathbf{h}\left(\mathbf{x}_t, \mathbf{m}_i\right)}{\partial \mathbf{m}_i} = \\ = \frac{\partial \mathbf{h}\left(\mathbf{M}'_i\right)}{\partial \mathbf{M}'_{k,i}} \frac{\partial \mathbf{M}'_{k,i}\left(\mathbf{M}_i\right)}{\partial \mathbf{M}_i} \frac{\partial \mathbf{M}_i\left(\mathbf{m}_i\right)}{\partial \mathbf{m}_i} \tag{35}$$

where we approached the complex projection of a landmark into an observation as a sequence of three, more handy steps: (1) $\mathbf{M}_i\left(\cdot\right)$ converts the set of inverse-depth parameters into their corresponding global coordinates, as shown in Equation 30; (2) $\mathbf{M}'_{k,i}\left(\cdot\right)$ is in charge of transforming them into local coordinates in the camera frame of reference as specified in Equation 27; and (3) the observation model $\mathbf{h}\left(\cdot\right)$ finally projects the landmark into pixel coordinates according to Equation 29. Recall that all these partial Jacobians, which are straightforward to obtain, must be evaluated at the latest estimated value of each of the involved variables.

It is worth mentioning that the advantages of the inverse-depth parameterization come only at the cost of employing six components for each landmark instead of the minimum of three as would correspond to a simple point in the space. Assuming an EKF implementation with a computational complexity cubic with the dimensionality, those extra parameters imply multiplying the execution time by eight. Therefore, it comes as no surprise the existence of proposals in the literature that recover the simpler $\left(x, y, z\right)$ parameterization as long as the reduction in the depth uncertainty makes it an acceptable approximation (Civera, Davison, & Montiel, 2008).

7. BAYESIAN ESTIMATION OF LANDMARK MAPS: RANGE-ONLY SENSORS

We finally arrive at the third type of landmark observations: those that only measure a range or distance from the sensor to one or a set of fixed points in the environments. As discussed in chapter 2 section 9, Range-Only (RO) sensors are commonly employed in submarine robotics but some practical applications have also appeared during the last decade for ground robots. In contrast to the *passive* nature of landmarks studied in previous sections, we deal here with observations that measure purposely-placed *active* devices in most cases, emitting radio or ultrasonic signals, thus, for differentiation we will refer to them as *beacons*. Beacons, landmarks, and features are all names of interchangeable entities in our present context of object-based mapping.

Disregarding the specific device utilized to obtain RO measurements, we will approach the problem in an abstract way focusing on the properties that all these sensors share. Firstly, they are typically able to detect several beacons simultaneously and to identify each one unequivocally by means of some sort of identification code, which is transmitted wirelessly: with RO sensors, we can avoid the hard problem of data association. Consequently, a RO observation for one beacon simply contains a range value, that is, $\mathbf{z}_k = \left(r_k\right)$.

Secondly, another fundamental characteristic of RO sensors is that they naturally lead to multiple hypotheses about the location of the beacons. To illustrate this point, refer to the example in Figure 11, where a robot moving in a straight line makes three range measurements for a beacon with location $\mathbf{m}_i$. In a two-dimensional approach, each range observation tells us that the beacon must lie around a circle centered at the robot position $\mathbf{x}_k$ with a radius of r_k. Under our probabilistic viewpoint, observations actually are assumed to be corrupted with an additive zero-mean Gaussian noise $n_k \sim N\left(0, \sigma_n^2\right)$, that is:

$$r_k = \left|\mathbf{m}_i - \mathbf{x}_k\right| + n_k \tag{36}$$

Figure 11. An example we have simulated of map building from range-only observations. (a) The first time the robot detects a landmark, it is introduced in the map via the inverse sensor model. In this case we employ a sum of Gaussians approximation to the actual, ring-shaped pdf (see text). (b) – (d) Subsequent observations further reduce the uncertainty in the landmark location. Notice how two potential locations for the landmark remain until the robot turns in (d) breaking the symmetry that existed up to that instant.

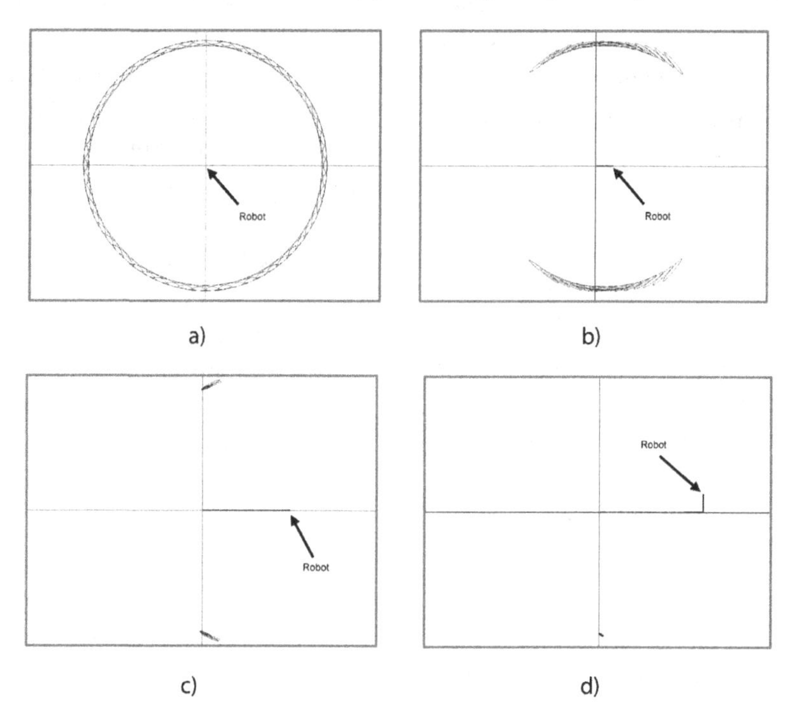

which suggests us to consider instead a "thick ring"-shaped pdf instead of a unidimensional circumference. One of such rings is depicted in the figure, centered at the robot pose where an observation was made. Intuitively speaking (the exact Bayesian approach is discussed below), the most likely locations at which the beacon may be actually located are those where the different rings intersect with each other.

Unlike range-bearing and bearing-only observations, in this case it is not only possible but quite common that the explained rings end up marking several separate locations as the likely location of the beacon. Therefore, probabilistic mapping with RO observations should be addressed with multimodal distributions, a requisite somewhat unique in mapping and SLAM.

The Inverse Sensor Model

Provided the sensor model of Equation 36 we find that just like with bearing-only observations, these ones also reduce the dimensionality of the observed beacon location from two (if we assume a planar map) to only one, the range measurement.

Therefore, we face again the non-existence of an inverse sensor function. A probabilistic inverse sensor model is easy to devise instead, since all we need is a pdf that assigns each potential beacon location $\mathbf{m}_i$ a likelihood according to Equation 36 and the known distribution of the additive n_t noise:

$$\underbrace{p\left(\mathbf{z}_k \middle| \mathbf{m}_i, \mathbf{x}_k\right)}_{\text{Inverse sensor model}} = \frac{1}{\sigma_n \sqrt{2\pi}} \exp\left\{ -\frac{1}{2} \left(\frac{\left|\mathbf{m}_i - \mathbf{x}_k\right| - r_k}{\sigma_n} \right)^2 \right\} \tag{37}$$

Notice that the thick ring displayed in figure 11a closely corresponds to this pdf, where the thickness is associated to the noise variance σ_n^2. The problem of this distribution is that it cannot be written as a Gaussian in terms of the (x, y) coordinates of the beacon $\mathbf{m}_i$.

Among others, two workarounds have been proposed in the literature for approximating the ring-shape distribution in Equation 37 while still being able to perform Bayesian filtering. The first one resembles the change of landmark parameterization explained above for bearing-only sensors. In this method, over-parameterization of the beacon position with the location where the robot made its first observation plus a pair of polar coordinates that identify a global direction and a distance from that robot pose to the beacon is used (Djugash, Singh, & Grocholsky, 2008). As illustrated in Figure 12, this idea allows representing each beacon with just one Gaussian over the five abovementioned parameters. However, in order to deal with multi-modality one would need to introduce additional heuristics to determine when and how to split the Gaussian into two or more pdf modes.

The second approach, introduced in (Blanco, González, & Fernández-Madrigal, 2008b) and already employed in the example of Figure 11, consists of approximating Equation 37 with a Sum Of Gaussians (SOG) over the ordinary Cartesian coordinates, such that:

Figure 12. An approximation of the inverse sensor model of RO observations by means of a single Gaussian, in the parameter space of range-bearing with respect a robot pose. In this example, three confidence intervals are shown, corresponding to 1σ, 2σ, and 3σ. The observation has a mean orientation of 45° and a mean distance of 9m, while the associated standard deviations is 50° and 0.2m, respectively. Notice how a large enough uncertainty in the orientation could approximate well a complete ring-like pdf.

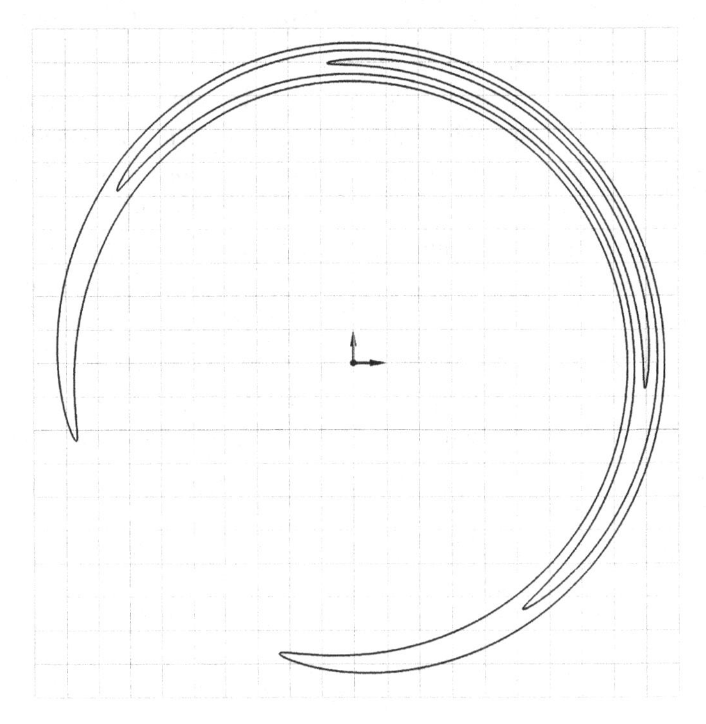

$$p\left(\mathbf{z}_k \middle| \mathbf{m}_i, \mathbf{x}_k\right) = \sum_j \omega_j N\left(\mathbf{m}_i; \bar{\mathbf{m}}_i^j, \mathbf{\Sigma}_{\mathbf{m}_i}^j\right),$$
$$\text{with} \sum_j \omega_j = 1 \tag{38}$$

where ω_j are the weights of each Gaussian mode. Upon its first observation, all the SOG modes have equal weights and are distributed in such a way that they approximate the thick ring-shape of the actual inverse sensor pdf. At the cost of maintaining several Gaussians for each beacon, the clear advantage of this method is its natural capability of representing an arbitrary number of modes just by readjusting the parameters of each SOG mode and their weights, which can be achieved by Bayesian filtering.

Recursive Bayesian Estimation

If beacons were modeled according to the first approach above, namely the over-parameterization that includes polar coordinates, we would have a single Gaussian for each beacon, which should be updated with subsequent observations. A natural implementation here is to employ the EKF as already explained above for the other types of landmarks. However, the strong non-linearities present in this parameterization would render the UKF, explained in chapter 7 section 3, a better candidate.

In the case that beacons are modeled as a weighted SOG, we would face a new problem not dealt with yet in this book. Nonetheless, it can be easily shown that updating a SOG with a new observation $\mathbf{z}_k = \left(r_k\right)$ can be easily realized through the following three steps:

1. The weights ω_j must be updated to reflect how well each Gaussian mode explains the observation. Quantitatively, this implies evaluating the likelihood of the observation against the prediction of the $j-th$ mode:

$$\omega_j \leftarrow \omega_j N\left(\mathbf{z}_k; \bar{h}_i^j, \sigma_i^{2j}\right)$$

with the mean obtained through the sensor model in Equation 36

$$\bar{\mathbf{h}}_i^j = \left|\bar{\mathbf{m}}_i^j - \mathbf{x}_k\right|$$

and the covariance got through first-order Taylorseries linearization (see chapter 3 section 8):

$$\sigma_i^{2j} = \mathbf{H}_i^j \mathbf{\Sigma}_{\mathbf{m}_i}^j \mathbf{H}_i^{jT} + \sigma_n^2 \tag{39}$$

2. Weights are then renormalized such that they sum the unity, in order to assure that the SOG is kept as a pdf.
3. The parameters of each Gaussian mode $N\left(\mathbf{m}_i; \bar{\mathbf{m}}_i^j, \mathbf{\Sigma}_{\mathbf{m}_i}^j\right)$ are updated following the standard EKF equations—refer for example to Equation 34. Unlike with the polar-coordinate parameterization, the more complex UKF is not required here since the uncertainty of each SOG mode is quite small in comparison to the entire ring-like pdf, and assuming linearity in the observation of one mode is perfectly acceptable.

Notice that this technique exploits the freedom, assumed in this chapter, in modeling of pdfs for each map element independently: each beacon may be represented by a different number of SOG modes. In practice, the weights of many modes will soon become negligible after a few observations. Thus, it becomes convenient to discard those SOG nodes that have weights below some certain threshold. After the robot moves around, the number of modes will reduce, dynamically adapting itself to the actual uncertainty in the beacon location, as shown with the example of Figure 11c.

8. OTHER MAP BUILDING ALGORITHMS

Previous sections have addressed the problem of initializing and updating the most common probabilistic representations of metric maps from sequences of sensor readings: grid maps and feature or landmark maps. In the following, we study other mapping algorithms, which either do not rely on a probabilistic foundation or are not so widely spread.

Point Maps

There exist two main families of sensors which are used to generate point maps: laser range scanners and 3D cameras—reviewed in chapter 2 sections 7 and 8, respectively. As already mentioned above, point maps are among the most "sub-symbolic" representations, in the sense that very little (or none) post-processing is required for the raw sensor readings to be merged within the maps.

In its most basic approach, maintaining a point map could be as simple as keeping a list of 2D or 3D point coordinates and appending new ones with each sensor reading, taking into account the corresponding geometrical transformation for the pose of the sensor within the global frame of reference. However, given that state-of-the-art range scanners provide dozens of completed scans per second, if we were to insert all of them in a point map, it would grow as rapidly as to render any localization or SLAM algorithm useless after a few seconds of operation. This is why all practical implementations discard most of the 2D or 3D range scans provided by the sensors. This may seem a waste of information, but, in fact, consecutive scans are highly redundant and hence provide very little new relevant information. One alternative for not dropping scans is to *fuse* points which are established to correspond to previously observed points, which can drastically reduce the number of points in a map at the expense of introducing the complexities and potential mistakes of data association.

Typical heuristics to determine which scans to keep are: (1) the usage of a fixed subsample rate (e.g., keeping only one scan out of ten) and (2) discarding all the captured scans until the robot moves or turns more than a certain threshold distance or angle, respectively. Selecting the parameters of any of these two heuristics is a critical step since they heavily condition the performance of any posterior localization or SLAM method, which has to work on the point map. Unfortunately, and to the best of our knowledge, this topic has not been properly addressed in the scientific literature, thus it requires the experience of the operator and some doses of manual tuning with trials and errors.

Continuous Markov Random Fields

Recall that we loosely defined a random field in section 2 as "a spatial domain where each point is associated a random variable." An example of a discrete random field has already been provided with occupancy grid maps. The property of interest at each location of the space was there the occupancy (or freeness), a discrete random variable with only two possible outcomes.

There exists, however, a variety of other properties we might be interested in while building a map with a mobile robot, and most of them are continuous magnitudes. For instance, we could map the height at each (x, y) position for an outdoor environment; or the temperature along the inside of an office building; or the different concentration of certain gases within a factory facility.

All these are examples of physical properties that vary from one point to the other—we will leave the dynamics aside and assume static environments, as usual. However, it is typical that the spatial variations of the magnitude of interest

Figure 13. Example of Kernel-based mapping with gas-concentration sensors mounted in a mobile robot of our lab. (a) The actual gas concentration in a simulation environment, from which a sequence of gas sensor observations have been simulated in order to build the gas map shown in (b). Notice that typical hurdles found in real sensors, such as a high response time, were taken into account in this simulation, hence the poor resolution attainable in the reconstruction, a common problem found in most kinds of gas maps (Images courtesy of Javier González Monroy, University of Málaga).

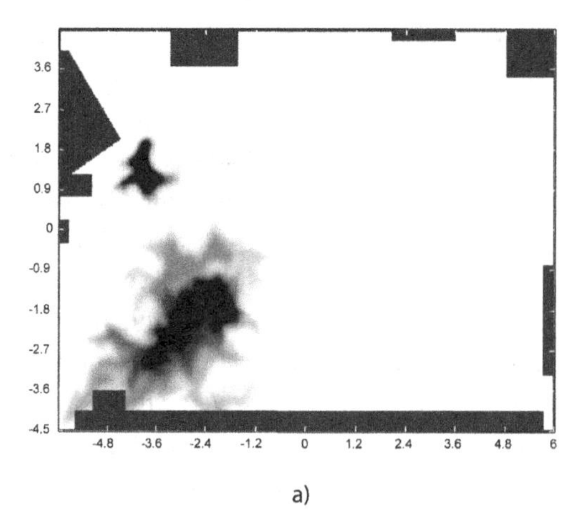

a)

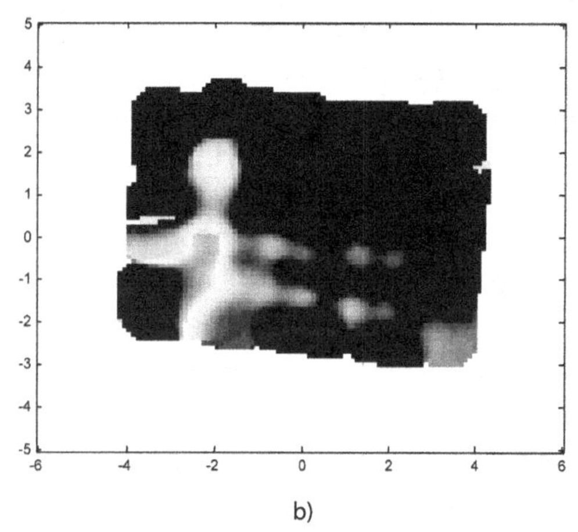

b)

are somewhat limited, with no abrupt changes between two nearby points, or, in more technical terms, a probability distribution could exist that, *conditioned* on any point of the environment, satisfactorily models the properties of its surroundings. In general, although the magnitude of interest for mapping may tend to remain constant around any given point, the variance of this prediction grows as we get farther from any central point. It is in those cases where the magnitude of interest satisfies these generic requirements when a map of the environment could be modeled as a Markov Random Field (MRF) (Winkler, 1995) (see Figure 13).

As already discussed in previous chapters, a MRF is a random field in which random variables can be organized on a discrete lattice and where the Markov condition holds between adjacent variables. This is in contrast to the common application of the Markov condition to sequences of r.v. that are consecutive in time. If we denote the r.v. for the property of interest at some arbitrary map coordinates (i, j) as $m_{i,j}$, and define the finite set of all adjacent coordinates as N, the *spatial* Markov property reads:

$$p\left(m_{i,j} \middle| \left\{m_p\right\}_{p \in N}, \left\{m_p\right\}_{p \notin N}\right) = p\left(m_{i,j} \middle| \left\{m_p\right\}_{p \in N}\right) \tag{40}$$

that is: a map element, conditioned on all its neighbors, is conditionally independent of the rest of the map. Moreover, if that conditional distribution can be properly modeled as a Gaussian, the MRF becomes a *Gaussian Markov Process*. Due to their general applicability, Gaussian Processes (GP) have received an immense attention in the research community (Rasmussen & Williams, 2006).

There exist different map building methods relying on the assumption of a MRF or GP-like map. We find classical references in the branch of geology and mining literature called *geostatistics*. In particular, dating back to the 1950s, we find the Kriging technique (Matheron, 1963) which applies what later on were to be called GP to interpolate geological measurements and therefore reconstruct non-measured areas. Closer to modern mobile robotics, we can find algorithms for mapping the concentration of gases relying on a simplified estimation method known as the Kernel method (Lilienthal & Duckett, 2004). A more direct application of GP to mapping was reported in Stachniss, Plagemann, Lilienthal, and Burgard (2008).

Pose Constraint Maps

As introduced in section 2, relational maps are all those representations that model some kind of relationship existing between the mapped elements instead of, for example, their global coordinates in some fixed frame of reference. Among this family of models, Pose Constraint Maps (PCMs) are probably the most representative nowadays due to their popularity.

Building a PCM implies creating and populating a directed graph, a computational abstract data type that has already been explained, instantiated such that its nodes represent robot poses (in an abstract sense, since the numeric values of those poses are unknowns) and its arcs represent the relative coordinates of poses with respect to each other. Typically, a PCM is built incrementally as the robot moves and explores its environment. New nodes (also called *keyframes*) are created wherever some heuristic is fulfilled, such as when it is reached a minimum distance from the last node. It is convenient to store (or *annotate*) the most recent robot observations (e.g. laser range scans, images from its cameras, etc.) within each node in order to provide it with some metrical information suitable to the determination of potential arcs. Indeed, for each newly created node, arcs should also be defined between nodes for which their relative spatial pose could be deduced from the annotated sensory data. The specific approach for obtaining such relative poses strongly depends on the employed sensors. As an example, the ICP matching algorithm, explained in chapter 6, could be used for the common case of working with laser range scanners. In order to cope with uncertainty, arcs often hold probability distributions of those relative poses, typically in the form of a Gaussian distribution.

Although the apparent simplicity of such a graph representation is appealing, its practical realization reveals two important hurdles, which are still an active area of research. Firstly, notice how global coordinates do not appear anywhere in the discussion above: all the map building process relies entirely on node-to-node relative coordinates. If for some reason the map is needed in a common frame of reference, computing the global coordinates for each node may become not an easy task at all. To illustrate the challenge, consider the map in Figure 5a which contains many closed loops in the graph topology. The global coordinates for that figure were obtained by arbitrarily fixing one of the nodes as the origin of coordinates, then creating a spanning tree from that node to all the others, e.g. using the Dijkstra's algorithm (Dijkstra, 1959). In a PCM with a tree structure, the global pose of each node is the simple accumulation of all the edges from the root node at the origin. However, there exist obvious mismatches or inconsistencies in the coordinates so obtained—which correspond to the arcs crossing wide gaps in the figure. The bunch of techniques generically dubbed as *Graph SLAM* (and reviewed in chapter 10 section 3) precisely address the problem of estimating a set of *consistent* global coordinates from PCMs with arbitrarily topologies that may include any number of loops.

The second problem of these relational maps is related to the creation of arcs when the robot closes a loop. This is the well-known *loop-closure* problem and consists in reliably detecting such a situation in order to define the corresponding arc with a relative pose obtained from the metrical registration of the observations in each node. Some approaches to efficiently detect loop closures include geometrical information, that is, firstly solving the graph for its global coordinates, then considering all the potential matches of the latest node with all its neighbors. When working with images, other authors propose purely topological methods where node-to-node pairings are established uniquely from the detection of similar visual features (Cummins & Newman, 2008). We will further explore the process of building maps of this kind in chapter 10.

REFERENCES

Anderson, M. L. (2003). Embodied cognition: A field guide. *Artificial Intelligence*, *149*, 91–130. doi:10.1016/S0004-3702(03)00054-7

Arras, K. O., Castellanos, J. A., Schilt, M., & Siegwart, R. (2003). Feature-based multi-hypothesis localization and tracking using geometric constraints. *Robotics and Autonomous Systems*, *44*, 41–53. doi:10.1016/S0921-8890(03)00009-5

Birk, A., & Carpin, S. (2006). Merging occupancy grid maps from multiple robots. *Proceedings of the IEEE*, *96*(7), 1384–1397. doi:10.1109/JPROC.2006.876965

Blanco, J. L., Fernández-Madrigal, J. A., & González, J. (2008). Towards a unified Bayesian approach to hybrid metric-topological SLAM. *IEEE Transactions on Robotics*, *24*(2), 259–270. doi:10.1109/TRO.2008.918049

Blanco, J. L., Fernández-Madrigal, J. A., & González, J. (2008b). Efficient probabilistic range-only SLAM. In *Proceedings of the IEEE/RSJ 2008 International Conference on Intelligent Robots and Systems*, (pp. 1017-1022). IEEE Press.

Blanco, J. L., González, J., & Fernández-Madrigal, J. A. (2009). Subjective local maps for hybrid metric-topological SLAM. *Robotics and Autonomous Systems*, *57*(1), 64–74. doi:10.1016/j.robot.2008.02.002

Brooks, R. A. (1982). Solving the find-path problem by good representation of free space. *IEEE Systems. Man and Cybernetics*, *13*, 190–197.

Brooks, R. A. (1991). Intelligence without representation. *Artificial Intelligence Journal*, *47*, 139–159. doi:10.1016/0004-3702(91)90053-M

Choset, H., & Burdick, J. W. (1995). Sensor based planning, part 1: The generalized Voronoi graph. In *Proceedings of the IEEE International Conference on Robotics and Automation*. IEEE Press.

Civera, J., Davison, A., & Montiel, J. (2008). Inverse depth parametrization for monocular SLAM. *IEEE Transactions on Robotics*, *24*(5), 932–945. doi:10.1109/TRO.2008.2003276

Coradeschi, S., & Saffiotti, A. (2003). An introduction to the anchoring problem. *Robotics and Autonomous Systems*, *43*(2-3), 85–96. doi:10.1016/S0921-8890(03)00021-6

Cummins, M., & Newman, P. (2008). Accelerated appearance-only SLAM. In *Proceedings of the IEEE International Conference on Robotics and Automation*, (pp. 1828-1833). IEEE Press.

Davison, A. (2003). Real-time simultaneous localization and mapping with a single camera. In *Proceedings of the International Conference on Computer Vision*, (pp. 1403-1410). IEEE.

Davison, A. J., Reid, I. D., Molton, N. D., & Stasse, O. (2007). MonoSLAM: Real-time single camera SLAM. *IEEE Transactions on Pattern Analysis and Machine Intelligence*, *29*(6), 1052–1067. doi:10.1109/TPAMI.2007.1049

Dijkstra, E. W. (1959). A note on two problems in connection with graphs. *Numerische Mathematik, 1*, 269–271. doi:10.1007/BF01386390

Djugash, J., Singh, S., & Grocholsky, B. (2008). Decentralized mapping of robot-aided sensor networks. In *Proceedings of the IEEE International Conference on Robotics and Automation*. IEEE Press.

Duda, R. O., & Hart, P. E. (1972). Use of the Hough transformation to detect lines and curves in pictures. *Communications of the ACM, 15*(1), 11–15. doi:10.1145/361237.361242

Fabrizi, E., & Saffiotti, A. (2000). Augmenting topology-based maps with geometric information. In *Proceedings of the 6th International Conference on Intelligent Autonomous Systems*. IEEE.

Fernández-Madrigal, J. A., & González, J. (2002). Multi-hierarchical representation of large-scale space: Applications to mobile robots. *Intelligent Systems, Control and Automation: Science and Engineering, 24*.

Fox, D. (2003). *The Intel lab dataset*. Retrieved Mar 1, 2012, from http://cres.usc.edu/radishrepository/view-one.php?name=intel_lab

Galindo, C., Fernández-Madrigal, J. A., & González, J. (2007). Multiple abstraction hierarchies for mobile robot operation in large environments. *Studies in Computational Intelligence, 68*.

Galindo, C., Fernández-Madrigal, J. A., González, J., & Saffiotti, A. (2008). Robot task planning using semantic maps. *Robotics and Autonomous Systems, 56*(11), 955–966. doi:10.1016/j.robot.2008.08.007

Grisetti, G., Grzonka, S., Stachniss, C., Pfaff, P., & Burgard, W. (2007). Efficient estimation of accurate maximum likelihood maps in 3D. In *Proceedings of the IEEE/RSJ International Conference on Intelligent Robots and Systems*, (pp. 3472-3478). IEEE Press.

Grisetti, G., Stachniss, C., Grzonka, S., & Burgard, W. (2007). A tree parameterization for efficiently computing maximum likelihood maps using gradient descent. In *Proceedings of Robotics: Science and Systems (RSS)*. IEEE.

Guivant, J., & Nebot, E. (2001). Optimization of the simultaneous localization and map building algorithm for real-time implementation. *IEEE Transactions on Robotics and Automation, 17*(3), 242–257. doi:10.1109/70.938382

Harnad, S. (1990). The symbol grounding problem. *Physica D. Nonlinear Phenomena, 42*, 335–346. doi:10.1016/0167-2789(90)90087-6

Hartley, R., & Zisserman, A. (2003). *Multiple view geometry in computer vision*. Cambridge, UK: Cambridge University Press.

Johnson, A. E., & Hebert, M. (1999). Using spin images for efficient object recognition in cluttered 3D scenes. *IEEE Transactions on Pattern Analysis and Machine Intelligence, 21*(5), 433–449. doi:10.1109/34.765655

Klein, G., & Murray, D. (2007) Parallel tracking and mapping for small AR workspaces. In *Proceedings of the 2007 6th IEEE and ACM International Symposium on Mixed and Augmented Reality*. IEEE Press.

Konolige, K. (2005). SLAM via variable reduction from constraint maps. In *Proceedings of the IEEE International Conference on Robotics and Automation*. IEEE Press.

Konolige, K., Bowman, J., Chen, J. D., Mihelich, P., Calonder, M., Lepetit, V., & Fua, P. (2010). View-based maps. *The International Journal of Robotics Research, 29*(8), 941–957. doi:10.1177/0278364910370376

Konolige, K., Fox, D., Limketkai, B., Ko, J., & Stewart, B. (2003). Map merging for distributed robot navigation. In *Proceedings of the IEEE/RSJ International Conference on Intelligent Robots and Systems*, (pp. 212-227). IEEE Press.

Kuipers, B. J. (1977). *Representing knowledge of large-scale space.* (Doctoral Dissertation). Massachusetts Institute of Technology. Cambridge, MA.

Kuipers, B. J. (1978). Modeling spatial knowledge. *Cognitive Science*, *2*, 129–153. doi:10.1207/s15516709cog0202_3

Kuipers, B. J. (2000). The spatial semantic hierarchy. *Artificial Intelligence*, *119*, 191–233. doi:10.1016/S0004-3702(00)00017-5

Latombe, J. C. (1991). *Robot motion planning.* Boston, MA: Kluwer Academic Publishers.

Leonard, J. J., & Durrant-Whyte, H. F. (1991). Mobile robot localization by tracking geometric beacon. *IEEE Transactions on Robotics and Automation*, *7*(3), 376–382. doi:10.1109/70.88147

Lilienthal, A., & Duckett, T. (2004). Building gas concentration gridmaps with a mobile robot. *Robotics and Autonomous Systems*, *48*(1), 3–16. doi:10.1016/j.robot.2004.05.002

Lowe, D. G. (2004). Distinctive image features from scale-invariant keypoints. *International Journal of Computer Vision*, *60*(2), 91–110. doi:10.1023/B:VISI.0000029664.99615.94

Lozano-Pérez, T., & Wesley, M. A. (1979). An algorithm for planning collision-free paths among polyhedral obstacles. *Communications of the ACM*, *22*(10), 560–570. doi:10.1145/359156.359164

Lu, F., & Milios, E. (1997). Globally consistent range scan alignment for environment mapping. *Autonomous Robots*, *4*(4), 333–349. doi:10.1023/A:1008854305733

Matheron, G. (1963). Principles of geostatistics. *Economic Geology and the Bulletin of the Society of Economic Geologists*, *58*(8), 1246. doi:10.2113/gsecongeo.58.8.1246

Microsoft. (2011). *Xbox Kinect web site*. Retrieved July 19, 2011, from http://www.xbox.com/en-US/kinect

Mikhail, E., Bethel, J., & McGlone, J. C. (2001). *Introduction to modern photogrammetry*. New York, NY: Wiley.

Mikolajczyk, K., & Schmid, C. A. (2005). Performance evaluation of local descriptors. *IEEE Transactions on Pattern Analysis and Machine Intelligence*, *27*(10), 1615–1630. doi:10.1109/TPAMI.2005.188

Moravec, H. P., & Elfes, A. (1985). High resolution maps from wide angle sonar. In *Proceedings of the IEEE International Conference on Robotics and Automation.* IEEE Press.

MRPT. (2011). *The mobile robot programming toolkit website*. Retrieved Mar 1, 2012, from http://www.mrpt.org/

Nixon, M., & Aguado, A. S. (2008). *Feature extraction and image processing*. New York, NY: Academic Press.

Nüchter, A. (2009). *3D robotic mapping: The simultaneous localization and mapping problem with six degrees of freedom*. Berlin, Germany: Springer Verlag.

Núñez, P., Vázquez-Martín, R., del Toro, J. C., Bandera, A., & Sandoval, F. (2006). Feature extraction from laser scan data based on curvature estimation for mobile robotics. In *Proceedings of the IEEE International Conference on Robotics and Automation*, (pp. 1167-1172). IEEE Press.

Piaget, J. (1948). *The child's conception of space*. London, UK: Routledge.

Ranganathan, A., & Dellaert, F. (2011). Online probabilistic topological mapping. *The International Journal of Robotics Research, 30*(6), 755–771. doi:10.1177/0278364910393287

Rasmussen, C. E., & Williams, C. K. I. (2006). *Gaussian processes for machine learning*. Cambridge, MA: The MIT Press.

Remolina, E., Fernández-Madrigal, J. A., Kuipers, B. J., & González-Jiménez, J. (1999). Formalizing regions in the spatial semantic hierarchy: An AH-graphs implementation approach. *Lecture Notes in Computer Science, 1661*, 109–124. doi:10.1007/3-540-48384-5_8

Rotwat, P. F. (1979). *Representing the spatial experience and solving spatial problems in a simulated robot environment.* (Doctoral Dissertation). University of British Columbia. British Columbia, Canada.

Rusu, R. B., Marton, Z. C., Blodow, N., Dolha, M., & Beetz, M. (2008). Towards 3D point cloud based object maps for household environments. *Robotics and Autonomous Systems, 56*(11), 927–941. doi:10.1016/j.robot.2008.08.005

Smith, M., Baldwin, I., Churchill, W., Paul, R., & Newman, P. (2009). The new college vision and laser data set. *The International Journal of Robotics Research, 28*(5), 595–599. doi:10.1177/0278364909103911

Stachniss, C., Plagemann, C., Lilienthal, A., & Burgard, W. (2008). Gas distribution modeling using sparse Gaussian process mixture models. In *Proceedings of Robotics: Science and Systems*. IEEE. doi:10.1007/s10514-009-9111-5

Strasdat, H., Montiel, J. M. M., & Davison, A. J. (2010). Scale-drift aware large scale monocular SLAM. In *Proceedings of Robotics Science and Sytems*. IEEE.

Tardós, J. D., Neira, J., Newman, P., & Leonard, J. (2002). Robust mapping and localization in indoor environments using sonar data. *The International Journal of Robotics Research, 21*(4), 311–330. doi:10.1177/027836402320556340

Thrun, S., Burgard, W., & Fox, D. (2005). *Probabilistic robotics*. Cambridge, MA: The MIT Press.

Tipaldi, G. D., & Arras, K. O. (2010). FLIRT – Interest regions for 2D range data. In *Proceedings of the IEEE International Conference on Robotics and Automation*. IEEE Press.

Tolman, E. C. (1948). Cognitive maps in rats and men. *Psychological Review, 55*(4), 189–208. doi:10.1037/h0061626

Trucco, E., & Verri, A. (1998). *Introductory techniques for 3-D computer vision*. Upper Saddle River, NJ: Prentice Hall.

Trucco, E., & Verri, A. (1998). *Introductory techniques for 3-D computer vision*. Upper Saddle River, NJ: Prentice Hall.

Trudeau, R. J. (1994). *Introduction to graph theory*. New York, NY: Dover Publications.

Winkler, G. (1995). *Image analysis, random fields and dynamic Monte Carlo methods*. Berlin, Germany: Springer.

Wurm, K. M., Hornung, A., Bennewitz, M., Stachniss, C., & Burgard, W. (2010). OctoMap: A probabilistic, flexible, and compact 3D map representation for robotic systems. In *Proceedings of the ICRA Workshop on Best Practice in 3D Perception and Modeling for Mobile Manipulation.* ICRA.

ENDNOTES

[1] Notice that denoting the "inverse observation function" as $\mathbf{h}^{-1}(\cdot)$ may be seen as an abuse of notation. Strictly speaking, if the observation function (disregarding the additive noise) is represented as the function $\mathbf{z}_k = \mathbf{h}_i(\mathbf{x}_k, \mathbf{m}_i)$, its mathematical inverse function should be $(\mathbf{x}_k, \mathbf{m}_i) = \mathbf{h}_i^{-1}(\mathbf{z}_k)$ instead of the commonly employed $\mathbf{m}_i = \mathbf{h}_i^{-1}(\mathbf{x}_k, \mathbf{z}_k)$, which is the function of our interest for map building and SLAM.

Chapter 9
The Bayesian Approach to SLAM

ABSTRACT

This is the second chapter of the third section. It deals with the situation arising when neither the environment nor the exact localization of a mobile robot are known, that is, when we face the hard problem of SLAM. It reviews the most common solutions to that problem found in literature, especially those based on statistical estimation. Both parametric and non-parametric filters are explained as practical solutions to this problem, including analysis of their advantages and weaknesses that must be both taken into account in order to design a robust SLAM system. Complete examples and algorithms for these filters are included.

CHAPTER GUIDELINE

- You will learn:
  - The graphical models of *full SLAM* and *marginalized* (*on-line*) *SLAM*, which lie behind all metrical solutions to SLAM.
  - What is the relationship between the Bayesian and the maximum likelihood estimation approaches to SLAM.
  - How do loop closures affect the covariance matrices of classical parametric filters.
  - A short glimpse at the most advanced algorithms for non-parametric Bayesian filtering.
  - An insight in the fundamental limitations of both parametric and non-parametric filters for metric SLAM.
- Provided tools:
  - A table classifying all existing approaches to metrical SLAM.
  - Pseudo-code descriptions and detailed formulation of the most important Bayesian algorithms to solve SLAM with different kinds of maps.
- Relation to other chapters:
  - Bayesian filters presented here are modified versions of those introduced in chapter 7 while addressing the localization-only problem.
  - Probabilistic motion and sensor models, described in chapters 5 and 6, respectively, are assumed to known in this chapter.

DOI: 10.4018/978-1-4666-2104-6.ch009

1. INTRODUCTION

In previous chapters, we have devised probabilistic approaches to two problems: (1) mobile robot localization and (2) map building, assuming that the map and the localization were already known, respectively. Now we turn our attention to the *simultaneous* estimation of both, robot localization and the map of the environment, called SLAM, or Simultaneous Localization and Mapping. Approaching SLAM as an estimation problem results in a much more challenging task due to its higher dimensionality and the interdependence of the observer positions and the reconstructed model of the environment.

The aim of this chapter is twofold. Firstly, this first section is devoted to providing a short historical outline of the research conducted on SLAM and an overview of the numerous proposed solutions; we will describe them under the unifying light of their associated graphical models. The rest of the chapter is devoted to analyze in detail the two most popular families of Bayesian solutions.

SLAM is of wide applicability; it naturally arises in many measuring and reconstruction issues. Hence, it comes at no surprise that, in different contexts, exactly the same estimation problem had been proposed under a variety of names. We can trace its probable origins back to the 50s of the last century, when the first least-squares methods were proposed in the photogrammetry and surveying communities (Golub & Plemmons, 1980) as efficient solutions to simultaneously estimating both the location of cameras and the observed points, leading to the so-called *Bundle Adjustment* (BA) approaches (Brown, 1958; Slama, 1980). In short, BA aims at minimizing the residual errors from reprojecting the observed points into the reconstructed camera positions, which is typically accomplished by iterative non-linear least-squares methods, e.g. the Gauss-Newton algorithm. The unprecedented accuracy that these methods achieved in aerial cartography was such that they helped detecting several imperfections in the construction of photographic cameras and films that previously had gone unnoticed.

Independently, the same goal was being pursued by *Structure From Motion* (SFM) methods since the 90s, also within the computer vision community (Tomasi & Kanade, 1992; Varga, 2007). However, with some exceptions such as (Forsyth, Ioffe, & Haddon, 1999), in SFM the focus shifts from least-squares optimization towards sequentially incorporating observations and updating the estimations making an intensive use of pure geometry: triangulation, epipolar lines, the fundamental matrix, etc. (Hartley & Zisserman, 2004).

As a third independent line of research, the idea of providing a probabilistic representation of the spatial relationships between a robot and a set of world objects was already present in the mobile robotics community by the end of the 80s, including the proposal of Kalman filters for incrementally learning those maps (Smith & Cheeseman, 1986; Durrant-Whyte, 1988). However, that framework remained in a second plane until 2001, when the seminal paper (Dissanayake, Newman, Clark, Durrant-Whyte, & Csorba, 2001) mathematically proved for the first time that, indeed, SLAM with Kalman filters was a valid approach that under certain conditions converges towards the real solution. Since then, SLAM became an area of frenetic research activity and other Bayesian estimation frameworks different from Kalman filters have been investigated.

It has not been until recently that all these three lines have finally converged: at present, BA, SFM and SLAM are all understood as a unique problem which can be addressed by a number of different estimation algorithms. Many modern methods have being recognized as reinventions—or even approximations—of older solutions, which were well established in other communities (Triggs, McLauchlan, Hartley, & Fitzgibbon, 2000). Ironically, some of the latest proposals in SLAM (reviewed in chapter 10) are closer to the least-squares photogrammetry methods of the 50s than to any other modern SLAM framework of the 2000s.

After this short historical perspective, we now define the BA, SFM, and SLAM problems in a unifying way. Alike with the localization-only problem explored in chapter 7, we will describe the problems by means of their graphical models—recall that chapter 3 section 10 provided an introduction to this mathematical tool. To begin with, and as is usual in the literature, we will assume a static world, thus the map can be represented by a variable $\mathbf{m}$, which does not change with time. Depending of the kind of map representation, this variable would be actually a set comprising any number of variables, one for each individual map element. For instance, for landmark maps we have $\mathbf{m} = \{\mathbf{m}_1, \mathbf{m}_2, \ldots \mathbf{m}_L\}$ with $\mathbf{m}_i$ being the parameterization of the *i*-th individual landmark. Following the notation of previous chapters, the sequence of robot poses, observations and actions will be denoted as $\{\mathbf{x}_0, \mathbf{x}_1, \ldots, \mathbf{x}_k\}$, $\{\mathbf{z}_1, \mathbf{z}_2, \ldots, \mathbf{z}_k\}$, and $\{\mathbf{u}_1, \mathbf{u}_2, \ldots, \mathbf{u}_k\}$, respectively. Then, the structure of the variables involved in SLAM is that shown in Figure 1a and dubbed *full SLAM* due to the inclusion of the complete (full) time series for each dynamic variable. From the perspective of graphical models, these variables form a partially observable DBN with known structure (Russell & Norvig, 2002), exactly like in the localization-only problem studied in chapter 7. Notice that observed variables have been represented as unshaded nodes (robot observations and actions), while the series of robot poses and the map elements are hidden variables, i.e. the unknowns we want to estimate.

The conditional independences encoded in this DBN allow us to factorize the joint pdf of all the variables involved in full SLAM as:

$$p\left(\mathbf{x}_{0:k}, \mathbf{m}_{1:L}, \mathbf{z}_{1:k}, \mathbf{u}_{1:k}\right) = \\ = p\left(\mathbf{x}_0\right)\prod_{i=1}^{L} p\left(\mathbf{m}_i\right)\prod_{i=1}^{k} p\left(\mathbf{x}_i \middle| \mathbf{x}_{i-1}, \mathbf{u}_i\right)\prod_{i=1}^{k} p\left(\mathbf{z}_i \middle| \mathbf{x}_i, \mathbf{m}_{O(i)}\right) \quad (1)$$

where $\mathbf{m}_{O(i)}$ represents the set of all map elements observed at the $i-th$ time step. Since the priors over the map elements, the $p\left(\mathbf{m}_i\right)$ terms, are commonly assumed to be uniform distributions in order to represent our lack of any a priori information, the joint becomes:

$$p\left(\mathbf{x}_{0:k}, \mathbf{m}_{1:L}, \mathbf{z}_{1:k}, \mathbf{u}_{1:k}\right) \propto \\ \propto p\left(\mathbf{x}_0\right)\prod_{i=1}^{k} p\left(\mathbf{x}_i \middle| \mathbf{x}_{i-1}, \mathbf{u}_i\right)\prod_{i=1}^{k} p\left(\mathbf{z}_i \middle| \mathbf{x}_i, \mathbf{m}_{O(i)}\right) \quad (2)$$

Regarding the hidden variables, all SLAM methods studied in this chapter estimate the map elements, but differences exist regarding which portion of the robot path is actually estimated. Methods that estimate the entire robot path $\mathbf{x}_{1:t}$ from the sequence of all the actions and observations until some time step t, are called *full SLAM* methods in reference to the abovementioned DBN model on which they rely. The goal in this approach is finding the joint distribution of *all* hidden variables conditioned on all observed data, which coincides with the right-hand side of the last expression (something which will always occur when conditioning on all hidden variables):

$$p\left(\mathbf{x}_{0:k}, \mathbf{m}_{1:L} \middle| \mathbf{z}_{1:k}, \mathbf{u}_{1:k}\right) \propto \\ \propto p\left(\mathbf{x}_0\right)\prod_{i=1}^{k} p\left(\mathbf{x}_i \middle| \mathbf{x}_{i-1}, \mathbf{u}_i\right)\prod_{i=1}^{k} p\left(\mathbf{z}_i \middle| \mathbf{x}_i, \mathbf{m}_{O(i)}\right) \quad (3)$$

Since robot poses and landmarks are typically parameterized using *global* coordinates, while actions and observations encode relative constraints only, the problem as expressed in

Figure 1. (a) The graphical model (dynamic Bayesian network) of the SLAM problem and (b) its corresponding graph of correlations after performing Bayesian inference from observed variables. Shaded and unshaded nodes represent hidden and observed variables, respectively.

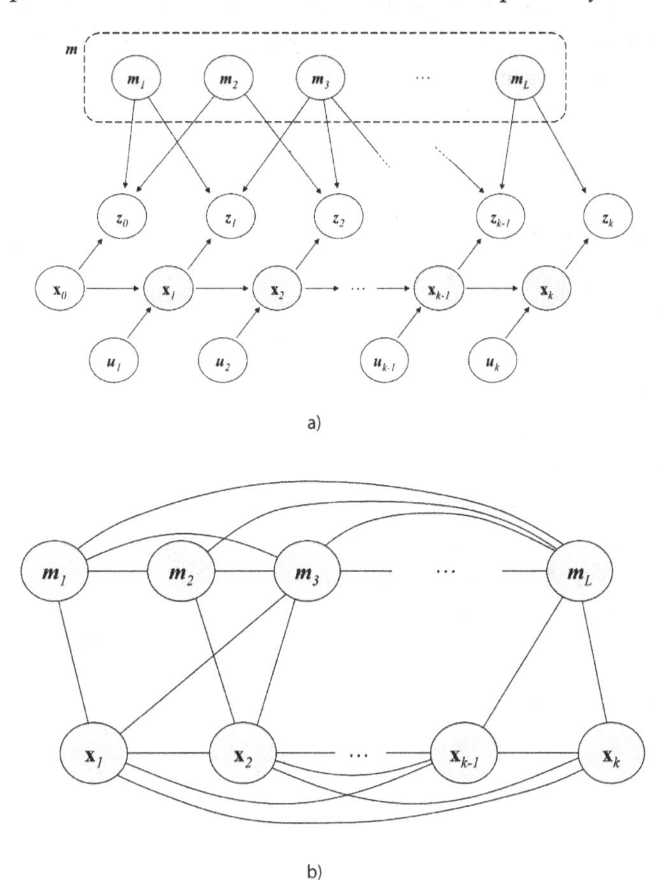

Equation 3 would be degenerate, since it would have an infinity number of equivalent solutions: any rigid rotation of a robot path and a map would lead to exactly the same likelihood values than the original. The common workaround is to set the pdf of the starting pose $\mathbf{x}_0$ to any arbitrary distribution, in practice assumed to be a Dirac delta distribution centered (arbitrarily) at the origin of coordinates. That is, it is assumed a perfect knowledge, without uncertainty, for the initial robot pose. The reason to eliminate this uncertainty is that it would otherwise impose an unnecessary bound to the accuracy at which the rest of the variables can be estimated, as was demonstrated in (Dissanayake, Newman, Clark, Durrant-Whyte, & Csorba, 2001).

It is interesting to find out the graph of correlations for full SLAM. Recall from chapter 3 section 10 that such a graph reflects which hidden variables are correlated to each other in the joint distribution obtained after doing inference given the values of the observed variables. Following the rules devised in that section, and from the DBN in Figure 1a, we end up with the graph of correlations shown in Figure 1b. The most important observation that can be stated from that graph is that the two groups of robot poses and map elements are internally fully connected, while the only uncorrelated variable pairs are those poses and map elements that were not connected by any observation in the DBN. If we pretended to represent the full SLAM conditional joint distribution as a multivariate Gaussian, the number of non-zero entries in its covariance matrix would therefore grow as $O\left(k^2 + L^2\right)$. We shall return to this point below when comparing full SLAM to other alternatives.

Even though the full SLAM DBN correctly reflects the underlying structure of all BA, SFM and SLAM problems, it is only the starting point from which the different approaches start to differentiate. To begin with, the least squares optimization algorithms typically applied in BA directly work with the joint conditional in Equation 3. But instead of pursuing a Bayesian *filtering* approach to estimate the unknowns, these methods can be interpreted as *batch* Maximum Likelihood Estimators (MLE) which obtain the values $\left\{\mathbf{x}_{1:t}^*, \mathbf{m}_{1:L}^*\right\}$ that best explain the actions and observations, that is, that maximize the joint conditional likelihood:

$$\left\{\mathbf{x}_{1:t}^*, \mathbf{m}_{1:L}^*\right\} = \arg\max_{\mathbf{x}_{1:t}, \mathbf{m}} p\left(\mathbf{x}_{0:k}, \mathbf{m}_{1:L} \middle| \mathbf{z}_{1:k}, \mathbf{u}_{1:k}\right) =$$

(due to the logarithm being monotonically increasing)

$$= \arg\min_{\mathbf{x}_{1:t}, \mathbf{m}} \left\{ -\log p\left(\mathbf{x}_{0:k}, \mathbf{m}_{1:L} \middle| \mathbf{z}_{1:k}, \mathbf{u}_{1:k}\right)\right\} =$$

$$= \arg\min_{\mathbf{x}_{1:t}, \mathbf{m}} \left\{ -\log \left(p\left(\mathbf{x}_0\right) \underbrace{\prod_{i=1}^{k} p\left(\mathbf{x}_i \middle| \mathbf{x}_{i-1}, \mathbf{u}_i\right)}_{\text{Motion model terms}} \underbrace{\prod_{i=1}^{k} p\left(\mathbf{z}_i \middle| \mathbf{x}_i, \mathbf{m}_{O(i)}\right)}_{\text{Observation likelihood terms}} \right) \right\} \tag{4}$$

Assuming Gaussian distributions for all the state (motion model) and observation predictions, this last expression becomes a sum of squared Mahalanobis distances or *squared weighted residuals*, since the logarithm transforms products into sums and when applied to the individual Gaussian pdfs yield only their exponents:

$$\left\{\mathbf{x}_{1:t}^*, \mathbf{m}_{1:L}^*\right\} =$$

$$= \arg\min_{\mathbf{x}_{1:t}, \mathbf{m}} \left\{ \underbrace{\sum_{i=1}^{k} \left\| \mathbf{x}_i - \hat{\mathbf{x}}_i \right\|_{\Sigma \hat{\mathbf{x}}_i}^2}_{\text{Residuals from motion model}} + \underbrace{\sum_{i=1}^{k} \left\| \mathbf{z}_i - \hat{\mathbf{z}}_i \right\|_{\Sigma \hat{\mathbf{z}}_i}^2}_{\text{Residuals from observation predictions}} \right\}$$

where the squared Mahalanobis distance appearing above is an abbreviation:

$$\left\| \mathbf{e} \right\|_{\Sigma}^2 = \mathbf{e}^T \Sigma^{-1} \mathbf{e}$$

and the predictions are given by:

$$\begin{aligned} \hat{\mathbf{x}}_i &= \mathbf{f}\left(\mathbf{x}_{i-1}, \mathbf{u}_i\right) + \mathbf{w}_i, \\ \mathbf{w}_i &\sim N\left(0, \mathbf{Q}_i\right) \quad - \quad \text{(Motion model)} \\ \hat{\mathbf{z}}_i &= \mathbf{h}\left(\mathbf{x}_i, \mathbf{m}_{O(i)}\right) + \mathbf{r}_i, \\ \mathbf{r}_i &\sim N\left(0, \mathbf{R}_i\right) \quad - \quad \text{(Observation model)} \end{aligned} \tag{5}$$

The covariance matrices $\Sigma_{\hat{\mathbf{x}}_i}$ and $\Sigma_{\hat{\mathbf{z}}_i}$ have the role of weighting the mismatch between the motion and observation predictions and the values $\mathbf{x}_i$ and $\mathbf{z}_i$ sought by the estimator, respectively. By identifying those two covariances with the covariance of noises in the motion and observation models, $\mathbf{Q}_i$ and $\mathbf{R}_i$ in the equation above, respectively, we are able to easily evaluate all the residuals. Non-linear least squares methods exist for iteratively solving this batch optimization problem, or its sliding window and on-line equivalents. These methods will be addressed in detail in chapter 10, but we should remark at this point that although they provide the maximum likelihood estimation of the hidden variables, in SLAM problems one should also evaluate its associated uncertainty in order to solve data association—a point which is not always stressed enough in the least-squares-based SLAM literature. The relevance of this uncertainty becomes clear by observing in the graph of correlations in Figure 1b how poses and map elements are correlated to each other, which implies that observation likelihoods should account for those cross-covariances. Section 2 will discuss how the cross-covariances may affect observation predictions used for data association, as illustrated with Figure 4.

A technique that can be regarded as an approximation to least-squares methods is the so-called *optimization by relaxation* (Press, Teukolsky, Vetterling, & Flannery, 1992). Here, the residuals for the robot actions and observations are considered independently one by one, computing at each iteration successive approximations towards the joint MLE of robot poses and map elements (Duckett, Marsland, & Shapiro, 2002). Despite of their simpler update equations, relaxation methods require many more iterations than least-squares methods, which consider all the constraints at once. Thus, it is not clear that they represent an advantage against smart implementations of least-square methods that exploit the sparseness of the matrices involved—refer to chapter 10 section 5.

Other approaches apart from least squares can be used to perform approximate inference with the full SLAM DBN. We turn now towards the Bayesian filters for doing inference with full SLAM, some of which were already presented in chapter 7. Parametric filters include the family of Kalman filters, where unknowns are estimated as multivariate Gaussians. As we saw above, the covariance matrix of the joint pdf of k robot poses and L map elements would require $O\left(k^2 + L^2\right)$ non-zero elements due to the correlation terms introduced by inference over the DBN (refer to chapter 4 section 10). This sustained quadratic growth with the simply passage of time—as k increases—even if no new landmarks are mapped, renders this approach impractical, and that is why Kalman filters are not applied to solve full SLAM. Alternatively, other parametric solutions exist that work with the *inverse* of the covariance matrix, the *information matrix*. In particular, the Square-Root Information Filters (SRIF) and Smoothers (SRIS) present two of the most efficient and numerically stable solutions to Bayesian inference with full SLAM. The advantage of the information matrix is that the number of non-zero elements typically grows like $O\left(k + L\right)$, thus they can be efficiently stored by means of sparse matrix representations. Although with moderate repercussion, SRIF and SRIS have been proposed as solutions to SLAM in the literature (Dellaert & Kaess, 2006). At present, these filters are better understood and classified as particular implementations of sparse least-squares methods.

Regarding the family of particle filtering algorithms (non-parametric filters), they would require a number of particles in the order of $O\left(c^{k+L}\right)$, with c an arbitrary constant, to maintain the state space well-sampled, although it can be reduced to $O\left(c^k\right)$ thanks to a technique known as Rao-Blackwellization Particle Filter (RBPF), previously discussed in chapter 4 and described for this case in section 4. It may seem surprising that, as we said above, Kalman filters are not a practical solution to full SLAM due to its quadratic growth while, in turn, RBPFs became extremely popular in despite of their theoretical need of an exponentially increasing size of the sample representing the posterior. There exist two reasons for this success. Firstly, a RBPF is an *approximate* estimator whose quality increases with the employed particle population but providing reasonable results even for very few particle counts. It turns out that, for some of the variations of the algorithm (those explained in section 5), even just *one particle* can provide an accurate result. In contrast, the dimensionality of the covariance in Kalman filters is firmly determined by the dimensionality of the hidden variables. Secondly, RBPF can work with observation likelihoods that are described only pointwise, while all the methods mentioned above for solving full SLAM (least-squares, relaxation, SRIF, etc.) requires a parametric observation likelihood and thus are much better suited for maps of landmarks. In other words, probabilistic SLAM with some kinds of maps and sensors (e.g. occupancy grid maps, range-only sensors) can be *only* approached rigorously with a RBPF.

Up to this point, we have discussed methods that work with the full SLAM DBN. A popular alternative consists of reducing the complexity of the estimation problem by removing, or *marginalizing out*, all the old robot poses but the most recent one, which corresponds, when solving SLAM in real-time, to the current robot pose. Following the rules for marginalization in a DBN that we devised in chapter 3 section 10, the graph for full SLAM in Figure 1a can be modified by successively removing the oldest robot pose until only $\mathbf{x}_t$ remains. The resulting network is shown in Figure 2a and will be called the *marginalized SLAM* or *on-line SLAM* DBN. Notice how the process of removing old robot poses introduces dependencies of the unique remaining pose $\mathbf{x}_t$ with the whole history of actions and observations. The associated graph of correlations for this problem, when performing inference from the observed data, is the one shown in Figure 2b. By contrasting this with the corresponding correlation graph for full SLAM in Figure 1b we find out that the last robot pose becomes now correlated with all the map elements instead of with just those actually observed at that time step. Representing the covariance matrix of the joint distribution for on-line SLAM would then involve a number of non-zero entries in the order of $O\left(L^2\right)$, in contrast to the $O\left(k^2 + L^2\right)$ for the full SLAM joint: on-line SLAM represents a significant storage (and computational) reduction. This explains the tremendous popularity of solutions based on the on-line SLAM model during the last decade, in particular, with the Kalman family of parametric filters. As stated above, an important limitation of such filters is the requirement for a parametric observation model, but that is not problem in landmark-based mapping. These solutions to SLAM have been so common that during the most part of the 2000s, talking about SLAM implicitly assumed referring to the KF-based solution to on-line SLAM with landmarks. However, experience has demonstrated that, just like all the other estimation methods discussed above, this approach has its own strengths and weaknesses. For instance, a recent thorough analysis reported in (Strasdat, Montiel, & Davison, 2010) compares the two alternative frameworks, full and on-line SLAM, in the context of visual SLAM with a single camera. Their findings are that, in general, the least-squares approach to full SLAM is capable of a superior efficiency and accuracy, whereas a Bayesian solution to on-line SLAM excels in its capability of modeling the high uncertainty in the depth of landmarks during their initial observations.

To summarize all the techniques mentioned in this section, we have classified them in the table shown in Figure 3. Firstly, there is a broad division between methods based on the full SLAM DBN and the on-line SLAM methods that marginalize out all but the latest robot pose. This latter group only includes a few variations of Kalman-like filters. However, the main body of SLAM methods lay in the part of full SLAM, thus we have further categorized them regarding three aspects: (1) their suitability for processing all the data at once (*batch* operation), or for processing data sequentially in order to estimate the latest robot pose (*filtering*) or the robot pose with a certain fixed delay (*fixed-lag smoothing*); (2) whether they are formulated under the perspective

Figure 2. (a) The DBN of marginalized (or "on-line") SLAM, obtained by marginalizing out all the robot poses but the latest one from the full SLAM DBN. Shaded and unshaded nodes represent hidden and observed variables, respectively. (b) The corresponding graph of correlations for this DBN.

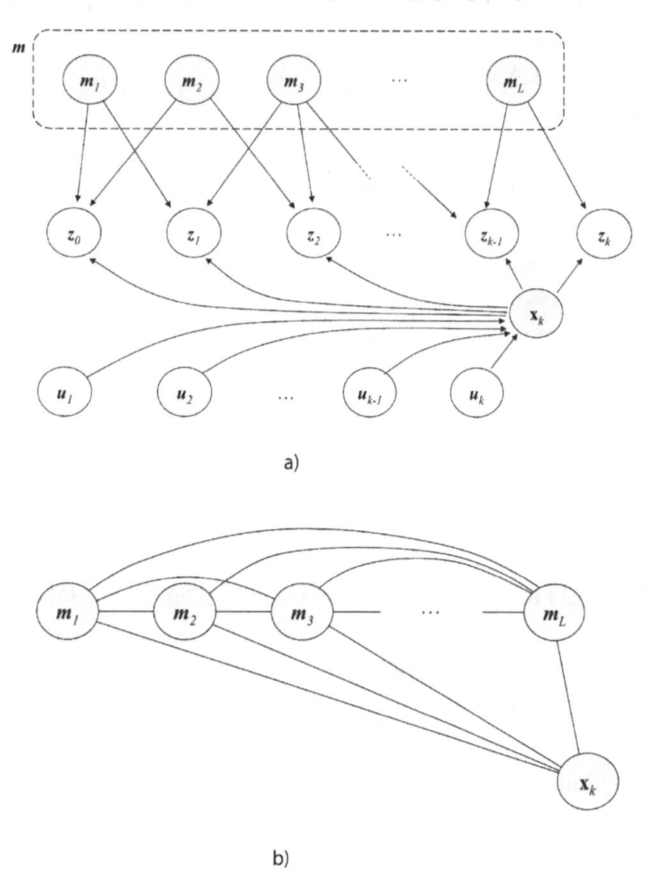

Figure 3. A classification of all the metrical SLAM methods mentioned in the text

Bayesian networks for metrical SLAM	Suitability for sequential operation			Bayesian estimators		Non-Bayesian estimators	
	Batch	Smoothing	Filtering	Non-parametric	Parametric		
Full SLAM 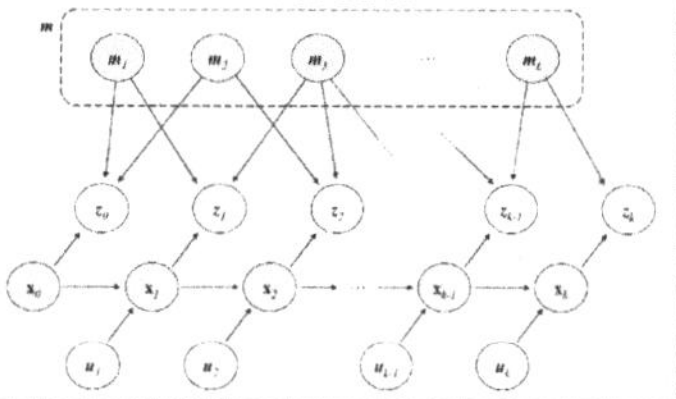	✓	✓	✓			✓	Least squares methods
			✓		✓		Square-Root Information Filter (SRIF)
	✓	✓			✓		Square-Root Information Smoothing (SRIS)
		✓		✓			RBPF-based smoothing
			✓	✓			RBPF-based SLAM filtering (FastSLAM, FastSLAM 2.0)
Marginalized ("online") SLAM 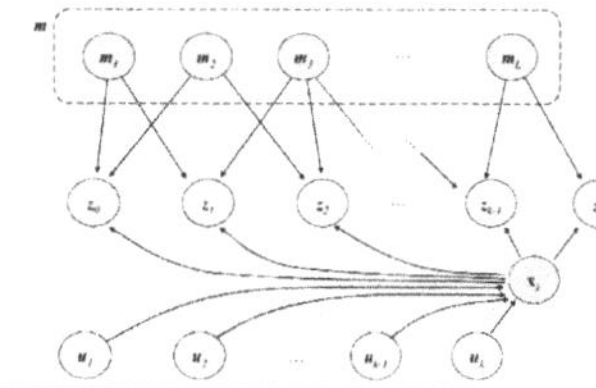			✓		✓		Sequential filtering with the family of Kalman filters (EKF, UKF, IKF, ...)

of Bayesian estimation or not; and (3) if they are Bayesian, whether they assume parameterized representations of probability distributions.

In next sections, the stress will be on the most relevant Bayesian solutions to SLAM, attending to their widespread usage in the robotics community during the last decade. We firstly describe the family of Kalman-like filters with landmark maps, that is, the "classical" solution to SLAM as reported in the influential (Dissanayake, Newman, Clark, Durrant-Whyte, & Csorba, 2001) and subsequent works. Next, we explore the most popular particle filtering solutions to SLAM with occupancy grid maps, as representative examples of non-parametric SLAM estimators.

2. ON-LINE SLAM: THE CLASSICAL EKF SOLUTION

The first solution to be studied here is the popular EKF-based sequential approach to SLAM for maps of landmarks. Recall from chapter 7 that the EKF is an extension of the standard Kalman filter to non-linear systems by means of first-order Taylor series approximation. Since the foundations of the filter were detailed in that chapter, we will focus here only on the specifics of its application to SLAM.

The aim then is to estimate the marginalized SLAM posterior, whose DBN was depicted in Figure 2a, conditioned on all the evidence gathered until some time step k, that is, to estimate:

$$bel\left(\mathbf{x}_k, \mathbf{m}_{1:L} \middle| \mathbf{z}_{0:k}, \mathbf{u}_{1:k}\right) \quad (6)$$

Since mobile robots need the output from the SLAM estimation process in real-time as they explore the environment, a sequential form (also called recursive) of the estimation process is in order. Next, it is rigorously demonstrated how to arrive at such a sequential form from Equation 6 by using only the fundamental probabilistic rules introduced in chapter 3:

By Application of the Bayes' rule over $\mathbf{z}_t$

$$\underbrace{bel\left(\mathbf{x}_k, \mathbf{m}_{1:L} \middle| \mathbf{z}_{0:k}, \mathbf{u}_{1:k}\right)}_{\text{Posterior pdf for time step } k} \propto$$
$$\propto p\left(\mathbf{x}_k, \mathbf{m}_{1:L} \middle| \mathbf{z}_{0:k-1}, \mathbf{u}_{1:k}\right) p\left(\mathbf{z}_k \middle| \mathbf{x}_k, \mathbf{m}_{1:L}, \mathbf{z}_{0:k-1}, \mathbf{u}_{1:k}\right) =$$

(by the conditional independence $\mathbf{z}_k \perp \mathbf{z}_{0:k-1}, \mathbf{u}_{1:k} \middle| \mathbf{x}_k, \mathbf{m}_{1:L}$)

$$= \underbrace{p\left(\mathbf{z}_k \middle| \mathbf{x}_k, \mathbf{m}_{1:L}\right)}_{\text{Observation likelihood}} p\left(\mathbf{x}_k, \mathbf{m}_{1:L} \middle| \mathbf{z}_{0:k-1}, \mathbf{u}_{1:k}\right) \quad (7)$$

where the conditional independence follows from the application of the rules given in chapter 3 section 10 to the DBN of Figure 2a. Now, we artificially introduce the variable $\mathbf{x}_{k-1}$ for the previous time step at the right-hand term by means of the law of total probability (notice that due to the assumption of a static map, the same trick has not to be done for the map). This can be interpreted as *undoing the marginalization* implied by the on-line SLAM model:

$$bel\left(\mathbf{x}_k, \mathbf{m}_{1:L} \middle| \mathbf{z}_{0:k}, \mathbf{u}_{1:k}\right) \propto p\left(\mathbf{z}_k \middle| \mathbf{x}_k, \mathbf{m}_{1:L}\right) \cdot$$
$$\int_{-\infty}^{\infty} p\left(\mathbf{x}_k, \mathbf{m}_{1:L} \middle| \mathbf{x}_{k-1}, \mathbf{z}_{0:k-1}, \mathbf{u}_{1:k}\right) p\left(\mathbf{x}_{k-1} \middle| \mathbf{z}_{0:k-1}, \mathbf{u}_{1:k}\right) d\mathbf{x}_{k-1} \quad (8)$$

Now, the first term inside the integral can be split in two due to the conditional independence of the robot pose and the map (which also follows from the DBN):

(by $\mathbf{x}_k \perp \mathbf{m}_{1:L} \middle| \mathbf{x}_{k-1}, \mathbf{u}_k$)

$$bel\left(\mathbf{x}_k, \mathbf{m}_{1:L} \middle| \mathbf{z}_{0:k}, \mathbf{u}_{1:k}\right) \propto p\left(\mathbf{z}_k \middle| \mathbf{x}_k, \mathbf{m}_{1:L}\right) \times$$
$$\int_{-\infty}^{\infty} p\left(\mathbf{x}_k \middle| \mathbf{x}_{k-1}, \mathbf{z}_{0:k-1}, \mathbf{u}_{1:k}\right) \underbrace{p\left(\mathbf{m}_{1:L} \middle| \mathbf{x}_{k-1}, \mathbf{z}_{0:k-1}, \mathbf{u}_{1:k}\right) p\left(\mathbf{x}_{k-1} \middle| \mathbf{z}_{0:k-1}, \mathbf{u}_{1:k}\right)}_{\text{These terms can be joined together}} d\mathbf{x}_{k-1} =$$

(since $p(a,b) = p(a)p(b \mid a)$)

$$= p\left(\mathbf{z}_k \middle| \mathbf{x}_k, \mathbf{m}_{1:L}\right) \int_{-\infty}^{\infty} p\left(\mathbf{x}_k \middle| \mathbf{x}_{k-1}, \mathbf{z}_{0:k-1}, \mathbf{u}_{1:k}\right) p\left(\mathbf{x}_{k-1}, \mathbf{m}_{1:L} \middle| \mathbf{z}_{0:k-1}, \mathbf{u}_{1:k}\right) d\mathbf{x}_{k-1}$$

(and due to the unconditional independence $\mathbf{x}_{k-1}, \mathbf{m}_{1:L} \perp \mathbf{u}_k$)

$$= p\left(\mathbf{z}_k \middle| \mathbf{x}_k, \mathbf{m}_{1:L}\right) \int_{-\infty}^{\infty} p\left(\mathbf{x}_k \middle| \mathbf{x}_{k-1}, \mathbf{z}_{0:k-1}, \mathbf{u}_{1:k}\right) \underbrace{bel\left(\mathbf{x}_{k-1}, \mathbf{m}_{1:L} \middle| \mathbf{z}_{0:k-1}, \mathbf{u}_{1:k-1}\right)}_{\text{Posterior pdf at } k\text{-}1} d\mathbf{x}_{k-1} \quad (9)$$

Notice how we have managed to turn the original identity in Equation 8 into a recursive expression: the filtered estimation for time step k now is given as a function of the estimation for $k-1$. Finally, the first distribution within the integral can be simplified due to another conditional independence from the DBN, arriving at the well-known recursive form of Bayesian estimation for on-line SLAM:

(using the conditional independence $\mathbf{x}_k \perp \mathbf{z}_{0:k-1}, \mathbf{u}_{1:k-1} \middle| \mathbf{x}_{k-1}, \mathbf{u}_k$)

$$\underbrace{bel\left(\mathbf{x}_k, \mathbf{m}_{1:L} \middle| \mathbf{z}_{0:k}, \mathbf{u}_{1:k}\right)}_{\text{Posterior pdf for } k} \propto \underbrace{p\left(\mathbf{z}_k \middle| \mathbf{x}_k, \mathbf{m}_{1:L}\right)}_{\text{Observation likelihood}} \int_{-\infty}^{\infty} \underbrace{p\left(\mathbf{x}_k \middle| \mathbf{x}_{k-1}, \mathbf{u}_k\right)}_{\text{Motion model}} \underbrace{bel\left(\mathbf{x}_{k-1}, \mathbf{m}_{1:L} \middle| \mathbf{z}_{0:k-1}, \mathbf{u}_{1:k-1}\right)}_{\text{Posterior pdf at } k\text{-}1} d\mathbf{x}_{k-1} \quad (10)$$

It is worth the reader reserves a moment for understanding the meaning of each term in this recursive formula, and the subtle differences with the similar expression in chapter 7's Equation 4 for mobile robot localization. Equation 10 is a mathematical relationship between generic probability distributions and likelihood functions, without making any assumptions about them; in order to implement this filter into a computer, a particular form for the equation must be firstly found.

In particular, we will adopt the Extended Kalman Filter (EKF) solution, an extension of the linear Kalman filter already introduced in chapter 7 section 3 which can cope with non-linearities in the motion and observation models. In this framework, the hidden variables of the DBN to be estimated are modeled as a multivariate Gaussian distribution. In the present case, those variables are the latest robot pose $\mathbf{x}_k$ and the location of all the landmarks in the map $\mathbf{m}_{1:L}$. We stack all these variables together into one state vector named $\mathbf{s}_k$:

$$bel\left(\underbrace{\mathbf{x}_k, \mathbf{m}_{1:L}}_{\mathbf{s}_k} \middle| \mathbf{z}_{1:k}, \mathbf{u}_{1:k}\right) = bel\left(\mathbf{s}_k \middle| \mathbf{z}_{1:k}, \mathbf{u}_{1:k}\right) = N(\mu_k, \Sigma_k)$$

with

$$\mathbf{s}_k = \begin{pmatrix} \mathbf{x}_k \\ \hline \mathbf{m}_1 \\ \vdots \\ \mathbf{m}_L \end{pmatrix} \qquad \mu_k = \begin{pmatrix} \mu_{x_k} \\ \hline \mu_{\mathbf{m}_1} \\ \vdots \\ \mu_{\mathbf{m}_L} \end{pmatrix} \quad (11)$$

Following the tradition in the SLAM literature, we will focus next on the two-dimensional problem, without loss of generality regarding all the aspects not related to the parameterization of the robot pose and the landmarks. Thus, the robot pose will be completely described as a point plus a heading, and each landmark location can be modeled as a single point:

$$\mathbf{x}_k = \begin{pmatrix} x_k \\ y_k \\ \theta_k \end{pmatrix} \qquad \mathbf{m}_i = \begin{pmatrix} m_{x_i} \\ m_{y_i} \end{pmatrix} \quad (12)$$

Since adding (or removing) landmarks to the map will require a clear understanding of the structure of the covariance matrix Σ_k of the posterior, it is worth deserving a moment to analyze it:

$$\boldsymbol{\Sigma}_k = \left(\begin{array}{c|ccc} \boldsymbol{\Sigma}_{\mathbf{x}_k} & \boldsymbol{\Sigma}_{\mathbf{x}_k\mathbf{m}_1} & \cdots & \boldsymbol{\Sigma}_{\mathbf{x}_k\mathbf{m}_L} \\ \hline \boldsymbol{\Sigma}_{\mathbf{m}_1\mathbf{x}_k} & \boldsymbol{\Sigma}_{\mathbf{m}_1} & \cdots & \boldsymbol{\Sigma}_{\mathbf{m}_1\mathbf{m}_L} \\ \vdots & \vdots & \ddots & \vdots \\ \boldsymbol{\Sigma}_{\mathbf{m}_L\mathbf{x}_k} & \boldsymbol{\Sigma}_{\mathbf{m}_L\mathbf{m}_1} & \cdots & \boldsymbol{\Sigma}_{\mathbf{m}_L} \end{array}\right) \quad (13)$$

Notice how this symmetric matrix has four clearly distinguished parts: the covariance of the robot pose $\boldsymbol{\Sigma}_{\mathbf{x}_k}$, the covariances $\boldsymbol{\Sigma}_{\mathbf{m}_i}$ and cross-covariances $\boldsymbol{\Sigma}_{\mathbf{m}_i\mathbf{m}_j}$ of all the landmarks and the robot-landmark cross-covariances $\boldsymbol{\Sigma}_{\mathbf{x}_k\mathbf{m}_i}$. Note as well that the matrix grows as new landmarks are incorporated into the map—the exact process for achieving this is discussed below.

We can now recall the graph of correlations of Figure 2b for on-line SLAM, whose meaning has important practical consequences at this point: the graph predicts that all map landmarks will be correlated to each other and with the robot pose, hence we can assure $\boldsymbol{\Sigma}_k$ will be densely populated, i.e., it will not be a sparse matrix. Since all the covariance blocks in $\boldsymbol{\Sigma}_k$ will tend to be non-zero (except for fortunate coincidences) we must store all of them. This lack of sparseness favors implementing $\boldsymbol{\Sigma}_k$ in computer programs as one contiguous, large matrix. In contrast, chapter 10 will provide examples of other solutions to SLAM heavily relying on sparse matrices.

Recall from chapter 7 that an EKF relies on a pair of functions which must be established for each specific problem, namely, the state transition and the observation models, denoted as $\mathbf{g}(\cdot)$ and $\mathbf{h}(\cdot)$, respectively:

$$\mathbf{s}_k = \mathbf{g}(\mathbf{s}_{k-1}, \mathbf{u}_k) + \varepsilon_k$$
$$\mathbf{z}_k = \mathbf{h}(\mathbf{s}_k) + \delta_k$$

where the additive Gaussian noises are:

$$\varepsilon_k \sim \mathrm{N}(\mathbf{s}_k; 0, \mathbf{R}_k)$$
$$\delta_k \sim \mathrm{N}(\mathbf{z}_k; 0, \mathbf{Q}_k)$$

and

$$\varepsilon_k \perp \delta_k \quad (14)$$

As long as we hold the static map assumption, the state transition function leaves the map part of $\mathbf{s}_k$ unmodified, that is:

$$\mathbf{s}_k = \mathbf{g}(\mathbf{s}_{k-1}, \mathbf{u}_k) + \varepsilon_k$$
$$\left(\begin{array}{c} \mathbf{x}_k \\ \hline \mathbf{m}_1 \\ \vdots \\ \mathbf{m}_L \end{array}\right) = \left(\begin{array}{c} \mathbf{g}'(\mathbf{x}_{k-1}, \mathbf{u}_k) + \varepsilon'_k \\ \hline \mathbf{m}_1 \\ \vdots \\ \mathbf{m}_L \end{array}\right)$$

with the process noise being:

$$\varepsilon'_k \sim \mathrm{N}(\mathbf{x}_k; 0, \mathbf{R}'_k)$$
$$\mathbf{R}_k = \left(\begin{array}{c|c} \mathbf{R}'_k & \mathbf{0} \\ \hline \mathbf{0} & \mathbf{0} \end{array}\right) \quad (15)$$

Any of the two-dimensional motion models studied in chapter 5 is suitable for playing the role of the transition function $\mathbf{g}'(\cdot)$ above.

Regarding observations, we will assume they belong to the range-bearing type, for that is the most common situation studied in the literature. Let $\mathbf{z}_k = \left(\mathbf{z}_{k,1}^T \; \cdots \; \mathbf{z}_{k,M}^T\right)^T$ denote the vector of feature observations, each one comprising a range and a bearing value:

$$\mathbf{z}_{k,i} = \begin{bmatrix} r_{k,i} \\ b_{k,i} \end{bmatrix} \quad (16)$$

Now we have set up all the elements needed for the estimation, which is discussed in the next subsections.

Algorithm Description

Once defined all the variables involved in the problem, we are in position of outlining in detail what takes place at each time step in range-bearing EKF-based SLAM. The method, summarized in Algorithm 1, takes as input the Gaussian parameters of the previous posterior $bel\left(\mathbf{s}_{k-1} \middle| \mathbf{z}_{1:k-1}, \mathbf{u}_{1:k-1}\right)$ and the new action ($\mathbf{u}_k$) and observation ($\mathbf{z}_k$), giving us the resulting parameters of the updated posterior. It includes some differences with the EKF algorithm explained in chapter 7 that will be explained below.

If compared to the description of EKF in chapter 7, one immediately identifies the state vector prediction and update stages in the first and third steps in the algorithm. In contrast, steps 2 and 4 are new and particular to SLAM, thus will require further discussion. We describe now the entire algorithm step by step.

The first step simply propagates the mean of the state vector using any appropriate motion

Algorithm 1. A summary of all the processes invoked during one time step in EKF-based SLAM

algorithm EKF_SLAM
Inputs: $\left(\mu_{k-1}, \Sigma_{k-1}, \mathbf{u}_k, \mathbf{z}_k\right)$
Outputs: $\left(\mu_k, \Sigma_k\right)$

Step 1: State vector prediction
1.1 Mean propagation: $\bar{\mu}_k \leftarrow \mathbf{g}\left(\mu_{k-1}, \mathbf{u}_k\right)$
1.2 Covariance prediction: $\bar{\Sigma}_k \leftarrow \mathbf{G}_k \Sigma_{k-1} \mathbf{G}_k^T + \mathbf{R}_k$, with $\mathbf{G}_k = \left.\frac{\partial \mathbf{g}\left(\mathbf{s}_{k-1}, \mathbf{u}_k\right)}{\partial \mathbf{s}_{k-1}}\right|_{\mathbf{s}_{k-1}=\mu_{k-1}}$

Step 2: Data association
2.1 Predict observations
2.1.1 Means: $\bar{\mu}_{z_k} \leftarrow \mathbf{h}\left(\bar{\mu}_k\right)$
2.1.2 Joint covariance: $\mathbf{S}_k \leftarrow \mathbf{H}_k \bar{\Sigma}_k \mathbf{H}_k^T + \mathbf{Q}_t$, with $\mathbf{H}_k = \left.\frac{d\mathbf{h}\left(\mathbf{s}_k\right)}{d\mathbf{s}_k}\right|_{\mathbf{s}_k=\bar{\mu}_k}$
2.2 Establish data associations: $\alpha_k \leftarrow data\_assoc\left(\mathbf{z}_k, \mathbf{S}_k, \bar{\mu}_{z_k}\right)$

Step 3: State vector update
3.1 Measurement residuals: $\tilde{\mathbf{y}}_k \leftarrow \mathbf{z}_k - \bar{\mu}_{z_k}\left(\alpha_k\right)$
3.2 Kalman gain matrix: $\mathbf{K}_k \leftarrow \bar{\Sigma}_k \mathbf{H}_k^T \mathbf{S}_k^{-1}$
3.3 Mean update: $\mu_k \leftarrow \bar{\mu}_k + \mathbf{K}_k \tilde{\mathbf{y}}_k$
3.4 State vector normalization (if needed): $\mu_k \leftarrow normalize\left(\mu_k\right)$
3.5 Covariance update: $\Sigma_k \leftarrow \left(\mathbf{I} - \mathbf{K}_k \mathbf{H}_k\right) \bar{\Sigma}_k$

Step 4: Creation of new landmarks
4.1 for each non-associated feature i in α_k
4.1.1 Augment mean of state vector: $\mu_k \leftarrow \left(\mu_k{}^T \quad \mathbf{h}^{-1}\left(\mathbf{z}_{k,i}\right)^T\right)^T$
4.1.2 Augment covariance matrix: see Equation 28.

model. As already stated in Equation 15 above, the assumption of a static map implies that only the first part of the state vector (the robot pose $\mathbf{x}_k$) is actually modified by the motion model function, that is:

$$\bar{\mu}_{\mathbf{x}_k} = \mathbf{g}'(\mu_{\mathbf{x}_{k-1}}, \mathbf{u}_k) \tag{17}$$

When updating the covariance (step 1.2 of the algorithm), the static map assumption again allows a simplification. From Equation 15 it follows that the Jacobian $\mathbf{G}_k$ can be decomposed into two parts, one of them being an identity matrix:

$$\left.\frac{\partial \mathbf{g}(\mathbf{x}_{k-1}, \mathbf{u}_k)}{\partial \mathbf{s}_{k-1}}\right|_{\mathbf{s}_{k-1}=\mu_{k-1}} = \left.\frac{\partial\{\mathbf{g}'(\mathbf{x}_{k-1}, \mathbf{u}_k), \mathbf{m}_1, \ldots, \mathbf{m}_L\}}{\partial \mathbf{s}_{k-1}}\right|_{\mathbf{s}_{k-1}=\mu_{k-1}} = \left(\begin{array}{c|c} \left.\frac{\partial \mathbf{g}'(\mathbf{x}_{k-1}, \mathbf{u}_k)}{\partial \mathbf{s}_{k-1}}\right|_{\mathbf{s}_{k-1}=\mu_{k-1}} & \mathbf{0}_{3\times 2L} \\ \hline \mathbf{0}_{2L\times 3} & \mathbf{I}_{2L} \end{array}\right) \tag{18}$$

which leads to the following simplified expression for the covariance of the prediction:

$$\bar{\Sigma}_k = \left.\frac{\partial \mathbf{g}(\mathbf{x}_{k-1}, \mathbf{u}_k)}{\partial \mathbf{s}_{k-1}}\right|_{\mu_{k-1}} \bar{\Sigma}_{k-1} \left.\frac{\partial \mathbf{g}(\mathbf{x}_{k-1}, \mathbf{u}_k)}{\partial \mathbf{s}_{k-1}}\right|_{\mu_{k-1}}^T + \mathbf{R}_k =$$

(by introducing the definition $\mathbf{G}' = \left.\frac{\partial \mathbf{g}'(\mathbf{x}_{k-1}, \mathbf{u}_k)}{\partial \mathbf{x}_{k-1}}\right|_{\mathbf{x}_{k-1}=\mu_{x_k-1}}$)

$$= \left(\begin{array}{c|c} \mathbf{G}' & \mathbf{0}_{3\times 2L} \\ \hline \mathbf{0}_{2L\times 3} & \mathbf{I}_{2L} \end{array}\right) \Sigma_{k-1} \left(\begin{array}{c|c} \mathbf{G}' & \mathbf{0}_{3\times 2L} \\ \hline \mathbf{0}_{2L\times 3} & \mathbf{I}_{2L} \end{array}\right)^T + \mathbf{R}_k =$$

(replacing $\mathbf{R}_k$ by its structure in Equation 15)

$$= \left(\begin{array}{c|ccc} \mathbf{G}'\Sigma_{\mathbf{x}_{k-1}}\mathbf{G}'^T + \mathbf{R}'_k & \mathbf{G}'\Sigma_{\mathbf{x}_{k-1}\mathbf{m}_1} & \cdots & \mathbf{G}'\Sigma_{\mathbf{x}_{k-1}\mathbf{m}_L} \\ \hline \Sigma_{\mathbf{m}_1\mathbf{x}_{k-1}}\mathbf{G}'^T & \Sigma_{\mathbf{m}_1} & \cdots & \Sigma_{\mathbf{m}_1\mathbf{m}_L} \\ \vdots & \vdots & \ddots & \vdots \\ \Sigma_{\mathbf{m}_L\mathbf{x}_{k-1}}\mathbf{G}'^T & \Sigma_{\mathbf{m}_L\mathbf{m}_1} & \cdots & \Sigma_{\mathbf{m}_L} \end{array}\right) \tag{19}$$

That is, the covariance of the robot pose (top-left block) is updated according to the probabilistic motion model, which also affects all the correlations of that pose with the map landmarks (the top-right and bottom-left blocks). The covariance of the map (right-bottom block) remains unmodified, as one would expect for a static map model.

The second step in the algorithm is Data Association (DA), and is mandatory for all sensors except for those which unequivocally identify landmarks in some way (for example, purposely placed visual landmarks with a unique design each). If we denote as M the number of feature observations stacked in $\mathbf{z}_k = \left(\mathbf{z}_{k,1}^T \ \cdots \ \mathbf{z}_{k,M}^T\right)^T$, the goal of this step is ending up with a vector α_k of length M whose $i-th$ entry identifies which map element corresponds to the observation $\mathbf{z}_{k,i}$. Two typical approaches to solve DA are the Nearest Neighbor (NN) and the Joint-Compatibility Branch and Bound (JCBB) algorithms, both already explained in chapter 6 section 4. Disregarding the particular method used to address DA, it is worth highlighting the point that DA occurs in the space of observations, that is, each observation $\mathbf{z}_{k,i}$ (or the entire set $\mathbf{z}_k$ in joint-compatibility methods) is compared to the predicted observations generated from each map element, and all this takes into account the uncertainty of the robot pose and the map, modeled up to this point as the Gaussian $\bar{\mathbf{s}}_k = N\left(\mathbf{s}_k; \bar{\mu}_k, \bar{\Sigma}_k\right)$. This is exactly the role of the step 2.1 in the algorithm: modeling the pdf of the observation predictions as another Gaussian $N\left(\mathbf{z}_k;\bar{\boldsymbol{\mu}}_{z_k},\mathbf{S}_k\right)$ whose parameters are approximated by first-order Taylor series expansion—recall chapter 3 section 8. The mean of those predictions are simply the propagation of the state vector mean through the observation function $\mathbf{h}(\cdot)$; its covariance matrix, called *innovation covariance*, already appeared in chapter 7 and condensates the uncertainty in both the state vector and the measurement process. It also describes the uncertainty of the residuals $\mathbf{z}_k - \mathbf{h}(\cdot)$, which follows from the fact that $\mathbf{z}_k$ is a constant. As stated in Algorithm 9.1, this covariance has the following form:

$$\mathbf{S}_k = \mathbf{H}_k\bar{\boldsymbol{\Sigma}}_k\mathbf{H}_k^T + \mathbf{Q}_k\text{, with }\ \mathbf{H}_k = \left.\frac{\partial\mathbf{h}\left(\mathbf{s}_k\right)}{\partial\mathbf{s}_k}\right|_{\mathbf{s}_k=\bar{\boldsymbol{\mu}}_k} \tag{20}$$

It is instructive to find out the expected structure of this matrix, which depends on that of the Jacobian $\mathbf{H}_k$ and the sensor noise $\mathbf{Q}_k$. We will aim at finding the sparsity of these block matrices, that is, which entries are non-zero and thus imply correlation between variables; for that purpose it will suffice to represent all non-zero submatrices as solid blocks (■).

Making the realistic assumption that measurement noises are uncorrelated between each landmark observation, it follows that $\mathbf{Q}_k$ has a diagonal block structure:

$$\mathbf{Q}_k = \begin{pmatrix} \mathbf{Q}_{k,1} & \mathbf{0} & \cdots & \mathbf{0} \\ \mathbf{0} & \mathbf{Q}_{k,2} & & \mathbf{0} \\ \vdots & & \ddots & \vdots \\ \mathbf{0} & \mathbf{0} & \cdots & \mathbf{Q}_{k,M} \end{pmatrix} = \begin{pmatrix} \blacksquare & & & \\ & \blacksquare & & \\ & & \ddots & \\ & & & \blacksquare \end{pmatrix} \tag{21}$$

and the observation Jacobian for the case of simultaneously observing M landmarks also has a sparse structure:

$$\frac{\partial\mathbf{h}\left(\mathbf{s}_k\right)}{\partial\mathbf{s}_k} = \left(\begin{array}{c|cccc} \frac{\partial\mathbf{h}_1}{\partial\mathbf{x}_k} & \frac{\partial\mathbf{h}_1}{\partial\mathbf{m}_1} & \cdots & & \frac{\partial\mathbf{h}_1}{\partial\mathbf{m}_L} \\ \vdots & \vdots & & \ddots & \vdots \\ \frac{\partial\mathbf{h}_M}{\partial\mathbf{x}_k} & \frac{\partial\mathbf{h}_M}{\partial\mathbf{m}_1} & \cdots & & \frac{\partial\mathbf{h}_M}{\partial\mathbf{m}_L} \end{array}\right) =$$

since $\frac{\partial\mathbf{h}_i\left(\mathbf{s}_k\right)}{\partial\mathbf{m}_j}$ is zero for all observations $\mathbf{z}_{k,i}$ not observing the $j-th$ landmark

$$= \left(\begin{array}{c|ccccc} \blacksquare & \blacksquare & & & & \\ \blacksquare & & \blacksquare & & & \\ \vdots & & & \cdots\ \blacksquare & & \\ \blacksquare & & & & \blacksquare & \\ \blacksquare & & & & & \blacksquare \end{array}\right) \tag{22}$$

Despite the sparsity of these matrices, when they are combined to form the innovation covariance we reach a dense matrix with all its entries being non-zero (even for the most advantageous starting point of $\bar{\boldsymbol{\Sigma}}_k$, when it is block diagonal):

$$\mathbf{S}_k = \mathbf{H}_k\bar{\boldsymbol{\Sigma}}_k\mathbf{H}_k^T + \mathbf{Q}_k =$$

$$= \left(\begin{array}{c|ccc} \blacksquare & \blacksquare & & \\ \vdots & & \cdots\ \blacksquare & \\ \blacksquare & & & \blacksquare \end{array}\right) \left(\quad \bar{\boldsymbol{\Sigma}}_k \quad \right) \left(\begin{array}{ccc} \blacksquare & \cdots & \blacksquare \\ \hline & & \blacksquare \\ & & \vdots \\ & \blacksquare & \\ \blacksquare & & \end{array}\right)$$

$$+ \begin{pmatrix} \blacksquare & & \\ & \ddots & \\ & & \blacksquare \end{pmatrix} =$$

$$= \begin{pmatrix} \blacksquare & \blacksquare & \cdots & \blacksquare \\ \blacksquare & \blacksquare & & \blacksquare \\ \vdots & & \ddots & \vdots \\ \blacksquare & \blacksquare & \cdots & \blacksquare \end{pmatrix} \tag{23}$$

Consequently, all observation predictions are correlated to each other. Two important lessons can be learned from the equation above. Firstly, that establishing correspondences between observations and predictions should not be addressed by considering individual observations isolated from each other. The joint-compatibility method for DA (recall chapter 6 section 4) is built upon this realization. Secondly, that uncertainty in observation space might (and usually does) look very different than uncertainty in the map $\left(\bar{\Sigma}_k\right)$, mainly because of the rearrangement due to the Jacobian but also for the correlations implied by $\bar{\Sigma}_k$.

It is important for beginners in SLAM to fully grasp the meaning of the last statement, thus a few words are in order to clarify the related concepts. Consider the situation depicted in Figure 4a, where a robot is close to a pair of landmarks #1 and #2 after describing a long path since the first time it observed #1. Observing #1 again after coming back to the same place is what in robotics is called *closing a loop*, an operation that, assuming data association is correctly solved, drastically reduces the robot localization uncertainty—which otherwise can only grow with time while exploring new areas. From the figure, it is clear that the uncertainty in the localization of the robot and landmark #2 are much higher than that of landmark #1. However, this is only because we are considering them in a *global* frame of coordinates. By evaluating now the observation predictions and its covariance $\mathbf{S}_k$, both illustrated in Figure 4b, we can now consider a change of coordinates such that the robot itself becomes the new origin of the coordinate system. After all, observations are how the world elements look like from the robot perspective. Then, we can notice in the figure how the prediction is much more accurate for landmark #2 than for landmark #1, exactly the contrary as one might naively expect from Figure 4a. The reason of this apparent contradiction comes from the higher correlation of the robot pose with landmark #2. In fact, that is why observing poorly correlated landmarks after closing a loop provides much richer information (with the subsequent reduction in robot uncertainty) than continuously observing the same landmarks once and over again.

Following with our explanation of the EKF-SLAM algorithm, step 3 comprises the standard EKF update equations for the state vector mean and covariance. The only novelty with respect to a generic EKF is the existence of an optional step (numbered as 3.4) for state vector *normalization*. Note that the update formula for the mean μ_k consists of adding a certain increment obtained from the Kalman gain and the measurement residuals, without taking into account any constraint regarding potential internal degrees of freedom within the state vector. For instance, in the case of a planar robot pose parameterized like $\mathbf{x}_k = \left(x_k \quad y_k \quad \theta_t\right)^T$, the updated angle $\theta_t \leftarrow \bar{\theta}_t + \Delta\theta_t$ has not to be restricted to $\left]-\pi, \pi\right]$ the usual range of valid angles, thus normalizing the state vector would imply here adding or subtracting an integer multiple of 2π. In practical applications for three-dimensional SLAM, where the robot pose has six degrees of freedom, it is common to parameterize the attitude angles as a unit quaternion, leading to an overparameterization of the 6D pose as seven values (Davison, Reid, Molton & Stasse, 2007). In that case, the internal constraint is the requirement of the quaternion to have a unit norm, which should be accounted for with this normalization step.

We finally arrive at the last step of the algorithm, related to the addition of new landmarks to the map. Recall from the data association stage that each landmark measurement $\mathbf{z}_{k,i}$ was assigned an index for the corresponding landmark in the map. There is nonetheless another possibility, which is that of measurements not actually having any plausible pairing in the map, becoming

Figure 4. Uncertainty can look very differently depending on the reference frame. The observation of two landmarks from one robot pose, seen globally in (a), gives very different results from those taken from another robot coordinates (b). The key to understand these effects is the correlation between the robot pose and all map landmarks, as further discussed in the text.

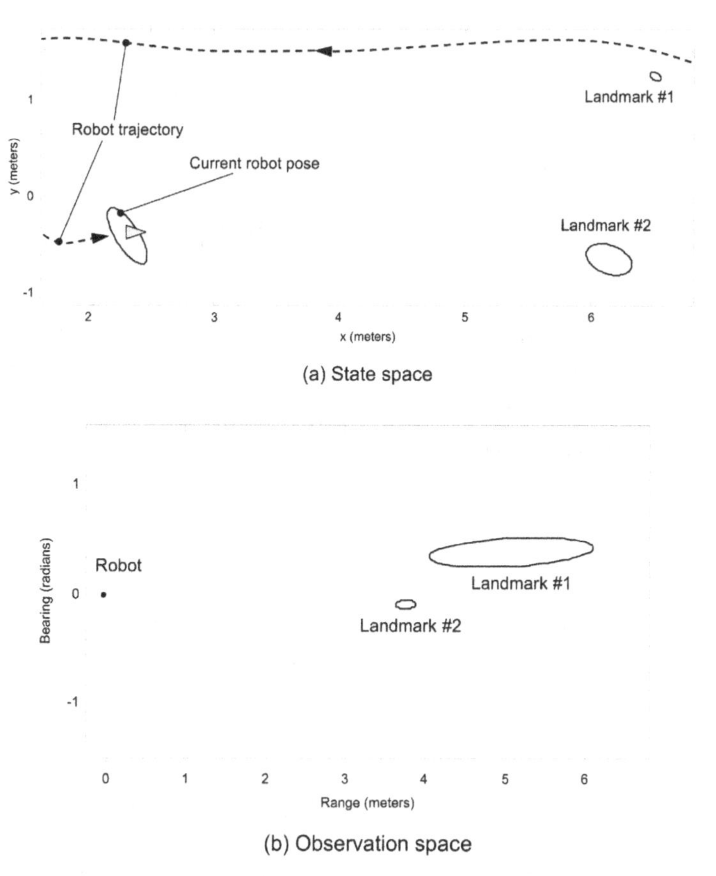

negative associations. For instance, all the observations within the first time step belong to this category simply because the map is initially empty.

Each of these measurements without association must be converted into a new landmark in the map by augmenting the state mean and covariance and making use of the inverse sensor model $\mathbf{h}^{-1}\left(\mathbf{x}_k, \mathbf{z}_{k,i}\right)$, the mathematical model that specifically tries to reconstruct a map element from its observations. The particular inverse sensor model could be one of the three possibilities explored in chapter 8 sections 5 – 7 for range-bearing, bearing-only and range-only observations, respectively. In this chapter we are assuming a range-bearing sensor and in consequence the function $\mathbf{h}^{-1}\left(\mathbf{x}_k, \mathbf{z}_{k,i}\right)$ exists and is well-defined. While in the previous chapter we assumed a perfectly known robot pose, that hypothesis does not hold in the context of EKF-based SLAM, thus the uncertainty in both the robot pose $\left(\mathbf{\Sigma}_{\mathbf{x}_k}\right)$ and the observation $\left(\mathbf{Q}_{k,i}\right)$ must be considered.

Let $\mathbf{a}\left(\mathbf{s}_k, \mathbf{z}_{k,i}\right)$ denote the function that augments the state vector $\mathbf{s}_k$ from having L mapped landmarks to the new state $\underset{\frown}{\mathbf{s}}_k$ with $L+1$ landmarks, by appending a new one from the non-associated observation $\mathbf{z}_{k,i}$, that is:

$$\underset{\frown}{\mathbf{s}}_k = \mathbf{a}\left(\mathbf{s}_k, \mathbf{z}_{k,i}\right) = \begin{pmatrix} \mathbf{x}_k \\ \hline \mathbf{m}_1 \\ \vdots \\ \mathbf{m}_L \\ \hline \mathbf{m}_{L+1} \end{pmatrix}$$

with

$$\mathbf{m}_{L+1} = \mathbf{h}^{-1}\left(\mathbf{x}_k, \mathbf{z}_{k,i}\right) \qquad (24)$$

Applying a Taylor series-based linear approximation, it is easily seen that the new state vector mean is built by simple concatenation of the old mean and the prediction from the inverse sensor model:

$$\underset{\frown}{\overline{\mathbf{s}}}_k = \begin{pmatrix} \overline{\mathbf{x}}_k \\ \hline \overline{\mathbf{m}}_1 \\ \vdots \\ \overline{\mathbf{m}}_L \\ \hline \mathbf{h}^{-1}\left(\overline{\mathbf{x}}_k, \mathbf{z}_{k,i}\right) \end{pmatrix} \qquad (25)$$

The new covariance $\underset{\frown}{\Sigma}_k$ demands a little more attention. It first requires devising the structure of the Jacobian of the augmentation function $\mathbf{a}\left(\mathbf{s}_k, \mathbf{z}_{k,i}\right)$ with respect to its inputs:

$$\frac{\partial \mathbf{a}\left(\mathbf{s}_k, \mathbf{z}_{k,i}\right)}{\partial\left\{\mathbf{s}_k, \mathbf{z}_{k,i}\right\}} = \frac{\partial\left\{\underset{\frown}{\mathbf{x}}_k, \mathbf{m}_1, \ldots, \mathbf{m}_L, \mathbf{m}_{L+1}\right\}}{\partial\left\{\mathbf{x}_k, \mathbf{m}_1, \ldots, \mathbf{m}_L, \mathbf{z}_{k,i}\right\}} = \left(\begin{array}{cc|c} \mathbf{I}_{3+2L} & & \mathbf{0}_{2L\times 2} \\ \hline \frac{\partial \mathbf{h}^{-1}}{\partial \mathbf{x}_k} & \mathbf{0}_{2\times 2L} & \frac{\partial \mathbf{h}^{-1}}{\partial \mathbf{z}_{k,i}} \end{array}\right) \qquad (26)$$

and secondly, the joint covariance matrix of the inputs $\mathbf{s}_k$ and $\mathbf{z}_{k,i}$, that is Σ_k and $\mathbf{Q}_{k,i}$. Certainly, assuming independence between both sets of variables is reasonable, thus we can proceed as follows:

$$\underset{\frown}{\Sigma}_k = \frac{\partial \mathbf{a}\left(\mathbf{s}_k, \mathbf{z}_{k,i}\right)}{\partial\left\{\mathbf{s}_k, \mathbf{z}_{k,i}\right\}} \left(\begin{array}{c|c} \Sigma_k & \mathbf{0}_{2L\times 2} \\ \hline \mathbf{0}_{2\times 2L} & \mathbf{Q}_{k,i} \end{array}\right) \frac{\partial \mathbf{a}\left(\mathbf{s}_k, \mathbf{z}_{k,i}\right)}{\partial\left\{\mathbf{s}_k, \mathbf{z}_{k,i}\right\}}^T =$$

$$= \left(\begin{array}{cc|c} \mathbf{I}_{3+2L} & & \mathbf{0}_{2L\times 2} \\ \hline \frac{\partial \mathbf{h}^{-1}}{\partial \mathbf{x}_k} & \mathbf{0}_{2\times 2L} & \frac{\partial \mathbf{h}^{-1}}{\partial \mathbf{z}_{k,i}} \end{array}\right) \left(\begin{array}{c|ccc|c} \Sigma_{\mathbf{x}_k} & \Sigma_{\mathbf{x}_k\mathbf{m}_1} & \cdots & \Sigma_{\mathbf{x}_k\mathbf{m}_L} & \\ \Sigma_{\mathbf{m}_1\mathbf{x}_k} & \Sigma_{\mathbf{m}_1} & \cdots & \Sigma_{\mathbf{m}_1\mathbf{m}_L} & \mathbf{0}_{2L\times 2} \\ \vdots & \vdots & \ddots & \vdots & \\ \Sigma_{\mathbf{m}_L\mathbf{x}_k} & \Sigma_{\mathbf{m}_L\mathbf{m}_1} & \cdots & \Sigma_{\mathbf{m}_L} & \\ \hline & \mathbf{0}_{2\times 2L} & & & \mathbf{Q}_{k,i} \end{array}\right) \left(\begin{array}{c|c} \mathbf{I}_{3+2L} & \frac{\partial \mathbf{h}^{-1}}{\partial \mathbf{x}_k}^T \\ & \mathbf{0}_{2L\times 2} \\ \hline \mathbf{0}_{2L\times 2} & \frac{\partial \mathbf{h}^{-1}}{\partial \mathbf{z}_{k,i}}^T \end{array}\right) \qquad (27)$$

Multiplying the matrices, we finally arrive at the expression for the covariance after the map augmentation, as shown in Box 1.

Notice that the process in Equations 24 through 28 must be repeated for each non-associated measurement until all the new landmarks have been added to the map.

Computational Complexity

Regarding the computational complexity of EKF-based on-line SLAM, we have summarized the cost of each algorithm step in Table 1. It can be compared to chapter 7 Table 4, where the complexity of EKF-based localization was also analyzed. Since in this SLAM solution the state vector grows with time as new landmarks are incorporated into the map, the complexity is clearly dominated by L, the number of landmarks, leading therefore to an overall $O\left(L^3\right)$ complexity for each time step. It must be remarked however, that an implementation that exploits the sparseness of the Jacobians would achieve an $O\left(L^2\right)$ complexity (Paz, Tardós, & Neira, 2008).

Box 1.

$$\underset{\frown}{\boldsymbol{\Sigma}}_k = \left(\begin{array}{cccc|c} & & & & \boldsymbol{\Sigma}_{\mathbf{x}_k} \frac{\partial \mathbf{h}^{-1}}{\partial \mathbf{x}_k}^T \\ & & \boldsymbol{\Sigma}_k & & \boldsymbol{\Sigma}_{\mathbf{m}_1\mathbf{x}_k} \frac{\partial \mathbf{h}^{-1}}{\partial \mathbf{x}_k}^T \\ & & & & \vdots \\ & & & & \boldsymbol{\Sigma}_{\mathbf{m}_L\mathbf{x}_k} \frac{\partial \mathbf{h}^{-1}}{\partial \mathbf{x}_k}^T \\ \hline \frac{\partial \mathbf{h}^{-1}}{\partial \mathbf{x}_k}\boldsymbol{\Sigma}_{\mathbf{x}_k} & \frac{\partial \mathbf{h}^{-1}}{\partial \mathbf{x}_k}\boldsymbol{\Sigma}_{\mathbf{m}_1\mathbf{x}_k} & \cdots & \frac{\partial \mathbf{h}^{-1}}{\partial \mathbf{x}_k}\boldsymbol{\Sigma}_{\mathbf{m}_L\mathbf{x}_k} & \frac{\partial \mathbf{h}^{-1}}{\partial \mathbf{x}_k}\boldsymbol{\Sigma}_{\mathbf{x}_k}\frac{\partial \mathbf{h}^{-1}}{\partial \mathbf{x}_k}^T + \frac{\partial \mathbf{h}^{-1}}{\partial \mathbf{z}_{k,i}}\mathbf{Q}_k\frac{\partial \mathbf{h}^{-1}}{\partial \mathbf{z}_{k,i}}^T \end{array}\right) \quad (28)$$

This is the reason EKF-based SLAM is well suited for building small maps (i.e., with not too many landmarks) but its performance quickly degrades with the size of the maps until a point where its application to real-time operation becomes unfeasible. Still, it remains being a reasonable approach for real-world mobile robotics if the size of the map is known in advance to be manageable.

Uncertainty and Loop Closures

Now that we have exposed the details of the EKF-based approach to on-line SLAM, it is enlightening to consider a small, simulated example of a robot exploring an environment that contains a loop. The final uncertainty in both the robot pose and all the landmarks is shown in Figure 5a as confidence ellipses. Notice how the robot location is such that it has just observed some landmarks already detected at the beginning of its track, which coincides with the origin of coordinates. Loop closures like this have two very evident effects. The first one is already clearly visible in the map: the uncertainty in the landmarks close to the loop closure is drastically reduced, in contrast with the unavoidable growth in the uncertainty of the robot pose as it explores new areas (and therefore, in landmarks seen for the first time from those uncertain locations), an effect clearly visible along the highlighted track of the robot in the figure.

The second place where loop closures are visible is a little more subtle, but not less important to learn. A quick look at the full state covariance matrix $\boldsymbol{\Sigma}_k$ in Figure 5b reveals what was already expected from the graph of correlations of the on-line SLAM model in Figure 2b, that is, that all variables (the robot pose and the map elements) are highly correlated to each other. Instead, the inverse of that matrix (the information matrix $\boldsymbol{\Sigma}_k^{-1}$), shown in Figure 5c, reveals better the sparse structure of the problem *constraints* (as opposed to the problem *correlations*): the map elements are highly correlated to those with which share covisibility from old robot poses, hence the dominant banded diagonal structure of the largest part of the matrix—recall Equation 13 for the structure of the covariance, with coincides with that of the information matrix. However, the highlighted nonzero entries off the diagonal band are an outcome of the loop closure, where new landmarks (bottom-right of the matrix) share covisibility with old ones (top-left). In EKF-based SLAM the sparsity of the information matrix is not relevant at all, but the importance of under-

Table 1. Computational complexity analysis of one time step in the EKF-based on-line SLAM algorithm. These symbols are in use: L is the number of landmarks in the map, M the number of simultaneously observed landmarks, $C \leq M$ the number of landmarks observed for the first time and K the dimensionality of the robot pose parameterization.

Step 1: State vector prediction	1.1 Mean propagation	typically $O(K^2)$
	1.2 Covariance propagation	$O(K^3 + K^2 L)$
Step 2: Data association	2.1.1 Predict observation mean	$O(L)$
	2.1.2 Predict observation covariance	$O(L^2)$
	2.2 Establish associations	typically $O(M\,L)$
Step 3: State vector update	3.1 Measurement residuals	$O(M)$
	3.2 Kalman gain matrix	$O(L^2\,M + L\,M^2)$
	3.3 Mean update	$O(L\,M)$
	3.4 Normalization	$O(K)$
	3.5 Covariance update	$O(M\,L^2 + L^3)$
Step 4: Creation of new landmarks	4.1.1 Augment mean	$O(C)$
	4.1.2 Augment covariance	$O(C\,L\,K^2)$
Overall complexity:		$O(L^3 + ML^2 + M^2L + K^3 + CLK^2)$

standing all the effects of loop closures is that other methods such as Bundle Adjustment (described in chapter 10) or Square-Root Information Filtering heavily rely on this sparse structure of the constraints in SLAM.

Another relevant reflection regarding loop closure in on-line SLAM is in order. It is worth stressing again that in its path before closing a loop, landmarks are created taking as reference robot poses, which thereafter are marginalized out. In turn, the relationship between all the history of robot poses and the latest one is kept in form of correlations in the dense covariance matrix—recall Figure 2b whose implications are visible in Figure 5b. Right after making observations that close a loop for the first time, the entire history of robot poses would have to be recalculated to correct the accumulated errors and, consequently, the positions of all the landmarks observed meanwhile. The "magic" of EKF-based on-line SLAM resides in avoiding this need to explicitly correct the robot path while still fixing the location of all the affected landmarks straight away. When this method was seriously approached in the beginning of the 2000s, other contemporary alternative algorithms for SLAM required explicit path recalculations after each loop closure; that became one of the reasons for its enthusiastic reception in the mobile robotics community.

Critical Analysis

Despite its suitability for moderately sized environments, EKF-based on-line SLAM as explained

Figure 5. Experimental results we have obtained from 2D range-bearing SLAM in a simulation environment. (a) The final map as estimated after finishing one loop within the synthetic environment. (b) The covariance Σ_k and (c) information (inverse covariance Σ_k^{-1}) matrices for the final robot-map state space depicted in (a). Lighter colors represent higher absolute values for each entry in the matrices. Notice how the dense covariance (b) reveals its sparseness when inverted in (c). For a better interpretation of both matrices (b) and (c), bear in mind that they exhibit the block structure described in equation 13

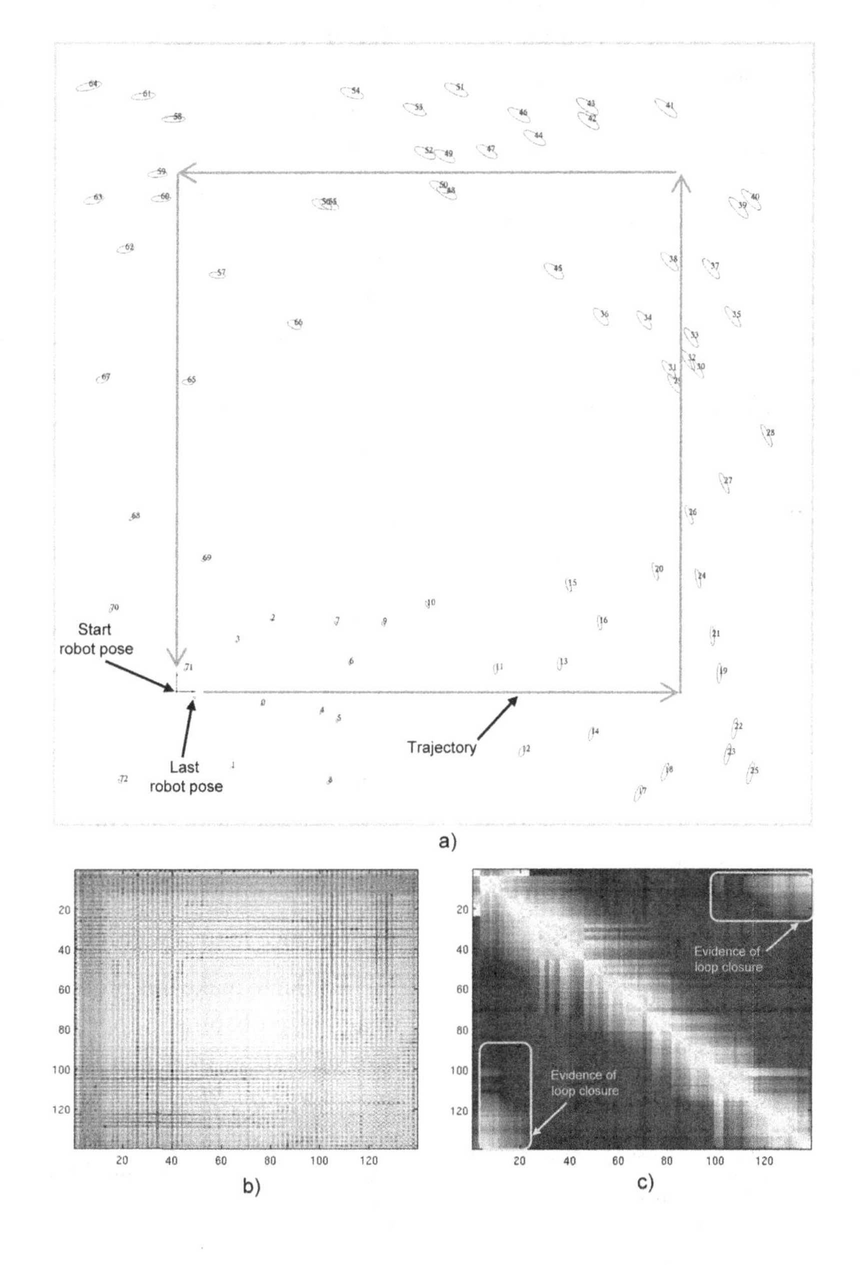

in this section has a few drawbacks and incompatibilities, which we summarize next.

First of all, this approach is only practical for building sparse maps of features. If for some reason it is preferred to construct dense occupancy grid maps, another (full SLAM) technique must be employed instead.

Secondly, one of the most unexpected lessons learnt from the recent merge of the Bundle Adjustment and SLAM communities was a honest reflection on the reasons behind the immense popularity of Kalman filtering and their derived forms, such as the EKF. It has been well-known for decades that the update stage in the EKF is exactly equivalent to *one iteration* of the non-linear least-squares Gauss-Newton algorithm (Bell & Cathey, 1993). With a quadratic convergence rate towards the optimum (Dennis & Schnabel, 1996), no one would ever think of executing only the first iteration of such algorithm; still, that is exactly what the KF or EKF do. Algorithms such as the Iterated EKF (IEKF) recover the iterative nature of Gauss-Newton to achieve much more accurate results (Jazwinsky, 1970). It would be always preferable to use the IEKF instead of the EKF despite its slightly higher computational burden but, for some reason, EKF has been being applied to a variety of problems without taking one step backwards and asking ourselves if that was the most optimal algorithm available. For all this, iterative methods are in recent years gaining popularity within the newly proposed SLAM frameworks.

Finally, the assumption of a fixed global frame of coordinates has a negative impact that limits the size of attainable maps with this technique. As the robot gets farther from the origin of coordinates, the theoretical posterior distribution of its pose become less and less similar to a Gaussian, while its uncertainty also gets larger. In practice, both effects invalidate the assumptions of the EKF algorithm, i.e. Gaussian distributions and lack of strong non-linearities. Solutions to overcome this problem can be divided into two branches: those which propose dividing large maps into a set of smaller ones (*submapping*, or hierarchical SLAM, as we shall see in the next chapter) and those attacking the problem of the non-linearities by changing the filtering algorithm, e.g. via the unscented transformation (recall chapter 7 section 3). Alternatively, one can switch to a totally different family of solutions, those based on the full SLAM model.

3. FULL SLAM: THE BASIC RBPF SOLUTION

We turn our attention now to a different way of approaching the SLAM problem via the full SLAM DBN, whose structure was depicted in Figure 1a. In contrast to the marginalized version of the problem (the so-called on-line SLAM), here the hidden variables are the map and the entire history of robot poses up to sometime step k, which are to be inferred from observations and actions. Therefore, the goal in probabilistic full SLAM is estimating the following distribution:

$$bel\left(\mathbf{x}_{0:k}, \mathbf{m}_{1:L} \middle| \mathbf{z}_{0:k}, \mathbf{u}_{1:k}\right) \tag{29}$$

Notice how the unique difference with Equation 6, the corresponding distribution for on-line SLAM, is the inclusion of all previous robot poses apart from the latest $\mathbf{x}_k$. This is not a minor change, since the dimensionality of the problem now increases with each time step. For instance, if the full joint in Equation 29 were to be represented as a multivariate Gaussian it would require a covariance matrix with $O\left(k^2 + L^2\right)$ elements.

In order to make this problem somewhat more treatable, probabilistic approaches to full SLAM take advantage of a marginalization technique known as *Rao-Blackwellization*, that can be applied when estimating (at least part of) the problem with importance sampling MonteCarlo estimators (Doucet, De Freitas, Murphy, & Russell, 2000). A variety of SLAM methods fall within this category under the common name of *Rao-Blackwellized Particle Filters* (RBPF). This section is devoted to an introduction of the simplest RBPF algorithm and to a thorough derivation of its well-known recursive equations; more efficient (and in turn slightly more complex) algorithms will be treated in section 4.

Given a DBN on which to perform inference, and an estimation problem $p(a,b \mid z)$, the RBPF approach consists of separating $p(a,b \mid z)$ into two inference processes $p(b \mid a,z)$ and $p(a \mid z)$ and then exploiting the fact that the conditional distribution $p(b \mid a,z)$ is available in closed form given a sample-based (*particle filter*) estimation of $p(a \mid z)$. For the particular case of full SLAM, and denoting the map as $\mathbf{m}$ (which may stand for any kind of metric map), the joint $p(a,b \mid z)$ factors as follows:

$$bel\left(\mathbf{x}_{0:k}, \mathbf{m} \middle| \mathbf{z}_{0:k}, \mathbf{u}_{1:k}\right) =$$

(using the chain rule of conditional probability, i.e., $p\left(a,b \middle| y\right) = p\left(a \middle| y\right) p\left(b \middle| a,y\right)$)

$$= bel\left(\mathbf{x}_{0:k} \middle| \mathbf{z}_{0:k}, \mathbf{u}_{1:k}\right) bel\left(\mathbf{m} \middle| \mathbf{x}_{0:k}, \mathbf{z}_{0:k}, \mathbf{u}_{1:k}\right) =$$

(and by the conditional independence $\mathbf{m} \perp \mathbf{u}_{1:k} \middle| \mathbf{x}_{0:k}, \mathbf{z}_{0:k}$ provided by the DBN)

$$= \underbrace{bel\left(\mathbf{x}_{0:k}, \mathbf{m} \middle| \mathbf{z}_{0:k}, \mathbf{u}_{1:k}\right)}_{\text{Full SLAM posterior estimation}} = \underbrace{bel\left(\mathbf{x}_{0:k} \middle| \mathbf{z}_{0:k}, \mathbf{u}_{1:k}\right)}_{\text{Robot path estimation}} \underbrace{bel\left(\mathbf{m} \middle| \mathbf{x}_{0:k}, \mathbf{z}_{0:k}\right)}_{\text{Map estimation}} \tag{30}$$

That is, the estimation of the robot path and the map is split into one process in charge of the robot poses and, taking the output from that estimator, another process that estimates the map from the robot path. The key advantage underlying this approach is the existence of closed-form Bayesian estimators of the map for the most common representations (i.e. occupancy grid maps, maps of landmarks). That estimation is conditioned on the robot path $\mathbf{x}_{0:k}$, thus we turn our attention to that process first.

The posterior distribution for the robot path,

$$bel\left(\mathbf{x}_{0:k} \middle| \mathbf{z}_{0:k}, \mathbf{u}_{1:k}\right) \tag{31}$$

can be estimated by means of a particle filter—except for the first time step $k = 0$, for which a fixed known distribution is assumed, typically a Dirac's delta at the origin of coordinates, just like in the EKF approach. As explained in chapter 7, particle filters approximate probability distributions non-parametrically as a set of *weighted observations* (a sample in the statistical terminology) in the corresponding state space of the robot paths $\mathbf{x}_{0:k}$ (the symbol $\triangleright$ stands for "is represented by"):

$$bel(\mathbf{x}_{0:k} \middle| \mathbf{z}_{0:k}, \mathbf{u}_{1:k}) \triangleright \left\{\left(w_k^{[i]}, \mathbf{x}_{0:k}^{[i]}\right)\right\} \quad \text{(indexed set)}$$

with

$$\sum_{i=1}^{N} w_k^{[i]} = 1 \tag{32}$$

where each $\mathbf{x}_{0:k}^{[i]}$ corresponds in this context to one hypothesis about the complete robot path (which we call a *particle*), and the $w_k^{[i]}$ are their corresponding importance factors or *weights*. The total number of hypotheses, denoted as N, is usually fixed and set heuristically—however, automatic methods for dynamically determining this number have been proposed in the SLAM literature (Blanco, González, & Fernandez-Madrigal, 2010).

Since robots typically require executing SLAM in real-time as new evidence is collected, the posterior of Equation 31 is estimated sequentially by computing the values of the particles $\mathbf{x}_{1:k}^{[i]}$

and their weights $w_k^{[i]}$ from the previous set $\left\{\left(w_{k-1}^{[i]}, \mathbf{x}_{0:k-1}^{[i]}\right)\right\}$. The algorithm in charge of these updates is summarized in Algorithm 2, but before proceeding further on we believe it is educational to provide a thorough derivation from the basic probabilistic principles.

We start by applying the Baye's rule on the latest observation, that is:

$$\underbrace{bel\left(\mathbf{x}_{0:k} \middle| \mathbf{z}_{0:k}, \mathbf{u}_{1:k}\right)}_{\text{Posterior pdf for time step } k} \propto$$
$$\propto p\left(\mathbf{z}_t \middle| \mathbf{x}_{0:k}, \mathbf{z}_{0:k-1}, \mathbf{u}_{1:k}\right) p\left(\mathbf{x}_{0:k} \middle| \mathbf{z}_{0:k-1}, \mathbf{u}_{1:k}\right) \quad (33)$$

where we can apply the law of total probability to the right-most term to make it recursive:

$$bel\left(\mathbf{x}_{0:k} \middle| \mathbf{z}_{0:k}, \mathbf{u}_{1:k}\right) \propto$$
$$\propto p\left(\mathbf{z}_t \middle| \mathbf{x}_{0:k}, \mathbf{z}_{0:k-1}, \mathbf{u}_{1:k}\right) p\left(\mathbf{x}_k, \mathbf{x}_{0:k-1} \middle| \mathbf{z}_{0:k-1}, \mathbf{u}_{1:k}\right) =$$

(by the law of total probability over $\mathbf{x}_{0:k-1}$)

$$= p\left(\mathbf{z}_t \middle| \mathbf{x}_{0:k}, \mathbf{z}_{0:k-1}, \mathbf{u}_{1:k}\right)$$
$$\int_{\mathbf{x}_{0:k-1}} p\left(\mathbf{x}_k, \mathbf{x}_{0:k-1} \middle| \mathbf{z}_{0:k-1}, \mathbf{u}_{1:k}, \mathbf{x}_{1:k-1}\right) p\left(\mathbf{x}_{0:k-1} \middle| \mathbf{z}_{0:k-1}, \mathbf{u}_{1:k}\right) d\mathbf{x}_{0:k-1} =$$

(due to the independence $\mathbf{x}_{0:k-1} \perp \mathbf{u}_k$)

$$= p\left(\mathbf{z}_t \middle| \mathbf{x}_{0:k}, \mathbf{z}_{0:k-1}, \mathbf{u}_{1:k}\right)$$
$$\int_{\mathbf{x}_{0:k-1}} p\left(\mathbf{x}_k, \mathbf{x}_{0:k-1} \middle| \mathbf{z}_{0:k-1}, \mathbf{u}_{1:k}, \mathbf{x}_{1:k-1}\right)$$
$$\underbrace{bel\left(\mathbf{x}_{0:k-1} \middle| \mathbf{z}_{0:k-1}, \mathbf{u}_{1:k-1}\right)}_{\text{Posterior at step } k-1} d\mathbf{x}_{0:k-1} =$$

(since $p\left(a, b \mid b\right) = p(a \mid b)$)

$$= p\left(\mathbf{z}_t \middle| \mathbf{x}_{1:k}, \mathbf{z}_{0:k-1}, \mathbf{u}_{1:k}\right)$$
$$\int_{\mathbf{x}_{0:k-1}} p\left(\mathbf{x}_k \middle| \mathbf{z}_{0:k-1}, \mathbf{u}_{1:k}, \mathbf{x}_{0:k-1}\right)$$
$$bel\left(\mathbf{x}_{0:k-1} \middle| \mathbf{z}_{0:k-1}, \mathbf{u}_{1:k-1}\right) d\mathbf{x}_{0:k-1} =$$

(and due to the conditional independence $\mathbf{x}_k \perp \mathbf{x}_{0:k-2}, \mathbf{u}_{1:k-1}, \mathbf{z}_{0:k-1} \middle| \mathbf{x}_{k-1}, \mathbf{u}_k$)

$$= p\left(\mathbf{z}_t \middle| \mathbf{x}_{0:k}, \mathbf{z}_{0:k-1}, \mathbf{u}_{1:k}\right)$$
$$\int_{\mathbf{x}_{0:k-1}} \underbrace{p\left(\mathbf{x}_k \middle| \mathbf{u}_k, \mathbf{x}_{k-1}\right)}_{\text{Motion model}} bel\left(\mathbf{x}_{0:k-1} \middle| \mathbf{z}_{0:k-1}, \mathbf{u}_{1:k-1}\right) d\mathbf{x}_{0:k-1} \quad (34)$$

Notice that from the two terms in the product, the integral represents what in Kalman filtering-based SLAM is called a robot pose *prediction*, since it only takes into account the motion model and the previous posterior pdf. In turn, the left-hand term that is out of the integral closely resembles an observation model $p\left(\mathbf{z}_t \middle| \mathbf{x}_k, \mathbf{m}\right)$, except for the omission of the map variable, which follows from the decoupling of the SLAM problem into two separate estimation problems in the RBPF approach. However, both estimation processes are actually interweaved, thus we can now introduce the map variables by applying the law of total probability, leading to:

(applying the law of total probability on $\mathbf{m}$)

$$p\left(\mathbf{z}_t \middle| \mathbf{x}_{0:k}, \mathbf{z}_{0:k-1}, \mathbf{u}_{1:k}\right) =$$
$$\int_{\mathbf{m}} p\left(\mathbf{z}_t \middle| \mathbf{x}_{0:k}, \mathbf{z}_{0:k-1}, \mathbf{u}_{1:k}, \mathbf{m}\right) p\left(\mathbf{m} \middle| \mathbf{x}_{0:k}, \mathbf{z}_{0:k-1}, \mathbf{u}_{1:k}\right) d\mathbf{m} =$$

(and due to the conditional independence $\mathbf{z}_k \perp \mathbf{x}_{0:k-1}, \mathbf{z}_{0:k-1}, \mathbf{u}_{1:k} \middle| \mathbf{x}_k, \mathbf{m}$)

$$= \int_{\mathbf{m}} \underbrace{p\left(\mathbf{z}_t \middle| \mathbf{x}_k, \mathbf{m}\right)}_{\text{Observation model}} p\left(\mathbf{m} \middle| \mathbf{x}_{0:k}, \mathbf{z}_{0:k-1}, \mathbf{u}_{1:k}\right) d\mathbf{m} =$$

(and due to the conditional independence $\mathbf{m} \perp \mathbf{x}_k, \mathbf{u}_{1:k} \big| \mathbf{x}_{0:k-1}, \mathbf{z}_{0:k-1}$)

$$= \int_{\mathbf{m}} p\left(\mathbf{z}_t \middle| \mathbf{x}_k, \mathbf{m}\right) \underbrace{p\left(\mathbf{m} \middle| \mathbf{x}_{0:k-1}, \mathbf{z}_{0:k-1}\right)}_{\text{Map estimation at step } k-1} d\mathbf{m} \tag{35}$$

Put in words, in order to account for the map in the likelihood function of the poses estimator of Equation 34 we need to integrate over all the possible map values. Notice how at this point the chicken-and-egg nature of SLAM reveals itself clearly by leading to a term for the map estimation, the $p\left(\mathbf{m} \middle| \mathbf{x}_{0:k-1}, \mathbf{z}_{0:k-1}\right)$ above, that depends on the (uncertain) estimated robot path until the previous time step $k-1\ldots$ which in turn depends on the previous map (see again Equation 34). To makes clearer this recursive dependence between path and map, we remind that the robot path is represented as a sample of certain size, thus the map estimation for the $i-th$ particle at time step $k-1$ is actually $p\left(\mathbf{m} \middle| \mathbf{x}^{[i]}_{0:k-1}, \mathbf{z}_{0:k-1}\right)$, therefore:

$$p\left(\mathbf{z}_t \middle| \mathbf{x}_{0:k}, \mathbf{z}_{0:k-1}, \mathbf{u}_{1:k}\right) =$$
$$= \int_{\mathbf{m}} p\left(\mathbf{z}_t \middle| \mathbf{x}_k, \mathbf{m}\right) \underbrace{p\left(\mathbf{m} \middle| \mathbf{x}^{[i]}_{0:k-1}, \mathbf{z}_{0:k-1}\right)}_{\substack{\text{Map estimation at step } k-1 \\ \text{in the } i-th \text{ particle}}} d\mathbf{m} \tag{36}$$

Since $\mathbf{x}^{[i]}_{0:k-1}$ is a particular value, as opposed to a pdf, this expression implies building a map from a sequence of known observations and robot poses, for which closed-form expressions exist. Chapter 8 was devoted to explaining these mapping methods, thus we shall not discuss them further here. In the robotics literature it is common to shorten the integral above with the introduction of the variables $\mathbf{m}^{[i]}$, denoting "the map for the $i-th$ particle," such that, probably in an abuse of notation, the observation likelihood can be written in a much more compact and expressive way like:

$$\underbrace{p\left(\mathbf{z}_t \middle| \mathbf{x}_k, \mathbf{m}^{[i]}\right)}_{\substack{\text{Observation likelihood} \\ \text{for the } i\text{-}th \text{ particle}}} \triangleq$$
$$\triangleq \int_{\mathbf{m}} p\left(\mathbf{z}_t \middle| \mathbf{x}_k, \mathbf{m}\right) p\left(\mathbf{m} \middle| \mathbf{x}^{[i]}_{0:k-1}, \mathbf{z}_{0:k-1}\right) d\mathbf{m} =$$
$$= p\left(\mathbf{z}_t \middle| \mathbf{x}_k, \mathbf{x}^{[i]}_{0:k-1}, \mathbf{z}_{0:k-1}, \mathbf{u}_{1:k}\right) \tag{37}$$

Putting all together, we arrive at the well-known recursive expression for sequential Bayesian estimation of the history of poses in a full SLAM framework with RBPF:

$$\underbrace{bel\left(\mathbf{x}_{0:k} \middle| \mathbf{z}_{0:k}, \mathbf{u}_{1:k}\right)}_{\substack{\text{Path posterior pdf} \\ \text{at time step } k}} \propto \underbrace{p\left(\mathbf{z}_t \middle| \mathbf{x}_k, \mathbf{m}^{[i]}\right)}_{\substack{\text{Observation likelihood} \\ \text{for the } i\text{-}th \text{ particle}}}$$
$$\int_{\mathbf{x}_{0:k-1}} \underbrace{p\left(\mathbf{x}_k \middle| \mathbf{u}_k, \mathbf{x}_{k-1}\right)}_{\text{Motion model}} bel\left(\mathbf{x}_{0:k-1} \middle| \mathbf{z}_{0:k-1}, \mathbf{u}_{1:k-1}\right) d\mathbf{x}_{0:k-1} \tag{38}$$

which in turn is fed into the map estimator (recall Equation 30), such that we have a probabilistic map for each particle. Furthermore, if the map is composed of several elements (e.g. landmarks) like $\mathbf{m} = \left\{\mathbf{m}_1, \ldots, \mathbf{m}_L\right\}$, the conditional independence of each element with respect to the robot path given the observations (refer to the DBN in Figure 1a allows further factoring, since the pdf of each map element can be estimated independently, that is:

$$\underbrace{bel\left(\mathbf{m}^{[i]} \middle| \mathbf{x}_{0:k}, \mathbf{z}_{1:k}\right)}_{\substack{\text{Map posterior pdf} \\ \text{for the } i\text{-}th \text{ particle}}} \triangleq$$
$$\triangleq bel\left(\mathbf{m} \middle| \mathbf{x}^{[i]}_{0:k}, \mathbf{z}_{0:k}\right) = \prod_{j=1\ldots L} bel\left(\mathbf{m}^{[i]}_j \middle| \mathbf{x}^{[i]}_{0:k}, \mathbf{z}_{0:k}\right) \tag{39}$$

This is one of the major advantages of RBPF-based SLAM, since we have the freedom to estimate each landmark with any appropriate Bayesian filter of our choice (e.g. particle filter, Kalman filter, evidente grid) and even to dynamically switch between the different representations at convenience; for example, collapsing a set of particles as a Gaussian when that seems appropriate.

After this introduction to the foundations of RBPF-based SLAM, we must remark that the final expression in Equation 38 is generic and does not specify how to compute the particles and their weights, but only the condition that the represented densities must fulfill. That is why different RBPF algorithms exist for implementing the same equation. We start next exposing the simplest of them.

Algorithm Description: RBPF with the Standard Proposal

The recursive Bayesian filtering problem in Equation 38 can be addressed using the SIR algorithm, already introduced in the context of robot localization in chapter 7 section 4. For convenience, we have summarized its application to the context of full SLAM in Algorithm 2.

Firstly, notice how the aim of the algorithm is updating a number N of weighted particles, each comprising a robot path $\mathbf{x}_{0:k}^{[i]}$ and a map denoted as $\mathbf{m}^{[i]}$ following the common convention in the SLAM literature. It must be kept in mind, however, that in spite of the superscript notation $\cdot^{[i]}$, this term is not a *value* (as $\mathbf{x}_{0:k}^{[i]}$ is) but stands instead for the entire probability distribution in Equation 39.

The first part of the algorithm, numbered as 1, is in charge of updating the robot path and their weights, that is, the part of the SLAM posterior represented as particles in the RBPF; then, the step 2 simply updates, one by one, the parametric or non-parametric Bayesian filter associated to every map element in each $\mathbf{m}^{[i]}$, assuming that the robot path is exactly $\mathbf{x}_{0:k}^{[i]}$ without uncertainty. For instance, if maps are being represented as occupancy grid maps, the grid associated to each particle can be updated using the log-odds technique explained in chapter 8 section 3. In the case of a map of landmarks (e.g. a range-bearing sensor) we can use one EKF to estimate a Gaussian pdf for each landmark, in contrast to the usage of one large Gaussian for the entire robot-map state vector employed in EKF-based SLAM. The impact of this fact on computational complexity is explored later on.

Regarding the RBPF part (step 1 in the algorithm), a new candidate pose at time step k is firstly drawn from the probabilistic motion model, that is, $\mathbf{x}_k^{[i]} \sim \mathbf{p}\left(\mathbf{x}_k \middle| \mathbf{x}_{k-1}^{[i]}, \mathbf{u}_k\right)$. Then, its weight is modified to account for how well this new pose explains the latest observation (after performing data association if we are dealing with maps of landmarks), by mean of multiplying the old weights by the observation likelihood $\mathbf{p}\left(\mathbf{z}_k \middle| \mathbf{x}_k^{[i]}, \mathbf{m}^{[i]}\right)$, after which all the weights are renormalized for they to sum up the unity. We insist at this point in our practical recommendation to implementors already given in chapter 7: keep the logarithm of weights (*log-weights*) instead of their natural values to avoid the common pitfall of falling into overflows and underflows between weight renormalizations steps. To end the review of the RBPF part, the impoverishment of the particle representation is watched via the *effective sample size* (or ESS, defined in chapter 7) and whenever this value falls below a certain threshold (typically half the number of particles), resampling is performed (step 1.3.1) by following any of the algorithms described in Appendix B.

We must remark three fundamental properties which characterize this RBPF approach to full SLAM. Firstly, notice how the history of past robot poses is totally immutable: once a candidate robot pose is drawn from the motion model it never changes. The only possible correction for robot poses that are incompatible with subsequent observations is a reduction of the weights for its associated path and, eventually, being removed in resampling. Compare this to the implicit path correction (e.g. after closing a loop) with each

Algorithm 2. The operations involved in RBPF-based SLAM, when using the standard proposal distribution

algorithm RBPF_SLAM_STANDARD

Inputs: $\left\{\left(\mathbf{x}_{0:k-1}^{[i]}, w_{k-1}^{[i]}\right), \mathbf{m}^{[i]}\right\}_{i=1}^{N}, \mathbf{u}_k, \mathbf{z}_k$

Outputs: $\left\{\left(\mathbf{x}_{0:k}^{[i]}, w_{k}^{[i]}\right), \mathbf{m}^{[i]}\right\}_{i=1}^{N}$

1 Robot path Bayesian estimate (SIR algorithm)

1.1 For each particle index $i \in \{1, \ldots, N\}$

1.1.1 Draw pose from motion model: $\mathbf{x}_k^{[i]} \sim \mathbf{p}\left(\mathbf{x}_k \middle| \mathbf{x}_{k-1}^{[i]}, \mathbf{u}_k\right)$

1.1.2 Data association (if applicable): $\alpha_k \leftarrow data\_assoc\left(\mathbf{z}_k, \mathbf{m}^{[i]}, \mathbf{x}_k^{[i]}\right)$

1.1.3 New weight from observation likelihood: $\hat{w}_k^{[i]} \leftarrow w_{k-1}^{[i]}\ \mathbf{p}\left(\mathbf{z}_k\left(\alpha_k\right) \middle| \mathbf{x}_k^{[i]}, \mathbf{m}^{[i]}\right)$

1.1.4 Augment robot path history: $\mathbf{x}_{0:k}^{[i]} \leftarrow \left\{\mathbf{x}_{0:k-1}^{[i]}, \mathbf{x}_k^{[i]}\right\}$

1.2 Normalize weights: $w_k^{[i]} \leftarrow \dfrac{\hat{w}_k^{[i]}}{\sum_{i=1\ldots M} \hat{w}_k^{[i]}}$

1.3 if $ESS\left(\left\{w_k^{[i]}\right\}_{i=1}^{N}\right) < \frac{1}{2}N$

1.3.1 Perform resampling: $\left\{\left(\mathbf{x}_{0:k-1}^{[i]}, w_{k-1}^{[i]}\right), \mathbf{m}^{[i]}\right\}_{i=1}^{N} \leftarrow resample\left\{\left(\mathbf{x}_{0:k-1}^{[i]}, w_{k-1}^{[i]}\right), \mathbf{m}^{[i]}\right\}_{i=1}^{N}$

2 Map Bayesian estimate

2.1 For each particle index $i \in \{1, \ldots, N\}$

2.1.1 For each map element $j \in \{1, \ldots, L\}$

2.1.1.1 Bayesian update: $\mathbf{m}_j^{[i]} \leftarrow map\_update\left(\mathbf{m}_j^{[i]}, \mathbf{z}_k\left(\alpha_k\right)\right)$

observation integrated into an EKF for on-line SLAM. Secondly, this version of the RBPF algorithm (named *standard proposal* RBPF for reasons that will become clear below) demands very little knowledge on the robot models. All we need is being able to *draw samples* from the motion model and to *evaluate pointwise* the sensor observation likelihood (steps 1.1.1 and 1.1.2, respectively). This contrasts with the need for a closed-form and (unimodal) Gaussian approximation of both models in EKF-based on-line SLAM. Therefore, a RBPF approach may be the unique way to approach SLAM for some combinations of motion models and sensors for which the EKF-SLAM requirements cannot be met. Finally, when building maps of landmarks it turns out that data association must be carried out for each particle *independently*. While this indeed represents a performance penalty if compared to the EKF-based solution to on-line SLAM where data association must be solved only once, the advantages of doing so can outweigh the inconveniences. In particular, data association can be ambiguous in cluttered environments with a high density of landmarks; the possibility of exploring the different potentially valid associations in each particle in such a scenario minimizes the catastrophic consequences of wrong associations to the scope of only those particles where data association took the incorrect choices. Moreover, and as proposed in (Blanco, González, & Fernández-Madrigal, 2012), RB-PFs offer a perfect opportunity for dealing with extremely noisy data association. This would be possible by splitting each particle into many in the next time step, one for each plausible association hypothesis, then weighting each of them according to the likelihood of its associations. As insightfully mentioned in (Frese, Larsson, & Duckett, 2005), SLAM comprises two qualitatively-different sources of uncertainty: continuous uncertainty regarding the noise and uncertainty in sensors and actuators, and the discrete uncertainty in establishing data association. Disregarding their drawbacks, RBPFs present a seamlessly integration of these two types of uncertainty into a unique Bayesian framework.

Concerning the computational complexity of this algorithm, we summarize the cost of each step in Table 2. It is worth mentioning that the selective resampling stage (step 1.3) comprises two separate operations, computing the ESS and performing the actual resampling, where the latter would be only executed occasionally. Therefore, the amortized complexity over large execution sequences becomes a weighted average of both costs.

Assuming a fixed number of particles N over time, which is a common practice in RBPF SLAM, it follows that the dominant complexity becomes linear with the number of map elements L. This is in strong contrast to the complexity of EKF-based SLAM, which exhibits a cubic execution time with the number of landmarks, as we saw in section 2. However, remember from chapter 7 that the number of particles may be high in some situations.

Critical Analysis

As stated above, the important advantages of a RBPF approach to full SLAM are a great flexibility in modeling the probability distributions for each map element (which become independent from each other) and a linear computational complexity with the size of the map. However, the method also suffers from two severe hurdles.

Firstly, representing the robot path pdf as a set of particles is only a good idea while the number of particles is high enough to provide a sufficient sample density in the interesting areas of the state space. However, the dimensionality of this state space ($\mathbf{x}_{0:k}$) grows with each time step, thus in theory the number of particles should grow exponentially with time if we want to maintain a

Table 2. Computational complexity analysis of one time step in the standard proposal RBPF algorithm. Note that N is the number of particles in use, L the number of map elements and that we have considered as $O(1)$ the execution times of drawing samples from the motion model, evaluating the observation likelihood and updating one map element. Most resampling algorithms can be implemented to achieve a complexity of $O(N)$ at least (refer to Appendix B).

Step 1: Robot path estimate	1.1 Drawing new values, weight update and path history augmentation	Without DA: $O(N)$ With DA: $O(N\,L)$
	1.2 Weight renormalization	$O(N)$
	1.3 Conditional resampling	Compute ESS: $O(N)$ Resampling (not in all iterations): $O(N)$
Step 2: Update of maps	2.1 Update all map elements	$O(N\,L)$
Overall complexity:		Worst case: $O(N\,L)$

properly sampled representation. Obviously, such an evolution in the number of particles would render RBPF-based SLAM prohibitive. In practice, however, using only a few particles or an increasing (slower than exponential) number of them, provides an acceptable approximation of the posterior pdf.

The second problem is partly a consequence of the former: if the number of particles does not grow exponentially with time, the state space becomes poorly sampled in some degree and after the resampling stage inherent to each loop closure it is common that only a very few path histories remain with non-negligible weights. In that case the pdf of the entire robot path collapses to only a few particles (or even just one), losing the probabilistic representation of $\mathbf{x}_{0:k}$. This problem is called *particle depletion,* and it was already explained in chapter 7. It becomes a serious hurdle when the robot explores an environment with nested loops. As illustrated in Figure 6, in RBPF SLAM the accuracy of the robot path estimation after closing a loop uniquely depends on the existence of a variety of path hypotheses, from which only the most likely will survive the resampling, but after closing an inner loop within a nested loop, the first resampling may discard most particles as exemplified in Figure 7, effectively rendering impossible for the outer loop to be closed accurately.

The two problems above are serious ones and inherent to the sampled-based nature of RBPF; thus in some sense, they are unavoidable and we should learn to live with them if employing RBPFs. Still, there exist advanced techniques aimed at minimizing those negative effects as much as possible, up to the point of turning RBPF-based SLAM into one of the most-widely used SLAM techniques in the robotics community—especially when working with 2D grid maps and laser scans. The next section is devoted to those improved RBPF algorithms.

4. FULL SLAM: IMPROVED RBPF SOLUTIONS

As we have seen above, the main drawbacks of a RBPF approach to SLAM are a consequence of a poor sampling of the history of poses state space, either for a reduced number of particles or an excessive resampling that discards too many path hypotheses. In this section we explore some of the most interesting algorithms that have been proposed in the literature in order to alleviate as much as possible these disadvantages.

About Importance Weights

The method presented in section 3 was called *standard proposal* RBPF. It is time now to explain the meaning of that name, and why the Algorithm 2 is only one of the possible ways to address sequential estimation with a RBPF.

We must get back to Equation 38, which establishes the sequential relationship existing between the posterior beliefs of consecutive time steps; we repeat it here for the convenience of the reader:

$$\underbrace{bel\left(\mathbf{x}_{0:k} \middle| \mathbf{z}_{0:k}, \mathbf{u}_{1:k}\right)}_{\substack{\text{Path posterior pdf} \\ \text{for time step } k}} \propto \underbrace{p\left(\mathbf{z}_t \middle| \mathbf{x}_k, \mathbf{m}^{[i]}\right)}_{\substack{\text{Observation likelihood} \\ \text{for the } i\text{-}th \text{ particle}}} \int_{-\infty}^{\infty} \underbrace{p\left(\mathbf{x}_k \middle| \mathbf{u}_k, \mathbf{x}_{k-1}\right)}_{\text{Motion model}} \underbrace{bel\left(\mathbf{x}_{0:k-1} \middle| \mathbf{z}_{0:k-1}, \mathbf{u}_{1:k-1}\right)}_{\text{Previous path posterior pdf}} d\mathbf{x}_{0:k-1} \tag{40}$$

One way to perform a Monte Carlo simulation of this equation in order to obtain a representation of the left-hand posterior is drawing samples from the pdf resulting from the integral. This part represents the possible poses of the robot given only our previous belief and the motion model. Clearly, this is exactly what the algorithm of the previous section achieves by drawing samples from $\mathbf{p}\left(\mathbf{x}_k \middle| \mathbf{x}_{k-1}^{[i]}, \mathbf{u}_k\right)$. In order to account for the term that is out of the integral, the weights were then

Figure 6. Experimental results we have obtained with one of our robots in our building for grid mapping with RBPF-based SLAM. The map and robot path hypotheses are shown for two different time steps (150 and 180), right before and after closing a loop. The loop closure is visible in the evolution of the filter ESS as the recovery of the maximum ESS value. You can observe that many of the path hypotheses have been lost in the resampling process.

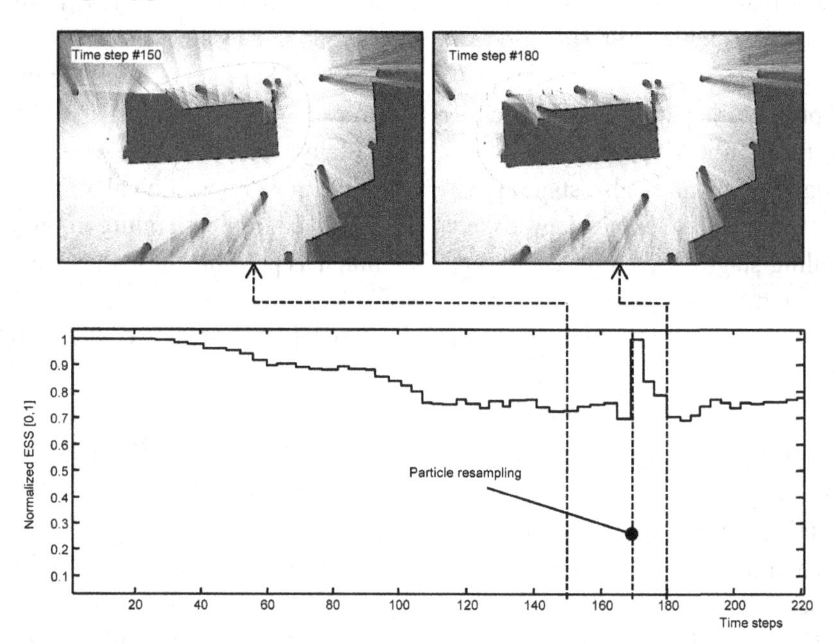

multiplied by their corresponding observation likelihood. These two operations correspond to steps 1.1.1 and 1.1.3 in Algorithm 2. It would be instructive for the reader to deserve a few moments to verify the exact correspondence between the analytic Equation 40 and its sampled-based simulation.

It can be shown however that the aforementioned method is only a particular version of a generic sample-based solution of Equation 40. As already discussed in chapter 7, the most generic method consists of drawing samples from an arbitrary pdf which is named the *proposal distribution* $\mathbf{q}(\cdot)$, and then updating the importance weights to account for the mismatch between the proposal and the latest observation $\mathbf{z}_k$. Mathematically, the algorithm reads (Doucet, Godsill, & Andrieu, 2000):

Draw sample:

$$\mathbf{x}_k^{[i]} \sim q\left(\mathbf{x}_k \middle| \mathbf{x}_{0:k-1}^{[i]}, \mathbf{z}_{0:k}, \mathbf{u}_{1:k}\right)$$

Weight update:

$$\hat{w}_k^{[i]} = w_k^{[i]} \frac{p\left(\mathbf{z}_k \middle| \mathbf{x}_{0:k}^{[i]}, \mathbf{z}_{0:k-1}, \mathbf{u}_{1:k}\right) p\left(\mathbf{x}_k \middle| \mathbf{x}_{k-1}^{[i]}, \mathbf{u}_k\right)}{q\left(\mathbf{x}_k \middle| \mathbf{x}_{0:k-1}^{[i]}, \mathbf{z}_{0:k}, \mathbf{u}_{1:k}\right)} \tag{41}$$

We call *standard* proposal to the particular choice $q\left(\mathbf{x}_k \middle| \mathbf{x}_{0:k-1}^{[i]}, \mathbf{z}_{0:k}, \mathbf{u}_{1:k}\right) = p\left(\mathbf{x}_k \middle| \mathbf{x}_{k-1}^{[i]}, \mathbf{u}_k\right)$, which if replaced in the weight update formula above leads to a simple:

$$\hat{w}_k^{[i]} = w_{k-1}^{[i]} \frac{p\left(\mathbf{z}_k \middle| \mathbf{x}_k, \mathbf{x}_{0:k-1}^{[i]}, \mathbf{z}_{0:k-1}, \mathbf{u}_{1:k}\right) p\left(\mathbf{x}_k \middle| \mathbf{x}_{k-1}^{[i]}, \mathbf{u}_k\right)}{p\left(\mathbf{x}_k \middle| \mathbf{x}_{k-1}^{[i]}, \mathbf{u}_k\right)} =$$

$$= w_{k-1}^{[i]}\, p\left(\mathbf{z}_k \middle| \mathbf{x}_k, \mathbf{x}_{0:k-1}^{[i]}, \mathbf{z}_{0:k-1}, \mathbf{u}_{1:k}\right) =$$

(using the notation introduced in Equation 37)

Figure 7. An example of the problems associated to particle impoverishment after closing the inner loop within an environment with nested loops. The evolution of the ESS (left-bottom graph) over the map building process reveals three resampling episodes: the first one (around time step 3100) occurs after closing the first loop, the one at left hand in the map, while the other two resampling events were caused by closing the largest loop. It can be seen in the map snapshot for time step 3400 how the inner nested loop is closed with an acceptable accuracy. However, the small diversity of survival particles makes harder to close the outer loop as becomes evident with the poor quality of the final map, where clear inconsistencies are visible at the right-bottom corner.

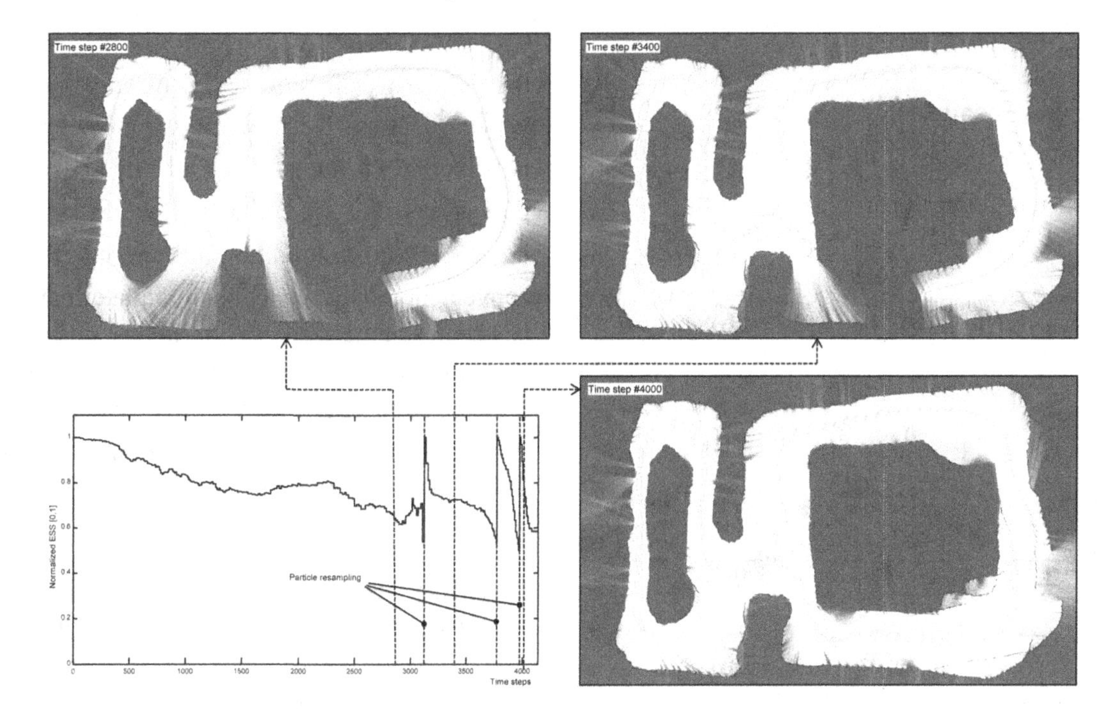

$$= w_{k-1}^{[i]} \, p\left(\mathbf{z}_k \middle| \mathbf{x}_k^{[i]}, \mathbf{m}^{[i]}\right) \tag{42}$$

Notice how this equation matches the one in step 1.1.3 of Algorithm 2, which can be now clearly seen as a particular case of RBPF for the standard proposal distribution.

However, in the context of SLAM this approach can be quite suboptimal due to the high uncertainty of the motion model (typically, from odometry) if compared to the observation model (e.g. a laser scanner). This mismatch causes most of the values to be drawn in areas of the state space to which the observation likelihood will immediately assign a negligible weight. As a consequence, most of the particles can be wasted since those with low weights are likely to be discarded in the next resampling, which impoverishes the diversity of path hypotheses and has fatal consequences for the accuracy of loop closures. Technically speaking, the optimality criterion we seek is to minimize the expected variance of the next weights $\left\{w_k^{[i]}\right\}_{i=1}^{N}$, as that would imply that all values represent equally well the pdf, thus delaying the need to resample.

In fact, an *optimal proposal distribution* $q\left(\cdot\right)$ was demonstrated to exist such that it minimizes the variance of the new importance weights, given by (Zaritskii, Svetnik, & Shimelevich, 1975; Doucet, Godsill, & Andrieu, 2000):

$$q_{\text{opt}}\left(\mathbf{x}_k \middle| \mathbf{x}^{[i]}_{0:k-1}, \mathbf{z}_{0:k}, \mathbf{u}_{1:k}\right) = p\left(\mathbf{x}_k \middle| \mathbf{x}^{[i]}_{0:k-1}, \mathbf{z}_{0:k}, \mathbf{u}_{1:k}\right) =$$

(using Bayes' rule over $\mathbf{z}_k$)

$$= \frac{p\left(\mathbf{z}_k \middle| \mathbf{x}_k, \mathbf{x}^{[i]}_{0:k-1}, \mathbf{z}_{0:k}, \mathbf{u}_{1:k}\right) p\left(\mathbf{x}_k \middle| \mathbf{x}^{[i]}_{0:k-1}, \mathbf{z}_{0:k-1}, \mathbf{u}_{1:k}\right)}{p\left(\mathbf{z}_k \middle| \mathbf{x}^{[i]}_{0:k-1}, \mathbf{z}_{0:k-1}, \mathbf{u}_{1:k}\right)} =$$

(since the denominator
$\mu_i = p\left(\mathbf{z}_k \middle| \mathbf{x}^{[i]}_{0:k-1}, \mathbf{z}_{0:k-1}, \mathbf{u}_{1:k}\right)$
is a constant for each particle)

$$= \frac{1}{\mu_i} p\left(\mathbf{z}_k \middle| \mathbf{x}_k, \mathbf{x}^{[i]}_{0:k-1}, \mathbf{z}_{0:k}, \mathbf{u}_{1:k}\right) p\left(\mathbf{x}_k \middle| \mathbf{x}^{[i]}_{0:k-1}, \mathbf{z}_{0:k-1}, \mathbf{u}_{1:k}\right) =$$

(due to $\mathbf{x}_k \perp \mathbf{x}^{[i]}_{0:k-2}, \mathbf{u}_{1:k-1}, \mathbf{z}_{0:k-1} \middle| \mathbf{x}^{[i]}_{k-1}, \mathbf{u}_k$)

$$= \frac{1}{\mu_i} \underbrace{p\left(\mathbf{z}_k \middle| \mathbf{x}_k, \mathbf{x}^{[i]}_{0:k-1}, \mathbf{z}_{0:k-1}, \mathbf{u}_{1:k}\right)}_{\substack{\text{Observation model for any} \\ \text{potential pose } \mathbf{x}_k}} \underbrace{p\left(\mathbf{x}_k \middle| \mathbf{x}^{[i]}_{k-1}, \mathbf{u}_k\right)}_{\text{Motion model}} \tag{43}$$

In words, the first line of this derivation means that sample must be drawn from the Bayes posterior density: the pdf of robot poses after incorporating the knowledge of the latest action $\mathbf{u}_k$ and observation $\mathbf{z}_k$. This may seem, after all, a trivial idea since that posterior is the *unknown* distribution we were trying to approximate with the particle filter in the first place. Though, the distribution appears here *conditioned* on the path hypotheses, which eases its approximation for each map type as will be shown later on. About the associated weight update, embedding Equation 43 into Equation 41 leads to:

$$\begin{aligned} w^{[i]}_k &\propto w^{[i]}_{k-1} \frac{p\left(\mathbf{z}_k \middle| \mathbf{x}_k, \mathbf{x}^{[i]}_{0:k-1}, \mathbf{z}_{0:k-1}, \mathbf{u}_{1:k}\right) p\left(\mathbf{x}_k \middle| \mathbf{x}^{[i]}_{k-1}, \mathbf{u}_k\right)}{q\left(\mathbf{x}_k \middle| \mathbf{x}^{[i]}_{0:k-1}, \mathbf{z}_{0:k}, \mathbf{u}_{1:k}\right)} \\ &= w^{[i]}_{k-1} \mu_i \frac{p\left(\mathbf{z}_k \middle| \mathbf{x}_k, \mathbf{x}^{[i]}_{0:k-1}, \mathbf{z}_{0:k-1}, \mathbf{u}_{1:k}\right) p\left(\mathbf{x}_k \middle| \mathbf{x}^{[i]}_{k-1}, \mathbf{u}_k\right)}{p\left(\mathbf{z}_k \middle| \mathbf{x}_k, \mathbf{x}^{[i]}_{0:k-1}, \mathbf{z}_{0:k-1}, \mathbf{u}_{1:k}\right) p\left(\mathbf{x}_k \middle| \mathbf{x}^{[i]}_{k-1}, \mathbf{u}_k\right)} \\ &= w^{[i]}_{k-1} \mu_i \\ &= w^{[i]}_{k-1} p\left(\mathbf{z}_k \middle| \mathbf{x}^{[i]}_{0:k-1}, \mathbf{z}_{0:k-1}, \mathbf{u}_{1:k}\right) \end{aligned} \tag{44}$$

that is, the weights are scaled by the term μ_i, which represents the *expectation* of the observation likelihood for $\mathbf{z}_k$ over all potential robot poses given only the action $\mathbf{u}_k$; notice the missing $\mathbf{x}_k$ among the conditioning variables. For the convenience of the reader, the complete approach has been summarized in Algorithm 3 in spite of the only differences with Algorithm 2 being the drawing of poses and the weight update formula. Next, we explore how to apply this optimal sampling scheme in practice within a RBPF-based SLAM framework.

Optimal Proposal Distribution with Landmark Maps ("FastSLAM 2.0")

In the robotics literature, the idea of applying the optimal sampling distribution to RBPF SLAM with landmarks was firstly introduced under the name *FastSLAM 2.0* by M. Montemerlo in (Montemerlo, Thrun, Koller, & Wegbreit, 2003), and further developed in his PhD thesis (Montemerlo, 2003).

This method relies on an analysis of the two terms in the product of Equation 43: the motion model and the observation likelihood, from whose product we must be able to draw random samples. In chapter 5, we discussed how the former can be approximated as a unimodal Gaussian, thus we can write:

$$p\left(\mathbf{x}_k \middle| \mathbf{x}^{[i]}_{k-1}, \mathbf{u}_k\right) = N\left(\mathbf{x}_k; \bar{\mathbf{x}}_{k|k-1}, \mathbf{\Sigma}_{\mathbf{x}_{k|k-1}}\right) \tag{45}$$

Algorithm 3. The operations involved in RBPF-based SLAM when applying the optimal proposal distribution

algorithm RBPF_SLAM_OPTIMAL

Inputs: $\left\{\left(\mathbf{x}_{0:k-1}^{[i]}, w_{k-1}^{[i]}\right), \mathbf{m}^{[i]}\right\}_{i=1}^{N}, \mathbf{u}_k, \mathbf{z}_k$

Outputs: $\left\{\left(\mathbf{x}_{0:k}^{[i]}, w_{k}^{[i]}\right), \mathbf{m}^{[i]}\right\}_{i=1}^{N}$

1 Robot path Bayesian estimate (SIR algorithm)

- 1.1 For each particle index $i \in \{1,\dots,N\}$
  - 1.1.1 Draw pose from proposal: $\mathbf{x}_k^{[i]} \sim p\left(\mathbf{x}_k \middle| \mathbf{x}_{0:k-1}^{[i]}, \mathbf{z}_{1:k}, \mathbf{u}_{1:k}\right)$
  - 1.1.2 Data association (if applicable): $\boldsymbol{\alpha}_k \leftarrow data\_assoc\left(\mathbf{z}_k, \mathbf{m}^{[i]}, \mathbf{x}_k^{[i]}\right)$
  - 1.1.3 Weight update: $\hat{w}_k^{[i]} \leftarrow w_{k-1}^{[i]}\ p\left(\mathbf{z}_k\left(\boldsymbol{\alpha}_k\right) \middle| \mathbf{x}_{0:k-1}^{[i]}, \mathbf{z}_{0:k-1}, \mathbf{u}_{1:k}\right)$
  - 1.1.4 Augment robot path history: $\mathbf{x}_{0:k}^{[i]} \leftarrow \left\{\mathbf{x}_{0:k-1}^{[i]}, \mathbf{x}_k^{[i]}\right\}$
- 1.2 Normalize weights: $w_k^{[i]} \leftarrow \dfrac{\hat{w}_k^{[i]}}{\sum_{i=1\dots M} \hat{w}_k^{[i]}}$
- 1.3 if $ESS\left(\left\{w_k^{[i]}\right\}_{i=1}^{N}\right) < \frac{1}{2}N$
  - 1.3.1 Perform resampling: $\left\{\left(\mathbf{x}_{0:k-1}^{[i]}, w_{k-1}^{[i]}\right), \mathbf{m}^{[i]}\right\}_{i=1}^{N} \leftarrow resample\left\{\left(\mathbf{x}_{0:k-1}^{[i]}, w_{k-1}^{[i]}\right), \mathbf{m}^{[i]}\right\}_{i=1}^{N}$

2 Map Bayesian estimate

- 2.1 For each particle index $i \in \{1,\dots,N\}$
  - 2.1.1 For each map element $j \in \{1,\dots,L\}$
    - 2.1.1.1 Bayesian update: $\mathbf{m}_j^{[i]} \leftarrow map\_update\left(\mathbf{m}_j^{[i]}, \mathbf{z}_k\left(\boldsymbol{\alpha}_k\right)\right)$

Regarding the observation likelihood $p\left(\mathbf{z}_k \middle| \mathbf{x}_k, \mathbf{m}^{[i]}\right)$, it can be always approximated as some sort of parameterized distribution (e.g. a unimodal Gaussian or a sum of Gaussians) when dealing with maps of landmarks. Keep in mind that the free variable in that likelihood is the robot pose $\mathbf{x}_k$, that is, its correct interpretation is that of the probability density of the robot being at each possible pose, given a particular sensor reading and a known distribution for the map.

This observation likelihood depends on the parameterization of landmarks and the observation type. Focusing on the common case of the range-bearing type of observations and being mapped landmarks directly parameterized by their Cartesian coordinates, a reasonable approximation of the observation likelihood as a Gaussian is possible. By factoring the observation likelihood into the product of its independent terms, we can rewrite the optimal proposal as:

$$
\begin{aligned}
& q_{\mathrm{opt}}\left(\mathbf{x}_k \middle| \mathbf{x}_{0:k-1}^{[i]}, \mathbf{z}_{0:k}, \mathbf{u}_{1:k}\right) = \\
& = p\left(\mathbf{x}_k \middle| \mathbf{x}_{0:k-1}^{[i]}, \mathbf{z}_{0:k}, \mathbf{u}_{1:k}\right) \propto \\
& \propto p\left(\mathbf{x}_k \middle| \mathbf{x}_{k-1}^{[i]}, \mathbf{u}_k\right) p\left(\mathbf{z}_k \middle| \mathbf{x}_k, \mathbf{x}_{0:k-1}^{[i]}, \mathbf{z}_{0:k-1}, \mathbf{u}_{1:k}\right) =
\end{aligned}
$$

(since $\mathbf{z}_{k,a} \perp \mathbf{z}_{k,b} \middle| \mathbf{x}_k$ from an analysis of the problem DBN)

$$
\begin{aligned}
& = \underbrace{p\left(\mathbf{x}_k \middle| \mathbf{x}_{k-1}^{[i]}, \mathbf{u}_k\right)}_{\substack{=N\left(\mathbf{x}_k;\bar{\mathbf{x}}_{k|k-1}, \Sigma_{\mathbf{x}_{k|k-1}}\right) \\ \text{Motion model}}} \prod_{j=1\dots M} \underbrace{p\left(\mathbf{z}_{k,j} \middle| \mathbf{x}_k, \mathbf{x}_{0:k-1}^{[i]}, \mathbf{z}_{0:k-1}, \mathbf{u}_{1:k}\right)}_{\substack{\text{Observation likelihood for each} \\ \text{of the } M \text{ observed landmarks}}} \cong \\
& \cong N\left(\mathbf{x}_k; \bar{\mathbf{x}}_k, \boldsymbol{\Sigma}_{\mathbf{x}_k}\right)
\end{aligned} \tag{46}
$$

where the first term in the product is a Gaussian obtained from the motion model. Notice that, for range-bearing sensors, the observation likelihood for each individual landmark peaks at all the areas of the robot pose state space from which a given range-bearing observation compatible with the readings could most likely have been taken. Those poses constitute a "ring-shaped" area around the landmark, with a different robot heading at each point along the "ring." The product of several such "ring-shaped" distributions (one per observed landmark) and the prediction from the observation model will be approximated as a unimodal Gaussian $N\left(\mathbf{x}_k; \overline{\mathbf{x}}_k, \Sigma_{\mathbf{x}_k}\right)$ in order to be able to easily draw random samples.

Following the notation of chapter 8 section 4, let each observation comprise a set of measurements of individual landmarks such that $\mathbf{z}_k = \left\{\mathbf{z}_{k,i_1}, \mathbf{z}_{k,i_2}, \ldots\right\}$ with $\left\{i_1, i_2, \ldots\right\}$ standing for the indices of the corresponding landmark already existing in the map, i.e. the result from data association, and let the additive zero-mean Gaussian noise of observations be $\mathbf{n}_{k,i} \sim N\left(\mathbf{z}_{j,i} \middle| \mathbf{0}, \mathbf{R}\right)$. Recall as well from Equation 39 that in a RBPF every landmark is uncorrelated to each other, thus the reconstructed map for the $i-th$ particle is factored as the product of L independent Gaussians:

$$bel\left(\mathbf{m}^{[i]} \middle| \mathbf{x}_{0:k}, \mathbf{z}_{0:k}\right) = \prod_{j=1\ldots L} bel\left(\mathbf{m}_j^{[i]} \middle| \mathbf{x}_{0:k}^{[i]}, \mathbf{z}_{0:k}\right) = \prod_{j=1\ldots L} N\left(\mathbf{m}_j; \overline{\mathbf{m}}_j, \boldsymbol{\Sigma}_{\mathbf{m}_j}\right) \quad (47)$$

We can now come back to the target distribution in Equation 46, whose negative logarithm will be considered for convenience in order to convert the product of a number of Gaussians into $r\left(\mathbf{x}_k\right)$, the sum of residuals, which has a quadratic form:

denoting a quadratic form $\mathbf{x}^T \boldsymbol{\Sigma}^{-1} \mathbf{x}$ as $\left\|\mathbf{x}\right\|_{\boldsymbol{\Sigma}}^2$

$$r\left(\mathbf{x}_k\right) \triangleq -\log q_{\text{opt}}\left(\mathbf{x}_k \middle| \mathbf{x}_{0:k-1}^{[i]}, \mathbf{z}_{0:k}, \mathbf{u}_{1:k}\right) = \underbrace{\eta}_{\text{Constant}} + \frac{1}{2}\left\|\mathbf{x}_k - \overline{\mathbf{x}}_{k|k-1}\right\|_{\boldsymbol{\Sigma}_{\mathbf{x}_{k|k-1}}}^2 + \sum_{j=1\ldots M} \frac{1}{2}\left\|h\left(\mathbf{x}_k, \overline{\mathbf{m}}_j\right) - \mathbf{z}_{k,j}\right\|_{\mathbf{S}_{k,j}}^2 \quad (48)$$

with the covariance $\mathbf{S}_{k,j}$ being the projection of the uncertainties in the landmarks plus the corresponding measurement noise into the observation space. Then, the sought mean $\overline{\mathbf{x}}_k$ of the optimal proposal coincides with the robot pose that minimizes the expression above, and its covariance $\boldsymbol{\Sigma}_{\mathbf{x}_k}$ is obtained as the inverse of the Hessian matrix, taking derivatives with respect to the robot pose. Replacing the non-linear function $h\left(\cdot\right)$, which was defined in chapter 8 section 4, by a first-order Taylor series approximation (see Appendix E):

$$h\left(\mathbf{x}_k, \overline{\mathbf{m}}_j\right) \cong \hat{\mathbf{z}}_{k,j} + \mathbf{H}_{k,j}\left(\mathbf{x}_k - \overline{\mathbf{x}}_{k|k-1}\right), \quad \text{with: } \begin{cases} \hat{\mathbf{z}}_{k,j} = h\left(\overline{\mathbf{x}}_{k|k-1}, \overline{\mathbf{m}}_j\right) \\ \mathbf{H}_{k,j} = \left.\dfrac{\partial h}{\partial \mathbf{x}_k}\right|_{\mathbf{x}_k = \overline{\mathbf{x}}_{k|k-1}, \mathbf{m}_j = \overline{\mathbf{m}}_j} \end{cases} \quad (49)$$

we arrive at a linearized version of the quadratic residuals:

$$r\left(\mathbf{x}_k\right) = \eta + \frac{1}{2}\left\|\mathbf{x}_k - \overline{\mathbf{x}}_{k|k-1}\right\|_{\boldsymbol{\Sigma}_{\mathbf{x}_{k|k-1}}}^2 + \sum_{j=1\ldots M} \frac{1}{2}\left\|\hat{\mathbf{z}}_{k,j} + \mathbf{H}_{k,j}\left(\mathbf{x}_k - \overline{\mathbf{x}}_{k|k-1}\right) - \mathbf{z}_{k,j}\right\|_{\mathbf{S}_{k,j}}^2 \quad (50)$$

for which the robot pose that minimizes the sum can be obtained by identifying the first derivative with zero:

Given that $\left\|\mathbf{x}\right\|_{\Sigma}^{2} = \mathbf{x}^{T}\mathbf{\Sigma}^{-1}\mathbf{x}$ and $\frac{\partial \mathbf{x}^{T}\mathbf{\Sigma}\mathbf{x}}{\partial \mathbf{x}} = \left(\mathbf{\Sigma} + \mathbf{\Sigma}^{T}\right)\mathbf{x} = 2\mathbf{\Sigma}\mathbf{x}$,

$$\begin{aligned}
&\frac{\partial r(\mathbf{x}_k)}{\partial \mathbf{x}_k} = \mathbf{\Sigma}_{\mathbf{x}_{k|k-1}}^{-1}\left(\mathbf{x}_k - \overline{\mathbf{x}}_{k|k-1}\right) + \\
&+ \sum_{j=1\ldots M} \mathbf{H}_{k,j}^{T}\mathbf{S}_{k,j}^{-1}\left(\mathbf{z}_{k,j} - \hat{\mathbf{z}}_{k,j} - \mathbf{H}_{k,j}\mathbf{x}_k + \mathbf{H}_{k,j}\overline{\mathbf{x}}_{k|k-1}\right) = 0 \\
&\underbrace{\left[\mathbf{\Sigma}_{\mathbf{x}_{k|k-1}}^{-1} - \sum_{j=1\ldots M} \mathbf{H}_{k,j}^{T}\mathbf{S}_{k,j}^{-1}\mathbf{H}_{k,j}\right]}_{\mathbf{P}_k}\mathbf{x}_k - \mathbf{\Sigma}_{\mathbf{x}_{k|k-1}}^{-1}\overline{\mathbf{x}}_{k|k-1} + \\
&+ \sum_{j=1\ldots M} \mathbf{H}_{k,j}^{T}\mathbf{S}_{k,j}^{-1}\left(\mathbf{z}_{k,j} - \hat{\mathbf{z}}_{k,j} + \mathbf{H}_{k,j}\overline{\mathbf{x}}_{k|k-1}\right) = 0 \\
&\rightarrow \mathbf{x}_k = \mathbf{P}_k^{-1}\left[\mathbf{\Sigma}_{\mathbf{x}_{k|k-1}}^{-1}\overline{\mathbf{x}}_{k|k-1} - \sum_{j=1\ldots M} \mathbf{H}_{k,j}^{T}\mathbf{S}_{k,j}^{-1}\left(\mathbf{z}_{k,j} - \hat{\mathbf{z}}_{k,j} + \mathbf{H}_{k,j}\overline{\mathbf{x}}_{k|k-1}\right)\right]
\end{aligned} \tag{51}$$

Notice that the matrix $\mathbf{P}_k$ defined above coincides with the Hessian of the observation, which can be verified by taking derivatives again with respect to the robot pose:

$$\begin{aligned}
\frac{\partial^2 r(\mathbf{x}_k)}{\partial^2 \mathbf{x}_k} &= \frac{\partial}{\partial \mathbf{x}_k}\Bigg[\mathbf{P}_k\mathbf{x}_k - \Sigma_{\mathbf{x}_{k|k-1}}^{-1}\overline{\mathbf{x}}_{k|k-1} + \\
&\quad + \sum_{j=1\ldots M} \mathbf{H}_{k,j}^{T}\mathbf{S}_{k,j}^{-1}\left(\mathbf{z}_{k,j} - \hat{\mathbf{z}}_{k,j} + \mathbf{H}_{k,j}\overline{\mathbf{x}}_{k|k-1}\right)\Bigg] = \\
&= \mathbf{P}_k
\end{aligned} \tag{52}$$

and, consequently, its inverse represents a good approximation to the covariance of the optimal proposal distribution (we defer the demonstration of this end until chapter 10 section 4):

$$\begin{aligned}
&q_{\text{opt}}\left(\mathbf{x}_k \middle| \mathbf{x}_{1:k-1}^{[i]}, \mathbf{z}_{1:k}, \mathbf{u}_{1:k}\right) = \\
&= N\left(\mathbf{x}_k; \overline{\mathbf{x}}_k, \Sigma_{\mathbf{x}_k}\right) \rightarrow \begin{cases} \overline{\mathbf{x}}_k \text{ given in eq. 51} \\ \Sigma_{\mathbf{x}_k} = \mathbf{P}_k^{-1} \end{cases}
\end{aligned} \tag{53}$$

Although this linearized solution was the original proposal introduced with FastSLAM 2.0 in (Montemerlo, 2003), the non-linear nature of the observation function $h(\cdot)$ would suggest the application of iterative non-linear least-squares methods like the Gauss-Newton algorithm (described in chapter 10) in order to further refine the mean of the proposal distribution.

To sum up, it turns out that the optimal proposal $q_{opt}(\cdot)$ for each particle becomes a unimodal Gaussian in the robot pose state space, from which we can easily draw a random sample—see the corresponding algorithm in Appendix C. By definition, the usage of the optimal proposal assures that the so-generated particle will be located in a region of the state space of the highest interest, i.e., that is simultaneously compatible with the odometry and with all the observations, and therefore the reconstructed path will be very close to the most-likely one.

Concerning the update of the importance weights, recall from Equation 44 that weights must be multiplied by the expectation of the observation likelihood over all potential robot poses given only the action $\mathbf{u}_k$:

$$\begin{aligned}
&w_k^{[i]} \propto w_{k-1}^{[i]} p\left(\mathbf{z}_k \middle| \mathbf{x}_{0:k-1}^{[i]}, \mathbf{z}_{0:k-1}, \mathbf{u}_{1:k}\right) = \\
&= w_{k-1}^{[i]} \int_{-\infty}^{\infty} \underbrace{p\left(\mathbf{z}_k \middle| \mathbf{x}_k, \mathbf{x}_{0:k-1}^{[i]}, \mathbf{z}_{0:k-1}, \mathbf{u}_{1:k}\right)}_{p\left(\mathbf{z}_k|\mathbf{x}_k,\mathbf{m}^{[i]}\right)} \underbrace{p\left(\mathbf{x}_k \middle| \mathbf{x}_{k-1}^{[i]}, \mathbf{u}_k\right)}_{\text{Motion model}} d\mathbf{x}_k
\end{aligned} \tag{54}$$

In order to determine this value, the predicted observation can be approximated as the Gaussian $\mathbf{z}_{k,j} \sim N\left(\overline{\mathbf{z}}_{k,j}, \mathbf{\Sigma}_{\mathbf{z}_{k,j}}\right)$, whose parameters can be determined, as usual, by linearizing the sensor model $\mathbf{z}_{k,j} = \mathbf{h}\left(\mathbf{m}_j, \mathbf{x}_{k|k-1}\right) + \mathbf{n}_{k,j}$ and projecting the uncertainty from the mapped landmarks and the pose prediction of Equation 45. Then, weight updates simply become the evaluation of the real measurement within the predicted Gaussians, that is, $w_k^{[i]} \propto w_{k-1}^{[i]} N\left(\mathbf{z}_{k,j}; \hat{\mathbf{z}}_{k,j}, \mathbf{S}_{k,j}\right)$. Note that the proportionality is due to the weight renormalization to be done once all particles in the filter have been updated—step 1.2 in Algorithm 3.

Optimal Proposal Distribution with Other Maps

When working with maps of landmarks, the existence of explicit data associations between observed features and those in the map provides us with a solid foundation for approximating the optimal proposal distribution as a parametric distribution starting from the Gaussian model of the sensor. However, not all map representations allow such a straight approach.

When constructing point maps or occupancy grids, the observations (typically scans from a laser range finder) do not have any well-defined correspondence with the map elements. For the specific case of a range scan observation, notice that each sensed point may not coincide exactly with any other previous point while the robot is moving. That is why this kind of maps does not provide any closed-form parametric observation likelihood. In turn, several approximations exist that provide pointwise-evaluable observation likelihoods $p\left(\mathbf{z}_k \middle| \mathbf{x}_k, \mathbf{x}_{1:k-1}^{[i]}, \mathbf{z}_{1:k-1}, \mathbf{u}_{1:k}\right)$, such as the beam model for grid maps already discussed in chapter 6 section 2—other alternatives exist, such as (Thrun, Burgard, & Fox, 2005; Plagemann, Kersting, Pfaff, & Burgard, 2007).

Let us assume therefore the existence of a pointwise-evaluable observation likelihood and a pointwise-evaluable motion models, both available for any arbitrary robot pose $\mathbf{x}_k$. Then, we will show how to approximate the optimal proposal distribution as a Gaussian from which to draw samples using a technique described in (Grisetti, Stachniss, & Burgard, 2007). In that work, the authors propose to firstly obtain a guess of the most likely robot pose $\mathbf{x}_k^*$ via a local optimizer, which for the case of working with laser scanners would be an ICP point cloud matcher, an algorithm already described in chapter 6 (Besl & McKay, 1992; Censi, 2008). Then, a fixed number of K auxiliary poses are drawn in the vicinity of the local optimum (within some heuristic radius Δ), at which the motion and observation models are evaluated pointwise:

(applying a local optimizer)

$$\mathbf{x}_k^* = \arg\max_{\mathbf{x}_k} p\left(\mathbf{z}_k \middle| \mathbf{x}_k, \mathbf{x}_{1:k-1}^{[i]}, \mathbf{z}_{1:k-1}, \mathbf{u}_{1:k}\right)$$

(drawing values within a vicinity of the optimum)

$$\tilde{\mathbf{x}}_j \sim \left\{\mathbf{x}_k \middle| \left|\mathbf{x}_k - \mathbf{x}_k^*\right| < \Delta\right\}, \text{ for } j = 1...K$$

(evaluating pointwise the product of the motion and the observation models)

$$\tilde{\omega}_j = p\left(\tilde{\mathbf{x}}_j \middle| \mathbf{x}_{k-1}^{[i]}, \mathbf{u}_k\right) p\left(\mathbf{z}_k \middle| \mathbf{x}_k^*, \mathbf{m}^{[i]}\right) \tag{55}$$

The optimal proposal can then be approximated by the unimodal Gaussian fit by those weighted values, that is:

$$\begin{aligned} q_{\text{opt}}\left(\mathbf{x}_k \middle| \mathbf{x}_{0:k-1}^{[i]}, \mathbf{z}_{0:k}, \mathbf{u}_{1:k}\right) &= p\left(\mathbf{x}_k \middle| \mathbf{x}_{0:k-1}^{[i]}, \mathbf{z}_{0:k}, \mathbf{u}_{1:k}\right) \propto \\ &\propto p\left(\mathbf{x}_k \middle| \mathbf{x}_{k-1}^{[i]}, \mathbf{u}_k\right) p\left(\mathbf{z}_k \middle| \mathbf{x}_k, \mathbf{x}_{0:k-1}^{[i]}, \mathbf{z}_{0:k-1}, \mathbf{u}_{1:k}\right) \propto \\ &\propto N\left(\mathbf{x}_k; \overline{\mathbf{x}}_k, \Sigma_k\right) \end{aligned}$$

$$\rightarrow \begin{cases} \overline{\mathbf{x}}_k = \dfrac{1}{\mu_i} \sum_{j=1...K} \tilde{\omega}_j \tilde{\mathbf{x}}_j \\ \mathbf{\Sigma}_k = \dfrac{1}{\mu_i} \sum_{j=1...K} \tilde{\omega}_j \left(\tilde{\mathbf{x}}_j - \overline{\mathbf{x}}_k\right)\left(\tilde{\mathbf{x}}_j - \overline{\mathbf{x}}_k\right)^T \end{cases}$$

with the normalization term defined as:

$$\mu_i = \sum_{j=1...K} \tilde{\omega}_j \tag{56}$$

Regarding the weight update, it turns out that the normalization term μ_i above coincides with the factor by which each particle weight must be multiplied, thus:

$$
\begin{aligned}
w_k^{[i]} &\propto w_{k-1}^{[i]} p\left(\mathbf{z}_k \middle| \mathbf{x}_{0:k-1}^{[i]}, \mathbf{z}_{0:k-1}, \mathbf{u}_{1:k}\right) = \\
&= w_{k-1}^{[i]} \int_{-\infty}^{\infty} \underbrace{p\left(\mathbf{z}_k \middle| \mathbf{x}_k, \mathbf{x}_{0:k-1}^{[i]}, \mathbf{z}_{0:k-1}, \mathbf{u}_{1:k}\right)}_{p\left(\mathbf{z}_k \middle| \mathbf{x}_k, \mathbf{m}^{[i]}\right)} \underbrace{p\left(\mathbf{x}_k \middle| \mathbf{x}_{k-1}^{[i]}, \mathbf{u}_k\right)}_{\text{Motion model}} d\mathbf{x}_k \cong \\
&\cong w_{k-1}^{[i]} \underbrace{\sum_{j=1\ldots K} \overbrace{p\left(\mathbf{z}_k \middle| \tilde{\mathbf{x}}_j, \mathbf{m}^{[i]}\right) p\left(\tilde{\mathbf{x}}_j \middle| \mathbf{x}_{k-1}^{[i]}, \mathbf{u}_k\right)}^{\tilde{\omega}_j}}_{\mu_i} = w_{k-1}^{[i]}\ \mu_i
\end{aligned} \tag{57}
$$

When using precise sensors (e.g. laser scanners), one can also drop the contribution of the motion model to the proposal without incurring in a significant loss, since the part of the state space where the observation likelihood peaks is much smaller than the extension where the motion model pdf has non-negligible values, and therefore the latter is considerably flat (approximately constant) in the intersection with the observation likelihood. A method following this idea was proposed in Grisetti, Stachniss, and Burgard (2005).

In the cases where a local optimizer is not a practical solution (which depends on the kind of sensor employed) or where the motion model easily allows drawing samples but not doing pointwise evaluation, a different method must be employed. In particular, the authors proposed in (Blanco, González, & Fernandez-Madrigal, 2010) a universal solution to approximate the optimal proposal distribution for any kind of motion and sensor model by means of an adaptive sample size and rejection sampling. As a result, we get a method whose accuracy can be arbitrarily improved as much as desired at the cost of increasing the computation time.

Finally, it is worth mentioning that the idea of using the proposal distribution for RBPF filtering has been extended by other authors by considering not just the last robot pose at each time step, but a window of the N most recent robot poses simultaneously. This is called fixed-lag "roughening" and is based on the optimal block proposal distribution. Interested readers may refer to existing literature for further details (Beevers & Huang, 2007; Doucet, Briers, & Sénécal, 2006).

REFERENCES

Beevers, K. R., & Huang, W. H. (2007). Fixed-lag sampling strategies for particle filtering SLAM. In *Proceedings of the IEEE International Conference on Robotics and Automation*, (pp. 2433-2438). IEEE Press.

Bell, B. M., & Cathey, F. W. (1993). The iterated Kalman filter update as a Gauss-Newton method. *IEEE Transactions on Automatic Control*, *38*(2), 294–297. doi:10.1109/9.250476

Besl, P. J., & McKay, N. D. (1992). A method for registration of 3D shapes. *IEEE Transactions on Pattern Analysis and Machine Intelligence*, *14*(2), 239–256. doi:10.1109/34.121791

Bierman, G. J. (1977). *Factorization methods for discrete sequential estimation*. New York, NY: Academic Press.

Blanco, J. L., González, J., & Fernández-Madrigal, J. A. (2010). Optimal filtering for non-parametric observation models: Applications to localization and SLAM. *The International Journal of Robotics Research*, *29*(14), 1726–1742. doi:10.1177/0278364910364165

Blanco, J. L., González, J., & Fernández-Madrigal, J. A. (2012). An alternative to the Mahalanobis distance for determining optimal correspondences in data association. *IEEE Transactions on Robotics*.

Brown, D. C. (1958). *A solution to the general problem of multiple station analytical stereotriangulation.* Technical Report RCA-MTP Data Reduction Technical Report No. 43 (or AFMTC TR 58-8). Patrick Airforce Base, FL: Patrick Airforce Base.

Censi, A. (2008). An ICP variant using a point-to-line metric. In *Proceedings of the IEEE International Conference on Robotics and Automation (ICRA)*. IEEE Press.

Davison, A. J., Reid, I. D., Molton, N. D., & Stasse, O. (2007). MonoSLAM: Real-time single camera SLAM. *IEEE Transactions on Pattern Analysis and Machine Intelligence, 29*(6), 1052–1067. doi:10.1109/TPAMI.2007.1049

Dellaert, F., & Kaess, M. (2006). Square root SAM: Simultaneous localization and mapping via square root information smoothing. *The International Journal of Robotics Research, 25*(12), 1181–1203. doi:10.1177/0278364906072768

Dennis, J. E., & Schnabel, R. B. (1996). *Numerical methods for unconstrained optimization and non-linear equations*. New York, NY: Society for Industrial Mathematics. doi:10.1137/1.9781611971200

Dissanayake, M. W. M. G., Newman, P., Clark, S., Durrant-Whyte, H. F., & Csorba, M. (2001). A solution to the simultaneous localization and map building (SLAM) problem. *IEEE Transactions on Robotics and Automation, 17*(3), 229–241. doi:10.1109/70.938381

Doucet, A., Briers, M., & Sénécal, S. (2006). Efficient block sampling strategies for sequential Monte Carlo methods. *Journal of Computational and Graphical Statistics, 15*(3), 693–711. doi:10.1198/106186006X142744

Doucet, A., De Freitas, N., Murphy, K., & Russell, S. (2000). Rao-Blackwellised particle filtering for dynamic Bayesian networks. In *Proceedings of the Sixteenth Conference on Uncertainty in Artificial Intelligence*, (pp. 176-183). IEEE.

Doucet, A., Godsill, S., & Andrieu, C. (2000). On sequential Monte Carlo sampling methods for Bayesian filtering. *Statistics and Computing, 10*(3), 197–208. doi:10.1023/A:1008935410038

Duckett, T., Marsland, S., & Shapiro, J. (2002). Fast, on-line learning of globally consistent maps. *Autonomous Robots, 12*(3), 287–300. doi:10.1023/A:1015269615729

Durrant-Whyte, H. F. (1988). Uncertain geometry in robotics. *IEEE Journal on Robotics and Automation, 4*(1), 23–31. doi:10.1109/56.768

Forsyth, D. A., Ioffe, S., & Haddon, J. (1999). Bayesian structure from motion. In *Proceedings of the IEEE International Conference of Computer Vision*, (pp. 660-665). IEEE Press.

Frese, U., Larsson, P., & Duckett, T. (2005). A multilevel relaxation algorithm for simultaneous localisation and mapping. *IEEE Transactions on Robotics, 21*(2), 1–12. doi:10.1109/TRO.2004.839220

Golub, G. H., & Plemmons, R. J. (1980). Large-scale geodetic least-squares adjustment by dissection and orthogonal decomposition. *Linear Algebra and Its Applications, 34*, 3–28. doi:10.1016/0024-3795(80)90156-1

Grisetti, G., Stachniss, C., & Burgard, W. (2005). Improving grid-based slam with Rao-Blackwellized particle filters by adaptive proposals and selective resampling. In *Proceedings of the 2005 IEEE International Conference on Robotics and Automation*, (pp. 2432-2437). IEEE Press.

Grisetti, G., Stachniss, C., & Burgard, W. (2007). Improved techniques for grid mapping with Rao-Blackwellized particle filters. *IEEE Transactions on Robotics, 23*(1), 34–46. doi:10.1109/TRO.2006.889486

Hartley, R. I., & Zisserman, A. (2004). *Multiple view geometry in computer vision*. Cambridge, UK: Cambridge University Press. doi:10.1017/CBO9780511811685

Jazwinsky, A. M. (1970). *Stochastic processes and filtering theory*. New York, NY: Academic Press.

Montemerlo, M. (2003). *FastSLAM: A factored solution to the simultaneous localization and mapping problem with unknown data association*. (Doctoral Dissertation). Carnegie Mellon University. Pittsburgh, PA.

Montemerlo, M., Thrun, S., Koller, D., & Wegbreit, B. (2003). FastSLAM 2.0: An improved particle filtering algorithm for simultaneous localization and mapping that provably converges. In *Proceedings of the International Joint Conference on Artificial Intelligence*, (vol 18), (pp. 1151-1156). IEEE.

Paz, L. M., Tardós, J. D., & Neira, J. (2008). Divide and conquer: EKF SLAM in O(n). *IEEE Transactions on Robotics*, *24*(5), 1107–1120. doi:10.1109/TRO.2008.2004639

Plagemann, C., Kersting, K., Pfaff, P., & Burgard, W. (2007). Gaussian beam processes: A nonparametric Bayesian measurement model for range finders. In *Proceedings of Robotics: Science and Systems Conference*. IEEE Press.

Russell, S., & Norvig, P. (2002). *Artificial intelligence: A modern approach*. Upper Saddle River, NJ: Prentice Hall.

Slama, C. C. (Ed.). (1980). *Manual of photogrammetry: American society of photogrammetry and remote sensing*. Falls Church, VA: American Society of Photogrammetry and Remote Sensing.

Smith, R. C., & Cheeseman, P. (1986). On the representation and estimation of spatial uncertainty. *The International Journal of Robotics Research*, *5*(4), 56–68. doi:10.1177/027836498600500404

Strasdat, H., Montiel, J. M. M., & Davison, A. J. (2010). Real-time monocular SLAM: Why filter? In *Proceedings of the IEEE International Conference on Robotics and Automation*, (pp. 2657-2664). IEEE Press.

Teukolsky, W. H., Vetterling, S. A., & Flannery, B. P. (1992). *Numerical recipes* (2nd ed). Cambridge, UK: Cambridge University Press.

Thrun, S., Burgard, W., & Fox, D. (2005). *Probabilistic robotics*. Cambridge, MA: The MIT Press.

Tomasi, C., & Kanade, T. (1992). Shape and motion from image streams under orthography: A factorization method. *International Journal of Computer Vision*, *9*(2), 137–154. doi:10.1007/BF00129684

Triggs, B., McLauchlan, P., Hartley, R., & Fitzgibbon, A. (2000). Bundle adjustment—A modern synthesis. *Vision Algorithms: Theory and Practice*, *1883*, 153–177.

Varga, M. (Ed.). (2007). *Practical image processing and computer vision*. New York, NY: Wiley.

Zaritskii, V. S., Svetnik, V. B., & Shimelevich, L. I. (1975). Monte Carlo technique in problems of optimal data processing. *Automation and Remote Control*, *12*, 95–103.

Chapter 10
Advanced SLAM Techniques

ABSTRACT

This chapter is the conclusion of the book. It is devoted to providing an overview of emerging paradigms that are appearing as outstanding the traditional approaches in scalability or efficiency, such as hierarchical sub-mapping, or hybrid metric-topological map models. Other techniques not based on Bayesian filtering, such as iterative sparse least-squares optimization (Graph-SLAM and Bundle adjustment), are also introduced due to their efficiency and increasing popularity.

CHAPTER GUIDELINE

- You will learn:
  - Why SLAM in 6D represents especial complications from a mathematical point of view.
  - How to modify classic parametric filters, like the EKF, to correctly deal with 6D robot poses.
  - A unifying view of *graph SLAM* and *Bundle Adjustment* under the perspective of graphical models.
  - What we mean with the *sparse structure* of the SLAM problem.
  - The close relation between abstract sparse algebra algorithms (mostly sparse Cholesky) and recent approaches to SLAM.
  - Least squares optimization algorithms: the Gauss-Newton and Levenberg-Marquardt methods.
  - What is the Schür complement of a matrix and why it is key to solve SLAM efficiently.
  - A broad discussion on approaches to SLAM alternative to the Bayesian recursive formulation, and potential future directions.
- Provided tools:
  - Detailed formulas for implementing (and understanding) the newest 6D SLAM algorithms.
  - Pseudo-code descriptions of most recent SLAM algorithms, including how to exploit the sparseness of the problem structure.

DOI: 10.4018/978-1-4666-2104-6.ch010

- Figures representing the most important sparse matrix operations in a visual way, for easily grasping how they work.
- A table summarizing robust kernels for least-squares optimizers.

- Relation to other chapters:
    - This final chapter presents alternative methods for situations that are especially complex, which should be put in contrast by the reader to the Bayesian recursive solutions of chapter 9.
    - Many of the detailed formulas required for 6D SLAM have been put separately in Appendix D, which the interested implementer should read.

1. INTRODUCTION

This book has explored in detail the most widely used approaches to *Bayesian* localization and SLAM in small to moderately sized scenarios, with the intention of revealing their mathematical foundations. But as the reader should have realized at this point, there exists no such thing as a *perfect* or "*magic*" approach to localization or SLAM that works in all cases for any kind of sensor and operation conditions. That limitation is more patent in SLAM than in localization, due to the more complex nature of the former estimation problem. While exposing each of the SLAM algorithms described in chapter 9, we stated the advantages of each method in contrast to the rest, but also insisted in their unique drawbacks. In this chapter we will reason further about those problems and will introduce different alternatives, out of the recursive Bayesian framework, that have been proposed in the literature to mitigate them in cases where the environment of the robot or the state-space of the problems are large or particularly complex. The objective is to offer the reader a wide perspective of the most relevant ideas present in the newest research, and also to serve as a complement to the rest of the book

In order to better realize the problems with all the methods for metric SLAM exposed so far when we augment the dimension or complexity of the mathematical setting, we could imagine what would be an "ideal," "perfect" solution for enabling SLAM in mid or large-sized complex environments over extended periods of time. We certainly believe that this goal should comprise:

- **Objective 1:** Allowing any arbitrary mix of sensors, with the intention that they complement each other. Notice that this includes handling different kinds of maps simultaneously.
- **Objective 2:** Robustly detecting loop closures, paying special attention to avoiding false positives.
- **Objective 3:** Being scalable for building large-scale maps.
- **Objective 4:** Being able to process arbitrary three-dimensional trajectories for the robot and its sensors, e.g. a hand-held camera, a laser range finder attached to a robotic arm, etc.

Next follows a discussion of the first three of these ambitious goals, stressing how they have been pursued in the research community in diverse ways. The fourth objective is the only one which can be considered as solved thanks to an increasingly popular approach which, due to its need for some extra mathematical background, will be discussed in section 2. Afterwards, a set of other selected ideas and especially advanced solutions aimed to overcome the difficulties of recursive Bayesian SLAM techniques in particularly difficult scenarios will be studied in sections 3 and on.

Objective 1: Seamless Multisensory Fusion

Regarding the first objective, it is clear that our perception of the environment as humans heavily relies on visual information, which is rich enough to eclipse the contributions of all our other senses when it comes about explaining the spatial structure that surrounds us. However, state-of-the-art computer vision techniques are still far from providing computers with the capability of interpreting images as efficiently and confidently as our brains. That is why the earliest approaches to SLAM relied on sensors that allowed for a much straightforward interpretation of the environment geometry (e.g. 2D laser scanners). However, visual SLAM has gained significant popularity in recent years. It is important to recall here that different sensors demand different types of maps; for instance, a wheeled robot equipped with a sonar and a camera would require an occupancy grid and a map of landmarks used jointly for localization; also, while doing SLAM both maps should be updated simultaneously.

There exists an important practical advantage in concurrently maintaining different metric representations of the world: wherever the observations from one sensor become ambiguous (i.e. provide little localization information), there is a chance of that deficiency being compensated by the readings from the other sensor, which are contrasted to a different map. Continuing with the example above, the landmark-based map used for localization with images from a camera may render useless when the robot faces a textureless wall; in contrast, sonar ranges would then still provide a good localization estimate from the occupancy grid. Notice that multisensory fusion must not only occur for qualitatively different sensors: the only requisite for it is that the sensors observe different aspects of the same environment. A clarifying example would be a robot equipped with several laser scanners, all placed to scan horizontally but at different heights. We could build one independent grid map or point map for each height since, in a sense, each sensor is in charge of exploring a different and independent aspect of the world, namely its "occupancy" at some given height.

Contrarily to what one may expect, it is actually quite straightforward to achieve multi-sensor fusion under most localization and SLAM frameworks if we assume statistical independence between the measurements coming from different sources. Such an assumption is plausible as long as each sensor reflects a different characteristic of the environment. If we stack together all the individual observations $\mathbf{z}_k^i$ (where i stands for the sensor index) into an augmented observation vector $\mathbf{z}_k = \left\{\mathbf{z}_k^1, \mathbf{z}_k^2, \ldots\right\}$, the particle filter algorithms described in chapter 9 can still be employed by just taking two modifications into account: (1) the evaluation of the observation likelihood factorizes now into the product of the individual likelihoods (where each sensor uses its corresponding map), and (2) each of the maps associated to the particles is updated taking into account only its corresponding observation. Fusion of sensors would be also possible in SLAM methods based on EKF-like filters, as long as each map-sensor pair provides a parametric observation model.

Objective 2: Robust Detection of Loop Closures

Any solution to SLAM pretending to cope with navigation in real-world scenarios must be able to correctly handle the revisiting problem, that is, to detect loop closures. Thinking of purely metrical approaches to SLAM, whenever a robot closes a loop the SLAM estimation must undergo two changes: firstly, the robot pose (or its entire path in full SLAM) is corrected, and, secondly, the entire map must be updated, including those parts not directly visible at that moment.

Loop closure stands out as one of the biggest challenges in any SLAM framework. The difficulty of correctly detecting loops increases with the size of maps, as it does the risk of falling into false positives, especially when significant perceptual aliasing (i.e. different places looking alike) exists in the environment. Furthermore, there exists no universal solution since the way in which it must be handled is strongly coupled with the specific SLAM algorithm and with the types of sensors and maps employed. In the following we will try however to classify the most popular loop closure methods according to the kind of maps on which they apply.

In the case of metric maps of landmarks or features it turns out that no special loop closing step is required: all what is needed is to correctly establish data association between sensed features and those stored in the map. However, this task may become an extremely challenging one. Since the uncertainty in the robot pose will be much larger at the moment of revisiting than the first time the landmarks were observed and added to the map, and this kind of maps are typically addressed with EKF-based SLAM, all the observation predictions of the revisited landmarks will have a large uncertainty and will be highly cross-correlated—recall chapter 9 section 2. In this context, the usage of the JCBB algorithm (Neira & Tardós, 2001) explained in chapter 6 for data association reveals as extremely handy in order to prune the large tree of potentials pairings, simultaneously considering all of them to assure the consistency of the associations. Alternatively, other authors have proposed to find the best matching between observations and the map under the perspective of representing landmarks and their spatial relationship as a graph. Then, data association can be seen as a particular instance of the problem of searching for the maximum common subgraph isomorphism between two graphs (Chen & Yun, 1998; Bailey, Nebot, Rosenblatt, & Durrant-Whyte, 2000), i.e., to find the largest portion of two graphs that they have in common. Independently of the algorithm employed to establish the most-likely pairings, it would be unpractical and senseless to consider all potential pairings between observations and all the landmarks in the map. In turn, only a small set of *individually compatible* pairings are considered for each observed feature, taking into account the current robot pose uncertainty and the known measurement noise. Obviously, finding the correct associations will be computationally easier as the total number of individually compatible pairings gets smaller. That is the reason behind the augmentation of map landmarks with *feature descriptors*, that is, a list of properties of each observed landmark such that it becomes (ideally) distinguishable from all the others. Using descriptors drastically reduces the number of individually compatible pairings by forcing us to only consider those between landmarks that simultaneously have similar descriptors and are spatially coincident. Descriptors have been largely employed for imaging sensors (Lowe, 2004; Bay, Tuytelaars, & Van Gool, 2006) but proposals have been also reported for raw laser scans (Vázquez-Martín, Núñez, Bandera, & Sandoval, 2009; Tipaldi & Arras, 2010).

Regarding map building with occupancy grid maps, we saw in chapter 9 that the lack of a parameterized observation model for these maps forces us to approach SLAM under the form of a RBPF. Recall that, in such a particle filtering solution, the joint SLAM posterior distribution is represented by a discrete set of path hypotheses, each carrying its corresponding version of the reconstructed map. In its most common version, grid mapping with a RBPF only adds new robot poses to the end of the path hypotheses, while the old part of paths never changes (Grisetti, Stachniss, & Burgard, 2007). Therefore, the only way in which a loop closure can improve the estimate of old portions of the path is by means of adjusting the weights of the existing particles, which typically leads to a resampling stage in which those hypotheses not consistent with the observations after the loop closure are discarded. Notice that for such a RBPF, the moment of a loop closure does not imply solving anything more complex than any other iteration (in contrast to the troubles of data association in an EKF). Nevertheless, the quality of the reconstructed path in a RBPF becomes strongly related to the number of particles employed, and therefore, with the amount of computational power available.

Finally, loop closure with topological maps represents its own challenges. On the one hand, some authors (Ranganathan, Menegatti, & Dellaert, 2006) have studied how probabilistic inference can be done on such maps by only examining the connections between nodes (i.e. the existence or not of arcs, without any metrical meaning). On the other hand, we find approaches where each node in the topology is associated with sensory information, thus loop closure becomes finding other nodes with observed data that look like they were gathered at approximately the same physical place. Under this paradigm, we can find the successful family of appearance-based localization works, based on visual descriptors (Cummins & Newman, 2008a, 2008b).

Objective 3: Scalability

The goal of achieving scalability means that the cost of maintaining or updating a map should remain *reasonably bounded* as the size of the mapped areas grows. Particularly, in landmark-based SLAM it has been proposed that a reasonable limit could be a linear update cost with the overall number of landmarks (Frese, 2006b). Speaking more concretely, the bound in computational complexity should be set such that the SLAM method can be executed in real-time under the expected operation conditions. However, being computational complexity often the most limiting factor, it is not the only one. *Consistency* is another major issue with many SLAM solutions (recall this property of estimators we discussed in chapter 4): the estimation of the SLAM posterior pdf may become overconfident, meaning that the real values of the map and the robot path may get a negligible likelihood. In other words, the SLAM method may get stuck onto an incorrect estimation without being able to recover; we then say that the estimation is inconsistent with the observed data, or that the estimation has "diverged." This problem typically arises in SLAM with EKF-like filters when the global coordinates of the robot are far from the origin. In such a situation, the actual uncertainty in the robot pose has little resemblance with a Gaussian, therefore invalidating the assumptions of the most common filtering algorithms and leading to a poor accuracy. Although typically relegated to the background, other scalability issues, which should be accounted for, are the storage requirements and the numerical stability of the required mathematical operations.

In the literature, we can find different viewpoints or strategies regarding how to pursue scalability in SLAM. We can establish a first group where they consist of *enhancing* existing methods so that, provided a fixed limit in the computational burden, they can be employed to build larger (or more accurate) maps than the original algorithms. Within this group we can find the memory and computational load optimizations for grid mapping with RBPF presented in (Grisetti, Stachniss, & Burgard, 2007), an efficient "divide and conquer" filtering algorithm (Paz, Tardós, & Neira, 2008) or the improved sparse solution to SLAM with information filters reported in (Walter, Eustice, & Leonard, 2007).

A second family of methods is constituted by those that address SLAM under a radically different perspective than the one presented for the SLAM filtering solutions explained in chapter 9. In this regard, two of the most influential lines of work are (1) approaching SLAM as a non-linear least-squares batch optimization problem, and (2) the proposal of hierarchical map models. Due to their importance in the newest research on very large environments or especially complex state-spaces, these two lines will be studied in sections 3 to 5, in spite of not fitting directly within the recursive Bayesian framework studied in the rest of this book. But before that we examine more in depth the particular issues related to the topology of the state-space.

2. ESTIMATION AS AN OPTIMIZATION PROBLEM: THE TOPOLOGY OF THE STATE-SPACE

Until very recently, the mobile robot community had not paid all the required attention to the mathematical issues related to the state space topology of the different estimation problems, mainly due to the good behavior of the methods explained in the rest of this book when applied to small to moderately sized planar environments. In this section, we provide an overview of the mathematical bases required to augment existing estimators to work with more complex spaces, as three-dimensional robot poses or landmark parameterizations that include spatial rotations.

Background

While performing localization with wheeled robots working on planar surfaces we can parameterize their pose as a three-component vector comprising the two-dimensional position plus a heading angle, as it has been seen in plenty of examples throughout this book. Properly speaking, such robot poses belong to a state-space with topological structure $\mathbb{R}^2 \times \mathbb{S}^1$, where $\mathbb{R}^D$ and $\mathbb{S}^N$ stand for the topology of a $D-$dimensional Euclidean space and an $N-$dimensional sphere, respectively. In particular, $\mathbb{S}^1$ represents the topology of a one-dimensional sphere, i.e., a circumference: orientations at 180 and -180 degrees actually stand for *exactly* the same point.

In the case of solving SLAM, this state-space also includes the parameterization of landmarks, which, in most cases, simply consists of their Euclidean coordinates. This gives us a SLAM problem with a topology $\mathbb{R}^2 \times \mathbb{S}^1 \times \mathbb{R}^{2L}$ for a bidimensional map of L landmarks. Note that other parameterizations such as the inverse depth approach described in chapter 8 section 4 may lead to more complex topologies. As will become evident below, such a set up has allowed us to work on bidimensional SLAM without concerning at all about topologies of the state-space. In contrast, they become an important issue when dealing with robots moving and rotating in three-dimensional space.

The relevance of the topology of a problem state-space comes from the fact that gradient descent, Gauss-Newton, Levenberg-Marquardt and the entire family of EKF-like algorithms share an important point at their core: they are non-linear optimization algorithms, that is, they all iteratively improve a state vector estimate $\mathbf{x}$ so that it minimizes a sum of (possibly covariance-weighted) square errors between some prediction and a vector of observed data $\mathbf{z}$ —notice that we did not mention here particle filters, for which all the following discussion does not apply. Independently of the algorithm employed to solve the estimation problem, what matters for the present discussion is that at some stage of the optimization, a prediction model $\mathbf{h}(\mathbf{x})$ (e.g. the robot motion and observation models) is used with the goal of minimizing the mismatch between observed data and the predictions of the current estimation. It can be shown that all the above-mentioned algorithms become a *least-squares minimization* over some target function which, in a generalized form, reads like this:

$$\mathrm{F}(\mathbf{x}) = \left(\mathbf{h}(\mathbf{x}) - \mathbf{z}\right)^T \left(\mathbf{h}(\mathbf{x}) - \mathbf{z}\right) = \underbrace{\mathbf{r}^T\mathbf{r}}_{\text{Square of residuals}} \tag{1}$$

In order to find the optimal $\mathbf{x}$ that minimizes this function, an initial value (e.g. the mean of the Gaussian distribution within an EKF) is updated iteratively by means of *small* increments:

$$\mathbf{x}_k \leftarrow \mathbf{x}_{k-1} + \boldsymbol{\delta}_k^* \tag{2}$$

where k stands for the iteration count—do not confuse with a time step index. For all the above-mentioned methods, the increments $\boldsymbol{\delta}_k^*$ can be shown to arise as a solution to the equation:

$$\boldsymbol{\delta}_k^* \leftarrow \left. \frac{\partial \mathrm{F}(\mathbf{x}_k + \boldsymbol{\delta})}{\partial \boldsymbol{\delta}} \right|_{\boldsymbol{\delta}=0} = 0 \tag{3}$$

because a null derivative corresponds to a critical point in the error function $\mathrm{F}(\mathbf{x})$, which can be proven to be a minimum—we will have much more to say about least-squares methods in next sections. Notice how the Jacobian is evaluated at $\boldsymbol{\delta} = 0$, that is, at the vicinity of the present estimation $\mathbf{x}_k$. Typically, Equations 2 and 3 are iterated for a fixed number of steps or until convergence. There exist some noticeably exceptions, like the non-iterated Kalman filter and its non-linear version of the EKF, which run the update step only once. As a side remark, it is worth stressing that in principle there is no reason to choose those filters instead of their iterated versions (apart from the obvious saving of computational cost), since the accuracy loss due to the linearization of the motion and observation models gets smaller with each iteration.

At this point, we can now raise the fundamental problem with the topology of the state space: all the optimization methods based on the principles described above assume that $\mathbf{x}_k$ is a vector belonging to a flat Euclidean state space. That is, a $\mathbb{R}^M$ state space is taken for granted. But then, how is it possible that in chapter 9 we introduced EKF-based SLAM as one of the most wide-spread solutions for a problem with a topology $\mathbb{R}^2 \times \mathbb{S}^1 \times \mathbb{R}^{2L}$, which is not flat due to the $\mathbb{S}^1$ component? The subtle answer is that, in fact, the EKF method has been always forcing us to consider the SLAM state $\mathbf{x}_k$ as a vector in the Euclidean flat space $\mathbb{R}^2 \times \mathbb{R}^1 \times \mathbb{R}^{2L}$ —note how the only change is the "flattening" of $\mathbb{S}^1$ into $\mathbb{R}^1$. In other words, the EKF (or any other estimator or optimizer that assumes a Euclidean space) cannot tell between angular and Euclidean coordinate dimensions, and will treat all of them equivalently. For a 2D robot, this approximation only implies having to take care of renormalizing the heading angle after each Kalman update to assure it remains within the valid range $\left]-\pi, \pi\right]$ (radians).

While not being a serious issue for 2D problems, things get more intricate when handling 3D robot poses, which comprise a 3D location plus three attitude angles—a total of 6 Degrees Of Freedom (DOFs). Mathematically speaking, the former family of planar translation and rotations defined the $\mathbf{SE}(2)$ group (named after *special Euclidean*), while the latter is represented by the group $\mathbf{SE}(3)$ of translations and rotations in the three-dimensional space—refer to Appendix A. To start with the complications, there exist a number of different ways to parameterize or represent poses in $\mathbf{SE}(3)$. The translational part is always described as a simple vector in $\mathbb{R}^3$, therefore the differences arise in the representation of rotations. Two of the most common parameterizations consist of Euler angles (standing for three sequential rotations around different axes, which vary among conventions) and unit quaternions (a unit-norm vector of four real numbers, which can be interpreted as a three-dimensional vector and a fourth value related to the angle of rotation around that vector). A third option is the direct usage of the 4×4 homogeneous matrices $M \in \mathbf{SE}(3)$ that define the linear equations of the coordinate transformations. As regards to their usage within the state vector of localization or SLAM, none of these three choices is an ideal solution, for different reasons:

- Euler angles (e.g. Tait-Bryan rotations, the so-called *yaw*, *pitch* and *roll* angles): This representation achieves a minimum storage requirement (six elements for a $\mathbf{SE}(3)$ pose which has six DOFs), but suffers from three important drawbacks: (1) the non-existence of (reasonably-simple) closed-form Jacobians for all the pose-to-pose and pose-to-point operations, (2) the need to renormalize the angles after each $\mathbf{x}_k \leftarrow \mathbf{x}_k + \boldsymbol{\delta}^*$ update, and (3) the risk of falling in a gimbal lock for an unfortunate combination of angles, which results in the loss of one DOF. Whenever a free DOF enters into any of the estimation algorithms named above, at some point we may face ill-conditioned matrices, which require a special treatment or even may render the estimation algorithm completely useless.

- Unit quaternions: Jacobians are always well defined under this model, but the representation of rotations by means of four real values is an over-parameterization that clearly introduces one extra DOF. Although this is not an issue as serious as the gimbal lock mentioned above, this DOF means an unnecessary cost (i.e. a larger state vector and larger covariance and Jacobian matrices) and introduces the need to renormalize the state vector after each incremental update. Notice, however, that this representation, not being perfect, behaves well enough to have allowed one of the most successful breakthroughs in real-time visual SLAM (Davison, Reid, Molton, & Stasse, 2007).
- Homogeneous matrices ($\mathbf{M} \in \mathbf{SE}(3) \subset \mathbf{GL}(4,\mathbb{R})$, where the last group is the *general linear* group of all 4×4 real matrices): Jacobians are always well-defined with this representation, but they introduce an unacceptable over-parameterization: for the rotational part (i.e. $\mathbf{SO}(3)$) we need $3 \times 3 = 9$ elements instead of the minimum of three components. Hence this representation is never used for encoding 3D transformations in optimization problems. In turn, it is often useful as an intermediary representation during geometric calculations.

In a few words: applying any optimizer or estimator *directly* to a state space comprising absolute values of poses in three-dimensional space (not a planar robot moving on a flat floor) is not a good idea, disregarding the choice for rotation parameterization.

Before proceeding with the proposed alternative, we need to briefly and informally present the concepts of *manifold*, *Lie algebra* and *exponential* and *logarithm maps*. A rigorous mathematical treatment of these concepts is out of the scope of this book, but the interested reader can check out the excellent introductions available elsewhere (Varadarajan, 1974; Gallier, 2001). In the case of our interest (a three-dimensional robot pose) we face a state space that belongs to $\mathbf{SE}(3)$, which is a 6-dimensional *manifold*. Informally, a D – dimensional manifold M embedded in $\mathbb{R}^N$ (with $N \geq D$) is any topological space where the vicinity of each point "looks like" a Euclidean space. A daily-life example is the surface of our Earth, which locally looks like a two-dimensional plane while on the large scale reveals itself as a curved surface in $\mathbb{R}^3$. We can define the *tangent space* at any non-singular point $\mathbf{x} \in M$ of the manifold as the vector space of all the possible "velocity vectors" of a particle at $\mathbf{x}$, if it was constrained to move along the manifold. Such tangent space is denoted as $T_{\mathbf{x}}M$ and has the same dimensionality as the manifold (D). Continuing with the Earth example, with $D = 2$ and $N = 3$, the tangent space would be any planar surface sufficiently small (e.g., a few kilometers squared) The reader is encouraged to refer to Figure 1, which illustrates these concepts.

Getting back to the group $\mathbf{SE}(3)$, it is a closed subgroup, under the group operation of matrix multiplication, of the more generic group of 4×4 matrices $\mathbf{GL}(4,\mathbb{R})$. Then, it follows from a theo-

Figure 1. A graphical illustration of the elements introduced in the text: an example of a two-dimensional manifold M (embedded in a 3D-space), a point on it $\mathbf{x} \in M$, the tangent space at $\mathbf{x}$, denoted as $T_{\mathbf{x}}M$ and the algebra $\mathfrak{m}$, which represents a vector base of that space.

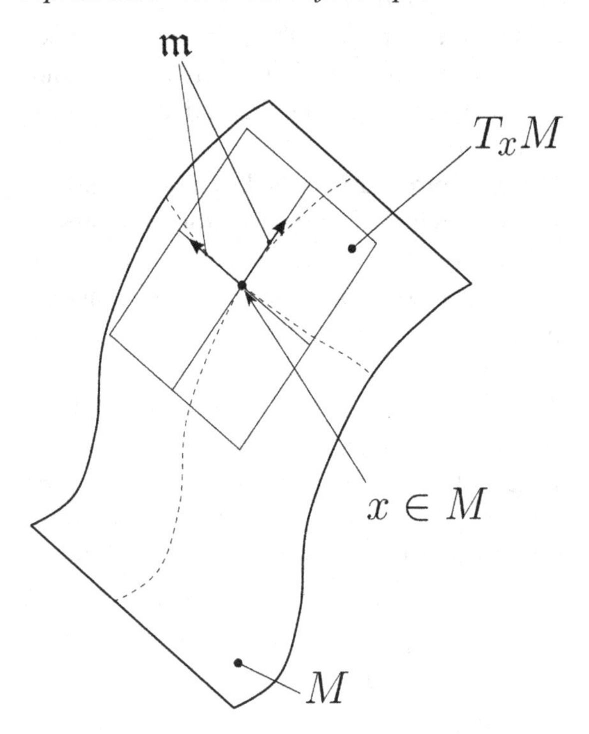

rem due to Von Newman and Cartan (Gallier, 2001, p. 397) that: (1) $\mathbf{SE}(3)$ is a linear Lie group; (2) it is also a smooth manifold in $\mathbb{R}^{4^2}$; (3) a vector base $\mathfrak{m}$ of the manifold is given by $T_{\mathbf{I}}\mathbf{SE}(3)$, the tangent space of $\mathbf{SE}(3)$ at the origin $\mathbf{I}$; and (4) $\mathfrak{m}$ is a *Lie algebra*. The reader must be advised not to getting confused by $\mathfrak{m}$ being named a "vector" base: the vector space we refer to is actually a space of matrices, so in practice $\mathfrak{m}$ consists of a set of *matrices:* each of them can be interpreted as a "unit vector" in the manifold of translations and rotations.

Finally, associated to each Lie group M and Lie algebra $\mathfrak{m}$ there exist two important mapping operators which convert vectors back and forth between M and $\mathfrak{m}$. They are the *exponential* and *logarithm maps* for the manifold:

$$\begin{aligned} \exp &: \mathfrak{m} \rightarrow M \\ \log &: M \rightarrow \mathfrak{m} \end{aligned} \qquad (4)$$

which play a central role in the method proposed below. The particular expressions for the cases of interest in mobile robotics and computer vision are provided in Appendix D.

An Elegant Solution to the Problem of the Topology of the State-Space: Optimizing on the Manifold

Although the idea is not new at all—refer to Gabay (1982) and Taylor and Kriegman (1994)—carrying out optimization directly on the manifold of $\mathbf{SE}(3)$ while keeping a unit quaternion or Euler angles parameterization in the state vector is a solution which has been gaining popularity in the robotics and computer vision community in recent years (Hertzberg, 2008; Strasdat, Montiel, & Davison, 2010a, 2010b; Küemmerle, Grisetti, Strasdat, Konolige, & Burgard, 2011). The reason for this popularity is that it avoids the inconveniences of both overparameterized representations and those that suffer from singularities.

Following the notation introduced in the work (Hertzberg, 2008), the only changes required in an estimation algorithm for working on a manifold are the following three replacements:

1st change: Take derivatives with respect to increments in the manifold

$$\begin{aligned} & \boldsymbol{\delta}_k^* \leftarrow \left. \frac{\partial \mathrm{F}(\mathbf{x}_k + \boldsymbol{\delta})}{\partial \boldsymbol{\delta}} \right|_{\boldsymbol{\delta}=0} = 0 \\ \Rightarrow \quad & \boldsymbol{\varepsilon}_k^* \leftarrow \left. \frac{\partial \mathrm{F}(\mathbf{x}_k \boxplus \boldsymbol{\varepsilon})}{\partial \boldsymbol{\varepsilon}} \right|_{\boldsymbol{\varepsilon}=0} = 0 \end{aligned}$$

2nd change: Incorporate manifold increments with an appropriate operator

$$\mathbf{x}_k \leftarrow \mathbf{x}_{k-1} + \boldsymbol{\delta}_k^* \Rightarrow \mathbf{x}_k \leftarrow \mathbf{x}_{k-1} \boxplus \boldsymbol{\varepsilon}_k^*$$

3rd change: Consider residuals (if they are poses) into the manifold

$$\mathbf{r}_k = \mathbf{h}\left(\mathbf{x}_k\right) - \mathbf{z} \Rightarrow \mathbf{r}_k = \mathbf{h}\left(\mathbf{x}_k\right) \boxminus \mathbf{z} \qquad (5)$$

Here $\mathbf{x}_k \in M$ is the state vector of the problem, which lies on some D – dimensional manifold M, $\boldsymbol{\varepsilon}_k^* \in \mathbb{R}^D$ is the optimal increment (to be determined) in the linearization of the manifold around $\mathbf{x}_k$ (using the Lie algebra of M as a vector base) and the operator $\boxplus : M \times \mathbb{R}^D \to M$ is a generalization of the common addition operator $+$ of Euclidean spaces. In the case that the observed data $\mathbf{z}$ also belong to a D'– dimensional manifold M' (which may or may not coincide with M), we must also introduce the replacement $\boxminus : M' \times M' \to \mathbb{R}^{D'}$ for the Euclidean subtraction operator: a new way in how differences between poses are measured which replaces the naive subtraction of their parameterizations.

There are two possible ways of implementing $\boxplus$, both of them perfectly valid. If we let $\mathbf{x}_k$ and $\mathbf{x}'_k$ denote two elements of the problem manifold and $\boldsymbol{\varepsilon} \in \mathbb{R}^D$ an increment in its linearized approximation, the two alternative conventions are shown in Box 1, with the multiplication of $e^{\boldsymbol{\varepsilon}}$ and $\mathbf{x}_{k-1}$ being the manifold group "product" operation (i.e., a multiplication of two 4×4 matrices) and $e^{\boldsymbol{\varepsilon}} = \exp\left(\boldsymbol{\varepsilon}\right)$ being the exponential map of the manifold. Regarding the $\boxminus$ operator in Equation 5, it corresponds to the logarithm map of the observation manifold M', applied to the "mismatch" between its two arguments; likewise with the $\boxplus$ operator, where one can choose among these two valid forms, as shown in Box 2.

It is important to highlight that the topological structure of the state vector $\mathbf{x}_k$ will most commonly be the product of many elemental topological substructures. For instance, representing a 3D robot pose and L spatial landmarks requires a $\mathbf{SE}\left(3\right) \times \mathbb{R}^{3L}$ space. Therefore, care must be taken to handle as explained above only the $\mathbf{SE}\left(3\right)$ parts, while treating as usual the pure Euclidean parts of the state vector, for which $\mathbf{x} \boxplus \boldsymbol{\varepsilon}$ and $\mathbf{h} \boxminus \mathbf{z}$ fall back to the classical $\mathbf{x} + \boldsymbol{\varepsilon}$ and $\mathbf{h} - \mathbf{z}$ respectively.

To sum up, the method for optimizing on manifolds consists of replacing the derivatives with respect to absolute 3D poses by the derivatives of the same pose modified by some small disturbance $\boldsymbol{\varepsilon}$, with respect to which we will take derivatives. The so-obtained optimal increment $\boldsymbol{\varepsilon}^*$ must be then added to the state vector through the $\boxplus$ operator.

Box 1.

$$\mathbf{x}_k = \mathbf{x}_{k-1} \boxplus \boldsymbol{\varepsilon} \Rightarrow \qquad (6)$$

$$\begin{array}{llcl} & \text{(In the } \oplus \text{ notation)} & & \text{(Poses as matrices)} \\ \left\{ \begin{array}{l} \text{Convention \#1:} \\ \\ \text{Convention \#2:} \end{array} \right. & \begin{array}{l} \mathbf{x}_k = \boldsymbol{\varepsilon} \oplus \mathbf{x}_{k-1} \\ \\ \mathbf{x}_k = \mathbf{x}_{k-1} \oplus \boldsymbol{\varepsilon} \end{array} & \begin{array}{c} \Leftrightarrow \\ \\ \Leftrightarrow \end{array} & \begin{array}{l} \mathbf{x}_k = e^{\boldsymbol{\varepsilon}} \mathbf{x}_{k-1} \\ \\ \mathbf{x}_k = \mathbf{x}_{k-1} e^{\boldsymbol{\varepsilon}} \end{array} \end{array}$$

Box 2.

$$\mathbf{r}_k = \mathbf{h}(\mathbf{x}_k) \boxminus \mathbf{z}_k \Rightarrow \tag{7}$$

$$\begin{array}{lcc} & \text{(In the } \ominus \text{ notation)} & & \text{(Poses as matrices)} \\ \left\{ \begin{array}{l} \text{Convention \#1:} \quad \mathbf{r}_k = \ln\left(\mathbf{z}_k \ominus \mathbf{h}(\mathbf{x}_k)\right) \\ \\ \text{Convention \#2:} \quad \mathbf{r}_k = \ln\left(\mathbf{h}(\mathbf{x}_k) \ominus \mathbf{z}_k\right) \end{array} \right. & \begin{array}{c} \Leftrightarrow \\ \\ \Leftrightarrow \end{array} & \begin{array}{l} \mathbf{r}_k = \ln\left(\mathbf{h}(\mathbf{x}_k)^{-1}\mathbf{z}_k\right) \\ \\ \mathbf{r}_k = \ln\left(\mathbf{z}_k^{-1}\mathbf{h}(\mathbf{x}_k)\right) \end{array} \end{array}$$

A Practical Example

Due to the abstract nature of the mathematics invoked above, it becomes instructive to study a short example of how to apply it to a practical situation.

Assume we are performing EKF-based SLAM, as described in chapter 9, section 2, with a state vector comprising the 3D robot pose and one landmark:

$$\mathbf{s}_k = \begin{pmatrix} \mathbf{x}_k \\ \mathbf{m}_1 \end{pmatrix} \tag{8}$$

where the rotational part of the pose $\mathbf{x}_k$ can be parameterized, in *any* form (e.g. unit quaternion, Tait-Bryan rotations or any other Euler angles), as a vector of length p.

The first relevant modification to the EKF method described in chapter 9's Algorithm 1 to allow it working with increments in the manifold $\mathbf{SE}(3)$ is related to the computation of the observation Jacobian:

$$\mathbf{H}_k = \left.\frac{d\mathbf{h}(\mathbf{s}_k)}{d\mathbf{s}_k}\right|_{\mathbf{s}_k=\bar{\boldsymbol{\mu}}_k} = \left.\frac{\partial \mathbf{h}(\mathbf{x}_k, \mathbf{m}_1)}{\partial\{\mathbf{x}_k, \mathbf{m}_1\}}\right|_{\mathbf{x}_k=\bar{\mathbf{x}}_k, \mathbf{m}_1=\bar{\mathbf{m}}_1} =$$

(splitting the Jacobian into two parts)

$$= \begin{pmatrix} \left.\dfrac{\partial \mathbf{h}(\mathbf{x}_k, \bar{\mathbf{m}}_1)}{\partial \mathbf{x}_k}\right|_{\mathbf{x}_k=\bar{\mathbf{x}}_k} \\ \left.\dfrac{\partial \mathbf{h}(\bar{\mathbf{x}}_k, \mathbf{m}_1)}{\partial \mathbf{m}_1}\right|_{\mathbf{m}_1=\bar{\mathbf{m}}_1} \end{pmatrix} = \begin{pmatrix} \mathbf{H}_k^{\mathbf{x}} \\ \mathbf{H}_k^{\mathbf{m}_1} \end{pmatrix} \tag{9}$$

Denoting the function that converts the parameterized robot pose into its associated 4×4 matrix as $\mathbf{M}(\mathbf{x}_k) : \mathbb{R}^p \to \mathbb{R}^{4^2}$, and adopting the pre-multiplication convention from the two legitimate options provided in Equation 6, we then take on the following change of variable:

$$\mathbf{x}_k = \boldsymbol{\varepsilon} \oplus \bar{\mathbf{x}}_k$$

with $\oplus$ being the pose composition operator; or in matrix form:

$$\underbrace{\mathbf{M}\left(\mathbf{x}_k\right)}_{\tilde{\mathbf{X}}} = e^{\varepsilon}\,\mathbf{M}\left(\bar{\mathbf{x}}_k\right), \quad \varepsilon \in \mathbb{R}^6 \tag{10}$$

Put in words: instead of working with the robot pose itself, we will deal now with small pose changes (ε) in the $\mathbf{SE}(3)$ manifold around the mean ($\bar{\mathbf{x}}_k$) of the robot pose Gaussian. In practice, this converts the Jacobian $\mathbf{H}_k$ above into an alternative version $\tilde{\mathbf{H}}_k$:

$$\tilde{\mathbf{H}}_k = \left.\frac{\partial \mathbf{h}\left(e^{\varepsilon}\,\mathbf{M}\left(\bar{\mathbf{x}}_k\right), \mathbf{m}_1\right)}{\partial\left\{\varepsilon, \mathbf{m}_1\right\}}\right|_{\varepsilon=0,\mathbf{m}_1=\bar{\mathbf{m}}_1} = \begin{pmatrix} \left.\dfrac{\partial \mathbf{h}\left(e^{\varepsilon}\,\mathbf{M}\left(\bar{\mathbf{x}}_k\right), \bar{\mathbf{m}}_1\right)}{\partial \varepsilon}\right|_{\varepsilon=0} \\ \left.\dfrac{\partial \mathbf{h}\left(\bar{\mathbf{x}}_k, \mathbf{m}_1\right)}{\partial \mathbf{m}_1}\right|_{\mathbf{m}_1=\bar{\mathbf{m}}_1} \end{pmatrix} = \begin{pmatrix} \tilde{\mathbf{H}}_k^{\varepsilon} \\ \mathbf{H}_k^{\mathbf{m}_1} \end{pmatrix} \tag{11}$$

where the bottom part of the Jacobian reduces to the common case when not working with on-manifold increments. It is worth spending a few moments grasping the meaning of the resulting Jacobian by comparing Equation 11 to Equation 9.

Now we address the partial Jacobian $\tilde{\mathbf{H}}_k^{\varepsilon}$, which will appear in any 3D localization or SLAM estimator. Its evaluation can be attacked by a step-by-step decomposition into known Jacobians via the chain rule of multivariate derivatives, as follows:

$$\tilde{\mathbf{H}}_k^{\varepsilon} = \left.\frac{\partial \mathbf{h}(\overbrace{e^{\varepsilon}\,\mathbf{M}(\bar{\mathbf{x}}_k)}^{\tilde{\mathbf{X}}}, \bar{\mathbf{m}}_1)}{\partial \varepsilon}\right|_{\varepsilon=0} = \left.\underbrace{\overbrace{\left.\frac{\partial \mathbf{h}\left(\tilde{\mathbf{X}}, \bar{\mathbf{m}}_1\right)}{\partial \tilde{\mathbf{X}}}\right|_{\tilde{\mathbf{X}}=\bar{\mathbf{x}}_k}}^{\mathbf{H}_k^{\tilde{\mathbf{X}}}}}_{\text{Jacobian of the common observation model}} \overbrace{\frac{\partial\, e^{\varepsilon}\,\mathbf{M}\left(\bar{\mathbf{x}}_k\right)}{\partial \varepsilon}}^{\tilde{\mathbf{X}}}\right|_{\varepsilon=0} \tag{12}$$

where it has to be highlighted that $\tilde{\mathbf{X}}$ stands for a 4×4 transformation matrix, not for the pose parameterization used in the state vector. Thus, the first Jacobian $\mathbf{H}_k^{\tilde{\mathbf{X}}}$ must be interpreted as a derivative with respect to each of the 12 relevant components of a transformation matrix. Regarding the second term in the product, it can be further decomposed as:

$$\left.\frac{\partial\, \overbrace{e^{\varepsilon}}^{\mathbf{A}}\,\mathbf{M}\left(\bar{\mathbf{x}}_k\right)}{\partial \varepsilon}\right|_{\varepsilon=0} = \underbrace{\left.\frac{\partial\, \mathbf{A}\,\mathbf{M}\left(\bar{\mathbf{x}}_k\right)}{\partial \mathbf{A}}\right|_{\mathbf{A}=e^{\varepsilon=0}=\mathbf{I}_4}}_{\text{This is simply the Jacobian of the multiplication of two matrices}} \left.\frac{d\,e^{\varepsilon}}{d\varepsilon}\right|_{\varepsilon=0} \tag{13}$$

where the last derivative has a known expression. Refer to Appendix D for further discussion on these and similar Jacobians.

Once the Jacobian has been evaluated, we can go on with the EKF-SLAM algorithm as usual, until we obtain the state-vector correction $\boldsymbol{\delta}_k = \mathbf{K}_k\tilde{\mathbf{y}}_k$ (step 3.3 in chapter 9's Algorithm 1). At this point, we must have in mind that such increment vector will be structured as follows:

$$\boldsymbol{\delta}_k = \begin{pmatrix} \varepsilon^* \\ \Delta\mathbf{m}_1 \end{pmatrix} \tag{14}$$

That is: the part to be added to the robot pose will be always retrieved as an increment in the linearized manifold (to be added via the $\boxplus$ operator), while the part corresponding to the landmark positions is still represented as a common Euclidean increment:

$$\bar{\mathbf{s}}_{k+1} = \begin{pmatrix} \bar{\mathbf{x}}_{k+1} \\ \bar{\mathbf{m}}_1 \end{pmatrix} \leftarrow \begin{pmatrix} \bar{\mathbf{x}}_{k+1} \boxplus \varepsilon^* \\ \bar{\mathbf{m}}_1 + \Delta\mathbf{m}_1 \end{pmatrix} \tag{15}$$

The process above illustrates a practical procedure for updating the mean vector in the EKF when optimizing on the manifold. An identical reasoning should be followed when working with least-squares methods, i.e. modification of the Jacobian, evaluation of it by repetitive application of the chain rule, and vector update with the $\boxplus$ operator.

When it comes about the covariance update in the EKF (or any other filter of the KF family), one more extra step must be accounted for: we must evaluate the Jacobian with respect to the particular pose parameterization ($\mathbf{H}_k^{\mathbf{x}}$) before using it in the covariance update formula (step 3.5 in chapter 9's Algorithm 1). However, we can easily derive this from the Jacobian $\mathbf{H}_k^{\tilde{\mathbf{X}}}$, already computed in Equation 12, as follows:

$$\mathbf{H}_k^{\mathbf{x}} = \left.\frac{\partial \mathbf{h}\left(\mathbf{x}_k, \bar{\mathbf{m}}_1\right)}{\partial \mathbf{x}_k}\right|_{\mathbf{x}_k=\bar{\mathbf{x}}_k} =$$

(by the variable change $\mathbf{x}_k = \varepsilon \oplus \bar{\mathbf{x}}_k$, or in matrix form $\tilde{\mathbf{X}} = e^{\varepsilon}\,\mathbf{M}(\bar{\mathbf{x}}_k)$, and applying the chain rule)

$$= \underbrace{\left.\frac{\partial \mathbf{h}\left(\tilde{\mathbf{X}}, \bar{\mathbf{m}}_1\right)}{\partial \tilde{\mathbf{X}}}\right|_{\tilde{\mathbf{X}}=e^{\varepsilon}\mathbf{M}(\bar{\mathbf{x}}_k)=\mathbf{M}(\bar{\mathbf{x}}_k)}}_{\mathbf{H}_k^{\tilde{\mathbf{X}}}} \left.\frac{\partial\, e^{\varepsilon}\,\mathbf{M}(\bar{\mathbf{x}}_k)}{\partial\, \mathbf{M}(\bar{\mathbf{x}}_k)}\right|_{\varepsilon=0} \left.\frac{d\,\mathbf{M}(\mathbf{x})}{d\mathbf{x}}\right|_{\mathbf{x}=\bar{\mathbf{x}}_k} =$$

(The second Jacobian reduces to $\left.\frac{\partial \mathbf{AB}}{\partial \mathbf{B}}\right|_{\mathbf{A}=\mathbf{I}} = \mathbf{I}$ thus):

$$= \mathbf{H}_k^{\tilde{\mathbf{X}}} \left.\frac{d\,\mathbf{M}(\mathbf{x})}{d\mathbf{x}}\right|_{\mathbf{x}=\bar{\mathbf{x}}_k} \qquad (16)$$

where the derivative in the last factor straightforwardly follows from the formula that gives us the 4×4 transformation matrix from the chosen parameterization in the state vector.

3. GRAPH SLAM: INTRODUCTION

The graphical representation of the variables involved in SLAM was presented in chapter 9 section 1 under the name of the *full SLAM* model. We saw there how its joint posterior pdf (or a marginalized version discarding old robot poses) can be computed sequentially by means of Bayesian estimators, namely the families of EKF-like filters and the RBPF algorithms. These approaches to SLAM stood as the *de facto* standard during the first half of the 2000s.

In spite of their suitability for successfully solving a number of realistic SLAM problems of moderate size, researchers however did not stop pushing the limits of SLAM, always pursuing building larger and more consistent maps as efficiently as possible. However, achieving something like closing a million-landmark loop (Frese & Schröder, 2006) with the classic EKF and RBPF-based methods has been shown as unfeasible in the last years, no matter how optimized an implementation is; sooner or later, the quadratic and exponential growth in computational complexity of EKF and RBPF, respectively, reveals as a limitation for estimating very large maps in a reasonable time.

Therefore, the solution for SLAM in the particular case of very large-scale scenarios came from a new paradigm, in which the structure of the problem is attacked from a new angle: by directly applying *maximum likelihood estimators* (MLE, recall chapter 4 section 7) over the *full SLAM* graphical model or its marginalized version without the map elements. This leads to the approaches known as *bundle adjustment* and *graph SLAM*, respectively. Although the bases on which both methods rely are almost identical, we analyze them separately in this and subsequent sections of the chapter in order to better study their particularities.

Framework Overview

Graph SLAM aims at obtaining a MLE of the history of robot poses. A direct estimation of the map is avoided by marginalizing the map elements out of the full SLAM DBN. There seems to exist some confusion about applying the name "graph SLAM" to methods where the map is also estimated directly along the poses; we will instead adhere to the largest part of the literature and classify those as instances of the Bundle Adjustment problem.

At first sight, it may look paradoxical to find a method for SLAM, which does not build any map! What is about the chicken-or-the-egg problem of robot poses and the map that we often mentioned throughout this book? In fact, the trick in graph SLAM is that map elements still are represented but only *implicitly* by means of the constraints imposed to the robot poses from which they are observed.

As we saw in chapter 3 section 10, marginalizing (the map) in a DBN has the effect of creating edges between previously unconnected nodes; in this case, new arcs will appear between robot poses from which the same map elements are observed. These new arcs represent constraints in the *relative* location of nodes with respect to each other. If we convert this graph into an undirected graphical model and assume that all joint and conditional pdfs are normally distributed we find the standard graphical model employed in graph SLAM: a Gaussian Markov Random Field (GMRF). The process by which the original full SLAM DBN becomes a GMRF is illustrated with an example in Figure 2a, b for a toy SLAM problem with only four robot poses. The reader should notice that despite the usage of an undirected graphical model for the underlying joint distribution, the edges still represent *relative* constraints with an implicit sense of direction in the relative poses.

A graph SLAM problem can be then represented as a graph whose nodes $\mathbf{x}_i$ stand for robot poses, or *frames*, and whose edges $\left\{\mathbf{z}_{i,j}\right\}_{\langle i,j\rangle\in\Upsilon}$ represent information about their relative positions, with Υ being the set of all edges. That is why the graphs used in graph SLAM are also called *pose graphs* or *graphs of pose constraints*. Edge constraints are all known data. Regarding nodes, from a total of $P = F + N$ poses $\mathbf{x}_{1:P}$ we have $F \geq 1$ whose global coordinates are fixed and known ($\mathbf{x}_{1:F} \triangleq \mathbf{x}_F$), while the other N are the unknowns to be determined ($\mathbf{x}_{F+1:P} \triangleq \mathbf{x}$). Typically, $F = 1$ and $\mathbf{x}_1$ is fixed arbitrarily to the origin of coordinates to avoid having infinite solutions[1]. Continuing the previous example, Figure 3a illustrates the geometrical representation of the GMRF problem. Here, we know $\mathbf{x}_0$ (that is the fixed pose in this case), as well as all the Gaussian parameters of the relative poses $\mathbf{z}_{i,j}$, and our aim is finding the *optimal* global poses of the free nodes $\mathbf{x}_{1:3}$.

In graph SLAM, optimal poses are those that maximize the likelihood of the observed data (relative poses in edges), that is, we are using a MLE. For that, we first need to establish the joint density $p\left(\mathbf{x}, \mathbf{x}_F, \left\{\mathbf{z}_{i,j}\right\}_\Upsilon\right)$ that is to be optimized. As explained in chapter 3 section 10, the pdf implied by a Markov random field can be factored into the product of potential functions (i.e. non-normalized pdfs) associated to the set of maximal cliques. By making the reasonable assumption that errors (the "uncertainty") in relative pose observations $\mathbf{z}_{i,j}$ are all independent with respect to each other, it follows[2] that the joint factors into the contributions from each individual edge, that is:

$$p(\mathbf{x}, \underbrace{\mathbf{x}_F, \{\mathbf{z}_{i,j}\}_\Upsilon}_{\text{Known data}}) \propto \prod_{\{i,j\}\in\Upsilon} \phi_{i,j}\left(\mathbf{x}_i, \mathbf{x}_j, \mathbf{z}_{i,j}\right) \quad (17)$$

where the proportionality factor does not depend on the unknowns, hence it can be simply ignored

Figure 2. (a) The DBN of a small full-SLAM problem and (b) the GMRF of its associated graph SLAM formulation. Notice that the map variables $\mathbf{m}$ *have been marginalized out, which introduces the arcs between the poses* $1 \leftrightarrow 3$ *due to the covisibility of* $\mathbf{m}_3$.

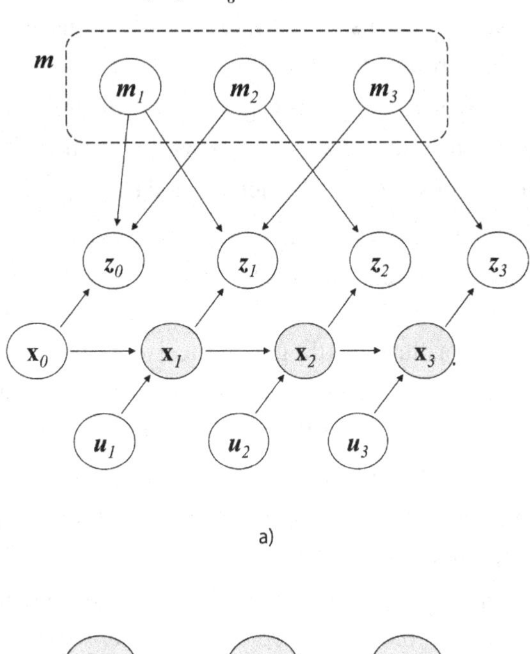

in the following. Each edge from node i to node j represents a measurement with a zero-mean Gaussian error, whose potential function reads:

$$\phi_{i,j}\left(\mathbf{x}_i, \mathbf{x}_j, \mathbf{z}_{i,j}\right) = \exp\left(-\frac{1}{2}\mathbf{r}_{i,j}^T\left(\mathbf{x}\right)\mathbf{\Lambda}_{i,j}\mathbf{r}_{i,j}\left(\mathbf{x}\right)\right) \tag{18}$$

where $\mathbf{r}_{i,j}\left(\mathbf{x}\right)$ is the vector of errors or *residuals*, a measure of the mismatch between the prediction and the observation, and $\mathbf{\Lambda}_{i,j}$ stands for the measurement *information* matrix, i.e. the inverse of the covariance matrix. This matrix must be interpreted as a measure of how much confident we are about the relative pose observation $\mathbf{z}_{i,j}$. Each edge may have a different information matrix and those with a large determinant (small covariance, or "strong links") will have more weight than edges with small determinants (large covariance, or "weak links") during optimization. In practice, edges derived from odometry and motion models (recall chapter 5) typically result in "weak links," whereas those obtained by accurate registration of precise sensors (e.g. ICP with point clouds) are "strong links."

In MLE we seek for the unknowns $\mathbf{x} = \mathbf{x}^*$ that maximize the product in Equation 17 or, more conveniently, that minimize its negative log likelihood. Replacing the potential functions by their values from Equation 18 we can define the target *cost function* as:

$$\begin{aligned}&\text{negative log likelihood}=\\&= -\log p(\mathbf{x}, \mathbf{x}_F, \{\mathbf{z}_{i,j}\}_\Upsilon) =\\&= \text{const} - \sum_{\{i,j\}\in\Upsilon} \log\phi_{i,j}\left(\mathbf{x}_i, \mathbf{x}_j, \mathbf{z}_{i,j}\right) =\\&= \text{const} + \frac{1}{2}\sum_{\{i,j\}\in\Upsilon}\mathbf{r}_{i,j}^T\left(\mathbf{x}\right)\mathbf{\Lambda}_{i,j}\mathbf{r}_{i,j}\left(\mathbf{x}\right) \rightarrow\end{aligned}$$

(discarding the constant term, since it does not change the optimization result):

$$\rightarrow \underbrace{\mathrm{F}\left(\mathbf{x}\right)}_{\text{Cost function}} = \frac{1}{2}\sum_{\{i,j\}\in\Upsilon}\mathbf{r}_{i,j}^T\left(\mathbf{x}\right)\mathbf{\Lambda}_{i,j}\mathbf{r}_{i,j}\left(\mathbf{x}\right) \tag{19}$$

and we can finally state graph SLAM as the following non-linear least-squares minimization problem:

$$\mathbf{x}^* = \arg\min_{\mathbf{x}} \mathrm{F}\left(\mathbf{x}\right) \tag{20}$$

Figure 3. (a) A representation of the problem of Figure 2 as a graph of pose constraints (the $\boldsymbol{\Delta}_{ij}$ variables) with uncertainty. (b) The resulting global poses after optimization. The goal of graph SLAM is finding these optimal poses that minimize the overall mismatch of all constraints. Note that at least one frame (here $\mathbf{x}_0$) has to be maintained fixed.

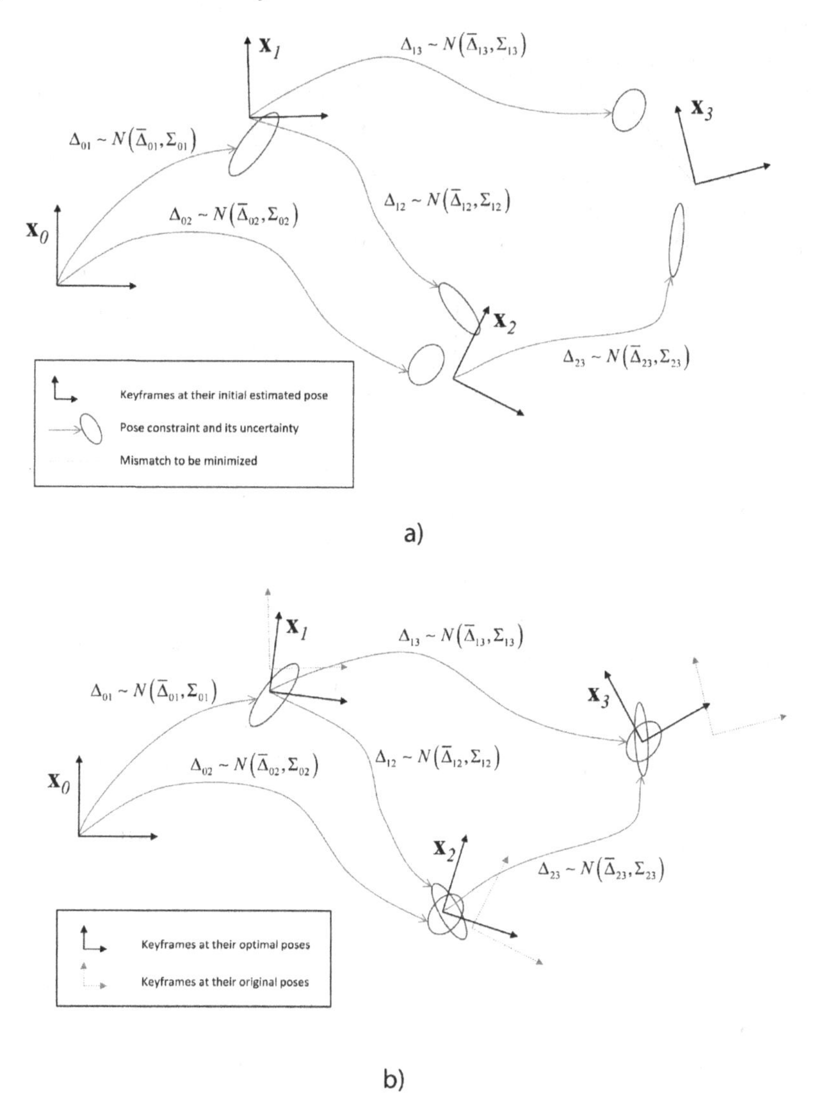

It only remains to be defined the form of the residuals $\mathbf{r}_{i,j}(\mathbf{x})$, which can be implemented as any function that measures the mismatch between the present estimate of the pose j, given by $\mathbf{x}_j$, and its prediction by concatenating the edge $i \rightarrow j$ to the pose i, that is, $\mathbf{x}_i \oplus \mathbf{z}_{i,j}$. Refer to Figure 3a for an illustration of these geometrical relationships. In the next section, we describe a common implementation of these functions.

The resulting cost function $\mathrm{F}(\mathbf{x})$ of Equation 19 has three insightful interpretations. On the one hand, we can focus on its probabilistic significance. Each edge contributes with a square error term $\mathbf{r}_{i,j}^T \boldsymbol{\Lambda}_{i,j} \mathbf{r}_{i,j}$, which can be seen as a chi-square residual; therefore the MLE is searching for the unknowns (robot poses) that minimize the overall chi-square statistic χ^2 of the joint pdf. Ideally, a χ^2 of zero would mean that the model (i.e. the global poses) perfectly predicts all the measurements (i.e. the relative poses in edges). We can also look at the equivalent geometrical interpretation of the equation. In this case, the goal is searching for the global poses with the *minimum mismatch* between incompatible pose reconstructions along the potential paths along the graph. For instance, in the previous example of Figure 3 the node 3 has three possible global coordinates (with respect to the origin at node 0), which correspond to the possible paths $0 \to 1 \to 3$, $0 \to 1 \to 2 \to 3$ and $0 \to 2 \to 3$. Obviously, the existence of imperfections in measuring relative poses means that different results will be obtained for each path. Since there will always exist such inconsistencies, all we can do is trying to minimize their combined effects. A third interpretation of the cost function comes from physics: if nodes are seen as analogous to masses and edges as linear springs, which store elastic energy if forced off their rest position, the optimal graph solution arises as the minimum energy state of the analogous system. The determinant of information matrices would then be related to the elastic constant (stiffness) of each spring.

It is worth stressing an important practical difference between graph SLAM and the classic Bayesian approaches described in the previous chapter. While EKF and RBPF-based SLAM take a *filtering* approach, sequentially incorporating new observations as they are gathered, graph SLAM fundamentally works *off-line* over the entire data history. Still, this does not imply that graph SLAM is not suited for addressing SLAM on a robot in real-time: with modern implementations, performance is typically good enough as for performing batch optimization of mid-sized graphs at the time they are built. Furthermore, frequently only a partial optimization of the graph is really needed since a small incremental change in the graph does not modify a large number of nodes, unless when dealing with loop closures.

This mode of operation is related to an increasingly popular view of SLAM as a two-fold process. Firstly, a SLAM *front-end* is in charge of processing sensory data in real-time, giving as output their interpretation in terms of which new frames and edges must be incorporated into the graph. Secondly, the SLAM *back-end* consists in any graph SLAM solver, as those introduced next, for which the sensor-specific nature of the pose constraints is irrelevant. An advantage of this viewpoint is that several alternative implementations for each of the two stages can be interchanged, easing the reuse of implementations between researchers. Note however that, as illustrated in Figure 4, the front-end may require access to the output of the back-end (the optimized global poses) in order to establish data association. Therefore, one must keep in mind that both processes, normally almost independent, may become strongly coupled during loop-closing situations. Practical applications often exploit the front-end/back-end division to parallelize SLAM by running each task concurrently in a different processing thread. A paramount example of this principle is the Parallel Tracking and Mapping (PTAM) framework reported in Klein and Murray (2007) for monocular visual SLAM. Front-ends for graph SLAM have been also reported for 2D laser scans in the so-called SPA2D framework (Konolige, Grisetti, Kümmerle, Burgard, Limketkai, & Vincent, 2010) and more recently for RGB and depth images captured with the novel 3D sensors from *PrimeSense* like the Microsoft's Kinect™ (Henry, Krainin, Herbst, Ren, & Fox, 2010; Engelhard, Endres, Hess, Sturm, & Burgard, 2011).

A Brief Historical Perspective

Before discussing the current solutions to graph SLAM, it is instructive to take a short review of their historical development. The seminal work in Lu and Milios (1997) can fairly be considered

Figure 4. Graph SLAM as a compound process composed of well-differentiated front-end and back-end algorithms

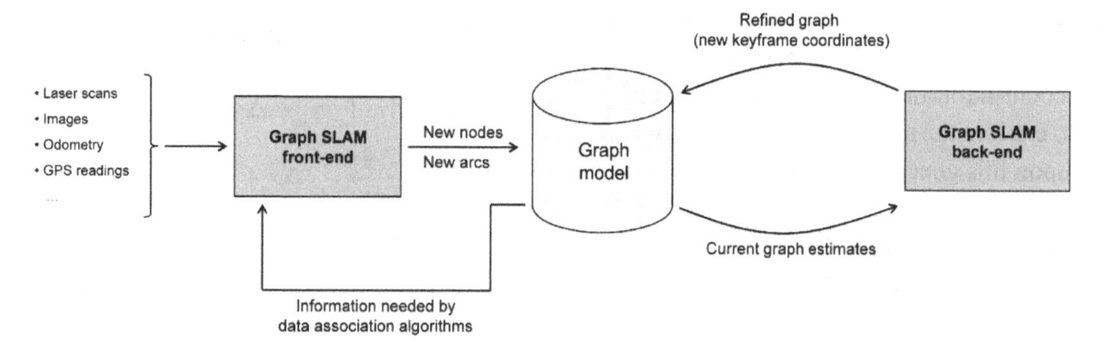

the first precursor of graph SLAM. In the late 90s, SLAM with 2D laser scanners was an active research area where most proposals included some sort of incremental growth of cumulative point maps, outside of any probabilistic framework. In that context, closing a loop typically led to maps, which were inconsistent due to the accumulation of incremental errors.

In order to deal with those errors and generate globally consistent maps, Lu and Milios proposed to construct a graph of pose constraints, exactly as in modern graph SLAM, but did not propose solving it with non-linear iterative optimization techniques like Gauss-Newton or Levenberg-Marquardt. An independent line of work was introduced with the PhD thesis of Tom Ducket and a subsequent series of works (Howard, Mataric, & Sukhatme, 2001; Duckett, Marsland, & Shapiro, 2002), where SLAM was also attacked as the optimization of pose graphs. In this case, the proposed solver was the Gauss-Seidel "relaxation" method (Demmel, 1997), in which constraints are considered only once at a time in order to iteratively approach the solution. Much faster convergence was afterward achieved by Udo Frese by means of a multilevel relaxation algorithm (Frese, Larsson, & Duckett, 2005), in which the graph of poses is first divided into hierarchical levels of decreasing detail in order to cope with a reduced number of constraints, each affecting a large group of nodes.

It would be not until Thrun and Montemerlo (2006) that the name "graph SLAM" is definitively established for grouping a growing body of work where the aim was solving graphs of pose constraints using non-linear least square methods. It was pretty clear by then that re-linearizing at each iteration of the least squares optimizer, a defining characteristic of graph SLAM, stands as a significant advantage in comparison to the single linearization step in EKF-based SLAM. When the starting point of the unknowns is far from the optimal solution, the non-linearities become very relevant; hence, this re-linearization is indispensable. Interested readers can find more reviews about graph SLAM in the recent research literature (Frese, Larsson, & Duckett, 2005; Thrun & Montemerlo, 2006; Dellaert, Carlson, Ila, Ni, & Thorpe, 2010).

4. GRAPH SLAM: OPTIMIZING ON MANIFOLDS

In this section we present the standard least-squares solution that lies at the core of modern graph SLAM back-ends, including the (now ubiquitous) application of manifolds that allows approaching 3D SLAM robustly. As will be clear soon, this solution basically consists of two alternating phases: (1) building a sparse linear system of equations and (2) solving it. In the first place we derive the equations of the most common sparse solvers, while a subsequent subsection analyzes how to efficiently build the linear system, taking into account the on-manifold optimization. Finally, we also explore the most recent proposals, which build upon this general framework.

On-Manifold Sparse Non-Linear Least Squares

We start by recalling that the target cost function to be optimized is defined as the following sum of quadratic forms (see Equation 19):

$$\mathrm{F}(\mathbf{x}) = \frac{1}{2} \sum_{\{i,j\} \in \Upsilon} \mathbf{r}_{i,j}^{T}(\mathbf{x}) \boldsymbol{\Lambda}_{i,j} \mathbf{r}_{i,j}(\mathbf{x}) \quad (21)$$

If the residuals $\mathbf{r}_{i,j}(\mathbf{x})$ were linear functions we could directly compute the optimal $\mathbf{x}^*$ that minimizes $\mathrm{F}(\mathbf{x})$. However, in graph SLAM the residuals are non-linear, thus no closed-form solution exists and we must rely on iterative methods, which linearize the problem around successively more accurate solutions until convergence. The exact implementation of the residuals functions $\mathbf{r}_{i,j}(\mathbf{x})$ is deferred until the next subsection; until that, we only has to keep in mind that each residual is a pose in either $\mathbf{SE}(2)$ or $\mathbf{SE}(3)$. Still, for the sake of clarity in the exposition, we will ignore in a first analysis their non-Euclidean nature, incorporating it afterwards.

Let $k = 1,2,3,\ldots$ be the iteration step in the optimization and $\mathbf{x}_k$ the corresponding estimate of the unknowns at each step—do not misread k as a time step index equivalent to the one of the Bayesian recursive approach. The starting point for the optimization, $\mathbf{x}_0$, is assumed to be known in advance. The cost function can then be approximated at the $k-th$ iteration by its second-order[3] Taylor series expansion $\hat{\mathrm{F}}_k(\cdot)$ in the vicinity of $\mathbf{x}_k$ (refer to Appendix E):

$$\begin{aligned}
&\mathrm{F}(\mathbf{x}_k + \Delta\mathbf{x}) \approx \hat{\mathrm{F}}_k(\mathbf{x}_k + \Delta\mathbf{x}) = \\
&= \mathrm{F}(\mathbf{x}_k) + \underbrace{\left.\frac{\partial \mathrm{F}}{\partial \mathbf{x}}\right|_{\mathbf{x}=\mathbf{x}_k}}_{\nabla_{\mathbf{x}} \mathrm{F}(\mathbf{x}_k)} \Delta\mathbf{x} + \frac{1}{2} \Delta\mathbf{x}^T \underbrace{\left.\frac{\partial^2 \mathrm{F}}{\partial \mathbf{x} \partial \mathbf{x}^T}\right|_{\mathbf{x}=\mathbf{x}_k}}_{\nabla^2_{\mathbf{x}} \mathrm{F}(\mathbf{x}_k)} \Delta\mathbf{x} = \\
&= \mathrm{F}(\mathbf{x}_k) + \underbrace{\nabla_{\mathbf{x}} \mathrm{F}(\mathbf{x}_k)}_{\mathbf{g}_k^T} \Delta\mathbf{x} + \frac{1}{2} \Delta\mathbf{x}^T \underbrace{\nabla^2_{\mathbf{x}} \mathrm{F}(\mathbf{x}_k)}_{\mathbf{H}_k} \Delta\mathbf{x} = \\
&= \mathrm{F}(\mathbf{x}_k) + \mathbf{g}_k^T \Delta\mathbf{x} + \frac{1}{2} \Delta\mathbf{x}^T \mathbf{H}_k \Delta\mathbf{x}
\end{aligned} \quad (22)$$

where we introduce the first and second-order derivatives of $\mathrm{F}(\mathbf{x})$, namely the gradient vector $\mathbf{g}_k = \nabla_{\mathbf{x}} \mathrm{F}(\mathbf{x})\big|_{\mathbf{x}=\mathbf{x}_{k-1}}^T$ and the Hessian matrix $\mathbf{H}_k = \nabla^2_{\mathbf{x}} \mathrm{F}(\mathbf{x}_k)\big|_{\mathbf{x}=\mathbf{x}_{k-1}}$, both evaluated at the central point $\mathbf{x}_{k-1}$—the next subsection will review how to evaluate these two terms. Geometrically speaking, the former defines the direction in the solution space in which the error function most quickly increases, while the latter contains the information about the curvature of the cost function.

By taking now derivatives with respect to an increment in the unknowns:

$$\begin{aligned}
&\frac{\partial \mathrm{F}(\mathbf{x}_k + \Delta\mathbf{x})}{\partial \Delta\mathbf{x}} \approx \frac{\partial \hat{\mathrm{F}}_k(\mathbf{x}_k + \Delta\mathbf{x})}{\partial \Delta\mathbf{x}} = \\
&= \underbrace{\frac{\partial}{\partial \Delta\mathbf{x}} \{\mathrm{F}(\mathbf{x}_k)\}}_{0} + \frac{\partial}{\partial \Delta\mathbf{x}} \{\mathbf{g}^T \Delta\mathbf{x}\} + \\
&+ \frac{\partial}{\partial \Delta\mathbf{x}} \left\{\frac{1}{2} \Delta\mathbf{x}^T \mathbf{H} \Delta\mathbf{x}\right\} =
\end{aligned}$$

(provided that $\frac{\partial \mathbf{a}^T \mathbf{M} \mathbf{a}}{\partial \mathbf{a}} = \left(\mathbf{M} + \mathbf{M}^T\right)\mathbf{a}$ and $\frac{\partial \mathbf{a}^T \mathbf{b}}{\partial \mathbf{b}} = \mathbf{a}$, for any vector $\mathbf{a}$ and $\mathbf{b}$ and any matrix $\mathbf{M}$, and since $\mathbf{H}^T = \mathbf{H}$)

$$\rightarrow \quad \frac{\partial \hat{\mathrm{F}}_k\left(\mathbf{x}_k + \mathbf{\Delta x}\right)}{\partial \mathbf{\Delta x}} = \mathbf{g} + \mathbf{H}\,\mathbf{\Delta x} \tag{23}$$

Identifying the derivative to zero, we can find the step $\mathbf{\Delta x}_k$ that minimizes our model $\hat{\mathrm{F}}_k\left(\mathbf{x}\right)$ of the error function[4]:

$$\left.\frac{\partial \hat{\mathrm{F}}_k\left(\mathbf{x}_k + \mathbf{\Delta x}\right)}{\partial \mathbf{\Delta x}}\right|_{\mathbf{\Delta x}=0} = 0 \rightarrow \mathbf{g}_k + \mathbf{H}_k\,\mathbf{\Delta x}_k^* = 0$$

Therefore, $\mathbf{\Delta x}_k^*$ is obtained by solving the linear system of the form $\mathbf{A}\,\mathbf{x} = \mathbf{b}$:

$$\mathbf{H}_k \mathbf{\Delta x}_k^* = -\mathbf{g}_k \tag{24}$$

The resulting increment is then used to compute the new estimate $\mathbf{x}_k = \mathbf{x}_{k-1} + \mathbf{\Delta x}_k^*$, around which the gradient and Hessian must be evaluated again, leading to a new linear system. The process is repeated until some convergence criterion is reached (Madsen, Nielsen, & Tingleff, 2004). The iterative nature of the solution arises as a consequence of optimizing a *model* of the cost function which only approximates its actual shape in the neighborhood of the central point. We have just described a technique named the *Newton method*, which represents the starting point from which many other second-order optimizers derive (Dennis & Schnabel, 1996, Kelley, 1999).

An outstanding characteristic of the Newton method is its quadratic rate of convergence, which means that for every step, the new increment $\mathbf{\Delta x}_k^*$ is in the order of $\mathbf{\Delta x}_{k-1}^{*\,2}$. Although this means that convergence may be reached in very few steps (e.g. the number of significant digits almost doubles with each iteration), in practice that is only true when the estimate $\mathbf{x}_k$ is in the proximity of the optimal, where the Hessian can approximate well the curvature of the error function. Otherwise, the Newton method has a very slow rate of convergence, especially for a poor initial guess $\mathbf{x}_0$.

That is why Levenberg-Marquardt (LM) algorithms are more often employed instead. In LM, a damping parameter λ_k is introduced in the linear system to modify the Hessian matrix:

$$\left(\mathbf{H}_k + \lambda_k \mathbf{D}\right)\mathbf{\Delta x}_k^* = -\mathbf{g}_k$$

with any positive definite matrix $\mathbf{D}$, typically the identity:

$$\underbrace{\left(\mathbf{H}_k + \lambda_k \mathbf{I}\right)}_{\mathbf{H}'_k}\mathbf{\Delta x}_k^* = -\mathbf{g}_k \tag{25}$$

The damping parameter controls the direction of the increments to something in between[5] the Newton method (for $\lambda_k = 0$) and the direction of the gradient (for $\lambda_k \rightarrow \infty$). These strategies are appropriate when being close or far from the optimal, respectively; that is the reason why λ_k is modified heuristically with each iteration in the LM algorithm.

Regarding the solution of the linear system itself, solving Equation 25 for $\mathbf{\Delta x}^*$ is possible by directly inverting the (modified) Hessian $\mathbf{H}'_k$ that is, $\mathbf{\Delta x}^* = -\mathbf{H}'^{-1}_k \mathbf{g}$. However, the sparse structure of $\mathbf{H}'_k$ makes advisable to exploit sparse linear algebra methods, such as decomposing it with a sparse Cholesky algorithm (Davis, 2006):

$$\mathbf{H}'_k = \mathbf{L}\mathbf{L}^T \tag{26}$$

which turns the original linear system into two triangular systems, much more efficient to solve by forward and back-substitution (Meyer, 2001), respectively:

$$\mathbf{L}\underbrace{\mathbf{L}^T\Delta\mathbf{x}^*}_{\mathbf{y}} = -\mathbf{g}$$
$$\rightarrow \begin{cases} 1^{st} \text{ step}: \mathbf{L}\,\mathbf{y} = -\mathbf{g} & \rightarrow \text{ Solve for } \mathbf{y} \\ 2^{nd} \text{ step}: \mathbf{L}^T\Delta\mathbf{x}^* = \mathbf{y} & \rightarrow \text{ Solve for } \Delta\mathbf{x}^* \end{cases} \tag{27}$$

In contrast to the cubic complexity of the naive Hessian inversion method, using a sparse Cholesky algorithm and solving the two sparse triangular systems may achieve, under ideal conditions, a nearly linear complexity with the number of graph nodes. Nevertheless, the actual performance strongly depends on the density of edges, their *ordering of appearance* in the Hessian and on the specific Cholesky implementation. One of the most relevant and efficient implementations at present is CHOLMOD, available as a C library and as a function within MATLAB (Chen, Davis, Hager, & Rajamanickam, 2008).

At this point, it should be clear that MLE in graph SLAM can be approached through an iterative process, involving the two alternating steps of constructing and solving a sparse linear system of equations. In the next section, we describe how to build that sparse system from the graph elements, but before that, we must address the issue of dealing with an optimization that does not take on a Euclidean space, but on a manifold. According to the method explained in section 2, there exist up to three required changes in our method - recall Equation 5 - from which only one has to be considered at this point (the other two will be applied in the next section):

$$\mathbf{x}_k \leftarrow \mathbf{x}_{k-1} + \Delta\mathbf{x}_k^* \;\Rightarrow\; \mathbf{x}_k \leftarrow \mathbf{x}_{k-1} \boxplus \varepsilon_k^* \tag{28}$$

where ε_k now represents increments in the linearized manifold around each frame pose. Arbitrarily choosing the pre-multiplication convention from the two possibilities in Equation 6 and denoting the global poses of nodes and their on-manifold corrections as $\mathbf{x}_k^i$ and ε_k^{i*} with $i = 1 \ldots N$, respectively, we end up with the following pose updates:

$$\mathbf{x}_k^i \leftarrow \mathbf{x}_{k-1}^i \boxplus \varepsilon_k^{i*} \;\Rightarrow\; \mathbf{x}_k^i \leftarrow \varepsilon_k^{i*} \oplus \mathbf{x}_{k-1}^i \tag{29}$$

To finish the description of the sparse LM solver we have summarized all the steps as pseudo-code in Algorithm 1.

Efficiently Building the Sparse Linear System

We have seen above how solving for the unknowns in graph SLAM requires providing the gradient $\mathbf{g}$ and the Hessian $\mathbf{H}$ of the cost function $\mathrm{F}(\mathbf{x})$ in order to solve the system of equations $\mathbf{H}\Delta\mathbf{x} = -\mathbf{g}$. Now we will see how to take advantage of the particular form of that cost function for efficiently computing $\mathbf{g}$ and building a sparse representation of $\mathbf{H}$.

It follows from the MLE interpretation of graph SLAM in Equation 19 that the cost function in the vicinity of any point $\mathbf{x}$, given a small displacement $\Delta\mathbf{x}$, becomes the sum of one quadratic residual for each edge, that is:

$$\mathrm{F}(\mathbf{x}+\Delta\mathbf{x}) = \frac{1}{2}\sum_{\{i,j\}\in\Upsilon} \mathbf{r}_{i,j}^T(\mathbf{x}+\Delta\mathbf{x})\,\mathbf{\Lambda}_{i,j}\,\mathbf{r}_{i,j}(\mathbf{x}+\Delta\mathbf{x}) \tag{30}$$

Since residuals $\mathbf{r}_{i,j}(\mathbf{x})$ vary non-linearly with $\mathbf{x}$ the overall cost is also non-linear. As explained above, the most interesting optimizers assume a second-order Taylor approximation of $\mathrm{F}(\mathbf{x})$. We show next how a first-order approximation of the residuals takes us to the parameters of the second-order approximation $\hat{\mathrm{F}}(\mathbf{x}+\Delta\mathbf{x}) \approx \mathrm{F}(\mathbf{x}+\Delta\mathbf{x})$.

Algorithm 1. The Levenberg-Marquardt algorithm for problems with a sparse structure and on-manifold unknowns. Given a starting point $\mathbf{x}_0$, the algorithm looks for the closest critical point of the target function $\mathrm{F}(\mathbf{x})$ assuming the availability of its gradient vector and its Hessian matrix—whose computation is given in Algorithm 2. Optimization ends when one of three termination criteria is reached: (1) a maximum number of iterations, (2) a value of the gradient infinity norm (i.e. the maximum of its element absolute values) is below a threshold, or (3) we get a negligible incremental correction (Madsen, Nielsen, & Tingleff, 2004).

algorithm sparse_Levenberg_Marquardt_on_manifold
Inputs: $\mathbf{x}_0$ (Initial guess of unknown poses $\mathbf{X}$), $\mathbf{z}_F$ (known, fixed poses), $\{\mathbf{z}_{i,j}\}_\Upsilon$ (the edges)
Outputs: $\mathbf{x}^*$ (Approximation to the optimal solution)

1: $k \leftarrow 1$ // Iteration counter
2: $\{\mathbf{g}_k, \mathbf{H}_k\} \leftarrow \mathrm{graph\_SLAM\_sparse\_linear\_system}(\mathbf{x}_{k-1}, \underbrace{\mathbf{x}_F, \{\mathbf{z}_{i,j}\}_\Upsilon}_{\text{Known data}})$
3: $\lambda \leftarrow \tau \max(\mathrm{diag}(\mathbf{H}_k))$ // Initial value for damping parameter
4: $\nu \leftarrow 2$ // Heuristic factor for step control
5: $\mathrm{done} \leftarrow \|\mathbf{g}_k\|_\infty < \tau_1$ // Termination criterion #1
5: **while** ($k < k_{max}$ and $\neg$done) // Termination criteria
  5.1: $\varepsilon_k^* \leftarrow \mathrm{sparse\_cholesky\_solver}((\mathbf{H}_k + \lambda\mathbf{I})\varepsilon_k^* = -\mathbf{g}_k)$ // Solve linear system
  5.2: **if** $(|\varepsilon_k^*| < \tau_2(|\varepsilon_k^*| + \tau_2))$ // Termination criterion #2
    5.2.1: done $\leftarrow$ *true*
    else
    5.2.2: $\mathbf{x}_k \leftarrow \mathbf{x}_{k-1} \boxplus \varepsilon_k^*$ // New tentative solution: on-manifold increments
    5.2.3: $l \leftarrow \dfrac{\mathrm{F}(x_{k-1}) - \mathrm{F}(x_k)}{\frac{1}{2}\varepsilon_k^{*T}(\lambda\varepsilon_k^* - \mathbf{g}_k)}$ // Discriminant
    5.2.4: **if** $(l > 0)$
      5.2.4.1: $k \leftarrow k + 1$ // Accept this new estimate
      5.2.4.2: $\{\mathbf{g}_k, \mathbf{H}_k\} \leftarrow \mathrm{graph\_SLAM\_sparse\_linear\_system}(\mathbf{x}_{k-1}, \underbrace{\mathbf{x}_F, \{\mathbf{z}_{i,j}\}_\Upsilon}_{\text{Known data}})$
      5.2.4.3: $\mathrm{done} \leftarrow \|\mathbf{g}_k\|_\infty < \tau_1$ // Termination criterion #1
      5.2.4.4: $\lambda \leftarrow \lambda \max\left(\frac{1}{3}, 1 - (2l-1)^3\right)$, $\nu \leftarrow 2$
      else
      5.2.4.5: $\mathbf{x}_k \leftarrow \mathbf{x}_{k-1}$ // Reject new estimate
      5.2.4.6: $\lambda \leftarrow \lambda\nu$, $\nu \leftarrow 2\nu$ // and retry with a larger λ
      5.2.4.7: $k \leftarrow k + 1$
      5.2.4.8: $\{\mathbf{g}_k, \mathbf{H}_k\} \leftarrow \{\mathbf{g}_{k-1}, \mathbf{H}_{k-1}\}$ // Reuse previous values
      end-if
    end-if
  end-while
6: Set output: $\mathbf{x}^* \leftarrow \mathbf{x}_k$

Again, the non-Euclidean nature of the residuals will be ignored for now for the sake of clarity.

Each edge $i \rightarrow j$ has an associated residual $\mathbf{r}_{i,j}(\mathbf{x})$ for which a linearized approximation $\hat{\mathbf{r}}_{i,j}(\mathbf{x})$ can be defined as:

$$\mathbf{r}_{i,j}(\mathbf{x}+\Delta\mathbf{x}) \approx \hat{\mathbf{r}}_{i,j}(\mathbf{x}+\Delta\mathbf{x}) = \mathbf{r}_{i,j}(\mathbf{x}) + \underbrace{\left.\frac{\partial \mathbf{r}_{i,j}(\breve{\mathbf{x}})}{\partial \breve{\mathbf{x}}}\right|_{\breve{\mathbf{x}}=\mathbf{x}}}_{\mathbf{J}_{i,j}} \Delta\mathbf{x} = \mathbf{r}_{i,j}(\mathbf{x}) + \mathbf{J}_{i,j}\Delta\mathbf{x} \quad (31)$$

The key observation that allows for a sparse construction of the problem is that $\mathbf{r}_{i,j}(\mathbf{x})$ does not depend on the entire vector $\mathbf{x}$, but only on the two poses $\mathbf{x}_i$ and $\mathbf{x}_j$; hence its Jacobian only contains, at most, two nonzero blocks:

$$\mathbf{J}_{i,j} = \begin{bmatrix} 0 & \cdots & 0 & \mathbf{J}_{i,j}^{i} & 0 & \cdots & 0 & \mathbf{J}_{i,j}^{j} & 0 & \cdots & 0 \end{bmatrix}$$

where:

$$\mathbf{J}_{i,j}^{i} = \left.\frac{\partial \mathbf{r}_{i,j}(\breve{\mathbf{x}})}{\partial \breve{\mathbf{x}}_i}\right|_{\breve{\mathbf{x}}=\mathbf{x}} \qquad \mathbf{J}_{i,j}^{j} = \left.\frac{\partial \mathbf{r}_{i,j}(\breve{\mathbf{x}})}{\partial \breve{\mathbf{x}}_j}\right|_{\breve{\mathbf{x}}=\mathbf{x}} \quad (32)$$

We said "at most" because if either $\mathbf{x}_i$ or $\mathbf{x}_j$ belongs to the set of fixed (known) poses, its corresponding Jacobian block ($\mathbf{J}_{i,j}^{i}$ or $\mathbf{J}_{i,j}^{j}$) will not appear in $\mathbf{J}_{i,j}$. By replacing the approximation of the residuals above into Equation 30 we find the approximate model $\hat{\mathrm{F}}(\mathbf{x}+\Delta\mathbf{x})$:

$$\mathrm{F}(\mathbf{x}+\Delta\mathbf{x}) \approx \hat{\mathrm{F}}(\mathbf{x}+\Delta\mathbf{x}) = = \frac{1}{2}\sum_{\{i,j\}\in\Upsilon} \hat{\mathbf{r}}_{i,j}^{T}(\mathbf{x}+\Delta\mathbf{x})\boldsymbol{\Lambda}_{i,j}\hat{\mathbf{r}}_{i,j}(\mathbf{x}+\Delta\mathbf{x}) = = \frac{1}{2}\sum_{\{i,j\}\in\Upsilon} \left(\mathbf{r}_{i,j}(\mathbf{x})+\mathbf{J}_{i,j}\Delta\mathbf{x}\right)^{T}\boldsymbol{\Lambda}_{i,j}\left(\mathbf{r}_{i,j}(\mathbf{x})+\mathbf{J}_{i,j}\Delta\mathbf{x}\right) =$$

(denoting $\mathbf{r}_{i,j} \triangleq \mathbf{r}_{i,j}(\mathbf{x})$ for clarity of notation)

$$= \frac{1}{2}\sum_{\{i,j\}\in\Upsilon} \mathbf{r}_{i,j}^{T}\boldsymbol{\Lambda}_{i,j}\mathbf{r}_{i,j} + \underbrace{\mathbf{r}_{i,j}^{T}\boldsymbol{\Lambda}_{i,j}\mathbf{J}_{i,j}\Delta\mathbf{x} + \Delta\mathbf{x}^{T}\mathbf{J}_{i,j}^{T}\boldsymbol{\Lambda}_{i,j}\mathbf{r}_{i,j}}_{\text{Both terms evaluate to the same scalar value}} + +\Delta\mathbf{x}^{T}\mathbf{J}_{i,j}^{T}\boldsymbol{\Lambda}_{i,j}\mathbf{J}_{i,j}\Delta\mathbf{x} =$$

$$= \sum_{\{i,j\}\in\Upsilon} \underbrace{\frac{1}{2}\mathbf{r}_{i,j}^{T}\boldsymbol{\Lambda}_{i,j}\mathbf{r}_{i,j}}_{\text{Constant term}} + \underbrace{\mathbf{r}_{i,j}^{T}\boldsymbol{\Lambda}_{i,j}\mathbf{J}_{i,j}}_{=\left(\mathbf{J}_{i,j}^{T}\boldsymbol{\Lambda}_{i,j}\mathbf{r}_{i,j}\right)^{T}}\Delta\mathbf{x} + +\frac{1}{2}\Delta\mathbf{x}^{T}\mathbf{J}_{i,j}^{T}\boldsymbol{\Lambda}_{i,j}\mathbf{J}_{i,j}\Delta\mathbf{x} \quad (33)$$

Comparing this expression to the second-order Taylor series expansion of the cost function in Equation 22 we easily identify the gradient and Hessian as formed by individual contributions from each edge:

$$\hat{\mathrm{F}}(\mathbf{x}+\Delta\mathbf{x}) = \mathrm{F}(\mathbf{x}) + \mathbf{g}^{T}\Delta\mathbf{x} + \frac{1}{2}\Delta\mathbf{x}^{T}\mathbf{H}\Delta\mathbf{x} = \sum_{\{i,j\}\in\Upsilon} \frac{1}{2}\mathbf{r}_{i,j}^{T}\boldsymbol{\Lambda}_{i,j}\mathbf{r}_{i,j} + \underbrace{\left(\mathbf{J}_{i,j}^{T}\boldsymbol{\Lambda}_{i,j}\mathbf{r}_{i,j}\right)^{T}}_{\mathbf{g}_{i,j}}\Delta\mathbf{x} + +\frac{1}{2}\Delta\mathbf{x}^{T}\underbrace{\mathbf{J}_{i,j}^{T}\boldsymbol{\Lambda}_{i,j}\mathbf{J}_{i,j}}_{\mathbf{H}_{i,j}}\Delta\mathbf{x}$$

$$\Rightarrow \begin{cases} \mathbf{g} = \sum_{\{i,j\}\in\Upsilon} \mathbf{g}_{i,j}, & \mathbf{g}_{i,j} = \mathbf{J}_{i,j}^{T}\boldsymbol{\Lambda}_{i,j}\mathbf{r}_{i,j} \\ \mathbf{H} = \sum_{\{i,j\}\in\Upsilon} \mathbf{H}_{i,j}, & \mathbf{H}_{i,j} = \mathbf{J}_{i,j}^{T}\boldsymbol{\Lambda}_{i,j}\mathbf{J}_{i,j} \end{cases} \quad (34)$$

We can further develop those expressions by taking into account the sparsity of the Jacobians, according to Equation 32. It is then found out that each edge contributes (at most) with two nonzero blocks to the gradient and four to the Hessian (of which two are the transpose of each other):

$$\mathbf{g}_{i,j} = \mathbf{J}_{i,j}^{T} \mathbf{\Lambda}_{i,j} \mathbf{r}_{i,j} = \begin{bmatrix} 0 \\ \vdots \\ 0 \\ \mathbf{J}_{i,j}^{i\ T} \mathbf{\Lambda}_{i,j} \mathbf{r}_{i,j} \\ 0 \\ \vdots \\ 0 \\ \mathbf{J}_{i,j}^{j\ T} \mathbf{\Lambda}_{i,j} \mathbf{r}_{i,j} \\ 0 \\ \vdots \\ 0 \end{bmatrix}$$

and Equation 35 (see Box 3).

It is helpful to reflect now a moment about the dimensions of the vectors and matrices involved in the present discussion. Let $|\Upsilon|$ denote the number of edges in the graph, N the number of free poses (the unknowns in $\mathbf{x}$) and p the dimensionality of each pose parameterization (i.e. 3 for 2D poses, 6 for Euler angles or 7 for unit quaternions). Then, the gradient can be seen as composed of N vectors of length p, while the Hessian comprises $N \times N = N^2$ blocks, each of size $p \times p$. However, of those N^2 blocks, all but $O(|\Upsilon|)$ are exactly zero, and since the number of edges in a typical mapping scenario grows linearly with the number of frames, it turns out that the number of non-zero blocks is $O(N)$. That is, by considering the sparseness of the problem we turn the construction of the system $\mathbf{H}\,\mathbf{\Delta x} = -\mathbf{g}$ into a problem of linear complexity with the size of the map (a really nice property!).

We must now focus on the fact that both the observation space and the state space are not Euclidean but pose manifolds. The reader should be aware of the involvement of two different pose parameterizations, which, in practice, may or may not coincide. For instance, the rotational part of the frame poses can be stored as Euler angles, while frame-to-frame constraints may be encoded as unit quaternions. For simplicity, we will assume the same parameterization of length p for both spaces, whose manifold dimensionality will be denoted as m. The only two possibilities

Box 3.

$$\mathbf{H}_{i,j} = \mathbf{J}_{i,j}^{T} \mathbf{\Lambda}_{i,j} \mathbf{J}_{i,j}$$

$$= \begin{bmatrix} 0 & & & & & \cdots & & & & 0 \\ & \ddots & & & & & & & & \\ & & 0 & & & & & & & \\ & & & \mathbf{J}_{i,j}^{i\ T} \mathbf{\Lambda}_{i,j} \mathbf{J}_{i,j}^{i} & 0 & \cdots & 0 & \mathbf{J}_{i,j}^{j\ T} \mathbf{\Lambda}_{i,j} \mathbf{J}_{i,j}^{i} & & \\ & & & 0 & 0 & & & 0 & & \\ \vdots & & & \vdots & & \ddots & & \vdots & & \vdots \\ & & & 0 & & & 0 & 0 & & \\ & & & \mathbf{J}_{i,j}^{i\ T} \mathbf{\Lambda}_{i,j} \mathbf{J}_{i,j}^{j} & 0 & \cdots & 0 & \mathbf{J}_{i,j}^{j\ T} \mathbf{\Lambda}_{i,j} \mathbf{J}_{i,j}^{j} & & \\ & & & & & & & & 0 & \\ & & & & & & & & & \ddots \\ 0 & & & & & \cdots & & & & 0 \end{bmatrix} \quad (35)$$

are working on $\mathbf{SE}(2)$ or $\mathbf{SE}(3)$ with $m = 3$ and $m = 6$, respectively.

The framework for on-manifold optimization presented in section 2 identified the need for three changes with respect to the Euclidean case. One of them was already applied in the last subsection; the other two replacements in Equation 5 must be considered at this point:

Optimization: take derivatives with respect to increments in the manifold

$$\Delta\mathbf{x}^* \leftarrow \left.\frac{\partial \hat{F}\left(\mathbf{x} + \Delta\mathbf{x}\right)}{\partial \Delta\mathbf{x}}\right|_{\Delta\mathbf{x}=0} = 0$$
$$\Rightarrow \quad \varepsilon^* \leftarrow \left.\frac{\partial \hat{F}\left(\mathbf{x} \boxplus \varepsilon\right)}{\partial \varepsilon}\right|_{\varepsilon=0} = 0$$

Residuals: consider errors measured on the manifold

$$\begin{aligned} &\mathbf{r}_{i,j}\left(\mathbf{x}\right) = \mathbf{h}_{i,j}\left(\mathbf{x}\right) \ominus \mathbf{z}_{i,j} \\ \Rightarrow \quad &\tilde{\mathbf{r}}_{i,j}\left(\mathbf{x}\right) = \mathbf{h}_{i,j}\left(\mathbf{x}\right) \boxminus \mathbf{z}_{i,j} \end{aligned} \tag{36}$$

The first change forces us to reconsider the approximation of the residuals to account for increments in the manifold ($\mathbf{x} \boxplus \varepsilon$), that is, instead of Equation 31 we now have:

$$\begin{aligned} \tilde{\mathbf{r}}_{i,j}\left(\mathbf{x} \boxplus \varepsilon\right) &\approx \hat{\tilde{\mathbf{r}}}_{i,j}\left(\mathbf{x} \boxplus \varepsilon\right) = \\ &= \tilde{\mathbf{r}}_{i,j}\left(\mathbf{x}\right) + \underbrace{\left.\frac{\partial \tilde{\mathbf{r}}_{i,j}\left(\mathbf{x} \boxplus \varepsilon\right)}{\partial \varepsilon}\right|_{\varepsilon=0}}_{\tilde{\mathbf{J}}_{i,j}} \varepsilon = \\ &= \tilde{\mathbf{r}}_{i,j}\left(\mathbf{x}\right) + \tilde{\mathbf{J}}_{i,j}\, \varepsilon \end{aligned} \tag{37}$$

Propagating this change throughout Equations 32, 33, 34, and 35, it can be seen that the expressions for the gradient and the Hessian only change in the usage of the modified Jacobians $\tilde{\mathbf{J}}_{i,j}$ on the manifold instead of the original $\mathbf{J}_{i,j}$, which depend on the parameterization of rotations in the state vector. We consider now the residuals on the manifold (the second replacement mentioned above), for which we must define first the observation model, i.e. the prediction of relative pose measurements for an edge $i \rightarrow j$:

$$\mathbf{h}_{i,j}\left(\mathbf{x}\right) \triangleq \mathbf{h}_{i,j}\left(\mathbf{x}_i, \mathbf{x}_j\right) = \mathbf{x}_j \ominus \mathbf{x}_i \tag{38}$$

This can be read as "the pose $\mathbf{x}_j$ as seen from $\mathbf{x}_i$"—refer to Appendix A. Then, the residuals on the manifold become:

$$\tilde{\mathbf{r}}_{i,j}\left(\mathbf{x}\right) = \mathbf{h}_{i,j}\left(\mathbf{x}_i, \mathbf{x}_j\right) \boxminus \mathbf{z}_{i,j} =$$

(by arbitrarily taking the second convention from Equation 7)

$$\begin{aligned} &= \ln\left(\mathbf{h}_{i,j}\left(\mathbf{x}_i, \mathbf{x}_j\right) \ominus \mathbf{z}_{i,j}\right) = \\ &= \ln\left(\left(\mathbf{x}_j \ominus \mathbf{x}_i\right) \ominus \mathbf{z}_{i,j}\right) = \ln\left(\mathbf{r}_{i,j}\left(\mathbf{x}\right)\right) \end{aligned} \tag{39}$$

Regarding the on-manifold Jacobian $\tilde{\mathbf{J}}_{i,j}$, it has the same sparse structure than the on-parameterization Jacobians $\mathbf{J}_{i,j}$, that is, $\tilde{\mathbf{J}}_{i,j}$ only comprises (at most) two nonzero blocks $\tilde{\mathbf{J}}^i_{i,j}$ and $\tilde{\mathbf{J}}^j_{i,j}$. Their evaluation can be simplified by putting them in terms of the original Jacobians $\mathbf{J}^i_{i,j}$ and $\mathbf{J}^j_{i,j}$ by means of the chain rule. If we denote the part of the manifold increment ε that affects the $i-th$ pose as ε_i, it follows that:

$$\tilde{\mathbf{J}}^i_{i,j} = \left.\frac{\partial \tilde{\mathbf{r}}_{i,j}\left(\mathbf{x} \boxplus \varepsilon_i\right)}{\partial \varepsilon_i}\right|_{\varepsilon_i=0} = \underbrace{\left.\frac{\partial \ln\left(\mathbf{y}\right)}{\partial \mathbf{y}}\right|_{\mathbf{y}=\mathbf{r}_{i,j}(\mathbf{x})}}_{\substack{\mathbf{L}(\mathbf{y}) \\ \text{Jacobian of the} \\ \text{logarithm map}}} \times \underbrace{\left.\frac{\partial \mathbf{r}_{i,j}\left(\breve{\mathbf{x}}_i, \mathbf{x}_j\right)}{\partial \breve{\mathbf{x}}_i}\right|_{\breve{\mathbf{x}}_i = \mathbf{x}_i \boxplus \varepsilon_i = \mathbf{x}_i}}_{\substack{\mathbf{J}^i_{i,j} \\ \text{The on-parameterization} \\ \text{Jacobian}}} \times \underbrace{\left.\frac{\partial \mathbf{x}_i \boxplus \varepsilon_i}{\partial \varepsilon_i}\right|_{\varepsilon_i=0}}_{\substack{\mathbf{E}(\mathbf{x}) \\ \text{Jacobian of manifold} \\ \text{increments}}} \quad (40)$$

thus:

$$\tilde{\mathbf{J}}^i_{i,j} = \underbrace{\mathbf{L}\left(\mathbf{r}_{i,j}\left(\mathbf{x}\right)\right)}_{m \times p} \underbrace{\mathbf{J}^i_{i,j}}_{p \times p} \underbrace{\mathbf{E}\left(\mathbf{x}_i\right)}_{p \times m}$$
$$and \quad \tilde{\mathbf{J}}^j_{i,j} = \underbrace{\mathbf{L}\left(\mathbf{r}_{i,j}\left(\mathbf{x}\right)\right)}_{m \times p} \underbrace{\mathbf{J}^j_{i,j}}_{p \times p} \underbrace{\mathbf{E}\left(\mathbf{x}_j\right)}_{p \times m} \quad (41)$$

Notice that for two-dimensional poses we have $m = p = 3$ and both $\mathbf{L}$ and $\mathbf{E}$ are constants (in fact, they are the identity $\mathbf{I}_3$). In contrast, in the three-dimensional case they depend on the chosen parameterization. For the convenience of the reader, we made available some of the most common Jacobians in Appendix D.

To summarize all the discussion above we provide a pseudo-code description of the final algorithm for building the gradient and sparse Hessian in Algorithm 2. Implementers should take care of not misinterpreting the term $\widetilde{\mathbf{\Lambda}}_{i,j}$, which stands for the precision matrix of its corresponding constraint *on the manifold* (i.e. its size is $m \times m$). It is related to the precision matrix in the parameterization space ($\mathbf{\Lambda}_{i,j}$) by means of:

$$\widetilde{\mathbf{\Lambda}}_{i,j} = \left(\mathbf{L}\left(\mathbf{r}_{i,j}\right) \mathbf{\Lambda}_{i,j}^{-1} \mathbf{L}\left(\mathbf{r}_{i,j}\right)^T\right)^{-1} \quad (42)$$

It is worth stressing a warning at this point: although one may be tempted of extending the formula for linear propagation of covariances $\mathbf{\Sigma}_y = \mathbf{J} \mathbf{\Sigma}_x \mathbf{J}^T$ for propagating information matrices, the only situation when it holds that $\mathbf{\Sigma}_y^{-1} = \mathbf{J} \mathbf{\Sigma}_x^{-1} \mathbf{J}^T$ is when $\mathbf{J}^{-1} = \mathbf{J}^T$, i.e. for $\mathbf{J}$ being an orthonormal matrix—refer to Appendix E. In the general case when Jacobians do not satisfy this condition, Equation 42 must be used instead.

Finally, a few words are in order about the *uncertainty* of graph SLAM results. As opposed to classic Bayesian filtering approaches to SLAM, here we have only addressed the MLE of the poses $\mathbf{x}^*_k$, which does not provide direct information of the uncertainty of the result. However, and although uncertainty is seldom considered in the graph SLAM literature, it is actually easy to show[6] that the Hessian matrix approximates the inverse covariance matrix of the estimated parameters. The most relevant reason why one may want to find the uncertainty in the MLE of robot poses is establishing potential loop closures by means of geometric information—note that pose uncertainty imposes a bound to how far one should look for potential loop closure candidates. However, the Hessian derived in this section models the uncertainty on the manifold. To retrieve the uncertainty estimation in the parameterization space, the Hessian should be computed again using the original Jacobians instead of their manifold counterparts, as well as the information matrix corresponding to parameterization space; that is, this alternative Hessian matrix would be obtained by replacing step 4.8 in the algorithm with:

$$\begin{aligned}
\mathbf{H}(i,i) &\leftarrow \mathbf{H}(i,i) + \mathbf{J}^{i\ T}_{i,j} \mathbf{\Lambda}_{i,j} \mathbf{J}^i_{i,j} \\
\mathbf{H}(i,j) &\leftarrow \mathbf{H}(i,j) + \mathbf{J}^{j\ T}_{i,j} \mathbf{\Lambda}_{i,j} \mathbf{J}^i_{i,j} \\
\mathbf{H}(j,i) &\leftarrow \mathbf{H}(j,i) + \mathbf{J}^{i\ T}_{i,j} \mathbf{\Lambda}_{i,j} \mathbf{J}^j_{i,j} \\
\mathbf{H}(j,j) &\leftarrow \mathbf{H}(j,j) + \mathbf{J}^{j\ T}_{i,j} \mathbf{\Lambda}_{i,j} \mathbf{J}^j_{i,j}
\end{aligned} \quad (43)$$

Algorithm 2. Algorithm for efficiently building the sparse linear system $\tilde{\mathbf{H}}\mathbf{x} = -\tilde{\mathbf{g}}$. Recall that $\mathbf{x}$ and $\mathbf{x}_F$ stand for the poses of unknown and fixed (known) frames, respectively, and that Υ symbolizes the set of all graph edges. We used $\tilde{\mathbf{g}}(i)$ to represent the $i-th$ vector of length m within the gradient and $\tilde{\mathbf{H}}(i,j)$ for the block at the $i-th$ row and $i-th$ column of the Hessian, if viewed as composed of $N \times N$ blocks of size $m \times m$ each.

Algorithm graph_SLAM_sparse_linear_system
Inputs: $\{\mathbf{x}, \mathbf{x}_F\}$ (global poses of nodes), $\{\mathbf{z}_{i,j}\}_\Upsilon$ (edge observations)
Outputs: $\tilde{\mathbf{g}}$, $\tilde{\mathbf{H}}$ (The gradient and Hessian for a manifold increment ε)

1: $\tilde{\mathbf{g}} \leftarrow \mathbf{0}_{Nm\times 1}$ // Initialize gradient to zero
2: $\tilde{\mathbf{H}} \leftarrow empty\_sparse\_matrix(Nm \times Nm)$ // Initialize sparse data structures
3: **for-each** node n appearing in Υ
  3.1: $\mathbf{E}_n \leftarrow \left.\frac{\partial \mathbf{x}_n \boxplus \varepsilon_n}{\partial \varepsilon_n}\right|_{\varepsilon_n=0}$ // Precompute Jacobians of manifold increments
end-for-each
4: **for-each** edge $\langle i,j \rangle \in \Upsilon$ // (which provides $\mathbf{z}_{i,j}$ and $\mathbf{\Lambda}_{i,j}$)
  4.1: $\mathbf{r}_{i,j} \leftarrow \mathbf{h}_{i,j}(\mathbf{x}_i, \mathbf{x}_j) \ominus \mathbf{z}_{i,j}$ // Residuals on parameterization space
  4.2: $\mathbf{L}_{i,j} \leftarrow \left.\frac{\partial \ln(\mathbf{y})}{\partial \mathbf{y}}\right|_{\mathbf{y}=\mathbf{r}_{i,j}}$ // Jacobian of residuals on the manifold
  4.3: $\mathbf{J}^i_{i,j} \leftarrow \left.\frac{\partial \mathbf{r}_{i,j}(\breve{\mathbf{x}}_i, \mathbf{x}_j)}{\partial \breve{\mathbf{x}}_i}\right|_{\breve{\mathbf{x}}_i=\mathbf{x}_i}$ $\mathbf{J}^j_{i,j} \leftarrow \left.\frac{\partial \mathbf{r}_{i,j}(\mathbf{x}_i, \breve{\mathbf{x}}_j)}{\partial \breve{\mathbf{x}}_j}\right|_{\breve{\mathbf{x}}_j=\mathbf{x}_j}$
  4.4: $\tilde{\mathbf{J}}^i_{i,j} \leftarrow \mathbf{L}_{i,j}\mathbf{J}^i_{i,j}\mathbf{E}_i$, $\tilde{\mathbf{J}}^j_{i,j} \leftarrow \mathbf{L}_{i,j}\mathbf{J}^j_{i,j}\mathbf{E}_j$ // Jacobians of on-manifold residuals
  4.5: $\tilde{\mathbf{r}}_{i,j} \leftarrow \ln(\mathbf{r}_{i,j})$ // Residuals on the manifold
  4.6: $\tilde{\mathbf{\Lambda}}_{i,j} \leftarrow \left(\mathbf{L}_{i,j}\mathbf{\Lambda}_{i,j}^{-1}\mathbf{L}^T_{i,j}\right)^{-1}$ // Observation uncertainty in the manifold
  4.7: // Update gradient:
    $\tilde{\mathbf{g}}(i) \leftarrow \tilde{\mathbf{g}}(i) + \tilde{\mathbf{J}}^{i\,T}_{i,j}\tilde{\mathbf{\Lambda}}_{i,j}\tilde{\mathbf{r}}_{i,j}$ $\tilde{\mathbf{g}}(j) \leftarrow \tilde{\mathbf{g}}(j) + \tilde{\mathbf{J}}^{j\,T}_{i,j}\tilde{\mathbf{\Lambda}}_{i,j}\tilde{\mathbf{r}}_{i,j}$
  4.8: // Update Hessian:
    $\tilde{\mathbf{H}}(i,i) \leftarrow \tilde{\mathbf{H}}(i,i) + \tilde{\mathbf{J}}^{i\,T}_{i,j}\tilde{\mathbf{\Lambda}}_{i,j}\tilde{\mathbf{J}}^i_{i,j}$ $\tilde{\mathbf{H}}(i,j) \leftarrow \tilde{\mathbf{H}}(i,j) + \tilde{\mathbf{J}}^{j\,T}_{i,j}\tilde{\mathbf{\Lambda}}_{i,j}\tilde{\mathbf{J}}^i_{i,j}$
    $\tilde{\mathbf{H}}(j,i) \leftarrow \tilde{\mathbf{H}}(j,i) + \tilde{\mathbf{J}}^{i\,T}_{i,j}\tilde{\mathbf{\Lambda}}_{i,j}\tilde{\mathbf{J}}^j_{i,j}$ $\tilde{\mathbf{H}}(j,j) \leftarrow \tilde{\mathbf{H}}(j,j) + \tilde{\mathbf{J}}^{j\,T}_{i,j}\tilde{\mathbf{\Lambda}}_{i,j}\tilde{\mathbf{J}}^j_{i,j}$
end-for-each

Related Methods and Recent Developments

At present, most newest proposals for robotic localization and mapping employ some variant of graph SLAM, and due to the frenetic research activity in the field, new advances are continuously being published as this book is written; hence the following discussion may probably soon become obsolete and should be complemented by contemporary state-of-the-art research articles by the interested reader.

We must firstly mention another family of SLAM methods: the information-theoretic incremental solutions. Those works adopt the point of view of filtering or smoothing (over a fixed window of time steps) much like in the Kalman filter, but relying on the canonical form of the Gaussian (briefly mentioned in Chapter 7 and Appendix E):

Canonical form:

$$N^{-1}\left(\mathbf{x};\eta,\mathbf{\Lambda}\right) \propto \exp\left(-\frac{1}{2}\mathbf{x}^T\mathbf{\Lambda}\mathbf{x} + \eta^T\mathbf{x}\right)$$

Standard form:

$$N\left(\mathbf{x};\mu,\mathbf{\Sigma}\right) \propto \exp\left(-\frac{1}{2}\left(\mathbf{x}-\mu\right)^T\mathbf{\Sigma}^{-1}\left(\mathbf{x}-\mu\right)\right)$$

thus:

$$\underbrace{\mathbf{\Lambda}}_{\substack{\text{Information}\\\text{matrix}}} = \mathbf{\Sigma}^{-1}, \quad \underbrace{\eta}_{\substack{\text{Information}\\\text{vector}}} = \mathbf{\Sigma}^{-1}\mu \qquad (44)$$

The equivalence to the least-squares framework explained in this chapter so far is obvious by noticing that the information vector and matrix correspond to the minus gradient and the Hessian matrix, respectively. Dual versions of the Kalman filter method exist for this canonical form, probably being the most prominent of them the Square Root Information Filtering (SRIF) and Smoothing (SRIS) algorithms (Bierman, 1977). These methods can be seen as incremental ("on-line") versions of the popular batch ("off-line") solutions to least-squares MLE discussed in this chapter. Exactly as in least squares methods, recovering the state vector in these filters still implies solving a sparse linear system $\left(\mathbf{\Sigma}^{-1}\mu = \eta\right)$. Information filters can then be understood as alternative algorithms for building the Hessian matrix ($\mathbf{\Lambda}$ or $\mathbf{\Sigma}^{-1}$) in problems where old robot poses are marginalized out of the problem at each time step (Walter, Eustice, & Leonard, 2007). As a final remark, the clear superiority of SRIF in comparison to the Kalman filter or EKF, notably regarding numerical stability, makes advisable to always use them instead of the latter (Dellaert & Kaess, 2006).

Getting back to the body of work on graph SLAM, we review next some of the most influential methods proposed in recent years which differ from the approach introduced in previous sections. In Olson, Leonard, and Teller (2006), the authors proposed to attack the problem by a first-order optimization technique known as stochastic gradient descent, in which only one edge (picked at random) is considered at each iteration. Instead of working on the state space of node global poses, an alternative "incremental" pose state-space is used, where the coordinates of each node are obtained by arithmetic addition of the poses in all the edges from the origin—notice the difference between relative poses $\mathbf{a} \ominus \mathbf{b}$ and "incremental" poses $\mathbf{a} - \mathbf{b}$, where the effects of rotations are ignored. Thanks to that approximation, Jacobians are enormously simplified. Incremental steps are taken proportionally to the gradient, that is, $\mathbf{\Delta x}_k = \alpha \mathbf{g}_k$; the authors also proposed there the application of Jacobi preconditioning, a technique which improves the numerical accuracy of the solutions by left-multiplying the step with a diagonal matrix, in particular, with the diagonal of the Hessian. The advantages of this approach are its efficiency and an exceptional robustness against wrong initial guesses.

An interesting work derived from Olson *et al.*'s solution is TORO (after "*Tree-based netwORk Optimizer*"), based on a tree parameterization of the problem (Giorgio, Grzonka, Stachniss, Pfaff, & Burgard, 2007; Giorgio, Stachniss, Grzonka, & Burgard, 2007). There, a spanning tree is built from the graph and the state space on which optimization takes places is now the sequence of "incremental" poses of each node relative to its parent in the tree. In this way, each update of the stochastic gradient descent algorithm optimizing an observation in the edge $i \rightarrow j$ only affects those nodes along the path between them, according to the tree. Notice that optimization is only possible when there exists *another* path between i and j apart from the edge being optimized.

It was not until 2010 that we could find the first method proposing the usage of on-manifold optimization as a way to robustly consider all the constraints simultaneously in three-dimensional problems (Grisetti, Küemmerle, Stachniss, Frese, & Hertzberg, 2010). Furthermore, the authors also proposed to compensate the larger computational burden of this technique, if compared to TORO where only one edge was considered at once, by building a hierarchical graph that reduces the overall number of nodes and edges. Another interesting recent work is (Küemmerle, Grisetti, Strasdat, Konolige, & Burgard, 2011), where a general programming framework is presented for on-manifold optimization of a variety of graph SLAM-like problems.

Finally, we must mention the efforts related to enable graph SLAM for on-line operation. This is still challenging with state-of-the-art methods and large maps, since continuously adding nodes impedes achieving a proper computational complexity as for real-time performance. A simple and practical solution is the so-called "continuable" Levenberg-Marquardt algorithm proposed in (Konolige, Grisetti, Kümmerle, Burgard, Limketkai, & Vincent, 2010), which only runs one iteration at once, saving the value of the damping parameter (the λ in Algorithm 1) between iterations and updating the graph with new sensory information in between. Since the incremental changes in a graph does not typically change much the shape of the overall cost function, by doing so we are saving the initial steps that it takes to tune λ for each iteration. Only in the case of loop closures we could need to spend a few iterations until λ gets adapted to the new situation. Another active area of research is reducing the number of frames using grounded (versus *ad hoc*) principles. In Kretzschmar, Stachniss, and Grisetti (2011) the authors develop an information-theoretic method to determine which nodes can be safely removed from the graph without affecting too much to the posterior reconstruction of an occupancy grid map. However, removing nodes should make the graph less sparse (recall the effects of marginalization on graphs discussed in chapter 3 section 10), and thus less efficient to solve. To avoid the subsequent loss of efficiency, an approximate tree is built to minimize the losses by removing some of the edges that, in theory, should be added after marginalization.

5. VISUAL SLAM WITH BUNDLE-ADJUSTMENT

Bundle Adjustment (BA) is a popular method in the computer vision community aimed at solving the problem of simultaneously estimating both a map of landmarks of certain environment and the poses from which those were observed. Therefore, BA is another solution to the *full SLAM* estimation problem described in chapter 9 section 1. The defining characteristics of BA that set it apart from other solutions are (1) the usage of least-squares optimization to obtain a MLE of the unknowns (in contrast to the sequential Bayesian algorithms in chapter 9, and (2) the inclusion in the problem of a map of discrete entities, typically point landmarks or features (as opposed to graph SLAM where the map was marginalized out).

Just like in graph SLAM, the variables involved in the problem can be modeled as a GMRF, a graphical representation where nodes correspond to either a pose or a landmark (the unknowns), while edges are observations (known data) that involve pairs of nodes. The original formulation of BA in (Brown, 1958) only considered pose-to-landmark observations in the context of visual reconstruction of cartographic maps from aerial images (Triggs, McLauchlan, Hartley & Fitzgibbon, 2000). That is where the name bundle adjustment comes from: each image feature can be seen as contained in a straight line (a "ray") that emerges from the unknown camera pose, thus by minimizing the problem constraints we are adjusting "bundles of rays" corresponding to the same feature so they all coincide as accurately as possible around one single point, which will be the reconstructed 3D location of the landmark —refer to Figure 5a. It is quite common to also introduce the camera calibration parameters, a series of parameters that establish the relationship existing between the 3D coordinates of a world feature and the image pixel coordinates on which it projects, as unknowns which are also determined by this MLE process.

In spite of the strong association of BA with computer vision and visual-based SLAM, the least-squares framework and the techniques employed in BA can be perfectly adapted to many other SLAM-like problems, and it has the option to accommodate other sources of information about the problem variables. All those problems would then be solvable by the same methods classically applied to BA for visual reconstruction; hence, this section is also devoted to analyze some well-known applications of BA techniques, even if they are not related at all with camera sensors. As an example of BA applied to a non-visual problem, consider Figure 5b where a mobile robot explores an environment and detects landmarks with a 2D laser scanner. Under this perspective, the solutions to BA described below may refer to any sensor from which relative constraints could be derived between the robot and any kind of map entity. Furthermore, it is not uncommon in BA set-ups to also consider pose-to-pose constraints (as edges between pairs of pose nodes in the graphical model), just like in the graph SLAM framework.

All in all, it would be appropriate to consider BA as a generalization of graph SLAM, being the latter a special instance of BA that can deal with cases that use no landmarks in the map. Under a mobile robotics perspective, pose-to-pose constraints can be derived, like in graph SLAM, from a probabilistic motion model, while pose-to-landmark constraints come from any range-bearing or bearing-only sensor—range-only sensors are especially troublesome for their need to multi-modal distributions, thus other SLAM solutions are recommended instead (recall chapter 8 section 4). Therefore, behind the vision-related name of Bundle Adjustment we find a quite generic technique that aims at performing MLE on a GMRF by exploiting some information about the *problem structure* rising from our knowledge about the connection patterns of edges in the graph. In the next subsection, we explore how this structure leads to optimized solutions.

The Structure of Bundle Adjustment Problems

As explained above, in BA the aim is performing MLE on the problem unknowns given a set of observations. In a graphical representation, the former are nodes and the latter are edges. The solution is fundamentally similar to that described in previous chapters for graph SLAM, that is: (1) build a system of linear equations in the form $\mathbf{Ax} = \mathbf{b}$ that represents a linearized approximation of all the problem constraints, and (2) solve it as efficiently as possible using sparse algebra techniques, repeating this process until convergence is reached. Thus, instead of focusing on the algorithmic details of these two stages, as we did for graph SLAM, the interesting part in BA lies on understanding the high level structure present in the linear system. In the past, BA has often (and incorrectly) been discarded as a practical solution to full SLAM due to its computational complexity, but those analyses were only the result of ignoring the problem sparsity. Since the highly sparse structure of BA reveals itself in the Jacobian and Hessian matrices, a good starting point is to analyze them.

In section 4 we learned that a least-squares problem can be optimized by finding successive increments $\mathbf{\Delta x}_k^*$ to be added to the problem state estimation $\mathbf{x}_k$ by solving one of these linear systems of equations:

$$\mathbf{H}_k \mathbf{\Delta x}_k^* = -\mathbf{g}_k \text{ (Gauss-Newton step)}$$

Figure 5. (a) Bundle adjustment as the process of looking for the best coincidence of "bundles of rays" emerging from aerial camera poses towards the spatial location of the photographed landmarks. (b) A mobile robot detects features (e.g., wall corners) with a laser scanner. This is another estimation problem, which can be seen as an especial instance of bundle adjustment.

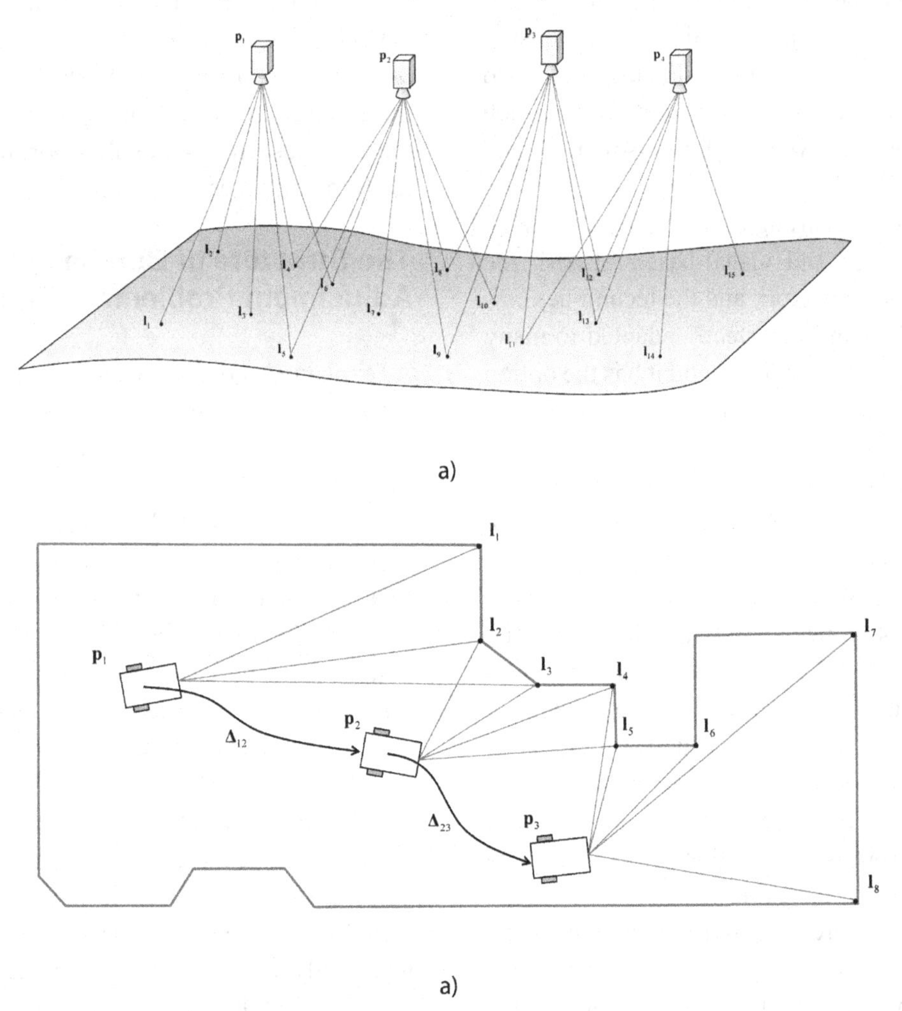

or

$$\left(\mathbf{H}_k + \lambda \mathbf{I}\right) \Delta \mathbf{x}_k^* = -\mathbf{g}_k$$

(Levenberg-Marquardt step) (45)

where

$\mathbf{g}_k = \nabla_{\mathbf{x}} \hat{\mathrm{F}}\left(\mathbf{x}\right)\Big|_{\mathbf{x}=\mathbf{x}_{k-1}}^{T}$ and $\mathbf{H}_k = \nabla_{\mathbf{x}}^2 \hat{\mathrm{F}}\left(\mathbf{x}\right)\Big|_{\mathbf{x}=\mathbf{x}_{k-1}}$ stand for the gradient and Hessian, respectively, of the second-order Taylor approximation $\hat{\mathrm{F}}\left(\mathbf{x}\right)$ of the cost function $\mathrm{F}\left(\mathbf{x}\right)$ that we want to optimize, representing the negative log likelihood of the GMRF graph.

In a typical BA problem[7], the variables $\mathbf{x}$ (i.e. the graph nodes) can be split into two categories: a set of N poses $\mathbf{p}_i$ and another set of M landmarks $\mathbf{l}_j$, that is, $\mathbf{x} = \left\{\mathbf{p}_{1:N}, \mathbf{l}_{1:M}\right\}$. For generality, let us assume a BA problem with two kinds of constraints: pose-to-landmark (i.e. landmark observations) and pose-to-pose (i.e. odometry, loop closures). By denoting these two sets of constraints as Υ and Θ, respectively, the associated cost function is then directly related to the full-SLAM model of chapter 9's Equation 5 and can be written down as:

$$\underbrace{\mathrm{F}(\mathbf{x})}_{\text{Cost function}} = \sum_{\{i,j\}\in\Upsilon} \underbrace{\mathrm{F}^{o}_{i,j}(\mathbf{x})}_{\substack{\text{Cost of each}\\ \text{observation}}} + \sum_{\{i,j\}\in\Theta} \underbrace{\mathrm{F}^{r}_{i,j}(\mathbf{x})}_{\substack{\text{Cost of relative}\\ \text{pose constraints}}} \tag{46}$$

which, assuming that all conditional probabilities in the MRF are Gaussians, is completed by the following definitions of the two types of individual cost functions, taking the appropriate one for each graph edge:

$$\begin{aligned} \mathrm{F}^{o}_{i,j}(\mathbf{x}) &\triangleq \mathrm{F}^{o}_{i,j}(\mathbf{p}_i,\mathbf{l}_j) = \\ &= \frac{1}{2}\sum_{\{i,j\}\in\Upsilon} \mathbf{r}^{o\,T}_{i,j}(\mathbf{p}_i,\mathbf{l}_j)\,\mathbf{\Lambda}^{o}_{i,j}\,\mathbf{r}^{o}_{i,j}(\mathbf{p}_i,\mathbf{l}_j) \\ \mathrm{F}^{r}_{i,j}(\mathbf{x}) &\triangleq \mathrm{F}^{r}_{i,j}(\mathbf{p}_i,\mathbf{p}_j) = \\ &= \frac{1}{2}\sum_{\{i,j\}\in\Theta} \mathbf{r}^{r\,T}_{i,j}(\mathbf{p}_i,\mathbf{p}_j)\,\mathbf{\Lambda}^{r}_{i,j}\,\mathbf{r}^{r}_{i,j}(\mathbf{p}_i,\mathbf{p}_j) \end{aligned} \tag{47}$$

Here, $\langle \mathbf{r}^{o}_{i,j}, \mathbf{\Lambda}^{o}_{i,j} \rangle$ and $\langle \mathbf{r}^{r}_{i,j}, \mathbf{\Lambda}^{r}_{i,j} \rangle$ stand for the *residuals* and their corresponding precision (information) matrices for each observation or relative pose constraint, respectively, that are associated to the edge $i \rightarrow j$. To illustrate this with an example, consider the BA problem depicted in Figure 6a, where there exist eight edges, all of them corresponding to pose-to-landmark observations. The state space $\mathbf{x}$ consists of the parameterization of three poses $\mathbf{p}_{1:3}$ (e.g., as quaternions) and four landmarks $\mathbf{l}_{1:4}$ (e.g., typically by their global Euclidean coordinates). As in graph SLAM, we must fix at least one pose in order to allow the problem to have one single solution, so we arbitrary assume $\mathbf{p}_1$ as the origin of coordinates. In this example, the overall cost function then reads (see Figure 6).

$$\begin{aligned} \mathrm{F}(\mathbf{x}) &= \sum_{\{i,j\}\in\Upsilon} \mathrm{F}^{o}_{i,j}(\mathbf{x}) \\ &= \mathrm{F}^{o}_{1,1}(\mathbf{p}_1,\mathbf{l}_1) + \mathrm{F}^{o}_{1,2}(\mathbf{p}_1,\mathbf{l}_2) + \mathrm{F}^{o}_{2,1}(\mathbf{p}_2,\mathbf{l}_1) + \mathrm{F}^{o}_{2,3}(\mathbf{p}_2,\mathbf{l}_3) + \\ &\quad \mathrm{F}^{o}_{2,3}(\mathbf{p}_2,\mathbf{l}_3) + \mathrm{F}^{o}_{3,2}(\mathbf{p}_3,\mathbf{l}_2) + \mathrm{F}^{o}_{3,3}(\mathbf{p}_3,\mathbf{l}_3) + \mathrm{F}^{o}_{3,4}(\mathbf{p}_3,\mathbf{l}_4) \end{aligned} \tag{49}$$

Also as with graph, SLAM cost functions (recall Equations 33, 34, and 35), we can easily demonstrate that each edge in the BA graph ends up contributing (at most) with two vector blocks to the gradient $\mathbf{g}_k$ and (at most) with four matrix blocks to the sparse Hessian $\mathbf{H}_k$. Again, the "at most" means that the rows and columns corresponding to variables whose values have been assumed to be known ($\mathbf{p}_1$ in the example) do not appear in the problem matrices.

A particularity found in BA is that we can exploit the division of variables into poses and landmarks to further factorize the problem. We can see this by analyzing the Jacobian structure for the cost function of the previous example (notice how we skip the column of derivatives with respect to the fixed variable $\mathbf{p}_1$):

$$\mathbf{J} = \frac{d\,\mathrm{F}(\mathbf{x})}{d\mathbf{x}} = \frac{d\,\mathrm{F}(\mathbf{p}_i,\mathbf{l}_j)}{d\{\mathbf{p}_i,\mathbf{l}_j\}} = \begin{array}{c|cccccc} & \frac{\partial\bullet}{\partial\mathbf{p}_2} & \frac{\partial\bullet}{\partial\mathbf{p}_3} & \frac{\partial\bullet}{\partial\mathbf{l}_1} & \frac{\partial\bullet}{\partial\mathbf{l}_2} & \frac{\partial\bullet}{\partial\mathbf{l}_3} & \frac{\partial\bullet}{\partial\mathbf{l}_4} \\ \hline \mathrm{F}^{o}_{\mathbf{p}_1,\mathbf{l}_1} & & & \blacksquare & & & \\ \mathrm{F}^{o}_{\mathbf{p}_1,\mathbf{l}_2} & & & & \blacksquare & & \\ \mathrm{F}^{o}_{\mathbf{p}_2,\mathbf{l}_1} & \blacksquare & & \blacksquare & & & \\ \mathrm{F}^{o}_{\mathbf{p}_2,\mathbf{l}_3} & \blacksquare & & & & \blacksquare & \\ \mathrm{F}^{o}_{\mathbf{p}_2,\mathbf{l}_4} & \blacksquare & & & & & \blacksquare \\ \mathrm{F}^{o}_{\mathbf{p}_3,\mathbf{l}_2} & & \blacksquare & & \blacksquare & & \\ \mathrm{F}^{o}_{\mathbf{p}_3,\mathbf{l}_3} & & \blacksquare & & & \blacksquare & \\ \mathrm{F}^{o}_{\mathbf{p}_3,\mathbf{l}_4} & & \blacksquare & & & & \blacksquare \end{array} \tag{48}$$

In a BA Jacobian we have as many rows of block matrices as observations (i.e. edges in the graph) and at most two nonzero entries will be found in each row, corresponding to the columns of the two involved variables (i.e. the graph nodes that the edge connects). Since fixed variables which do not enter in the optimization process do not appear in the Jacobian, the corresponding observations will have only one non-zero block in their rows. If we now derive the associated structure of the approximated Hessian $\mathbf{H} = \mathbf{J}^T \mathbf{\Lambda} \mathbf{J}$, and assume independence between observations, it follows that $\mathbf{\Lambda}$ is block diagonal and thus the Hessian structure is identical to that of $\mathbf{J}^T \mathbf{J}$. Box 4 follows the previous example.

We should spend a moment analyzing the resulting structure, because it will be found in all BA problems. To start with, it is clear that if we order all the problem variables like $\mathbf{x} = \left\{\mathbf{p}_{1:N}, \mathbf{l}_{1:M}\right\}$ so that all poses come first, the resulting Hessian could be always split in four parts: $\mathbf{H}_\mathrm{p}$, related to poses only; $\mathbf{H}_\mathrm{l}$, to landmarks only; and $\mathbf{H}_\mathrm{pl}$, to both. If we now consider the structure of the observations and their associated Jacobian matrix, we can make further statements about each of those blocks:

- From the assumption of independence in the errors of each landmark observation it follows that the Hessian of landmarks $\mathbf{H}_\mathrm{l}$ will be, in all cases, *exactly* block diagonal.
- In a pure BA problem, i.e. with pose-to-landmark observations, it turns out that the Hessian of poses $\mathbf{H}_\mathrm{p}$ is also *exactly* block diagonal.
- In a BA problem where pose-to-pose constraints are also present, the Hessian of poses $\mathbf{H}_\mathrm{p}$ *tends* to be banded or, at least, there exists a reordering of variables that converts it into an approximately banded matrix. Nonzero blocks outside of the diagonal band may exist only due to loop closures. This was already illustrated in chapter 9 Figure 5 with the information matrix of a map where a loop closure was included, built with EKF in chapter 9 section 2—recall that the Hessian matrix is the equivalent to the information matrix of the unknowns.
- The pose-landmark Hessian submatrix $\mathbf{H}_\mathrm{pl}$ has nonzero entries only in those rows and columns of pairs of variables linked by one observation—the reader can compare the graph in Figure 6a to the nonzero blocks of $\mathbf{H}_\mathrm{pl}$ in the equation above.

Examples of typical Hessian matrices for the two kinds of problems, pure BA and BA with pose constraints, are represented in Figure 6b and Figure 7b. It is this preeminently diagonal structure of $\mathbf{H}_\mathrm{p}$ and $\mathbf{H}_\mathrm{l}$ what is called the *primary structure* of the BA problem, whereas we refer to the frequently sparse nature of $\mathbf{H}_\mathrm{pl}$ as the *secondary structure* (Triggs, McLauchlan, Hartley, & Fitzgibbon, 2000). We describe next how to smartly exploit this knowledge of the problem structure to reduce the computational burden of finding its MLE solution.

Consider the following linear system for a Gauss-Newton algorithm, where the step in the unknowns $\mathbf{\Delta x}$ must be solved within the iterative estimator (in a Levenberg-Marquardt method, only the diagonal of $\mathbf{H}$ changes and the entire subsequent discussion applies unmodified):

$$\mathbf{H}\,\mathbf{\Delta x} = -\mathbf{g} \tag{51}$$

By replacing the Hessian by its four parts from Equation 50 and by splitting the gradient into two according to the partition of variables into poses and landmarks, the linear system can be written in terms of two matrix equations:

Figure 6. (a) A small BA problem consisting of only three poses, four landmarks, and eight observations. (b) The sparse structure of this problem becomes clear in the corresponding Jacobian and Hessian matrices, which exhibit characteristic patterns. Solid blocks stand for nonzero submatrices in this sparse matrix representation. Note that Λ *is assumed to be block diagonal and that pose* $\mathbf{p}_1$ *has been considered fixed in both problems.*

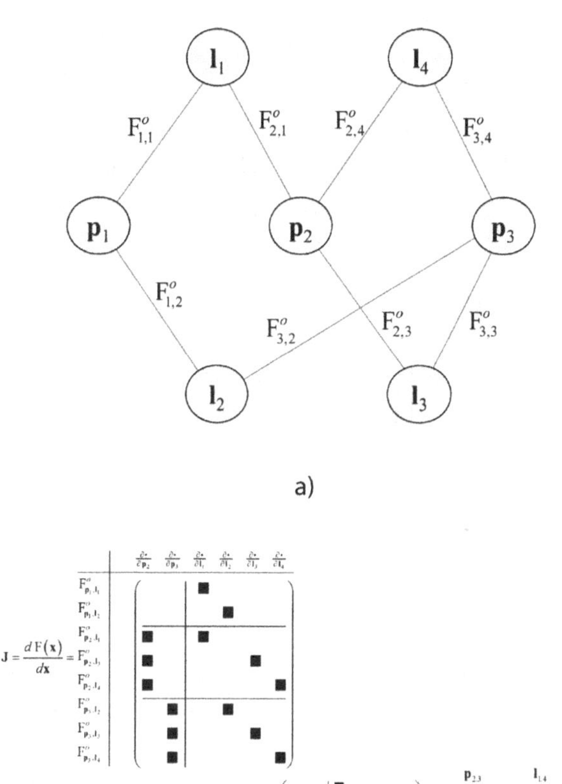

b)

$$\begin{pmatrix} \mathbf{H}_{\mathrm{p}} & \mathbf{H}_{\mathrm{pl}} \\ \mathbf{H}_{\mathrm{pl}}^{T} & \mathbf{H}_{\mathrm{l}} \end{pmatrix} \begin{pmatrix} \Delta\mathbf{x}_{\mathrm{p}} \\ \Delta\mathbf{x}_{\mathrm{l}} \end{pmatrix} = \begin{pmatrix} -\mathbf{g}_{\mathrm{p}} \\ -\mathbf{g}_{\mathrm{l}} \end{pmatrix} \tag{52}$$

which can then be attacked by firstly building a *reduced linear system* which only involves one type of unknowns (either poses or landmarks), solving then for the optimal step in those variables and finally solving for the step in the rest of unknowns by back substitution. Obviously, the idea is that the combined computational burden of solving those two sub-problems should present an advantage in comparison to solve at once the original system; this is assured by the special structure of the Hessian submatrices, as we see next.

The reduced linear system is got by Gaussian elimination (or *marginalization*, in probabilistic parlance) of one of the two blocks of variables. From the two possibilities (i.e., eliminating all

Box 4.

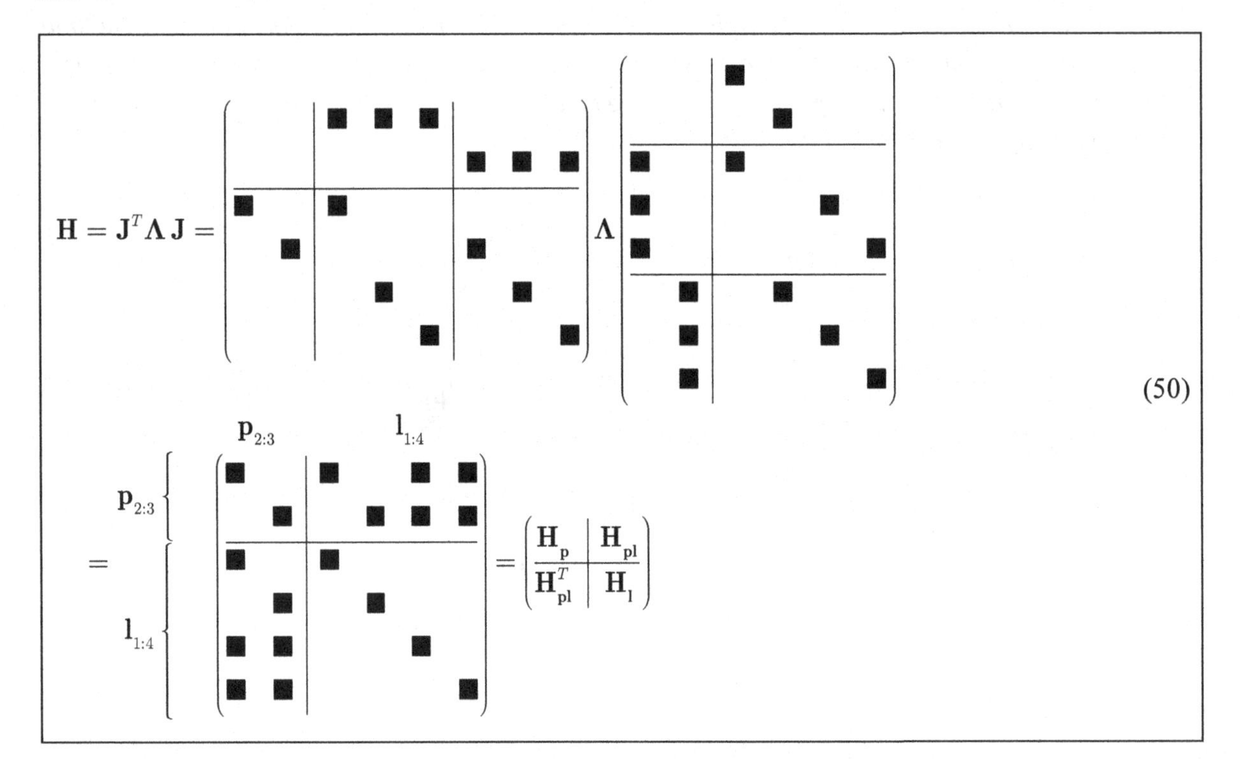

robot poses or all the landmarks) we shall be normally interested in minimizing the size of the reduced system. In practice, the number of landmarks M typically exceeds by far the number of poses N, thus we will remove them in order to obtain a pose-only reduced system from which to get the optimal step $\mathbf{\Delta p}^*$. Notice that all the constraints implied by the eliminated variables will be still present in the reduced system: that is, marginalizing out all the landmarks makes all their observations to be implicit embedded in the resulting equations.

It can be shown that left multiplying both sides of Equation 52 by the appropriate matrix we can obtain the poses-only or landmarks-only reduced system:

Poses-only reduced system:

$$\begin{pmatrix} \mathbf{I} & -\mathbf{H}_{pl}\mathbf{H}_l^{-1} \\ \mathbf{0} & \mathbf{I} \end{pmatrix}\begin{pmatrix} \mathbf{H}_p & \mathbf{H}_{pl} \\ \mathbf{H}_{pl}^T & \mathbf{H}_l \end{pmatrix}\begin{pmatrix} \mathbf{\Delta x}_p \\ \mathbf{\Delta x}_l \end{pmatrix} = \begin{pmatrix} \mathbf{I} & -\mathbf{H}_{pl}\mathbf{H}_l^{-1} \\ \mathbf{0} & \mathbf{I} \end{pmatrix}\begin{pmatrix} -\mathbf{g}_p \\ -\mathbf{g}_l \end{pmatrix}$$
$$\begin{pmatrix} \overbrace{\mathbf{H}_p - \mathbf{H}_{pl}\mathbf{H}_l^{-1}\mathbf{H}_{pl}^T}^{\bar{\mathbf{H}}_p} & \mathbf{0} \\ \mathbf{H}_{pl}^T & \mathbf{H}_l \end{pmatrix}\begin{pmatrix} \mathbf{\Delta x}_p \\ \mathbf{\Delta x}_l \end{pmatrix} = \begin{pmatrix} \overbrace{-\mathbf{g}_p + \mathbf{H}_{pl}\mathbf{H}_l^{-1}\mathbf{g}_l}^{-\bar{\mathbf{g}}_p} \\ -\mathbf{g}_l \end{pmatrix}$$
$$\bar{\mathbf{H}}_p \mathbf{\Delta x}_p = -\bar{\mathbf{g}}_p$$

Landmarks-only reduced system:

$$\begin{pmatrix} \mathbf{I} & \mathbf{0} \\ -\mathbf{H}_{pl}^T\mathbf{H}_p^{-1} & \mathbf{I} \end{pmatrix}\begin{pmatrix} \mathbf{H}_p & \mathbf{H}_{pl} \\ \mathbf{H}_{pl}^T & \mathbf{H}_l \end{pmatrix}\begin{pmatrix} \mathbf{\Delta x}_p \\ \mathbf{\Delta x}_l \end{pmatrix} = \begin{pmatrix} \mathbf{I} & \mathbf{0} \\ -\mathbf{H}_{pl}^T\mathbf{H}_p^{-1} & \mathbf{I} \end{pmatrix}\begin{pmatrix} -\mathbf{g}_p \\ -\mathbf{g}_l \end{pmatrix}$$
$$\begin{pmatrix} \mathbf{H}_p & \mathbf{H}_{pl} \\ \mathbf{0} & \underbrace{\mathbf{H}_l - \mathbf{H}_{pl}^T\mathbf{H}_p^{-1}\mathbf{H}_{pl}}_{\bar{\mathbf{H}}_l} \end{pmatrix}\begin{pmatrix} \mathbf{\Delta x}_p \\ \mathbf{\Delta x}_l \end{pmatrix} = \begin{pmatrix} -\mathbf{g}_p \\ \underbrace{-\mathbf{g}_l + \mathbf{H}_{pl}^T\mathbf{H}_p^{-1}\mathbf{g}_p}_{-\bar{\mathbf{g}}_l} \end{pmatrix}$$
$$\bar{\mathbf{H}}_l \mathbf{\Delta x}_l = -\bar{\mathbf{g}}_l \quad (53)$$

where the barred Hessian submatrices $\bar{\mathbf{H}}_p$ and $\bar{\mathbf{H}}_l$ are known as the *Schur complements* of their

Figure 7. (a) The same problem of Figure 6a, where edges also exist between poses. (b) The sparse structure of the modified problem. Compare to Figure 6b.

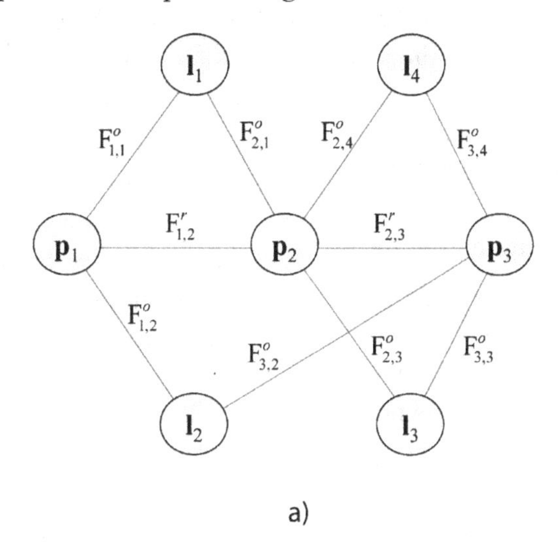

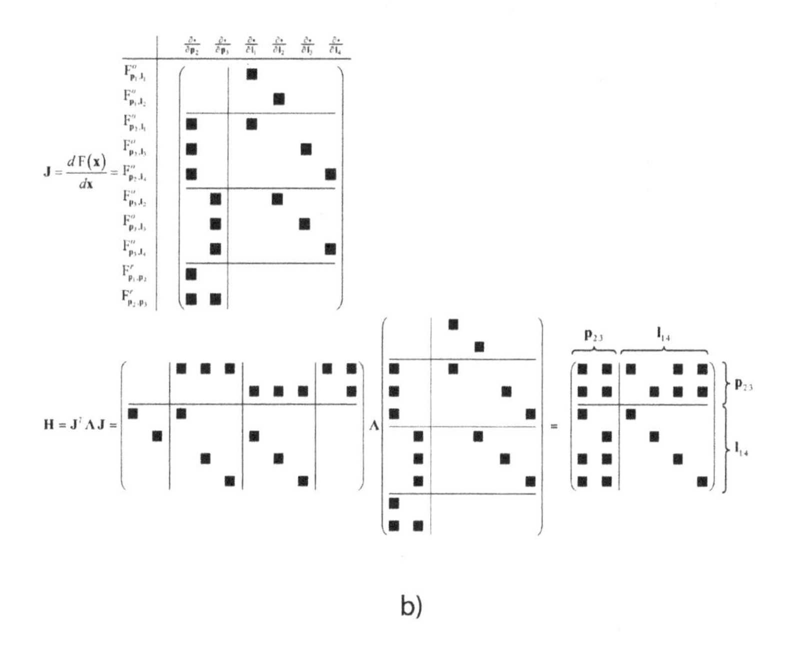

unbarred versions $\mathbf{H}_p$ and $\mathbf{H}_l$ within the complete Hessian. As it can be seen in their expressions above, both Schur complements require inverting the Hessian of the other set of variables. The key to understand how these reduced systems can be built efficiently becomes clear by realizing of the (typically) block diagonal structure of the matrices to be inverted. In particular, the poses-only system $\bar{\mathbf{H}}_p \Delta \mathbf{x}_p = -\bar{\mathbf{g}}_p$ requires evaluating the inverse $\mathbf{H}_l^{-1}$, but we saw above that the Hessian of the landmarks is always *exactly* block diagonal, thus its inverse is simply computed by inverting each diagonal block $\mathbf{H}_{\mathrm{l}_k}^{-1}$ independently, a property of any block diagonal matrix, which has a linear complexity with the number of landmarks. For the sake of clarity, the whole process from the Jacobian matrix up to computing the reduced system by means of the Schur complement has been illustrated in Figure 8.

As a side note, we must remark how solving for only the poses $\Delta\mathbf{x}_{\mathrm{p}}$ actually turns the first stage of BA into a graph SLAM problem, where the nonzero blocks outside the diagonal of the reduced Hessian $\bar{\mathbf{H}}_{\mathrm{p}}$ can be interpreted as pose-to-pose constraints, whose uncertainty subsumes all the information of the marginalized-out landmarks. Sometimes, reducing BA to this graph of pose constraints and forgetting about recovering the map of landmarks is a practical SLAM solution for a mobile robot (Konolige & Agrawal, 2008). This is why we presented BA as a variation of the graph SLAM framework.

The so-called *secondary structure* of BA, i.e. the sparsity of $\mathbf{H}_{\mathrm{pl}}$, can be exploited to efficiently compute a sparse reduced Hessian $\bar{\mathbf{H}}_{\mathrm{p}}$ which, in turn, can be used as the input to sparse solvers of the system $\bar{\mathbf{H}}_{\mathrm{p}}\Delta\mathbf{x}_{\mathrm{p}} = -\bar{\mathbf{g}}_{\mathrm{p}}$, just like we saw in Equations 26 and 27. Expanding the terms of the reduced Hessian and gradient according to Equation 52 we arrive at these expressions for efficiently computing their nonzero blocks:

$$\begin{cases} \bar{\mathbf{H}}_{\mathrm{p}} = \mathbf{H}_{\mathrm{p}} - \mathbf{H}_{\mathrm{pl}}\mathbf{H}_{\mathrm{l}}^{-1}\mathbf{H}_{\mathrm{pl}}^{T} \\ \rightarrow \bar{\mathbf{H}}_{\mathrm{p}_{i,j}} = \mathbf{H}_{\mathrm{p}_{i,j}} - \sum_{k\in\mathbf{V}_{i,j}} \mathbf{H}_{\mathrm{p}_i\mathrm{l}_k}\mathbf{H}_{\mathrm{l}_k}^{-1}\mathbf{H}_{\mathrm{p}_j\mathrm{l}_k}^{T} \qquad i,j\in 1\ldots N \\ \bar{\mathbf{g}}_{\mathrm{p}} = \mathbf{g}_{\mathrm{p}} - \mathbf{H}_{\mathrm{pl}}\mathbf{H}_{\mathrm{l}}^{-1}\mathbf{g}_{\mathrm{l}} \\ \rightarrow \bar{\mathbf{g}}_{\mathrm{p}_i} = \mathbf{g}_{\mathrm{p}_i} - \sum_{k\in\mathbf{V}_i} \mathbf{H}_{\mathrm{p}_i\mathrm{l}_k}\mathbf{H}_{\mathrm{l}_k}^{-1}\mathbf{g}_{\mathrm{l}_k} \qquad i\in 1\ldots N \end{cases} \tag{54}$$

where the sets $\mathbf{V}_i$ stand for the indices of all those landmarks observed from a pose $\mathbf{p}_i$ and the $\mathbf{V}_{i,j}$ are those landmarks that are visible from any pair of poses, that is, $\mathbf{V}_{i,j} = \mathbf{V}_i \cap \mathbf{V}_j$. Caution must be paid to the possibility of finding $\mathbf{H}_{\mathrm{l}_k}$ blocks that are rank-deficient, which may occur if the parallax between all observations to a landmark is too small. In those cases, an inverse of $\mathbf{H}_{\mathrm{l}_k}$ does not exist, thus one possibility is to ignore the contributions of that landmark to the reduced system of equations. Ideally, one should have had present this issue from the very beginning (i.e. image capturing and front-end processing) and do not include such ill-conditioned observations by means of a proper evaluation of whether a given keyframe is suitable to be included in the BA problem or not.

Once an optimal step $\Delta\mathbf{x}_{\mathrm{p}}^{*}$ is available we can also compute the corresponding correction for the global location of landmarks. By replacing the known value of $\Delta\mathbf{x}_{\mathrm{p}}^{*}$ into Equation 53 we arrive at another equation of $\mathbf{A}\mathbf{x} = \mathbf{b}$ form:

$$\begin{aligned} \mathbf{H}_{\mathrm{l}}\Delta\mathbf{x}_{\mathrm{l}} &= -\mathbf{g}_{\mathrm{l}} - \mathbf{H}_{\mathrm{pl}}^{T}\Delta\mathbf{x}_{\mathrm{p}}^{*} \\ \Delta\mathbf{x}_{\mathrm{l}} &= \mathbf{H}_{\mathrm{l}}^{-1}\left(-\mathbf{g}_{\mathrm{l}} - \mathbf{H}_{\mathrm{pl}}^{T}\Delta\mathbf{x}_{\mathrm{p}}^{*}\right) \end{aligned} \tag{55}$$

which has an especially efficient solution, as follows by observing that $\mathbf{H}_{\mathrm{l}}$ is block diagonal and, therefore, its inverse is also block diagonal and can be computed by simply inverting each of those blocks one by one. By doing so and exploiting the sparse structure of $\mathbf{H}_{\mathrm{pl}}$ we obtain:

$$\Delta\mathbf{x}_{\mathrm{l}_k}^{*} = \mathbf{H}_{\mathrm{l}_k}^{-1}\left(-\mathbf{g}_{\mathrm{l}_k} - \sum_{i=\mathbf{B}_k}\mathbf{H}_{\mathrm{p}_i\mathrm{l}_k}^{T}\Delta\mathbf{x}_{\mathrm{p}_i}^{*}\right) \qquad k = 1\ldots M \tag{56}$$

where the sets $\mathbf{B}_k$ stand for all those pose indices from which $\mathbf{l}_k$ is visible. As in graph SLAM, the optimal steps $\Delta\mathbf{x}_{\mathrm{p}}^{*}$ and $\Delta\mathbf{x}_{\mathrm{l}_k}^{*}$ must be computed and added to the current approximation of the MLE within the iterative optimizer until some convergence criterion is reached. Notice as well that when dealing with $\mathbf{SE}(3)$ poses—the most

Figure 8. A summary of the conceptual steps involved in building the pose-only reduced Hessian for a BA problem. (a) Computation of the entire problem Hessian from the Jacobians. (b) The four parts in which the Hessian matrix is clearly divided. $\mathbf{H}_l$ *will be always block diagonal. Only if there existed no pose-to-pose constraints would* $\mathbf{H}_p$ *be also block diagonal. (c) How the Schur complement* $\bar{\mathbf{H}}_p$ *is computed by manipulating the different parts of the Hessian. Notice that all the depicted operations should take into account the sparsity of the matrices to achieve a reasonable performance.*

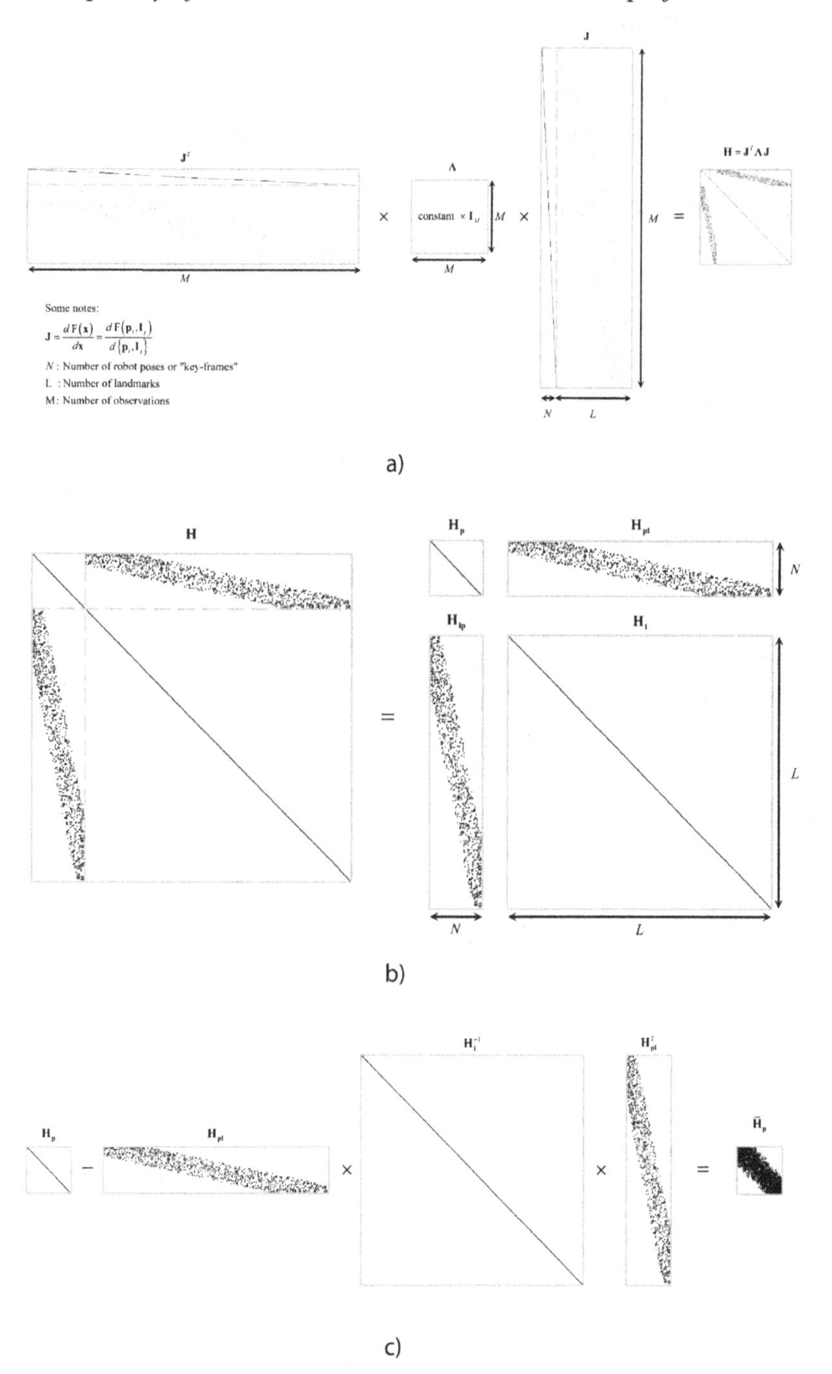

common situation in BA—the on-manifold optimization procedure introduced in section 2 should be also applied, although it has not been mentioned throughout the discussion above for the sake of conciseness.

Robustness against Outliers

One important difference of BA with respect to graph SLAM is the existence of pose-to-landmark edges in the associated problem graph. Those edges imply *identifying* which landmarks are observed from each pose. That is, in BA we must have solved the *data association* problem like in any other SLAM framework of those seen in chapter 9. And due to the difficulty of correctly performing data association with visual features, we must live with the fact that at least a small fraction of the associations may be possibly wrong. Furthermore, when working with images, the observation data themselves consist of pixel coordinates, which are subject to a variety of large errors under troublesome conditions, such as wrong tracking of visual features, partial occlusions, etc. In summary, in BA we have certain observations whose errors are much larger than the predictions (i.e. the $\mathbf{\Lambda}_{i,j}$ precision matrices) from the sensor model. Those are called *outliers*, while correct measurements are called *inliers*. Ideally, we would like to discard all the outliers so they do not get into the optimization, but unfortunately telling them from inliers is not easy *before* optimizing.

A powerful workaround is the employment of *robust estimators*, a technique proposed in (Huber, 1964). The basic idea consists of replacing the common cost function used for MLE by a modified, sometimes called *robustified*, cost function:

$$\begin{aligned} F(\mathbf{x}) &= \underbrace{\sum_i \frac{1}{2} \mathbf{r}_i^T \mathbf{r}_i}_{\text{Least-squares cost function}} = \sum_i \frac{1}{2} \left\|\mathbf{r}_i\right\|^2 \\ \rightarrow \quad F(\mathbf{x}) &= \underbrace{\sum_i \frac{1}{2} \rho\left(\left\|\mathbf{r}_i\right\|^2\right)}_{\text{Robustified cost function}} \end{aligned} \tag{57}$$

where $\rho(s): \mathbb{R}^{0,+} \rightarrow \mathbb{R}^{0,+}$ is a *kernel function* that can modify the weight of each residual according to its norm $s = \left\|\mathbf{r}_i\right\|^2$. Common least-squares can then be seen as a particular instance of a robust estimator with kernel $\rho(s) = s$. To understand how introducing this kernel may help us with ourliers we must take a step backward and recall that the role of cost functions $F(\mathbf{x})$ is to stand as approximations of the negative logarithm of the problem joint distribution. The particular case of least squares is the outcome of modeling all the conditional distributions as Gaussians—as we saw in chapter 9 section 1 and also in section 3. Among the remarkable properties of Gaussian distributions we find the central limit theorem (recall chapter 4 section 2), which serves us to allege their application to all kind of complex *measuring* processes, such as pose-to-landmark observations in BA. Outliers, however, completely invalidate this view, since errors are not caused by *small* defects or imperfections but can be orders or magnitude larger than the errors in inliers.

The key is then trying to design a probabilistic observation model that accounts for both possibilities (inliers and outliers), something similar to what we already faced while discussing models for laser scanners in chapter 6 section 2. For instance, we could think of a probabilistic sensor model with a main Gaussian mode representing inliers, plus a uniform distribution, which models the possibility of outliers. By denoting the norm of residuals as $r = \left\|\mathbf{r}\right\|$ we can write down this model as follows:

$$p(r) = \begin{cases} \alpha N(r; 0, \sigma^2) + (1-\alpha)\frac{1}{2D} & \text{, for } |r| < D \\ 0 & \text{, otherwise} \end{cases} \tag{58}$$

(for a mix coefficient $a \in [0,1]$, and D being a sufficiently large value for residuals)

The negative logarithm of this model, evaluated in the working region $|r| < D$, gives us the corresponding kernel function:

$$\begin{aligned} \rho(r) &= -\log p(r)_{|r|<D} = \\ &= -\log\left(\exp\left(-\frac{1}{2}\frac{r^2}{\sigma^2}\right) + \eta\right) \\ &\rightarrow \begin{cases} \cong \eta' \, r^2 & \text{, for } |r^2| \ll |\sigma^2| \\ \cong \eta'' & \text{, for } |r^2| \gg |\sigma^2| \end{cases} \end{aligned} \tag{59}$$

where η, η' and η'' are all different constants or put in words: for "*small*" residuals the Gaussian pdf dominates and the cost becomes quadratic, while for "*large*" residuals the cost becomes practically constant. In all kernel functions, the criterion to classify residuals as "small" or "large" is determined by some threshold, in this particular case represented by σ.

The intention of using residual reweighting kernels is that outliers, characterized by large residuals, are automatically assigned smaller weights than in a pure least-squares approach. It is obvious that, since the optimization algorithm will minimize the overall cost from all the problem constraints, the negative impact of outliers can be drastically reduced or even completely avoided, depending on the choice for a robust kernel and its threshold. As a consequence, a robustified cost function will "turn the focus" of any optimizer on minimizing the error in the interesting observations (the inliers), while that in outliers will be almost ignored.

In summary, the squared error kernel, $\rho(s) = s$, implicitly assumed in all least-squares MLE of previous sections, is the simplest choice but has the dangerous property that all observations have equal weights. If we cannot say for sure that our observations are free of outliers, a robust kernel *must* be used instead, for a single outlier can have catastrophic effects in the accuracy of least-squares estimators.

Several kernels have been proposed in the robust estimation literature, but two of the most widely used are the *Huber* and the *pseudo-Huber* kernels (Huber, 1964; Hartley & Zisserman, 2004), which are heuristics rather than derived from a pdf as in the example elaborated above. Please refer to Figure 9 for a graphical comparison of these and other common kernels. Notice that not all are *convex* functions, a property with profound implications in MLE methods: since the sum of convex functions is also convex, the existence of only one global minimum in the final cost function can be only assured by convex models. Therefore, it is a good advice in general to avoid non-convex kernels.

Understandably, using robust kernels for the individual cost functions, as shown in Equation 57, modifies up to some extent the shape of the global cost $F(\mathbf{x})$. As a consequence, we should adapt the optimizers searching for its minimum. By approximating the kernel $\rho(s)$ with a truncated Taylor series, it can be shown (Triggs, McLauchlan, Hartley, & Fitzgibbon, 2000) that only slight modifications are required with respect to the non-robust generic form, which we saw in Equation 34. In particular, a first-order approximation that can suffice in practice only implies this change:

$$\mathbf{g}_{i,j} = \mathbf{J}_{i,j}^T \mathbf{\Lambda}_{i,j} \mathbf{r}_{i,j} \quad \xrightarrow{\text{Robustified version}} \quad \mathbf{g}_{i,j} = \rho'_{i,j} \mathbf{J}_{i,j}^T \mathbf{\Lambda}_{i,j} \mathbf{r}_{i,j}$$

with:

$$\rho'_{i,j} \triangleq \left. \frac{d\sqrt{2\rho(s)}}{ds} \right|_{s=\mathbf{r}_{i,j}{}^T \mathbf{\Lambda}_{i,j} \mathbf{r}_{i,j}} \quad (60)$$

Although robust kernels are essential to find accurate solutions in the presence of outliers, we must also warn the reader about the need to wisely set thresholds according to the expected residuals of true inliers. A too conservative threshold could lead to wrongly treat many inliers as outliers, making their contribution to the joint likelihood to be totally or almost (depending on the particular kernel) neglected. The iterative optimizer may still be able to find an accurate MLE solution with an overconfident robust kernel, but it may takes 10 to 100 times more iterations, ruining the overall performance. In a few words: robust kernels are *a must* in practice, but have to be applied judiciously.

Other Advanced Techniques in Bundle Adjustment

Up to this point, we have seen how BA is basically a least-squares geometry optimization problem whose variables happen to exhibit a particularly sparse structure. We now turn our attention to a pair of ideas, which implementors can use to improve the efficiency of BA solutions.

The first one is related to on-line operation. Just like in graph SLAM, sparse solvers for BA are fundamentally designed for batch, off-line operation over the entire sequence of data. But let us consider for a moment what is needed to adapt those methods for incrementally incorporating observations as they are gathered. We know that solving BA comprises an alternation between (1) a stage where the sparse linear system $\mathbf{H}\Delta\mathbf{x} = -\mathbf{g}$ is built and (2) another stage where it is solved to determine the state space corrections $\Delta\mathbf{x}$ (as described earlier in this section, this latter stage in turn comprises two phases to exploit the primary structure of the problem). The alternation ends upon convergence when the minimum of the cost function is reached. If more observations or nodes (either poses or landmarks) are then added to the problem, we can naively restart the entire process rebuilding from scratch the new linear system, i.e. the new Hessian $\mathbf{H}$ and gradient $\mathbf{g}$. However, it makes sense to wonder what previous calculations can be reused to reduce such a computational burden.

We can identify three potential spots subject to optimized implementations in incremental BA. Firstly, the update of the sparse Hessian matrix: adding new poses or landmarks simply turns into extending the size of the matrix by the appropriate number of rows and columns, as many as needed by the pose or landmark parameterization. In most sparse matrix implementations, a size change has a constant (and negligible) time complexity, but only if the rows and columns are appended at the end of the existing matrix. Since we need poses and landmarks sorted and grouped together to exploit the so-called primary structure, a straightforward solution is to explicitly maintaining the three sparse submatrices $\mathbf{H}_\mathrm{p}$, $\mathbf{H}_\mathrm{pl}$, and $\mathbf{H}_\mathrm{l}$, separately:

$$\mathbf{H} = \begin{pmatrix} \mathbf{H}_\mathrm{p} & \mathbf{H}_\mathrm{pl} \\ \mathbf{H}_\mathrm{pl}^T & \mathbf{H}_\mathrm{l} \end{pmatrix} \quad (61)$$

In fact, $\mathbf{H}_\mathrm{p}$ and $\mathbf{H}_\mathrm{l}$, as well as the Hessian of any reduced system (see Equation 53) will be always sparse *symmetric* matrices, so in practice any efficient implementation should store them as triangular matrices. Moreover, implementations of factorization methods for symmetric definite (i.e. Cholesky) or semidefinite positive matrices typically expect the input matrices to be in this format. Note that the Hessian submatrix $\mathbf{H}_\mathrm{pl}$ will typically be sparse for large problems, but not symmetric; therefore, all nonzero entries must be stored in that case.

Secondly, if the secondary structure is exploited it means that the involved matrices have

Figure 9. Some possibilities for the residual weighting kernels, along their characteristics. The parameters used in the example graphs are $\alpha = 2$, $\varepsilon = 0.1$, *and* $b = 2$.

	Convex	Parameters	Cost kernel function Plots show: $\frac{1}{2}\rho(r^2)$	Probability density function (pdf) with: $s = r^2$
Standard least-squares	Yes	(none)	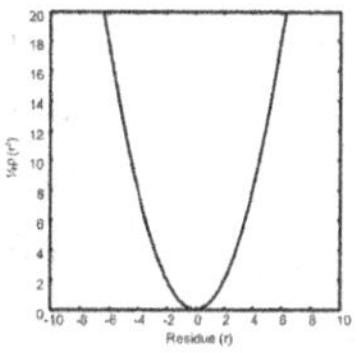 $\rho(s) = s$	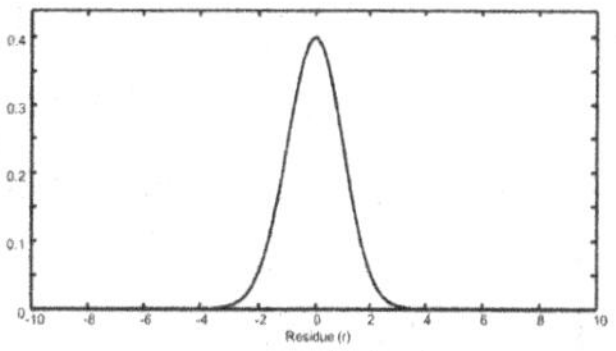 $p(r) \propto \exp\left\{-\frac{1}{2}r^2\right\}$
Original Blake-Zisserman	No	α	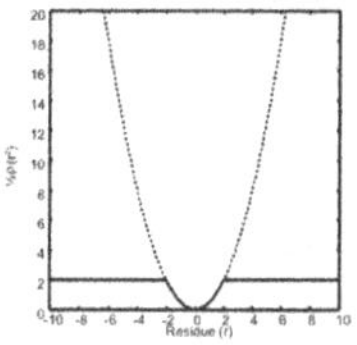 $\rho(s) = \min(s, \alpha^2)$	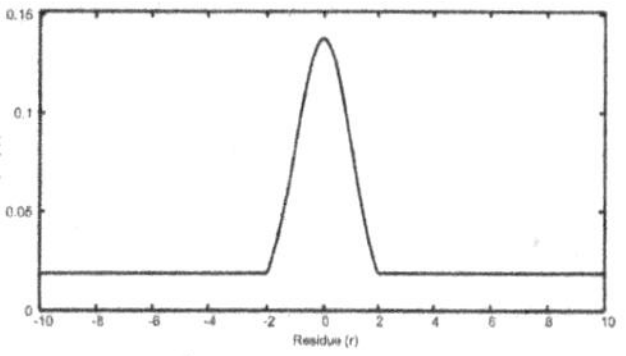 $p(r) \propto \begin{cases} \exp\left\{-\frac{1}{2}r^2\right\} & \text{, if } \lvert r \rvert < \alpha \\ \exp\left\{-\frac{1}{2}\alpha^2\right\} & \text{, otherwise} \end{cases}$
Modified Blake-Zisserman	No	ε	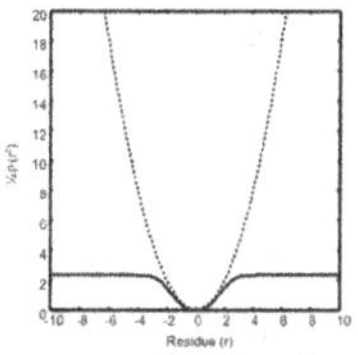 $\rho(r) = -2\log\left(\exp\left\{-\frac{1}{2}s\right\} + \varepsilon\right)$	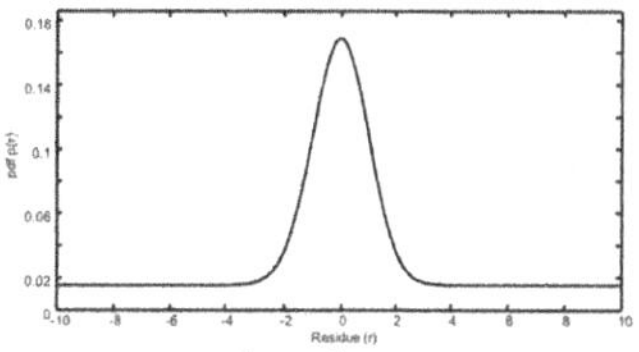 $p(r) \propto \exp\left\{-\frac{1}{2}r^2\right\} + \varepsilon$ (Not a real pdf !)
Huber	Yes	b	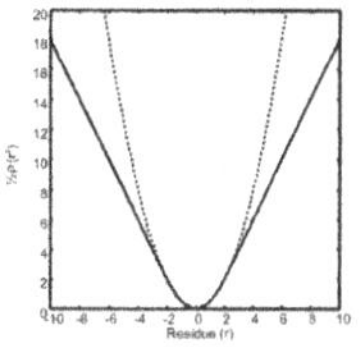 $\rho(s) = \begin{cases} s & \text{, if s} < b^2 \\ 2b\sqrt{s} - b^2 & \text{, otherwise} \end{cases}$	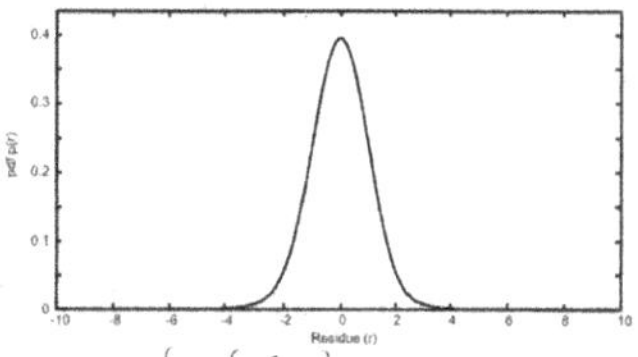 $p(r) \propto \begin{cases} \exp\left\{-\frac{1}{2}r^2\right\} & \text{, if } \lvert r \rvert < b \\ \exp\left\{-b\lvert r \rvert + \frac{1}{2}b^2\right\} & \text{, otherwise} \end{cases}$
Pseudo-Huber	Yes	b	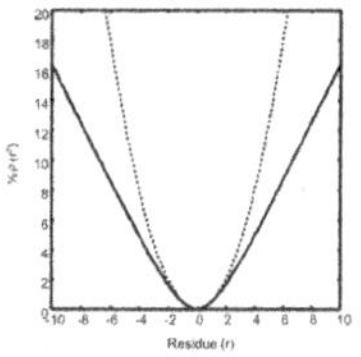 $\rho(s) = 2b^2\left[\sqrt{1 + \frac{s}{b^2}} - 1\right]$	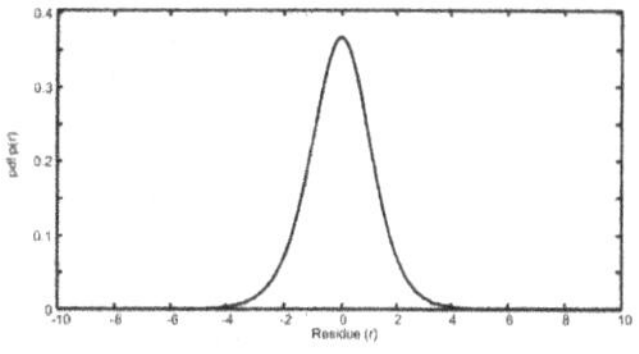 $p(r) \propto \exp\left\{-b^2\left[\sqrt{1 + \left(\frac{r}{b}\right)^2} - 1\right]\right\}$

been factorized, typically as $\mathbf{LL}^T$ (the Cholesky factorization) or $\mathbf{LDL}^T$ (the special case of **LDU** factorization for symmetric matrices), with **L** and **D** standing for lower triangular and diagonal matrices, respectively—recall Equation 27 for the reason why such factorizations help us to solve linear systems more efficiently. Computing those factorizations is typically one of the most costly operations within the entire BA (and graph SLAM) process, thus taking advantage of previous factorizations to speed up the new ones would be desirable. Unfortunately, algorithms for updating factorizations given a small set of changes in the original matrices are complex and an active area of research at present. One of the most popular choices is the CHOLMOD algorithm, presented in (Chen, Davis, Hager, & Rajamanickam, 2008).

Finally, in the third place, we must highlight that the inverse of the BA Hessian, which approximates the covariance of the estimated parameters (i.e. their uncertainty), may be needed in a typical on-line framework to help with the establishment of data association. Inverting a *sparse* Hessian always results in a *dense* covariance matrix, thus this reveals as another costly operation. The methods for updating factorizations are related to the update of the covariance matrix without a full recalculation. All these operations are out of the scope of the present book, hence we recommend consulting (Kaess, Johannsson, Roberts, Ila, Leonard, & Dellaert, 2011) and the work (Triggs, McLauchlan, Hartley, & Fitzgibbon, 2000), widely accepted as the most-influential BA reference for the last decade, for further discussions on the topic.

The second idea that we mentioned above for enhancing the performance of BA consists of a profound change in the problem statement: instead of reconstructing the poses and landmarks in a single *global* frame of coordinates, we could switch our state space to one with *relative* coordinates only. This is the essence of a recent proposal named *Relative Bundle Adjustment* (RBA), introduced by Sibley *et al.* in a series of works (Sibley, Mei, Reid, & Newman, 2010; Mei, Sibley, Cummins, Newman, & Reid, 2010). In this scheme, landmark coordinates are referenced to the first pose from which they are observed, and each pose coordinates are relative to its preceding pose in the sequence of data acquisition. Maintaining local coordinates is a recurrent topic in mobile robotics and has been defended in the context of other SLAM frameworks, as in EKF-based SLAM (Castellanos, Martinez-Cantin, Tardós, & Neira, 2007). Its principal advantage for large BA problems can be easily understood by realizing that each incremental observation will only affect a few neighboring relative poses. This is in contrast with global coordinates, when each single observation has the potential to affect *all* the variables in the problem—e.g., think of loop closures. Therefore, RBA allows the fixation of a maximum number of variables to update with each time step. In other words: RBA is a SLAM algorithm with a *constant time* complexity (that is *really* great news!). It should come at no surprise to the reader that there exists a high price to pay for this extraordinary property: the lack of global metric information hardens the determination of data association with features far down the chain of relative poses. Also, if for some reason global coordinates are required for all or part of the pose nodes, obtaining them actually turns into an iterative estimation problem in itself; in fact, into an independent graph SLAM problem. In spite of this, RBA becomes a practical and efficient solution if data association is possible without global metric information, something doable by means of recognizing revisitations with visual appearance techniques, which may be as efficient as linear with the number of previously visited locations (Cummins & Newman, 2008).

6. TOWARDS LIFELONG SLAM

We began this chapter by asking ourselves about what would be needed for an "ideal" solution to SLAM. The techniques explained so far have the potential for getting close to such a perfect solution by handling robots moving in a three-dimensional world with computational costs that, in the best cases, are nearly linear with the size of maps. Such an amazing finale may incite the reader to think that, at present, "SLAM is solved." Indeed, the approaches explained along this book seem capable of solving localization and mapping problems in several practical robotic applications, especially those where the robot is physically constrained to a relatively small workspace.

Still, state-of-the-art techniques are far from bringing to reality the enviable autonomy of robots we accustom to see in the sci-fi literature or films. The ultimate goal for SLAM clearly consists of enabling a robot to freely learn about its complex, real-world environment during long periods of time; ideally, in *lifelong operation*. All existing methods at present fail to achieve that for one reason or another, as we shall see below[8]. In the rest of this section, we briefly explore two major challenges of mobile robotics, which, up to this date, are open problems. In our opinion, these points should be seriously addressed by future research to help making lifelong operation a reality.

Stability vs. Plasticity

Firstly, we have the plain fact that there is no such thing as *the* representation of an environment. Every place undergoes changes in a wide range of timescales, forcing any realistic representation or map to be *dynamic*. For example, imagine an open square in a town: we could find people around, whose location changes with every second, parked vehicles, which stay static for periods of minutes or hours, temporary facilities that remain fixed for days. Over the course of years, major construction works can alter scene elements that would otherwise be considered as static, e.g. sidewalks, surrounding buildings, a bridge, etc. Any system facing the challenge of *learning* a model for such a changing reality will have to deal with two contradictory tendencies: on the one hand, fixing the model over time (*stability*) and on the other hand the necessity of change itself as a reaction to real variations in the external world (*plasticity*). This is the essence of the *stability-plasticity dilemma*, a term coined in (Grossberg, 1988) within the context of understanding how the brain learns.

Although not always made explicit—we do in this book—virtually all SLAM algorithms take the so-called "static world assumption," as briefly introduced in chapter 1. Such an approximation is assumable when the sensed information mostly reflects static aspects of the environment, either for the environment really being static or because the sensor ignores the dynamic elements (for example, a planar 2D laser scanner placed at a height of 3 meters in a crowded room still makes the static world assumption right). Otherwise, the SLAM method will be confronting two contradictory pieces of sensory data: one for the static environment and another for those parts that changed. As a result, the accuracy of the robot positioning and of the map can be jeopardized. Since this is the case of virtually all state-of-the-art SLAM algorithms at present, we must insist in the relevance that this problem, which will deserve a large research activity in the future.

Regarding the few existing approaches for handling dynamic environments, we can find a first idea related to what in Bayesian inference parlance is called the *transition function* of the state vector (recall, for example, the EKF equations for robot localization). Here, the static world assumption turns into an identity transition function: if $f(\cdot)$ stands for the function predicting new maps $\mathbf{m}_k$ from old maps $\mathbf{m}_{k-1}$, we would have $f\left(\mathbf{m}_{k-1}\right) = \mathbf{m}_{k-1}$. The omnipresence of this hypothesis is the reason why we dropped time step indices for map variables throughout this book, i.e. $\mathbf{m} \equiv \mathbf{m}_1 = \mathbf{m}_2 = \ldots = \mathbf{m}_k$. When it comes to designing transition functions $f\left(\cdot\right)$ for particular cases of dynamic maps, a powerful proposal is state augmentation, where dynamic information (e.g., the estimated velocity of elements in the map) is stored for each item for usage within the transition function. Alike to probabilistic robot motion models explored in chapter 5, these transition models can (and should) increase the uncertainty of the estimated parameters of every map element as a mean to model the unknowns of the system dynamics. Notice that this method is general enough to be applicable to maps of discrete elements (e.g. maps of landmarks) or to random fields (e.g. occupancy grids, gas concentration grid maps, etc.). In its simplest form, a "no change" probabilistic evolution model can be always used (a dual of the "no-motion motion model" of chapter 5), where no state augmentation is needed and the only map change consists of increasing the uncertainty of all map elements—e.g., in EKF-based SLAM this turns out to summing a positive constant to the diagonal of the map covariance submatrix. As an example of a work demonstrating state augmentation of map landmarks can be found in (Wang, Thorpe, Thrun, Hebert, & Durrant-Whyte, 2007), where the sensory inputs are segmented into static or dynamic features. Dynamic features are tracked and their motion parameters estimated by means of separate Bayesian filters, one per dynamic feature.

A second idea to handle dynamic environments consists of trying to discern the static and dynamic parts of an environment, so each gets mapped into an independent map. Notice that this implies reusing common map building algorithms for the static part, for which the static world assumption holds. This idea was reported applied to grid mapping in Wolf and Sukhatme (2005), where the authors propose the maintenance of two separate occupancy grid maps: one for the "static world" and another for dynamic obstacles. Avoiding the problem of classifying which portions of the sensory data belong to each category is possible by creating several parallel maps in different stability-plasticity "time scales" (Biber & Duckett, 2005). Here, all maps are updated simultaneously, but map parameters are tuned so some are more moldable by quickly changing readings than others.

The Vastness and Complexity of the World

In comparison to current intelligent machines, we humans have a remarkable capacity for spatial self-awareness, which endures throughout a lifetime, no matter how many different places we visit. Thus, we propose the following thought experiments as a didactic way to introduce some clues on how we deal with mental models of our vast world, with the intention of extrapolating what we will learn to mobile robotics.

Imagine visiting a friend's apartment for the first time. After leaving, you would probably have a quite detailed idea about the structure of the building common areas, the outline of the different rooms and perhaps even lots of details on objects and furniture in them. Imagine now instead driving in the same city towards some destination in a neighborhood you never visited before. All the information which is really needed to build a mental reconstruction of your way is a simplified representation of streets with their approximate spatial arrangement, the placement of traffic lights, etc. Everything else off the road (i.e. sidewalks, buildings, parks, etc.) becomes largely irrelevant, thus any detailed description for them is totally out of order for your goals. You may even pass in front of your friend's apartment without noticing. Finally, imagine you do not repeat the visit to the apartment for one year; in your next visit, it is probable that you still remember the outline of the apartment before seeing it again but, almost for sure, many of the furniture and decoration details would seem new again after having been forgotten.

From these examples, we can gain a significant insight about how a lifelong learning system should operate regarding the construction of world models, even without pretending a serious analysis from the point of view of psychological sciences. In particular, four essential insights can be nailed down here. Firstly, mental models exhibit a strong *hierarchical organization*, since we dynamically select only those parts of a *world model* that are really needed at each instant (this can be interpreted as an "attention-focusing mechanism"). Secondly, there exists a long-term tendency to *forget* pieces of information about learned models, especially those of little relevance, which typically are those, which we take into account rarely for our operations. Thirdly, and as a consequence of the two previous points, we benefit from a seemingly unlimited capability for *effortlessly* learning models for new places during an entire lifetime. This clearly contrasts to what happens with *all* existing SLAM techniques for mobile robots, where the effort required to create models of new places and to update old ones increases as more and more new locations got mapped. Finally, a fourth lesson we can learn, for which we have provided a hint already, is that the decision about what portion of the world model is "active" at each instant of time critically depends on the specific *task* being performed. Put in other words, there exists no such thing as relevant and irrelevant information about an environment: that classification only makes sense under the light of performing some particular task or operation.

By translating these five principles into mobile robotics parlance, we arrive at the following desiderata for lifelong SLAM:

1. Usage of *hierarchical maps* such as the ones described in chapter 8,
2. Existence of *forgetting-like mechanisms* for map maintenance,
3. Unlimited *scalability*, that is, candidate algorithms should demonstrate a (probably amortized) constant time computational complexity,
4. Ideally, *multiple hierarchies*—also explained in chapter 8—should coexist, each one being useful for one particular robotic task, and finally,
5. Maps should also comprise *semantic* information to allow reasoning on them.

There exists a clear interest in hierarchical maps in the mobile robotics community, with a number of proposals having been made over the years. However, we must warn the reader that the term "hierarchical" has been inconsistently applied in the SLAM literature to denote a wide range of dissimilar techniques. In the case of topological maps—some hierarchical maps are pure metrical—all the proposed frameworks agree on the general idea of summarizing the information of large maps by means of *graph* representations which less elements than the original maps. However, they disagree in everything else, including the semantics of the graph nodes and edges.

The most promising of the hierarchical proposals divide the map of the world into pieces, which are called *local maps,* and then build a topological map where nodes represent those local maps. Each local map gets its own reference frame for local coordinates, thus the location of objects within them are totally independent of the location of the local map with respect to other maps. Furthermore, in this paradigm the concept of global coordinates seems to lose its usefulness since all we may be interested in are the relative positions of different local maps. This idea looks closer to the way we humans conceptualize models of the world than the classic SLAM approach of employing one single reference frame. Since this model consists of a graph representing a topology of a set of local metric maps, this kind of representations are called Hybrid Metric-Topological (HMT) maps, as it was explained in chapter 8 (Blanco, Fernández-Madrigal, & González, 2008). Notice as well that it looks like we are describing a one-level hierarchy, but the concept of graph abstraction can be applied recursively, ending up with graph representations with multiple levels of abstraction (Fernández-Madrigal & González, 2002), as illustrated with an example in Figure 10. Interest for these ideas seems to have been renewed recently with their introduction in the context of graph SLAM (Frese, 2006; Grisetti, Küemmerle, Stachniss, Frese, & Hertzberg, 2010).

Apart from the psychological-based motivation for hierarchies introduced above, there are many other technical aspects that recommend splitting a global model of the world into smaller pieces. The first important reason is that by doing so we set an upper bound to the computational complexity of the SLAM algorithm in charge of maintaining local maps, as long as their size can be assured not to grow indefinitely. Another justification for local maps comes from the convenient representation of uncertainties as Gaussians, which becomes a really bad approximation to the actual uncertainty when using a single frame of coordinates in large-scale maps. In a different context, we must point out that HMT maps allow interconnecting metric map SLAM and artificial intelligence reasoning carried out over the corresponding topologies.

From a probabilistic point of view, we can justify the splitting of the world model into local maps from the realization that robot observations have a negligible effect in the estimation of random variables, either robot poses or landmarks, that are far way from the place observed (obviously, this does not hold when using global coordinates and closing a loop but in hierarchical maps we use local coordinates instead). Think of the information matrix of the SLAM unknowns, which corresponds to the Hessian matrix described earlier in this chapter. If we could rearrange the order of the variables such that the Hessian becomes block diagonal, it would mean the sets of variables corresponding to each block can be estimated independently of the rest. Each diagonal block corresponds to the ideal local map within an HMT map, but unfortunately, the sparseness of the information matrix does not typically allow partitioning large maps into several maps of convenient sizes. There exist graph-partitioning methods to decide how to cluster variables together for their information matrix becoming as close as possible to block diagonal. In any case, the partitioning into local maps becomes an approximation where some information (entries outside of the diagonal blocks) will be lost. In practice, we can set a threshold to the amount of information loss, which can be tolerated without problems and only divide local maps when that condition holds (Blanco, González, & Fernández-Madrigal, 2009).

In despite of the advantages of HMT mapping, this paradigm introduces its own new challenges. The most relevant is the need to detect topological loop closures, that is, whether two local maps correspond in fact to the same place. Existing solutions depend on the specific type of local maps used and on the sensors employed by the robot. For example, in case of representing local maps as occupancy grids we could rely on grid matching algorithms. For robots equipped with cameras, the approaches that have demonstrated the best performance and reliability are those based on advanced appearance-based place recognition (Cummins & Newman, 2008a, 2008b).

The second desideratum for lifelong SLAM that we mentioned previously is the ability to *forget*. Within a framework for HMT maps with multi-level hierarchies as the one we just depicted, forgetting could be realized in two ways. Firstly, by considering at each instant the highest abstraction level suitable for establishing the relationship of the robot with its environment and disregarding ("forgetting") all the low-level information (remember from chapter 8 that information at the highest levels of the hierarchy is most robust

Figure 10. An illustration of the concept of multiple-level hierarchies for hybrid metric-topological maps

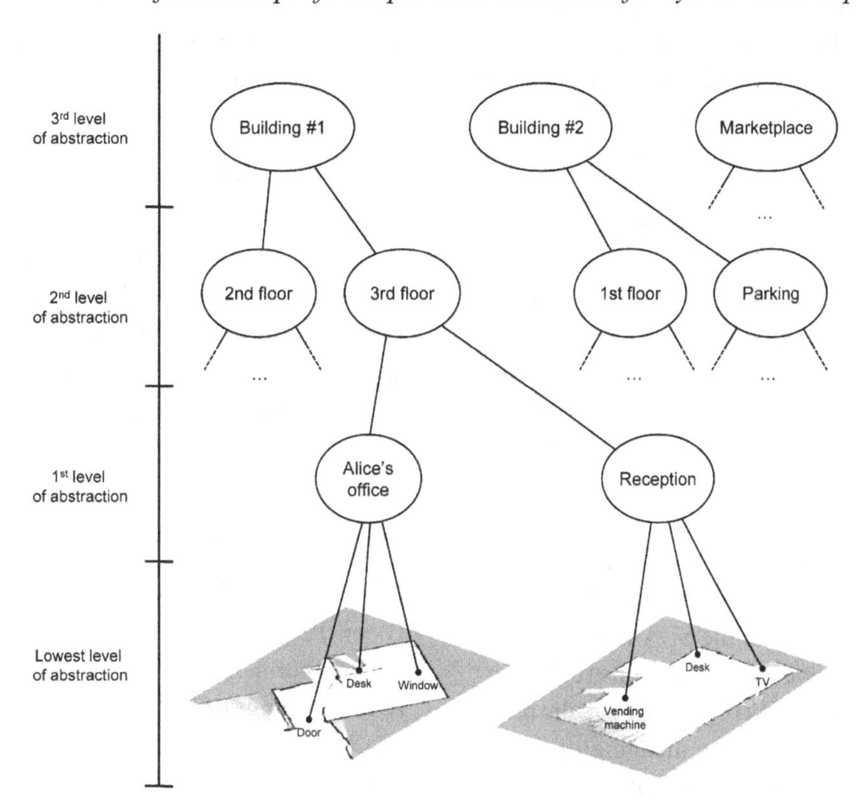

and less sensitive to changes in reality, i.e., more stable). Secondly, we can think of an actual permanent removal of information from maps as a mechanism to assure bounded storage requirements for lifelong SLAM. Methods in this line have been already proposed in the community of bundle adjustment and graph SLAM, exploring the delicate issues of how to forget variables (marginalize out, in statistical terms) while still preserving a minimalist representation of the environment (Konolige & Agrawal, 2008).

The third requirement for lifelong SLAM we mentioned before was unlimited scalability, or in other words, constant computational complexity. As one could expect, most existing SLAM methods have something in common: updating the estimate of the unknowns gets more computationally expensive as the map grows. The unique exceptions are RBPF algorithms, whose complexity is determined by the number of samples, which in some cases is kept constant. However, as we discussed in the previous chapter, RBPFs do not scale well for large maps due to a number of shortcomings that cannot be avoided without using an exponential number of particles as the map grows. The rest of algorithms have better complexities: EFK-based methods have quadratic complexity and MLE methods (graph SLAM and bundle adjustment) can approach linear complexity by exploiting sparsity. Furthermore, recent advances as the introduction of a combined Kalman-Information filter in Cadena and Neira (2010) make possible further reducing the complexity down to logarithmic with the number of landmarks. Nevertheless, devising an (amortized) constant-time mapping algorithm still remains as the unreachable "holy grail" of SLAM.

Regarding the fourth desideratum (the one about multiple hierarchies), we must clearly differentiate between several levels of abstraction in hierarchical maps and the existence of more than one hierarchy which can coexist in parallel. Multiple hierarchies were introduced in Fernández-Madrigal and González (2002b) as a powerful abstraction tool which resembles the way in which we human have different representations of the world depending on our intentions, i.e., on the task being carried out at each instant. It has been shown that this multihierarchical scheme allows optimizing each hierarchy for each robot task independently, and thus obtaining a better optimization of the robot operation (Galindo, Fernández-Madrigal, & González, 2004). As a clarifying example, consider the split of the environment into local map within a hierarchical HMT map. Those divisions are calculated to minimize a very specific measure (i.e. the elements outside of the diagonal blocks of the SLAM information matrix), thus a hierarchy can be constructed observing that criterion. On the other hand, the robot may need to understand the meaning of places such as "David's office" or "40 Holywell Street," which correspond to the way in which we humans divide the world. This latter organization of the world into rooms, buildings, street addresses, etc. would then correspond to another independent hierarchy in the robot world model.

Last but not least, we have the goal of augmenting metric maps with semantic information. This introduces new challenges on its own, as the so-called *anchoring* problem of how raw sensory data is robustly "anchored" to its corresponding symbol in the robot world model (Galindo, Fernández-Madrigal, González, & Saffiotti, 2008). In turn, semantic information opens the door to a wide range of potential applications related to high-level reasoning which, among other utilities, can help with robot localization.

To end this section, in which we provided a glimpse at how lifelong SLAM could be achieved, we can sum up all the discussion into one central idea: the only practical and realistic way in which robots could, someday, achieve a lifelong autonomous operation seems to require finding a point of convergence between estimation theory, computer vision and symbolic reasoning—artificial intelligence.

REFERENCES

Bailey, T., Nebot, E. M., Rosenblatt, J. K., & Durrant-Whyte, H. F. (2000). Data association for mobile robot navigation: A graph theoretic approach. *IEEE International Conference on Robotics and Automation, 3*, 2512-2517.

Bay, H., Tuytelaars, T., & Van Gool, L. (2006). Surf: Speeded up robust features. In *Proceedings of the European Conference on Computer Vision (ECCV)*, (pp. 404-417). ECCV.

Biber, P., & Duckett, T. (2005). Dynamic maps for long-term operation of mobile service robots. In *Proceedings of Robotics: Science and Systems (RSS)*, (pp. 17-24). RSS.

Bierman, G. J. (1977). *Factorization methods for discrete sequential estimation*. New York, NY: Academic Press.

Blanco, J. L., Fernández-Madrigal, J. A., & González, J. (2008). Towards a unified Bayesian approach to hybrid metric-topological SLAM. *IEEE Transactions on Robotics, 24*(2), 259–270. doi:10.1109/TRO.2008.918049

Blanco, J. L., González, J., & Fernández-Madrigal, J. A. (2009). Subjective local maps for hybrid metric-topological SLAM. *Robotics and Autonomous Systems, 57*(1), 64–74. doi:10.1016/j.robot.2008.02.002

Cadena, C., & Neira, J. (2010). SLAM in O (log n) with the combined Kalman-information filter. *Robotics and Autonomous Systems, 58*(11), 1207–1219. doi:10.1016/j.robot.2010.08.003

Castellanos, J. A., Martínez-Cantin, R., Tardós, J. D., & Neira, J. (2007). Robocentric map joining: Improving the consistency of EKF-SLAM. *Robotics and Autonomous Systems*, *55*(1), 21–29. doi:10.1016/j.robot.2006.06.005

Chen, C. K., & Yun, D. Y. Y. (1998). Unifying graph matching problems with a practical solution. In *Proceedings of the International Conference on Systems, Signals, Control, Computers*. IEEE.

Chen, Y., Davis, T. A., Hager, W. W., & Rajamanickam, S. (2008). Algorithm 887: CHOLMOD, supernodal sparse Cholesky factorization and update/downdate. *ACM Transactions on Mathematical Software*, *35*(3).

Cummins, M., & Newman, P. (2008). Accelerated appearance-only SLAM. In *Proceedings of the IEEE International Conference on Robotics and Automation*, (pp. 1828-1833). IEEE Press.

Cummins, M., & Newman, P. (2008b). FAB-MAP: Probabilistic localization and mapping in the space of appearance. *The International Journal of Robotics Research*, *27*(6), 647–665. doi:10.1177/0278364908090961

Davis, T. A. (2006). *Direct methods for sparse linear systems*. New York, NY: Society for Industrial Mathematics. doi:10.1137/1.9780898718881

Davison, A. J., Reid, I. D., Molton, N. D., & Stasse, O. (2007). MonoSLAM: Real-time single camera SLAM. *IEEE Transactions on Pattern Analysis and Machine Intelligence*, *29*(6), 1052–1067. doi:10.1109/TPAMI.2007.1049

Dellaert, F., Carlson, J., Ila, V., Ni, K., & Thorpe, C. E. (2010). Subgraph-preconditioned conjugate gradients for large scale SLAM. In *Proceedings of the IEEE/RSJ International Conference on Intelligent Robots and Systems (IROS)*, (pp. 2566-2571). IEEE Press.

Dellaert, F., & Kaess, M. (2006). Square root SAM: Simultaneous localization and mapping via square root information smoothing. *The International Journal of Robotics Research*, *25*(12), 1181–1203. doi:10.1177/0278364906072768

Demmel, J. W. (1997). *Applied numerical linear algebra*. Philadelphia, PA: Society for Industrial and Applied Mathematics. doi:10.1137/1.9781611971446

Dennis, J. E., & Schnabel, R. B. (1996). *Numerical methods for unconstrained optimization and nonlinear equations*. New York, NY: Society for Industrial Mathematics. doi:10.1137/1.9781611971200

Duckett, T., Marsland, S., & Shapiro, J. (2002). Fast, on-line learning of globally consistent maps. *Autonomous Robots*, *12*(3), 287–300. doi:10.1023/A:1015269615729

Engelhard, N., Endres, F., Hess, J., Sturm, J., & Burgard, W. (2011). Real-time 3D visual SLAM with a hand-held RGB-D camera. In *Proceedings of the RGB-D Workshop on 3D Perception in Robotics at the European Robotics Forum*. RGB-D.

Fernández-Madrigal, J. A., & González, J. (2002). Multi-hierarchical representation of large-scale space: Applications to mobile robots. *Intelligent Systems, Control and Automation: Science and Engineering, 24*.

Fernández-Madrigal, J. A., & González, J. (2002b). Multihierarchical graph search. *IEEE Transactions on Pattern Analysis and Machine Intelligence*, *24*(1), 103–113. doi:10.1109/34.982887

Frese, U. (2006). Treemap: An O (log n) algorithm for indoor simultaneous localization and mapping. *Autonomous Robots*, *21*(2), 103–122. doi:10.1007/s10514-006-9043-2

Frese, U. (2006b). A discussion of simultaneous localization and mapping. *Autonomous Robots*, *20*(1), 25–42. doi:10.1007/s10514-006-5735-x

Frese, U., Larsson, P., & Duckett, T. (2005). A multilevel relaxation algorithm for simultaneous localisation and mapping. *IEEE Transactions on Robotics, 21*(2), 1–12. doi:10.1109/TRO.2004.839220

Frese, U., & Schröder, L. (2006). Closing a million-landmarks loop. In *Proceedings of the IEEE/RSJ International Conference on Intelligent Robots and Systems (IROS)*, (pp. 5032-5039). IEEE Press.

Gabay, D. (1982). Minimizing a differentiable function over a differential manifold. *Journal of Optimization Theory and Applications, 37*(2), 177–219. doi:10.1007/BF00934767

Galindo, C., Fernández-Madrigal, J. A., & González, J. (2004). Improving efficiency in mobile robot task planning through world abstraction. *IEEE Transactions on Robotics, 20*(4), 677–690. doi:10.1109/TRO.2004.829480

Galindo, C., Fernández-Madrigal, J. A., González, J., & Saffiotti, A. (2008). Robot task planning using semantic maps. *Robotics and Autonomous Systems, 56*(11), 955–966. doi:10.1016/j.robot.2008.08.007

Gallier, J. H. (2001). *Geometric methods and applications: For computer science and engineering*. Berlin, Germany: Springer Verlag.

Grisetti, G., Grzonka, S., Stachniss, C., Pfaff, P., & Burgard, W. (2007). Efficient estimation of accurate maximum likelihood maps in 3D. In *Proceedings of the IEEE/RSJ International Conference on Intelligent Robots and Systems (IROS)*, (pp. 3472-3478). IEEE Press.

Grisetti, G., Küemmerle, R., Stachniss, C., Frese, U., & Hertzberg, C. (2010). Hierarchical optimization on manifolds for online 2D and 3D mapping. In *Proceedings of the IEEE International Conference on Robotics and Automation*. IEEE Press.

Grisetti, G., Stachniss, C., Grzonka, S., & Burgard, W. (2007). A tree parameterization for efficiently computing maximum likelihood maps using gradient descent. In *Proceedings of Robotics: Science and Systems (RSS)*. RSS.

Grisetti, G., Stachniss, C., Grzonka, S., & Burgard, W. (2007). A tree parameterization for efficiently computing maximum likelihood maps using gradient descent. In *Proceedings of Robotics: Science and Systems (RSS)*. RSS.

Grossberg, S. (1988). *The adaptive brain*. London, UK: Elsevier.

Hartley, R., & Zisserman, A. (2003). *Multiple view geometry in computer vision*. Cambridge, UK: Cambridge University Press.

Henry, P., Krainin, M., Herbst, E., Ren, X., & Fox, D. (2010). RGB-D mapping: Using depth cameras for dense 3D modeling of indoor environments. In *Proceedings of the 12th International Symposium on Experimental Robotics (ISER)*. ISER.

Hertzberg, C. (2008). *A framework for sparse, non-linear least squares problems on manifolds*. (Master's Thesis). Universität Bremen. Bremen, Germany.

Howard, A., Mataric, M., & Sukhatme, G. (2001). Relaxation on a mesh: A formalism for generalized localization. In *Proceedings of the IEEE/RSJ International Conference on Intelligent Robots and Systems (IROS)*, (pp. 1055-1060). IEEE Press.

Huber, P. J. (1964). Robust estimation of a location parameter. *Annals of Mathematical Statistics, 35*, 73–101. doi:10.1214/aoms/1177703732

Kaess, M., Johannsson, H., Roberts, R., Ila, V., Leonard, J.J.,& Dellaert, F. (2011). iSAM2: Incremental Smoothing and Mapping with Fluid Relinearization and Incremental Variable Reordering. *IEEE International Conference on Robotics and Automation*, 3281-3288.

Kelley, C.T. (1999). *Iterative methods for optimization*. Philadelphia: Society for Industrial Mathematics.

Klein, G., & Murray, D. (2007). Parallel tracking and mapping for small AR workspaces. In *Proceedings of the 2007 6th IEEE and ACM International Symposium on Mixed and Augmented Reality.* IEEE Press.

Konolige, K., & Agrawal, M. (2008). FrameSLAM: From bundle adjustment to real-time visual mapping. *IEEE Transactions on Robotics*, *24*(5), 1066–1077. doi:10.1109/TRO.2008.2004832

Konolige, K., Grisetti, G., Kümmerle, R., Burgard, W., Limketkai, B., & Vincent, R. (2010). Sparse pose adjustment for 2D mapping. In *Proceedings of the IEEE/RSJ International Conference on Intelligent Robots and Systems (IROS)*. IEEE Press.

Kretzschmar, H., Stachniss, C., & Grisetti, G. (2011). Efficient information-theoretic graph pruning for graph-based SLAM with laser range finders. In *Proceedings of the IEEE/RSJ International Conference on Intelligent Robots and Systems (IROS)*. IEEE Press.

Küemmerle, R., Grisetti, G., Strasdat, H., Konolige, K., & Burgard, W. (2011). g2o: A general framework for graph optimization. In *Proceedings of the IEEE International Conference on Robotics and Automation (ICRA)*. IEEE Press.

Lowe, D. G. (2004). Distinctive image features from scale-invariant keypoints. *International Journal of Computer Vision*, *60*(2), 91–110. doi:10.1023/B:VISI.0000029664.99615.94

Lu, F., & Milios, E. (1997). Globally consistent range scan alignment for environment mapping. *Autonomous Robots*, *4*(4), 333–349. doi:10.1023/A:1008854305733

Madsen, K., Nielsen, H., & Tingleff, O. (2004). *Methods for non-linear least squares problems*. Copenhagen, Denmark: Technical University of Denmark.

Mei, C., Sibley, G., Cummins, M., Newman, P., & Reid, I. (2010). RSLAM: A system for large-scale mapping in constant-time using stereo. *International Journal of Computer Vision*, *94*(2), 198–214. doi:10.1007/s11263-010-0361-7

Meyer, C. D. (2001). *Matrix analysis and applied linear algebra*. Retrieved Mar 1, 2012, from http://www.matrixanalysis.com/

Neira, J., & Tardós, J. D. (2001). Data association in stochastic mapping using the joint compatibility test. *IEEE Transactions on Robotics and Automation*, *17*(6), 890–897. doi:10.1109/70.976019

Olson, E., Leonard, J. J., & Teller, S. (2006). Fast iterative optimization of pose graphs with poor initial estimates. In *Proceedings of the IEEE International Conference on Robotics & Automation*, (pp. 2262-2269). IEEE Press.

Paz, L. M., Tardós, J. D., & Neira, J. (2008). Divide and conquer: EKF SLAM in O(n). *IEEE Transactions on Robotics*, *24*(5), 1107–1120. doi:10.1109/TRO.2008.2004639

Ranganathan, A., Menegatti, E., & Dellaert, F. (2006). Bayesian inference in the space of topological maps. *IEEE Transactions on Robotics*, *22*(1), 92–107. doi:10.1109/TRO.2005.861457

Sibley, G., Mei, C., Reid, I., & Newman, P. (2010). Vast scale outdoor navigation using adaptive relative bundle adjustment. *The International Journal of Robotics Research*, *29*(8), 958–980. doi:10.1177/0278364910369268

Strasdat, H., Montiel, J. M. M., & Davison, A. J. (2010a). Scale-drift aware large scale monocular SLAM. In *Proceedings of Robotics Science and Sytems*. IEEE.

Strasdat, H., Montiel, J. M. M., & Davison, A. J. (2010b). Real-time monocular SLAM: Why filter? In *Proceedings of the IEEE International Conference on Robotics and Automation*, (pp. 2657-2664). IEEE Press.

Taylor, C. J., & Kriegman, D. J. (1994). *Minimization on the lie group SO(3) and related manifolds*. Technical Report 9405. New Haven, CT: Yale University.

Thrun, S., & Montemerlo, M. (2006). The graph SLAM algorithm with applications to large-scale mapping of urban structures. *The International Journal of Robotics Research*, *25*(5-6), 403–429. doi:10.1177/0278364906065387

Tipaldi, G. D., & Arras, K. O. (2010). FLIRT – Interest regions for 2D range data. In *Proceedings of the IEEE International Conference on Robotics and Automation*. IEEE Press.

Triggs, B., McLauchlan, P., Hartley, R., & Fitzgibbon, A. (2000). Bundle adjustment—A modern synthesis. *Vision Algorithms: Theory and Practice*, *1883*, 153–177.

Varadarajan, V. S. (1974). *Lie groups, lie algebras, and their representations*. Upper Saddle River, NJ: Prentice-Hall.

Vázquez-Martín, R., Núñez, P., Bandera, A., & Sandoval, F. (2009). Curvature-based environment description for robot navigation using laser range sensors. *Sensors (Basel, Switzerland)*, *9*(8), 5894–5918. doi:10.3390/s90805894

Walter, M. R., Eustice, R. M., & Leonard, J. J. (2007). Exactly sparse extended information filters for feature-based SLAM. *The International Journal of Robotics Research*, *26*(4), 335–359. doi:10.1177/0278364906075026

Wang, C. C., Thorpe, C., Thrun, S., Hebert, M., & Durrant-Whyte, H. (2007). Simultaneous localization, mapping and moving object tracking. *The International Journal of Robotics Research*, *26*(9), 889–916. doi:10.1177/0278364907081229

Wolf, D. F., & Sukhatme, G. S. (2005). Mobile robot simultaneous localization and mapping in dynamic environments. *Autonomous Robots*, *19*(1), 53–65. doi:10.1007/s10514-005-0606-4

ENDNOTES

1. If we had $F = 0$ we would be facing a graph with relative constraints only, which is a degenerated problem where any rigid transformation of an optimal solution is also an optimal solution, i.e., there exist infinite solutions.
2. The set of all maximal cliques in graph SLAM comprises, by definition, either pairwise edges or sets with three or more fully connected nodes. In this latter case, the independence assumption allows the factorization of those cliques into the product of all their pairwise constituents. Hence, we can always factor the joint pdf into the product of the potential functions of the individual edges.
3. Optimizing with first-order Taylor approximations (i.e. linearization) is also possible, as with the gradient descent method (Dennis & Schnabel, 1996). Although in most practical cases the second-order methods are preferable due to their much faster convergence, each iteration in first-order methods may be considerably cheaper to compute. An interesting discussion about this trade-off can be found in Triggs, McLauchlan, Hartley, and Fitzgibbon (2000).
4. In fact, a null derivative only implies a critical point in the cost function which could be a minimum, a maximum or a saddle point. For our multivariate function $\mathrm{F}\left(\mathbf{x}\right): \mathbb{R}^N \rightarrow \mathbb{R}$, we can tell between the three cases by checking whether the matrix of second derivatives (the Hessian $\mathbf{H}$) is definite positive, definite negative or indefinite, respectively. Thus, in order to localize a minimum we must assure a definite positive Hessian, that is, $\mathrm{rank}\left(\mathbf{H}\right) = N$. Since in practice the Hessian is always replaced by the Gauss-Newton approximation $\mathbf{H} \cong \mathbf{J}^T \boldsymbol{\Lambda} \mathbf{J}$ (as discussed in the text), it is

assured that $\mathbf{H}$ will be symmetric and, at least, semi-definitive positive. Given that $\mathbf{\Lambda}$ is definite positive by hypothesis and $\operatorname{rank}\left(\mathbf{J}^{T}\mathbf{\Lambda}\,\mathbf{J}\right) = \operatorname{rank}\left(\mathbf{J}\right)$, the condition for the positive definiteness of $\mathbf{H}$ reduces to $\operatorname{rank}\left(\mathbf{J}\right) = N$. This translates into the requirement for a minimum number of (linearly independent) constraints (i.e. edges) in any graph SLAM problem.

5 Evidently, a value of $\lambda_k = 0$ turns the Levenberg-Marquardt system $\left(\mathbf{H}_k + \lambda_k \mathbf{I}\right)\mathbf{\Delta x}_k^* = -\mathbf{g}_k^T$ into $\mathbf{H}_k \mathbf{\Delta x}_k^* = -\mathbf{g}_k^T$, which coincides with the Newton method. In contrast, when λ_k is large (in particular, much larger than the diagonal values of the Hessian) the linear system approximately becomes $\mathbf{\Delta x}_k^* = -\lambda_k^{-1}\,\mathbf{g}_k^T$, which corresponds to the steps from a first-order method known as steepest (or gradient) descent.

6 Assume that the uncertainty in the unknown parameters $\mathbf{x}$ of a cost function $\mathrm{F}\left(\mathbf{x}\right)$ can be properly modeled as a Gaussian with mean $\bar{\mathbf{x}}$ and covariance $\mathbf{\Sigma}$. The negative logarithm of such a multivariate pdf would vary with $\mathbf{x}$ like $\frac{1}{2}\left(\mathbf{x}-\bar{\mathbf{x}}\right)^T \mathbf{\Sigma}^{-1}\left(\mathbf{x}-\bar{\mathbf{x}}\right)$. On the other hand, the cost function itself is the negative log-likelihood of the unknown parameters. In the vicinity of the optimal solution $\mathbf{x}^*$ this function is well approximated by its second-order Taylor series expansion (recall Equation 22), which varies with the unknowns like $\mathbf{g}\left(\mathbf{x}-\mathbf{x}^*\right)+\frac{1}{2}\left(\mathbf{x}-\mathbf{x}^*\right)^T \mathbf{H}\left(\mathbf{x}-\mathbf{x}^*\right)$. By definition, if $\mathbf{x}^*$ is the exact minimum of $\mathrm{F}\left(\mathbf{x}\right)$ then its gradient $\mathbf{g}$ evaluates to zero there, so it becomes $\frac{1}{2}\left(\mathbf{x}-\mathbf{x}^*\right)^T \mathbf{H}\left(\mathbf{x}-\mathbf{x}^*\right)$ By identifying the optimal solution $\mathbf{x}^*$ with the mean $\bar{\mathbf{x}}$ of the tentative Gaussian distribution, it is obvious that the Hessian $\mathbf{H}$ plays the role of the inverse covariance matrix $\mathbf{\Sigma}^{-1}$ in the negative log-likelihood of a multivariate pdf, as we wanted to show. It must be noted however that in practice we will typically only have the Gauss-Newton approximation of the Hessian $\mathbf{H} \approx \mathbf{J}^T\mathbf{\Lambda}\,\mathbf{J}$. This stands out as a very good approximation when second-order derivatives (ignored in that approximation) do not have a significant weight in the Hessian diagonal, which typically occurs when all the residuals are small, i.e., there are no outliers.

7 In the photogrammetry literature it is also common to include the set of camera calibration parameters (or several sets if more than one camera were used to capture the images) so they are automatically optimized among the rest of unknowns. Since that is not common in the mobile robotics literature, we do not mention this possibility in the text and refer the interested reader to Triggs, McLauchlan, Hartley, and Fitzgibbon (2000).

8 Notice that the discussion focuses on lifelong *SLAM*. Performing *localization* (without mapping) for extended periods of time can, indeed, be considered a "solved problem" nowadays. For example, localization with particle filters has demonstrated to be extremely robust and reliable… as long as the uncertainty of all the involved processes is properly modeled. However, limiting ourselves to localization implicitly constrains the application of mobile robotics to restricted areas of mostly static environments, which furthermore have to be mapped in advance before deploying the robot for operation.

Appendix A: Common SE (2) and SE (3) Geometric Operations

Dealing with mobile robots necessarily implies dealing with geometric problems. Studying their kinematic models or their position and attitude in three-dimensional space, for example, requires us to handle spatial relationships. This appendix provides a summary of the basic mathematical concepts from 2D and 3D geometry that are needed for solving most mobile robotic problems. Some other mathematically more intricate concepts will be deferred until Appendix D.

1. ABOUT GEOMETRIC OPERATIONS AND THEIR NOTATION

The geometric operations we will discuss here work with two elements: *spatial locations* and *spatial transformations*. Locations are simply *points* in a two or three-dimensional Euclidean space, which we will describe with the vector of their (two or three) coordinates with respect to some reference frame, that is:

$$\begin{aligned} \mathbf{a}_2 &= \begin{bmatrix} x \\ y \end{bmatrix} \qquad \text{(a 2D point)} \\ \mathbf{a}_3 &= \begin{bmatrix} x \\ y \\ z \end{bmatrix} \qquad \text{(a 3D point)} \end{aligned} \tag{1}$$

Regarding spatial transformations, we firstly find *pure rotations*. It can be shown that pure rigid rotations in 2D and 3D form mathematical *groups* whose elements are 2×2 and 3×3 matrices, respectively, with unit determinants, and where the group inner operation is the standard matrix multiplication. These groups are named $\mathbf{SO}(2)$ and $\mathbf{SO}(3)$, after "*special orthogonal* group." We will denote its elements (matrices) as:

$$\begin{aligned} \mathbf{R}_{2\times2} &\in \mathbf{SO}(2) \qquad \text{(a pure rotation in 2D)} \\ \mathbf{R}_{3\times3} &\in \mathbf{SO}(3) \qquad \text{(a pure rotation in 3D)} \end{aligned} \tag{2}$$

Spatial points or transformations are never *absolute* in the strict sense of the word: they only make sense with respect to some frame of coordinates. A frame of coordinates can be visualized as a set of two or three orthogonal axes (for 2D or 3D, respectively) fixed at some arbitrary placement. Sometimes such frames are informally referred to as *corners*. We can therefore visually imagine a $\mathbf{SO}(n)$ transformation as an arbitrary rotation of those axes, without translation.

A broader class of spatial transformation also allows for spatial translations apart from rotation. It is common in robotics to name *poses* to those transformations. Thus, a pose can be visualized as a corner placed and rotated arbitrarily with respect to any other reference corner. Mathematically, poses form the "*special Euclidean groups*" $\mathbf{SE}(2)$ and $\mathbf{SE}(3)$, for 2D and 3D, respectively. The group elements are 3×3 and 4×4 matrices, respectively, and the group operation is also here the standard matrix multiplication. We will denote pose matrices with a capitalized $\mathbf{P}$, that is:

$$\begin{aligned} &\mathbf{P}_{3\times3} \in \mathbf{SE}(2) && \text{(a generic pose in 2D)} \\ &\mathbf{P}_{4\times4} \in \mathbf{SE}(3) && \text{(a generic pose in 3D)} \end{aligned} \tag{3}$$

Since pure rotations typically have less practical utility in mobile robotics than the more generic concept of poses (i.e., translations plus pure rotations), in most problems we will deal only with the latter.

As we will see in the next sections, poses are usually represented in a *parameterized form* instead of their matrix forms $\mathbf{P}$. The main reason is that pose matrices, while perfectly representing the spatial transformation, have far more DOFs than the actual poses. The number of independent values needed to completely specify a spatial pose is 3 or 6 (in 2D or 3D space, respectively), but the number of values in the matrices is much higher: $\mathbf{SE}(2)$ and $\mathbf{SE}(3)$ have matrices with $3 \times 3 = 9$ and $4 \times 4 = 16$ elements, respectively.

In general, we will denote pose parameterizations as *vectors* represented with an uncapitalized letter $\mathbf{p}$, whose length will depend on the particular parameterization —but which can never be smaller than the number of spatial DOFs. We will often need to transform between the vector of a pose parameters and its matrix form, an operation which we will denoted by means of the function $\mathbf{M}(\cdot)$, specific for each parameterization:

$$\underbrace{\mathbf{P}}_{\substack{\text{Pose in} \\ \text{matrix form}}} = \mathbf{M} \underbrace{(\mathbf{p})}_{\substack{\text{Pose} \\ \text{parameterization}}}$$

where the inverse function is defined as:

$$\underbrace{\mathbf{p}}_{\substack{\text{Pose} \\ \text{parameterization}}} = \mathbf{M}^{-1} \underbrace{(\mathbf{P})}_{\substack{\text{Pose in} \\ \text{matrix form}}} \tag{4}$$

Formally, only the matrices $\mathbf{P}$ belong to the $\mathbf{SE}(n)$ groups. However, in an abuse of notation we can also say that a pose parameterization $\mathbf{p}$ is a member of the groups if we interpret such a statement as equivalent to $\mathbf{M}(\mathbf{p}) \in \mathbf{SE}(n)$.

As shown below, mixing spatial transformations (*poses*) and spatial locations (*points*) requires working with pose $\mathbf{P}$ matrices and point vectors of the *correct length*: although 2D and 3D points are described by vectors of length 2 and 3, the matrices of 2D and 3D poses have, as said above, sizes of 3×3 and

4×4, respectively. For each point vector $\mathbf{a}$ we can define its associated *extended* vector $\mathbf{A}$ by means of a function $\mathbf{V}(\cdot)$ which simply appends an extra unit element. That is:

For 2D:

$$\mathbf{V}: \mathbb{R}^2 \to \mathbb{R}^3 \qquad \mathbf{A} = \mathbf{V}(\mathbf{a}) \quad \Rightarrow \mathbf{V}\begin{pmatrix} a_x \\ a_y \end{pmatrix} = \begin{pmatrix} a_x \\ a_y \\ 1 \end{pmatrix}$$

For 3D:

$$\mathbf{V}: \mathbb{R}^3 \to \mathbb{R}^4 \qquad \mathbf{A} = \mathbf{V}(\mathbf{a}) \quad \Rightarrow \mathbf{V}\begin{pmatrix} a_x \\ a_y \\ a_z \end{pmatrix} = \begin{pmatrix} a_x \\ a_y \\ a_z \\ 1 \end{pmatrix} \tag{5}$$

After all these preliminary definitions, we are ready to describe the four fundamental geometric operations regarding poses and points, illustrated in Figure 1:

1. Composition of two poses $\mathbf{p}_1$ and $\mathbf{p}_2$: the resulting pose is $\mathbf{p}_2$ as if it was expressed with respect to $\mathbf{p}_1$. That is, as if $\mathbf{p}_1$ was the new origin of coordinates for $\mathbf{p}_2$.
2. Inverse composition of two poses $\mathbf{p}_1$ and $\mathbf{p}_2$: the resulting pose is $\mathbf{p}_1$ "as seen from" $\mathbf{p}_2$.
3. Composition of a pose $\mathbf{p}$ and a point $\mathbf{a}$: the resulting point is $\mathbf{a}$ "as if" it was expressed in the coordinate frame defined by $\mathbf{p}$.
4. Inverse composition of a pose $\mathbf{p}$ and a point $\mathbf{a}$: the resulting point is $\mathbf{a}$ "as seen from" $\mathbf{p}$.

It is worth carefully matching these four definitions with their representations in Figure 1c-f in order to unambiguously grasp their geometrical meaning.

In order to denote all these operations rigorously, it is common to find in the literature the so-called "o plus" notation for pose operations, which employs the operators $\oplus$ and $\ominus$ (Smith, Self, & Cheeseman, 1988; Thrun, Burgard, & Fox, 2005). Under this notation, the four operations above and their matrix equivalents (with $\mathbf{P} = \mathbf{M}(\mathbf{p})$, $\mathbf{A}' = \mathbf{V}(\mathbf{a})$, etc.) can be shown to become matrix multiplications, such that:

Operation	"o plus" notation		Matrix notation
1. Pose-pose composition	$\mathbf{p} = \mathbf{p}_1 \oplus \mathbf{p}_2$	$\rightarrow$	$\mathbf{P} = \mathbf{P}_1 \mathbf{P}_2$
2. Pose-pose inverse composition	$\mathbf{p} = \mathbf{p}_1 \ominus \mathbf{p}_2$	$\rightarrow$	$\mathbf{P} = \mathbf{P}_2^{-1}\mathbf{P}_1$
3. Pose-point composition	$\mathbf{a}' = \mathbf{p} \oplus \mathbf{a}$	$\rightarrow$	$\mathbf{A}' = \mathbf{P}\mathbf{A}$
4. Pose-point inverse composition	$\mathbf{a}' = \mathbf{a} \ominus \mathbf{p}$	$\rightarrow$	$\mathbf{A}' = \mathbf{P}^{-1}\mathbf{A}$

Figure 1. Illustration of the four fundamental geometric operations with poses and points. (a) – (b) Two arbitrary poses $\mathbf{p}_1$ and $\mathbf{p}_2$, and their composition in (c). We also show (d) the pose inverse composition, (e) the pose-point composition, and (f) the inverse pose-point composition.

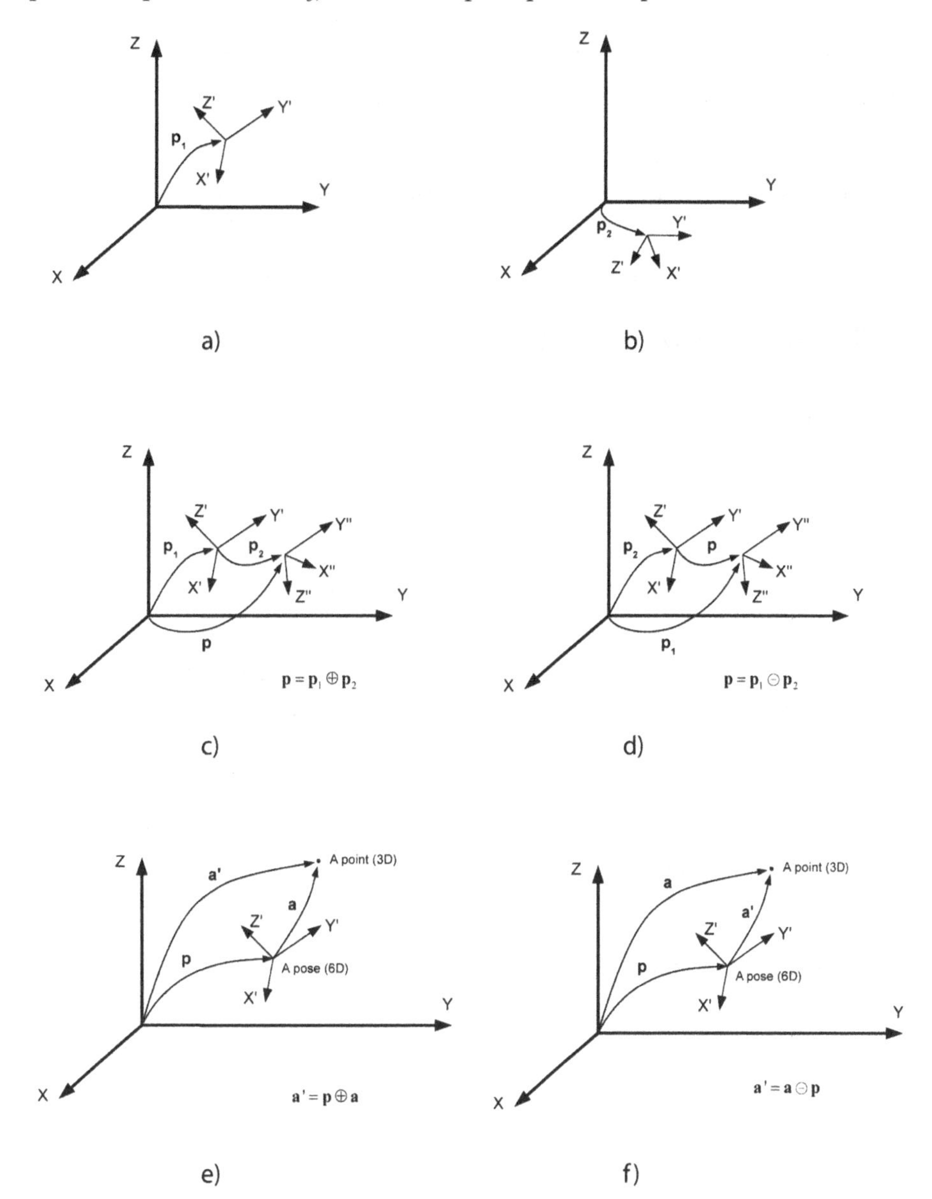

An extension to this basic notation is the *unary* $\ominus$ operator to denote pose inversion. It can be seen as a special case of pose inverse composition where the pose implicitly assumed at the left of the operator is the *identity element* of the $\mathbf{SE}(n)$ group, that is, the origin of coordinates:

$$\text{Pose inversion:} \quad \mathbf{p}' = \ominus\mathbf{p} \equiv \underbrace{\mathbf{p}_0}_{\substack{\text{Pose at the}\\ \text{origin of}\\ \text{coordinates}}} \ominus\, \mathbf{p} \quad \rightarrow \quad \mathbf{P}' = \mathbf{P}^{-1} \underbrace{\mathbf{P}_0}_{\mathbf{I}} = \mathbf{P}^{-1} \tag{6}$$

The usage of this unary operator allows us to restate an alternative formulation of the two operations involving inverse pose composition (the operations numbered as 2 and 4 above), now in terms of "normal" (not inverse) pose compositions:

$$\begin{array}{lll} \text{2) Pose-pose inverse composition} & \mathbf{p} = \mathbf{p}_1 \ominus \mathbf{p}_2 \equiv (\ominus \mathbf{p}_2) \oplus \mathbf{p}_1 \\ \text{4) Pose-point inverse composition} & \mathbf{a}' = \mathbf{a} \ominus \mathbf{p} \equiv (\ominus \mathbf{p}) \oplus \mathbf{a} \end{array} \tag{7}$$

In the next sections, we explore how all these operations can be implemented in practice for the cases of 2D and 3D geometry.

2. OPERATIONS WITH SE(2) POSES

A $\mathbf{SE}(2)$ pose is easily parameterized as a translation in 2D and a rotation, such as:

$$\mathbf{p} = \begin{pmatrix} x \\ y \\ \theta \end{pmatrix} \tag{8}$$

Under the convention of positive counterclockwise rotations, the corresponding matrix form reads as follows in order to satisfy the definitions of the operators $\oplus$ and $\ominus$ above:

$$\mathbf{P} = \mathbf{M}(\mathbf{p}) \equiv \begin{pmatrix} \cos\theta & -\sin\theta & x \\ \sin\theta & \cos\theta & y \\ 0 & 0 & 1 \end{pmatrix} \tag{9}$$

The inverse conversion is straightforward in this case, since:

$$\begin{array}{l} \mathbf{p} = \mathbf{M}^{-1}(\mathbf{P}), \quad \textit{with } \mathbf{P} = \begin{pmatrix} m_{11} & m_{12} & m_{13} \\ m_{21} & m_{22} & m_{23} \\ 0 & 0 & 1 \end{pmatrix} \\ \rightarrow \mathbf{M}^{-1}(\mathbf{P}) = \begin{pmatrix} m_{13} \\ m_{23} \\ \theta \end{pmatrix} \quad , \textit{with } \theta = \tan^{-1}(m_{21}/m_{11}) \quad \textit{or} \quad \theta = \operatorname{atan2}(m_{21}, m_{11}) \end{array} \tag{10}$$

Given a 2D pose as a matrix we can easily carry out any of the four pose-point operations in matrix form. However, since in practice we may be normally interested in the parameterized representations instead of matrices, it would be valuable to have expressions that directly give us the parameters of the final poses and points, skipping the explicit operations with the intermediary matrices. Such expressions are quite simple for 2D geometry, thus we provide all of them below. Due to their utility in several robotics operations (e.g. uncertainty propagation in an EKF) we also provide the corresponding Jacobian matrices with respect to each argument.

Composition of two poses:

$$\begin{pmatrix} x \\ y \\ \theta \end{pmatrix} = \mathbf{p} = \mathbf{p}_1 \oplus \mathbf{p}_2 = \mathbf{f}_c(\mathbf{p}_1, \mathbf{p}_2) = \begin{pmatrix} x_1 + x_2\cos\theta_1 - y_2\sin\theta_1 \\ y_1 + x_2\sin\theta_1 + y_2\cos\theta_1 \\ \theta_1 + \theta_2 \end{pmatrix}$$

and its Jacobians:

$$\left.\frac{\partial \mathbf{f}_c(\mathbf{p}_1, \mathbf{p}_2)}{\partial \mathbf{p}_1}\right|_{3\times3} = \begin{pmatrix} 1 & 0 & -x_2\sin\theta_1 - y_2\cos\theta_1 \\ 0 & 1 & x_2\cos\theta_1 - y_2\sin\theta_1 \\ 0 & 0 & 1 \end{pmatrix}$$
$$\left.\frac{\partial \mathbf{f}_c(\mathbf{p}_1, \mathbf{p}_2)}{\partial \mathbf{p}_2}\right|_{3\times3} = \begin{pmatrix} \cos\theta_1 & -\sin\theta_1 & 0 \\ \sin\theta_1 & \cos\theta_1 & 0 \\ 0 & 0 & 1 \end{pmatrix} \tag{11}$$

Inverse composition of two poses:

$$\begin{pmatrix} x \\ y \\ \theta \end{pmatrix} = \mathbf{p} = \mathbf{p}_1 \ominus \mathbf{p}_2 = \mathbf{f}_i(\mathbf{p}_1, \mathbf{p}_2) = \begin{pmatrix} (x_1 - x_2)\cos\theta_2 + (y_1 - y_2)\sin\theta_2 \\ -(x_1 - x_2)\sin\theta_2 + (y_1 - y_2)\cos\theta_2 \\ \theta_1 - \theta_2 \end{pmatrix}$$

and its Jacobians:

$$\left.\frac{\partial \mathbf{f}_i(\mathbf{p}_1, \mathbf{p}_2)}{\partial \mathbf{p}_1}\right|_{3\times3} = \begin{pmatrix} \cos\theta_2 & \sin\theta_2 & 0 \\ -\sin\theta_2 & \cos\theta_2 & 0 \\ 0 & 0 & 1 \end{pmatrix}$$
$$\left.\frac{\partial \mathbf{f}_i(\mathbf{p}_1, \mathbf{p}_2)}{\partial \mathbf{p}_2}\right|_{3\times3} = \begin{pmatrix} -\cos\theta_2 & -\sin\theta_2 & -(x_1 - x_2)\sin\theta_2 + (y_1 - y_2)\cos\theta_2 \\ \sin\theta_2 & -\cos\theta_2 & -(x_1 - x_2)\cos\theta_2 - (y_1 - y_2)\sin\theta_2 \\ 0 & 0 & -1 \end{pmatrix} \tag{12}$$

Composition of a pose and a point:

$$\begin{pmatrix} a_x' \\ a_y' \end{pmatrix} = \mathbf{a}' = \mathbf{p} \oplus \mathbf{a} = \mathbf{f}_{pc}(\mathbf{p}, \mathbf{a}) = \begin{pmatrix} x + a_x\cos\theta - a_y\sin\theta \\ y + a_x\sin\theta + a_y\cos\theta \end{pmatrix}$$

and its Jacobians:

$$\left.\frac{\partial \mathbf{f}_{pc}(\mathbf{p},\mathbf{a})}{\partial \mathbf{p}}\right|_{2\times3} = \begin{pmatrix} 1 & 0 & -a_x \sin\theta - a_y \cos\theta \\ 0 & 1 & a_x \cos\theta - a_y \sin\theta \end{pmatrix}$$
$$\left.\frac{\partial \mathbf{f}_{pc}(\mathbf{p},\mathbf{a})}{\partial \mathbf{a}}\right|_{2\times2} = \begin{pmatrix} \cos\theta & -\sin\theta \\ \sin\theta & \cos\theta \end{pmatrix} \tag{13}$$

Inverse composition of a pose and a point:

$$\begin{pmatrix} a_x' \\ a_y' \end{pmatrix} = \mathbf{a}' = \mathbf{a} \ominus \mathbf{p} = \mathbf{f}_{pi}(\mathbf{a},\mathbf{p}) = \begin{pmatrix} (a_x - x)\cos\theta + (a_y - y)\sin\theta \\ -(a_x - x)\sin\theta + (a_y - y)\cos\theta \end{pmatrix}$$

and its Jacobians:

$$\left.\frac{\partial \mathbf{f}_{pi}(\mathbf{a},\mathbf{p})}{\partial \mathbf{a}}\right|_{2\times2} = \begin{pmatrix} \cos\theta & \sin\theta \\ -\sin\theta & \cos\theta \end{pmatrix}$$
$$\left.\frac{\partial \mathbf{f}_{pi}(\mathbf{a},\mathbf{p})}{\partial \mathbf{p}}\right|_{2\times3} = \begin{pmatrix} -\cos\theta & -\sin\theta & -(a_x - x)\sin\theta + (a_y - y)\cos\theta \\ \sin\theta & -\cos\theta & -(a_x - x)\cos\theta - (a_y - y)\sin\theta \end{pmatrix} \tag{14}$$

Regarding the computation of the inverse of a pose, $\ominus\mathbf{p}$, it can be easily done by applying equation (6) and then equation (12).

3. OPERATIONS WITH SE(3) POSES

A $\mathbf{SE}(3)$ pose comprises a pure translation and a pure rotation, the latter belonging to $\mathbf{SO}(3)$. There exist two main families of parameterizations for this rotational part: triplets of Euler angles and the unit quaternion. We describe both of them next.

An important remark regarding Euler angles that is barely mentioned in the literature is the existence of *12 different such parameterizations* depending on the order in which the three rotations are applied to arrive at the desired attitude (Diebel, 2006). Therefore, it becomes crucial to always clearly state the chosen order, since a reader will not be able to unambiguously guess it. In the following, we will adopt the so-called *yaw-pitch-roll* representation, where the names of each rotation follow from their usage in airplane navigation. If we denote as ϕ (yaw), χ (pitch) and ψ (roll) the angles of these consecutive rotations (i.e. each rotation actuates around the already-rotated axes), which are applied in that same order as sketched in Figure 2a, then we can parameterize a 3D pose as:

Figure 2. (a) The particular convention adopted here for an Euler angles parameterization of 3D rotations. (b) A geometric interpretation of the unit quaternion, where the rotation θ relates to the quaternion parameters by $\theta = \cos^{-1}(2q_r)$.

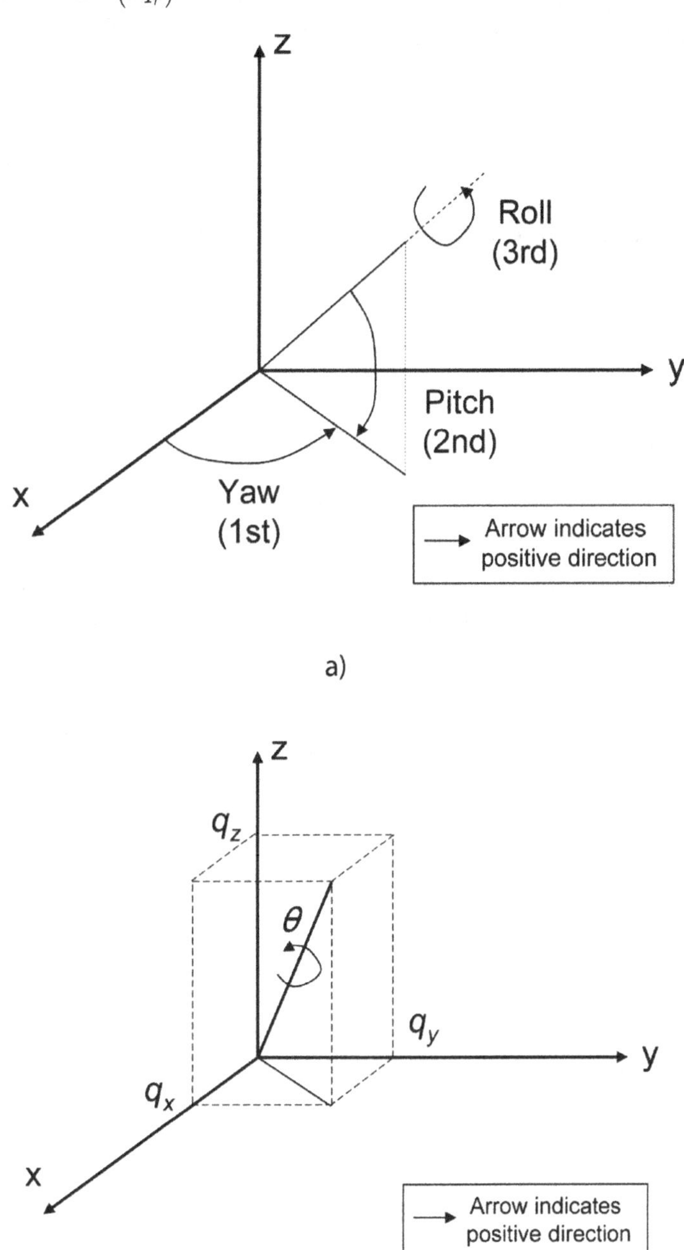

$$\mathbf{p} = \begin{pmatrix} x \\ y \\ z \\ \hline \phi \\ \chi \\ \psi \end{pmatrix} \qquad \text{(yaw-pitch-roll parameterization)} \tag{15}$$

The corresponding matrix form of such a pose has the structure:

$$\mathbf{P} = \mathbf{M}(\mathbf{p}) \equiv \left(\begin{array}{ccc|c} & & & x \\ & \mathbf{R}(\phi, \chi, \psi) & & y \\ & & & z \\ \hline 0 & 0 & 0 & 1 \end{array}\right) \tag{16}$$

where the 3×3 rotation matrix $\mathbf{R}(\phi, \chi, \psi)$ can be easily found by concatenating successive rotations of ϕ, χ and ψ radians around the z, y and x axes, respectively. Notice that rotations apply over the successively transformed axes (more on this below), which means that we must use right-hand matrix multiplications. Therefore, we have:

$$\mathbf{R}_{ypr}(\phi, \chi, \psi) = \underbrace{\mathbf{R}_z(\phi)}_{1^{st}:\text{yaw}} \underbrace{\mathbf{R}_y(\chi)}_{2^{nd}:\text{pitch}} \underbrace{\mathbf{R}_x(\psi)}_{3^{rd}:\text{ roll}}$$

with:

$$\mathbf{R}_z(\phi) = \begin{pmatrix} \cos\phi & -\sin\phi & 0 \\ \sin\phi & \cos\phi & 0 \\ 0 & 0 & 1 \end{pmatrix} = \begin{pmatrix} \mathrm{c}\,\phi & -\mathrm{s}\,\phi & 0 \\ \mathrm{s}\,\phi & \mathrm{c}\,\phi & 0 \\ 0 & 0 & 1 \end{pmatrix}$$

$$\mathbf{R}_y(\chi) = \begin{pmatrix} \cos\chi & 0 & \sin\chi \\ 0 & 1 & 0 \\ -\sin\chi & 0 & \cos\chi \end{pmatrix} = \begin{pmatrix} \mathrm{c}\,\chi & 0 & \mathrm{s}\,\chi \\ 0 & 1 & 0 \\ -\mathrm{s}\,\chi & 0 & \mathrm{c}\,\chi \end{pmatrix}$$

$$\mathbf{R}_x(\psi) = \begin{pmatrix} 1 & 0 & 0 \\ 0 & \cos\psi & -\sin\psi \\ 0 & \sin\psi & \cos\psi \end{pmatrix} = \begin{pmatrix} 1 & 0 & 0 \\ 0 & \mathrm{c}\,\psi & -\mathrm{s}\,\psi \\ 0 & \mathrm{s}\,\psi & \mathrm{c}\,\psi \end{pmatrix}$$

such that:

$$\mathbf{R}(\phi,\chi,\psi)=\begin{pmatrix} c\phi c\chi & c\phi s\chi s\psi - s\phi c\psi & c\phi s\chi c\psi + s\phi s\psi \\ s\phi c\chi & s\phi s\chi s\psi + c\phi c\psi & s\phi s\chi c\psi - c\phi s\psi \\ -s\chi & c\chi s\psi & c\chi c\psi \end{pmatrix} \tag{17}$$

It is common (and frustrating for students) to find different and apparently incompatible definitions for what are the *yaw-pitch-roll* angles. We just mentioned that they are rotations around the ("*dynamic*") z, y, and x axes, with "dynamic" meaning that successive rotations take into account how the axes were transformed by previous rotations. An alternative definition states that *roll-pitch-yaw* angles (notice the reverse order) are defined as rotations around the global (or "*fixed*") x, y, and z axes. In spite of the apparent contradictory definitions, if we realize that rotating around global axes is achieved by left-hand matrix multiplication, it turns out that the roll-pitch-yaw parameterization defines this rotation matrix:

$$\mathbf{R}_{rpy}(\psi,\chi,\phi)=\underbrace{\mathbf{R}_z(\phi)}_{3^{rd}:\text{yaw}}\underbrace{\mathbf{R}_y(\chi)}_{2^{nd}:\text{pitch}}\underbrace{\mathbf{R}_x(\psi)}_{1^{st}:\text{ roll}} \tag{18}$$

which coincides with the previous rotation in equation (17). To make it clear: any rotation has exactly the same *yaw, pitch,* and *roll* parameters, disregarding whether it is measured under the "dynamic" axes yaw-pitch-roll convention or under the "fixed" axes roll-pitch-roll convention. The trick here is that rotations are applied in reverse order in the two conventions but the difference between "dynamic" and "fixed" axes modifies the side on which rotation matrices accumulate, hence finally we obtain exactly the same rotation matrix.

Once we have addressed this probable source of confusion, we must mention one of the problematic aspects of the yaw-pitch-roll parameterization: the degeneration of one degree of freedom when the pitch (χ) approaches $\pm 90^{\circ}$—the so-called *gimbal lock.* Indeed, if $\chi = \pm 90^{\circ}$ we have $\cos\chi = 0$ and $\sin\chi = \pm 1$, which leads to this degenerated rotation matrix:

$$\mathbf{R}(\phi,\pm 90^{\circ},\psi)=\begin{pmatrix} 0 & \pm c\phi s\psi - s\phi c\psi & \pm c\phi c\psi + s\phi s\psi \\ 0 & \pm s\phi s\psi + c\phi c\psi & \pm s\phi c\psi - c\phi s\psi \\ \mp 1 & 0 & 0 \end{pmatrix}=$$

(using well-known trigonometric expressions)

$$=\begin{pmatrix} 0 & -\sin(\phi\mp\psi) & \cos(\phi\mp\psi) \\ 0 & \cos(\phi\mp\psi) & \sin(\phi\mp\psi) \\ \mp 1 & 0 & 0 \end{pmatrix} \overset{\alpha=\phi\mp\psi}{=} \begin{pmatrix} 0 & -\sin\alpha & \cos\alpha \\ 0 & \cos\alpha & \sin\alpha \\ \mp 1 & 0 & 0 \end{pmatrix} \tag{19}$$

where the other two angles (yaw and roll) do not represent independent rotations anymore. Another important inconvenient of this parameterization, derived from the gimbal lock problem, is the lack of a unique inverse function for the matrix function in equation (16). It can be shown that the matrix-to-parameterization function $M^{-1}(\cdot)$ becomes in this case:

$$\mathbf{p} = \mathbf{M}^{-1}(\mathbf{P}), \quad with\ \mathbf{P} = \begin{pmatrix} m_{11} & m_{12} & m_{13} & m_{14} \\ m_{21} & m_{22} & m_{23} & m_{24} \\ m_{31} & m_{32} & m_{33} & m_{34} \\ 0 & 0 & 0 & 1 \end{pmatrix} \rightarrow \mathbf{M}^{-1}(\mathbf{P}) = \begin{pmatrix} x \\ y \\ z \\ \hline \phi \\ \chi \\ \psi \end{pmatrix}$$

with

$$\begin{cases} x = m_{14} \\ y = m_{24} \\ z = m_{34} \end{cases} \quad and \quad \begin{cases} \chi = \operatorname{atan2}\left(-m_{31}, \sqrt{m_{11}^2 + m_{21}^2}\right) \\ \text{if } \chi = -90^\circ \rightarrow \begin{cases} \phi = \operatorname{atan2}(-m_{23}, -m_{13}) \\ \psi = 0 \end{cases} \\ \text{if } |\chi| \neq 90^\circ \rightarrow \begin{cases} \phi = \operatorname{atan2}(-m_{21}, -m_{11}) \\ \psi = \operatorname{atan2}(-m_{32}, -m_{33}) \end{cases} \\ \text{if } \chi = +90^\circ \rightarrow \begin{cases} \phi = \operatorname{atan2}(m_{23}, m_{13}) \\ \psi = 0 \end{cases} \end{cases} \tag{20}$$

To end with our treatment of Euler angles, we must mention that inverting a pose, $\ominus\mathbf{p}$, is more easily performed by first computing the pose in matrix form, $\mathbf{P} = \mathbf{M}(\mathbf{p})$, then inverting that matrix and then applying equation (20) to retrieve the inverse pose parameters. It must be noticed that inverting $\mathbf{P} \in \mathbf{SE}(3)$ matrices can be achieved without actually performing the costly matrix inversion: it can be shown that inverting $\mathbf{P} \in \mathbf{SE}(3)$ is equivalent to transposing the 3×3 rotational part and using the following expression for the translational part:

$$\mathbf{P}^{-1} = \begin{pmatrix} \mathbf{i} & \mathbf{j} & \mathbf{k} & \mathbf{t} \\ 0 & 0 & 0 & 1 \end{pmatrix}^{-1} = \begin{pmatrix} i_1 & j_1 & k_1 & x \\ i_2 & j_2 & k_2 & y \\ i_3 & j_3 & k_3 & z \\ 0 & 0 & 0 & 1 \end{pmatrix}^{-1} = \begin{pmatrix} i_1 & i_2 & i_3 & -\mathbf{i}\cdot\mathbf{t} \\ j_1 & j_2 & j_3 & -\mathbf{j}\cdot\mathbf{t} \\ k_1 & k_2 & k_3 & -\mathbf{k}\cdot\mathbf{t} \\ 0 & 0 & 0 & 1 \end{pmatrix}$$

(with $\mathbf{a}\cdot\mathbf{b}$ standing for the dot product) (21)

The complexity of the equations involved in every operation with Euler angles parameterizations prevents us from obtaining simple and closed-form equations for directly computing the *parameters* of the poses resulting from geometric operations, as we did in the previous section for $\mathbf{SE}(2)$ poses. However, this is still possible with the unit quaternion representation, which we address next. Notice that, in spite of its defects, the yaw-pitch-roll parameterization is widely used for the highly intuitive meaning of its parameters.

Another popular parameterization of 3D poses is by means of a 3D translation plus a unit quaternion for the attitude, such that:

$$\mathbf{p} = \begin{pmatrix} x \\ y \\ z \\ \hline q_r \\ q_x \\ q_y \\ q_z \end{pmatrix} \quad \text{(unit quaternion parameterization)} \tag{22}$$

$$\text{(with } q_r^2 + q_x^2 + q_y^2 + q_z^2 = 1\text{)}$$

For better grasping the geometry of quaternions it reveals as more convenient to consider its four elements as two differentiated parts (Horn, 2001): the scalar q_r and the vector $\begin{pmatrix} q_x & q_y & q_z \end{pmatrix}^T$. Any arbitrary rotation in the three-dimensional space can be interpreted as one single rotation (of magnitude θ radians) around a conveniently-chosen axis of rotation, say, a unitary vector $\mathbf{v} = \begin{pmatrix} v_x & v_y & v_z \end{pmatrix}^T$. It can be shown that the q_r component of a unit quaternion is related to the magnitude of the rotation, while the vector part indicates the rotation axis—see Figure 2b. More concretely:

$$\begin{aligned} q_r &= \cos\frac{\theta}{2} \\ \begin{pmatrix} q_x \\ q_y \\ q_z \end{pmatrix} &= \sin\frac{\theta}{2}\begin{pmatrix} v_x \\ v_y \\ v_z \end{pmatrix} \end{aligned} \tag{23}$$

The matrix form of the pose represented as a unit quaternion can be obtained as:

$$\mathbf{P} = \mathbf{M}(\mathbf{p}) \equiv \begin{pmatrix} q_r^2 + q_x^2 - q_y^2 - q_z^2 & 2(q_x q_y - q_r q_z) & 2(q_z q_x + q_r q_y) & x \\ 2(q_x q_y + q_r q_z) & q_r^2 - q_x^2 + q_y^2 - q_z^2 & 2(q_y q_z - q_r q_x) & y \\ 2(q_z q_x - q_r q_y) & 2(q_y q_z + q_r q_x) & q_r^2 - q_x^2 - q_y^2 + q_z^2 & z \\ 0 & 0 & 0 & 1 \end{pmatrix} \tag{24}$$

As with the case of the Euler angles, it is not easy to invert this function. Some methods based on eigendecomposition have been proposed in the literature (Bar-Itzhack, 2000), but probably the easiest way to retrieve the quaternion parameters is to firstly obtain the yaw (ϕ), pitch (χ) and roll (ψ) parameters as in equation (20), then applying the following equivalence relations existing between both parameterizations:

$$
\begin{aligned}
q_r &= \cos\frac{\psi}{2}\cos\frac{\chi}{2}\cos\frac{\phi}{2} + \sin\frac{\psi}{2}\sin\frac{\chi}{2}\sin\frac{\phi}{2} \\
q_x &= \sin\frac{\psi}{2}\cos\frac{\chi}{2}\cos\frac{\phi}{2} - \cos\frac{\psi}{2}\sin\frac{\chi}{2}\sin\frac{\phi}{2} \\
q_y &= \cos\frac{\psi}{2}\sin\frac{\chi}{2}\cos\frac{\phi}{2} + \sin\frac{\psi}{2}\cos\frac{\chi}{2}\sin\frac{\phi}{2} \\
q_z &= \cos\frac{\psi}{2}\cos\frac{\chi}{2}\sin\frac{\phi}{2} - \sin\frac{\psi}{2}\sin\frac{\chi}{2}\cos\frac{\phi}{2}
\end{aligned}
\tag{25}
$$

An advantage of quaternions is the simplicity of performing some operations with them. For example, it is easy to compute the inverse of a quaternion, that is, $\ominus\mathbf{p}$. The rotational part is inverted be simply inverting the vector formed by $\left(q_x \; q_y \; q_z\right)^T$. It might seem more reasonable to inverse the sign of q_r instead, but notice that the actual rotation angle θ is related to q_r by $\theta = \cos^{-1}\left(2q_r\right)$, as follows from equation (23) above. Therefore, θ is limited to the range of nonnegative values $\left[0, \pi\right]$, and the sign of q_r would be ignored. The common criterion is to always employ nonnegative values for q_r. Regarding the inversion of the translational part of the pose $\mathbf{p}$, it involves the function $\mathbf{f}_{pi}\left(\cdot\right)$ introduced in equation (30). To sum up, we end up with:

$$
\mathbf{p}' = \ominus\mathbf{p} = \begin{pmatrix} \mathbf{f}_{pi}\left(\left[0 \; 0 \; 0\right]^T, \mathbf{p}\right) \\ q_r \\ -q_x \\ -q_y \\ -q_z \end{pmatrix}
\tag{26}
$$

Finally, an operation which is specific to this particular parameterization is the *quaternion normalization*. Its aim is to assure that $q_r^2 + q_x^2 + q_y^2 + q_z^2$ equals the unity, that is:

$$
\begin{pmatrix} q_r' \\ q_x' \\ q_y' \\ q_z' \end{pmatrix} = \mathbf{f}_{\mathrm{qn}}(\mathbf{q}) = \frac{\mathbf{q}}{|\,\mathbf{q}\,|} = \frac{1}{(q_r^2 + q_x^2 + q_y^2 + q_z^2)^{1/2}} \begin{pmatrix} q_r \\ q_x \\ q_y \\ q_z \end{pmatrix}
\tag{27}
$$

This function, which *must* be employed after obtaining quaternion estimates from an EKF or any other least-squares estimation algorithm that do not respect the unit-length constraint, should be also applied when directly working with quaternions in order to eliminate potential numerical inaccuracies.

We can now address the implementation of the four fundamental geometric operations, which we will describe for the unit-quaternion parameterization only since it leads to relatively simple expressions, in comparison to Euler angles.

Composition of two poses:

$$\begin{pmatrix} x \\ y \\ z \\ q_r \\ q_x \\ q_y \\ q_z \end{pmatrix} = \mathbf{p} = \mathbf{p}_1 \oplus \mathbf{p}_2$$

$$\mathbf{p} = \mathbf{f}_{qn}\left(\mathbf{f}_c(\mathbf{p}_1, \mathbf{p}_2)\right), \quad with \quad \mathbf{f}_c(\mathbf{p}_1, \mathbf{p}_2) = \begin{pmatrix} \mathbf{f}_{pc}(\mathbf{p}_1, [x_2\; y_2\; z_2]^T) \\ q_{r1}q_{r2} - q_{x1}q_{x2} - q_{y1}q_{y2} - q_{z1}q_{z2} \\ q_{r1}q_{x2} + q_{r2}q_{x1} + q_{y1}q_{z2} - q_{y2}q_{z1} \\ q_{r1}q_{y2} + q_{r2}q_{y1} + q_{z1}q_{x2} - q_{z2}q_{x1} \\ q_{r1}q_{z2} + q_{r2}q_{z1} + q_{x1}q_{y2} - q_{x2}q_{y1} \end{pmatrix}$$

and its Jacobians:

$$\left.\frac{\partial \mathbf{f}_c(\mathbf{p}_1, \mathbf{p}_2)}{\partial \mathbf{p}_1}\right|_{7\times 7} = \begin{pmatrix} \multicolumn{5}{c}{\left.\dfrac{\partial \mathbf{f}_{pc}(\mathbf{p}_1, [x_2\; y_2\; z_2]^T)}{\partial \mathbf{p}_1}\right|_{3\times 7}} \\ & q_{r2} & -q_{x2} & -q_{y2} & -q_{z2} \\ \mathbf{0}_{4\times 3} & q_{x2} & q_{r2} & q_{z2} & -q_{y2} \\ & q_{y2} & -q_{z2} & q_{r2} & q_{x2} \\ & q_{z2} & q_{y2} & -q_{x2} & q_{r2} \end{pmatrix}$$

$$\left.\frac{\partial \mathbf{f}_c(\mathbf{p}_1, \mathbf{p}_2)}{\partial \mathbf{p}_2}\right|_{7\times 7} = \begin{pmatrix} \left.\dfrac{\partial \mathbf{f}_{pc}(\mathbf{p}_1, [x_2\; y_2\; z_2]^T)}{\partial [x_2\; y_2\; z_2]^{\top}}\right|_{3\times 3} & \multicolumn{4}{c}{\mathbf{0}_{3\times 4}} \\ & q_{r1} & -q_{x1} & -q_{y1} & -q_{z1} \\ \mathbf{0}_{4\times 3} & q_{x1} & q_{r1} & -q_{z1} & q_{y1} \\ & q_{y1} & q_{z1} & q_{r1} & -q_{x1} \\ & q_{z1} & -q_{y1} & q_{x1} & q_{r1} \end{pmatrix} \tag{28}$$

where $\mathbf{f}_{pc}$ and the corresponding Jacobian submatrix will be defined in equation (29) below.

2. Inverse composition of two poses: In this case it is more convenient to employ the equivalence $\mathbf{p} = \mathbf{p}_1 \ominus \mathbf{p}_2 \equiv (\ominus \mathbf{p}_2) \oplus \mathbf{p}_1$, with $\ominus \mathbf{p}_2$ evaluated as shown in equation (26) and the pose composition performed as just described above.

Composition of a pose and a point:

$$\begin{pmatrix} a_x' \\ a_y' \\ a_z' \end{pmatrix} = \mathbf{a}' = \mathbf{p} \oplus \mathbf{a}$$

$$\mathbf{a}' = \mathbf{f}_{pc}(\mathbf{p}, \mathbf{a}) = \begin{pmatrix} x + a_x + 2\left[-(q_y^2 + q_z^2)a_x + (q_x q_y - q_r q_z)a_y + (q_r q_y + q_x q_z)a_z\right] \\ y + a_y + 2\left[(q_r q_z + q_x q_y)a_x - (q_x^2 + q_z^2)a_y + (q_y q_z - q_r q_x)a_z\right] \\ z + a_z + 2\left[(q_x q_z - q_r q_y)a_x + (q_r q_x + q_y q_z)a_y - (q_x^2 + q_y^2)a_z\right] \end{pmatrix}$$

and its Jacobians:

$$\left.\frac{\partial \mathbf{f}_{pc}(\mathbf{p}, \mathbf{a})}{\partial \mathbf{p}}\right|_{3\times 7} = \begin{pmatrix} 1 & 0 & 0 & \\ 0 & 1 & 0 & \dfrac{\partial \mathbf{f}_{pc}(\mathbf{p}, \mathbf{a})}{\partial [qr\, qx\, qy\, qz]} \\ 0 & 0 & 1 & \end{pmatrix}$$

with

$$\frac{\partial \mathbf{f}_{pc}(\mathbf{p}, \mathbf{a})}{\partial [qr\, qx\, qy\, qz]} =$$

$$2\begin{pmatrix} -q_z a_y + q_y a_z & q_y a_y + q_z a_z & -2q_y a_x + q_x a_y + q_r a_z & -2q_z a_x - q_r a_y + q_x a_z \\ q_z a_x - q_x a_z & q_y a_x - 2q_x a_y - q_r a_z & q_x a_x + q_z a_z & q_r a_x - 2q_z a_y + q_y a_z \\ -q_y a_x + q_x a_y & q_z a_x + q_r a_y - 2q_x a_z & -q_r a_x + q_z a_y - 2q_y a_z & q_x a_x + q_y a_y \end{pmatrix}$$

and

$$\left.\frac{\partial \mathbf{f}_{pc}(\mathbf{p}, \mathbf{a})}{\partial \mathbf{a}}\right|_{3\times 3} = 2\begin{pmatrix} \frac{1}{2} - q_y^2 - q_z^2 & q_x q_y - q_r q_z & q_r q_y + q_x q_z \\ q_r q_z + q_x q_y & \frac{1}{2} - q_x^2 - q_z^2 & q_y q_z - q_r q_x \\ q_x q_z - q_r q_y & q_r q_x + q_y q_z & \frac{1}{2} - q_x^2 - q_y^2 \end{pmatrix} \tag{29}$$

Inverse composition of a pose and a point:

$$\begin{pmatrix} a_x' \\ a_y' \\ a_z' \end{pmatrix} = \mathbf{a}' = \mathbf{a} \ominus \mathbf{p}$$

$$\mathbf{a}' = \mathbf{f}_{pi}\left(\mathbf{a}, \mathbf{p}\right) =$$

$$\begin{pmatrix} (a_x - x) + 2\left[-(q_y^2 + q_z^2)(a_x - x) + (q_x q_y - q_r q_z)(a_y - y) + (q_r q_y + q_x q_z)(a_z - z)\right] \\ (a_y - y) + 2\left[(q_r q_z + q_x q_y)(a_x - x) - (q_x^2 + q_z^2)(a_y - y) + (q_y q_z - q_r q_x)(a_z - z)\right] \\ (a_z - z) + 2\left[(q_x q_z - q_r q_y)(a_x - x) + (q_r q_x + q_y q_z)(a_y - y) - (q_x^2 + q_y^2)(a_z - z)\right] \end{pmatrix}$$

and its Jacobians:

$$\frac{\partial \mathbf{f}_{pi}(\mathbf{a}, \mathbf{p})}{\partial \mathbf{a}} = \begin{pmatrix} 1 - 2(q_y^2 + q_z^2) & 2q_x q_y + 2qrq_z & -2qrq_y + 2q_x q_z \\ -2qrq_z + 2q_x q_y & 1 - 2(q_x^2 + q_z^2) & 2q_y q_z + 2qrq_x \\ 2q_x q_z + 2qrq_y & -2qrq_x + 2q_y q_z & 1 - 2(q_x^2 + q_y^2) \end{pmatrix}$$

$$\frac{\partial \mathbf{f}_{pi}(\mathbf{a}, \mathbf{p})}{\partial \mathbf{p}} =$$

$$\begin{pmatrix} 2q_y^2 + 2q_z^2 - 1 & -2q_r q_z - 2q_x q_y & 2q_r q_y - 2q_x q_z & \\ 2q_r q_z - 2q_x q_y & 2q_x^2 + 2q_z^2 - 1 & -2q_r q_x - 2q_y q_z & \dfrac{\partial \mathbf{f}_{pi}(\mathbf{a}, \mathbf{p})}{\partial \left\{q_r, q_x, q_y, q_z\right\}} \\ -2q_r q_y - 2q_x q_z & 2q_r q_x - 2q_y q_z & 2q_x^2 + 2q_y^2 - 1 & \end{pmatrix}$$

with

$$\frac{\partial \mathbf{f}_{pi}(\mathbf{a}, \mathbf{p})}{\partial \left\{q_r, q_x, q_y, q_z\right\}} =$$

$$\begin{pmatrix} 2q_y \Delta z - 2q_z \Delta y & 2q_y \Delta y + 2q_z \Delta z & 2q_x \Delta y - 4q_y \Delta x + 2qr\Delta z & 2q_x \Delta z - 2qr\Delta y - 4q_z \Delta x \\ 2q_z \Delta x - 2q_x \Delta z & 2q_y \Delta x - 4q_x \Delta y - 2q_r \Delta z & 2q_x \Delta x + 2q_z \Delta z & 2q_r \Delta x - 4q_z \Delta y + 2q_y \Delta z \\ 2q_x \Delta y - 2q_y \Delta x & 2q_z \Delta x + 2qr\Delta y - 4q_x \Delta z & 2q_z \Delta y - 2qr\Delta x - 4q_y \Delta z & 2q_x \Delta x + 2q_y \Delta y \end{pmatrix} \quad (30)$$

where for the sake of readability we introduced these replacements:

$$\begin{aligned} \Delta x &= a_x - x \\ \Delta y &= a_y - y \\ \Delta z &= a_z - z \end{aligned} \quad (31)$$

As a final note to this appendix, we can mention that all the operations described here are readily available in the MRPT C++ libraries. More details about these geometry transformations and many others can be found in the report (Blanco, 2010).

REFERENCES

Bar-Itzhack, I. Y. (2000). New method for extracting the quaternion from a rotation matrix. *Journal of Guidance, Control, and Dynamics, 23*(6), 1085–1087. doi:10.2514/2.4654

Blanco, J. L. (2010). *A tutorial on SE(3) transformation parameterizations and on-manifold optimization*. University of Málaga. Retrieved Mar 1, 2012, from http://citeseerx.ist.psu.edu/viewdoc/summary?doi=10.1.1.172.7103

Diebel, J. (2006). *Representing attitude: Euler angles, unit quaternions, and rotation vectors*. University of Stanford. Retrieved Mar 1, 2012, from http://citeseerx.ist.psu.edu/viewdoc/summary?doi=10.1.1.110.5134

Horn, B. K. P. (2001). *Some notes on unit quaternions and rotation*. Massachusetts Institute of Technology. Retrieved Mar 1, 2012, from http://people.csail.mit.edu/bkph/articles/Quaternions.pdf

Smith, R. C., & Cheeseman, P. (1986). On the representation and estimation of spatial uncertainty. *The International Journal of Robotics Research, 5*(4), 56–68. doi:10.1177/027836498600500404

Thrun, S., Burgard, W., & Fox, D. (2005). *Probabilistic robotics*. Cambridge, MA: The MIT Press.

Appendix B: Resampling Algorithms

A common problem of all particle filters is the degeneracy of weights, which consists of the unbounded increase of the variance of the importance weights $\omega^{[i]}$ of the particles with time. The term "variance of the weights" must be understood as the potential variability of the weights among the possible different executions of the particle filter. In order to prevent this growth of variance, which entails a loss of particle diversity, one of a set of *resampling methods* must be employed, as it was explained in chapter 7.

The aim of resampling is to replace an old set of N particles by a new one, typically with the same population size, but where particles have been duplicated or removed according to their weights. More specifically, the expected duplication count of the $i-th$ particle, denoted by N_i, must tend to $N\omega^{[i]}$. After resampling, all the weights become equal to preserve the importance sampling of the target pdf. Deciding whether to perform resampling or not is most commonly done by monitoring the Effective Sample Size (ESS). As mentioned in chapter 7, the ESS provides a measure of the variance of the particle weights, e.g. the ESS tends to 1 when one single particle carries the largest weight and the rest have negligible weights in comparison. In the following we review the most common resampling algorithms.

1. REVIEW OF RESAMPLING ALGORITHMS

This section describes four different strategies for resampling a set of particles whose normalized weights are given by $\omega^{[i]}$, for $i=1,...,N$. All the methods will be explained using a visual analogy with a "wheel" whose perimeter is assigned to the different particles in such a way that the length of the perimeter associated to each particle is proportional to its weight. Therefore, picking a random direction in this "wheel" implies choosing a particle with a probability proportional to its weight. For a more formal description of the methods, please refer to the excellent reviews in (Arumlampalam, Maskell, Gordon, & Clapp, 2002; Douc, Capp, & Moulines, 2005). The four methods described here have $O(N)$ implementations, that is, their execution times can be made to be linear with the number of particles (Carpenter, Clifford, & Fearnhead, 1999; Arumlampalam, Maskell, Gordon, & Clapp, 2002).

Multinomial resampling: It is the most straightforward resampling method, where N independent random numbers are generated to pick a particle from the old set. In the "wheel" analogy, illustrated in Figure 1, this method consists of picking N independent random directions from the center of the wheel and taking the pointed particle. This method is named after the fact that the probability mass function for the duplication counts N_i is a multinomial distribution with the weights as parameters. A naïve implementation would have a time complexity of $O(N \log N)$, but applying the method of simulating order statistics (Carpenter, Clifford, & Fearnhead, 1999), it can be implemented in $O(N)$.

Figure 1. The multinomial resampling algorithm

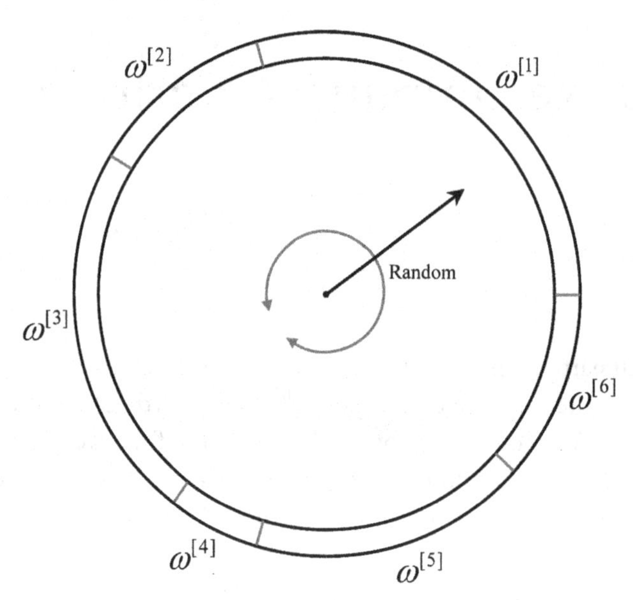

Figure 2. The residual resampling algorithm. The shaded areas represent the integer parts of $\omega^{[i]}/(1/N)$. The residual parts of the weights, subtracting these areas, are taken as the modified weights $\tilde{\omega}^{[i]}$.

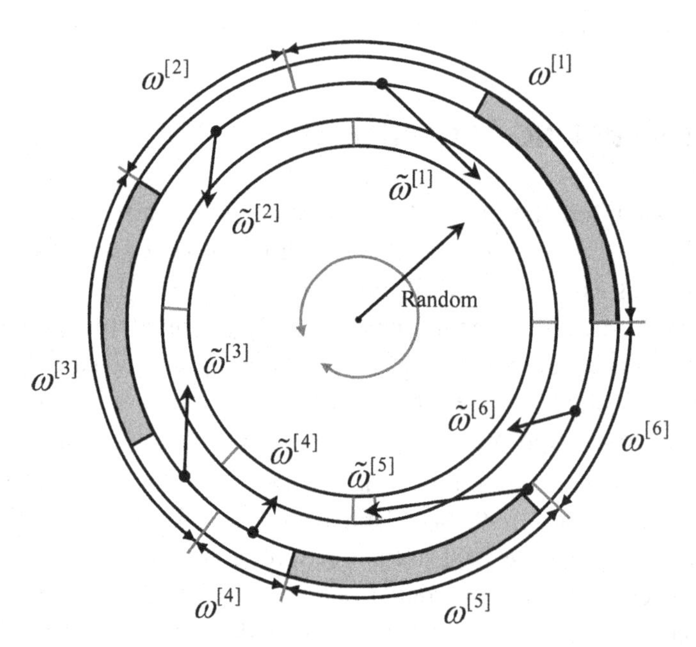

Figure 3. The stratified resampling algorithm. The entire circumference is divided into N equal parts, represented as the N circular sectors of $1/N$ perimeter lengths each.

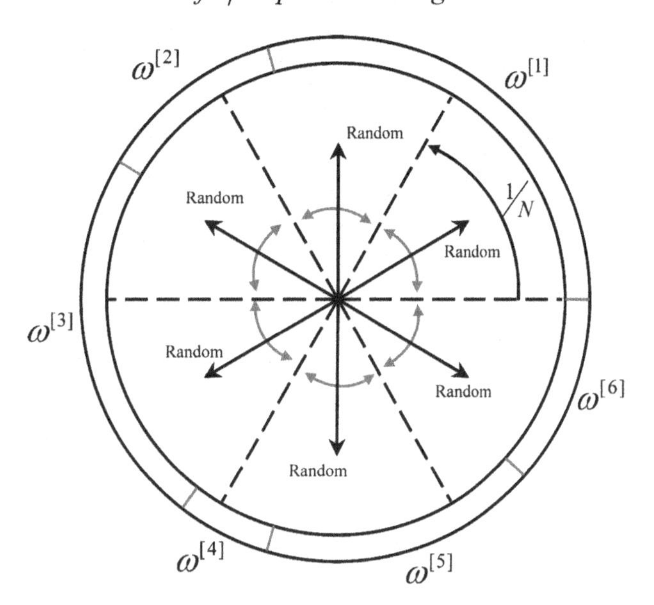

Figure 4. The systematic resampling algorithm

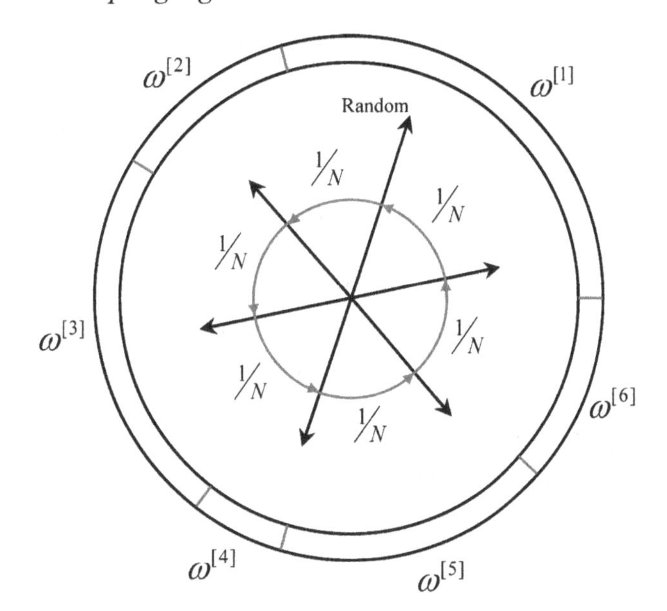

Figure 5. A simple benchmark to measure the loss of hypothesis diversity with time in an RBPF for the four different resampling techniques discussed in this appendix. The multinomial method clearly emerges as the worst choice.

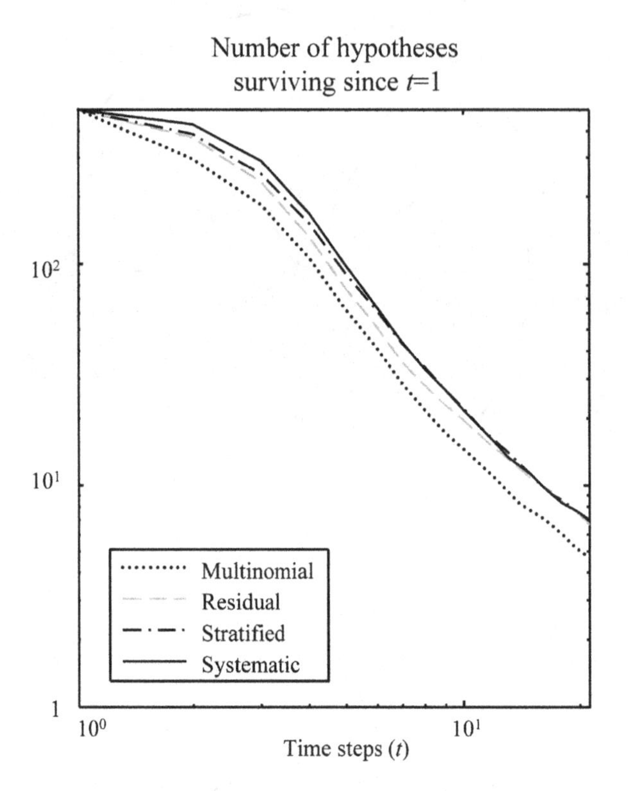

Residual resampling: This method comprises two stages, as can be seen in Figure 1. Firstly, particles are resampled deterministically by picking $N_i = \lfloor N\omega^{[i]} \rfloor$ copies of the $i-th$ particle—where $\lfloor x \rfloor$ stands for the floor of x, the largest integer above or equal to x. Then, multinomial sampling is performed with the residual weights: $\tilde{\omega}^{[i]} = \omega^{[i]} - N_i / N$ (see Figure 1-4).

Stratified resampling: In this method, the "wheel" representing the old set of particles is divided into N equally-sized segments, as represented in Figure 3. Then, N numbers are independently generated from a uniform distribution like in multinomial sampling, but instead of mapping each draw to the entire circumference, they are mapped within its corresponding partition out of the N ones.

Systematic resampling: Also called *universal sampling*, this popular technique draws only one random number, i.e., one direction in the "wheel," with the others $N-1$ directions being fixed at $1/N$ increments from that randomly picked direction.

2. COMPARISON OF THE DIFFERENT METHODS

In the context of Rao-Blackwellized Particle Filters (RBPF), where each particle carries a hypothesis of the complete history of the system state evolution, resampling becomes a crucial operation that reduces the diversity of the PF estimate for past states. We saw the application of those filters to SLAM in chapter 9.

In order to evaluate the impact of the resampling strategy on this loss, the four different resampling methods discussed above have been evaluated in a benchmark that measures the diversity of different states remaining after t time steps, assuming all the states were initially different. The results, displayed in Figure 5, agree with the theoretical conclusions in Douc, Capp, and Moulines (2005), stating that multinomial resampling is the worst of the four methods in terms of variance of the sample weights. Therefore, due to its simple implementation and good results, the systematic method is recommended when using a static number of particles in all the iterations. If a dynamic number of samples is desired, things get more involved and it is recommended to switch to a specific particle filter algorithm which simultaneously takes into account this particularity while also aiming at optimal sampling (Blanco, González, & Fernandez-Madrigal, 2010).

REFERENCES

Arumlampalam, M. S., Maskell, S., Gordon, N., & Clapp, T. (2002). A tutorial on particle filters for online nonlinear/non-Gaussian Bayesian tracking. *IEEE Transactions on Signal Processing, 50*(2), 174–188. doi:10.1109/78.978374

Blanco, J. L., González, J., & Fernández-Madrigal, J. A. (2010). Optimal filtering for non-parametric observation models: Applications to localization and SLAM. *The International Journal of Robotics Research, 29*(14), 1726–1742. doi:10.1177/0278364910364165

Carpenter, J., Clifford, P., & Fearnhead, P. (1999). Improved particle filter for nonlinear problems. *IEEE Proceedings on Radar. Sonar and Navigation, 146*(1), 2–7. doi:10.1049/ip-rsn:19990255

Douc, R., Capp, O., & Moulines, E. (2005). Comparison of resampling schemes for particle filtering. In *Proceedings of the 4th International Symposium on Image and Signal Processing and Analysis,* (pp. 64–69). IEEE.

Appendix C: Generation of Pseudo-Random Numbers

Computers are deterministic machines: if fed with exactly the same input data, a program will always arrive at exactly the same results. Still, there exist certain families of algorithms, which require some sort of randomness. The most important cases studied in this book are the different kinds of Monte Carlo sequential filters, or particle filters, applied to mobile robot localization and SLAM. Other practical applications of randomness in mobile robotics include randomized path-planning methods and the generation of noise and errors in simulations. In all these cases, our goal is being able to *draw samples* from some given (discrete or continuous) probability distribution.

The closest to real randomness that we can achieve with a computer program are the so called *Pseudo-Random* Number Generators (PRNG). The design of such algorithms is a complicated issue, which requires both a solid mathematical ground and some doses of art and creativity. Unfortunately, it seems that the importance of choosing a *"good"* PRNG has been often overlooked in the past, sometimes leading to disastrous results as was the case of the RANDU algorithm, designed for the IBM System/360 and widely used in the 60s-70s (Press, Teukolsky, Vetterling, & Flannery, 1992).

Since all PRNG methods output (alleged) "random" numbers, it may seem strange at a first glance the claim that some PRNGs are of *better quality* than others. To understand this, we must firstly focus on what all PRNG methods have in common. All PRNGs consist of a sequence of mathematical operations which are applied to some internal state every time a new pseudo-random number is required. As a result, we obtain one random sample and get the PRNG's internal state modified. Since the operations are (typically) the same for each new sample, the evolution of the internal state over time is the only reason why each sample differs from the previous one—PRNGs are *deterministic* algorithms! The initial state of a PRNG is set by means of the so called *seed* of the algorithm, consisting of one or more numbers. As one would expect, feeding the same seed to the same PRNG algorithm and requesting an arbitrary number of random samples will always gives us *exactly* the same sequence of pseudo-random numbers.

The quality of a PRNG depends on certain statistical characteristics of the so generated sequences. Two of the most important measures of the a PRNG "real randomness" are: (1) its period, i.e. how many samples can be generated before the exact sequence commences to repeat itself over and over again, and (2) the statistical correlation between each sample and the preceding or following ones. An ideal PRNG would have an infinite period and a correlation of exactly zero for any given pairs of samples in the sequences associated to any arbitrary seed. Existing implementations successfully achieve these goals up to different degrees.

From a practical perspective, the reader interested in generating random samples will do so in the context of some particular programming language. At present, C and C++ maintain their positions among the most widely used languages (Tiobe, 2012). Even if a user does not directly use them, most modern languages inherit the basic syntax of C for sequential programming, and the implementations of many popular languages rely on C and its standard library under the hood. Unfortunately, the C and C++ language standards do *not* specify what algorithm should be behind the PRNG functions, which in these languages are rand() and random(). Most C library vendors implement both based on a Linear Congruential Generator (LCG) which, as will be discussed below, is not the best choice. Furthermore, another reason to discourage employing those two standard functions is that there exist no guarantees that the same program will behave exactly the same under different operating systems or even if it is built with different compilers. The implementation of pseudo-random numbers in MATLAB follows a totally different approach (Marsaglia, 1968) and can be considered as of the highest quality.

In the following, we will describe algorithms for generating high-quality uniformly distributed numbers from which we will see how to generate other common distributions. The algorithms described here can be found as part of the C++ MRPT libraries (MRPT, 2011). Additionally, some of them have been recently approved by the corresponding ISO standardization committee as part of the latest C++ language standard (ISO/IEC 14882:2011), under the namespace std::tr1.

1. SAMPLING FROM A UNIFORM DISTRIBUTION

We start with the most basic type of PRNG: the one producing *integer* numbers following a discrete uniform distribution. For convenience, assume that the support of the probability distribution is the range $[0, m-1] \subset \mathbb{N}$. If we start with a *seed* value of i_0 and denote the $k-th$ sample returned by our PRNG as i_k, we can express our goal as:

$$i_k \sim U(i; 0, m-1) \ , \ \forall k \geq 1 \qquad \text{(discrete pmf)} \tag{1}$$

Since PRNGs for *all* other probability distributions (e.g. continuous uniform pdf, Gaussians, etc.) can be derived from a discrete uniform PRNG, it comes at no surprise that this type of generators had received a huge attention by researchers during the last two decades.

Without doubt, the most popular such PRNG methods belong to the family of Linear Congruential Generators (LCGs), which have been employed inside programming language libraries since the 60s.

Algorithm 1. The generic LCG algorithm. Note that the index k only has meaning for the invoker of the algorithm and is not used at all internally.

algorithm draw_uniform_LCG
Inputs: none
Outputs: i_{k+1} (a pseudo-random sample, as an unsigned integer)

Internal state: i_k (an unsigned integer)
1: if (this is the first call) // Do we have to initialize from seed?
  1.1: $i_k \leftarrow seed$
2: $i_{k+1} \leftarrow (a\, i_k + c) \bmod m$
3: $i_k \leftarrow i_{k+1}$ // Save state for the next call

Their popularity follows from their simplicity: as it can be seen in Algorithm 1, they only involve one multiplication, one addition and one modulus calculation (i.e. "wrapping" numbers above the given limit).

Different LCG implementations only differ in the choice of its parameters: the multiplier a, the constant c and the modulus m. The quality of the resulting random numbers vitally depends on a careful election of them. Some of the best combinations attainable in practice by an LCG were reported in Park and Miller (1988) to be $c = 0$, $m = 2^{31} - 1$ and a equaling either 16807, 48271, or 69621.

However, LCG algorithms in general (no matter what parameters you use) should be avoided if high-quality random numbers are desired, e.g., when performing a Monte Carlo simulation with tens of millions of random samples. There exist a variety of reasons that conspire to make LCGs undesirable: the important correlation existing between consecutive numbers for many choices of the parameters, a negligence in the ANSI C specification which *might* make standard library PRNG implementations (i.e. random() and rand()) to have periods as short as 2^{15}, etc. (Press, Teukolsky, Vetterling, & Flannery, 1992).

Instead, we strongly recommend using other PRNG algorithms. A good candidate is the *Mersenne twister* (Matsumoto & Nishimura, 1998), whose popular implementation known as MT19937 is sketched in Algorithm 2. The method is named after its *extremely large* period of $2^{19937} - 1$, or roughly $4.315 \cdot 10^{6001}$. As can be seen in the pseudocode, the method actually generates random numbers in blocks of 624 samples, then outputs them one by one until all its elements have been used; then a new block is computed. The resulting natural numbers approximately follow the discrete uniform distribution $U\left(0, 2^{32} - 1\right)$.

Up to this point we have seen how to draw samples from a *discrete* uniform distribution $i_k \sim U\left(i; 0, m-1\right)$. If we needed instead samples from a *continuous* uniform distribution, such as $x_k \sim U\left(x; x_{min}, x_{max}\right)$, we would easily generate the latter from the former as shown in Algorithm 3. Notice that the so obtained real numbers will be all spaced at intervals determined by Δ, which for a typical situation (32bit PSRG algorithm and a 64bit type for floating point numbers) will be several orders of magnitude larger than the machine precision or "epsilon" (i.e. the smallest representable number larger than zero). However, this should not be seen as an inconvenience since the accuracy will be actually determined by the ratio between the size of the pdf domain ($x_{max} - x_{min}$) and this smallest step (Δ). This ratio is given by $m-1$, which is high enough (typically $2^{32} - 1$) as to assure an excellent approximation of a continuous pdf.

2. SAMPLING FROM A 1-DIMENSIONAL GAUSSIAN

Although uniformly distributed numbers may find its utility in mobile robotics, Gaussian distributions hold the undisputed first place in the list of continuous distributions regarding their number of practical applications. We address here the unidimensional case, leaving the more complex multivariate Gaussian distribution for the next section.

The generic unidimensional Gaussian has an arbitrary mean μ and variance σ^2, such that a sequence of numbers $y_1, y_2, \ldots$ drawn from that pdf can be represented as:

$$y_k \sim N\left(y; \mu, \sigma^2\right) \quad (2)$$

Algorithm 2. The 32bit version of the MT19937 algorithm for generating high-quality pseudo-random integer numbers. Note that the operation $X >> N$ stands for a right shift of X by N bits, padding with zeros, $X \oplus Y$ is the bitwise exclusive or (xor) operation and $X \,\&\, Y$ is the bitwise and operator. The constants employed in the algorithm are: $N = 624$ (state length), $M = 397$ (a period), $L = 1812433253$ (an arbitrarily-chosen multiplier for the auxiliary LCG), $A = 2567483615$ (from the matrix involved in the underlying linear recurrent formula), $B = 2636928640$, $C = 4022730752$, $u = 11$, $s = 7$, $l = 18$, and $t = 15$. For further details, please refer to Matsumoto and Nishimura (1998).

algorithm draw_uniform_MT19937_uint32
Inputs: none
Outputs: i_{k+1} (a pseudo-random sample in the range $[0, 2^{32} - 1] \subset \mathbb{N}$)
Internal state: $b_{0 \ldots N-1}$ (a vector of N 32bit unsigned integers)
j (index for next output number from b)

1: if (this is the first call) // Do we have to initialize from seed?
1.1: $j \leftarrow 0$
1.2: $b_0 \leftarrow seed$
1.3: $b_j \leftarrow$ lowest 32bits of $\left[j + L\left(b_{j-1} \oplus \left(b_{j-1} >> 30\right)\right)\right]$ // An auxiliary LCG
2: if ($j \geq N$) // Need to generate the vector b?
2.1: $j \leftarrow 0$ // Reset index
2.2: for each $i \in [0, N-1]$
2.2.1: $y_i \leftarrow \begin{cases} \text{bit } 31 & \leftarrow \text{most significant bit of } b_i \\ \text{bits } 0 \ldots 30 & \leftarrow 31 \text{ least significant bits of } b_{(i+1) \bmod N} \end{cases}$
2.2.2: $b_i \leftarrow b_{(i+M) \bmod N} \oplus (y_i >> 1)$
2.2.3: only if y_i is odd: $b_i \leftarrow b_i \oplus A$
3: $y \leftarrow b_k \oplus (b_k >> u)$
4: $y \leftarrow y \oplus ((y << s) \,\&\, B)$
5: $y \leftarrow y \oplus ((y << t) \,\&\, C)$
6: $i_{k+1} \leftarrow y \oplus (y >> l)$ // The output random sample
7: $j \leftarrow j + 1$ // Increment index for the next call

Algorithm 3. The wrapper for the MT19937 algorithm in case of generating pseudo-random numbers approximating a continuous uniform distribution. Here, m stand for the amount of different integer values provided by Algorithm 2, that is, $m = 2^{32}$.

algorithm draw_uniform_MT19937_real
Inputs: x_{min}, x_{max} (the limits of the uniform distribution domain)
Outputs: x_k (a pseudo-random sample in the range $[x_{\min}, x_{\max}]$)

Internal state: (none)
1: $i_k \leftarrow \text{draw\_uniform\_MT19937\_uint32}(x_{max}, x_{min})$
2: $\Delta = \dfrac{x_{max} - x_{min}}{m - 1}$
3: $x_k = x_{min} + \Delta \cdot i_k$

Using the rules for linear transformations of r.v.s described in chapter 3 section 8 ("Scaling and Offsetting a Continuous r.v."), it is easily shown that drawing samples from an arbitrary Gaussian distribution can be achieved by sampling a new r.v. z_k from a standard normal distribution (with a mean of zero and unit variance) and then applying a linear transformation:

$$z_k \sim N\left(z;0,1\right),\ y_k = \mu + \sigma z_k \rightarrow y_k \sim N\left(y;\mu,\sigma^2\right) \tag{3}$$

Therefore, we can focus on generating samples from the standard normal distribution $N\left(0,1\right)$. There exist several proposed algorithms to convert samples (x) from a uniform distribution, obtained as shown in the previous section, into samples (z) from a standard normal pdf. Since most of them rely on the fundamental rule for r.v. transformation already introduced in chapter 3, we repeat it here again for convenience.

Assume we have a (let it be multivariate) r.v. $\mathbf{x}$ which is transformed into a different r.v. $\mathbf{z}$ by means of a function $\mathbf{z} = \mathbf{g}\left(\mathbf{x}\right)$. If we denote their density functions as $f_{\mathbf{x}}\left(\mathbf{x}\right)$ and $f_{\mathbf{z}}\left(\mathbf{z}\right)$, and the inverse of the transformation as $\mathbf{x} = \mathbf{g}^{-1}\left(\mathbf{z}\right)$ respectively, it can be shown that in some usual cases both pdfs are related by (refer to chapter 3 section 8):

$$f_{\mathbf{z}}(\mathbf{z}) = f_{\mathbf{x}}(\mathbf{g}^{-1}(\mathbf{z}))\left|\det\left(\mathbf{J}_{\mathbf{g}^{-1}}\left(\mathbf{z}\right)\right)\right| \tag{4}$$

where $\mathbf{J}_{\mathbf{g}^{-1}}$ stands for the Jacobian matrix of the inverse transformation. This expression has an insightful geometrical interpretation, as it was represented in chapter 3 Figure 8 for the particular case of a scalar function.

One of the most widely-spread methods for generating Gaussian samples for its simplicity is the Box-Muller transform in polar coordinates (Devroye, 1986; Press, Teukolsky, Vetterling, & Flannery, 1992), which we will describe here—for an extensive review of other 15 different methods the reader can refer to Thomas, Leong, Luk, and Villaseñor (2007). This technique takes as input a pair of random samples from a uniform distribution and generates a pair of independent (uncorrelated) samples that follow a standard normal pdf. It is therefore convenient to approach the method as a transformation of multivariate r.v.s of dimension two. Let $\mathbf{v}$ denote a vector comprising:

$$\mathbf{v} = \begin{pmatrix} \rho' \\ \theta \end{pmatrix} \quad \text{, such as} \begin{cases} \rho' \sim U\left(0,1\right) \\ \theta \sim U\left(0,2\pi\right) \end{cases} \tag{5}$$

We will interpret $\rho = \sqrt{\rho'}$ (that is, $\rho' = \rho^2$) and θ as the polar coordinates (distance and angle, respectively) of a point in the plane, whose position therefore is constrained to the unit circle centered at the origin. The Box-Muller transformation proposes the following change og variables:

$$\mathbf{y} = \mathbf{g}(\mathbf{v})$$
$$\begin{pmatrix} y_1 \\ y_2 \end{pmatrix} = \begin{pmatrix} \sqrt{-2\log\rho'}\cos\theta \\ \sqrt{-2\log\rho'}\sin\theta \end{pmatrix} \tag{6}$$

which can be demonstrated to leave two Gaussian samples in the vector $\mathbf{y}$. To demonstrate this, we can apply equation (4). By dividing and squaring and summing the two equations of $\mathbf{g}(\mathbf{v})$ above, we can easily arrive at the inverse transformation function:

$$\mathbf{v} = \mathbf{g}^{-1}(\mathbf{y})$$
$$\begin{pmatrix} \rho' \\ \theta \end{pmatrix} = \begin{pmatrix} e^{-\frac{1}{2}\left(y_1^2+y_2^2\right)} \\ \frac{1}{2\pi}\tan^{-1}\left(\frac{y_2}{y_1}\right) \end{pmatrix} \tag{7}$$

whose Jacobian, and the absolute value of its determinant, read:

$$\mathbf{J}_{\mathbf{g}^{-1}} = \begin{pmatrix} \frac{d\rho'}{dy_1} & \frac{d\rho'}{dy_2} \\ \frac{d\theta}{dy_1} & \frac{d\theta}{dy_2} \end{pmatrix} = \begin{pmatrix} -y_1 e^{-\frac{1}{2}\left(y_1^2+y_2^2\right)} & -y_2 e^{-\frac{1}{2}\left(y_1^2+y_2^2\right)} \\ \frac{1}{2\pi}\frac{-y_2}{y_1^2+y_2^2} & \frac{1}{2\pi}\frac{-y_1}{y_1^2+y_2^2} \end{pmatrix}$$
$$\left|\det\left(\mathbf{J}_{\mathbf{g}^{-1}}\right)\right| = \left|-\frac{1}{2\pi}\frac{y_1^2+y_2^2}{y_1^2+y_2^2}e^{-\frac{1}{2}\left(y_1^2+y_2^2\right)}\right| = \frac{1}{2\pi}e^{-\frac{1}{2}\left(y_1^2+y_2^2\right)} \tag{8}$$

and by replacing this result into equation (4) we arrive at the pdf of the transformed variable $\mathbf{y}$:

$$f_{\mathbf{y}}(\mathbf{y}) = \underbrace{f_{\mathbf{x}}(\mathbf{g}^{-1}(\mathbf{y}))}_{\substack{\text{Uniform distribution} \\ \text{where } \mathbf{g}^{-1} \text{ is defined}}} \left|\det\left(\mathbf{J}_{\mathbf{g}^{-1}}(\mathbf{z})\right)\right| = \frac{1}{2\pi}e^{-\frac{1}{2}\left(y_1^2+y_2^2\right)} \quad , \forall \mathbf{y} \in \text{unit circle} \tag{9}$$

Realize how there exists no term where the two components of $\mathbf{y}$ appear multiplying to each other, thus we can easily deduce that both are uncorrelated and that they follow an exact standard normal distribution, which was our goal:

$$f_{\mathbf{y}}(\mathbf{y}) = \frac{1}{2\pi}e^{-\frac{1}{2}\left(y_1^2+y_2^2\right)} = \left(\frac{1}{\sqrt{2\pi}}e^{-\frac{1}{2}y_1^2}\right)\left(\frac{1}{\sqrt{2\pi}}e^{-\frac{1}{2}y_2^2}\right)$$
$$\rightarrow \begin{cases} y_1 \sim N(0,1) \\ y_2 \sim N(0,1) \end{cases} \tag{10}$$

Figure 1. (a) Two-dimensional random samples uniformly distributed on the unit circle. (b) – (c) If we transform those samples by means of the Box-Muller transformation $\mathbf{y} = \mathbf{g}(\mathbf{v})$ and plot the histograms for each of the two output components independently, we can verify how they follow a standard normal distribution. The theoretical pdf of $N(0,1)$ has been overlaid for comparison. The mismatch between experimental histograms and the theoretical pdf is only due to the reduced number of samples (5000) employed here for illustrative purposes.

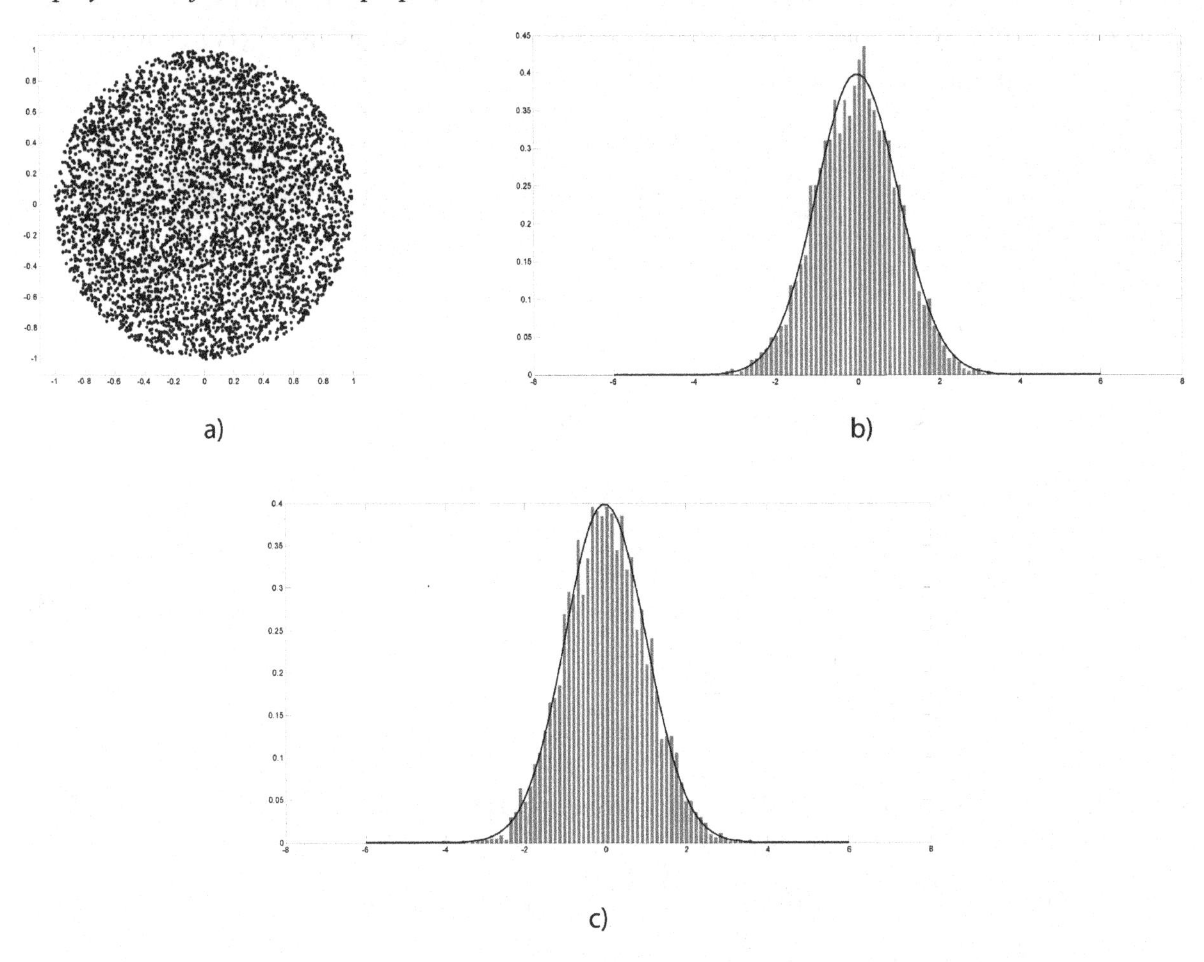

To illustrate this method, we have depicted in Figure 1 a number of random samples for the original r.v.s and the histograms of the transformed ones, which clearly match the expected theoretical pdf.

Finally, we must address one optimization that is employed in virtually all implementations. In order to avoid evaluating the trigonometric functions of equation (6) another change of variables is introduced: instead of starting from uniform samples of ρ' and θ, we generate instead pairs of variables x_1 and x_2 such that, interpreted as two dimensional coordinates (in the x and y axes), are samples *uniformly* drawn from the unit circle—as shown in Figure 1a. One easy way to achieve this is by *rejection sampling*: first, we generate samples for x_1 and x_2 in the square region $[-1,1]\times[-1,1]$ using any of the uniform PRNGs introduced above, and then the samples are accepted only if they fall within the unit circle; otherwise, they are thrown away and the process is repeated. Notice that about 21.46% of the samples will be discarded in this procedure, as follows from the areas of the square and the circle, i.e.,

Algorithm 4. An implementation of a PRNG for the standard normal distribution using the Box-Muller transformation (based on Devroye, 1986; Press, Teukolsky, Vetterling, & Flannery, 1992)

algorithm draw_standard_Gaussian
Inputs: none
Outputs: y (a pseudo-random sample from $N(0,1)$, a real number)
Internal state: y' (cached sample, a real number)
b (flag for cached sample, boolean)

1: if (this is the first call)
 1.1: $b \leftarrow false$
2: if $(b = true)$
 2.2: $y \leftarrow y'$ // Output the cached sample
 2.3: $b \leftarrow false$
 else
 2.4: repeat // Rejection sampling loop
  2.4.1: $x_1 \leftarrow \text{draw\_uniform\_MT19937\_real}(-1,1)$ // Draw two uniform samples
  2.4.2: $x_2 \leftarrow \text{draw\_uniform\_MT19937\_real}(-1,1)$ // in the interval [-1,1]
  2.4.3: $\rho' \leftarrow x_1^2 + x_2^2$
until $(\rho' > 0 \ AND \ \rho' < 1)$
 2.5: $y \leftarrow \sqrt{\frac{-2\log(\rho')}{\rho'}} x_1$ // Output one sample
 2.6: $y' \leftarrow \sqrt{\frac{-2\log(\rho')}{\rho'}} x_2$ // and save another one for the next call
 2.7: $b \leftarrow true$

$(2^2 - \pi 1^2) / 2^2 = 0.2146$. Once we have a valid sample within the unit circle, we apply the transformation $\rho' = x_1^2 + x_2^2$ and $\theta = \tan^{-1}(x_2 / x_1)$, from which follows:

$$\cos\theta = \frac{x_1}{x_1^2 + x_2^2} = \frac{x_1}{\sqrt{\rho'}}$$
$$\sin\theta = \frac{x_2}{x_1^2 + x_2^2} = \frac{x_2}{\sqrt{\rho'}} \tag{11}$$

It can be shown that by doing so, both ρ' and θ follow uniform distributions as required initially by the algorithm. The complete procedure has been summarized in Algorithm 4.

3. SAMPLING FROM AN N-DIMENSIONAL GAUSSIAN

After all the definitions in previous sections, we are finally ready to address the distribution with the most practical applications in probabilistic robotics. Our aim here will be drawing samples from an $n-$dimensional multivariate Gaussian distribution such as:

$$\mathbf{y}_k \sim N(\mathbf{y}; \boldsymbol{\mu}, \boldsymbol{\Sigma}) \tag{12}$$

As an example of the applicability of this operation, in chapter 5 we analyzed several motion models whose uncertainty can be approximated as a multivariate Gaussian. When employing those models within a particle filter (either for localization or for SLAM), one needs to draw samples from those distributions just like in the equation above.

Our first step will be to realize that, thanks to the properties of uncertainty propagation of Gaussians through linear transformations, we can simplify the problem by drawing samples from a different variable $\mathbf{z}$ which has identical covariance matrix than $\mathbf{y}$ but a mean of zero. That is:

$$\mathbf{z}_k \sim N\left(\mathbf{z};\mathbf{0},\boldsymbol{\Sigma}\right),\ \mathbf{y}_k = \boldsymbol{\mu} + \mathbf{z}_k \rightarrow \mathbf{y}_k \sim N\left(\mathbf{y};\boldsymbol{\mu},\boldsymbol{\Sigma}\right) \quad (13)$$

Our approach to generate samples for $\mathbf{z}$ will be quite simple: finding a linear change of variables from another auxiliary r.v., that we will denote as $\mathbf{x}$, from which we already know how to draw samples. Since in the previous section we learned how to draw samples from a standard normal distribution $N(0,1)$, the ideal situation would be that all the n components of $\mathbf{x}$ had a mean of zero and a unit variance and that they were all uncorrelated to each other. Put mathematically, we want $\mathbf{x}$ to follow this distribution:

$$\mathbf{x}_k \sim N\left(\mathbf{x};\mathbf{0},\mathbf{I}_n\right) \quad (14)$$

where $\mathbf{I}_n$ is the identity matrix of size $n \times n$. By hypothesis, the relationship between $\mathbf{x}$ and $\mathbf{z}$ is linear, thus we denote as $\mathbf{M}$ the corresponding matrix:

$$\mathbf{z} = \mathbf{M}\mathbf{x} \quad (15)$$

It is important to realize that all we need at this point is the value of $\mathbf{M}$, since we already know how to draw samples for each individual component of $\mathbf{x}$, which we could then stack into a vector, premultiply by $\mathbf{M}$ and finally add the mean vector $\boldsymbol{\mu}$ to obtain a sample of $\mathbf{y}$, our original r.v.

There exist two different $\mathbf{M}$ matrices, which can serve us for our purposes. They are related to the eigendecomposition and the Cholesky decomposition of the covariance matrix $\boldsymbol{\Sigma}$, respectively. While the latter is more numerically stable and is recommended in general, we will firstly explain the eigendecomposition approach since its geometrical meaning gives it a great didactic value.

As we saw in chapter 3, a linear transformation of r.v.s as that in equation (15) leads to a well known expression for the covariance of the target variable (which we want to equal $\boldsymbol{\Sigma}$) in terms of that of the source variable ($\mathbf{I}_n$), which is:

$$\boldsymbol{\Sigma} = \mathbf{M}\mathbf{I}_n\mathbf{M}^T$$

(since the identity matrix is superfluous)

$$= \mathbf{M}\mathbf{M}^T \quad (16)$$

One particular way to factor a square symmetric matrix such as $\mathbf{\Sigma}$ is by its eigendecomposition, which is defined as:

$$\mathbf{\Sigma} = \mathbf{V}\mathbf{D}\mathbf{V}^T \tag{17}$$

with $\mathbf{V}$ being a square matrix where each column contains an eigenvector of $\mathbf{\Sigma}$ and $\mathbf{D}$ is a diagonal matrix whose entries correspond to the covariance eigenvalues (in the same order than the columns in $\mathbf{V}$). If $\mathbf{\Sigma}$ is positive definite, which is always desirable, the eigenvalues are all positive, whereas for positive-semidefinite matrices some eigenvalues are exactly zero. All the eigenvectors are orthogonal and of unit length, thus $\mathbf{V}$ is called an *orthonormal* matrix. It is also said that $\mathbf{D}$ is the *canonical form* of $\mathbf{\Sigma}$ because the latter is just a "rotated version" of the former, as we will see immediately.

Geometrically speaking, it is convenient to visualize covariance matrices by means of their corresponding confidence ellipses (for bidimensional variables) or ellipsoids (for higher dimensions)—refer to the examples at the right hand of Figure 2. The eigendecomposition of a covariance matrix has a direct relation with this geometrical viewpoint: each eigenvector provides the direction for an axis of the ellipse, while eigenvalues state their lengths. For instance, if all the eigenvalues were equal, the ellipsoid would become a sphere, disregarding the particular value of the eigenvectors. When some of the eigenvalues are much larger than the others, it means that uncertainty is more prominent in some particular directions—those of the corresponding eigenvectors. In some degenerate cases, we may find positive-semidefinite covariance matrices, where null eigenvalues imply the loss of one spatial degree of freedom for the r.v., i.e., some axes of the ellipsoid have a null length. A full understanding of all these geometrical concepts is of paramount importance when facing the interpretation of results of statistical problems as those discussed in this book.

If we now wish to determine the value of $\mathbf{M}$ according to equations (16) – (17) we can proceed as follows:

$$\begin{aligned} \mathbf{\Sigma} &= \mathbf{M}\mathbf{M}^T \\ &= \mathbf{V}\mathbf{D}\mathbf{V}^T \\ &= \mathbf{V}\mathbf{D}^{\frac{1}{2}}\mathbf{D}^{\frac{1}{2}}\mathbf{V}^T \\ &= \mathbf{V}\mathbf{D}^{\frac{1}{2}}\left(\mathbf{V}\mathbf{D}^{\frac{1}{2}}\right)^T \end{aligned} \tag{18}$$

where we have used these facts: (1) the square root of a diagonal matrix gives us another diagonal matrix with the square root of each of the original entries, and (2) the transpose of a diagonal matrix is identical to itself. Therefore:

$$\mathbf{M} = \mathbf{V}\mathbf{D}^{\frac{1}{2}}$$

(First version: based on eigendecomposition) (19)

The linear transformation $\mathbf{z} = \mathbf{M}\mathbf{x}$ then adopts an extremely intuitive form: we firstly draw *independent* standard normal samples "for each axis of the ellipsoid" that represent the target Gaussian uncertainty and then *rotate* those samples according to a change of coordinates where the new axes

Algorithm 5. A Cholesky decomposition-based implementation of a PRNG for multivariate Gaussian distributions. Only applicable to positive-definite covariance matrices.

algorithm draw_multivariate_Gaussian
Inputs: μ (the mean vector)
Σ (covariance matrix)
Outputs: $\mathbf{y}$ (a pseudo-random sample from $N(\mu, \Sigma)$)
Internal state: (none)

1: $\mathbf{L} \leftarrow cholesky(\Sigma)$ // Such as $\Sigma = \mathbf{L}\mathbf{L}^T$
2: for $i = 1...n$ // n being the dimensionality of μ and Σ
2.1: $x_i \leftarrow \text{draw\_standard\_Gaussian}()$
3: $\mathbf{z} \leftarrow \mathbf{L}\mathbf{x}$
4: $\mathbf{y} \leftarrow \mu + \mathbf{z}$ // Output sample

Figure 2. An example of the process for generating pseudo-random samples for (a) 2D and (b) 3D multivariate Gaussian distributions. The ellipse (2D) and ellipsoid (3D) represent the 95% confidence intervals of each Gaussian. The two (scaled) eigenvectors of the 2D covariance matrix have been represented in the right-hand graph of (a) as thick lines. Observe how they coincide, by definition, with the axes of the ellipse.

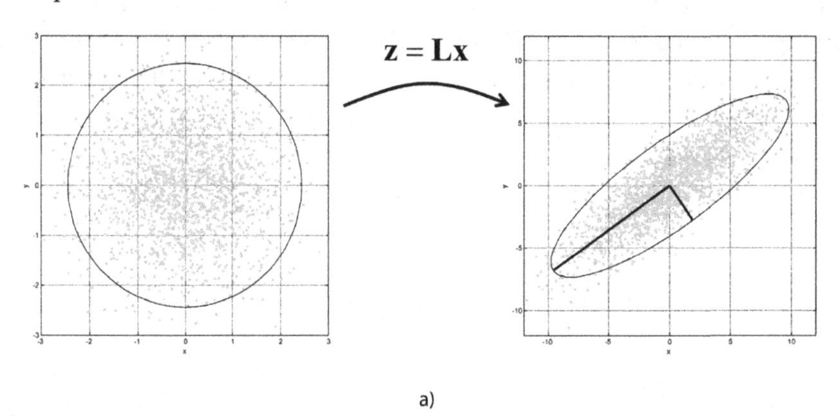

a)

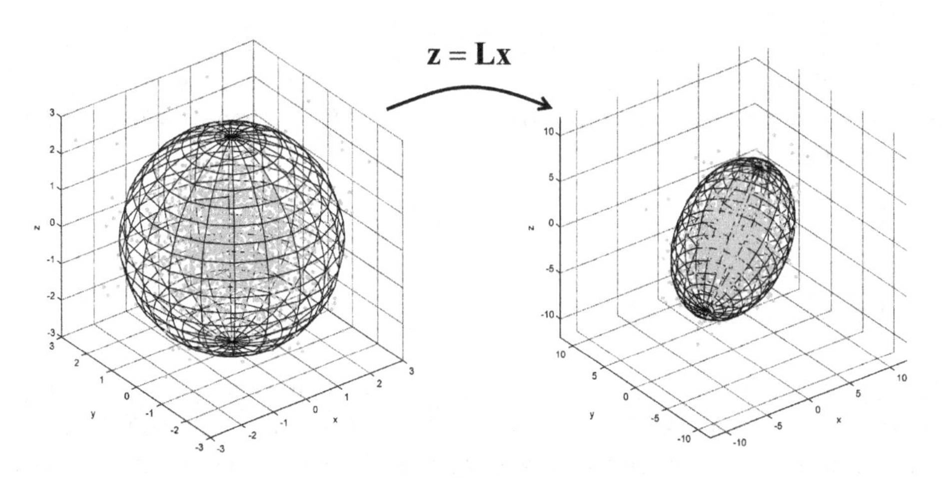

b)

coincide with the eigenvectors (which, recall, are the ellipsoid axes). The scale introduced by the square root of the eigenvalues enlarges or reduces the uncertainty in each direction according to the real uncertainty encoded by the covariance matrix.

Despite its didactic value, in practice it is advisable to employ instead the Cholesky decomposition of the covariance, for its simplicity and more efficient implementations. In this case, we have:

$$\begin{aligned}\mathbf{\Sigma} &= \mathbf{M}\mathbf{M}^T \\ &= \mathbf{L}\mathbf{L}^T\end{aligned} \tag{20}$$

where the Cholesky factorization is, by definition, $\mathbf{\Sigma} = \mathbf{L}\mathbf{L}^T$, thus obviously:

$$\mathbf{M} = \mathbf{L}$$

(Second version: based on Cholesky decomposition) (21)

To sum up, all the steps for drawing samples from a multivariate Gaussian distribution have been enumerated in Algorithm 5. Notice that the usage of the Cholesky factorization assumes a positive-definite covariance. In situations where the appearance of semidefinite-positive matrices cannot be ruled out, either the $\mathbf{LDL}^T$ decomposition (where we would find out that $\mathbf{M} = \mathbf{L}\mathbf{D}^{\frac{1}{2}}$) or the eigendecomposition described above should be employed instead.

REFERENCES

Devroye, L. (1986). *Non-uniform random variate generation*. New York, NY: Springer-Verlag.

Marsaglia, G. (1968). Random numbers fall mainly in the planes. *Proceedings of the National Academy of Sciences of the United States of America*, *61*, 25–28. doi:10.1073/pnas.61.1.25

Matsumoto, M., & Nishimura, T. (1998). Mersenne twister: A 623-dimensionally equidistributed uniform pseudo-random number generator. *ACM Transactions on Modeling and Computer Simulation*, *8*(1), 3–30. doi:10.1145/272991.272995

MRPT. (2011). *The mobile robot programming toolkit website*. Retrieved Mar 1, 2012, from http://www.mrpt.org/

Park, S. K., & Miller, K. W. (1988). Random number generators: Good ones are hard to find. *Communications of the ACM*, *31*, 1192–1201. doi:10.1145/63039.63042

Press, W. H., Teukolsky, S. A., Vetterling, W. T., & Flannery, B. P. (1992). *Numerical recipes in C: The art of scientific programming* (2nd ed.). Cambridge, UK: Cambridge University Press.

Thomas, D. B., Leong, P. H. W., Luk, W., & Villaseñor, J. D. (2007). Gaussian random number generators. *ACM Computing Surveys*, *39*(4). doi:10.1145/1287620.1287622

Tiobe Software. (2012). *Tiobe programming community index*. Retrieved April 11, 2012, from http://www.tiobe.com/index.php/content/paperinfo/tpci/index.html

Appendix D: Manifold Maps for SO(n) and SE(n)

As we saw in chapter 10, recent SLAM implementations that operate with three-dimensional poses often make use of on-manifold linearization of pose increments to avoid the shortcomings of directly optimizing in pose parameterization spaces. This appendix is devoted to providing the reader a detailed account of the mathematical tools required to understand all the expressions involved in on-manifold optimization problems. The presented contents will, hopefully, also serve as a solid base for bootstrapping the reader's own solutions.

1. OPERATOR DEFINITIONS

In the following, we will make use of some vector and matrix operators, which are rather uncommon in mobile robotics literature. Since they have not been employed throughout this book until this point, it is in order to define them here.

The "*vector to skew-symmetric matrix*" operator: A skew-symmetric matrix is any square matrix $\mathbf{A}$ such that $\mathbf{A} = -\mathbf{A}^T$. This implies that diagonal elements must be all zeros and off-diagonal entries the negative of their symmetric counterparts. It can be easily seen that any 2×2 or 3×3 skew-symmetric matrix only has 1 or 3 degrees of freedom (i.e. only 1 or 3 independent numbers appear in such matrices), respectively, thus it makes sense to parameterize them as a *vector*. Generating skew-symmetric matrices from such vectors is performed by means of the $[\cdot]_\times$ operator, defined as:

$$
\begin{aligned}
2\times 2: \quad & [\mathbf{v}]_\times = [(x)]_\times \equiv \begin{pmatrix} 0 & -x \\ x & 0 \end{pmatrix} \\
3\times 3: \quad & [\mathbf{v}]_\times = \left[\begin{pmatrix} x \\ y \\ z \end{pmatrix}\right]_\times \equiv \begin{pmatrix} 0 & -z & y \\ z & 0 & -x \\ -y & x & 0 \end{pmatrix}
\end{aligned}
\tag{1}
$$

The origin of the symbol $\times$ in this operator follows from its application to converting a cross product of two 3D vectors ($\mathbf{x} \times \mathbf{y}$) into a matrix-vector multiplication ($[\mathbf{x}]_\times \mathbf{y}$).

The "*skew-symmetric matrix to vector*" operator: The inverse of the $[\cdot]_\times$ operator will be denoted in the following as $[\cdot]_\nabla$, that is:

$$2\times 2: \quad \begin{pmatrix} 0 & -x \\ x & 0 \end{pmatrix}_{\nabla} \equiv (x)$$
$$3\times 3: \quad \begin{pmatrix} 0 & -z & y \\ z & 0 & -x \\ -y & x & 0 \end{pmatrix}_{\nabla} \equiv \begin{pmatrix} x \\ y \\ z \end{pmatrix} \tag{2}$$

The $vec(\cdot)$ operator: it stacks all the columns of an $M \times N$ matrix to form a $MN \times 1$ vector. For example:

$$vec\left(\begin{bmatrix} 1 & 2 & 3 \\ 4 & 5 & 6 \end{bmatrix}\right) = \begin{pmatrix} 1 \\ 4 \\ 2 \\ 5 \\ 3 \\ 6 \end{pmatrix} \tag{3}$$

The Kronecker operator (also called matrix *direct product*): Denoted as $\mathbf{A} \otimes \mathbf{B}$ for any two matrices $\mathbf{A}$ and $\mathbf{B}$ of dimensions $M_A \times N_A$ and $M_B \times N_B$, respectively, it gives us a tensor product of the matrices as an $M_A M_B \times N_A N_B$ matrix. That is:

$$\mathbf{A} \otimes \mathbf{B} = \begin{pmatrix} a_{11}\mathbf{B} & a_{12}\mathbf{B} & a_{13}\mathbf{B} & \ldots \\ a_{21}\mathbf{B} & a_{22}\mathbf{B} & a_{23}\mathbf{B} & \ldots \\ & \ldots & & \end{pmatrix} \tag{4}$$

2. LIE GROUPS AND LIE ALGEBRAS

Chapter 10 section 2 provided a brief mathematical definition for the concepts of manifold, Lie group and Lie algebra. For our purposes in this appendix, it will be enough to keep in mind these points:

1. All $\mathbf{SO}(n)$ and $\mathbf{SE}(n)$ groups—refer to Appendix A for their definitions—are Lie groups, with the main implication of this for us being that:
2. they are also smooth manifolds embedded in $\mathbb{R}^{m^2}$ —where m does not have to coincide with n.
3. Their tangent spaces at the identity matrix $\mathbf{I}$ (the "origin or coordinates" for both groups) are denoted as $T_{\mathbf{I}}\mathbf{SO}(n)$ and $T_{\mathbf{I}}\mathbf{SE}(n)$, respectively. The Lie algebras associated to those spaces provide us the space vector bases $\mathfrak{so}(n)$ and $\mathfrak{se}(n)$, respectively.

We have summarized the main properties of the groups in which we are interested in Table 1.where $\mathbf{GL}(n,\mathbb{R})$ stands for the *general linear* group of $n \times n$ real matrices. Informally, two spaces are dif-

Table 1. Main properties of the groups

	Closed subgroup of:	Manifold dimensionality (number of DOFs):	Is a manifold embedded in:	Diffeomorphic to:
$\mathbf{SO}(2)$	$\mathbf{GL}(2,\mathbb{R})$	1	$\mathbb{R}^{2^2}=\mathbb{R}^4$	–
$\mathbf{SO}(3)$	$\mathbf{GL}(3,\mathbb{R})$	3	$\mathbb{R}^{3^2}=\mathbb{R}^9$	–
$\mathbf{SE}(2)$	$\mathbf{GL}(3,\mathbb{R})$	3	$\mathbb{R}^{3^2}=\mathbb{R}^9$	$\mathbf{SO}(2)\times\mathbb{R}^2(2\times 2+2=6$ coordinates)
$\mathbf{SE}(3)$	$\mathbf{GL}(4,\mathbb{R})$	6	$\mathbb{R}^{4^2}=\mathbb{R}^{16}$	$\mathbf{SO}(3)\times\mathbb{R}^3(3\times 3+3=12$ coordinates)

feomorphic if there exists a one-to-one smooth correspondence between all its elements. In this case, the mathematical equivalence between groups allows us to treat robot poses (in $\mathbf{SE}(n)$) as a vector of coordinates with two *separate* parts: (1) the elements of the rotation matrix (from $\mathbf{SE}(n)$) and (2) the translational part (a simple vector in $\mathbb{R}^n$). We will see how to exploit such a representation in section 5.

Regarding the Lie algebras of these manifolds, they are nothing more that a vector base: a set of linearly independent elements (as many as the number of DOFs) such that any element in the manifold can be decomposed in a linear combination of them. Since the manifold elements are matrices, every component of a Lie algebra is also a matrix, i.e., instead of a *vector base* we have a *matrix base*.

The particular elements of the Lie algebra for $\mathbf{SE}(2)$, denoted as $\mathfrak{se}(2)$, are the following three matrices:

$$\mathfrak{se}(2)=\{\mathbf{G}_i^{\mathfrak{se}(2)}\}_{i=1,2,3}$$

$$\mathbf{G}_1^{\mathfrak{se}(2)}=\left(\begin{array}{cc|c}0&0&1\\0&0&0\\\hline 0&0&0\end{array}\right)\qquad \mathbf{G}_2^{\mathfrak{se}(2)}=\left(\begin{array}{cc|c}0&0&0\\0&0&1\\\hline 0&0&0\end{array}\right)\qquad \mathbf{G}_3^{\mathfrak{se}(2)}=\left(\begin{array}{cc|c}0&-1&0\\1&0&0\\\hline 0&0&0\end{array}\right)\tag{5}$$

while for $\mathbf{SE}(3)$ the Lie algebra $\mathfrak{se}(3)$ comprises these six matrices:

$$\mathfrak{se}(3) = \{\mathbf{G}_i^{\mathfrak{se}(3)}\}_{i=1\ldots6}$$

$$\mathbf{G}_1^{\mathfrak{se}(3)} = \left(\begin{array}{ccc|c} 0 & 0 & 0 & 1 \\ 0 & 0 & 0 & 0 \\ 0 & 0 & 0 & 0 \\ \hline 0 & 0 & 0 & 0 \end{array}\right) \quad \mathbf{G}_2^{\mathfrak{se}(3)} = \left(\begin{array}{ccc|c} 0 & 0 & 0 & 0 \\ 0 & 0 & 0 & 1 \\ 0 & 0 & 0 & 0 \\ \hline 0 & 0 & 0 & 0 \end{array}\right) \quad \mathbf{G}_3^{\mathfrak{se}(3)} = \left(\begin{array}{ccc|c} 0 & 0 & 0 & 0 \\ 0 & 0 & 0 & 0 \\ 0 & 0 & 0 & 1 \\ \hline 0 & 0 & 0 & 0 \end{array}\right)$$

$$\mathbf{G}_4^{\mathfrak{se}(3)} = \left(\begin{array}{ccc|c} & \begin{bmatrix} 1 \\ 0 \\ 0 \end{bmatrix}_\times & & \begin{matrix} 0 \\ 0 \\ 0 \end{matrix} \\ \hline 0 & 0 & 0 & 0 \end{array}\right) = \left(\begin{array}{ccc|c} 0 & 0 & 0 & 0 \\ 0 & 0 & -1 & 0 \\ 0 & 1 & 0 & 0 \\ \hline 0 & 0 & 0 & 0 \end{array}\right)$$

$$\mathbf{G}_5^{\mathfrak{se}(3)} = \left(\begin{array}{ccc|c} & \begin{bmatrix} 0 \\ 1 \\ 0 \end{bmatrix}_\times & & \begin{matrix} 0 \\ 0 \\ 0 \end{matrix} \\ \hline 0 & 0 & 0 & 0 \end{array}\right) = \left(\begin{array}{ccc|c} 0 & 0 & 1 & 0 \\ 0 & 0 & 0 & 0 \\ -1 & 0 & 0 & 0 \\ \hline 0 & 0 & 0 & 0 \end{array}\right) \tag{6}$$

$$\mathbf{G}_6^{\mathfrak{se}(3)} = \left(\begin{array}{ccc|c} & \begin{bmatrix} 0 \\ 0 \\ 1 \end{bmatrix}_\times & & \begin{matrix} 0 \\ 0 \\ 0 \end{matrix} \\ \hline 0 & 0 & 0 & 0 \end{array}\right) = \left(\begin{array}{ccc|c} 0 & -1 & 0 & 0 \\ 1 & 0 & 0 & 0 \\ 0 & 0 & 0 & 0 \\ \hline 0 & 0 & 0 & 0 \end{array}\right)$$

All these basis matrices of Lie algebras have a clear geometrical interpretation: they represent the directions of "infinitesimal transformations" along each of the different DOFs. Those "infinitesimal transformations" are actually the derivatives (the *tangent directions*), at the identity element $\mathbf{I}$ of the corresponding $\mathbf{SE}(n)$ manifold, with respect to each DOF. For example, consider the derivative of a $\mathbf{SE}(2)$ pose (a 3×3 matrix) with respect to the rotation parameter θ:

$$\frac{\partial}{\partial\theta}\begin{pmatrix} \cos\theta & -\sin\theta & x \\ \sin\theta & \cos\theta & y \\ 0 & 0 & 1 \end{pmatrix} = \begin{pmatrix} -\sin\theta & -\cos\theta & 0 \\ \cos\theta & -\sin\theta & 0 \\ 0 & 0 & 0 \end{pmatrix} \tag{7}$$

By evaluating this matrix at the identity $\mathbf{I}$, i.e. at $(x\ y\ \theta) = (0\ 0\ 0)$, we have:

$$\begin{pmatrix} 0 & -1 & 0 \\ 1 & 0 & 0 \\ 0 & 0 & 0 \end{pmatrix} \tag{8}$$

which exactly matches $\mathbf{G}_3^{\mathfrak{se}(2)}$ in equation (5). All the other Lie algebra basis matrices are obtained by the same procedure.

Regarding the Lie algebra bases of the pure rotation groups, it can be shown that $\mathbf{SO}(2)$ only has one basis matrix $\mathbf{G}_1^{\mathfrak{so}(2)}$ which coincides with the top-left submatrix of $\mathbf{G}_3^{\mathfrak{se}(2)}$ in equation (5), while the three bases $\mathbf{G}_{1,2,3}^{\mathfrak{so}(3)}$ of $\mathbf{SO}(3)$ are the top-left submatrices of $\mathbf{G}_4^{\mathfrak{se}(3)}$, $\mathbf{G}_5^{\mathfrak{se}(3)}$, and $\mathbf{G}_6^{\mathfrak{se}(3)}$ in equation (6), respectively.

3. EXPONENTIAL AND LOGARITHM MAPS

Each Lie group M has two associated mapping operators which convert matrices between M and its Lie algebra $\mathfrak{m}$. They are called the exponential and logarithm maps.

For any manifold, the exponential map is simply the *matrix exponential* function, such as:

$$\begin{array}{rlcl} \mathbf{exp}_{\mathfrak{m}}: & \mathfrak{m} & \rightarrow & M \\ & \boldsymbol{\omega} & \rightarrow & \mathbf{P} \end{array} \qquad \mathbf{P} = e^{\boldsymbol{\omega}} \tag{9}$$

Interestingly, the exponential function for matrices is defined as the sum of the infinite series:

$$e^{\mathbf{A}} = \mathbf{I} + \sum_{i=1}^{\infty} \frac{\mathbf{A}^i}{i!} \tag{10}$$

which coincides with the Taylor series expansion of the scalar exponential function and is always well-defined (i.e. the series converges) for any square matrix $\mathbf{A}$. Moreover, it turns out that the exponential of *any* skew-symmetric matrix is a well-defined rotation matrix (Gallier, 2001). This is in complete agreement with our purpose of converting elements from the Lie algebras $\mathfrak{so}(n)$ into rotation matrices, since any linear combination of the basis skew-symmetric matrices will still be skew-symmetric and, in consequence, will generate a rotation matrix when mapped through the exponential function. The conversion from $\mathfrak{se}(n)$ requires a little further analysis regarding the matrix structure, as we show shortly.

The universal definition of the exponential map is that one provided in equations (9)-(10), but in practice the matrix exponential leads to particular closed-form expressions for each of the manifolds of our interest. Next, we provide a complete summary of the explicit equations for all the interesting exponential maps. Some expressions can be found in the literature (Gallier, 2001), while the rest have been derived by the authors for completeness. For a manifold M with k DOFs we will denote as $\mathbf{v} = \{v_1, \ldots, v_k\}$ the vector of all the coordinates of a matrix $\boldsymbol{\Omega}$ belonging to its Lie algebra $\mathfrak{m}$. This means that such matrix is composed as $\boldsymbol{\Omega} = \sum_{i=1}^{k} v_i \mathbf{G}_i^{\mathfrak{m}}$, with the $\mathbf{G}_i^{\mathfrak{m}}$ matrices given in the last section.

Notice that the vector of parameters $\mathbf{v}$ stands as the minimum-DOF representation of a value in the linearized manifold; hence it is the form in use in all robotics optimization problems where exponential maps are employed. That is also the why we will provide the expression of the manifold maps as functions of their vectors of parameters ($\mathbf{v}$), not only their associated matrices ($\sum_{i=1}^{k} v_i \mathbf{G}_i^{\mathfrak{m}}$).

In $\mathbf{SO}(2)$, the unique Lie algebra coordinate θ represents the rotation in radians and the exponential map has this form:

$$\begin{array}{llll} \mathbf{exp}_{\mathfrak{so}(2)}: & \mathfrak{so}(2) & \to & \mathbf{SO}(2) \\ & \boldsymbol{\Omega}_{2\times 2}(\mathbf{v}) & \to & \mathbf{R}_{2\times 2} \end{array} \qquad \mathbf{R} = e^{\Omega}$$

$$\begin{aligned} &\boldsymbol{\Omega} = \left[(\theta)\right]_{\times} = \begin{pmatrix} 0 & -\theta \\ \theta & 0 \end{pmatrix} \qquad \text{Parameters: } \mathbf{v} = (\theta) \\ &\mathbf{R} = \mathbf{exp}_{\mathfrak{so}(2)}(\mathbf{v}) \equiv e^{\Omega} = \begin{pmatrix} \cos\theta & -\sin\theta \\ \sin\theta & \cos\theta \end{pmatrix} \end{aligned} \tag{11}$$

In $\mathbf{SO}(3)$ we have three Lie algebra coordinates $\mathbf{v} = (\omega_1\ \omega_2\ \omega_3)^T$ which determine the 3D rotation by means of its modulus (related to the rotation angle) and its direction (the rotation axis). In this case, the exponential map employs the well-known Rodrigues' formula:

$$\begin{array}{llll} \mathbf{exp}_{\mathfrak{so}(3)}: & \mathfrak{so}(3) & \to & \mathbf{SO}(3) \\ & \boldsymbol{\Omega}_{3\times 3}(\mathbf{v}) & \to & \mathbf{R}_{3\times 3} \end{array} \qquad \mathbf{R} = e^{\Omega}$$

$$\begin{aligned} &\boldsymbol{\Omega} = \left[\mathbf{v}\right]_{\times} = \begin{bmatrix} \begin{pmatrix} \omega_1 \\ \omega_2 \\ \omega_3 \end{pmatrix} \end{bmatrix}_{\times} = \begin{pmatrix} 0 & -\omega_3 & \omega_2 \\ \omega_3 & 0 & -\omega_1 \\ -\omega_2 & \omega_1 & 0 \end{pmatrix} \qquad \text{Parameters: } \mathbf{v} = \begin{pmatrix} \omega_1 \\ \omega_2 \\ \omega_3 \end{pmatrix} \\ &\mathbf{R} = \mathbf{exp}_{\mathfrak{so}(3)}(\mathbf{v}) \equiv e^{\Omega} = \begin{cases} \mathbf{I}_3 & , \text{if } |\mathbf{v}| = 0 \\ \mathbf{I}_3 + \dfrac{\sin|\mathbf{v}|}{|\mathbf{v}|}[\mathbf{v}]_{\times} + \dfrac{1-\cos|\mathbf{v}|}{|\mathbf{v}|^2}[\mathbf{v}]_{\times}^2 & , \text{otherwise} \end{cases} \end{aligned} \tag{12}$$

In $\mathbf{SE}(2)$, the Lie algebra has three coordinates: θ which represents the rotation in radians, and $\mathbf{t}' = (x'\ y')^T$ which *is related to* the spatial translation. Note that $(x'\ y')^T$ is *not* the spatial translation of the corresponding pose in $\mathbf{SE}(2)$, which turns out to be $\mathbf{t} = \mathbf{V}_{\mathfrak{se}(2)}\mathbf{t}'$. The exponential map here becomes:

$$\begin{array}{llll} \mathbf{exp}_{\mathfrak{se}(2)}: & \mathfrak{se}(2) & \to & \mathbf{SE}(2) \\ & \boldsymbol{\Psi}_{3\times 3}(\mathbf{v}) & \to & \mathbf{P}_{3\times 3} \\ & \left(\begin{array}{cc|c} \multicolumn{2}{c|}{[\theta]_{\times}} & \mathbf{t}' \\ \hline 0 & 0 & 0 \end{array}\right) & \to & \left(\begin{array}{cc|c} \multicolumn{2}{c|}{\mathbf{R}} & \mathbf{t} \\ \hline 0 & 0 & 1 \end{array}\right) \end{array} \qquad \mathbf{P} = e^{\Psi}$$

$$\boldsymbol{\Psi} = \left(\begin{array}{cc|c} 0 & -\theta & x' \\ \theta & 0 & y' \\ \hline 0 & 0 & 0 \end{array}\right) \qquad \text{Parameters: } \mathbf{v} = \begin{pmatrix} \mathbf{t}' \\ \theta \end{pmatrix} = \begin{pmatrix} x' \\ y' \\ \theta \end{pmatrix}$$

$$\mathbf{P} = \mathbf{exp}_{\mathfrak{se}(2)}(\mathbf{v}) \equiv e^{\boldsymbol{\Psi}} = \left(\begin{array}{c|c} e^{[\theta]_\times} & \mathbf{V}_{\mathfrak{se}(2)}\mathbf{t}' \\ \hline 0\ 0 & 1 \end{array}\right) = \left(\begin{array}{cc|c} \cos\theta & -\sin\theta & \mathbf{V}_{\mathfrak{se}(2)}\mathbf{t}' \\ \sin\theta & \cos\theta & \\ \hline 0 & 0 & 1 \end{array}\right)$$

with:

$$\mathbf{V}_{\mathfrak{se}(2)} = \begin{cases} \mathbf{I}_2 & \text{, if } \theta = 0 \\ \begin{pmatrix} \dfrac{\sin\theta}{\theta} & \dfrac{\cos\theta - 1}{\theta} \\ \dfrac{1-\cos\theta}{\theta} & \dfrac{\sin\theta}{\theta} \end{pmatrix} & \text{, otherwise} \end{cases} \tag{13}$$

Finally, the Lie algebra of $\mathbf{SE}(3)$ has six coordinates: $\boldsymbol{\omega} = \left(\omega_1\ \omega_2\ \omega_3\right)^T$ which parameterize the 3D rotation exactly like described above for $\mathbf{SO}(3)$, and $\mathbf{t}' = \left(x'\ y'\ z'\right)^T$ which is *related* to the spatial translation. Again, we must stress that this translation vector is *not* directly equal to the translation $\mathbf{t}$ of the pose. The corresponding exponential map is:

$$\begin{array}{llcl} \mathbf{exp}_{\mathfrak{se}(3)}: & \mathfrak{se}(3) & \rightarrow & \mathbf{SE}(3) \\ & \boldsymbol{\Psi}_{4\times4}(\mathbf{v}) & \rightarrow & \mathbf{P}_{4\times4} \\ & \left(\begin{array}{ccc|c} & [\boldsymbol{\omega}]_\times & & \mathbf{t}' \\ \hline 0 & 0 & 0 & 0 \end{array}\right) & \rightarrow & \left(\begin{array}{ccc|c} & \mathbf{R} & & \mathbf{t} \\ \hline 0 & 0 & 0 & 1 \end{array}\right) \end{array} \qquad \mathbf{P} = e^{\boldsymbol{\Psi}}$$

$$\boldsymbol{\Psi} = \left(\begin{array}{ccc|c} 0 & -\omega_3 & \omega_2 & x' \\ \omega_3 & 0 & -\omega_1 & y' \\ -\omega_2 & \omega_1 & 0 & z' \\ \hline 0 & 0 & 0 & 0 \end{array}\right) \qquad \text{Parameters: } \mathbf{v} = \begin{pmatrix} \mathbf{t}' \\ \boldsymbol{\omega} \end{pmatrix} = \begin{pmatrix} x' \\ y' \\ z' \\ \hline \omega_1 \\ \omega_2 \\ \omega_3 \end{pmatrix}$$

$$\mathbf{P} = \mathbf{exp}_{\mathfrak{se}(3)}(\mathbf{v}) \equiv e^{\boldsymbol{\Psi}} = \left(\begin{array}{c|c} e^{[\boldsymbol{\omega}]_\times} & \mathbf{V}_{\mathfrak{se}(3)}\mathbf{t}' \\ \hline 0\ 0\ 0 & 1 \end{array}\right)$$

with:

$$\mathbf{V}_{\mathfrak{se}(3)} = \begin{cases} \mathbf{I}_3 & , \text{if } \left|\boldsymbol{\omega}\right| = 0 \\ \mathbf{I}_3 + \dfrac{1-\cos\left|\boldsymbol{\omega}\right|}{\left|\boldsymbol{\omega}\right|^2}\left[\boldsymbol{\omega}\right]_\times + \dfrac{\left|\boldsymbol{\omega}\right| - \sin\left|\boldsymbol{\omega}\right|}{\left|\boldsymbol{\omega}\right|^3}\left[\boldsymbol{\omega}\right]_\times^2 & , \text{otherwise} \end{cases} \tag{14}$$

Once we defined the exponential maps for all the manifolds of our interest, we turn now to the corresponding *logarithm maps*. The goal of this function is to provide a mapping between matrices in the manifold M and in its Lie algebra $\mathfrak{m}$, that is:

$$\begin{aligned} \mathbf{log}_{\mathfrak{m}} : \; & M \rightarrow \mathfrak{m} \\ & \mathbf{P} \rightarrow \boldsymbol{\omega} \end{aligned} \qquad \boldsymbol{\omega} = \mathbf{log}\left(\mathbf{P}\right) \tag{15}$$

This is clearly the inverse function of the exponential map defined in equation (9), thus it comes at no surprise that this operation also corresponds to a standard function called *matrix logarithm*, the inverse of equation (10).

Iterative algorithms exist for numerically determining matrix logarithms of arbitrary matrices (Davies & Higham, 2010). Fortunately, efficient closed-form solutions are also available for the matrices of our interest. Notice that the logarithm of a matrix (in the manifold) is another matrix of the same size (in the Lie algebra), but in subsequent equations we will put the stress on recovering the *coordinates* (the vector $\mathbf{v}$) of the latter matrix in the Lie algebra bases.

For $\mathbf{SO}(2)$, the coordinate $\mathbf{v}$ only comprises one coordinate (the rotation θ). The corresponding logarithm map reads:

$$\begin{aligned} \mathbf{log}_{\mathfrak{so}(2)} : \quad & \mathbf{SO}(2) \rightarrow \mathfrak{so}(2) \\ & \mathbf{R}_{2\times 2} \rightarrow \boldsymbol{\Omega}_{2\times 2} \\ & \begin{pmatrix} R_{11} & R_{12} \\ R_{21} & R_{22} \end{pmatrix} \rightarrow \left[\mathbf{v}\right]_\times \end{aligned} \qquad \boldsymbol{\Omega} = \mathbf{log}\left(\mathbf{R}\right)$$

with:

$$\mathbf{v} = \left(\theta\right) = \left[\mathbf{log}_{\mathfrak{so}(2)}\left(\mathbf{R}\right)\right]_\nabla = \left[\mathbf{log}\left(\mathbf{R}\right)\right]_\nabla = \operatorname{atan2}\left(R_{21}, R_{11}\right) \tag{16}$$

In the logarithm of a $\mathbf{SO}(3)$ matrix our aim is to find out the three parameters $\mathbf{v} = \left(\omega_1 \; \omega_2 \; \omega_3\right)^T$ that determine the 3D rotation. In this case:

$$\begin{aligned} \mathbf{log}_{\mathfrak{so}(3)} : \; & \mathbf{SO}(3) \rightarrow \mathfrak{so}(3) \\ & \mathbf{R}_{3\times 3} \rightarrow \underbrace{\boldsymbol{\Omega}_{3\times 3}}_{\left[\mathbf{v}\right]_\times} \end{aligned} \qquad \boldsymbol{\Omega} = \mathbf{log}\left(\mathbf{R}\right)$$

$$\mathbf{\Omega} = [\mathbf{v}]_\times \Leftrightarrow \mathbf{v} = [\mathbf{\Omega}]_\nabla \qquad \rightarrow \mathbf{v} = [\log \mathbf{R}]_\nabla$$

with:

$$\log(\mathbf{R}) = \frac{\theta}{2\sin\theta}(\mathbf{R} - \mathbf{R}^T)$$

(where the rotation angle θ is computed as):

$$\theta = \cos^{-1}\left(\frac{tr(\mathbf{R}) - 1}{2}\right) \tag{17}$$

As happened with the exponential map, the logarithm map of $\mathbf{SE}(2)$ forces us to tell between the translation parameters $\mathbf{t}' = (x' \; y')^T$ in the Lie algebra, and the actual translation $\mathbf{t} = (x \; y)^T$. In this case, the logarithm map can be shown to be:

$$\begin{array}{llll} \mathbf{log}_{\mathfrak{se}(2)}: & \mathbf{SE}(2) & \rightarrow & \mathfrak{se}(2) \\ & \mathbf{P}_{3\times3} & \rightarrow & \mathbf{\Psi}_{3\times3} \qquad \mathbf{\Psi} = \log(\mathbf{P}) \\ & \left(\begin{array}{cc|c} \multicolumn{2}{c|}{\mathbf{R}} & \mathbf{t} \\ \hline 0 & 0 & 1 \end{array}\right) & \rightarrow & \left(\begin{array}{cc|c} \multicolumn{2}{c|}{[\theta]_\times} & \mathbf{t}' \\ \hline 0 & 0 & 1 \end{array}\right) \end{array}$$

with parameters:

$$\mathbf{v} = \begin{pmatrix} \mathbf{t}' \\ \theta \end{pmatrix} = \begin{pmatrix} x' \\ y' \\ \theta \end{pmatrix}$$

$$\rightarrow \begin{cases} \theta = [\mathbf{log}_{\mathfrak{so}(2)}(\mathbf{R})]_\nabla \\ \mathbf{t}' = \mathbf{V}_{\mathfrak{se}(2)}^{-1}\mathbf{t} \end{cases}$$

(where $\mathbf{log}_{\mathfrak{so}(2)}(\cdot)$ and $\mathbf{V}_{\mathfrak{se}(2)}$ are given in equation (16) and equation (13), respectively, and)

$$\mathbf{V}_{\mathfrak{se}(2)}^{-1} = \begin{cases} \mathbf{I}_2 & \text{, if } \theta = 0 \\ \dfrac{\theta}{2}\begin{pmatrix} \dfrac{\sin\theta}{1-\cos\theta} & 1 \\ -1 & \dfrac{\sin\theta}{1-\cos\theta} \end{pmatrix} & \text{, otherwise} \end{cases} \tag{18}$$

And finally, the logarithm for $\mathbf{SE}(3)$ takes this form:

$$\begin{array}{llll} \mathbf{log}_{\mathfrak{se}(3)}: & \mathbf{SE}(3) & \rightarrow & \mathfrak{se}(3) \\ & \mathbf{P}_{4\times 4} & \rightarrow & \mathbf{\Psi}_{4\times 4} \qquad \mathbf{\Psi} = \log(\mathbf{P}) \\ & \left(\begin{array}{ccc|c} & \mathbf{R} & & \mathbf{t} \\ \hline 0 & 0 & 0 & 1 \end{array}\right) & \rightarrow & \left(\begin{array}{ccc|c} & [\boldsymbol{\omega}]_\times & & \mathbf{t}' \\ \hline 0 & 0 & 0 & 1 \end{array}\right) \end{array}$$

with parameters:

$$\mathbf{v} = \begin{pmatrix} \mathbf{t}' \\ \boldsymbol{\omega} \end{pmatrix} = \begin{pmatrix} x' \\ y' \\ z' \\ \hline \omega_1 \\ \omega_2 \\ \omega_3 \end{pmatrix}$$

$$\rightarrow \begin{cases} \boldsymbol{\omega} = \left[\mathbf{log}_{\mathfrak{so}(3)}(\mathbf{R})\right]_\nabla \\ \mathbf{t}' = \mathbf{V}_{\mathfrak{se}(3)}^{-1}\,\mathbf{t} \end{cases}$$

(where $\mathbf{log}_{\mathfrak{so}(3)}(\cdot)$ and $\mathbf{V}_{\mathfrak{se}(3)}$ are given in equation [17] and equation [14], respectively) (19)

4. PSEUDO-EXPONENTIAL AND PSEUDO-LOGARITHM MAPS

If the reader has carefully studied recent literature about on-manifold optimization, he or she may have noticed that the equations employed there for the different manifold maps are *almost* exactly those introduced in the previous section. In particular, the unique difference between the commonly used formulas and those above are related to the treatment of the *translation* vectors in $\mathbf{SE}(n)$ groups, where the distinction between the vectors of translations in the pose ($\mathbf{t}$) and in the Lie-algebra ($\mathbf{t}'$) is ignored.

We must highlight that the mathematically correct exponential and logarithm maps for $\mathbf{SE}(n)$ groups are, indeed, those reported in the previous section. As can be seen in equations (13)–(14) and equations (18)–(19), in these maps the two translation vectors are not equivalent since they are related to each other by $\mathbf{t} = \mathbf{V}_{\mathfrak{se}(n)}\mathbf{t}'$. However, it can be shown that we can safely replace the manifold maps with alternative versions where the translations in the manifold are identified with those of the real pose (i.e., $\mathbf{t}' = \mathbf{t}$) and still perform optimizations as described in chapter 10 section 2 without varying the final results, i.e. the same minimum of the cost function will be reached. For the sake of rigorousness, we will name those alternative maps the *pseudo-exponential* ($\mathbf{pexp}_m$) and the *pseudo-logarithm* ($\mathbf{plog}_m$).

Their practical usefulness is the obvious simplification of dropping the $\mathbf{V}_{\mathfrak{se}(n)}$ terms in all the transformations and, consequently, in all the Jacobian matrices involved in the optimization problem.

Regarding the pseudo-exponential functions, for $\mathbf{SE}(2)$ we have:

$$\begin{array}{llcll} \mathbf{pexp}_{\mathfrak{se}(2)}: & \mathfrak{se}(2) & \rightarrow & \mathbf{SE}(2) & (\text{Note: } \mathbf{P} \neq e^{\Psi}) \\ & \Psi_{3\times 3}(\mathbf{v}) & \rightarrow & \mathbf{P}_{3\times 3} & \\ & \left(\begin{array}{cc|c} \multicolumn{2}{c|}{[\theta]_\times} & \mathbf{t} \\ \hline 0 & 0 & 0 \end{array}\right) & \rightarrow & \left(\begin{array}{cc|c} \multicolumn{2}{c|}{\mathbf{R}} & \mathbf{t} \\ \hline 0 & 0 & 1 \end{array}\right) & \text{Parameters: } \mathbf{v} = (\theta) \end{array}$$

$$\mathbf{R} = e^{[\theta]_\times} = \begin{pmatrix} \cos\theta & -\sin\theta \\ \sin\theta & \cos\theta \end{pmatrix} \tag{20}$$

while for $\mathbf{SE}(3)$:

$$\begin{array}{llcll} \mathbf{pexp}_{\mathfrak{se}(3)}: & \mathfrak{se}(3) & \rightarrow & \mathbf{SE}(3) & (\text{Note: } \mathbf{P} \neq e^{\Psi}) \\ & \Psi_{4\times 4}(\mathbf{v}) & \rightarrow & \mathbf{P}_{4\times 4} & \\ & \left(\begin{array}{ccc|c} \multicolumn{3}{c|}{[\omega]_\times} & \mathbf{t} \\ \hline 0 & 0 & 0 & 0 \end{array}\right) & \rightarrow & \left(\begin{array}{ccc|c} \multicolumn{3}{c|}{\mathbf{R}} & \mathbf{t} \\ \hline 0 & 0 & 0 & 1 \end{array}\right) & \text{Parameters: } \mathbf{v} = \begin{pmatrix} \mathbf{t} \\ \omega \end{pmatrix} \end{array}$$

$$\begin{array}{l} \mathbf{R} = e^{[\omega]_\times} = \mathbf{exp}_{\mathfrak{so}(3)}(\omega) \\ \text{Parameters: } \mathbf{v} = \begin{pmatrix} \mathbf{t} \\ \omega \end{pmatrix} \end{array} \tag{21}$$

with $\mathbf{exp}_{\mathfrak{so}(3)}(\cdot)$ defined in equation (12).

The pseudo-logarithm for $\mathbf{SE}(2)$ becomes:

$$\begin{array}{llcll} \mathbf{plog}_{\mathfrak{se}(2)}: & \mathbf{SE}(2) & \rightarrow & \mathfrak{se}(2) & (\text{Note: } \Psi \neq \mathbf{log}(\mathbf{P})) \\ & \mathbf{P}_{3\times 3} & \rightarrow & \Psi_{3\times 3}(\mathbf{v}) & \\ & \left(\begin{array}{cc|c} \multicolumn{2}{c|}{\mathbf{R}} & \mathbf{t} \\ \hline 0 & 0 & 1 \end{array}\right) & \rightarrow & \left(\begin{array}{cc|c} \multicolumn{2}{c|}{[\theta]_\times} & \mathbf{t} \\ \hline 0 & 0 & 1 \end{array}\right) & \text{Parameters: } \mathbf{v} = (\theta) \end{array}$$

with

$$\theta = \left[\mathbf{log}_{\mathfrak{so}(2)}(\mathbf{R})\right]_\nabla \tag{22}$$

and for the $\mathbf{SE}(3)$ group we have:

$$
\begin{array}{llll}
\mathbf{plog}_{\mathfrak{se}(3)}: & \mathbf{SE}(3) & \rightarrow & \mathfrak{se}(3) & \left(\text{Note: } \mathbf{\Psi} \neq \mathbf{log}(\mathbf{P})\right) \\
& \mathbf{P}_{4\times4} & \rightarrow & \mathbf{\Psi}_{4\times4}(\mathbf{v}) & \\
& \left(\begin{array}{ccc|c} & \mathbf{R} & & \mathbf{t} \\ \hline 0 & 0 & 0 & 1 \end{array}\right) & \rightarrow & \left(\begin{array}{ccc|c} & [\boldsymbol{\omega}]_\times & & \mathbf{t} \\ \hline 0 & 0 & 0 & 1 \end{array}\right) & \text{Parameters: } \mathbf{v} = \begin{pmatrix} \mathbf{t} \\ \boldsymbol{\omega} \end{pmatrix}
\end{array}
$$

with

$$
\boldsymbol{\omega} = \left[\mathbf{log}_{\mathfrak{so}(3)}(\mathbf{R})\right]_\nabla \tag{23}
$$

with the definition of $\mathbf{log}_{\mathfrak{so}(3)}(\cdot)$ already provided in equation (17).

5. ABOUT DERIVATIVES OF POSE MATRICES

One of the goals of this appendix is providing the expressions for a set of useful Jacobians, which are reported in the next section. The most useful Jacobians in robotics applications (e.g. graph SLAM, bundle adjustment) can be split by means of the chain rule in a series of smaller Jacobians, of which many of them will be often related to geometry transformations. In particular, some of them may involve taking derivatives *with respect to matrices*. This topic has not been addressed anywhere else in this book, thus we devote the present section to introduce the related notation.

Let us focus now exclusively on $\mathbf{SE}(3)$ poses. We know that any pose $\mathbf{P} \in \mathbf{SE}(3)$ has the structure:

$$
\mathbf{P} = \left(\begin{array}{ccc|c} & & & x \\ & \mathbf{R}_{3\times3} & & y \\ & & & z \\ \hline 0 & 0 & 0 & 1 \end{array}\right) \tag{24}
$$

and that it belongs to a manifold which is embedded in $\mathbb{R}^{4^2}$ and is diffeomorphic to $\mathbf{SO}(3) \times \mathbb{R}^3$ (see section 1). Thus, the manifold has a dimensionality of 12: nine coordinates for the 3×3 rotation matrix plus other three for the translation vector.

Since we will be interested here in expressions involving *derivatives* of functions of poses, we need to define a clear notation for what a derivative of a matrix actually means. As an example, consider an arbitrary function, e.g. the map of pairs of poses $\mathbf{P}_1$ and $\mathbf{P}_2$ to their composition $\mathbf{P}_1 \oplus \mathbf{P}_2$, that is, $\mathbf{f}_\oplus : \mathbf{SE}(3) \times \mathbf{SE}(3) \rightarrow \mathbf{SE}(3)$. Then, what does the expression:

$$
\frac{\partial \mathbf{f}_\oplus(\mathbf{P}_1, \mathbf{P}_2)}{\partial \mathbf{P}_1} \tag{25}
$$

means? If $\mathbf{P}_i$ were vectors the expression above would be interpreted as a Jacobian matrix without further complications. But since they are poses, in the first place we must make explicit in which parameterization we are describing them. One possibility is to interpret that poses are given as the vectors of their parameters, in which case equation (25) would be a Jacobian. Indeed, that was the assumption followed in Appendix A and we were able to provide the corresponding Jacobian of all the relevant geometric operations treated there.

However, one must observe that, interpreted in this way, the geometry functions: (1) are typically non-linear, which entails inaccuracies when using their Jacobians for optimization, and (2) may become somewhat complicated—see for example, Appendix A equation (30). Therefore, it makes sense to employ an alternative: to parameterize poses directly with coordinates in their diffeomorphic spaces. Put simple, this means that a $\mathbf{SE}(3)$ pose will be represented with 12 scalars, i.e. the three first rows in its corresponding 4×4 matrix. Although this implies a clear over-parameterization of an entity with 6 DOFs, it turns out that many important operations become linear under this representation, enabling us to obtain exact derivatives in an efficient way. Observe how we are over-parameterizing poses only while evaluating Jacobians with respect to them, which does not have the adverse effects of employing over-parameterized pose representations within state spaces—as discussed in chapter 10 section 2.

Recovering the example in equation (25), we can now say that the derivative will be a 12×12 matrix. It is illustrative to further work on this example: using the standard notation for denoting matrix elements, that is:

$$\mathbf{M} = \begin{pmatrix} m_{11} & m_{12} & m_{13} & m_{14} \\ m_{21} & m_{22} & m_{23} & m_{24} \\ m_{31} & m_{32} & m_{33} & m_{34} \\ m_{41} & m_{42} & m_{43} & m_{44} \end{pmatrix} \tag{26}$$

and denoting the resulting matrix from $\mathbf{f}_{\oplus}(\mathbf{P}_1, \mathbf{P}_2)$ as $\mathbf{F}$, we can unroll the derivative in equation (25) as follows:

$$\begin{aligned} \frac{\partial \mathbf{f}_{\oplus}(\mathbf{p}, \mathbf{q})}{\partial \mathbf{P}} &\overset{\mathbf{F}=\mathbf{PQ}}{=} \frac{\partial \mathbf{F}}{\partial \mathbf{P}} \\ &= \frac{\partial vec(\mathbf{F})}{\partial vec(\mathbf{P})} \\ &= \frac{\partial [f_{11} f_{21} f_{31} f_{12} f_{22} \cdots f_{33} f_{14} f_{24} f_{34}]}{\partial [p_{11} p_{21} p_{31} p_{12} p_{22} \cdots p_{33} p_{14} p_{24} p_{34}]} = \begin{pmatrix} \frac{\partial f_{11}}{\partial p_{11}} & \frac{\partial f_{11}}{\partial p_{21}} & \cdots & \frac{\partial f_{11}}{\partial p_{34}} \\ \cdots & \cdots & \cdots & \cdots \\ \frac{\partial f_{34}}{\partial p_{11}} & \frac{\partial f_{34}}{\partial p_{21}} & \cdots & \frac{\partial f_{34}}{\partial p_{34}} \end{pmatrix}_{12 \times 12} \end{aligned} \tag{27}$$

where we have employed the $vec(\cdot)$ operator (see section 1) to reshape the top 3×4 portion of its arguments as 12×1 vectors.

While reading the following section the reader should keep in mind that each derivative taken *with respect to a pose matrix* should be interpreted as we have just described, i.e., they become $n \times 12$ Jacobian matrices.

6. SOME USEFUL JACOBIANS

We provide in the following a set of closed-form expressions which may be useful while designing optimization algorithms that involve 3D poses. Some of the Jacobians below already were employed while discussing graph SLAM and Bundle Adjustment methods in chapter 10.

Notice that Jacobians for purely geometric operations are also provided here since they are intermediary results required while applying the chain rule within more complex (and more useful) functions. Those geometry functions differ from those already studied in Appendix A in the adoption of a direct matrix parameterization of poses, for reasons explained in section 5.

Jacobian of the SE(3) Pseudo-Exponential Map e^{ε}

This is the most basic Jacobian that will be found in all on-manifold optimization problems, since the term e^{ε} will always appear—see chapter 10 section 2. Notice that we focus on the *pseudo-exponential* version instead of the actual exponential map for its simplicity. Therefore, we must assume the following replacement (which is not explicitly stated in graph SLAM literature):

$$e^{\varepsilon} \Rightarrow \mathbf{pexp}(\varepsilon) \tag{28}$$

Furthermore, we will take derivatives at the Lie algebra coordinates $\varepsilon = \mathbf{0}$ since our derivation is aimed at being used within the context of chapter 10's equation (5). Proceeding so, and given the definition of the pseudo-exponential in equation (21), we obtain:

$$\left.\frac{d\,\mathbf{pexp}(\varepsilon)}{d\,\varepsilon}\right|_{\varepsilon=\mathbf{0}} = \begin{pmatrix} \mathbf{0}_{3\times3} & -[\mathbf{e}_1]_\times \\ \mathbf{0}_{3\times3} & -[\mathbf{e}_2]_\times \\ \mathbf{0}_{3\times3} & -[\mathbf{e}_3]_\times \\ \mathbf{I}_3 & \mathbf{0}_{3\times3} \end{pmatrix} \qquad (\text{A } 12\times6 \text{ Jacobian}) \tag{29}$$

with $\mathbf{e}_1 = [1\ 0\ 0]^T$, $\mathbf{e}_2 = [0\ 1\ 0]^T$ and $\mathbf{e}_3 = [0\ 0\ 1]^T$. Notice that the resulting Jacobian is for the ordering convention of $\mathfrak{se}(3)$ coordinates in equation (21), which are denoted there as $\mathbf{v}$ instead of ε.

Jacobian of $\mathbf{a} \oplus \mathbf{b}$

Let $\mathbf{f}_{\oplus} : \mathbf{SE}(3) \times \mathbf{SE}(3) \to \mathbf{SE}(3)$ denote the pose composition operation, such that $\mathbf{f}_{\oplus}(\mathbf{a}, \mathbf{b}) = \mathbf{a} \oplus \mathbf{b}$. Then we can take derivatives of $\mathbf{f}_{\oplus}(\mathbf{a}, \mathbf{b})$ with respect to the two poses involved. If the 4×4 transformation matrix associated to a pose $\mathbf{x}$ is denoted as:

$$\mathbf{X} = \begin{pmatrix} & \mathbf{R}_X & & \mathbf{t}_X \\ 0 & 0 & 0 & 1 \end{pmatrix} \tag{30}$$

then the matrix of the resulting pose becomes the product $\mathbf{A}\,\mathbf{B}$ which, if we expand element by element and rearrange the resulting terms, leads us to:

$$\frac{\partial \mathbf{f}_\oplus(\mathbf{a},\mathbf{b})}{\partial \mathbf{a}} = \frac{\partial \mathbf{AB}}{\partial \mathbf{A}} = \mathbf{B}^T \otimes \mathbf{I}_3 \qquad \text{(A } 12\times12 \text{ Jacobian)} \tag{31}$$

$$\frac{\partial \mathbf{f}_\oplus(\mathbf{a},\mathbf{b})}{\partial \mathbf{b}} = \frac{\partial \mathbf{AB}}{\partial \mathbf{B}} = \mathbf{I}_4 \otimes \mathbf{R}_A \qquad \text{(A } 12\times12 \text{ Jacobian)} \tag{32}$$

Jacobian of a ⊕ p

Let $\mathbf{g}_\oplus : \mathbf{SE}(3)\times\mathbb{R}^3 \rightarrow \mathbb{R}^3$ denote the pose-point composition operation such that $\mathbf{g}_\oplus(\mathbf{a},\mathbf{p}) = \mathbf{a}\oplus\mathbf{p}$. Then we can take derivatives of $\mathbf{g}_\oplus(\mathbf{a},\mathbf{p})$ with respect to either the pose matrix $\mathbf{A}$ or the point $\mathbf{p}$. Using the same notation that in equation (30), we obtain in this case:

$$\frac{\partial \mathbf{g}_\oplus(\mathbf{a},\mathbf{p})}{\partial \mathbf{p}} = \frac{\partial \mathbf{Ap}}{\partial \mathbf{p}} = \frac{\partial(\mathbf{R}_A\mathbf{p}+\mathbf{t}_A)}{\partial \mathbf{p}} = \mathbf{R}_A \qquad \text{(A } 3\times3 \text{ Jacobian)} \tag{33}$$

$$\frac{\partial \mathbf{g}_\oplus(\mathbf{a},\mathbf{p})}{\partial \mathbf{A}} = \frac{\partial \mathbf{Ap}}{\partial \mathbf{A}} = \left(\mathbf{p}^T \;\; 1\right)\otimes\mathbf{I}_3 \qquad \text{(A } 3\times12 \text{ Jacobian)} \tag{34}$$

Jacobian of $e^\varepsilon \oplus d$

Let $\mathbf{d}$ be a $\mathbf{SE}(3)$ pose with an associated 4×4 matrix:

$$\mathbf{D} = \begin{pmatrix} \mathbf{d}_{c1} & \mathbf{d}_{c2} & \mathbf{d}_{c3} & \mathbf{d}_t \\ 0 & 0 & 0 & 1 \end{pmatrix} \tag{35}$$

Following the convention of left-composition for the incremental pose $\mathbf{pexp}(\varepsilon)$ described in chapter 10 section 2 (see chapter 10's equation **[6]**), we are interested in the derivative of $\mathbf{pexp}(\varepsilon)\oplus\mathbf{D}$ with respect to the increment ε in the manifold linearization. Applying the chain rule:

$$\left.\frac{\partial\,\mathbf{pexp}(\varepsilon)\oplus\mathbf{d}}{\partial\varepsilon}\right|_{\varepsilon=0} = \left.\frac{\partial\,\mathbf{pexp}(\varepsilon)\,\mathbf{D}}{\partial\varepsilon}\right|_{\varepsilon=0}$$

$$= \left.\frac{\partial\mathbf{AD}}{\partial\mathbf{A}}\right|_{\mathbf{A}=\mathbf{I}_4=\mathbf{pexp}(\mathbf{0})} \left.\frac{d\,\mathbf{pexp}(\varepsilon)}{d\,\varepsilon}\right|_{\varepsilon=0}$$

(using equation **[31]**)

$$= \left[\mathbf{T}(\mathbf{D})^{\top}\otimes\mathbf{I}_3\right]\left.\frac{d\,\mathbf{pexp}(\varepsilon)}{d\,\varepsilon}\right|_{\varepsilon=0}$$

(replacing equation **[29]** and rearranging)

$$= \begin{pmatrix} \mathbf{0}_{3\times3} & -[\mathbf{d}_{\mathbf{c1}}]_{\times} \\ \mathbf{0}_{3\times3} & -[\mathbf{d}_{\mathbf{c2}}]_{\times} \\ \mathbf{0}_{3\times3} & -[\mathbf{d}_{\mathbf{c3}}]_{\times} \\ \mathbf{I}_3 & -[\mathbf{d}_{\mathbf{t}}]_{\times} \end{pmatrix} \qquad \text{(A } 12\times6 \text{ Jacobian)} \tag{36}$$

Jacobian of $\mathbf{d}\oplus\mathbf{e}^{\varepsilon}$

Let $\mathbf{d}$ be a $\mathbf{SE}(3)$ pose with an associated 4×4 matrix:

$$\mathbf{D} = \begin{pmatrix} \mathbf{d}_{\mathbf{c1}} & \mathbf{d}_{\mathbf{c2}} & \mathbf{d}_{\mathbf{c3}} & \mathbf{t}_{\mathrm{D}} \\ 0 & 0 & 0 & 1 \end{pmatrix} = \begin{pmatrix} & \mathbf{R}_{\mathrm{D}} & & \mathbf{t}_{\mathrm{D}} \\ 0 & 0 & 0 & 1 \end{pmatrix} \tag{37}$$

The derivative of $\mathbf{d}\oplus e^{\varepsilon}$ with respect to the increment ε can be obtained as follows:

$$\left.\frac{\partial\,\mathbf{d}\oplus\mathbf{pexp}(\varepsilon)}{\partial\varepsilon}\right|_{\varepsilon=0} = \left.\frac{\partial\,\mathbf{Dpexp}(\varepsilon)}{\partial\varepsilon}\right|_{\varepsilon=0}$$

$$= \left.\frac{\partial\mathbf{AB}}{\partial\mathbf{A}}\right|_{\mathbf{A}=\mathbf{D},\mathbf{B}=\mathbf{I}_4=\mathbf{pexp}(\mathbf{0})} \left.\frac{d\,\mathbf{pexp}(\varepsilon)}{d\,\varepsilon}\right|_{\varepsilon=0}$$

(using equation **[32]** and equation **[29]**)

$$= \left[\mathbf{I}_4 \otimes \mathbf{R}_D \right] \begin{pmatrix} \mathbf{0}_{3\times3} & -[\mathbf{e}_1]_\times \\ \mathbf{0}_{3\times3} & -[\mathbf{e}_2]_\times \\ \mathbf{0}_{3\times3} & -[\mathbf{e}_3]_\times \\ \mathbf{I}_3 & \mathbf{0}_{3\times3} \end{pmatrix}$$

(doing the math and rearranging)

$$= \begin{pmatrix} & \mathbf{0}_{3\times1} & -\mathbf{d}_{c3} & \mathbf{d}_{c2} \\ \mathbf{0}_{9\times3} & \mathbf{d}_{c3} & \mathbf{0}_{3\times1} & -\mathbf{d}_{c1} \\ & -\mathbf{d}_{c2} & \mathbf{d}_{c1} & \mathbf{0}_{3\times1} \\ \mathbf{R}_D & & \mathbf{0}_{3\times3} & \end{pmatrix} \quad \text{(a } 12\times6 \text{ Jacobian)} \tag{38}$$

Jacobian of $\mathbf{e}^{\varepsilon} \oplus \mathbf{d} \oplus \mathbf{p}$

Let $\mathbf{p}$ be a point in $\mathbb{R}^3$ and $\mathbf{d}$ a $\mathbf{SE}(3)$ pose with an associated matrix:

$$\mathbf{D} = \begin{pmatrix} d_{11} & d_{12} & d_{13} & d_{tx} \\ d_{21} & d_{22} & d_{23} & d_{ty} \\ d_{31} & d_{32} & d_{33} & d_{tz} \\ 0 & 0 & 0 & 1 \end{pmatrix} = \begin{pmatrix} \mathbf{d}_{c1} & \mathbf{d}_{c2} & \mathbf{d}_{c3} & \mathbf{d}_t \\ 0 & 0 & 0 & 1 \end{pmatrix} = \begin{pmatrix} & \mathbf{R}_D & & \mathbf{d}_t \\ 0 & 0 & 0 & 1 \end{pmatrix} \tag{39}$$

The derivative of $\mathbf{pexp}(\varepsilon) \oplus \mathbf{d} \oplus \mathbf{p}$ with respect to the increment ε is an operation needed, for example, in Bundle Adjustment (see chapter 10) while optimizing the camera poses—when using the common convention of $\mathbf{d}$ representing the *inverse* of a camera pose, such as $\mathbf{d} \oplus \mathbf{p}$ represents the relative location of a landmark $\mathbf{p}$ with respect to that camera. We can do:

$$\left. \frac{\partial\, \mathbf{pexp}(\varepsilon) \oplus \mathbf{d} \oplus \mathbf{p}}{\partial \varepsilon} \right|_{\varepsilon=0} = \left. \frac{\partial \mathbf{A} \oplus \mathbf{p}}{\partial \mathbf{A}} \right|_{\mathbf{A}=\mathbf{pexp}(\mathbf{0})\mathbf{D}=\mathbf{D}} \left. \frac{\partial\, \mathbf{exp}(\varepsilon)\mathbf{D}}{\partial \varepsilon} \right|_{\varepsilon=0}$$

(using equation **[34]** and equation **[29]**)

$$= \left(\begin{pmatrix} \mathbf{p}^T & 1 \end{pmatrix} \otimes \mathbf{I}_3 \right) \begin{pmatrix} \mathbf{0}_{3\times3} & -[\mathbf{d}_{c1}]_\times \\ \mathbf{0}_{3\times3} & -[\mathbf{d}_{c2}]_\times \\ \mathbf{0}_{3\times3} & -[\mathbf{d}_{c3}]_\times \\ \mathbf{I}_3 & -[\mathbf{d}_t]_\times \end{pmatrix}$$

(developing and rearranging)

$$= \begin{pmatrix} \mathbf{I}_3 & -[\mathbf{D} \oplus \mathbf{p}]_\times \end{pmatrix} \qquad \text{(a } 3\times 6 \text{ Jacobian)} \tag{40}$$

Jacobian of $\mathbf{a} \oplus \mathbf{e}^{\varepsilon} \oplus \mathbf{d} \oplus \mathbf{p}$

This expression appears in problems such as relative Bundle Adjustment (Sibley, Mei, Reid, & Newman, 2010). Let $\mathbf{p} \in \mathbb{R}^3$ be a 3D point (a landmark in relative coordinates) and $\mathbf{a}, \mathbf{d} \in \mathbf{SE}(3)$ be two poses, so that $\mathbf{R}_\mathbf{A}$ is the 3×3 rotation matrix associated to $\mathbf{a}$. We will denote the rows and columns of the matrix associated to $\mathbf{d}$ as:

$$\mathbf{D} = \begin{pmatrix} \mathbf{d}_{c1} & \mathbf{d}_{c2} & \mathbf{d}_{c3} & \mathbf{d}_{t} \\ 0 & 0 & 0 & 1 \end{pmatrix} = \begin{pmatrix} \mathbf{d}_{r1}^T & d_{tx} \\ \mathbf{d}_{r2}^T & d_{ty} \\ \mathbf{d}_{r3}^T & d_{tz} \\ 0\ 0\ 0 & 1 \end{pmatrix} \tag{41}$$

Then, the Jacobian of the chained poses-point composition with respect to the increment in the linearized manifold can be shown to be:

$$\left.\frac{\partial\, \mathbf{a} \oplus \mathbf{pexp}(\varepsilon) \oplus \mathbf{d} \oplus \mathbf{p}}{\partial \varepsilon}\right|_{\varepsilon=0} = \left.\frac{\partial \mathbf{A} \oplus \mathbf{pexp}(\varepsilon) \oplus \mathbf{D} \oplus \mathbf{p}}{\partial \varepsilon}\right|_{\varepsilon=0}$$

(using equation **[31]**, equation **[34]** and equation **[29]**, and rearranging)

$$= \mathbf{R}_\mathbf{A} \left(\mathbf{I}_3 \;\; \begin{matrix} 0 & \mathbf{p}\cdot\mathbf{d}_{r3} + d_{tz} & -(\mathbf{p}\cdot\mathbf{d}_{r2} + d_{ty}) \\ -(\mathbf{p}\cdot\mathbf{d}_{r3} + d_{tz}) & 0 & \mathbf{p}\cdot\mathbf{d}_{r1} + d_{tx} \\ \mathbf{p}\cdot\mathbf{d}_{r2} + d_{ty} & -(\mathbf{p}\cdot\mathbf{d}_{r1} + d_{tx}) & 0 \end{matrix} \right) \qquad \text{(A } 3\times 6 \text{ Jacobian)} \tag{42}$$

where $\mathbf{a} \cdot \mathbf{b}$ stands for the scalar product of two vectors.

Analyzing the expression above we can observe that an approximation can be used when both $\mathbf{a}$ and $\mathbf{d}$ represent small pose increments. In that case:

$$\left.\frac{\partial\, \mathbf{a} \oplus \mathbf{pexp}(\varepsilon) \oplus \mathbf{d} \oplus \mathbf{p}}{\partial \varepsilon}\right|_{\varepsilon=0} \approx \begin{pmatrix} \mathbf{I}_3 & -[\mathbf{p} + \mathbf{d}_t]_\times \end{pmatrix} \qquad \text{(A } 3\times 6 \text{ Jacobian)} \tag{43}$$

REFERENCES

Davies, P. I., & Higham, N. J. (2010). A Schur-Parlett algorithm for computing matrix functions. *SIAM Journal on Matrix Analysis and Applications*, *25*(2), 464–485. doi:10.1137/S0895479802410815

Gallier, J. H. (2001). *Geometric methods and applications: For computer science and engineering*. Berlin, Germany: Springer Verlag.

Sibley, G., Mei, C., Reid, I., & Newman, P. (2010). Vast scale outdoor navigation using adaptive relative bundle adjustment. *The International Journal of Robotics Research*, *29*(8), 958–980. doi:10.1177/0278364910369268

Appendix E: Basic Calculus and Algebra Concepts

One of the aims of this text is provide the reader with as much self-contained expositions as possible, even of the most involved concepts. Since absolute self-containment cannot be achieved in practice, this appendix provides a brief review of some theoretical concepts and tools of calculus and matrix algebra that have wide-spread applicability throughout this book. If the reader wishes to delve into these issues more thoroughly, we recommend consulting Meyer (2001) or Apostol (1967). An extensive repository of matrix formulas and identities, without demonstrations, can be found in Petersen and Pedersen (2008).

1. BASIC MATRIX ALGEBRA

A matrix is said to be an $n \times m$ matrix if it has n rows and m columns. Two matrices $\mathbf{A}$ and $\mathbf{B}$ can be added or subtracted only if they have exactly the same size:

$$\underbrace{\mathbf{R}}_{n\times m} = \underbrace{\mathbf{A}}_{n\times m} \pm \underbrace{\mathbf{B}}_{p\times q} \qquad \text{is defined iff} \quad n = p,\ m = q \tag{1}$$

Matrix addition and subtraction are commutative:

$$\mathbf{A} + \mathbf{B} = \mathbf{B} + \mathbf{A} \tag{2}$$

and associative:

$$\mathbf{A} + \mathbf{B} + \mathbf{C} = (\mathbf{A} + \mathbf{B}) + \mathbf{C} = \mathbf{A} + (\mathbf{B} + \mathbf{C}) \tag{3}$$

Two matrices $\mathbf{A}$ and $\mathbf{B}$ can be multiplied only if they are *conformant* matrices, which means that the number of columns in the former matches the number of rows in the latter:

$$\underbrace{\mathbf{R}}_{n\times m} = \underbrace{\mathbf{A}}_{n\times p}\,\underbrace{\mathbf{B}}_{p\times m} \tag{4}$$

Matrix multiplication is not commutative, thus left-multiplying and right-multiplying by a matrix $\mathbf{M}$, assuming that in both cases the matrices are conformant, give us different results:

$$\mathbf{M}\,\mathbf{A} \neq \mathbf{A}\mathbf{M} \quad \text{(in general)} \tag{5}$$

In turn, multiplication is distributive:

$$\begin{aligned} \mathbf{A}(\mathbf{B}+\mathbf{C}) &= \mathbf{AB}+\mathbf{AC} \\ (\mathbf{A}+\mathbf{B})\mathbf{C} &= \mathbf{AC}+\mathbf{BC} \end{aligned} \tag{6}$$

and associative:

$$\mathbf{ABC} = (\mathbf{AB})\mathbf{C} = \mathbf{A}(\mathbf{BC}) \tag{7}$$

The transpose of an $n \times m$ matrix $\mathbf{A}$ is an $m \times n$ matrix denoted as $\mathbf{A}^T$ whose columns are the rows of the original matrix $\mathbf{A}$. Transposing twice gives us the original matrix:

$$\left(\mathbf{A}^T\right)^T = \mathbf{A} \tag{8}$$

and for any symmetric matrix $\mathbf{S}$, we have:

$$\mathbf{S}^T = \mathbf{S} \tag{9}$$

Only for $n \times n$ square matrices $\mathbf{A}$ we can define its inverse matrix $\mathbf{A}^{-1}$ as that one fulfilling:

$$\mathbf{A}\mathbf{A}^{-1} = \mathbf{I}_n \tag{10}$$

with $\mathbf{I}_n$ the identity matrix of size $n \times n$, with all entries zeros but its diagonal which only has ones. Not all matrices have an inverse, thus those that have one are called invertible matrices or non-singular matrices.

A square $n \times n$ matrix $\mathbf{A}$ is said to be an orthogonal matrix (or sometimes orthonormal), if:

$$\mathbf{A}\mathbf{A}^T = \mathbf{I}_n \tag{11}$$

which implies, from equation (10), that the inverse of any orthogonal matrix is always its transpose:

$$\mathbf{A}^T = \mathbf{A}^{-1} \qquad \text{(for } \mathbf{A} \text{ orthogonal)} \tag{12}$$

which can be exploited while working with rotation matrices, always orthogonal.

In general, the order of matrix transposition and inversion can be always exchanged:

$$\left(\mathbf{A}^T\right)^{-1} = \left(\mathbf{A}^{-1}\right)^T \tag{13}$$

Transposing a sum of matrices becomes the sum of their transposed versions:

$$\begin{aligned}\left(\mathbf{A}_1 + \mathbf{A}_2\right)^T &= \mathbf{A}_1^T + \mathbf{A}_2^T \\ &\vdots \\ \left(\mathbf{A}_1 + \mathbf{A}_2 + \ldots + \mathbf{A}_n\right)^T &= \mathbf{A}_1^T + \mathbf{A}_2^T + \ldots + \mathbf{A}_n^T\end{aligned} \tag{14}$$

while transposing a product of matrices becomes the product, in reverse order, of the transposed versions:

$$\begin{aligned}\left(\mathbf{A}_1\mathbf{A}_2\right)^T &= \mathbf{A}_2^T\mathbf{A}_1^T \\ &\vdots \\ \left(\mathbf{A}_1\mathbf{A}_2\cdots\mathbf{A}_n\right)^T &= \mathbf{A}_n^T\cdots\mathbf{A}_2^T\mathbf{A}_1^T\end{aligned} \tag{15}$$

The inverse of a sum of matrices cannot be further simplified, in general. Do *not* assume that it equals the sum of the inverse of each matrix. In turn, inverting a product of conformant matrices can be converted into the product, in reverse order, of the inverted matrices:

$$\begin{aligned}\left(\mathbf{A}_1\mathbf{A}_2\right)^{-1} &= \mathbf{A}_2^{-1}\mathbf{A}_1^{-1} \\ &\vdots \\ \left(\mathbf{A}_1\mathbf{A}_2\cdots\mathbf{A}_n\right)^{-1} &= \mathbf{A}_n^{-1}\cdots\mathbf{A}_2^{-1}\mathbf{A}_1^{-1}\end{aligned} \tag{16}$$

Given a square, real and symmetric $n \times n$ matrix $\mathbf{A}$, it will always have n eigenvectors $\mathbf{v}_i$ and n real eigenvalues λ_i, which are defined as the solutions to the system of linear equations:

$$\mathbf{A}\mathbf{v}_i = \lambda_i\mathbf{v}_i \qquad \text{, for } i = 1, 2, \ldots, n \tag{17}$$

Since for each eigenvalue λ_i there exist infinite possible eigenvectors fulfilling equation (17), it is convention to impose the additional restriction that all eigenvectors must have a unit norm, i.e., $\left\|\mathbf{v}_i\right\| = 1$. If a certain eigenvalue appears more than once, we say it is a degenerate eigenvalue and it implies the lost of additional degrees of freedom while determining the corresponding eigenvectors.

Then, we define the eigen decomposition of the *real* symmetric matrix $\mathbf{A}$ as its factorization as the product of these three matrices:

$$\mathbf{A} = \mathbf{V}\mathbf{D}\mathbf{V}^{-1}$$

(and, since when $\mathbf{A}$ is real we have $\mathbf{V}^{-1} = \mathbf{V}^T$)

$$= \mathbf{V}\mathbf{D}\mathbf{V}^T \tag{18}$$

with:

$$\mathbf{V} = \begin{pmatrix} | & | & & | \\ \mathbf{v}_1 & \mathbf{v}_2 & \cdots & \mathbf{v}_n \\ | & | & & | \end{pmatrix} \qquad \mathbf{D} = \begin{pmatrix} \lambda_1 & 0 & \cdots & 0 \\ 0 & \lambda_2 & & 0 \\ \vdots & & \ddots & \vdots \\ 0 & 0 & \cdots & \lambda_n \end{pmatrix} \tag{19}$$

Matrix factorizations find numerous applications in simplifying the solution of numerical problems. For instance, the inverse of a real symmetric matrix can be computed as:

$$\begin{aligned} \mathbf{A}^{-1} &= \left(\mathbf{V}\mathbf{D}\mathbf{V}^T\right)^{-1} \\ &= \left(\mathbf{V}^T\right)^{-1}\mathbf{D}^{-1}\mathbf{V}^{-1} \\ &= \mathbf{V}\mathbf{D}^{-1}\mathbf{V}^{-1} \end{aligned} \tag{20}$$

which only involves the trivial inversion of a diagonal matrix (i.e. inverting one by one its diagonal entries). Another useful factorization is the Cholesky decomposition, which is addressed in section 3.

An $n \times n$ matrix $\mathbf{A}$ is said to be a positive-semidefinite matrix if it fulfills:

$$\mathbf{x}^T\mathbf{A}\mathbf{x} \geq 0 \quad , \forall \mathbf{x} \in \mathbb{R}^n \quad (\text{excepting } \mathbf{x} = 0) \tag{21}$$

or a positive-definite matrix if it fulfills instead the more restrictive condition:

$$\mathbf{x}^T\mathbf{A}\mathbf{x} > 0 \quad , \forall \mathbf{x} \in \mathbb{R}^n \quad (\text{excepting } \mathbf{x} = 0) \tag{22}$$

where the terms $\mathbf{x}^T\mathbf{A}\mathbf{x}$ are called quadratic forms.

Two completely equivalent definitions are saying that positive-definite matrices only have positive eigenvalues while positive-semidefinite matrices have nonnegative eigenvalues, a weaker condition since one or more null eigenvalues are permitted. Any positive-semidefinite matrix has a determinant of zero and is, therefore, noninvertible.

Covariance matrices, introduced in chapter 3, are especial matrices because they are always real symmetric matrices and usually positive-definite—although, occasionally, can be positive-semidefinite. If a covariance matrix $\mathbf{\Sigma}$ is positive-definite, its inverse will always exist and will be also positive-definite and symmetric:

$$\left(\mathbf{\Sigma}^T\right)^{-1} = \left(\mathbf{\Sigma}^{-1}\right)^T = \mathbf{\Sigma}^{-1} \quad (\text{with } \mathbf{\Sigma} \text{ definite-positive}) \tag{23}$$

Dense matrices are defined in opposition to *sparse matrices*. A sparse matrix is one whose ratio of nonzero entries is somewhat reduced, typically a 10% or less of the entire matrix. Efficient storage of sparse matrices can be achieved by only keeping the nonzero entries, i.e. we assume that all non-stored elements are zeros. A popular storage format for sparse matrices is the Column-Compressed Sparse (CCS) matrix form, available in C/C++ via the set of ubiquitous libraries *SuiteSparse*, by Timoty Davis (2006), and in MATLAB via the *sparse()* function—which internally relies on *SuiteSparse*. Sparse matrices require especial algorithms for replacing the most common operations such as addition or multiplication but, in turn, they can dramatically increase the efficiency of solving certain mathematical problems, as we explored in chapter 10.

2. THE MATRIX INVERSION LEMMA

A useful result which we need for the derivation of the Kalman filter in chapter 7 is the equality called the *matrix inversion lemma*:

$$\left(\mathbf{E}+\mathbf{FGH}\right)^{-1}=\mathbf{E}^{-1}-\mathbf{E}^{-1}\mathbf{F}\left(\mathbf{G}^{-1}+\mathbf{HE}^{-1}\mathbf{F}\right)\mathbf{HE}^{-1} \tag{24}$$

That this equality holds can be demonstrated by using only the basic concepts described above:(post-multiplying by $\left(\mathbf{E}+\mathbf{FGH}\right)$ both sides of the equality, which can be done since it is not zero—otherwise it would not appear inverted in the original expression)

$$\underbrace{\left(\mathbf{E}+\mathbf{FGH}\right)^{-1}\left(\mathbf{E}+\mathbf{FGH}\right)}_{\mathbf{I}}=\left(\mathbf{E}^{-1}-\mathbf{E}^{-1}\mathbf{F}\left(\mathbf{G}^{-1}+\mathbf{HE}^{-1}\mathbf{F}\right)^{-1}\mathbf{HE}^{-1}\right)\left(\mathbf{E}+\mathbf{FGH}\right)$$

$$\mathbf{I}=\left(\mathbf{E}^{-1}-\mathbf{E}^{-1}\mathbf{F}\left(\mathbf{G}^{-1}+\mathbf{HE}^{-1}\mathbf{F}\right)^{-1}\mathbf{HE}^{-1}\right)\underline{\left(\mathbf{E}+\mathbf{FGH}\right)}$$

(by associativity with the second factor of the right-hand side)

$$\mathbf{I}=\underbrace{\mathbf{E}^{-1}\mathbf{E}}_{\mathbf{I}}+\mathbf{E}^{-1}\mathbf{FGH}-\mathbf{E}^{-1}\mathbf{F}\left(\mathbf{G}^{-1}+\mathbf{HE}^{-1}\mathbf{F}\right)^{-1}\mathbf{HE}^{-1}\left(\mathbf{E}+\mathbf{FGH}\right)$$

$$\cancel{\mathbf{I}}=\cancel{\mathbf{I}}+\mathbf{E}^{-1}\mathbf{FGH}-\underline{\mathbf{E}^{-1}\mathbf{F}\left(\mathbf{G}^{-1}+\mathbf{HE}^{-1}\mathbf{F}\right)^{-1}\mathbf{HE}^{-1}\left(\mathbf{E}+\mathbf{FGH}\right)}$$

(canceling the two identity matrices and by associativity in the underlined term)

$$\mathbf{0}=\mathbf{E}^{-1}\mathbf{FGH}-\mathbf{E}^{-1}\mathbf{F}\left(\mathbf{G}^{-1}+\mathbf{HE}^{-1}\mathbf{F}\right)^{-1}\mathbf{H}\underbrace{\mathbf{E}^{-1}\mathbf{E}}_{\mathbf{I}}-\mathbf{E}^{-1}\mathbf{F}\left(\mathbf{G}^{-1}+\mathbf{HE}^{-1}\mathbf{F}\right)^{-1}\mathbf{HE}^{-1}\mathbf{FGH}$$

$$0 = \underline{\mathbf{E}^{-1}\mathbf{F}}\mathbf{GH} - \underline{\mathbf{E}^{-1}\mathbf{F}}\left(\mathbf{G}^{-1} + \mathbf{HE}^{-1}\mathbf{F}\right)^{-1}\mathbf{H} - \underline{\mathbf{E}^{-1}\mathbf{F}}\left(\mathbf{G}^{-1} + \mathbf{HE}^{-1}\mathbf{F}\right)^{-1}\mathbf{HE}^{-1}\mathbf{FGH}$$

(using associativity with the underlined factor)

$$0 = \mathbf{E}^{-1}\mathbf{F}\left(\mathbf{GH} - \left(\mathbf{G}^{-1} + \mathbf{HE}^{-1}\mathbf{F}\right)^{-1}\mathbf{H} - \left(\mathbf{G}^{-1} + \mathbf{HE}^{-1}\mathbf{F}\right)^{-1}\mathbf{HE}^{-1}\mathbf{FGH}\right)$$

(inserting the factor $\mathbf{G}^{-1}\mathbf{G}$ the result is not altered, since $\mathbf{G}^{-1}\mathbf{G} = \mathbf{I}$ and we know that $\mathbf{G}^{-1}$ exists for it appears in the initial expression)

$$0 = \mathbf{E}^{-1}\mathbf{F}\left(\mathbf{GH} - \left(\mathbf{G}^{-1} + \mathbf{HE}^{-1}\mathbf{F}\right)^{-1}\underline{\mathbf{G}^{-1}\mathbf{G}}\mathbf{H} - \left(\mathbf{G}^{-1} + \mathbf{HE}^{-1}\mathbf{F}\right)^{-1}\mathbf{HE}^{-1}\mathbf{FGH}\right)$$

(taking out the common factor $\mathbf{GH}$)

$$0 = \mathbf{E}^{-1}\mathbf{F}\left(\mathbf{GH} - \underline{\left(\left(\mathbf{G}^{-1} + \mathbf{HE}^{-1}\mathbf{F}\right)^{-1}\mathbf{G}^{-1} + \left(\mathbf{G}^{-1} + \mathbf{HE}^{-1}\mathbf{F}\right)^{-1}\mathbf{HE}^{-1}\mathbf{F}\right)}\mathbf{GH}\right)$$

(by associativity in the underlined term)

$$0 = \mathbf{E}^{-1}\mathbf{F}\left(\mathbf{GH} - \underbrace{\left(\mathbf{G}^{-1} + \mathbf{HE}^{-1}\mathbf{F}\right)^{-1}\left(\mathbf{G}^{-1} + \mathbf{HE}^{-1}\mathbf{F}\right)}_{\mathbf{I}}\mathbf{GH}\right)$$

$$0 = \mathbf{E}^{-1}\mathbf{F}\underbrace{\left(\mathbf{GH} - \mathbf{GH}\right)}_{0}$$

$$0 = 0 \tag{25}$$

3. CHOLESKY DECOMPOSITION

Some of the algorithms presented in this text require decomposing a positive-definite matrix $\mathbf{A}$ (e.g. a covariance or an information matrix) into the product of two triangular matrices such that:

$$\mathbf{A} = \mathbf{LL}^T \tag{26}$$

with $\mathbf{L}$ being a lower triangular matrix (i.e., all the entries above the main diagonal are zero) with strictly positive diagonal entries. This is the Cholesky decomposition of a matrix, sometimes simply called a $\mathbf{LL}^T$ decomposition. Let us denote the $k-th$ column of a $d \times d$ matrix $\mathbf{A}$ as $\left[\mathbf{A}\right]_{*,k}$, and the entry at the $j-th$ row and $k-th$ column as $\left[\mathbf{A}\right]_{j,k}$. Then, we can refer to the elements of this lower triangular matrix as follows:

$$\mathbf{L} = \begin{pmatrix} \left[\mathbf{L}\right]_{1,1} & 0 & \cdots & 0 \\ \left[\mathbf{L}\right]_{2,1} & \left[\mathbf{L}\right]_{2,2} & \cdots & 0 \\ \vdots & \vdots & \ddots & \vdots \\ \left[\mathbf{L}\right]_{d,1} & \left[\mathbf{L}\right]_{d,2} & \cdots & \left[\mathbf{L}\right]_{d,d} \end{pmatrix}, \qquad i = 1, 2, \ldots, d \tag{27}$$

There are several ways of computing the matrix $\mathbf{L}$. One that is specially concise and clear is the Cholesky–Banachiewicz algorithm, which finds the elements of $\mathbf{L}$ going from top to bottom and from left to right:

$$\begin{aligned} \left[\mathbf{L}\right]_{i,i} &= \sqrt{\left[\mathbf{A}\right]_{i,i} - \sum_{k=1}^{i-1} \left[\mathbf{L}\right]_{i,k}^{2}} \\ \left[\mathbf{L}\right]_{i,j} &= \frac{1}{\left[\mathbf{L}\right]_{j,j}} \left(\left[\mathbf{A}\right]_{i,j} - \sum_{k=1}^{j-1} \left[\mathbf{L}\right]_{i,k} \left[\mathbf{L}\right]_{j,k} \right) \quad , \text{ for } \; i > j \end{aligned} \tag{28}$$

This algorithm is appropriate when dealing with dense covariance matrices, as those found in chapters 7 and 9. In case of handling sparse real symmetric matrices, we should turn to specialized algorithms which efficiently exploit the sparse structure of such matrices. One such specialized Cholesky decomposition methods is CHOLMOD, available as a C library and as a function within MATLAB (Chen, Davis, Hager, & Rajamanickam, 2008). This algorithm finds its applicability in the advanced SLAM methods discussed in chapter 10.

Positive-semidefinite matrices (either dense or sparse) cannot be decomposed by means of the Cholesky factorization $\mathbf{LL}^T$, but require an $\mathbf{LDL}^T$ factorization instead, which includes an extra diagonal matrix $\mathbf{D}$ and, usually, a permutation matrix to reorder the terms such that only the latest elements in the diagonal are zeros. Refer to *Eigen* (Guennebaud & Jacob, 2010), a popular C++ library, for an efficient implementation of $\mathbf{LDL}^T$ algorithms.

Finally, let us derive an auxiliary result regarding the Cholesky factorization $\mathbf{A} = \mathbf{LL}^T$, which reveals useful in chapter 7. We start by evaluating the entry at the $j-th$ row and $k-th$ column of $\mathbf{A}$:

$$\mathbf{A} = \mathbf{LL}^T$$

(by the definition of matrix multiplication)

$$\left[\mathbf{A}\right]_{j,k} = \left[\mathbf{LL}^T\right]_{j,k} = \sum_{i=1}^{d} \left[\mathbf{L}\right]_{j,i} \left[\mathbf{L}^T\right]_{i,k} =$$

(since $[\mathbf{X}]_{j,i}[\mathbf{Y}]_{i,k} = \left[[\mathbf{X}]_{*,i}[\mathbf{Y}]_{i,*}\right]_{j,k}$)

$$= \sum_{i=1}^{d}\left[[\mathbf{L}]_{*,i}\left[\mathbf{L}^{T}\right]_{i,*}\right]_{j,k} =$$

(changing the scope of the transpose)

$$= \sum_{i=1}^{d}\left[[\mathbf{L}]_{*,i}[\mathbf{L}]_{*,i}^{T}\right]_{j,k} =$$

(since the sum does not alter the position of elements in the resulting matrix)

$$= \left[\sum_{i=1}^{d}[\mathbf{L}]_{*,i}[\mathbf{L}]_{*,i}^{T}\right]_{j,k} \quad (29)$$

And therefore, by generalization over all the entries of the $\mathbf{A}$ matrix:

$$\mathbf{A} = \sum_{i=1}^{d}[\mathbf{L}]_{*,i}[\mathbf{L}]_{*,i}^{T} \quad (30)$$

4. THE GAUSSIAN CANONICAL FORM

In chapter 3, we introduced the *standard form* of a Gaussian pdf—not to be confused with the standard normal distribution, which is a normal distribution with zero mean and unit variance. There exists, however, another way to parameterize a multivariate Gaussian pdf: the *canonical form*. Among other applications, this formulation becomes useful during our derivation of the Kalman filter in chapter 7. We have included it in the present appendix since its definition consists almost entirely on the application of elementary algebra transformations to the standard form.

We start repeating here for convenience the standard form of a multivariate Gaussian pdf:

$$p(\mathbf{x};\boldsymbol{\mu},\boldsymbol{\Sigma}) = \frac{1}{\sqrt{(2\pi)^{d}\det(\boldsymbol{\Sigma})}}\exp\left(-\frac{1}{2}(\mathbf{x}-\boldsymbol{\mu})^{T}\boldsymbol{\Sigma}^{-1}(\mathbf{x}-\boldsymbol{\mu})\right) \quad (31)$$

where both the column vector $\mathbf{x}$ (i.e. the point at which we evaluate the pdf) and the mean vector $\boldsymbol{\mu}$ have d elements, while $\boldsymbol{\Sigma}$ is a $d \times d$ covariance matrix.

The *canonical form* of a Gaussian pdf assumes instead the following alternative parameterization of the same pdf (Wu, 2005):

$$p\left(\mathbf{x};\boldsymbol{\mu},\boldsymbol{\Sigma}\right) \propto \exp\left(-\frac{1}{2}\mathbf{x}^T\boldsymbol{\Lambda}\mathbf{x} + \boldsymbol{\eta}^T\mathbf{x} + \delta\right) \tag{32}$$

Here, $\boldsymbol{\Lambda}$ is called the *information* matrix or *precision* matrix and $\boldsymbol{\eta}$ is the *information* vector. Both the standard and the canonical representations of a Gaussian density distribution function are equivalent, and thus they can be derived from each other.

Actually, such derivation can be carried out in a more general form than the standard form. Consider the following function, which we will call the *generalized standard form* of a Gaussian pdf, slightly more general than the exponential of a standard form Gaussian, where $\mathbf{E}$ is any $d \times d$ matrix (we have dropped the constant factor of the pdf):

$$\exp\left(\alpha\right) = \exp\left(-\frac{1}{2}\left(\mathbf{E}\mathbf{x} - \boldsymbol{\mu}\right)^T \boldsymbol{\Sigma}^{-1}\left(\mathbf{E}\mathbf{x} - \boldsymbol{\mu}\right)\right) \tag{33}$$

We can also derive a canonical form for this exponential by expanding this exponent, using just the basic algebra concepts mentioned above:

$$\begin{aligned}
\alpha &= -\frac{1}{2}(\mathbf{E}\mathbf{x} - \boldsymbol{\mu})^T\boldsymbol{\Sigma}^{-1}(\mathbf{E}\mathbf{x} - \boldsymbol{\mu}) \\
&= -\frac{1}{2}(\mathbf{x}^T\mathbf{E}^T - \boldsymbol{\mu}^T)\boldsymbol{\Sigma}^{-1}(\mathbf{E}\mathbf{x} - \boldsymbol{\mu}) \\
&= -\frac{1}{2}(\mathbf{x}^T\mathbf{E}^T\boldsymbol{\Sigma}^{-1} - \boldsymbol{\mu}^T\boldsymbol{\Sigma}^{-1})(\mathbf{E}\mathbf{x} - \boldsymbol{\mu}) \\
&= -\frac{1}{2}(\mathbf{x}^T\mathbf{E}^T\boldsymbol{\Sigma}^{-1}\mathbf{E}\mathbf{x} \underbrace{-\boldsymbol{\mu}^T\boldsymbol{\Sigma}^{-1}\mathbf{E}\mathbf{x} - \mathbf{x}^T\mathbf{E}^T\boldsymbol{\Sigma}^{-1}\boldsymbol{\mu}}_{\substack{\text{one is the transpose of the other} \\ \text{and both are scalars}}} + \boldsymbol{\mu}^T\boldsymbol{\Sigma}^{-1}\boldsymbol{\mu}) \\
&= -\frac{1}{2}(\mathbf{x}^T \underbrace{\mathbf{E}^T\boldsymbol{\Sigma}^{-1}\mathbf{E}}_{\Lambda}\mathbf{x} - 2\underbrace{\boldsymbol{\mu}^T\boldsymbol{\Sigma}^{-1}\mathbf{E}}_{\eta^T}\mathbf{x} + \underbrace{\boldsymbol{\mu}^T\boldsymbol{\Sigma}^{-1}\boldsymbol{\mu}}_{-2\delta}) \\
&= -\frac{1}{2}\mathbf{x}^T\boldsymbol{\Lambda}\mathbf{x} + \boldsymbol{\eta}^T\mathbf{x} + \delta
\end{aligned} \tag{34}$$

where we have established these identities:

$$\begin{aligned}
\boldsymbol{\Lambda} &= \mathbf{E}^T\boldsymbol{\Sigma}^{-1}\mathbf{E} \\
\boldsymbol{\eta} &= \mathbf{E}^T\boldsymbol{\Sigma}^{-1}\boldsymbol{\mu} \\
\delta &= -\frac{1}{2}\boldsymbol{\mu}^T\boldsymbol{\Sigma}^{-1}\boldsymbol{\mu}
\end{aligned} \tag{35}$$

Therefore, we have demonstrated a how to pass from generalized standard exponents to canonical form exponents. Now, by setting $\mathbf{E} = \mathbf{I}_d$, with $\mathbf{I}_d$ the $d \times d$ identity matrix, we obtain the formulas for passing from standard form exponents to canonical exponents:

$$\begin{aligned} \Lambda &= \Sigma^{-1} \\ \eta &= \Sigma^{-1}\mu \\ \delta &= -\frac{1}{2}\mu^T\Sigma^{-1}\mu \end{aligned} \tag{36}$$

Additionally, we can easily find from the equation above the reverse transformation, that is, passing from a canonical form exponent to a standard form exponent:

$$\begin{aligned} \Sigma &= \Lambda^{-1} \\ \mu &= \Sigma\eta = \Lambda^{-1}\eta \end{aligned} \tag{37}$$

where the inverse of Λ exists iff Σ is positive-definite and, therefore, non-singular. Anyway, if Σ was positive-semidefinite (i.e., it has at least one null eigenvalue), we would not have any valid finite representation of Λ.

Note that when doing this transformation for going back from a canonical to a standard form exponent one only needs the two mentioned equations for calculating Σ and μ, but a few words are in order about the parameter δ, which must also be determined. It is common to find a canonical form that presents this structure:

$$p(\mathbf{x};\mu,\Sigma) = \exp\left(-\frac{1}{2}\mathbf{x}^T\Lambda\mathbf{x} + \eta^T\mathbf{x} + \delta + \varepsilon\right) \tag{38}$$

with ε being an additional nonzero term independent of the variable $\mathbf{x}$. That exponent can be considered as actually comprising two exponents:

- One corresponding to an exact canonical form, $-\frac{1}{2}\mathbf{x}^T\Lambda\mathbf{x} + \eta^T\mathbf{x} + \delta$,
- and another one corresponding to the additional term ε.

Once the former is transformed into the exponent of a Gaussian in standard form by computing the Σ and μ parameters as indicated above, we must account for the extra term in the canonical exponent: it will represent a term that multiplies the standard Gaussian form outside the exponential. The complete transformation in this case will produce the scaled standard form:

$$p(\mathbf{x};\mu,\Sigma) = \exp(\varepsilon)\exp\left(-\frac{1}{2}(\mathbf{x}-\mu)^T\Sigma^{-1}(\mathbf{x}-\mu)\right) \tag{39}$$

Since both, this density, and the standard form in equation (31) must integrate up to one to be valid pdfs it becomes clear that this extra constant $\exp(\varepsilon)$ must coincide with the constant term in equation (31), thus it must hold that:

$$\exp(\varepsilon) = \frac{1}{\sqrt{(2\pi)^d \det(\mathbf{\Sigma})}} \tag{40}$$

5. JACOBIAN AND HESSIAN OF A FUNCTION

The Jacobian is a natural extension to the concept of derivative of a function for the case of multivariate functions. Let $\mathbf{f}(\mathbf{x}) : \mathbb{R}^n \to \mathbb{R}^m$ denote an arbitrary vector function, with $n \geq 1$ and $m \geq 1$. Since it generates vectors of m real numbers we can consider it instead comprising m individual scalar functions $f_i : \mathbb{R}^n \to \mathbb{R}$, such that:

$$\mathbf{f}(\mathbf{x}) = \begin{bmatrix} f_1(\mathbf{x}) \\ f_2(\mathbf{x}) \\ \vdots \\ f_m(\mathbf{x}) \end{bmatrix} \in \mathbb{R}^m \qquad \text{, with } \mathbf{x} = \begin{bmatrix} x_1 \\ x_2 \\ \vdots \\ x_n \end{bmatrix} \in \mathbb{R}^n \tag{41}$$

Then, we denote the *Jacobian matrix* (or simply, the *Jacobian*) of $\mathbf{f}(\mathbf{x})$ as either $\mathbf{J}_{\mathbf{f},\mathbf{x}}(\mathbf{x})$ or $\nabla_{\mathbf{x}}\mathbf{f}(\mathbf{x})$, and define it to be the following $m \times n$ matrix:

$$\mathbf{J}_{\mathbf{f},\mathbf{x}}(\mathbf{x}) = \nabla_{\mathbf{x}}\mathbf{f}(\mathbf{x}) \triangleq \begin{pmatrix} \frac{\partial f_1(\mathbf{x})}{\partial x_1} & \cdots & \frac{\partial f_1(\mathbf{x})}{\partial x_n} \\ \vdots & \ddots & \vdots \\ \frac{\partial f_m(\mathbf{x})}{\partial x_1} & \cdots & \frac{\partial f_m(\mathbf{x})}{\partial x_n} \end{pmatrix} \tag{42}$$

Sometimes we may be interested in the Jacobian of a function with respect to a given subset of its parameters $\mathbf{z} \subset \mathbf{x}$. In those cases, the Jacobian is defined exactly the same but replacing the parameter $\mathbf{x}$ above by $\mathbf{z}$ and considering the other parameters as constants with regard to derivatives. Naturally, the resulting Jacobian will have less columns that the full Jacobian with respect to all the parameters. As an example, consider the following division of a Jacobin into two parts when we split its parameters $\mathbf{x}$ into two disjoint sets of variables $\mathbf{y}, \mathbf{z} \subseteq \mathbf{x}$ (such that $\mathbf{x} = \mathbf{y} \cup \mathbf{z}$):

$$\nabla_{\mathbf{y},\mathbf{z}}\mathbf{f}(\mathbf{y},\mathbf{z}) = \begin{pmatrix} \nabla_{\mathbf{y}}\mathbf{f} & \nabla_{\mathbf{z}}\mathbf{f} \end{pmatrix}$$

with:

$$\nabla_{\mathbf{y}}\mathbf{f}(\mathbf{y},\mathbf{z}) \triangleq \begin{pmatrix} \frac{\partial f_1(\mathbf{y},\mathbf{z})}{\partial y_1} & \cdots & \frac{\partial f_1(\mathbf{y},\mathbf{z})}{\partial y_p} \\ \vdots & \ddots & \vdots \\ \frac{\partial f_m(\mathbf{y},\mathbf{z})}{\partial y_1} & \cdots & \frac{\partial f_m(\mathbf{y},\mathbf{z})}{\partial y_p} \end{pmatrix} \quad \nabla_{\mathbf{z}}\mathbf{f}(\mathbf{y},\mathbf{z}) \triangleq \begin{pmatrix} \frac{\partial f_1(\mathbf{y},\mathbf{z})}{\partial z_1} & \cdots & \frac{\partial f_1(\mathbf{y},\mathbf{z})}{\partial z_q} \\ \vdots & \ddots & \vdots \\ \frac{\partial f_m(\mathbf{y},\mathbf{z})}{\partial z_1} & \cdots & \frac{\partial f_m(\mathbf{y},\mathbf{z})}{\partial z_q} \end{pmatrix}$$

and:

$$\mathbf{y} = \begin{bmatrix} y_1 \\ y_2 \\ \vdots \\ y_p \end{bmatrix} \in \mathbb{R}^p \qquad \mathbf{z} = \begin{bmatrix} z_1 \\ z_2 \\ \vdots \\ z_q \end{bmatrix} \in \mathbb{R}^q \tag{43}$$

For scalar functions, $f(\mathbf{x}) : \mathbb{R}^n \to \mathbb{R}$, the Jacobian becomes a row vector, which is the transposed gradient of the scalar field defined by that function:

$$\underbrace{\mathbf{g}}_{\text{Gradient vector}} = \nabla_{\mathbf{x}} f(\mathbf{x})^T \tag{44}$$

Therefore, the Jacobian reflects all the first-order derivatives for a vector function. The next higher-order derivative equivalent for multivariate functions is the *Hessian matrix*. In this case we only address the case of scalar functions, which are the simplest to formulate and the most useful for mobile robotics. Thus, given a scalar function, $f(\mathbf{x}) : \mathbb{R}^n \to \mathbb{R}$, its Hessian will be always a square $n \times n$ matrix containing all the second-order derivatives with respect to the parameters $\mathbf{x}$:

$$\nabla^2_{\mathbf{x}} f(\mathbf{x}) \triangleq \begin{pmatrix} \frac{\partial^2 f(\mathbf{x})}{\partial x_1^2} & \cdots & \frac{\partial^2 f(\mathbf{x})}{\partial x_1 \partial x_n} \\ \vdots & \ddots & \vdots \\ \frac{\partial^2 f(\mathbf{x})}{\partial x_n \partial x_1} & \cdots & \frac{\partial^2 f(\mathbf{x})}{\partial x_n^2} \end{pmatrix} \tag{45}$$

In the context of localization and SLAM, the Hessian will mostly appear while working with least-squares optimization (refer to chapter 10). In that case, if the evaluation point $\mathbf{x}$ is close to a minimum of the function $f(\mathbf{x})$ it can be shown (Triggs, McLauchlan, Hartley, & Fitzgibbon, 2000) that the Hessian can be accurately approximated without evaluating second-order derivatives as:

$$\nabla^2_{\mathbf{x}} f(\mathbf{x}) \approx \left[\nabla_{\mathbf{x}} f(\mathbf{x})\right]^T \nabla_{\mathbf{x}} f(\mathbf{x}) \tag{46}$$

which only requires computing the simpler Jacobian $\nabla_{\mathbf{x}} f(\mathbf{x})$. This is called the Gauss-Newton approximation to the Hessian and is at the core of the most relevant optimization algorithms.

6. TAYLOR SERIES EXPANSIONS

The Taylor series expansion of a function $f(x)$ is a tool developed by the English mathematician Brooks Taylor in the 18th century and consists of another function $\hat{f}(x)$, which approximates the original one in the vicinity of a given point $x = a$ (called the *linearization* point when using a linearized, first-order Taylor series expansion).

The Taylor series is based on the infinite derivatives of the function $f(x)$ at the linearization point—obviously, assuming that the function is infinitely differentiable at that point. More concretely, the Taylor series expansion for a scalar function is defined as:

$$
\begin{aligned}
f(x) &= f(a + \Delta x) \\
&\approx \hat{f}(a + \Delta x) \\
&= \sum_{i=0}^{n} \frac{1}{i!} \left. \frac{d^i f(x)}{dx^i} \right|_{x=a} (\Delta x)^i \\
&= \underbrace{f(a)}_{i=0} + \underbrace{\frac{1}{1!} \left. \frac{df(x)}{dx} \right|_{x=a} \Delta x}_{i=1} + \underbrace{\frac{1}{2!} \left. \frac{d^2 f(x)}{dx^2} \right|_{x=a} \Delta x^2}_{i=2} + \ldots
\end{aligned}
\tag{47}
$$

If the original function can be expressed as a convergent sum of infinite power terms, that is, making $n \to \infty$ in the equation above, it is then called an *analytic* function. The maximum index n included in the series is called the order of the series expansion, and as one could expect, the larger the order, the better will be the approximation of the function at points far from $x = a$. Taylor expansions are prominently used throughout this book for linearization of non-linear functions, mostly corresponding to first order ($n = 1$) expansions, although second order ($n = 2$) approximations are also touched while discussing least-squares methods in chapter 10.

In the case of multivariate functions, the same principle applies by replacing derivatives with Jacobian matrices. For example, a second-order Taylor series expansion of both, a univariate $f(\mathbf{x})$ and a multivariate function $\mathbf{F}(\mathbf{x})$, read:

$$
f(x) = f(a + \Delta x) \approx f(a) + \left. \frac{df(x)}{dx} \right|_{x=a} \Delta x + \frac{1}{2} \left. \frac{d^2 f(x)}{dx^2} \right|_{x=a} \Delta x^2
$$

$$\begin{aligned} F(\mathbf{x}) = F(\mathbf{a}+\Delta\mathbf{x}) \approx F(\mathbf{a}) + \left.\frac{dF(\mathbf{x})}{d\mathbf{x}}\right|_{\mathbf{x}=\mathbf{a}} \Delta\mathbf{x} + \frac{1}{2}\Delta\mathbf{x}^T \left.\frac{d^2F(\mathbf{x})}{d\mathbf{x}\,d\mathbf{x}^T}\right|_{\mathbf{x}=\mathbf{a}} \Delta\mathbf{x} \\ = F(\mathbf{a}) + \nabla_{\mathbf{x}} F\Big|_{\mathbf{x}=\mathbf{a}} \Delta\mathbf{x} + \frac{1}{2}\Delta\mathbf{x}^T \nabla_{\mathbf{x}}^2 F\Big|_{\mathbf{x}=\mathbf{a}} \Delta\mathbf{x} \end{aligned} \tag{48}$$

In order to illustrate how a scalar function can be approximated by Taylor series of increasingly higher orders, please refer to the example in Figure 1.

Figure 1. Example of Taylor expansion of the function corresponding to the pdf of an exponentially distributed r.v. (i.e., $f(x) = \lambda e^{-\lambda x}$ with mean 1). The more terms considered in the Taylor series expansion of that function, the better the approximation in the neighborhood of the point x=1.

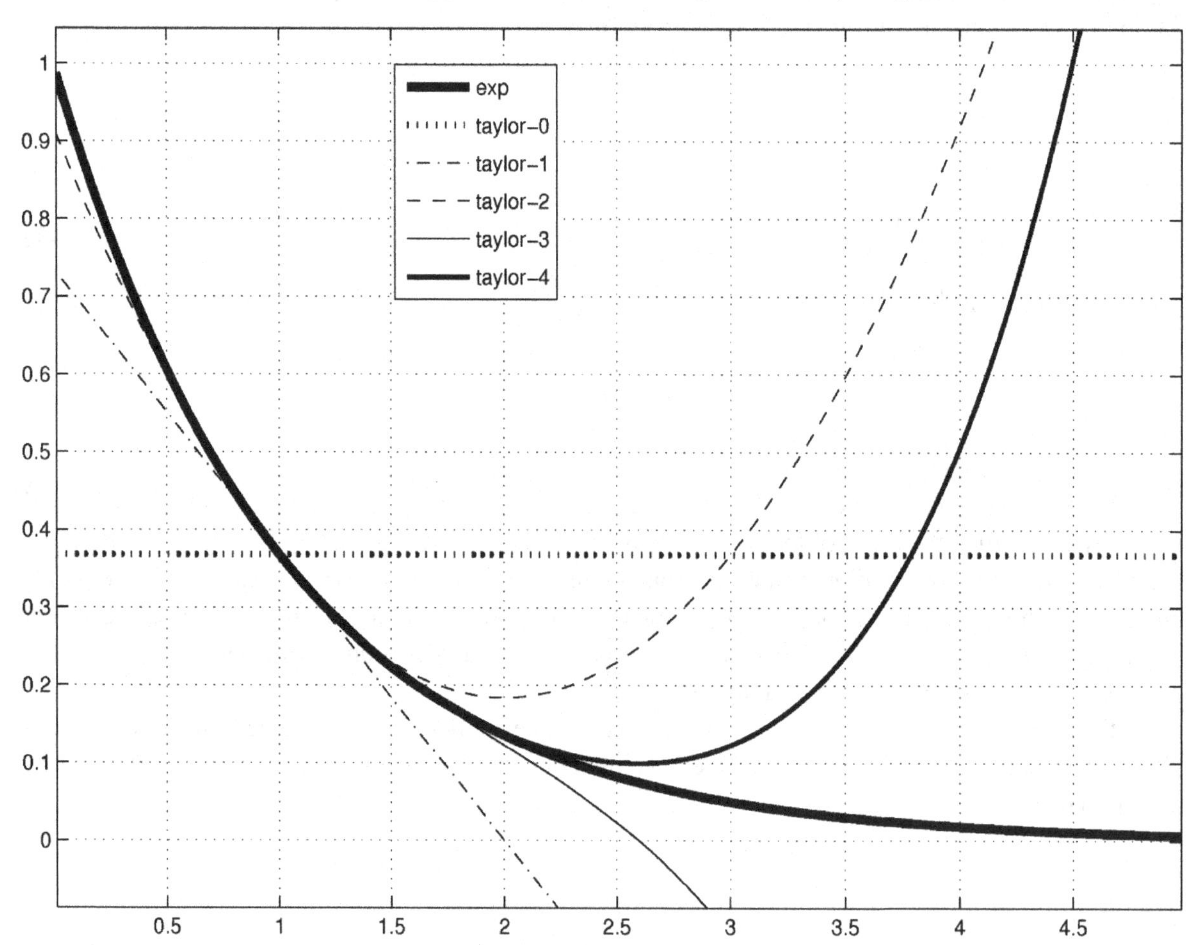

REFERENCES

Apostol, T. M. (1967). *Calculus* (*Vol. 1*). New York, NY: Wiley.

Chen, Y., Davis, T. A., Hager, W. W., & Rajamanickam, S. (2008). Algorithm 887: CHOLMOD, supernodal sparse Cholesky factorization and update/downdate. *ACM Transactions on Mathematical Software*, *35*(3).

Davis, T. A. (2006). *Direct methods for sparse linear systems*. New York, NY: Society for Industrial Mathematics. doi:10.1137/1.9780898718881

Guennebaud, G., & Jacob, B. (2010). *Eigen v3*. Retrieved Mar 1, 2012, from http://eigen.tuxfamily.org

Meyer, C. D. (2001). *Matrix analysis and applied linear algebra*. Retrieved Mar 1, 2012, from http://www.matrixanalysis.com/

Petersen, K. B., & Pedersen, M. S. (2008). *The matrix cookbook*. Retrieved Mar 1, 2012, from http://citeseer.ist.psu.edu/viewdoc/summary?doi=10.1.1.113.6244

Triggs, B., McLauchlan, P., Hartley, R., & Fitzgibbon, A. (2000). Bundle adjustment—A modern synthesis. *Vision Algorithms: Theory and Practice*, *1883*, 153–177.

Appendix F: Table Notation

Table 1. Summary of Notation

c, k, ...	Scalar values
$\mathbf{x}$, $\mathbf{v}$, ...	Vectors
$\mathbf{A}$, $\mathbf{B}$, ...	Matrices
$\mathbf{I}_n$	The $n \times n$ identity matrix
$\left[x_1\ x_2\ x_3\ \cdots\ x_n\right]^T = \left[x_1, x_2, x_3, \cdots, x_n\right]$	Equivalent definitions for vectors, as $n \times 1$ column matrices
$\lvert c \rvert$	Absolute value
$\lvert \mathbf{A} \rvert$, $\det(\mathbf{A})$	Determinant of a square matrix
$\lVert \mathbf{x} \rVert$	Norm of a vector
$f(x)$, $f(\mathbf{x})$	Scalar functions, whose range is one-dimensional
$\mathbf{f}(x)$, $\mathbf{f}(\mathbf{x})$	Vector functions, whose range is n-dimensional
$\left.\frac{d\mathbf{f}}{d\mathbf{x}}\right\rvert_{\mathbf{x}=\mathbf{a}} \equiv \nabla_{\mathbf{x}}\mathbf{f}(\mathbf{x})\rvert_{\mathbf{x}=\mathbf{a}} \equiv \mathbf{J}_{\mathbf{f}}(\mathbf{a})$	Jacobian matrix of a function $\mathbf{f}(\mathbf{x})$ evaluated at point $\mathbf{x} = \mathbf{a}$
$\left.\frac{\partial\mathbf{f}}{\partial\mathbf{x}}\right\rvert_{\substack{\mathbf{x}=\mathbf{a}\\ \mathbf{y}=\mathbf{b}}} \equiv \nabla_{\mathbf{x}}\mathbf{f}(\mathbf{x},\mathbf{y})\rvert_{\substack{\mathbf{x}=\mathbf{a}\\ \mathbf{y}=\mathbf{b}}}$	Jacobian matrix of $\mathbf{f}(\mathbf{x},\mathbf{y})$ evaluated at $\mathbf{x} = \mathbf{a}$, $\mathbf{y} = \mathbf{b}$
$\nabla_{\mathbf{x}}^2 f(\mathbf{x})\rvert_{\mathbf{x}=\mathbf{a}}$	Hessian matrix of $f(\mathbf{x})$ evaluated at point $\mathbf{x} = \mathbf{a}$
$p(x)$	Probability density function (pdf)
$P(x)$	Probability mass function (pmf)
$P[event]$	Probability operator
$E[x]$	Expectation operator
$V[x]$	Variance operator

$x \sim pdf$	x follows a distribution with the given pdf, or x is an observation drawn from a distribution with that pdf
(A,B,C)	Tuples
$\langle a,b,c,\ldots\rangle$	Sequences
$\mathbf{x}_{1:n} \triangleq \{x_1, x_2, \ldots, x_n\} = \{x_i\}_{i=1\ldots n}$	Sets of variables, interpreted as a stacked vector, sequences if the elements are sub-indexed by natural numbers
$a \gg b$ or $a \ll b$	a is much greater (or less) than b
$a >> n$ or $a << n$	Bitwise logical right (or left) shift of a by n bits
$\triangleq$, $\approx$, $\propto$	"is defined as", "is approximately equal to", "is proportional to"
$\leftrightarrow$, $\rightarrow$	"if and only if" ("equivalent to"), "then" ("implies")

Table 2. Summary of Common Abbreviations

iid	Independent and identically distributed
iff	If and only if
KF, EKF, …	Kalman filter, Extended Kalman filter, …
pdf	Probability density function
PF, RBPF	Particle filter, Rao-Blackwellized particle filter
pmf	Mass distribution function
r.v.(s)	Random variable(s)

Compilation of References

Access, A. I. (2011). *Glossary of data modelling*. Retrieved March 15, 2011, from http://www.aiaccess.net/e_gm.htm

Adams, J. L. (1961). *Remote control* with *long transmission delays.* (Doctoral Dissertation). Stanford University. Palo Alto, CA.

Aho, A. V., Ullman, J. D., & Hopcroft, J. E. (1983). *Data structures and algorithms*. Reading, MA: Addison-Wesley.

Ambrose, R. O., Aldridge, H., Askew, R. S., Burridge, R. R., Bluethmann, W., Diftler, M., ... Rehnmark, F. (2002). Robonaut: NASA's space humanoid. *IEEE Intelligent Systems and their Applications, 15*(4), 57-63.

Anderson, H. L. (1986). Metropolis, Monte Carlo and the MANIAC. *Los Alamos Science*, *14*, 96–108.

Anderson, M. L. (2003). Embodied cognition: A field guide. *Artificial Intelligence*, *149*, 91–130. doi:10.1016/S0004-3702(03)00054-7

Andreasson, H., Treptow, A., & Duckett, T. (2007). Self-localization in non-stationary environments using omni-directional vision. *Robotics and Autonomous Systems*, *55*(7), 541–551. doi:10.1016/j.robot.2007.02.002

Ankishan, H., & Efe, M. (2011). Adaptive neuro fuzzy supported Kalman filter approach for simultaneous localization and mapping. In *Proceedings of the IEEE 19th Conference on Signal Processing and Communications Applications (SIU 2011),* (pp. 266-270). IEEE Press.

Aoi, S., & Tsuchiya, K. (2011). Generation of bipedal walking through interactions among the robot dynamics, the oscillator dynamics, and the environment: Stability characteristics of a five-link planar biped robot. *Autonomous Robots*, *30*(2), 123–141. doi:10.1007/s10514-010-9209-9

Apostol, T. M. (1967). *Calculus* (2nd ed., *Vol. 1*). New York, NY: John Wiley & Sons.

Apostol, T. M. (1969). *Calculus* (2nd ed., *Vol. 2*). New York, NY: John Wiley & Sons.

Arras, K. O. (1998). *An introduction to error propagation: Derivation, meaning and examples of equation* $\mathbf{C}_Y = \mathbf{F}_X \mathbf{C}_X \mathbf{F}_X^T$. Technical report no. EPFL-ASL-TR-98-01 R3. Geneva, Switzerland: Swiss Federal Institute of Technology.

Arras, K. O., Castellanos, J. A., Schilt, M., & Siegwart, R. (2003). Feature-based multi-hypothesis localization and tracking using geometric constraints. *Robotics and Autonomous Systems*, *44*, 41–53. doi:10.1016/S0921-8890(03)00009-5

Arumlampalam, M. S., Maskell, S., Gordon, N., & Clapp, T. (2002). A tutorial on particle filters for on-line nonlinear/non-Gaussian Bayesian tracking. *IEEE Transactions on Signal Processing*, *50*(2), 174–188. doi:10.1109/78.978374

Ash, R. B. (1999). *Probability and measure theory*. New York, NY: Academic Press.

Asimov, I. (1941, May). Liar!. *Astounding Science Fiction*.

Bagrow, L., & Skelton, R. A. (2009). *History of cartography* (2nd ed.). Piscataway, NJ: Transactions Publishers.

Bailey, T., Nebot, E. M., Rosenblatt, J. K., & Durrant-Whyte, H. F. (2000). Data association for mobile robot navigation: A graph theoretic approach. *IEEE International Conference on Robotics and Automation, 3*, 2512-2517.

Baron, R. J. (1981). Mechanisms of human facial recognition. *International Journal of Man-Machine Studies*, *15*(2), 137–178. doi:10.1016/S0020-7373(81)80001-6

Bar-Shalom, Y., & Fortmann, T. E. (1988). *Tracking and data association*. San Diego, CA: Academic Press Professional, Inc.

Basseville, M., & Nikiforov, I. V. (1993). *Detection of abrupt changes: Theory and application.* Upper Saddle River, NJ: Prentice-Hall.

Bay, H., Tuytelaars, T., & Van Gool, L. (2006). Surf: Speeded up robust features. In *Proceedings of the European Conference on Computer Vision (ECCV)*, (pp. 404-417). ECCV.

Beevers, K. R., & Huang, W. H. (2007). Fixed-lag sampling strategies for particle filtering SLAM. In *Proceedings of the IEEE International Conference on Robotics and Automation*, (pp. 2433-2438). IEEE Press.

Bell, B. M., & Cathey, F. W. (1993). The iterated Kalman filter update as a Gauss-Newton method. *IEEE Transactions on Automatic Control*, *38*(2), 294–297. doi:10.1109/9.250476

Besl, P. J., & McKay, N. D. (1992). A method for registration of 3D shapes. *IEEE Transactions on Pattern Analysis and Machine Intelligence*, *14*(2), 239–256. doi:10.1109/34.121791

Bessière, P., Laugier, C., & Siegwart, R. (Eds.). (2008). *Probabilistic reasoning and decision making in sensory-motor systems*. Berlin, Germany: Springer. doi:10.1007/978-3-540-79007-5

Biber, P., & Duckett, T. (2005). Dynamic maps for long-term operation of mobile service robots. In *Proceedings of Robotics: Science and Systems (RSS)*, (pp. 17-24). RSS.

Bierman, G. J. (1977). *Factorization methods for discrete sequential estimation*. New York, NY: Academic Press.

Birk, A., & Carpin, S. (2006). Merging occupancy grid maps from multiple robots. *Proceedings of the IEEE*, *96*(7), 1384–1397. doi:10.1109/JPROC.2006.876965

Bishop, C. M. (2006). *Pattern recognition and machine learning*. New York, NY: Springer.

Blackwell, D. (1947). Conditional expectation and unbiased sequential estimation. *Annals of Mathematical Statistics*, *18*(1), 105–110. doi:10.1214/aoms/1177730497

Blanco, J. L. (2009). *Contributions to localization, mapping and navigation in mobile robotics.* (Doctoral Dissertation). University of Málaga. Malaga, Spain.

Blanco, J. L., Fernández-Madrigal, J. A., & González, J. (2008). Efficient probabilistic range-only slam. In *Proceedings of the IEEE/RSJ International Conference on Intelligent Robots and Systems*, (pp. 1017-1022). IEEE Press.

Blanco, J. L., González, J., & Fernández-Madrigal, J. A. (2007). A consensus-based approach for estimating the observation likelihood of accurate range sensors. In *Proceedings of the IEEE International Conference on Robotics and Automation*. IEEE Press.

Blanco, J. L., González, J., & Fernández-Madrigal, J. A. (2012). An alternative to the Mahalanobis distance for determining optimal correspondences in data association. *IEEE Transactions on Robotics*, *28*(4), 980-986.

Blanco, J. L., Fernández-Madrigal, J. A., & González, J. (2008). Towards a unified Bayesian approach to hybrid metric-topological SLAM. *IEEE Transactions on Robotics*, *24*(2), 259–270. doi:10.1109/TRO.2008.918049

Blanco, J. L., González, J., & Fernández-Madrigal, J. A. (2008). Extending obstacle avoidance methods through multiple parameter-space transformations. *Autonomous Robots*, *24*(1), 29–48. doi:10.1007/s10514-007-9062-7

Blanco, J. L., González, J., & Fernández-Madrigal, J. A. (2009). Subjective local maps for hybrid metric-topological SLAM. *Robotics and Autonomous Systems*, *57*(1), 64–74. doi:10.1016/j.robot.2008.02.002

Blanco, J. L., González, J., & Fernández-Madrigal, J. A. (2009). Subjective local maps for hybrid metric-topological SLAM. *Robotics and Autonomous Systems*, *57*(1), 64–74. doi:10.1016/j.robot.2008.02.002

Blanco, J. L., González, J., & Fernández-Madrigal, J. A. (2010). Optimal filtering for non-parametric observation models: Applications to localization and SLAM. *The International Journal of Robotics Research*, *29*(14), 1726–1742. doi:10.1177/0278364910364165

Borenstein, J. (1998). Experimental results from internal odometry error correction with the OmniMate mobile robot. *IEEE Transactions on Robotics and Automation*, *14*(6), 963–969. doi:10.1109/70.736779

Borenstein, J., Everett, B., & Feng, L. (1996). *Navigating mobile robots: Systems and techniques*. Wellesley, MA: A. K. Peters, Ltd.

Borenstein, J., Everett, H. R., & Feng, L. (1996). *Where am I? Sensors and methods for mobile robot positioning. Technical Report*. Ann Arbor, MI: University of Michigan.

Boston Dynamics. (2010). *Petman - BigDog gets a big brother*. Retrieved Mar 1, 2012, from http://www.bostondynamics.com/robot_petman.html

Bowditch, N. (2010). *The American practical navigator*. Arcata, CA: Paradise Cay Publications.

Brooks, R. A. (1982). Solving the find-path problem by good representation of free space. *IEEE Systems. Man and Cybernetics*, *13*, 190–197.

Brooks, R. A. (1991). Intelligence without representation. *Artificial Intelligence Journal*, *47*, 139–159. doi:10.1016/0004-3702(91)90053-M

Brown, D. C. (1958). *A solution to the general problem of multiple station analytical stereotriangulation*. Technical Report RCA-MTP Data Reduction Technical Report No. 43 (or AFMTC TR 58-8). Patrick Airforce Base, FL: Patrick Airforce Base.

Bucy, R. S. (1965). Nonlinear filtering theory. *IEEE Transactions on Automatic Control*, *10*(2), 198. doi:10.1109/TAC.1965.1098109

Caballero, F., Merino, L., Ferruz, J., & Ollero, A. (2009). *Vision-based odometry and SLAM for medium and high altitude flying UAVs*. Retrieved from http://grvc.us.es/aware/papers/aware_paper_11.pdf

Cadena, C., & Neira, J. (2010). SLAM in O (log n) with the combined Kalman-information filter. *Robotics and Autonomous Systems*, *58*(11), 1207–1219. doi:10.1016/j.robot.2010.08.003

Capek, K. (1921). *RUR (Rossum's universal robots)*. New York, NY: Penguin Books.

Casacuberta, D., Ayala, S., & Vallverdú, J. (2011). Embodying cognition: A morphological perspective. In *Machine Learning: Concepts, Methodologies, Tools and Applications*. Hershey, PA: IGI Global. doi:10.4018/978-1-60960-818-7.ch707

Castellanos, J. A., Martínez-Cantin, R., Tardós, J. D., & Neira, J. (2007). Robocentric map joining: Improving the consistency of EKF-SLAM. *Robotics and Autonomous Systems*, *55*(1), 21–29. doi:10.1016/j.robot.2006.06.005

Censi, A. (2008). An ICP variant using a point-to-line metric. In *Proceedings of the IEEE International Conference on Robotics and Automation (ICRA)*, (pp. 19-25). IEEE Press.

Ceruzzi, P. E. (2003). *A history of modern computing*. Cambridge, MA: MIT Press.

Chatterjee, A., & Matsuno, F. (2007). A neuro-fuzzy assisted extended Kalman filter-based approach for simultaneous localization and mapping (SLAM) problems. *IEEE Transactions on Fuzzy Systems*, *15*(5), 984–997. doi:10.1109/TFUZZ.2007.894972

Chen, C. K., & Yun, D. Y. Y. (1998). Unifying graph matching problems with a practical solution. In *Proceedings of the International Conference on Systems, Signals, Control, Computers*. IEEE.

Chen, R., & Liu, J. S. (2000). Mixture Kalman filters. *Journal of the Royal Statistical Society. Series B. Methodological*, *62*(3), 493–508. doi:10.1111/1467-9868.00246

Chen, Y., Davis, T. A., Hager, W. W., & Rajamanickam, S. (2008). Algorithm 887: CHOLMOD, supernodal sparse Cholesky factorization and update/downdate. *ACM Transactions on Mathematical Software*, *35*(3).

Choset, H., & Burdick, J. W. (1995). Sensor based planning, part 1: The generalized Voronoi graph. In *Proceedings of the IEEE International Conference on Robotics and Automation*. IEEE Press.

Ciocarlie, M., Hsiao, K., Jones, E. G., Chitta, S., Rusu, R. B., & Sucan, I. A. (2008). Towards reliable grasping and manipulation in household environments. *Robotics and Autonomous Systems*, *56*(1), 54–65.

Civera, J., Davison, A., & Montiel, J. (2008). Inverse depth parametrization for monocular SLAM. *IEEE Transactions on Robotics*, *24*(5), 932–945. doi:10.1109/TRO.2008.2003276

Coppersmith, D., & Winograd, S. (1990). Matrix multiplication via arithmetic progressions. *Journal of Symbolic Computation*, *9*(3), 251–280. doi:10.1016/S0747-7171(08)80013-2

Coradeschi, S., & Saffiotti, A. (2003). An introduction to the anchoring problem. *Robotics and Autonomous Systems*, *43*(2-3), 85–96. doi:10.1016/S0921-8890(03)00021-6

Cover, T. A., & Thomas, J. A. (2006). *Elements of information theory*. New York, NY: Wiley-Interscience.

Cox, I. J., Wilfong, G. T., & Lozano-Pérez, T. (1990). *Autonomous robot vehicles*. Berlin, Germany: Springer. doi:10.1007/978-1-4613-8997-2

Craig, J. J. (1989). *Introduction to robotics mechanics and control* (2nd ed.). Reading, MA: Addison-Wesley.

Cramér, H. (1946). *Mathematical methods of statistics*. Princeton, NJ: Princeton University Press.

Crowley, J. L. (1985). Navigation for an intelligent mobile robot. *IEEE Journal on Robotics and Automation, 1*(1). doi:10.1109/JRA.1985.1087002

Cruz-Martín, A., Fernández-Madrigal, J. A., Galindo, C., González-Jiménez, J., Stockmans-Daou, C., & Blanco, J. L. (2012). A lego mindstorms NXT approach for teaching at data acquisition, control systems engineering and real-time systems undergraduate courses. *Computers & Education*. doi:10.1016/j.compedu.2012.03.026

Cummins, M., & Newman, P. (2008). Accelerated appearance-only SLAM. In *Proceedings of the IEEE International Conference on Robotics and Automation*, (pp. 1828-1833). IEEE Press.

Cummins, M., & Newman, P. (2008). FAB-MAP: Probabilistic localization and mapping in the space of appearance. *The International Journal of Robotics Research, 27*(6), 647–665. doi:10.1177/0278364908090961

Davison, A. (2003). Real-time simultaneous localization and mapping with a single camera. In *Proceedings of the International Conference on Computer Vision*, (pp. 1403-1410). IEEE.

Davison, A. J. (1998). *Mobile robot navigation using active vision.* (Doctoral Dissertation). University of Oxford. Oxford, UK.

Davison, A. J., Reid, I. D., Molton, N. D., & Stasse, O. (2007). MonoSLAM: Real-time single camera SLAM. *IEEE Transactions on Pattern Analysis and Machine Intelligence, 29*(6), 1052–1067. doi:10.1109/TPAMI.2007.1049

Davis, T. A. (2006). *Direct methods for sparse linear systems*. New York, NY: Society for Industrial Mathematics. doi:10.1137/1.9780898718881

Dellaert, F., Carlson, J., Ila, V., Ni, K., & Thorpe, C. E. (2010). Subgraph-preconditioned conjugate gradients for large scale SLAM. In *Proceedings of the IEEE/RSJ International Conference on Intelligent Robots and Systems (IROS)*, (pp. 2566-2571). IEEE Press.

Dellaert, F., & Kaess, M. (2006). Square root SAM: Simultaneous localization and mapping via square root information smoothing. *The International Journal of Robotics Research, 25*(12), 1181–1203. doi:10.1177/0278364906072768

Demmel, J. W. (1997). *Applied numerical linear algebra*. Philadelphia, PA: Society for Industrial and Applied Mathematics. doi:10.1137/1.9781611971446

Dempster, A. P., Laird, N. M., & Rubin, D. B. (1977). Maximum likelihood from incomplete data via the EM algorithm. *Journal of the Royal Statistical Society. Series B. Methodological, 39*(1), 1–38.

Dennis, J. E., & Schnabel, R. B. (1996). *Numerical methods for unconstrained optimization and nonlinear equations*. New York, NY: Society for Industrial Mathematics. doi:10.1137/1.9781611971200

Desai, J. P., Ostrowski, J., & Kumar, V. (1998). Controlling formations of multiple mobile robots. In *Proceedings of the IEEE International Conference on Robotics and Automation*, (pp. 2864-2869). IEEE Press.

Devol, G. C. (1961). *Programmed article transfer*. US Patent no. 2988237. Washington, DC: US Patent Office.

Diamond, J. (2005). *Guns, germs, and steel: The fates of human societies*. New York, NY: W. W. Norton and Co.

Dijkstra, E. W. (1959). A note on two problems in connection with graphs. *Numerische Mathematik, 1*, 269–271. doi:10.1007/BF01386390

Dissanayake, M. W. M. G., Newman, P., Clark, S., Durrant-Whyte, H. F., & Csorba, M. (2001). A solution to the simultaneous localization and map building (SLAM) problem. *IEEE Transactions on Robotics and Automation, 17*(3), 229–241. doi:10.1109/70.938381

Djugash, J., Singh, S., & Grocholsky, B. (2008). Decentralized mapping of robot-aided sensor networks. In *Proceedings of the IEEE International Conference on Robotics and Automation*. IEEE Press.

Doh, N., Choset, H., & Chung, W. K. (2003). Accurate relative localization using odometry. In *Proceedings of the IEEE International Conference on Robotics and Automation*, (vol 2), (pp. 1606-1612). IEEE Press.

Douc, R., Capp, O., & Moulines, E. (2005). Comparison of resampling schemes for particle filtering. In *Proceedings of the 4th International Symposium on Image and Signal Processing and Analysis*, (pp. 64-69). IEEE.

Doucet, A., de Freitas, N., Murphy, K., & Russell, S. (2000). Rao-Blackwellised particle filtering for dynamic Bayesian networks. In *Proceedings of the Conference on Uncertainty in Artificial Intelligence (UAI)*. UAI.

Doucet, A., Briers, M., & Sénécal, S. (2006). Efficient block sampling strategies for sequential Monte Carlo methods. *Journal of Computational and Graphical Statistics*, *15*(3), 693–711. doi:10.1198/106186006X142744

Doucet, A., Godsill, S., & Andrieu, C. (2000). On sequential Monte Carlo sampling methods for Bayesian filtering. *Statistics and Computing*, *10*(3), 197–208. doi:10.1023/A:1008935410038

Doucet, A., & Johansen, A. M. (2008). *A tutorial on particle filters and smoothing: Fifteen years later*. British Columbia, Canada: University of British Columbia.

Doyle, J. C., Francis, B. A., & Tannenbaum, A. (1992). *Feedback control theory*. New York, NY: Dover Publications.

Dubois, D., & Prade, H. (2001). Possibility theory, probability theory and multiple-valued logics: A clarification. *Annals of Mathematics and Artificial Intelligence*, *32*, 35–66. doi:10.1023/A:1016740830286

Duckett, T., Marsland, S., & Shapiro, J. (2002). Fast, online learning of globally consistent maps. *Autonomous Robots*, *12*(3), 287–300. doi:10.1023/A:1015269615729

Duda, R. O., & Hart, P. E. (1972). Use of the Hough transformation to detect lines and curves in pictures. *Communications of the ACM*, *15*(1), 11–15. doi:10.1145/361237.361242

Durrant-Whyte, H. F. (1988). Uncertain geometry in robotics. *IEEE Journal on Robotics and Automation*, *4*(1), 23–31. doi:10.1109/56.768

Elfes, A. (1990). *Occupancy grids: A stochastic spatial representation for active robot perception*. Paper presented at the 6th Conference on Uncertainty and AI. Cambridge, MA.

Endo, T., Kawasaki, H., Mouri, T., Yoshida, T., Ishigure, Y., & Shimomura, H. … Koketsu, K. (2009). Five-fingered haptic interface robot: HIRO III. In *Proceedings of the Third Joint EuroHaptics Conference and Symposium on Haptic Interfaces for Virtual Environment and Teleoperator Systems (World Haptics Conference)*, (pp. 458-463). World Haptics Conference.

Engelhard, N., Endres, F., Hess, J., Sturm, J., & Burgard, W. (2011). Real-time 3D visual SLAM with a hand-held RGB-D camera. In *Proceedings of the RGB-D Workshop on 3D Perception in Robotics at the European Robotics Forum*. RGB-D.

Engelson, S. P., & McDermott, D. V. (1992). Error correction in mobile robot map learning. In *Proceedings of the International Conference on Robotics and Automation*. IEEE.

Espiau, B., Chaumette, F., & Rives, P. (1993). A new approach to visual servoing in robotics. *Geometric Reasoning for Perception and Action*, 106-136.

Everett, H. R. (1995). *Sensors for mobile robots*. Boca Raton, FL: CRC Press.

Fabrizi, E., & Saffiotti, A. (2000). Augmenting topology-based maps with geometric information. In *Proceedings of the 6th International Conference on Intelligent Autonomous Systems*. IEEE.

Fernández-Madrigal, J. A., & González, J. (2002). Multihierarchical representation of large-scale space: Applications to mobile robots. *Intelligent Systems, Control and Automation: Science and Engineering, 24*.

Fernández-Madrigal, J. A., & González, J. (2002b). Multihierarchical graph search. *IEEE Transactions on Pattern Analysis and Machine Intelligence*, *24*(1), 103–113. doi:10.1109/34.982887

Forsyth, D. A., Ioffe, S., & Haddon, J. (1999). Bayesian structure from motion. In *Proceedings of the IEEE International Conference of Computer Vision*, (pp. 660-665). IEEE Press.

Fox, D. (2003). *The Intel lab dataset*. Retrieved Mar 1, 2012, from http://cres.usc.edu/radishrepository/view-one.php?name=intel_lab

Fox, D., Burgard, W., Dellaert, F., & Thrun, S. (1999). Monte Carlo localization: Efficient position estimation for mobile robots. In *Proceedings of the National Conference on Artificial Intelligence (AAAI)*. AAAI.

Fox, D. (2003). Adapting the sample size in particle filters through KLD-sampling. *The International Journal of Robotics Research*, *22*(12), 985–1003. doi:10.1177/0278364903022012001

Fox, D., Burgard, W., & Thrun, S. (1999). Markov localization for mobile robots in dynamic environments. *Journal of Artificial Intelligence Research*, *11*, 391–427.

Franklin, J. (2001). *The science of conjecture: Evidence and probability before Pascal*. Baltimore, MD: The Johns Hopkins University Press. doi:10.1007/BF02985402

Frese, U., & Schröder, L. (2006). Closing a million-landmarks loop. In *Proceedings of the IEEE/RSJ International Conference on Intelligent Robots and Systems (IROS)*, (pp. 5032-5039). IEEE Press.

Frese, U. (2006). Treemap: An O (log n) algorithm for indoor simultaneous localization and mapping. *Autonomous Robots*, *21*(2), 103–122. doi:10.1007/s10514-006-9043-2

Frese, U., Larsson, P., & Duckett, T. (2005). A multilevel relaxation algorithm for simultaneous localisation and mapping. *IEEE Transactions on Robotics*, *21*(2), 1–12. doi:10.1109/TRO.2004.839220

Gabay, D. (1982). Minimizing a differentiable function over a differential manifold. *Journal of Optimization Theory and Applications*, *37*(2), 177–219. doi:10.1007/BF00934767

Gago, A., Fernández-Madrigal, J. A., Galindo, C., & Cruz-Martín, A. (2010). *Statistical characterization of the time-delay for Web-based networked telerobots*. Paper presented at the 5th International Workshop in Applied Probability (IWAP 2010). Madrid, Spain.

Galindo, C., Fernández-Madrigal, J. A., & González, J. (2007). Multiple abstraction hierarchies for mobile robot operation in large environments. *Studies in Computational Intelligence, 68*.

Galindo, C., Fernández-Madrigal, J. A., & González, J. (2004). Improving efficiency in mobile robot task planning through world abstraction. *IEEE Transactions on Robotics*, *20*(4), 677–690. doi:10.1109/TRO.2004.829480

Galindo, C., Fernández-Madrigal, J. A., González, J., & Saffiotti, A. (2008). Robot task planning using semantic maps. *Robotics and Autonomous Systems*, *56*(11), 955–966. doi:10.1016/j.robot.2008.08.007

Galindo, C., Fernández-Madrigal, J. A., González-Jiménez, J., & Saffiotti, A. (2008). Robot task planning using semantic maps. *Robotics and Autonomous Systems*, *56*(11), 955–966. doi:10.1016/j.robot.2008.08.007

Gallier, J. H. (2001). *Geometric methods and applications: For computer science and engineering*. Berlin, Germany: Springer Verlag.

Garey, M. R., & Johnson, D. S. (1979). *Computers and intractability: A guide to the theory of NP-completeness*. New York, NY: W. H. Freeman.

Gelb, A. (Ed.). (1974). *Applied optimal estimation*. Cambridge, MA: The MIT Press.

Georgakis, C. (1994, February). A note on the Gaussian integral. *Mathematics Magazine*, 47. doi:10.2307/2690556

Gezici, S., Tian, Z., Giannakis, G. B., Kobayashi, H., Molisch, A. F., Poor, H. V., & Sahinoglu, Z. (2005). Localization via ultra-wideband radios: A look at positioning aspects for future sensor networks. *IEEE Signal Processing Magazine*, *22*(4), 70–84. doi:10.1109/MSP.2005.1458289

Ghahramani, Z. (2001). An introduction to hidden Markov models and Bayesian networks. *International Journal of Pattern Recognition and Artificial Intelligence*, *15*(1), 9–42. doi:10.1142/S0218001401000836

Golub, G. H., & Plemmons, R. J. (1980). Large-scale geodetic least-squares adjustment by dissection and orthogonal decomposition. *Linear Algebra and Its Applications*, *34*, 3–28. doi:10.1016/0024-3795(80)90156-1

González, J., Blanco, J. L., Galindo, C., Ortiz-de-Galisteo, A., Fernández-Madrigal, J. A., Moreno, F. A., & Martinez, J. L. (2009). Mobile robot localization based on ultra-wide-band ranging: A particle filter approach. *Robotics and Autonomous Systems*, *57*(5), 496–507. doi:10.1016/j.robot.2008.10.022

González-Jiménez, J., Blanco, J. L., Galindo, C., Ortiz-de-Galisteo, A., Fernández-Madrigal, J. A., Moreno, F., & Martínez, J. L. (2009). Mobile robot localization based on ultra-wide-band ranging: A particle filter approach. *Robotics and Autonomous Systems*, *57*(5), 496–507. doi:10.1016/j.robot.2008.10.022

Gordon, N. J., Salmond, D. J., & Smith, A. F. M. (1993). Novel approach to nonlinear/non-Gaussian Bayesian state estimation. *IEE Proceedings. Part F. Communications, Radar and Signal Processing*, *140*(2), 107–113. doi:10.1049/ip-f-2.1993.0015

Gozick, B., Subbu, K. P., Dantu, R., & Maeshiro, T. (2011). Magnetic maps for indoor navigation. *IEEE Transactions on Instrumentation and Measurement*, *60*(12), 3883–3891. doi:10.1109/TIM.2011.2147690

Grimson, W. E. L. (1991). *Object recognition by computer: The role of geometric constraints*. Cambridge, MA: The MIT Press.

Grimson, W. E. L., & Lozano-Pérez, T. (1987). Localizing overlapping parts by searching the interpretation tree. *IEEE Transactions on Pattern Analysis and Machine Intelligence*, *9*(4), 469–482. doi:10.1109/TPAMI.1987.4767935

Grisetti, G., Grzonka, S., Stachniss, C., Pfaff, P., & Burgard, W. (2007). Efficient estimation of accurate maximum likelihood maps in 3D. In *Proceedings of the IEEE/RSJ International Conference on Intelligent Robots and Systems (IROS)*, (pp. 3472-3478). IEEE Press.

Grisetti, G., Küemmerle, R., Stachniss, C., Frese, U., & Hertzberg, C. (2010). Hierarchical optimization on manifolds for online 2D and 3D mapping. In *Proceedings of the IEEE International Conference on Robotics and Automation*. IEEE Press.

Grisetti, G., Stachniss, C., & Burgard, W. (2005). Improving grid-based slam with Rao-Blackwellized particle filters by adaptive proposals and selective resampling. In *Proceedings of the 2005 IEEE International Conference on Robotics and Automation*, (pp. 2432-2437). IEEE Press.

Grisetti, G., Stachniss, C., Grzonka, S., & Burgard, W. (2007). A tree parameterization for efficiently computing maximum likelihood maps using gradient descent. In *Proceedings of Robotics: Science and Systems (RSS)*. RSS.

Grisetti, G., Stachniss, C., & Burgard, W. (2007). Improved techniques for grid mapping with Rao-Blackwellized particle filters. *IEEE Transactions on Robotics*, *23*(1), 34–46. doi:10.1109/TRO.2006.889486

Grisetti, G., Stachniss, C., Grzonka, S., & Burgard, W. (2007). A tree parameterization for efficiently computing maximum likelihood maps using gradient descent. In *Proceedings of Robotics: Science and Systems (RSS)*. IEEE.

Grossberg, S. (1988). *The adaptive brain*. London, UK: Elsevier.

Guivant, J., & Nebot, E. (2001). Optimization of the simultaneous localization and map building algorithm for real-time implementation. *IEEE Transactions on Robotics and Automation*, *17*(3), 242–257. doi:10.1109/70.938382

Gustafsson, F. (2000). *Adaptive filtering and change detection*. New York, NY: John Wiley & Sons.

Harnad, S. (1990). The symbol grounding problem. *Physica D. Nonlinear Phenomena*, *42*, 335–346. doi:10.1016/0167-2789(90)90087-6

Hartley, R. I., & Zisserman, A. (2004). *Multiple view geometry in computer vision*. Cambridge, UK: Cambridge University Press. doi:10.1017/CBO9780511811685

Hartley, R., & Zisserman, A. (2003). *Multiple view geometry in computer vision*. Cambridge, UK: Cambridge University Press.

Havangi, R., Teshnehlab, M., Nekoui, M. A., & Taghirad, H. (2011). An adaptive neuro-fuzzy Rao-Blackwellized particle filter for SLAM. In *Proceedings of the IEEE International Conference on Mechatronics (ICM 2011)*, (pp. 487-492). IEEE Press.

Hazewinkel, M. (2002). *Encyclopaedia of mathematics*. Berlin, Germany: Springer-Verlag. Retrieved Mar 1, 2012, from http://www.encyclopediaofmath.org/

Henry, P., Krainin, M., Herbst, E., Ren, X., & Fox, D. (2010). RGB-D mapping: Using depth cameras for dense 3D modeling of indoor environments. In *Proceedings of the 12th International Symposium on Experimental Robotics (ISER)*. ISER.

Herman, W. A. (1923). *Founders of oceanography and their work: An introduction to the science of the sea.* London, UK: Edward Arnold & Co. Retrieved Mar 1, 2012, from http://www.archive.org/details/foundersofoceano1923herd

Hertzberg, C. (2008). *A framework for sparse, non-linear least squares problems on manifolds.* (Master's Thesis). Universität Bremen. Bremen, Germany.

Hirose, R., & Takenaka, T. (2001). Development of the humanoid robot ASIMO. *Honda R&D Technical Review, 13*(1), 1-6.

Hoffmann, G. M., Tomlin, C. J., Montemerlo, M., & Thrun, S. (2007). Autonomous automobile trajectory tracking for off-road driving: Controller design, experimental validation and racing. In *Proceedings of the American Control Conference (ACC)*, (pp. 2296-2301). ACC.

Hokuyo. (2011). *Hokuyo Automatic Co. Ltd. corporate website.* Retrieved Mar 1, 2012, from http://www.hokuyo-aut.jp/

Howard, A., Mataric, M., & Sukhatme, G. (2001). Relaxation on a mesh: A formalism for generalized localization. In *Proceedings of the IEEE/RSJ International Conference on Intelligent Robots and Systems (IROS)*, (pp. 1055-1060). IEEE Press.

Huang, Q., Yokoi, K., Kajita, S., Kaneko, K., Arai, H., Koyachi, N., & Tanie, K. (2001). Planning walking patterns for a biped robot. *IEEE Transactions on Robotics and Automation, 17*(3), 180–189.

Huber, P. J. (1964). Robust estimation of a location parameter. *Annals of Mathematical Statistics, 35*, 73–101. doi:10.1214/aoms/1177703732

Idel, M. (1990). *Golem: Jewish magical and mystical traditions on the artificial anthropoid.* Albany, NY: State University of New York Press.

Ishida, H., Suetsugu, K., Nakamoto, T., & Moriizumi, T. (1994). Study of autonomous mobile sensing system for localization of odor source using gas sensors and anemometric sensors. *Sensors and Actuators. A, Physical, 45*(2), 153–157. doi:10.1016/0924-4247(94)00829-9

JAVAD. (2011). *JAVAD GNSS Inc. corporate website.* Retrieved Mar 1, 2012, from http://www.javad.com/

Jazwinsky, A. H. (1970). *Stochastic processes and filtering theory.* New York, NY: Academic Press.

Jefferies, M., & Yeap, W. K. (Eds.). (2008). *Robotics and cognitive approaches to spatial mapping.* Berlin, Germany: Springer. doi:10.1007/978-3-540-75388-9

Jenkins, D. (2003). The western wool textile industry in the nineteenth century. In *The Cambridge History of Western Textiles.* Cambridge, UK: Cambridge University Press.

Johnson, A. E., & Hebert, M. (1999). Using spin images for efficient object recognition in cluttered 3D scenes. *IEEE Transactions on Pattern Analysis and Machine Intelligence, 21*(5), 433–449. doi:10.1109/34.765655

Julier, S. J., & Uhlmann, J. K. (1997). A new extension of the Kalman filter to nonlinear systems. In *Proceedings of the International Symposium Aerospace/Defense Sensing, Simulation and Controls,* (vol 3068), (pp. 182-193). Berlin, Germany: Springer.

Kallenberg, O. (1997). *Foundations of modern probability.* New York, NY: Springer-Verlag.

Kalman, R. E. (1960). A new approach to linear filtering and prediction problems. *Transactions of the ASME Journal of Basic Engineering, 82*(D), 35-45.

Kalman, R. E., & Bucy, R. S. (1961). New results in linear filtering and prediction theory. *Transactions of the ASME Journal of Basic Engineering, 83*(D), 95-108.

Kalman, R. E. (1960). A new approach to linear filtering and prediction problems. *Transactions of the ASME Journal of Basic Engineering, 82*, 35–45. doi:10.1115/1.3662552

Kanwal, R. P. (1998). *Generalized functions, theory and technique* (2nd ed.). Berlin, Germany: Birkhäuser.

Kassam, S. A. (1988). *Signal detection in non-Gaussian noise.* Berlin, Germany: Springer-Verlag.

Kay, S. M. (1993). Fundamentals of statistical signal processing: *Vol. 1. Estimation theory.* Upper Saddle River, NJ: Prentice-Hall.

Kindermann, R., & Snell, J. L. (1980). *Markov random fields and their applications.* Providence, RI: American Mathematical Society.

Klein, G., & Murray, D. (2007) Parallel tracking and mapping for small AR workspaces. In *Proceedings of the 2007 6th IEEE and ACM International Symposium on Mixed and Augmented Reality*. IEEE Press.

Klir, G. J. (2006). *Uncertainty and information: Foundations of generalized information theory*. New York, NY: John Wiley & Sons.

Kneip, L., Tache, F., Caprari, G., & Siegwart, R. (2009). Characterization of the compact Hokuyo URG-04LX 2D laser range scanner. In *Proceedings of the IEEE International Conference on Robotics and Automation*, (pp. 1447-1454). IEEE Press.

Koller, D., & Friedman, N. (2009). *Probabilistic graphical models: Principles and techniques*. Cambridge, MA: The MIT Press.

Konolige, K. (1999). Markov localization using correlation. In *Proceedings of the International Joint Conference on Artificial Intelligence*. IEEE.

Konolige, K. (2005). SLAM via variable reduction from constraint maps. In *Proceedings of the IEEE International Conference on Robotics and Automation*. IEEE Press.

Konolige, K., Fox, D., Limketkai, B., Ko, J., & Stewart, B. (2003). Map merging for distributed robot navigation. In *Proceedings of the IEEE/RSJ International Conference on Intelligent Robots and Systems*, (pp. 212-227). IEEE Press.

Konolige, K., Grisetti, G., Kümmerle, R., Burgard, W., Limketkai, B., & Vincent, R. (2010). Sparse pose adjustment for 2D mapping. In *Proceedings of the IEEE/RSJ International Conference on Intelligent Robots and Systems (IROS)*. IEEE Press.

Konolige, K., & Agrawal, M. (2008). FrameSLAM: From bundle adjustment to real-time visual mapping. *IEEE Transactions on Robotics*, *24*(5), 1066–1077. doi:10.1109/TRO.2008.2004832

Konolige, K., Bowman, J., Chen, J. D., Mihelich, P., Calonder, M., Lepetit, V., & Fua, P. (2010). View-based maps. *The International Journal of Robotics Research*, *29*(8), 941–957. doi:10.1177/0278364910370376

Kretzschmar, H., Stachniss, C., & Grisetti, G. (2011). Efficient information-theoretic graph pruning for graph-based SLAM with laser range finders. In *Proceedings of the IEEE/RSJ International Conference on Intelligent Robots and Systems (IROS)*. IEEE Press.

Krzysztof, K. (Ed.). (2006). Robot motion and control: Recent developments. *Lecture Notes in Control and Information Sciences, 335*.

Küemmerle, R., Grisetti, G., Strasdat, H., Konolige, K., & Burgard, W. (2011). g2o: A general framework for graph optimization. In *Proceedings of the IEEE International Conference on Robotics and Automation (ICRA)*. IEEE Press.

Kuipers, B. J. (1977). *Representing knowledge of large-scale space*. (Doctoral Dissertation). Massachusetts Institute of Technology. Cambridge, MA.

Kuipers, B. J. (1978). Modeling spatial knowledge. *Cognitive Science*, *2*, 129–153. doi:10.1207/s15516709cog0202_3

Kuipers, B. J. (1983). *The cognitive map: could it have been any other way? Spatial Orientation: Theory, Research, and Application* (pp. 345–359). New York, NY: Plenum Press.

Kuipers, B. J. (2000). The spatial semantic hierarchy. *Artificial Intelligence*, *119*, 191–233. doi:10.1016/S0004-3702(00)00017-5

Kuipers, B. J. (2008). *An intellectual history of the spatial semantic hierarchy*. Berlin, Germany: Springer. doi:10.1007/978-3-540-75388-9_15

Kurfess, T. R. (Ed.). (2005). *Robotics and automation handbook*. Boca Raton, FL: CRC Press.

Landes, J. B. (2007). The anatomy of artificial life: An eighteenth-century perspective. In Riskin, J. (Ed.), *Genesis Redux: Essays in the History and Philosophy of Artificial Life*. Chicago, IL: The University of Chicago Press.

Larsen, M. B. (2000). High performance doppler-inertial navigation-experimental results. In *Proceedings of the MTS/IEEE Conference and Exhibition OCEANS 2000*, (vol 2), (pp. 1449-1456). IEEE Press.

Latombe, J. C. (1991). *Robot motion planning*. Boston, MA: Kluwer Academic Publishers.

Lee, K. H., & Ehsani, R. (2008). Comparison of two 2D laser scanners for sensing object distances, shapes, and surface patterns. *Computers and Electronics in Agriculture*, *60*(2), 250–262. doi:10.1016/j.compag.2007.08.007

Lee, M. H., Meng, Q., & Chao, F. (2007). Developmental learning for autonomous robots. *Robotics and Autonomous Systems*, *55*, 750–759. doi:10.1016/j.robot.2007.05.002

Lego. (2011). *Lego mindstorms robots homepage*. Retrieved July 12, 2011, from http://mindstorms.lego.com/en-us/Default.aspx

Lehman, M. M. (1990). Uncertainty in computer application. *Communications of the ACM*, *33*(5).

Lehmann, E. L., & Casella, G. (1998). *Theory of point estimation* (2nd ed.). Berlin, Germany: Springer.

Leonard, J. J., & Durrant-Whyte, H. F. (1991). Mobile robot localization by tracking geometric beacon. *IEEE Transactions on Robotics and Automation*, *7*(3), 376–382. doi:10.1109/70.88147

Lilienthal, A., & Duckett, T. (2004). Building gas concentration gridmaps with a mobile robot. *Robotics and Autonomous Systems*, *48*(1), 3–16. doi:10.1016/j.robot.2004.05.002

Li, T. S. (2003). *On exponentially weighted recursive least squares for estimating time-varying parameters*. Hawthorne, NY: IBM. doi:10.1080/15598608.2008.10411879

Liu, C. (2011). *Foundations of MEMS* (2nd ed.). Upper Saddle River, NJ: Prentice Hall.

Lowe, D. G. (2004). Distinctive image features from scale-invariant keypoints. *International Journal of Computer Vision*, *60*(2), 91–110. doi:10.1023/B:VISI.0000029664.99615.94

Lozano-Pérez, T., & Wesley, M. A. (1979). An algorithm for planning collision-free paths among obstacles. *Communications of the ACM*, *22*, 560–570. doi:10.1145/359156.359164

Lu, F., & Milios, E. (1997). Globally consistent range scan alignment for environment mapping. *Autonomous Robots*, *4*(4), 333–349. doi:10.1023/A:1008854305733

Lu, F., & Milios, E. (1997). Robot pose estimation in unknown environments by matching 2d range scans. *Journal of Intelligent & Robotic Systems*, *18*, 249–275. doi:10.1023/A:1007957421070

Lungarella, M., Metta, G., Pfeifer, R., & Sandini, G. (2003). Developmental robotics: A survey. *Connection Science*, *15*(4), 151–190. doi:10.1080/09540090310001655110

Madsen, K., Nielsen, H., & Tingleff, O. (2004). *Methods for non-linear least squares problems*. Copenhagen, Denmark: Technical University of Denmark.

Mamen, R. (2003). Applying space technologies for human benefit: The Canadian experience and global trends. In *Proceedings of the International Conference on Recent Advances in Space Technologies (RAST 2003)*. RAST.

Mandow, A., Gómez-de-Gabriel, J. M., Martínez, J. L., Muñoz, V. F., Ollero, A., & García-Cerezo, A. (1996). The autonomous mobile robot AURORA for greenhouse operation. *IEEE Robotics & Automation Magazine*, *3*(4), 18–28. doi:10.1109/100.556479

Manolakis, D. G., Ingle, V. K., & Kogon, S. M. (2005). *Statistical and adaptive signal processing*. London, UK: Artech House.

Manoonpong, P., Geng, T., Kulvicius, T., Porr, B., & Wörgötter, F. (2007). Adaptive, fast walking in a biped robot under neuronal control and learning. *PLoS Computational Biology*, *3*(7), 134. doi:10.1371/journal.pcbi.0030134

Martin, M. C., & Moravec, H. P. (1996). *Robot evidence grids*. CMU Technical Report CMU-RI-TR-96-06. Pittsburgh, PA: Carnegie Mellon University.

Martínez, M. A., González, J., & Martínez, J. L. (1997). The DSP multi-frequency sonar configuration of the RAM-2 mobile robot. In *Proceedings of the Second EUROMICRO Workshop on Advanced Mobile Robots*. EUROMICRO.

Martín-Fernández, M. (2004). Fundamentals on estimation theory. Boston, MA: Harvard Medical School. Retrieved March 10, 2011, from http://lmi.bwh.harvard.edu/papers/pdfs/2004/martin-fernandezCOURSE04b.pdf

Matheron, G. (1963). Principles of geostatistics. *Economic Geology and the Bulletin of the Society of Economic Geologists*, *58*(8), 1246. doi:10.2113/gsecongeo.58.8.1246

McDowall, J. (2000). Conventional battery technologies-present and future. In *Proceedings of the IEEE 2000 Power Engineering Society Summer Meeting,* (vol 3), (pp. 1538-1540). IEEE Press. Mesa. (2011). *MESA imaging corporate website*. Retrieved Mar 1, 2012, from http://www.mesa-imaging.ch/

Meester, R. (2008). *A natural introduction to probability theory* (2nd ed.). Berlin, Germany: Birkhäuser Verlag.

Mei, C., Sibley, G., Cummins, M., Newman, P., & Reid, I. (2010). RSLAM: A system for large-scale mapping in constant-time using stereo. *International Journal of Computer Vision, 94*(2), 198–214. doi:10.1007/s11263-010-0361-7

Merrian Webster Inc. (2011). *Merrian-Webster on-line dictionary and thesaurus*. Retrieved Mar 1, 2012, from http://www.merriam-webster.com/

Meyer, C. D. (2001). *Matrix analysis and applied linear algebra*. Retrieved Mar 1, 2012, from http://www.matrixanalysis.com/

Michael, N., Mellinger, D., Lindsey, Q., & Kumar, V. (2010). The GRASP multiple micro-UAV testbed. *IEEE Robotics & Automation Magazine, 17*(3), 56–65. doi:10.1109/MRA.2010.937855

Microsoft. (2011). *Xbox kinect web site*. Retrieved July 19, 2011, from http://www.xbox.com/en-US/kinect

Mikhail, E., Bethel, J., & McGlone, J. C. (2001). *Introduction to modern photogrammetry*. New York, NY: Wiley.

Mikolajczyk, K., & Schmid, C. A. (2005). Performance evaluation of local descriptors. *IEEE Transactions on Pattern Analysis and Machine Intelligence, 27*(10), 1615–1630. doi:10.1109/TPAMI.2005.188

Minguez, J., & Montano, L. (2008). Extending collision avoidance methods to consider the vehicle shape, kinematics, and dynamics of a mobile robot. *IEEE Transactions on Robotics, 25*(2), 367–381. doi:10.1109/TRO.2009.2011526

Minguez, J., Montesano, L., & Lamiraux, F. (2006). Metric-based iterative closest point scan matching for sensor displacement estimation. *IEEE Transactions on Robotics, 22*(5), 1047–1054. doi:10.1109/TRO.2006.878961

Mobile Robotics. (2011). *Mobile Robotics Inc. corporate website*. Retrieved Mar 1, 2012, from http://www.mobilerobots.com/

Montemerlo, M. (2003). *FastSLAM: A factored solution to the simultaneous localization and mapping problem with unknown data association.* (Doctoral Dissertation). Carnegie Mellon University. Pittsburgh, PA.

Montemerlo, M., Thrun, S., Koller, D., & Wegbreit, B. (2002). FastSLAM: A factored solution to the simultaneous localization and mapping problem. In *Proceedings of the AAAI National Conference on Artificial Intelligence*. AAAI.

Montemerlo, M., Thrun, S., Koller, D., & Wegbreit, B. (2003). FastSLAM 2.0: An improved particle filtering algorithm for simultaneous localization and mapping that provably converges. In *Proceedings of the International Joint Conference on Artificial Intelligence,* (vol 18), (pp. 1151-1156). IEEE.

Montemerlo, M., & Thrun, S. (2007). *FastSLAM: A scalable method for the simultaneous localization and mapping problem in robotics*. Berlin, Germany: Springer.

Montgomery, D. C. (2005). *Introduction to statistical quality control*. New York, NY: John Wiley & Sons.

Moravec, H. (1980). *Obstacle avoidance and navigation in the real world by a seeing robot rover.* (Doctoral Dissertation). Stanford University. Palo Alto, CA. Retrieved Mar 1, 2012, from http://www.frc.ri.cmu.edu/~hpm/hpm.pubs.html

Moravec, H. P., & Elfes, A. (1985). High resolution maps from wide angle sonar. In *Proceedings of the IEEE International Conference on Robotics and Automation.* IEEE Press.

Morgan, D. (1992). *Numerical methods: Real-time and embedded systems programming*. Evansville, IN: M&T Publishing.

Morris, W., Dryanovski, I., & Xiao, J. (2010). 3D indoor mapping for micro-UAVs using hybrid range finders and multi-volume occupancy grids. In *Proceedings of the RSS 2010 Workshop on RGB-D: Advanced Reasoning with Depth Cameras*. RSS.

Morris, A. S. (2005). *Measurement and instrumentation principles*. London, UK: Elsevier.

MRPT. (2011). *The mobile robot programming toolkit website*. Retrieved Mar 1, 2012, from http://www.mrpt.org/

Neira, J., & Tardós, J. D. (2001). Data association in stochastic mapping using the joint compatibility test. *IEEE Transactions on Robotics and Automation, 17*(6), 890–897. doi:10.1109/70.976019

Newman, P., & Leonard, J. (2003). Pure range-only sub-sea SLAM. In *Proceedings of the IEEE International Conference on Robotics and Automation*, (vol 2), (pp. 1921-1926). IEEE Press.

Newman, P., Leonard, J., & Rikoski, R. J. (2005). Towards constant-time SLAM on an autonomous underwater vehicle using synthetic aperture sonar. In *Robotics Research* (pp. 409–420). Berlin, Germany: Springer.

Niku, S. B. (2010). *Introduction to robotics: Analysis, control, applications*. New York, NY: John Wiley & Sons.

Nilsson, N. (1988). *Shakey the robot. Tech. Note 323*. Palo Alto, CA: Artificial Intelligence Center, SRI International.

Nilsson, N. (1998). *Artificial intelligence: A new synthesis*. San Francisco, CA: Morgan Kaufmann Publishers Inc.

Nistér, D., Naroditsky, O., & Bergen, J. (2006). Visual odometry for ground vehicle applications. *Journal of Field Robotics, 23*(1), 3–20. doi:10.1002/rob.20103

Nixon, M., & Aguado, A. S. (2008). *Feature extraction and image processing*. New York, NY: Academic Press.

Nüchter, A. (2009). *3D robotic mapping: The simultaneous localization and mapping problem with six degrees of freedom*. Berlin, Germany: Springer Verlag.

Núñez, P., Vázquez-Martín, R., del Toro, J. C., Bandera, A., & Sandoval, F. (2006). Feature extraction from laser scan data based on curvature estimation for mobile robotics. In *Proceedings of the IEEE International Conference on Robotics and Automation*, (pp. 1167-1172). IEEE Press.

Ollero, A., & Merino, L. (2004). Control and perception techniques for aerial robotics. *Annual Reviews in Control, 28*(2), 167–178. doi:10.1016/j.arcontrol.2004.05.003

Olson, E., Leonard, J. J., & Teller, S. (2006). Fast iterative optimization of pose graphs with poor initial estimates. In *Proceedings of the IEEE International Conference on Robotics & Automation*, (pp. 2262-2269). IEEE Press.

Pal Robotics. (2011). *Corporate website*. Retrieved Mar 1, 2012, from http://www.pal-robotics.com/

Parallax. (2011). *Corporate website*. Retrieved Mar 1, 2012, from http://www.parallax.com/

Paz, L. M., Tardós, J. D., & Neira, J. (2008). Divide and conquer: EKF SLAM in O(n). *IEEE Transactions on Robotics, 24*(5), 1107–1120. doi:10.1109/TRO.2008.2004639

Piaget, J. (1948). *The child's conception of space*. London, UK: Routledge.

Plagemann, C., Kersting, K., Pfaff, P., & Burgard, W. (2007). Gaussian beam processes: A nonparametric Bayesian measurement model for range finders. In *Proceedings of Robotics: Science and Systems Conference*. IEEE. Press.

Poor, H. V., & Hadjiliadis, O. (2009). *Quickest detection*. Cambridge, UK: Cambridge University Press.

Prešeren, P. (Ed.). (2004). *World's oldest wheel*. Slovenia News.

Qiu, D., May, S., & Nüchter, A. (2009). GPU-accelerated nearest neighbor search for 3D registration. In *Proceedings of the 7th International Conference on Computer Vision Systems: Computer Vision Systems*, (pp. 194-203). IEEE.

Quinlan, S., & Khatib, O. (1993). Elastic bands: Connecting path planning and control. In *Proceedings of the IEEE International Conference on Robotics and Automation*, (pp. 802-807). IEEE Press.

Raibert, M. H., & Tello, E. R. (1986). Legged robots that balance. *IEEE Expert, 1*(4), 89. doi:10.1109/MEX.1986.4307016

Raiffa, H., & Schlaifer, R. (2000). *Applied statistical decision theory*. New York, NY: Wiley-Interscience.

Ranganathan, A., & Dellaert, F. (2011). Online probabilistic topological mapping. *The International Journal of Robotics Research, 30*(6), 755–771. doi:10.1177/0278364910393287

Ranganathan, A., Menegatti, E., & Dellaert, F. (2006). Bayesian inference in the space of topological maps. *IEEE Transactions on Robotics*, *22*(1), 92–107. doi:10.1109/TRO.2005.861457

Rao, C. R. (1945). Information and the accuracy attainable in the estimation of statistical parameters. *Bulletin of the Calcutta Mathematical Society*, *37*, 81–89.

Rao, C. R. (1965). *Linear statistical inference and its applications* (2nd ed.). New York, NY: Wiley-Interscience.

Rasmussen, C. E., & Williams, C. K. I. (2006). *Gaussian processes for machine learning*. Cambridge, MA: The MIT Press.

Real World Interface. (2011). *Corporate website*. Retrieved Mar 1, 2012, from http://www.rwii.com/

Remolina, E., Fernández-Madrigal, J. A., Kuipers, B. J., & González-Jiménez, J. (1999). Formalizing regions in the spatial semantic hierarchy: An AH-graphs implementation approach. *Lecture Notes in Computer Science*, *1661*, 109–124. doi:10.1007/3-540-48384-5_8

Riegl. (2011). *RIEGL GmbH corporate website*. Retrieved Mar 1, 2012, from http://www.riegl.com/

Robot Electronics. (2011). *Corporate website*. Retrieved Mar 1, 2012, from http://www.robot-electronics.co.uk/

Rodwell, E. J., & Cloud, M. J. (2012). Automatic error analysis using intervals. *IEEE Transactions on Education*, *55*(1), 9–15. doi:10.1109/TE.2011.2109722

Roth, S. D. (1982). Ray casting for modeling solids. *Computer Graphics and Image Processing*, *18*, 109–144. doi:10.1016/0146-664X(82)90169-1

Rotwat, P. F. (1979). *Representing the spatial experience and solving spatial problems in a simulated robot environment.* (Doctoral Dissertation). University of British Columbia. British Columbia, Canada.

Rubin, D. B. (1987). A noniterative sampling/importance resampling alternative to the data augmentation algorithm for creating a few imputations when fractions of missing information are modest: The SIR algorithm. *Journal of the American Statistical Association*, *82*(398), 543–546. doi:10.2307/2289460

Russell, S., & Norvig, P. (2002). *Artificial intelligence: A modern approach*. Upper Saddle River, NJ: Prentice Hall.

Rusu, R. B., Marton, Z. C., Blodow, N., Dolha, M., & Beetz, M. (2008). Towards 3D point cloud based object maps for household environments. *Robotics and Autonomous Systems*, *56*(11), 927–941. doi:10.1016/j.robot.2008.08.005

Salicone, S. (2007). *Measurement uncertainty. An approach via the mathematical theory of evidence*. Berlin, Germany: Springer.

Sandler, B. Z. (1999). *Robotics: Designing the mechanisms for automated machinery* (2nd ed.). New York, NY: Academic Press.

Saripalli, S., Montgomery, J. F., & Sukhatme, G. S. (2003). Visually guided landing of an unmanned aerial vehicle. *IEEE Transactions on Robotics and Automation*, *19*(3), 371–380. doi:10.1109/TRA.2003.810239

Scheinman, V. (1969). *Design of a computer controlled manipulator. Tech Report*. Palo Alto, CA: University of Stanford.

Schön, T. B. (2006). *Estimation of non-linear dynamic systems: Theory and applications.* (Doctoral Dissertation). Linköpings Universitet. Linköpings, Sweden.

Schraer, W. D., & Stoltze, H. J. (1999). *Biology: The study of life*. Upper Saddle River, NJ: Prentice Hall.

Schultz, A. C., & Parker, L. E. (Eds.). (2010). *Multi-robot systems: From swarms to intelligent automata*. Berlin, Germany: Springer.

Scuro, S. R. (2004). *Introduction to error theory. Technical Report*. College Station, TX: Texas A&M University.

Selig, J. M. (1992). *Introductory robotics*. Upper Saddle River, NJ: Prentice-Hall.

Serway, R. A., & Jewett, J. W. (2009). *Physics for scientists and engineers*. New York, NY: Brooks Cole.

Shachter, R. D. (1998). Bayes-ball: The rational pastime (for determining irrelevance and requisite information in belief networks and influence diagrams). In *Proceedings of the Fourteenth Conference in Uncertainty in Artificial Intelligence*, (pp. 480-487). IEEE.

Shaffer, G., & Vovk, V. (2005). *The origins and legacy of Kolmogorov's Grundbegriffe*. Working Paper. Retrieved February 2, 2010 from http://www.probabilityandfinance.com/articles/04.pdf

Sharp. (2011). *Sharp microelectronics of the Americas Inc. corporate website*. Retrieved Mar 1, 2012, from http://www.sharpsma.com/

Shimon, Y. N. (1999). *Handbook of industrial robotics* (2nd ed.). New York, NY: John Wiley & Sons.

Sibley, G., Mei, C., Reid, I., & Newman, P. (2010). Vast scale outdoor navigation using adaptive relative bundle adjustment. *The International Journal of Robotics Research*, *29*(8), 958–980. doi:10.1177/0278364910369268

Siciliano, B., & Khatib, O. (Eds.). (2008). *Handbook of robotics*. Berlin, Germany: Springer. doi:10.1007/978-3-540-30301-5

Sick. (2011). *SICK AG corporate website*. Retrieved Mar 1, 2012, from http://www.sick.com/

Siegwart, R., & Nourbakhsh, I. R. (2004). *Introduction to autonomous mobile robots: Intelligent robotics and autonomous agents*. Cambridge, MA: MIT Press.

Slama, C. C. (Ed.). (1980). *Manual of photogrammetry: American society of photogrammetry and remote sensing*. Falls Church, VA: American Society of Photogrammetry and Remote Sensing.

Sleeswyk, A. W. (1981, October). Vitruvius' odometer. *Scientific American*.

Smith, M., Baldwin, I., Churchill, W., Paul, R., & Newman, P. (2009). The new college vision and laser data set. *The International Journal of Robotics Research*, *28*(5), 595–599. doi:10.1177/0278364909103911

Smith, R. C., & Cheeseman, P. (1986). On the representation and estimation of spatial uncertainty. *The International Journal of Robotics Research*, *5*(4), 56–68. doi:10.1177/027836498600500404

Smith, R. C., Self, M., & Cheeseman, P. (1990). Estimating uncertain spatial relationships in robotics. *Autonomous Robot Vehicles*, *1*, 167–193. doi:10.1007/978-1-4613-8997-2_14

Snyman, J. A. (2005). *Practical mathematical optimization: An introduction to basic optimization theory and classical and new gradient-based algorithms*. Berlin, Germany: Springer Publishing.

Sonardyne. (2011). *Sonardyne corporate website*. Retrieved Mar 1, 2012, from http://www.sonardyne.com/

Stachniss, C. (2009). *Robotic mapping and exploration*. Berlin, Germany: Springer.

Stachniss, C., Plagemann, C., Lilienthal, A., & Burgard, W. (2008). Gas distribution modeling using sparse Gaussian process mixture models. In *Proceedings of Robotics: Science and Systems*. IEEE. doi:10.1007/s10514-009-9111-5

Stoll, E., Letschnik, J., Walter, U., Artigas, J., Kremer, P., Preusche, C., & Hirzinger, G. (2009). On-orbit servicing. *IEEE Robotics & Automation Magazine*, *6*(4), 29–33. doi:10.1109/MRA.2009.934819

Strasdat, H., Montiel, J. M. M., & Davison, A. J. (2010). Real-time monocular SLAM: Why filter? In *Proceedings of the IEEE International Conference on Robotics and Automation*, (pp. 2657-2664). IEEE Press.

Strasdat, H., Montiel, J. M. M., & Davison, A. J. (2010). Scale-drift aware large scale monocular SLAM. In *Proceedings of Robotics Science and Sytems*. IEEE.

Strassen, V. (1969). Gaussian elimination is not optimal. *Numerische Mathematik*, *13*(4), 354–356. doi:10.1007/BF02165411

Sturm, J., Magnenat, S., Engelhard, N., Pomerleau, F., Colas, F., & Burgard, W. … Siegwart, R. (2011). Towards a benchmark for RGB-D SLAM evaluation. In *Proceedings of the RGB-D Workshop on Advanced Reasoning with Depth Cameras at Robotics: Science and Systems Conference*. RGB-D.

Tabak, J. (2004). *Probability and statistics: The science of uncertainty*. New York, NY: Facts on File.

Tardós, J. D., Neira, J., Newman, P., & Leonard, J. (2002). Robust mapping and localization in indoor environments using sonar data. *The International Journal of Robotics Research*, *21*(4), 311–330. doi:10.1177/027836402320556340

Taylor, C. J., & Kriegman, D. J. (1994). *Minimization on the lie group SO(3) and related manifolds*. Technical Report 9405. New Haven, CT: Yale University.

Taylor, J. R. (1997). *An introduction to error analysis* (2nd ed.). New York, NY: University Science Books.

Tellez, R., Ferro, F., Mora, D., Pinyol, D., & Faconti, D. (2008). Autonomous humanoid navigation using laser and odometry data. In *Proceedings of the 8th IEEE-RAS International Conference on Humanoid Robots*, (pp. 500-506). IEEE Press.

Tester, J. W. (2005). *Sustainable energy: Choosing among options*. Cambridge, MA: The MIT Press.

Thomas, D. B., Leong, P. H. W., Luk, W., & Villaseñor, J. D. (2007). Gaussian random number generators. *ACM Computing Surveys*, *39*(4). doi:10.1145/1287620.1287622

Thorpe, C. E. (1984). *Path relaxation: Path planning for a mobile robot.* Technical Report CMU-RI-TR-84-5. Pittsburgh, PA: Carnegie Mellon University.

Thrun, S. (2001). A probabilistic on-line mapping algorithm for teams of mobile robots. *The International Journal of Robotics Research*, *20*(5), 335. doi:10.1177/02783640122067435

Thrun, S. (2001). Is robotics going statistics? The field of probabilistic robotics. *Communications of the ACM*, *45*(3), 1–8.

Thrun, S., Burgard, W., & Fox, D. (2005). *Probabilistic robotics*. Cambridge, MA: MIT Press.

Thrun, S., Fox, D., Burgard, W., & Dellaert, F. (2000). Robust Monte Carlo localization for mobile robots. *Artificial Intelligence*, *128*(1-2).

Thrun, S., & Montemerlo, M. (2006). The graph SLAM algorithm with applications to large-scale mapping of urban structures. *The International Journal of Robotics Research*, *25*(5-6), 403–429. doi:10.1177/0278364906065387

Tipaldi, G. D., & Arras, K. O. (2010). FLIRT – Interest regions for 2D range data. In *Proceedings of the IEEE International Conference on Robotics and Automation*. IEEE Press.

Tolman, E. C. (1948). Cognitive maps in rats and men. *Psychological Review*, *55*(4), 189–208. doi:10.1037/h0061626

Tomasi, C., & Kanade, T. (1992). Shape and motion from image streams under orthography: A factorization method. *International Journal of Computer Vision*, *9*(2), 137–154. doi:10.1007/BF00129684

Triggs, B., McLauchlan, P., Hartley, R., & Fitzgibbon, A. (2000). Bundle adjustment—A modern synthesis. *Vision Algorithms: Theory and Practice*, *1883*, 153–177.

Trucco, E., & Verri, A. (1998). *Introductory techniques for 3-D computer vision*. Upper Saddle River, NJ: Prentice Hall.

Trudeau, R. J. (1994). *Introduction to graph theory*. New York, NY: Dover Publications.

Turing, A. (1948). *Intelligent machinery.* Retrieved Feb 1, 2012, from http://www.turingarchive.org/browse.php/C/11

uBlox. (2011). *u-Blox AG corporate website*. Retrieved Mar 1, 2012, from http://www.u-blox.com/

Ulrich, I., & Nourbakhsh, I. (2000). Appearance-based place recognition for topological localization. In *Proceedings of the IEEE International Conference on Robotics and Automation*, (pp. 1023-1029). IEEE Press.

Varadarajan, V. S. (1974). *Lie groups, lie algebras, and their representations*. Upper Saddle River, NJ: Prentice-Hall.

Varga, M. (Ed.). (2007). *Practical image processing and computer vision.* New York, NY: Wiley.

Vázquez-Martín, R., Núñez, P., Bandera, A., & Sandoval, F. (2009). Curvature-based environment description for robot navigation using laser range sensors. *Sensors (Basel, Switzerland)*, *9*(8), 5894–5918. doi:10.3390/s90805894

Velodyne. (2011). *Velodyne lidar corporate website*. Retrieved Mar 1, 2012, from http://velodynelidar.com/

Waldron, K. J. (1985). Mobility and controllability characteristics of mobile robotic platforms. In *Proceedings of the IEEE International Conference on Robotics and Automation*, (pp. 237-243). IEEE Press.

Walpole, R. E., Myers, R. L., & Myers, S. H. (1997). *Probability and statistics for engineers and scientists*. Upper Saddle River, NJ: Prentice Hall. doi:10.2307/2530629

Walter, M. R., Eustice, R. M., & Leonard, J. J. (2007). Exactly sparse extended information filters for feature-based SLAM. *The International Journal of Robotics Research*, *26*(4), 335–359. doi:10.1177/0278364906075026

Walter, W. G. (1963). *The living brain*. New York, NY: W. W. Norton and Co.

Wan, E., & Van der Merwe, R. (2000). The unscented Kalman filter for nonlinear estimation. In *Proceedings of Symposium on Adaptive Systems for Signal Processing, Communication and Control (AS-SPCC)*. AS-SPCC.

Wang, C. C., Thorpe, C., Thrun, S., Hebert, M., & Durrant-Whyte, H. (2007). Simultaneous localization, mapping and moving object tracking. *The International Journal of Robotics Research*, *26*(9), 889–916. doi:10.1177/0278364907081229

Watanabe, K., Pathiranage, C. D., & Izumi, K. (2009). T-S fuzzy model adopted SLAM algorithm with linear programming based data association for mobile robots. In *Proceedings of the IEEE International Symposium on Industrial Electronics (ISIE 2009)*, (pp. 244-249). IEEE Press.

Web, E. K. F. (2012). *Kalman filters website of the MRPT project*. Retrieved Mar 1, 2012, from http://www.mrpt.org/Kalman_Filters

Web, P. F. (2012). *Application:pf-localization website of the MRPT project*. Retrieved Mar 1, 2012, from http://www.mrpt.org/Application:pf-localization

Weisstein, E. (2011). *Wolfram mathworld: On-line mathematical encyclopedia*. Retrieved Mar 1, 2012, from http://mathworld.wolfram.com

Welch, G., & Bishop, G. (2001). An introduction to the Kalman filter. In *Proceedings of SIGGRAPH 2001*. ACM Press.

Wetherill, G. B. (1961). Bayesian sequential analysis. *Biometrika*, *48*, 281–292.

Wiener, N. (1965). *Cybernetics or the control and communication in the animal and the machine*. Cambridge, MA: MIT Press. doi:10.1037/13140-000

Wilcox, R. R. (2005). *Introduction to robust estimation and hypothesis testing* (2nd ed.). London, UK: Elsevier.

Willow Garage. (2012). *Willow garage corporate website*. Retrieved Mar 1, 2012, from http://www.willowgarage.com/

Winkler, G. (1995). *Image analysis, random fields and dynamic Monte Carlo methods*. Berlin, Germany: Springer.

Wolf, D. F., & Sukhatme, G. S. (2005). Mobile robot simultaneous localization and mapping in dynamic environments. *Autonomous Robots*, *19*(1), 53–65. doi:10.1007/s10514-005-0606-4

Wu, J. (2005). *Some properties of the Normal distribution*. Atlanta, GA: Georgia Institute of Technology.

Wurm, K. M., Hornung, A., Bennewitz, M., Stachniss, C., & Burgard, W. (2010). OctoMap: A probabilistic, flexible, and compact 3D map representation for robotic systems. In *Proceedings of the ICRA Workshop on Best Practice in 3D Perception and Modeling for Mobile Manipulation*. ICRA.

Yuh, J. (2000). Design and control of autonomous underwater robots: A survey. *Autonomous Robots*, *8*(1), 7–24. doi:10.1023/A:1008984701078

Zadeh, L. A. (1965). Fuzzy sets. *Control*, *8*, 338–353. doi:10.1016/S0019-9958(65)90241-X

Zadeh, L. A. (2005). Toward a generalized theory of uncertainty (GTU)—An outline. *Information Sciences*, *172*, 1–40. doi:10.1016/j.ins.2005.01.017

Zaloga, S., & Laurier, J. (2005). *V-1 flying bomb 1942-52: Hitler's infamous 'doodlebug*. London, UK: Osprey Publishing.

Zaritskii, V. S., Svetnik, V. B., & Shimelevich, L. I. (1975). Monte Carlo technique in problems of optimal data processing. *Automation and Remote Control*, *12*, 95–103.

Zhao, H., Chiba, M., Shibasaki, R., Shao, X., Cui, J., & Zha, H. (2009). A laser-scanner-based approach toward driving safety and traffic data collection. *IEEE Transactions on Intelligent Transportation Systems*, *10*(3), 534–546. doi:10.1109/TITS.2009.2026450

Zheng, Y. F., & Zheng, Y. F. (1994). *Recent trends in mobile robots*. New York, NY: World Scientific Pub Co Inc.

Ziparo, V. A., Kleiner, A., Nebel, B., & Nardi, D. (2007). RFID-based exploration for large robot teams. In *Proceedings of the IEEE International Conference on Robotics and Automation*, (pp. 4606-4613). IEEE Press.

Zorich, Z. (2011). Paleolithic tools - Plakias, Crete. *Archaeology Magazine, 64*(1).

About the Authors

Juan-Antonio Fernández-Madrigal holds a PhD in Computer Science and is tenured Associate Professor in the University of Málaga (Spain). He has been teaching since 1998 in graduate and post-graduate courses on real-time systems, control engineering, and robotics. He has supervised PhD theses on cognitive robotics and probabilistic localization and mapping for mobile robots, and a relevant number of BSc and MSc theses on very diverse subjects. His research work has been developed mainly on different aspects of the modeling of the environment for mobile robots, cognitive robotics, and robotic software development. He has three books published internationally and nearly 80 scientific papers on these and other topics. He has been involved in different roles in regional, national, and European research projects, and is co-inventor of several patents. Regarding his personal interests, they currently include programming, drawing, and writing, having authored five sci-fi books in Spanish and more than a hundred short stories.

José Luis Blanco Claraco is a Lecturer in the University of Málaga (Spain), where he has taught in graduate courses on mechatronics and material science. He obtained a PhD in Mobile Robotics from the same university in 2009. His research interests include estimation theory, mobile robot navigation, large-scale map building, and computer vision. Since 2005, he has participated in several national and European research projects, as well as research collaborations with private companies. As a result, he has published more than 40 scientific papers and is co-inventor of three patents and one utility model. He is also an active supporter of Open Source initiatives, having participated in more than 10 software projects, most remarkably in the Mobile Robot Programming Toolkit (MRPT), which he started in 2005 and still actively maintains at present. Among his non-academic interests, he collaborates in popular science blogs aimed at the Spanish speaking community.

Index

3D cameras 48, 290

A

accelerometers 37-39, 53
Ackermann model 161
Ackerman steering 34, 35
active beacons 12, 19
active exploration 13
active vision 13, 24
Actuator dynamics 63
adaptive tessellation 259
anchoring problem 267, 293, 384
ancillary statistic 121, 122
appearance-based approaches 235
appearance-based place recognition 252, 382
approximate nearest neighbors (ANN) 198
asymptotically unbiased estimator 124
Autonomous Underwater Vehicles (AUVs) 33

B

Baron's cross-correlation coefficient 195-197, 202
Bayes, Thomas 66
Bayes Filter
    Discrete Bayes Filter (DBF) 234
Bayesian estimators 22, 110, 111, 129, 130, 133, 169, 171, 175, 203, 256, 319, 348
Bayesian network
    dynamic 204, 270
Bayesian networks (BNs) 99
Bayesian paradigm 13, 71, 96, 97
Bayes' Rule 19, 60, 97, 98, 135, 244, 270-272, 276, 277, 306, 328
Bayes' theorem 66
beam model 176-179, 226, 273, 332
bearings 16, 49, 51, 176, 180-182, 185, 198, 274, 275, 278, 285, 308
bearing-only observation 281, 284
behavior-based control architectures 17
Bernoulli distribution 257
bias of an estimator 124
biped robots 31, 32
Branch and Bound (BB) strategy 190
Bundle Adjustment (BA) 299, 364

C

cartography 15, 24, 299
casters wheels 33
central limit theorem 67, 110, 118, 142, 207, 374
central moments 82
Chebyshev's inequality 85, 86, 115, 116
chi-squared distribution 60, 81, 95, 118, 148, 188, 189
chi-squared probability distribution 188
    degrees of freedom 188
chi-squared test 68, 189
CHOLMOD algorithm 356, 378, 385
clique 101
cluttered environments 10, 323
codomain 75-78, 85, 91, 108, 190
cognitive maps 19, 25, 255, 269
complete estimator 121, 122
conditional expectation 98, 99, 106, 125, 126, 130, 131
conditional independence 98-102, 105, 175, 205, 270, 306, 307, 319-321
conditional probability 60, 64, 96-99, 122, 236, 237, 245, 319
conditional probability chain rule 97
conjugate distributions 130, 207
consistent estimator 123
constraint maps 264, 292, 294
contact sensors 17, 39
continuous random variables 75
convolution 93
countable additivity 71, 72
covariance matrix 88-91, 96, 104, 105, 109, 146,

147, 154, 155, 159, 165, 168, 169, 172, 181, 185-188, 201, 208, 217, 219, 229, 280, 284, 302-304, 308, 311, 314-318, 350, 361, 378, 389
Cramér-Rao Lower Bound (CRLB) 125
cross-correlation 93, 180, 195-197, 202
cross-covariance matrix 90, 219
curse of dimensionality 243, 259
curve fitting 68

D

Data Association (DA) 263, 310
data association problem (DA) 185, 263
datasets 66-68, 111, 250
dead-reckoning 16, 19, 33, 39, 167
dead-zones 63
Decision processes 63
Deliberative robot control architectures 17
Descriptive statistics 111
developmental robotics 19, 25, 256
Differential drive 34, 156, 161
Dirac's delta 240-242, 246, 319
Discrete Bayes Filter (DBF) 234
discrete random variables 75, 234
domain 75, 76, 89, 101, 108, 190, 223, 235, 240, 277, 290
Doppler effect 38
Doppler Velocity Log (DVL) 33
drive wheels 33-35, 173
d-separation 61, 102, 276
Dynamic Bayesian Network (DBN) 204, 270
dynamic environments 11, 24, 251, 379, 380, 388
dynamic stochastic process 133

E

Effective Sample Size (ESS) 247, 322
efficient estimator 125, 126
embodied cognition 18, 23, 255, 293
engineered environments 11, 12
environmental sensors 12, 36, 52
estimator 110, 115-135, 141, 175, 222, 229, 239, 240, 302, 303, 319, 321, 342, 343, 347, 368, 374
expectation of a matrix of r.v.s 91
Exponentially Weighted Recursive Least Squares (EW-RLS) 208
exteroceptive sensors 12, 36, 39, 40, 45, 48-51, 142, 169

F

feature descriptor 281
Field Of View (FOV) 42, 46
first-order ancillary 121, 122
Fisher information matrix 125-128
Flying Robots 32, 33, 44, 55, 143
fuzzy theory 64

G

gas sensing 51
gauge freedom 36
Gaussian Markov Random Field (GMRF) 349
Gauss-Newton algorithm 179, 299, 318, 331, 333, 336, 341, 353, 365, 368, 388, 389
Generalized Voronoi Graphs 260, 261
General Theory of Uncertainty (GTU) 64
geometrical maps 6, 93
global localization 14, 20, 205, 222, 250
Global Navigation Satellite Systems (GNSS) 12, 53
Global Positioning System 53
good old fashioned AI (GOFAI) 18
graphs 8, 61, 99-105, 108, 201, 265, 267, 276, 292, 293, 296, 301-305, 308, 315, 327, 336, 339, 348-356, 359-368, 372-378, 381-389
graphical models 60, 61, 64, 99, 135, 251, 298-300, 336
graph of correlations 61, 104, 105, 301-305, 308, 315
graph theory 99, 265, 296
greedy algorithm 187
Grid maps 39, 193-199, 202, 255-261, 264, 269, 272, 273, 290, 293, 304, 306, 318, 319, 322, 325, 332, 339, 380
ground-truth 111
gyroscopes 37-39, 171

H

Hessian matrix 330, 354-357, 361, 363, 368, 373, 376, 382
hierarchy of abstraction 268, 269
Histogram Filter (HF) 235, 238, 239
Holonomic mobile robots 151
human cognitive map 19, 269
Humanoids 31
hybrid metric-topological (HMT) maps 8, 265, 382
hybrid metric-topological SLAM 259, 293, 384
hysteresis 63

I

importance sampling 69, 240, 243, 244, 318
inertial sensors 11, 12, 16, 33, 36-38
inference problem with Bayesian Networks 101
Infinite Impulse Response filter (IIR) 208
Information Filter (IF) 209
infrared (IR) 39
innovation covariance 219, 220, 232, 233, 311
interpretation tree 186, 187, 191, 200, 202
Iterated Extended Kalman Filter 226
Iterative Closest Point (ICP) algorithms 198

J

Jacobian structure 367
joint compatibility 191
Joint-Compatibility Branch and Bound (JCBB) 310
joint probability density functions 88
joint probability distribution 87, 103
joint probability mass function 88

K

Kalman Filter (KF) 19, 20, 69, 105, 171, 207-211, 215, 217, 220-223, 226, 231, 251, 252, 280, 285, 306, 307, 321, 333, 342, 362, 363
  computational complexity 222
  Extended Kalman Filter (EKF) 207, 223, 280, 307
  Extended Mixture Kalman Filter (EMKF) 226
  prediction stage 209, 215, 231
  residual error 222
  unscented 226, 252
  update stage 209, 217
Kalman gain 219-224, 232, 280, 285, 312
KD-trees 198
kidnapped robot problem 14, 133
Kinect™ 49, 50, 352
kinematic designs for wheeled robots 33
Kinematic Equations 33, 143, 151, 156, 160, 163-169, 173
Kullback-Leibler divergence (KLD) 250
kurtosis 84

L

landmark maps 254-257, 261, 263, 274, 278, 280, 286, 290, 300, 306, 328
landmarks 11, 14, 19, 179-184, 257-263, 274-280, 283-286, 289, 300, 303-316, 319-324, 328-332, 338-341, 345, 364-383
large-scale robotic mapping 19
large-scale space 25, 252, 257, 265, 294, 295, 385
laser scanners 20, 32, 46-48, 51, 57, 176, 332, 333, 338, 353, 374
law of large numbers 66, 110, 111, 115, 117, 123, 128, 131, 132, 175, 239, 240
learning problems with Bayesian Networks 101
least squares estimator (LSE) 128
least-squares minimization 341
Legged Robots 31, 58
Lehmann-Scheffé theorem 126
Levenberg-Marquardt (LM) algorithms 336, 341, 353-357, 364-368, 389
lifelong operation 379
likelihood field 178, 199
Limit Laws 115
linear correlation coefficient 90, 197
linearization point 95, 96, 146
line of sight 50, 177, 178, 193, 273
lithium-ion polymer (Li-Po) 55
localization 1-16, 19-36, 39-42, 45-51, 57, 60-70, 80, 87, 96, 105, 106, 110-112, 120, 129, 133, 140-144, 149, 155, 156, 167-179, 184, 185, 195, 198-207, 222-228, 234-236, 239-245, 248-258, 261, 263, 269, 273, 280, 284, 290, 293-299, 307, 312, 314, 322, 333-342, 347, 362, 379, 384-389
local maps 174, 193-197, 293, 381-384
log-likelihood 128, 197, 199, 389
loop closure 257, 315, 316, 325, 326, 338-340, 361, 368

M

machine perception and intelligent robotics group (MAPIR) 3
Mahalanobis distance 185-192, 199, 201, 302, 333
  joint 191
Manifold 149, 343-347, 356, 359-361, 386
map matching 19, 90, 193-198, 259
mapping 1-16, 19-32, 35, 36, 39-42, 45-48, 52, 58-69, 76, 80, 91, 96, 105, 106, 110-112, 120, 129, 140, 141, 172-179, 182-186, 198-203, 207, 251, 254-259, 263, 269, 272-275, 280, 281, 284-287, 290-296, 299, 304, 321, 326, 334, 335, 339, 340, 344, 352, 359, 362, 379-389
Maritime navigation 15, 16
Markov chains 67
Markovian assumption 204
Markov localization 24, 200, 235, 251, 257
Markov property 99, 135, 291

Markov random field (MRF) 99, 100, 291
Markov's inequality 85, 86
Matching Likelihood (ML) 185, 192, 198
mathematical graph 265
maximal clique 101
Maximum A Posteriori Estimator (MAP) 131
Maximum Likelihood Estimators (MLE) 302
mean squared error (MSE) 124
measure-theoretic definition of probability 69
Median Estimator (MED) 132
method of least squares 67
method of moments 110, 128
Micro Air Vehicles (MAVs) 32
MicroElectroMechanical Systems (MEMS) 12
minimal sufficient estimator 122
Minimum Mean Squared Error Estimator (MMSE) 130
minimum mismatch 352
minimum variance unbiased estimator (MVUE) 125, 222
modified environments 11
moments of a probability distribution 82
Monte Carlo localization 27, 106, 239, 251
Monte Carlo methods 20, 68, 69, 172, 239, 243, 296, 334
motion control 29
MRPT toolkit 226
multimodal distributions 80, 84, 118, 222, 287
multinomial resampling 246
Multisensory Fusion 338

N

navigation 4, 5, 10-17, 20, 24, 26, 33, 39, 53, 57, 59, 63, 64, 73, 111, 208, 255, 258-260, 264, 265, 273, 295, 338, 384, 387, 388
Nearest Neighbor DA Algorithm (NN) 187
Newton method 355, 389
non-holonomic restrictions 157
Non-parametric filters 22, 203, 207, 234, 257, 298, 303
non-repeatability of sensors 62
normal curve 66
Nouvelle AI 18

O

observation likelihood 141, 142, 175, 176, 179, 180, 193-199, 207, 217, 234-237, 241, 246, 304, 321-333, 338
occupancy grids 20, 24, 58, 177, 193, 254, 257, 258, 332, 380, 382
octomap 259, 296
octree 259
Odometry 5, 16, 24, 33-36, 39, 56-59, 140-143, 152-173, 217, 226, 248, 327, 331, 350, 366
one-to-one function 74, 185
on-manifold optimization 354, 360, 364, 374
on-to function 75
optimal estimator 124
optimal proposal distribution 244, 327-333
optimization by relaxation 303
Outdoor scenarios 8, 10, 261
over-parameterization 284, 288, 289, 343

P

paradigms of probability 66
parametric distribution 172, 206, 332
Parametric filters 206, 207, 234, 243, 298, 303, 304, 336
partially observable 36
particle depletion problem 246
Particle Filter (PF) 239
    vanilla 246
particle impoverishment 246, 327
partition function 101
passive beacons 19
paths 8, 12, 25, 33, 102-105, 186, 187, 191, 192, 295, 319, 339, 352
Perception processes 62
phase aliasing 44
pinhole projective model 283
point clouds 50, 198, 226, 248, 259, 260, 350
point maps 193, 198, 259-261, 290, 332, 353
pose 2, 6, 36, 37, 48, 67, 70, 73-75, 87, 89, 97, 111, 123, 130, 133, 135, 141-160, 163-172, 175-186, 189, 192-195, 198-201, 205-208, 217-228, 231-240, 243-250, 255, 264, 269, 270, 276-293, 301, 304-322, 329-333, 338-356, 359-369, 372-378, 387
pose constraint map 264
positive-definite matrix 91, 229
positive-semidefinite matrix 91, 109, 229
possibility theory 64, 106
prediction 11, 25, 106, 111, 135, 172, 181, 183, 201, 206, 209, 210, 215-218, 222, 224, 231-234, 239, 246, 251, 280, 289, 291, 309-314, 320, 330, 331, 341, 350, 351, 360
prehistoric maps 15
Probabilistic Motion Model 142-145, 150-153, 159, 165, 168-170, 236, 310, 322, 365

probability density 60, 64, 79, 88, 92, 122, 127, 129, 175, 237, 240, 241, 329
probability function 70-72, 75, 76, 80, 91, 97, 111, 133
probability space 69-78, 85, 87, 97, 111, 120-122, 138
proprioceptive sensors 11, 12, 36, 140, 142, 169, 171
pseudo-Huber kernels 375
Pulsed-Signal Time of Flight (P-ToF) 42
p-value 188, 189

Q

quaternions 281

R

RADAR 42, 251
Random errors 61, 62
random events 70-76, 89, 96, 98, 240
random field 99, 100, 257, 290, 291, 349
range 12, 20, 22, 28, 29, 32, 39-51, 55-58, 75, 117, 144, 147, 170, 176-181, 185, 194-200, 235, 244, 247-250, 257-263, 270-275, 278, 286, 287, 290, 292, 295, 296, 308, 312, 332, 335, 337, 342, 379, 381, 384, 387, 388
range-bearing sensors 12, 51, 176, 179, 278, 279, 283, 330
rangefinders 40-45, 48
Range-Only Sensors 49-51, 254, 286, 304, 365
Rao-Blackwellization, general theory of 69, 126, 303, 318
Rao-Blackwellization Particle Filter (RBPF) 182, 318, 334
Rao-Blackwell theorem 126
raw moments 82
ray-casting 176, 177
Real-time Kinematics (RTK) 53
Real World Interface (RWI) 156
reduced linear system 369
reflectivity sensors 39, 40
Relative Bundle Adjustment (RBA) 378
RFID tags 53
right-tail of a pdf 189
rigid transformation 19, 193, 388
risk of an estimator 124
robot actions 22, 29, 30, 35, 140, 141, 175, 303
robot awakening problem 14, 222, 234
robotic data set 13
robotic manipulators 1-4, 16
robust estimator 123, 374
robustified cost function 375

S

sample space 70-78, 85, 91, 108
sample variance 110, 118, 129
Sampling Importance 246
saturation 272
Scalability 23, 336, 340, 381, 383
Schur complements 370, 371
semantic maps 24, 265, 268, 294, 386
sensor fusion 171
sensor linearity 61
sensor model
    inverse 272-274, 278, 279, 282-284, 287, 288, 313, 314
sensor resolution 61, 63
sequential Bayesian filtering 135
sequential estimation 106, 130, 169, 250, 325, 333, 384
Sequential Importance Sampling (SIS) algorithm 243
Sequential Monte Carlo (SMC) estimation 239
Settling times 63
sigma-point filtering 228
sigma-points 229-232
simultaneity of random events 70, 71
simultaneous localization and mapping (SLAM) 35, 48, 106, 112
    back-end 352
    EKF-based 309, 313, 315, 322, 324, 339, 342, 346, 353, 378, 380
    front-end 352
    full 298-305, 318-325, 338, 348, 349, 364, 365
    graph 292, 336, 348-356, 361-367, 372-378, 382, 383, 388, 389
    graph-SLAM approaches 264
    hybrid metric-topological 259, 293, 384
    large-scale 20
    on-line 304-308, 314-318, 323
skewness 84
SLAM problem 14, 180, 187, 193, 255, 276, 301, 318, 320, 336, 341, 349, 372, 378, 389
sonars 42-45, 176
sources of uncertainty 61, 63, 177, 178, 324
sparse Cholesky algorithm 355, 356
Spatial Semantic Hierarchy 25, 269, 295, 296
squared Mahalanobis distance (SMD) 185
squared weighted residuals 302
Square Root Information Filtering (SRIF) 363
Square Root Information Smoothing 363

stability-plasticity dilemma 379
standardized moments 82
standard proposal distribution 245, 323, 327
State Space Model (SSM) 208
state space topology 341
static workspaces 11
statistical inference 29, 95, 107, 111
Statistics 6, 19-22, 26, 60, 64-68, 90, 106, 107, 110-117, 120, 137, 171, 189, 251, 269, 334, 386
stereo camera 51, 263
stochastic process 67-76, 108, 111-120, 126, 133, 136
strong law of large numbers 117
strongly consistent estimator 123
Structured-light cameras 49
structured scenarios 8
Structure From Motion (SFM) 299
submaps 193
Submarine Robots 33
sub-symbolic maps 256, 264
sufficient estimator 122
sum of Gaussians (SOG) 288
symbol grounding problem 256, 265, 267, 294
symbolic maps 256, 265-269
Synchro drive 35
Systematic errors 61, 142, 152

T

Taylor series expansion 95, 146, 168, 181, 311, 354, 358, 389
theorem of total probability 97
theory of probability 64, 65, 71
Time-of-Flight (ToF) 40
topological map 8, 75, 264, 265, 268, 381
    hybrid metrical 8, 265
topological relations 256, 264, 269
tracking problem 13, 14
Tree-based network optimizer (TORO) 261, 295, 363, 364
tricycle model 34, 161, 162

U

Ultra Wide Band (UWB) 50
unbiaseness 116
uncertaintly inflation 175
    overestimation 175
    underestimation 175
unimodal distributions 80
unmanned aerial vehicles (UAV) 32
Unscented Kalman Filter 226, 252
Unscented Transformation 96, 318
Unscented Transform (UT) 228

V

Valve-Regulated Lead-Acid (VRLA) 55
Vented Lead-Acid (VLA) 55
view-based maps 264, 294
visual servoing 29
voxels 257

W

weak law of large numbers 115
weakly consistent estimator 123
wheeled-robot odometry 33
Wheeled robots 22, 32-35, 140, 341
working ranges 61, 63